RF/MICROWAVE CIRCUIT DESIGN FOR WIRELESS APPLICATIONS

TP13002480
ENG

RF/MICROWAVE CIRCUIT DESIGN FOR WIRELESS APPLICATIONS

SECOND EDITION

Ulrich L. Rohde

Brandenburg University of Technology
Cottbus, Germany
Synergy Microwave Corporation
Paterson, New Jersey

Matthias Rudolph

Brandenburg University of Technology
Cottbus, Germany

A JOHN WILEY & SONS, INC., PUBLICATION

Library of Congress Cataloging-in-Publication Data:

Rohde, Ulrich L.
  RF/microwave circuit design for wireless applications / Ulrich L. Rohde, Matthias Rudolph. – 2nd ed.
    p. cm.
    ISBN 978-0-470-90181-6 (hardback)
  1. Microwave integrated circuits–Computer-aided design. 2. Wireless communication
systems–Equipment and supplies–Design and construction. 3. Semiconductors–Computer-aided design.
4. Microwave circuits–Design and construction. 5. Radio frequency integrated circuits–Design and
construction. I. Rudolph, Matthias, 1969- II. Title.
    TK7876.R65 2013
    621.381'32–dc23

                                                                        2012013020

Printed in the United States of America

10 9 8 7 6 5 4 3 2 1

To Professor Vittorio Rizzoli

CONTENTS

Foreword   **xiii**

Preface   **xv**

1   Introduction to Wireless Circuit Design   **1**

   1.1   Introduction / 1
   1.2   System Functions / 3
   1.3   The Radio Channel and Modulation Requirements / 5
         1.3.1   Introduction / 5
         1.3.2   Channel Impulse Response / 7
         1.3.3   Doppler Effect / 12
         1.3.4   Transfer Function / 13
         1.3.5   Time Response of Channel Impulse Response
                 and Transfer Function / 14
         1.3.6   Lessons Learned / 16
         1.3.7   Wireless Signal Example: The TDMA System in GSM / 17
         1.3.8   From GSM to UMTS to LTE / 28
   1.4   About Bits, Symbols, and Waveforms / 29
         1.4.1   Introduction / 29
         1.4.2   Some Fundamentals of Digital Modulation Techniques / 37
   1.5   Analysis of Wireless Systems / 50
         1.5.1   Analog and Digital Receiver Designs / 50
         1.5.2   Transmitters / 57
   1.6   Building Blocks / 78
   1.7   System Specifications and Their Relationship to Circuit Design / 79
         1.7.1   System Noise and Noise Floor / 80
         1.7.2   System Amplitude and Phase Behavior / 89
   1.8   Testing / 108
         1.8.1   Introduction / 108
         1.8.2   Transmission and Reception Quality / 112
         1.8.3   Base Station Simulation / 113
         1.8.4   GSM / 122
         1.8.5   DECT / 122
   1.9   Converting C/N or SNR to E_B/N_0 / 123
   References / 124
   Further Reading / 125

2 Models for Active Devices   **127**

2.1   Diodes / 128
    2.1.1   Large-Signal Diode Model / 129
    2.1.2   Mixer and Detector Diodes / 133
    2.1.3   PIN Diodes / 141
    2.1.4   Tuning Diodes / 158
2.2   Bipolar Transistors / 203
    2.2.1   Transistor Structure Types / 203
    2.2.2   Large-Signal Behavior of Bipolar Transistors / 204
    2.2.3   Large-Signal Transistors in the Forward-Active Region / 218
    2.2.4   Improving RF Performance by Means of Heterostructures / 225
    2.2.5   Effects of Collector Voltage on Large-Signal Characteristics in
          the Forward-Active Region of BJTs / 226
    2.2.6   Effects of Collector Current and Voltage on Large-Signal
          Characteristics in the Forward-Active Region of HBTs / 229
    2.2.7   Saturation and Inverse Active Regions / 231
    2.2.8   Self-Heating / 236
    2.2.9   Small-Signal Models of Bipolar Transistors / 239
2.3   Field-Effect Transistors / 239
2.4   Large-Signal Behavior of JFETs / 248
    2.4.1   Small-Signal Behavior of JFETs / 251
    2.4.2   Large-Signal Behavior of MOSFETs / 256
    2.4.3   Small-Signal Model of the MOS Transistor in Saturation / 263
    2.4.4   Short-Channel Effects in FETs / 267
    2.4.5   Small-Signal Models of MOSFETs / 272
    2.4.6   III–V MESFETs and HEMTs / 289
    2.4.7   Small-Signal GaAs MESFET and HEMT Model / 298
2.5   Parameter Extraction of Active Devices / 324
    2.5.1   Introduction / 324
    2.5.2   Typical SPICE Parameters / 324
    2.5.3   Noise Modeling / 326
    2.5.4   Scalable Device Models / 334
    2.5.5   Generating a Databank for Parameter Extraction / 334
    2.5.6   Conclusions / 345
    2.5.7   Device Libraries / 347
    2.5.8   Physics-Based MESFET Modeling / 348
    2.5.9   Example: Improving the BFR193W Model / 351
References / 355
Further Reading / 357

3 Amplifier Design with BJTs and FETs   **359**

3.1   Properties of Amplifiers / 359
    3.1.1   Introduction / 359
    3.1.2   Gain / 364
    3.1.3   Noise Figure (NF) / 369
    3.1.4   Linearity / 397
    3.1.5   AGC / 413
    3.1.6   Bias and Power Voltage and Current (Power Consumption) / 417

3.2   Amplifier Gain, Stability, and Matching / 423
    3.2.1   Scattering Parameter Relationships / 423
    3.2.2   Low-Noise Amplifiers / 429
    3.2.3   High-Gain Amplifiers / 467
    3.2.4   Low-Voltage Open-Collector Design / 476
3.3   Single-Stage Feedback Amplifiers / 484
    3.3.1   Lossless or Noiseless Feedback / 489
    3.3.2   Broadband Matching / 490
3.4   Two-Stage Amplifiers / 490
3.5   Amplifiers with Three or More Stages / 499
    3.5.1   Stability of Multistage Amplifiers / 504
3.6   A Novel Approach to Voltage-Controlled Tuned Filters Including CAD
    Validation / 505
    3.6.1   Diode Performance / 505
    3.6.2   A VHF Example / 508
    3.6.3   An HF/VHF Voltage-Controlled Filter / 508
    3.6.4   Improving the VHF Filter / 513
    3.6.5   Conclusion / 514
3.7   Differential Amplifiers / 514
3.8   Frequency Doublers / 518
3.9   Multistage Amplifiers with Automatic Gain Control (AGC) / 524
3.10   Biasing / 524
    3.10.1   RF Biasing / 535
    3.10.2   dc Biasing / 535
    3.10.3   dc Biasing of IC-Type Amplifiers / 539
3.11   Push–Pull/Parallel Amplifiers / 539
3.12   Power Amplifiers / 542
    3.12.1   Example 1: 7-W Class C BJT Amplifier for 1.6 GHz / 552
    3.12.2   Example: A Highly Efficient 3.5 GHz Inverse Class-F GaN
            HEMT Power Amplifier / 565
    3.12.3   Linear Amplifier Systems / 575
    3.12.4   Impedance Matching Networks Applied to RF Power
            Transistors / 585
    3.12.5   Example 2: Low-Noise Amplifier Using Distributed
            Elements / 606
    3.12.6   Example 3: 1-W Amplifier Using the CLY15 / 612
    3.12.7   Example 4: 90-W Push–Pull BJT Amplifier at 430 MHz / 617
    3.12.8   Quasiparallel Transistors for Improved Linearity / 621
    3.12.9   Distribution Amplifiers / 622
    3.12.10 Stability Analysis of a Power Amplifier / 623
References / 631
Further Reading / 635

4   **Mixer Design**   **637**

4.1   Introduction / 637
4.2   Properties of Mixers / 640
    4.2.1   Conversion Gain/Loss / 640
    4.2.2   Noise Figure / 642

4.2.3　Linearity / 650
4.2.4　LO Drive Level / 652
4.2.5　Interport Isolation / 652
4.2.6　Port VSWR / 652
4.2.7　dc Offset / 654
4.2.8　dc Polarity / 654
4.2.9　Power Consumption / 654
4.3　Diode Mixers / 654
4.3.1　Single-Diode Mixer / 655
4.3.2　Single-Balanced Mixer / 664
4.3.3　Diode-Ring Mixer / 666
4.4　Transistor Mixers / 685
4.4.1　BJT Gilbert Cell / 686
4.4.2　BJT Gilbert Cell with Feedback / 689
4.4.3　FET Mixers / 697
4.4.4　MOSFET Gilbert Cell / 700
4.4.5　GaAsFET Single-Gate Switch—Resistive Mixer / 701
References / 725
Further Reading / 726

5　RF/Wireless Oscillators　　　　　　　　　　　　　　　　　　**727**

5.1　Introduction of Frequency Control / 727
5.2　Background / 727
5.3　Oscillator Design / 728
5.3.1　Basics of Oscillators / 730
5.4　Oscillator Circuits / 744
5.4.1　Hartley / 744
5.4.2　Colpitts / 745
5.4.3　Clapp–Gouriet / 745
5.5　Design of RF Oscillators / 746
5.5.1　General Thoughts on Transistor Oscillators / 746
5.5.2　Two-Port Microwave/RF Oscillator Design / 749
5.5.3　Ceramic-Resonator Oscillators / 754
5.5.4　Using a Microstrip Inductor as the Oscillator Resonator / 757
5.5.5　Hartley Microstrip Resonator Oscillator / 763
5.5.6　Crystal Oscillators / 763
5.5.7　Voltage-Controlled Oscillators / 768
5.5.8　Diode-Tuned Resonant Circuits / 770
5.5.9　Practical Circuits / 775
5.6　Noise in Oscillators / 781
5.6.1　Linear Approach to the Calculation of Oscillator Phase Noise / 781
5.6.2　Phase-Noise Analysis Based on the Feedback Model / 789
5.6.3　AM-to-PM Conversion / 792
5.6.4　Numerically Optimized Oscillators / 800
5.7　Oscillators in Practice / 803
5.7.1　Oscillator Specifications / 803
5.7.2　More Practical Circuits / 808

5.8   Phase-Noise Improvements of Integrated RF and Millimeterwave
      Oscillators / 814
     5.8.1   Abstract / 814
     5.8.2   Review of Noise Analysis / 815
     5.8.3   Workarounds / 818
     5.8.4   Reduction of Flicker Noise / 819
     5.8.5   Applications to Integrated Oscillators / 820
     5.8.6   Summary / 826
References / 826
Interesting Patents / 828
Further Reading / 828

6 Wireless Synthesizers   **831**

6.1   Introduction / 831
6.2   Phase-Locked Loops / 831
     6.2.1   PLL Basics / 831
     6.2.2   Phase-Frequency Comparators / 834
     6.2.3   Filters for Phase Detectors Providing Voltage Output / 845
     6.2.4   Charge-Pump-Based Phase-Locked Loops / 850
6.3   How to Do a Practical PLL Design Using CAD / 859
6.4   Fractional-N-Division PLL Synthesis / 864
     6.4.1   The Fractional-N Principle / 864
     6.4.2   Spur-Suppression Techniques / 866
6.5   Direct Digital Synthesis / 871
References / 879
Interesting Patents / 880
Further Reading / 881

Index   **883**

FOREWORD

Twelve years have passed since the publication of the first edition of *RF/Microwave Circuit Design for Wireless Applications*. Year 2000 was still the dawn of the wireless era; mobile telephony was about to change from highly expensive business equipment to a common ubiquitous consumer article. Chipsets for WiFi and GPS were available or under development but still to be deployed on the mass market. To remember those days, we kept the statement on the first page of the first chapter, saying that we concluded that 30% of the passengers waiting at an airport were on the air. Today, I would expect this number to be well beyond 100%, accounting for people being connected to the Internet by multiple devices while talking on their cell phone.

Other changes are also obvious requiring the book to be enhanced and partly rewritten. Semiconductor technology and device modeling have advanced rapidly. The first edition still discussed GaAs MESFET as an important device, and stated that CMOS is still too slow for wireless applications. Therefore, this edition had to account for these advances, and now discusses CMOS and CMOS circuits, BiCMOS, and HBTs on GaAs and SiGe, as well as GaN HEMTs.

It also happened in the last 15 years that the semiconductor branches of the major technology companies became independent companies. Just to name a few: chip-makers Siemens, Motorola, Philips, and Hewlett-Packard are now Infineon, Freescale, NXP, and Avago. On the other hand, most of the circuit design principles did not change much. We decided in many cases to rely on the same commercial circuit examples that were already discussed in the first edition. The majority of the example products are still on the market.

This book also discusses the GSM concept in detail. Important highlights in this new edition are power amplifiers with linearization. (This topic is discussed in great detail.) In addition to this, nonlinear noise in mixers and oscillators are new topics as is the treatment of nonlinear noise in cross-coupled oscillators that play an important role, and these have been covered as outlined above. The majority of the chapters have been updated, but the topics listed above are all new and fully covered.

Last but not the least, there was a change in the authors of this book. Author David P. Newkirk, whose valuable language skills as a technical writer ensured the high quality of the first edition, was unavailable for the second edition. However, I was fortunate to have Dr.-Ing. Matthias Rudolph join me in the revision of this book. He had a solid background in microwave engineering at Ferdinand Braun Institut in Berlin, Germany where he was in a management position responsible for device modeling and low-noise components. In the fall of 2009, he was appointed the Ulrich-L.-Rohde tenure-tracking Professor for RF and Microwave Techniques at Brandenburg University of Technology, Cottbus, Germany. His experience and dedication to this effort gives this new edition of the book an exceptionally wide base.

Finally, I would like to appreciate the support from the numerous companies and individuals that allowed us to use their images, datasheets, and technical papers as examples. Special thanks also to our publisher John Wiley & Sons, especially to George Telecki for his ongoing support and his patience.

ULRICH L. ROHDE

Marco Island, Florida
Fall 2012

PREFACE

When I started 2 years ago to write a book on wireless technology—specifically, circuit design—I had hoped that the explosion of the technology had stabilized. To my surprise, however, the technology is far from settled, and I found myself in a constant chase to catch up with the latest developments. Such a chase requires a fast engine like the Concorde.

In the case of this somewhat older technology, speed still has not been surpassed by any other commercial approach. This tells us there is a lot of design technology that needs to be understood or modified to handle today's needs. Because of the very demanding calculation effort for the circuits, this book makes heavy use of the most modern CAD tools. Hewlett-Packard[1] was kind enough to provide us with a copy of their advanced design system (ADS), which also comes with matching synthesis and a wideband CDMA library. Unfortunately, some of the mechanics of getting us started on the software collided with the already delayed schedule of this book, and we were only in a position to reference their advanced capability

---

[1] Now Agilent Technologies.

and not really demonstrate it. The use of this software, including the one from Eagleware, which was also provided to us, needed to be deferred to the next edition of this book. To meet time constraints and give a consistent presentation, we decided to stay with the Ansoft tools. One of the most time-consuming efforts was the actual modeling job, since we wanted to make sure all circuits would work properly. There are too many publications showing incomplete or nonworking designs.

On the positive side, trade journals give valuable insight in state-of-the-art designs, and I would recommend to all engineers to get a free subscription to them. Some of the major ones include *Applied Microwave & Wireless, Electronic Design, Electronic Engineering Europe, Microwave Journal, Microwaves & RF, Microwave Product Digest (MPD), RF Design, and Wireless Systems Design.*

There are also several conferences that have excellent proceedings, which can be obtained either in book or CD form: GaAs IC Symposium (annual; sponsored by IEEE-EDS, IEEE-MTT), IEEE International Solid-State Circuits Conference (annual), and IEEE MTT-S International Microwave Symposium (annual).

There may be other useful conferences along these lines that are being announced in the trade journals mentioned above, such as in England, Holland, and Germany, and workshops associated with conferences, such as the recent "Designing RF Receivers for Wireless Systems" associated with the IEEE MTT-S.

Other useful tools include courses such as *Introduction to RF/MW Design*, a four-day short course offered by Besser Associates.

Wireless design can be split into the digital part, which has to do with the various modulation and demodulation capabilities, advantages and disadvantages, and many analog technologies, of which most of this book is composed.

The analog part is complicated by the fact that we have three competing technologies. Given the fact that cost, space, and power consumption are issues for hand-held and battery-operated applications, CMOS has been a strong contestant in the area of cordless telephones because of the relaxed signal-to-noise-ratio specifications compared with cellular telephones. CMOS is much noisier than bipolar and GaAs technologies. One of the problems then is the input/output stage at UHF/SHF frequencies. Here, we find a fierce battle between silicon-germanium (SiGe) transistors and GaAs technology. Most prescalers are bipolar and most power amplifiers are based on GaAs FETs or LDMOS transistors for base stations. The most competitive technologies are the SiGe transistors and, of course, GaAs, the latter being the most expensive of the three mentioned. In the silicon-germanium area, IBM and Maxim seem to be the leaders, with many trying to catch up.

Another important issue is how to differentiate between hand-held or battery-operated applications and base stations. Most designers, who are tasked to look into battery-operated devices, ultimately resort to using available integrated circuits, which seem to change every 6–9 months, and new offerings occur. Given the multiple choices, we have not yet seen a systematic approach of how to select the proper IC families and their members. We have therefore decided to show some guidelines for the design applications of the ICs, mainly focusing on high-performance applications. In the case of high-performance applications, low power consumption is not that big an issue; dynamic range in its various forms tends to be more important. Most of these circuits are designed in discrete portions or use discrete parts. Anyone who has a reasonable antenna and has a line of sight to New York City, with the antenna connected to a spectrum analyzer, will immediately understand this. Between telephones, both cordless and cellular, high-powered pagers, and other services, the spectrum analyzer will be overwhelmed by these signals. IC applications for handsets and other applications already value their parts as "good." Their third-order intercept points are better than −10 dBm, while the real professional having to design a fixed station is looking for at least +10 dBm, if not more. This applies not only to amplifiers but also to mixer and oscillator performances. We, therefore, decided to give examples of this dynamic range. The following brief survey of current ICs has been assembled for the purpose of showing typical specifications that have been assembled to show the practical needs. It is useful that large companies make both the cellular telephones and integrated circuits or their discrete implementation for base stations. We strongly believe that the circuits selected by us will be useful for all applications.

Chapter 1, as mentioned, is an introduction to the digital modulations that form the foundation of wireless radiocommunication and its performance evaluation. We decided to leave the information regarding actual implementation to more qualified individuals. Since the standards for these modulations are still in a state of flux, we felt that it would not be possible to attack all angles. Chapter 1 contains some very nice material from various sources including tutorial material from my German company, Rohde & Schwarz, in Munich—specifically, from the digital modulation portion of their 1998 *Introductory Training for Sales Engineers* CD. *Note:* On a few rare equations, we have used either a picture or an equation more than once so that the reader need not refer to a previous chapter for full understanding of a discussion.

Chapter 2 is a comprehensive introduction into the various semiconductor technologies that enables the designer to make an educated decision. Relevant material such as PIN diodes has also been covered. In many applications, the transistors are being used close to their electrical limits, such as a combination of low voltage and low current. The f_T dependency, noise figure, and large-signal performance have to be evaluated. Another important application for diodes is their use as switches, as well as variable capacitances frequently referred to as tuning diodes. In order to better understand what the various parameters of semiconductors mean, we have included a variety of datasheets and some small applications showing which technology is best for what application. In linear applications, noise figure is extremely important; in nonlinear applications, the distortion products need to be known. Therefore, this chapter also includes not only the linear performance of semiconductors but also their nonlinear behavior, including even some details on parameter extraction. Given the number of choices the designer has today and the frequent lack of complete data from manufacturers, these are also important issues.

Chapter 3, the longest chapter, has the most detailed analysis and guidelines for discrete and integrated amplifiers providing deep insight into the semiconductor performance and circuitry necessary to get the best results from the devices. We deal with the properties of the amplifiers, gain stability, and matching, evaluated one-, two-, and three-stage amplifiers with internal dc coupling and feedback as are frequently found in integrated circuits. In doing so, we also provide examples of ICs currently in the market, knowing that every six months more sophisticated devices will appear. Another important topic in this chapter is the choice of bias point and matching for digital signal handling, and we provide insight into such complex issues as the adjacent channel power ratio, which is related to a form of distortion caused by the amplifier in its particular operating mode. To connect these amplifiers, impedance matching is a big issue, and we evaluate some useful couplers and broadband matching circuits useful to these high frequencies. Finally, we provide a tracking filter as preselector, using tuning diodes.

Chapter 4 is a detailed analysis of the available mixer circuits that are applicable to the wireless frequency range. The design also is supplied with the necessary mathematics to calculate the difference between insertion loss and noise figure, and receives insight into the differences between passive and active mixers, additive and multiplicative mixers, and other useful hints. We have also added some very clever circuits from companies such as Motorola and Siemens, as they are available as ICs.

Chapter 5, the oscillator section, is a logical next step to be considered, as many amplifiers turn out to oscillate. After a brief introduction explaining why voltage-controlled oscillators (VCOs) are needed, we cover the necessary conditions for oscillation and its resulting phase noise for various configurations, including microwave oscillators and the very important ceramic-resonator-based oscillator. This chapter walks the reader through the various noise-contributing factors and the performance differences between discrete and integrated oscillators and their performance. Here too, a large number of novel circuits are covered.

Chapter 6 deals with the frequency synthesizer, which depends heavily on the oscillators as shown in Chapter 5, and different system configurations to obtain the best performance. All components of a synthesizer, such as loop filters and phase frequency discriminators, and their actual performance are evaluated. Included are further applications for commercial synthesizer chips, and, of course, the direct digital frequency synthesizer as well as the fractional-N-division synthesizer principles are covered. The fractional-N-division synthesizer is probably one of the most exciting implementations of synthesizers, and we added interesting patents for those interested in coming up with their own design.

I would like to thank my co-author, David Newkirk, for the enormous effort he put into making this project possible. Not only does he have a wealth of information as to practical applications, but he has also worked as a Professional Editor for many years and was really the key factor in putting this book together. Finally, I would like to thank the many engineers from Ansoft, Alpha Industries, Motorola, National Semiconductor, Philips, Rohde & Schwarz, and Siemens (now Infineon) for providing current information and being able to get permission to reproduce some of the excellent material.

In the area of permissions, National Semiconductor has specifically asked us to include the following passage, which applies to all their permissions.

LIFE SUPPORT POLICY

NATIONAL'S PRODUCTS ARE NOT AUTHORIZED FOR USE AS CRITICAL COMPONENTS IN LIFE SUPPORT DEVICES OR SYSTEMS WITHOUT THE EXPRESS WRITTEN APPROVAL OF THE PRESIDENT OF NATIONAL SEMICONDUCTOR CORPORATION.

As used herein:

1. Life support devices or systems are devices or systems which, (a) are intended for surgical implant into the body, or (b) support or sustain life, and whose failure to perform, when properly used in accordance with instructions for use provided in the labeling, can be reasonably expected to result in a significant injury to the user.
2. A critical component is any component of a life support device or system whose failure to perform can be reasonably expected to cause the failure of the life support device or system, or to affect its safety or effectiveness.

I am also grateful to John Wiley & Sons, specifically, George Telecki, for tolerating the several slips in schedule, which were the result of the complexity of this effort.

Finally, I would like to dedicate this book to Professor Vittorio Rizzoli, who has been instrumental in the development of the powerful harmonic-balance analysis tool, specifically Microwave Harmonica, which is part of Ansoft's Serenade Design Environment. Most of the success had by Compact Software, now part of Ansoft, continues to be based on his far-reaching contributions.

ULRICH L. ROHDE

Upper Saddle River, New Jersey
March 2000

1

INTRODUCTION TO WIRELESS CIRCUIT DESIGN

1.1 INTRODUCTION

Wireless circuits are not that different from commonly known two-way radio, television, and broadcast arrangements. Some of them require high linearity in modulation (TV picture); some work via relay stations (two-way radio). The real differences lie in the fact that the cell sizes are much smaller, and that in most cases we attempt multiple channel use (reuse) using time-division multiplex, spread spectrum, or some other efficient means of reducing the bandwidth required for communication. One can argue that the wireless circuits include simple devices, such as garage-door openers and wireless keys for automobiles (we have seen many cases where strong interfering signals prevented the car owners from reclaiming their cars until the interfering signal disappeared). Another longtime favorite is cordless telephones: initially, 50-MHz models with essentially no privacy protection; later, more sophisticated models that operate at 900 MHz; and now, dual-band designs that use 900 MHz and 2.4 GHz.

The largest wireless growth area is probably the cellular telephones. The two major applications are the handsets, common referred to as cell phones or occasionally as "handies," and the base stations. The base stations have many more problems with large-signal-handling linearity at high power, although handset users may run into similar problems. An example of this is the waiting area of an airport, where many travelers are trying to conduct last-minute business; in one instance, we concluded that about 30% of all the people present were on the air! It would have been fun to evaluate this receiver-hostile environment with a spectrum analyzer.

From such use comes anxiety factors, the lesser of which is "When will my battery die?"—a spare battery tends to help—and the greater of which the ongoing question, "Will this cell-phone transmitter harm my body?" A brief comment for the self-proclaimed experts in this area: a 50–100-kW TV transmitter, specifically its video or picture portion, connected

RF/Microwave Circuit Design for Wireless Applications, Second Edition. Ulrich L. Rohde and Matthias Rudolph.
© 2013 John Wiley & Sons, Inc. Published 2013 by John Wiley & Sons, Inc.

to a high-gain antenna, emits levels of energy in line-of-sight paths that by far exceed the pulsed energy from a cell phone. Specifically, the duration of energy is significantly smaller, and the absolute energy is more than a thousandfold higher, than the RF supposedly harming us from the cellular phone. Hand-held two-way radios have been used for the last 30 years or so by police and other security interests, operating in the frequency range from 50 to 900 MHz with antennas close to the users' heads, and there are no known cases of cancer or any other illnesses caused by these handheld radios. Recent studies in England, debatably or not, showed that the reaction-time level of people using cell phones drastically actually increased—but then there are always the skeptics and politically motivated who ignore the facts, try to influence the media, and have their 15 min of fame (as Andy Warhol used to say).

The question if cellphone radiation is harmful for the user is a question of ongoing research. However, the issue is quite complicated, and the results of different studies are even contradictory, or hard to reproduce. Luckily, Professor James C. Lin, from the University of Illinois-Chicago, takes the effort to write review articles on recent studies in his series *Health Effects* in the *IEEE Microwave Magazine* since 2001. This series provides us with the respective information and it is not only comprehensive but also written for engineers. From the vast number of articles, we just cite the first [1] and the one on the multinational study of the possible relation of tumors and cell phone use [2]. The other articles are easily found by a IEEEXplore database search.

Concerning radiation, Figure 1.1 shows the simulated near-field radiation of a Motorola mobile phone. The antenna is hidden inside the phone for optical reasons, and it is most likely a radiating structure that looks quite different from the linear antennas that were used in the past. The field is mainly concentrated at the top of the phone, where it is unlikely that

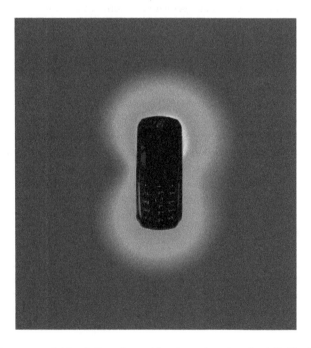

Figure 1.1 Antenna near-field radiation of a mobile phone (courtesy Prof. D. Manteuffel, Univ. Kiel, Germany).

the user's hand attenuates the transmit and receive power, and it is also directed away from the head. If the user will find a "warm" sensation, it will have more to do with the efficiency of the RF power amplifiers heating up the case than the effect of radiation, especially when many frontends are in use, like GSM, GPS, and WiFi.

With this introduction in place, we will first take a look at a typical UHF/SHF transceiver and explain the path from the microphone to the antenna and back. After this, we will inspect the radio channel and its effect on various methods of digital modulation. Analysis of wireless receivers and transmitters will be next, followed by a look at available building blocks and how they affect the overall system. To validate proper system operation, a fairly large number of measurements and tests must be performed, and conveying their purpose and importance will necessitate the definition of a number of system characteristics and concepts, such as dynamic range. Finally, after this is done, we will look at the issue of wireless system testing. Again, we intend to give guidance applicable to battery-operated, hand-held operation as well as high-powered base stations.

1.2 SYSTEM FUNCTIONS

A cellular telephone is a hybrid between a double-sideband and FM (PM) transceiver. The actual transmission is not continuous, but is pulsed, and because of the pulse spectrum there is a signal bandwidth concern due to keying transients, not unlike intermodulation products of an SSB transceiver cluttering up adjacent channels. The cellular telephone is also a linear transceiver in the sense that its signal-handling circuitry must be sufficiently amplitude- and phase-linear to preserve the modulation characteristics of the AM/PM hybrid emissions it transmits and receives. Containing such an emission's spectral regrowth, which affects operation on adjacent channels, is not unlike the linearity requirements we encounter in single-sideband (SSB) transceivers—requirements so stringent that amplifiers must be run nearly in Class A to meet them. The time division multiple access (TDMA) operating mode, which allows many stations to use the same frequency through the use of short, precisely timed transmissions, requires a system that transmits with a small duty cycle, putting much less thermal stress on a power amplifier than continuous operation. Power management, including a sleep mode, is another important issue in handset design.

Figure 1.2 shows the block diagram of a hand-held transceiver. This example shows an example chipset for a mobile phone that can operate in four 2G GSM (800/900 and 1800/1900 MHz) bands and in one 3G WCDMA (UMTS) band. Many functionalities are integrated into a few chips.

For those not too familiar with transceivers, here is a "walk" through the block diagram. The RF signal is intercepted by the antenna is fed to a switch that selects whether the GSM or WCDMA path is in use.

If we follow the GSM path first, the signal is fed to a diplexer/switch (called switchplexer in the figure). Since GSM operates transmits in bursts, the RF front end is switched between transmit and receive. Then follows a band-selection filter for each of the four GSM bands, and low-noise amplifiers. The signal is then directly downconverted to baseband by a quadrature demodulator using a local oscillator signal generated on chip. The baseband signal is low-pass filtered, amplified, and converted to the digital domain for further processing.

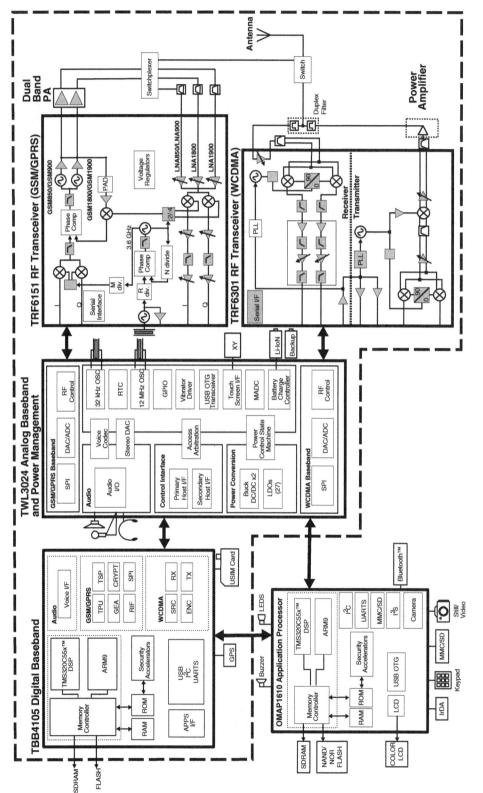

Figure 1.2 Block diagram of a handheld cellular telephone transceiver (courtesy Texas Instruments).

4

The WCDMA path is different from the GSM path at first sight since it requires a duplex filter instead of the switch/duplexer used in GSM. The reason is that in UMTS, the transmitter is in full-duplex operation and transmits and receives at the same time. The duplex filter therefore has to separate the two paths, especially it is has to prevent crosstalk at the transmit frequency that could overload the input or the receiver. Again there is a combination of band-selection filter and low-noise amplifier before the received signal is downconverted into the baseband, and its I and Q channels are converted to the digital domain.

A nice overview about DSP in "readable" form is Ref. [3].

The GSM transmit portion consists of an on-chip synthesizer that is modulated. Both receive and transmit frequencies are controlled by a miniature temperature-compensated crystal oscillator (TCXO). The output of the voltage-controlled oscillator (VCO) is then amplified and fed to the antenna through the same switch/duplexer as the receive portion.

The WCDMA transmit branch relies on the I–Q modulator architecture. In contrast to GSM the uses the GMSK coding scheme with constant envelope, WCDMA signals show high peak-to-average ratios and need good control over amplitude and phase of the RF signal.

Figure 1.2 also reveals which part of the functionality can be integrated in CMOS today. In fact, it is almost everything, so let us talk about what is still off-chip and what will not be integrated in CMOS anytime soon.

- The antenna, for obvious reasons. Unlike early mobile phones, radiating parts are integrated inside the phone that do not even look like classical antennas.
- Band-selection filters need very low insertion loss and high selectivity. The low loss is required to obtain high sensitivity. This type of high-Q filters are commonly realized as surface-acoustic wave (SAW) or bulk acoustic resonance (BAR) filters on piezo crystals.
- The antenna switch must be able to carry the high transmit power, provide low insertion loss, and high isolation. Besides of pin diodes, GaAs HEMT devices are often used.
- The power amplifiers need to provide high powers, high linearity, and a maximum of power-added efficiency. The technology of choice to date is InGaP/GaAs HBTs.

A mobile phone transmitter is more involved than, say, a WLAN or DECT transmitter, as it requires to operate at multiple frequencies according to different standards. And since it needs to operate over much wider distances and at higher powers. Less critical transceivers might be fully integrated in CMOS or in BiCMOS. However, an in-depth discussion of different transceiver architectures is beyond the scope of this book, and we stop at this point. A nice overview of different architectures is found in Ref. [4].

1.3 THE RADIO CHANNEL AND MODULATION REQUIREMENTS

1.3.1 Introduction

The transmission of information from a fixed station to a mobile is considerably influenced by the characteristics of the radio channel. The RF signal arrives at the receiving antenna not only on the direct path but is normally reflected by natural and artificial obstacles in its way. Consequently, the signal arrives at the receiver several times in the form of echoes that are

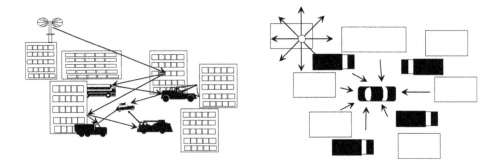

Figure 1.3 Mobile receiver affected by fading.

superimposed on the direct signal (Figure 1.3). This superposition may be an advantage as the energy received in this case is greater than in single-path reception. This feature is made use of in the DAB single-frequency network. However, this characteristic may be a disadvantage when the different waves cancel each other under unfavorable phase conditions. In conventional car radio reception, this effect is known as fading. It is particularly annoying when the vehicle stops in an area where the field strength is reduced because of fading (e.g., at traffic lights). Additional difficulties arise when digital signals are transmitted. If strong echo signals (compared to the directly received signal) arrive at the receiver with a delay in the order of a symbol period or more, time-adjacent symbols interfere with each other. In addition, the receive frequency may be falsified at high vehicle speeds because of the Doppler effect so that the receiver may have problems to estimate the instantaneous phase in the case of angle-modulated carriers. Both effects lead to a high symbol error rate even if the field strength is sufficiently high. Radio broadcasting systems using conventional frequency modulation are hardly affected by these interfering effects. If an analog system is replaced by a digital one that is expected to offer advantages over the previous system, it has to be ensured that these advantages—for example, better AF S/N and the possibility to offer supplementary services to the subscriber—are not at the expense of reception in hilly terrain or at high vehicle speeds because of extreme fading.

For this reason, a modulation method combined with suitable error protection has to be found for mobile reception in a typical radio channel, which is immune to fading, echo, and Doppler effects.

With a view to this, more detailed information on the radio channel is required. The channel can be described by means of a model. In the worst case, which may be the case for reception in built-up areas, it can be assumed that the mobile receives the signal on several indirect paths but not on a direct one. The signals are reflected, for example, by large buildings; the resulting signal delays are relatively long. In the vicinity of the receiver, these paths are split up into a great number of subpaths; the delays of these signals are relatively short. These signals may again be reflected by buildings but also by other vehicles or natural obstacles like trees. Assuming the subpaths being statistically independent of each other, the superimposed signals at the antenna input cause considerable time- and position-dependent field-strength variations with an amplitude obeying the Rayleigh distribution (Figures 1.4 and 1.5).

If a direct path is received in addition, the distribution changes to the Rice distribution and finally, when the direct path becomes dominant, the distribution follows the Gaussian distribution with the field strength of the direct path being used as the center value.

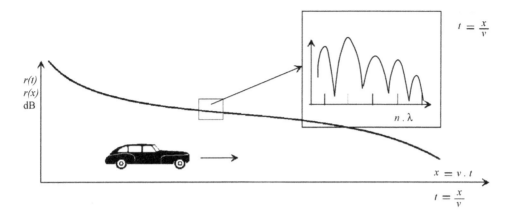

Figure 1.4 Receive signal as a function of time or position.

In a Rayleigh channel, the bit error rate increases dramatically compared to the BER in an additive white Gaussian noise (AWGN) channel produces (Figure 1.6).

1.3.2 Channel Impulse Response

This scenario can be demonstrated by means of the channel impulse response. Let us assume that a very short pulse of extremely high amplitude [in the ideal case a Dirac pulse $\delta(t)$] is sent by the transmitting antenna at a time $t_0 = 0$. This pulse arrives at the receiving antenna direct and in the form of reflections with different delays τ_i and different amplitudes because of path losses. The impulse response of the radio channel is the sum of all received pulses

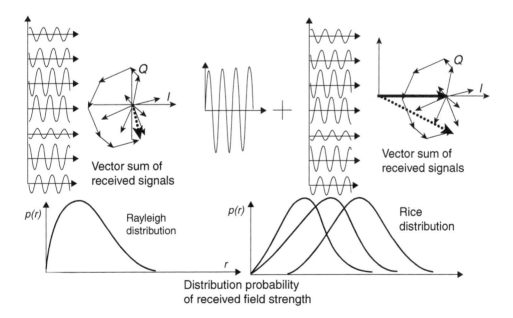

Figure 1.5 Rayleigh and Rice distribution.

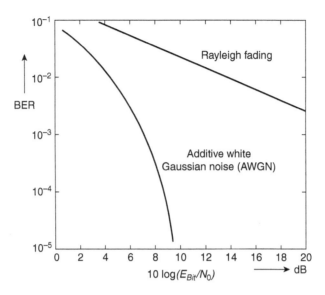

Figure 1.6   BER in a Rayleigh channel.

(Figure 1.7). Since the mobile receiver and also some of the reflecting objects are moving, the channel impulse response is a function of time and of delays τ_i, that is, it corresponds to

$$h(t, \tau) = \sum_N a_i \delta(t - \tau_i) \tag{1.1}$$

This shows that delta functions sent at different times t cause different reactions in the radio channel.

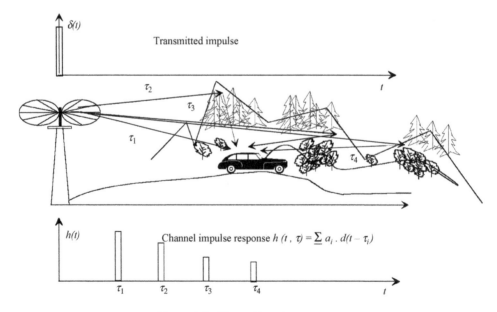

Figure 1.7   Channel impulse response.

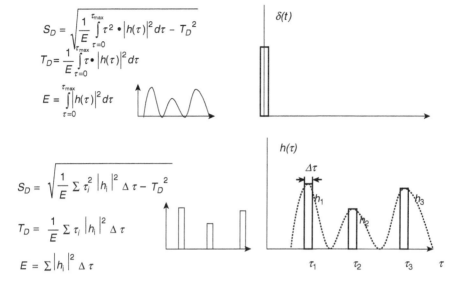

$$S_D = \sqrt{\frac{1}{E} \int_{\tau=0}^{\tau_{max}} \tau^2 \cdot |h(\tau)|^2 \, d\tau - T_D^2}$$

$$T_D = \frac{1}{E} \int_{\tau=0}^{\tau_{max}} \tau \cdot |h(\tau)|^2 \, d\tau$$

$$E = \int_{\tau=0}^{\tau_{max}} |h(\tau)|^2 \, d\tau$$

$$S_D = \sqrt{\frac{1}{E} \sum \tau_i^2 \, |h_i|^2 \, \Delta\tau - T_D^2}$$

$$T_D = \frac{1}{E} \sum \tau_i \, |h_i|^2 \, \Delta\tau$$

$$E = \sum |h_i|^2 \, \Delta\tau$$

Figure 1.8 Calculation of delay spread.

In many experimental investigations, different landscape models with typical echo profiles were created.

The most important are

- rural area (RA),
- typical urban area (TU),
- bad urban area (BA), and
- hilly terrain (HT).

The channel impulse response informs on how the received power is distributed to the individual echoes. A parameter, the "delay spread" can be calculated from the channel impulse response, permitting an approximate description of typical landscape models (Figure 1.8).

The delay spread also roughly informs on the modulation parameters carrier frequency, symbol period, and duration of guard interval, which have to be selected in relation to each other. If the receiver is located in an area with a high delay spread (e.g., in hilly terrain), echoes of the symbols sent at different times are superimposed when broadband modulation methods with a short symbol period are used. In the case of DAB, this problem is aggravated by the use of single-frequency networks. An adjacent transmitter emitting the same information on the same frequency has the effect of an artificial echo (Figure 1.9).

A constructive superposition of echoes is only possible if the symbol period is much greater than the delay spread. The following holds:

$$T_s > 10T_d \tag{1.2}$$

This has the consequence that relatively narrowband modulation methods have to be used. If this is not possible, channel equalizing is required.

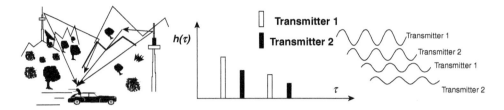

Figure 1.9 Artificial and natural echoes in the single-frequency network.

For the channel equalizing, a continuous estimation of the radio channel is necessary. The estimation is performed with the aid of a periodic transmission of data known to the receiver. In networks according to the GSA standards, a midamble consisting of 26 bits—the training sequence—is transmitted with every burst. The training sequence corresponds to a characteristic pattern of *I/Q* signals that is kept in a memory in the receiver. The baseband signals of every received training sequence are correlated with the stored ones. From this correlation, the channel can be estimated, the properties of the estimated channel will then be fed to the equalizer (Figure 1.10).

The equalizer uses the Viterbi algorithm (maximum sequence likelihood estimation) for the estimation of the phases that most likely have been sent at the sampling times. From these phases, the information bits are calculated (Figure 1.11). A well-designed equalizer then will superimpose the energies of the single echoes constructively, so that the result in an area, where the echoes are not to much delayed, delay times up to 16 μs have to be tolerated by a receiver, are better than in an area with no significant echoes (Figure 1.12).

Remaining bit errors are eliminated using another Viterbi decoder at the transmitter convolutionally encoded data sequences.

The ability of a mobile receiver to work in an hostile environment such as the radio channel with echoes must be proven. The test is performed with the aid of a fading simulator.

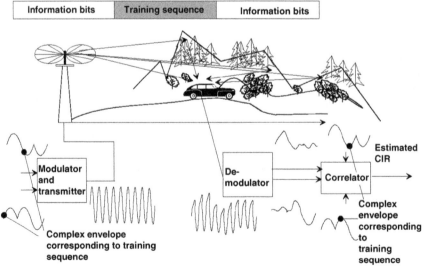

Figure 1.10 Channel estimation.

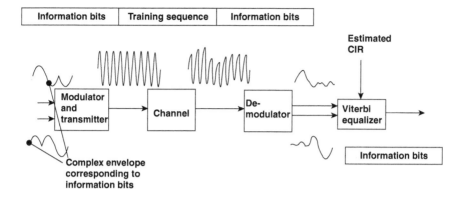

| Information bits | Training sequence | Information bits |

Figure 1.11   Channel equalization.

The fading simulator simulates different scenarios with different delay times and different Doppler profiles. A signal generator generates undistorted *I/Q* modulated RF signals that are downconverted into the baseband. Here, the *I/Q* signals are digitized and split into different channels where they are delayed and attenuated and where Doppler effects are superimposed. After combination of these distorted signals at the output of the baseband section of the simulator, these signals modulate the RF carrier that is the test signal for the receiver under test (Figure 1.13).

To make the tests comparable, GSM recommends typical profiles, for example

- rural area (RAx),
- typical urban (TUx), and
- hilly terrain (HTx).

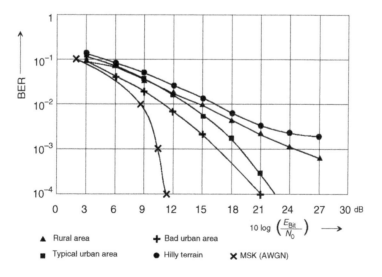

Figure 1.12   BERs after the channel equalizer in different areas.

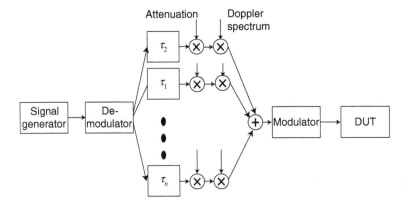

Figure 1.13 Fading simulator.

where number and strengths of the echoes and the Doppler spectra are prescribed (Figure 1.14).

1.3.3 Doppler Effect

Since the mobile receiver and some of the reflecting objects are in motion, the receive frequency is shifted because of the Doppler effect. In the case of single-path reception, this shift is calculated as follows:

$$f_d = \frac{v}{c} f_c \cos \alpha \qquad (1.3)$$

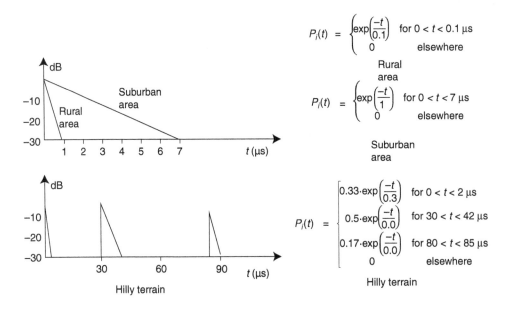

Figure 1.14 Typical landscape profiles.

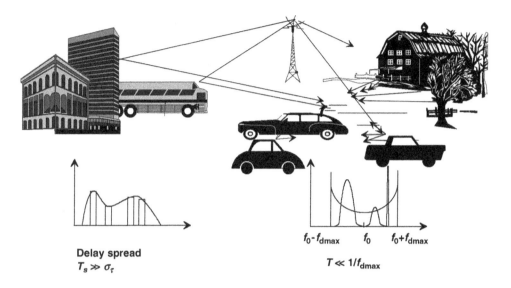

Figure 1.15 Doppler spread.

where v = speed of vehicle, c = speed of light, f = carrier frequency, and α = angle between v and the line connecting transmitter and receiver.

In the case of multipath reception, the signals on the individual paths arrive at the receiving antenna with different Doppler shifts because of the different angles α_i, and the receive spectrum is spread. Assuming an equal distribution of the angles of incidence, the power density spectrum can be calculated as follows:

$$P(f) = \frac{1}{\pi} \frac{1}{\sqrt{f_d^2 - f^2}} \text{ for } |f| < |f_d| \tag{1.4}$$

where f_d = maximum Doppler frequency.

Of course, other Doppler spectra are possible in addition to the pure Doppler shift described above, for example, spectra with a Gaussian distribution using one or several maxima. A Doppler spread can be calculated from the Doppler spectrum analogously to the delay spread (Figure 1.15).

1.3.4 Transfer Function

The FFT value of the channel impulse response is the transfer function $H(f, t)$ of the radio channel, which is also time dependent. The transfer function describes the attenuation of frequencies in the transmission channel. When examining the frequency dependence, it will be evident that the influence of the transmission channel on two sine-wave signals of different frequency becomes greater with increasing frequency difference. This behavior can be adequately described by the coherence bandwidth that is approximately equal to the reciprocal delay spread, that is

$$(\Delta f)_c = \frac{1}{T_d} \tag{1.5}$$

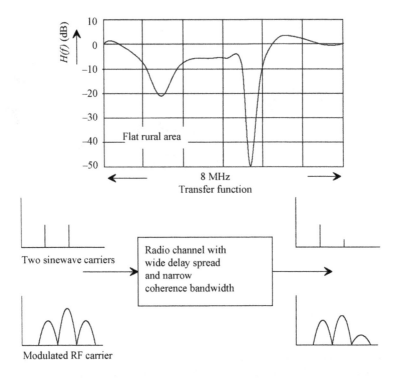

Figure 1.16 Effect of transfer function on modulated RF signals.

If the coherence bandwidth is sufficiently wide and, consequently, the associated delay spread is small, the channel is not frequency selective. This means that all frequencies are subject to the same fading. If the coherence bandwidth is narrow and the associated delay spread wide, even very close adjacent frequencies are attenuated differently by the channel. The effect on a broadband-modulated carrier with respect to the coherence bandwidth is obvious. The sidebands important for the transmitted information are attenuated to a different degree. The result is a considerable distortion of the receive signal combined with a high bit error rate even if the received field strength is high. This characteristic of the radio channel again speaks for the use of narrowband modulation methods (Figure 1.16).

1.3.5 Time Response of Channel Impulse Response and Transfer Function

The time response of the radio channel can be derived from the Doppler spread. It is assumed that the channel rapidly varies at high vehicle speeds. The time variation of the radio channel can be described by a figure, the coherence time, which is analogous to the coherence bandwidth. This calculated value is the reciprocal bandwidth of the Doppler spectrum. A wide Doppler spectrum therefore indicates that the channel impulse response and the transfer function vary rapidly with time (Figure 1.17). If the Doppler spread is reduced to a single line, the channel is time invariant. In other words, if the vehicle has stopped or moves at a constant speed in a terrain without reflecting objects, the channel impulse response and the transfer function measured at different times are the same.

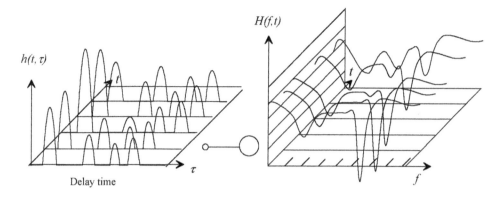

Figure 1.17   Channel impulse response and transfer function as a function of time.

The effect on information transmission will be illustrated in an example. In the case of MPSK modulation using hard keying, the transmitter holds the carrier phase for a certain period of time, that is, for the symbol period T. In the case of soft keying with low-pass-filtered baseband signals for limiting the modulated RF carrier, the nominal phase is reached at a specific time, the sampling time. In both cases, the phase error $\phi_f = f_d T_S$ is superimposed onto the nominal phase angle, which yields a phase uncertainty of $\Delta\phi = 2\phi_f$ at the receiver. The longer the symbol period the greater the angle deviation (Figure 1.18). Considering this characteristic of the transmission channel, a short symbol period of $T_s \ll (\Delta t)c$ should be used. However, this requires broadband modulation methods.

Figure 1.19 shows the field strength or power arriving at the mobile receiver if the vehicle moves in a Rayleigh distribution channel. Since the phase depends on the vehicle position, the receiver moves through positions of considerably differing field strength at different times (time dependence of radio channel). In the case of frequency-selective channels, this applies to one frequency only, that is, to a receiver using a narrowband IF filter for narrowband emissions. As Figure 1.19 shows, this effect can be reduced by increasing the bandwidth of the emitted signal and consequently the receiver bandwidth.

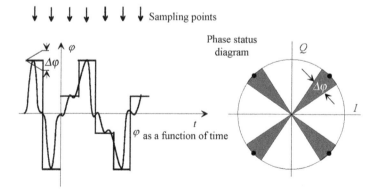

Figure 1.18   Phase uncertainty caused by Doppler effect.

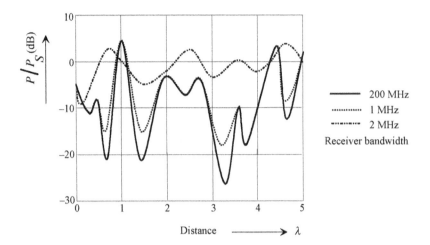

Figure 1.19   Effect of bandwidth on fading.

1.3.6 Lessons Learned

The strongly frequency-selective channel causing inadmissible distortion of the broadband-modulated carrier and channel impulse responses like those expected in a hilly terrain speak in favor of narrowband modulation methods with long symbol periods. In hilly terrain, extensively delayed echoes cause intersymbol interference when broadband modulation with short symbol periods is used. On the other hand, narrowband modulation has the disadvantage that the signals arrive at the receiver considerably attenuated and reception may be interrupted for an indefinite period of time. The burst errors occurring in digital information transmission cannot be corrected even with the most elaborate error protection methods.

These transmission interruptions can be avoided by using broadband modulation methods, but these are sensitive to strongly frequency-selective channels. Since broadband modulation is obtained through the use of short symbol periods, broadband modulation is unsuitable when greatly delayed echoes are expected.

It remains to be defined when a signal is considered a narrowband and when a broadband signal. This question shall be answered with the aid of an example. Apart from extremely narrowband analog modulation methods as are used for sound broadcasting in the longwave, mediumwave, and shortwave bands, FM sound broadcasting transmissions in the VHF bands are narrowband. In the case of digital modulation, this means that transmissions with a rate of 400 kbit/s modulated onto a carrier with a bandwidth efficiency of 1.5 (bit/s)Hz over a bandwidth of approximately 300 kHz can be regarded as narrowband transmissions according to the definition above. Consequently, a DAB signal with this gross bit rate would be a narrowband signal and suitable for transmission on the radio channel with restrictions only. Based on the experience with conventional FM broadcasting systems in large cities and hilly terrain, this was obvious from the very beginning.

Consequently, it is necessary to find ways for spreading the band artificially without reducing the bandwidth efficiency. This means that a large band must be available for the transmission of several programs, the full bandwidth being used by all the programs without mutual interference.

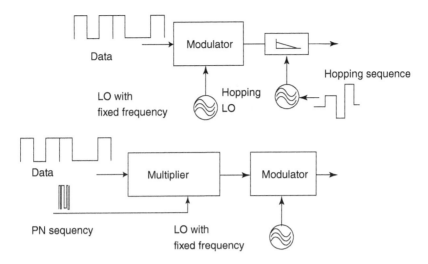

Figure 1.20   Band spreading by frequency hopping and CDMA.

Several approaches can be adopted to tackle the problem (Figure 1.20). One way would be a continuous change of the transmit and receive frequency according to a defined pattern (frequency hopping). This method is used in mobile radio, for instance, but only marginal investigations have been made in this respect for DAB.

Another possibility is to multiply the symbols of the individual programs with digital signals (pseudonoise function) using a much higher bit rate so that a higher symbol rate is obtained. In this case, the different programs are assigned different functions that must be orthogonal to each other (code division multiple access, CDMA). The "chopped " bit streams of the individual programs are modulated onto carriers of identical frequency and the modulated carriers are added. A correlation receiver knowing the pseudonoise function divides the incoming CDMA signal into the individual programs. The disadvantage is obvious. Symbol periods are very short and elaborate means will be required for compensating the intersymbol interference.

A different approach has been chosen for DAB, which does not involve continuous and elaborate channel measurements, so that it would be possible to use favorably priced receivers as are demanded in the field of consumer electronics. The method used for DAB is a multicarrier method where the information to be transmitted is spread onto many carriers using time and frequency interleaving. The terms time and frequency interleaving will be explained in the course of this discussion. The result is a broadband transmission method with long symbol periods. However, certain limitations caused by the Doppler effect will have to be accepted particularly at high carrier frequencies.

1.3.7   Wireless Signal Example: The TDMA System in GSM

1.3.7.1   Frequency Division Multiple Access (FDMA)

In analog radio systems, the trend has always been toward a more efficient utilization of the available frequency spectrum by reducing the channel spacing. The number of radio channels obtained at a channel spacing of 12.5 kHz is of course twice that obtained at 25 kHz. However, any improvement brings about its disadvantage: the narrower the channel spacing,

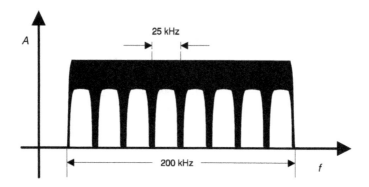

Figure 1.21 Channel spacing in broadband/narrowband systems.

the higher the required frequency accuracy and the lower the possible maximum deviation of the frequency modulation. The latter leads to a poorer transmission quality due to the lower S/N ratio. Furthermore, the gaps between the channels, which must be a number of kilohertz wide for safety reasons, also reduce the available system bandwidth (see Figures 1.21 and 1.22).

The use of an available system spectrum divided into individual frequency channels enables the user to simultaneously access a multitude of different frequencies. This multiple access is called frequency-division multiple access (FDMA). Consequently, all radio systems with a spectrum divided into channels are FDMA systems. At present, the technically useful limit is reached with a channel spacing of 10–12.5 kHz.

Advantages of FDMA.

- Simultaneous access to a given bandwidth by many subscribers.
- Increase in the number of channels through reduction of channel spacing.

Disadvantages of FDMA.

- Higher frequency accuracy required.
- Transmission quality decreasing with reduction of channel bandwidth.
- Better rejection filters required.
- One transmitter/receiver required per channel.

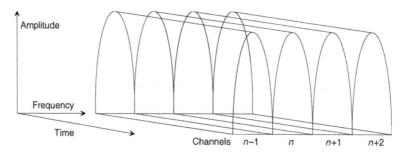

Figure 1.22 Frequency-division multiple access (FDMA).

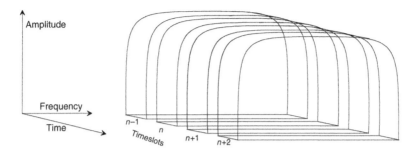

Figure 1.23   Time-division multiple access (TDMA).

1.3.7.2   Time-Division Multiple Access (TDMA)

With TDMA systems, the available bandwidth is divided into considerably fewer, and therefore wider, channels than in FDMA systems. Each of these channels is available to several subscribers quasi-simultaneously (see Figure 1.23). However, a given subscriber can use the whole channel for a very short period (timeslot) only, for the rest of the time, they have no access. This serial access of several users is repeated within a fixed time frame.

Advantages of TDMA.

- Simultaneous use of a specific bandwidth by a great number of subscribers.
- Depending on the number of available timeslots, several subscribers can be served by one transmitter/receiver.
- Transmitter and receiver are not permanently switched on (saves battery power).
- The RF section may carry out other tasks in the intervals between transmission and reception.
- Reduced susceptibility to frequency-selective fading in the case of larger channel bandwidths.

Disadvantages of TDMA.

- Accurate time synchronization of subscribers is required.
- Higher processor capacity is required.
- Broadband modulators are required.

1.3.7.3   Code-Division Multiple Access (CDMA)

The increasing use of low-priced and powerful signal processors allows a less common technique of multiple access to be employed in mass communication systems. In the case of code-division multiple access (CDMA), the whole system bandwidth is available to all subscribers at any time; that is, all send and receive simultaneously, with each using a specific code (Figure 1.24).

Logic "1" represents a certain bit sequence, logic "0" is the inversion of this sequence. The different signals are distinguished in the receiver by means of a cross-correlation of the received signal, which comprises a great number of codes, with the bit sequence expected so that the desired transmission signal can be detected.

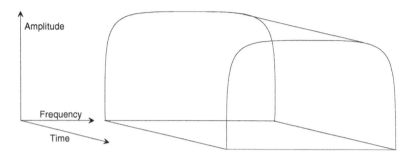

Figure 1.24   Code-division multiple access (CDMA).

Advantages of CDMA.

- Simultaneous use of a specific channel or subband many subscribers.
- Several signals can be received simultaneously by one receiver.
- Reduced susceptibility to frequency-selective fading in the case of large channel bandwidths.
- More subscribers can be served.
- Reduced costs for radio network planning.

Disadvantages of CDMA.

- Accurate time synchronization of subscribers required.
- Fast transmitter power control over a wide dynamic range.

1.3.7.4   TDMA in GSM

RF Data   In spite of the competition with other mobile radio systems, a common frequency band could be defined for GSM worldwide. All operators who signed the GSM memorandum of understanding committed themselves to install their GSM systems within the standardized frequency range. The competition for frequencies mainly affects countries using NMT900, the frequency range of which corresponds to the GSM P band. TACS also partly overlaps the GSM P band; the G1 band is completely within the TACS range. Cordless telephones operating in accordance with the CT1 standard also use the upper end of the GSM P band. CT1+ telephones, which had been assigned a frequency range below the P band years ago to protect them against GSM, have now been ousted by the G1 band. See Table 1.1.

Since each frequency channel is divided into eight timeslots, transmitter and receiver operate in an intermittent mode, with the receive and transmit time in the upper and lower channels shifted by three timeslots (Figure 1.25). Although this alternative sending and receiving scheme operates in half-duplex rather than full-duplex operation, the received signal sounds continuous to the user.

1.3.7.5   TDMA Structure

Frame and Multiframe   All GSM radio channels are organized in frames of approximately 4.62 ms duration. The frames are continuously repeated. Each frame is divided into

Table 1.1 RF Data for GSM900 and GSM1800

	GSM900		GSM1800
	P band	G1 band	
Frequency range			
Uplink (MHz)	890–915	880–890	1710–1785
(MS transmitting)			
Downlink (MHz)	935–960	925–935	1805–1880
(BTS transmitting)			
Duplex spacing (MHz)	45	45	95
Spectrum (MHz)	2×25	2×10	2×75
Frequency channels	124	49	374
Channel numbers	1–124	975–1023	512–885
(ARFCN)			
Channel spacing		200 kHz	
Modulation		GMSK with $B \times T = 0.3$	
Data transmission rate		270.833 kbit/s	
Bit duration		3.69 /μs	

eight timeslots of approximately 577 μs each. A timeslot contains an information packet, the burst. Twenty-six-type multiframes are used on all timeslots containing a traffic channel (voice and/or data), 51-type multiframes on all timeslots reserved for control channels. See Figure 1.26.

The TDMA structure uses other frame types above the multiframe level, as shown in Table 1.2.

Table 1.2 Superframes and Hyperframes

Superframe	$= 51 \times 26$ frames
	$= 1326$ frames
	$= 6.12$ s
Hyperframe	$= 2048$ superframes
	$= 2.715.648$ frames
	$= 3$ h, 29 min, 3.5 s

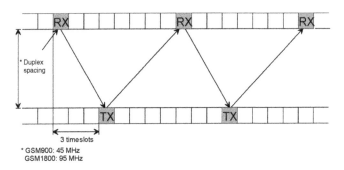

Figure 1.25 Duplex spacing of transmission and reception.

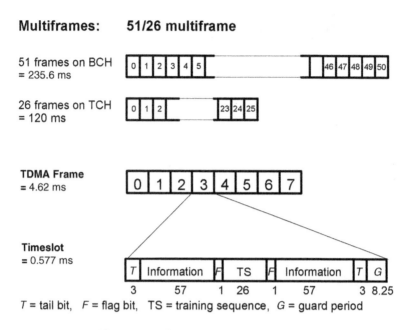

Figure 1.26   Timeslot, frame, and multiframe.

TDMA Timers   The frame number within the hyperframe is counted continually so that counting of the TDMA clock restarts after approximately 3.5 h. The frame number therefore represents a time unit in the GSM system. Similar to time counting, where the seconds are combined into minutes, hours, and days, GSM does not count the absolute frame numbers but uses timers instead. These timers are structured as follows as shown in Table 1.3.

The absolute frame number is obtained by a multiplication of the three timers. However, on certain occasions, a short version of the timers is used.

Burst Structures   Information between base station and mobile is sent in the timeslots. In each slot, a certain amount of information—that is, a burst—can be transmitted. Normally, the timeslot is occupied by the normal burst (Figure 1.27), which is used for signaling as well as for voice and data transmission.

Each part of the burst serves a specific purpose as described below.

Information Bits   The normal burst is able to transmit 2×57 information bits. Since this information can be replaced every 4.62 ms during an ongoing call, the average theoretical

Table 1.3   TDMA Timers

T1 = FN div. (26×51)	Range	0–2047
T2 = FN mod 26	Range	0–25
T3 = FN mod 51	Range	0–50
FN (Frame number)	Range	0....2715647
$FN_{max} = 51 \times 26 \times 2048 - 1$		

T	Information	F	Training sequence	F	Information	T	G
Bits 3	57	1	26	1	57	3	8.25

T = Tailbits   F = Flagbits   G = Guardtime

Figure 1.27   Normal burst.

transmission rate is

$$114\,\text{bits} \div 4.62\,\text{ms} \approx 24.7\,\text{kbit/s} \tag{1.6}$$

At the same time, this rate also represents the maximum transmission rate that can be obtained in the GSM system using one timeslot per transmission per time frame. Consequently, the transmission rate could be increased only if more than one time slot is used for the transmission.

The bit rate is much lower in the control channels; that is, the above transmission rate is only attained by the mobile station if a traffic channel has been set up. In this case, the base station and the mobile station use a signaling channel, which also uses up capacity, in addition to the voice or data channel. Table 1.4 shows the assignment of the theoretically available capacity with a traffic channel set up.

Training Sequence   In the middle of the normal burst, a 26-bit training sequence , the bit sequence of which is known to the receiver, is sent. The training-sequence code (TSC) can be one of eight different sequences. These sequences are stored in all receivers, and at the beginning of a transmission, the base transceiver station (BTS) decides on the TSC to be used. The training sequence serves two main purposes: bit synchronization and estimation of channel impulse response.

1.3.7.6 Bit Synchronization

Data transmitted via the air interface are in the asynchronous mode; that is, the receiver has to regenerate the bit clock from the data stream. To enable synchronization in the receiver, the transmitters adds synchronization bits to the information stream. Therefore, in normal data transmission, data telegrams start with a "...10101010..." sequence so that the receiver can regenerate the bit clock. A predefined bit word informs the receiver when the actual information (block synchronization) starts. A receiver synchronized in this way is able

Table 1.4   Transmission Bit Rates

	Information	Error Protection	Total
Traffic channel			22.8 kbit/s
Voice (full-rate)	13.0 kbit/s	9.8 kbit/s	
Data	2.4 kbit/s	20.4 kbit/s	
	4.8 kbit/s	18.0 kbit/s	
	9.6 kbit/s	13.2 kbit/s	
Control channels			0.95 kbit/s
Idle frame			0.95 kbit/s
Total			24.7 kbit/s

to decode the data stream online. The training sequence of the burst has to assume both synchronization tasks. Since it is in the middle of the burst, direct decoding is not possible.

Each burst must first be stored in the receiver and then decoded by postprocessing. The reason for using this method is the second task of the training sequence. Synchronization itself is carried out by means of cross-correlation; that is, the expected training sequence is compared (correlated) to the center of the received burst and to the beginning and end of the training sequence so that the bit clock is also known. A burst containing other than the expected training sequence cannot be synchronized and decoded.

1.3.7.7 Compensation of Multipath Reception
The signal from the transmitter (in Figure 1.28, BTS → MS; the same applies also in the opposite direction) arrives at the receiver not only along the direct path but also via various other paths as a result of reflection and diffraction caused by obstacles in the signal path.

Propagation conditions on these additional paths differ from those on the direct path. For instance, we can expect signals traveling via additional paths to exhibit

- longer travel times because of increased path length,
- various strengths, and
- different Doppler shifts.

Because of the different travel times, the signals arrive with a different phase at the receiving antenna. Depending on this phase, components may be canceled—that is, they may totally disappear—or added so that a high-quality signal is received for only a short period of time. RF-level variations are statistically distributed; level shifts due to fading may be as great as 40 dB.

In addition to RF-level fading, another annoying effect is encountered that, uncompensated, would make correct signal decoding rather difficult. Because of the additional distance, the signal travels via the indirect path, the signal arriving at the receiving antenna exhibits time delay of its modulation in addition to variable phase shifts. The total of all channel responses to a single transmitted pulse is called channel impulse response (CIR,

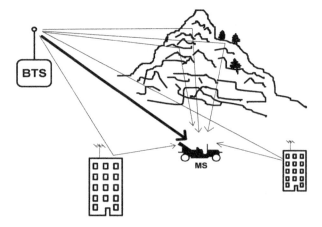

Figure 1.28 Multipath reception due to reflection and diffraction. The base transceiver station (BTS) is transmitting to the mobile station (MS).

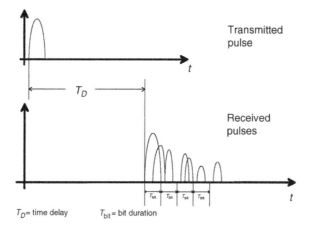

Figure 1.29   Channel impulse response.

Figure 1.29). If the indirect path is only 1 km longer, the GSM echo bit reaches the receiver later than the directly received bit and thus interferes with the next bit received. This intersymbol interference may occur over several bits in succession. With delays of up to 15 μs, differentiating the desired-signal components from echoes becomes more and more difficult. This problem can also be solved with the aid of the training sequence. The echoes on the delayed paths also contain the training sequence. The correlation used for detecting the original training sequence may also be used for detecting the training-sequence echoes as well as their delay and loss. With the aid of this information, the received signal can be corrected by a channel equalizer.

Guard Period   Transmission in each timeslot is terminated with a guard period (Figure 1.30) of 8.25 bit periods ($\approx$30 μs). During this time, the level of the burst must be reduced from nominal to a minimum value (by up to 70 dB) and the burst is modulated with so-called dummy bits (logic 1), that is, no information is transmitted. The user of the next

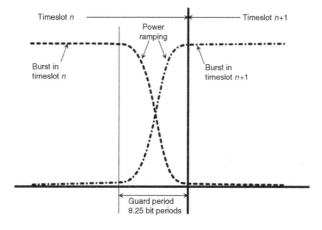

Figure 1.30   Guard period at the end of each timeslot.

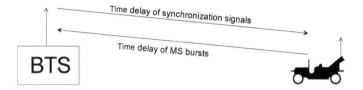

Figure 1.31 Signal delay and its effect.

timeslot should start sending during this guard period so that his burst has reached nominal power when the actual transmission in the timeslot starts. This means that the switching time that cannot be used for transmitting information is used twice.

Delay Correction The integrity of a timeslot depends on whether the subscribers send only during the period assigned to them and otherwise keep quiet. This is only possible when all subscribers are accurately synchronized. For practical reasons, the clock signal is generated by the BTS and the mobile stations synchronize to it. Conflicts with adjacent timeslots may occur in the uplink where several subscribers have to share the same channel. Although several subscribers are addressed in the downlink, signals can only be transmitted by the BTS.

We will now examine the effect of a distance of 10 km between an MS and a BTS (Figure 1.31). Synchronization of the MS is as follows.

Delay over 10 km = 33.3 μs (distance ÷ velocity of light).
The sync signals from the BTS requires this time for transmission.
→ MS is synchronized by 33.3 μs too late.
MS sends a burst at the correct time from its point of view.
→ The burst is sent 33.3 μs too late.

Signal delay over a distance of 10 km → the burst requires another 33.3 μs. From the point of view of the BTS, the burst arrives with a delay of twice the delay time. Transmission cannot be made in the assigned timeslot and interferes with the next one.

The guard period at the end of each burst is only approximately 30 μs long and fully used in the example above. The greater the distance between MS and BTS, the greater the effect of the signal delay. The only way to solve this problem is to make the MS send the burst at an earlier time. To do so, the distance between MS and BTS must be known. The BTS determines the distance by means of a delay measurement and informs the MS of the actual delay. As a result, the MS corrects its transmission time so that the signals again arrive time synchronized with the other mobiles at the BTS antenna.

In GSM, this procedure is called timing advance (TA) and is carried out continuously for all active mobile stations. Every 480 ms, a new TA value is sent to all active mobiles. A few limiting values of the GSM system can be deduced from the TA.

The TA is transmitted as a 6-bit word. Numerals from 0 to 63 can be represented with 6 bits. For instance, a TA of 10 means that for an MS from which bursts arrive with a delay of 10 bit periods, transmission must be advanced by this amount. Since the TA is transmitted as an absolute value, a maximum delay of 232.5 μs (63 × 3.69 μs) can be signaled and corrected. This maximum double delay corresponds to a distance of approximately 34.9 km.

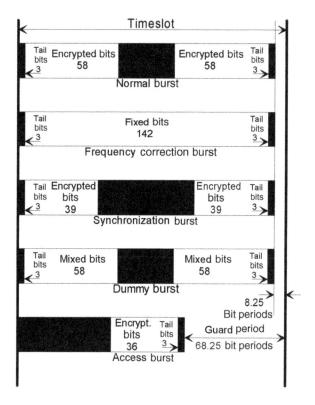

Figure 1.32 Burst types.

With the aid of the TA, the distance between the MS and the BTS can be determined. Since the smallest delay increment is a bit period, the resolution for a distance of one half of the delay corresponds to one bit period and therefore to approximately 550 m.

This synchronization scheme cannot solve an inherent problem. The delay of the very first burst sent by an MS cannot be corrected because, with contact between the MS and BTS yet to be established, the MS has not yet received an appropriate TA value from the BTS. Yet, if the mobile sends a normally timed burst, it may spill over into the next time slot and interfere with transmission from another MS. To avoid causing interference in such cases, the mobile uses a special burst at the beginning of a transmission or when no valid TA has been received. Called the access burst, it is considerably shorter than a normal burst. The next section describes the access burst and other additional burst types in greater detail.

Burst Types   In addition to the normal burst described before, other bursts, including the access burst, are available for special purposes (Figure 1.32).

Frequency Correction Burst. The 142 "fixed bits" of the frequency correction burst (FCB) are all set to logic 0. With Gaussian minimum shift keying (GMSK), the type of modulation used in GSM, a stationary carrier frequency deviation, in this case approximately 67.7 kHz, is generated with this burst. The FCB is sent by the BTS only and used by the mobile for synchronization to the carrier frequency and for compensating a possible Doppler shift. It is sent by the BTS every 10 frames (approximately every 46 ms), but only in the timeslot 0 and only on a single (the C0) carrier.

Synchronization Burst. One frame after the frequency correction burst, but also in timeslot 0, the synchronization burst (SB) is transmitted. This burst is sent only by the BTS, and only on the C0 carrier. A particular difference between it and the normal burst is its considerably longer training sequence. Like the 26-bit training sequence of a normal burst, the SB's training sequence is also used for bit synchronization. Due to the great length of the sequence, synchronization can be more exact.

The twice 39 encrypted bits comprise timers T1, T2, and T3 in a coded form, and also the base-station identification code (BSIC). When this "GSM time" is received, the MS is synchronized to the BTS.

Dummy Burst. A BTS must send continuously—that is, in all timeslots—on its C0 carrier because this carrier is used by the MS to find the nearest BTS and for evaluating reception quality. If no normal burst is available for transmission in a timeslot, the BTS sends dummy bursts instead, as the carrier cannot be transmitted without a modulation signal. Only the BTS sends dummy bursts, and only on the C0 carrier.

Access Burst. As already pointed out, the access burst is sent when the MS first calls the BTS as a means of minimizing interference to other MTs while initiating a delay measurement and the determination of the TA. In most cases, the access burst is also sent on the C0 carrier in the uplink direction, but in the case of a handover it may be transmitted on any carrier.

1.3.8 From GSM to UMTS to LTE

With GSM, as with the other second generation (2G) systems, mobile communications became digital. This system was of a completely new quality as compared to its analog predecessors. As an early system, this standard is less complex than newer systems, and easier to comprehend. It still is a good introductory example in a book like this, since it is the basis from which modern standards were developed.

When GSM was developed, computing power was limited, especially regarding mobile devices. Designing the required RFICs, too, was all but an easy task. The standard therefore aimed at taking full advantage of the new possibilities that were offered by the digital world, like privacy protection through encryption, secure billing, and squeezing as many connections into a physical radio channel as possible. But it still is mainly a telephone system, as it offers one dedicated line for each call. Packet-switched data connections were not in the focus—in absence of reasonable mobile computer power and display technology, and considering the low data rates and high costs: a reasonable approach. However, short-message texting (SMS) was build into the system from the start. A low data-rate package-switched signaling channel was used to transfer these few bits.

On the positive side, in the view of an RF designer, GSM is a nice standard. Relying on the GSMK modulation scheme (see Section 1.4.2), the transmitted RF signal is always of constant amplitude. The information is encoded in the phase shift from symbol to symbol. Due to the narrowband phase modulation, the signal did not consume more bandwidth than in AM. These signals were sent out in bursts, as discussed in the previous section.

The advancement of the semiconductor technology enables much more complex systems nowadays. The driving force today is mobile data transmission, mainly access to the internet, e-mail, and social media, but also to entertainment such as stream video. Considering data throughput, voice is becoming negligible.

The third generation (3G) successor of GSM is UMTS. With this system, any connection between the mobile device and the base station is treated as a data session. Instead of defining a fixed number of possible channels, defined in bandwidth and time slots, it uses a more flexible scheme, based on wideband code-division multiple access, which will be defined in a subsequent section.

In future mobile systems, like the long-term evolution (3GPP-LTE), a standard that is currently being deployed, adds numerous new features on the system level. For example, variable bandwidth, enhanced localization, spatial multiplexing, and multiple-input–multiple-output (MIMO) schemes that use multiple receive and transmit antennas in order to enhance data throughput, or maximum ratio combining schemes that reduce bit error rate relying on multiple transmit antennas and a single receive antenna.

An outline addressing most of the details of current and emerging standards beyond GSM are way beyond the scope of this book, and the reader is referred to the literature. Fortunately, a number of good introductory application notes are available, for example, [6–8, 10, 16–18].

1.4 ABOUT BITS, SYMBOLS, AND WAVEFORMS

1.4.1 Introduction

Digital modulation of an RF carrier is the allocation of physically existing RF waveforms to the single elements of an alphabet of logical symbols where the number of allowed waveforms is equal to the number of logical elements of the alphabet (Figure 1.33). The most common alphabet is the binary one with the two logical symbols "0" and "1", but we will also deal with quaternary, octernary, and hexadecimal alphabets or more generally with M-ary alphabets comprising many more elements when discussing the signal generation with signal generators and dedicated software packages. The waveforms representing these symbols differ from each other by their parameters amplitude $a(t)$, their frequency $f(t)$, and their phase $\phi(t)$.

A modulator therefore is nothing more than a device by which this allocation is performed (Figure 1.34). From a coder it receives the logical symbols and emits at its output the corresponding waveforms $s_i(t)$. The waveform generation may be done by using a set of distinct generators (e.g., two oscillators to generate two signals with different frequencies in the case of binary frequency shift keying), by classical amplitude or frequency modulators or by more sophisticated equipment such as I/Q-modulators for M-ary modulations.

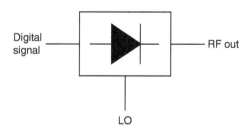

Figure 1.33 At base, digital modulation involves frequency-shifting a baseband digital signal to RF. In practice, the process is more complicated than this because of bandwidth constraints on the resulting RF signal.

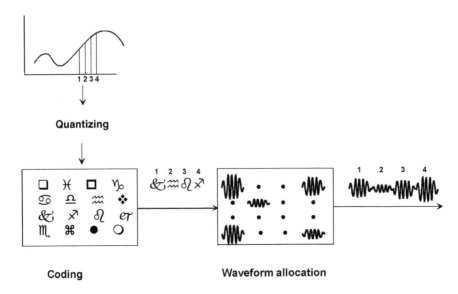

Figure 1.34 Digital modulator.

On their way across the RF channel from the transmitter to the receiver, these waveforms are distorted by noise and other disturbing properties of the RF channel.

The task of the receiver is to interpret the received waveforms $r_i(t)$ and to reallocate the proper logical symbols to them. For this purpose, it is not necessary to reconstruct the original waveforms from the distorted ones (Figure 1.35). The important thing is to find out

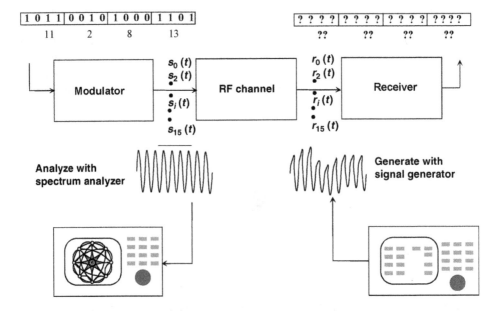

Figure 1.35 The information channel.

which symbol has most probably been sent when a certain signal $r_i(t)$ has been received, a process that is known as maximum likelihood estimation.

For meaningful receiver tests therefore waveforms have to be generated that mimic real, distorted signals to prove the ability of a receiver to tolerate waveform distortions to a certain extent.

1.4.1.1 Representation of a Modulated RF Carrier

The waveform of a modulated RF carrier can be expressed as

$$s(t) = a(t)\cos[2\pi f_c(t)t + \phi(t)] \tag{1.7}$$

and is defined by its amplitude $a(t)$, its carrier frequency $f_c(t)$, and its phase $\phi(t)$. All the three parameters are time variant and may be altered to generate different waveforms to represent logical symbols. If the occupied bandwidth of this modulated carrier is narrow compared to the carrier frequency f_c, we call this signal the RF-bandpass signal (Figure 1.36).

As any frequency variation causes a phase variation and vice versa a phase variation always causes a frequency variation, we can replace any frequency modulation by a corresponding phase modulation. Therefore, we simplify the above equation to

$$s(t) = a(t)\cos[2\pi f_c t + \phi(t)] \tag{1.8}$$

that is, we consider the carrier frequency as a constant and concentrate all frequency and phase variations into the parameter $\phi(t)$.

For our purposes, another representation is more suitable, we will have to look up some trigonometric identities and our formula processor finds that

$$a(t)\cos[2\pi f_c t + \phi(t)] = \cos[\phi(t)]a(t)\cos(2\pi f_c t)$$
$$- \sin[\phi(t)]a(t)\sin(2\pi f_c t) \tag{1.9}$$

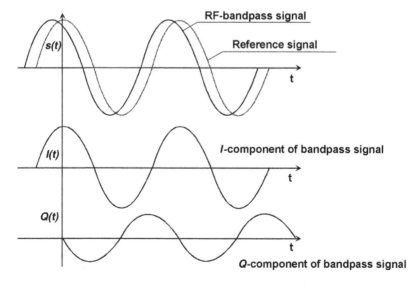

Figure 1.36 The bandpass signal and the *I/Q* representation of a carrier.

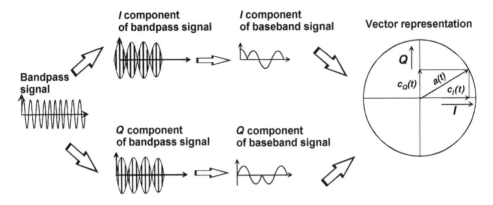

Figure 1.37 Different forms of signal representation.

which we call the *I/Q* representation of the RF-signal. *I/Q* means that we have an *I* (in phase) signal, namely, $\cos[\phi(t)]a(t)\cos(2\pi f_c t)$, and a *Q* (quadrature) signal, namely, $-\sin[\phi(t)]a(t)\sin(2\pi f_c t)$. These equations help us a lot in understanding an *I/Q* modulator. Because of the phase difference of 90° between the two carrier components, these are said to be orthogonal to each other.

All the information about the (modulated) carrier with the carrier frequency f_c is contained in the terms

$$c_I(t) = a(t)\cos[\phi(t)] \tag{1.10}$$

$$c_Q(t) = a(t)\sin[\phi(t)] \tag{1.11}$$

and, lazy as we are, we therefore disregard the terms $\cos(2\pi f_c t)$ and $-\sin(2\pi f_c t)$ for further considerations and denote the above signals $c_I(t)$ and $c_Q(t)$ as the components of the complex baseband waveform or baseband signal.

This leads us immediately to the vector representation of the signal, where we consider the two components $c_I(t)$ and $c_Q(t)$ of the complex baseband signal as the time-variant components of a time variant vector with the vector length $a(t)$ and the angle to the *I*-axis $\phi(t)$. We also get

$$a(t) = \sqrt{c_I^2(t) + c_Q^2(t)} \tag{1.12}$$

$$\phi(t) = \arctan\left(\frac{c_Q}{c_I}\right) \tag{1.13}$$

The vector can be depicted in the *I/Q* area (Figure 1.37).

Generation of the Modulated Carrier Once we have realized that the modulated carrier can be represented as the sum of its *I* and *Q* components, which are the product of the two baseband components with two orthogonal RF carriers of the same frequency, it is easy to understand the hardware of the modulator (Figure 1.38). An unmodulated RF carrier is split up into two equal oscillations $\cos(2\pi f_t)$, one of the two is then is shifted by $0.5\,\pi$, and therefore is described by $-\sin(2\pi f_t)$. The component $\cos(2\pi f_t)$ is multiplied with the *I* component of the baseband signal $c_I(t)$, the other one, $-\sin(2\pi f_t)$, is multiplied with the

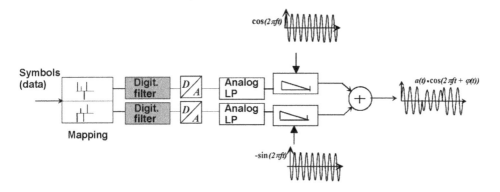

Figure 1.38 Principle of a digital *I/Q* modulator.

Q component $c_Q(t)$ of the baseband signal. Each multiplication may be performed using a double-balanced mixer. Afterward, the two RF-components are added in a simple power combiner. As it is difficult to shift the carrier by $90°$ over a broad frequency range, the modulated carrier is generated at an intermediate frequency and then upconverted to the wanted output frequency in a second mixer stage.

The baseband signals are generated by mapping every digital symbol into a pair of digital pulses that are fed to digital baseband filters. The output signal of these filters is D/A converted and smoothened by analog low-pass filters.

Figure 1.39 shows another example where, for a given modulation (MSK or GMSK), the instantaneous phase and then the corresponding cosinusoid and sinusoid, which modulate the two carrier components, are calculated from the data signal.

Digital designs of the modulator also exist, in which the IF-carrier generation, the time-variant phase shift, the multiplication with the baseband signals, and the sum of the components are calculated in a digital signal processor, the output of which is D/A-converted and upconverted to the output frequency in the classical way. A further possibility is the generation of the modulated carrier with direct digital synthesis (DDS), as it is used in the Rohde & Schwarz SME signal generator.

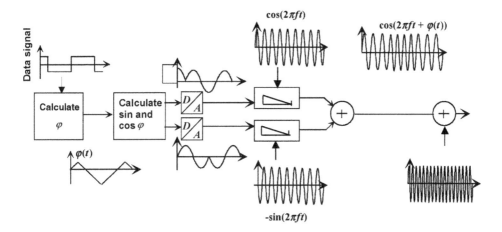

Figure 1.39 *I/Q* modulation (MSK and GMSK).

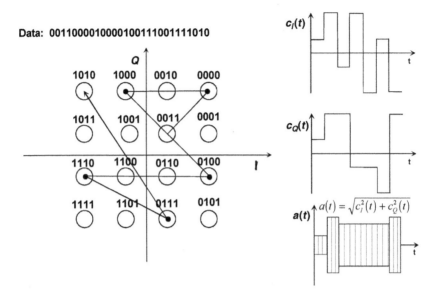

Figure 1.40   Constellation diagram, and baseband and RF signals of 16 QAM.

Mapping the Data into the Baseband Waveforms   The next question is, "How do we generate the baseband waveforms $c_I(t)$ and $c_Q(t)$?" There is no general answer to this question, as the generation of the baseband waveforms depends on the type of modulation. The following short descriptions will suffice for the moment.

Linear modulations (all kinds of amplitude and phase-shift keying, and M-ary QAM).

- For binary amplitude and phase shift keying (ASK and BPSK), the data signal itself represented as a unipolar (ASK) or bipolar (BPSK) nonreturn-to-zero (NRZ) signal is the baseband waveform $c_I(t)$; the component $c_Q(t)$ does not exist.

- For M-ary phase shift keying and M-ary quadrature amplitude modulation, N bits are combined to form new symbols that are elements of an alphabet with $M = 2N$ elements. In the simplest case, every symbol is allocated an I and a Q amplitude during the symbol duration, which is N times the bit duration. The modulating signals $c_I(t)$ and $c_Q(t)$ are then staircase functions, and the modulated carrier has a time-varying envelope with the instantaneous amplitude $a(t)$ (Figure 1.40). Because the steps of the envelope cause unwanted side lobes of the RF spectrum, the baseband signals are filtered to smooth the shape of the RF envelope and reduce the occupied bandwidth of the modulated RF signal.

Nonlinear modulations (frequency-shift keying, minimum shift keying and Gaussian minimum shift keying).

- Despite M-ary frequency shift keying (FSK) could be performed using an I/Q modulator, for this type of modulation much simpler equipment such as a voltage-controlled oscillator is used as a frequency modulator. Figure 1.41 shows an example of quaternary frequency shift keying (4FSK), which also is known as 4 PAM/FM. This term indicates that every two bits are combined to a dibit that is mapped into a baseband

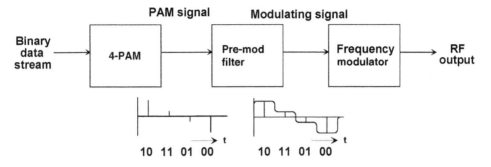

Figure 1.41   4PAM/FSK.

pulse with an amplitude taking on one of four possible levels. The pulse than is shaped by a baseband filter before being fed to the frequency modulator.

- If more precise modulations are required (e.g., MSK and GMSK, which also turn out to be frequency modulations), first the instantaneous phase of the modulated RF carrier is calculated from the data. The corresponding sine and cosine values that form the modulating baseband signals $c_I(t)$ and $c_Q(t)$ are determined from a look-up table. This operation is the reason for the fact that frequency modulation is called a nonlinear modulation.

1.4.1.2 The Spectrum of a Digitally Modulated Carrier

It is the task of any transmission process to occupy as little bandwidth as possible. The absolute lower limit in the baseband is half the symbol rate of the baseband signal, where for *M*-ary modulation the symbol rate r_{Symbol} is equal to the bit rate divided by $l_d(M)$. This lower limit is only theoretical as ideal rectangular filters that cannot be realized were necessary. Therefore, in practice, a minimum baseband bandwidth of about $0.75\, r_{\text{Symbol}}$ has to be taken into account.

With linear modulation, the occupied bandwidth in the RF range is twice the occupied baseband bandwidth. This follows from the lag theorem, according to which the double-sided spectrum of a time function is shifted from $f = 0$ to the frequency $f = f_c$ when the time function is multiplied with $\cos\left(2\pi f_c t\right)$ (Figure 1.42).

Expressing this with formulas, we find

$$c(t) \qquad \circ\!\!-\!\!\bullet \qquad C(f) \tag{1.14}$$

$$e^{j2\pi f_o t} c(t) \qquad \circ\!\!-\!\!\bullet \qquad C(f - f_0) \tag{1.15}$$

Therefore, if the baseband spectrum is limited by a low-pass filter, the RF spectrum is limited as if it was filtered by an RF bandpass filter with twice the bandwidth of the baseband filter.

Demodulation of Digitally Modulated Carriers   The demodulation process is an estimation process. The receiver compares the received waveform with all the possible waveforms

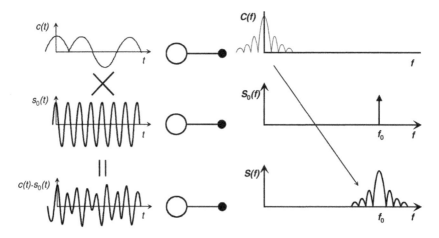

Figure 1.42 Occupied bandwidth in the baseband and the RF range.

and decides which symbol has been received. The waveform allocated is the one that is most similar to the received signal. This process is also called maximum likelihood estimation.

Figure 1.43 shows a possible demodulator. The received signals are cross-correlated with the stored ones in M correlators. High output from a given correlator indicates a match between its stored symbol and a symbol in the incoming signal.

As can easily be seen, such a receiver is quite complex. Therefore, other principles of receivers have been studied. The most often used receiver is based on an *I/Q* demodulator, the principle of which is shown in Figure 1.44.

The received signal is split into two equal components. One component is multiplied with the reconstructed carrier signal, the other with its orthogonal counterpart, that is, the $\pi/2$-shifted reconstructed carrier. After low-pass filtering, this multiplication delivers the

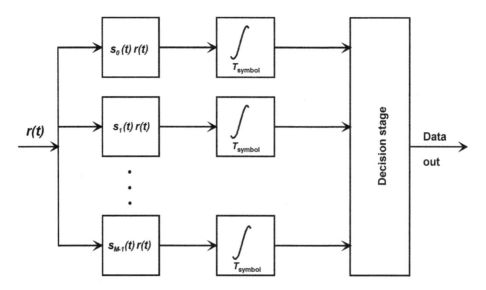

Figure 1.43 Correlation receiver.

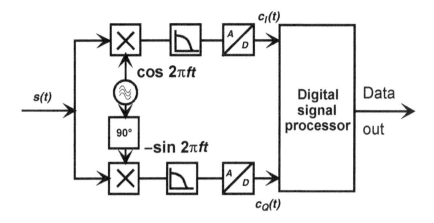

Figure 1.44 Coherent demodulation.

I and the Q components of the baseband signal, which is more or less distorted by interference, multipath, and noise in the radio channel. These demodulated baseband signals then can be A/D converted and treated using simple or sophisticated algorithms to decide which symbols were originally sent. This type of demodulation requires an exact reconstruction of the unmodulated carrier, not only with respect to frequency but also to phase, and is called coherent demodulation. While the frequency recovery is in principle quite simple, the acquisition and tracking of the signal phase requires complex signal processing of the baseband signal.

Coherent demodulation and its accompanying costly signal acquisition and tracking is necessary for all types of phase and quadrature amplitude modulation in which any part of the transmitted information is coded in the absolute phase of the carrier. It only can be avoided when the information is coded in the phase difference of two sequential waveforms. Such modulation schemes are called differential phase modulation. Examples of such modulation include differential binary phase-shift keying (DBPSK) and differential quadriphase-shift keying (DQPSK) and their derivatives.

The following list shows common digital modulation types and the telecom systems that use them.

BPSK—Satellite radio links, GPS, Inmarsat

QPSK—Satellite radio links, GPS, IS-95

OQPSK—IS-95 (reverse link)

$\pi/4$-QPSK—NADC, PHS, PDC. TETRA

GMSK—GSM, DCS 1800

GFSK (Gaussian FSK)—DECT

COFDM (coded orthogonal frequency-division multiplexing)—DAB, DVB.

1.4.2 Some Fundamentals of Digital Modulation Techniques

The first waveforms that were used for "digital modulation" were amplitude-shift keying (ASK), similar to on–off-keyed Morse code, and phase-shift keying (similar to FM).

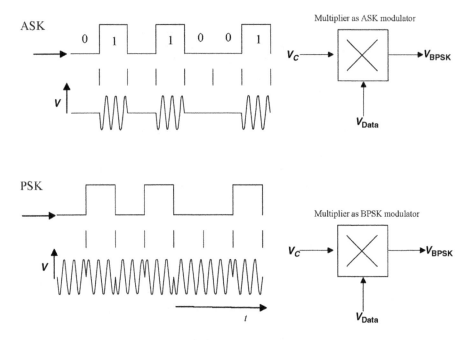

Figure 1.45 Amplitude-shift keying (ASK) and phase-shift keying (PSK) in the time domain.

Figures 1.45 shows the resulting waveforms in the time domain. It is also useful to look at them in the the phase domain—the in-phase *(I)* and quadrature *(Q)* planes. From Figure 1.46, we see that ASK has only two states: no signal and signal at *I* (in-phase), while phase-shift keying can have two states: $+I$ and $-I$. In the frequency domain, ASK and BPSK have the output spectra shown in Figure 1.47. These output spectra, with their $(\sin x)/x$ appearance, depend on the rise and decay times and duty cycle of the modulating signal.

BPSK sends one data bit per signal state. Another way to put this is that each of BPSK's two signal states is a symbol that stands for just one bit—0 or 1. Transmitting more than one bit per symbol would allow us to improve our data throughput per unit of time, and this is exactly what is done in quadrature PSK (QPSK), which uses four possible signal states as symbols for two-bit sequences called dibits (Figure 1.48). Figure 1.49 shows a QPSK modulator and the QPSK constellation diagram. Figure 1.50 shows the result of QPSK in the time and frequency domains. Figure 1.51 shows a spectrogram of an actual QPSK emission.

Moving from ASK to BPSK or QPSK results in increased resistance to noise. Figure 1.52 illustrates this by graphing bit error rate versus signal-to-noise ratio (here expressed as $10 \log [E_{\text{bit}} \div N_0]$, where E_{bit} is energy per bit and N_0 is the noise [42]). Figure 1.53 compares the maximum interference voltages for BPSK and QPSK.

Figure 1.54 shows bandwidth requirements and constellation diagrams for BPSK and QPSK. Unfiltered BPSK and QPSK are purely angle-modulated emissions; for each, all symbols are transmitted at the same amplitude. As we will see, however, bandwidth limiting a BPSK or QPSK signal to minimize adjacent-channel interference results in envelope variations that must be preserved through careful circuit design if bit errors are to be minimized.

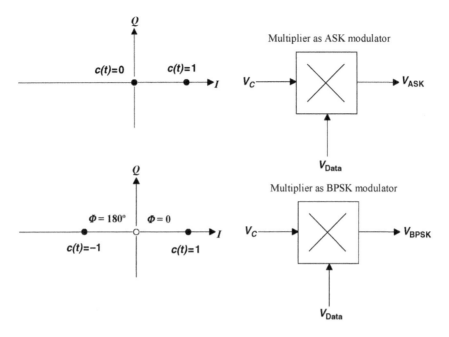

Figure 1.46 ASK and PSK in the I/Q plane.

Further increases in bit rate per symbol are possible with phase-modulation schemes that divide the 360° of each carrier cycle into increasingly smaller segments. Transmitting three bits per symbol, with each symbol spaced from its neighbors by 45°, gives 8-PSK; transmitting four bits per symbol, with each symbol spaced from its neighbors by 22.5°,

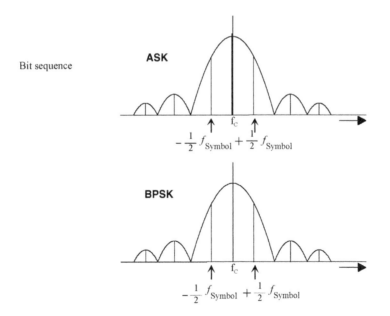

Figure 1.47 ASK and BPSK in the frequency domain.

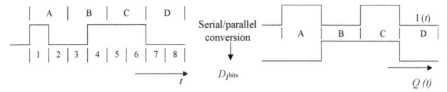

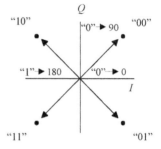

| Bit sequence | | $D_{j\text{bits}}$ | | Transmitted |
1	2	I	Q	phase
0	0	0	0	45° (D)
1	0	1	0	135° (A)
1	1	1	1	225° (C)
0	1	0	1	315° (B)

Figure 1.48 Quadrature PSK (QPSK) modulator (*I/Q* modulator).

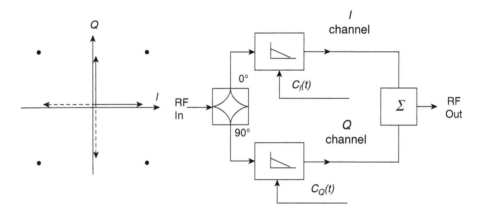

Figure 1.49 QPSK constellation diagram (left) and modulator (right).

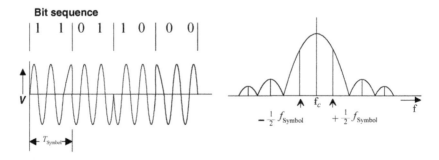

Figure 1.50 Result of QPSK modulation in the time and frequency domains.

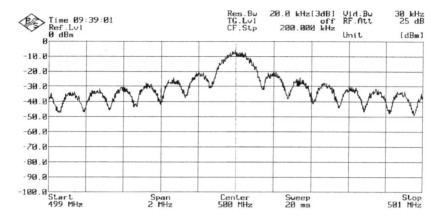

Figure 1.51   QPSK spectrum.

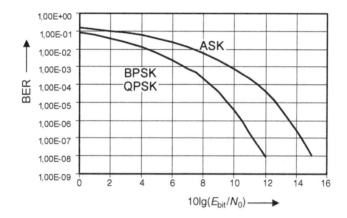

Figure 1.52   Bit error rate (BER) in terms of E_{bit}/N_0 for BPSK, QPSK, and ASK.

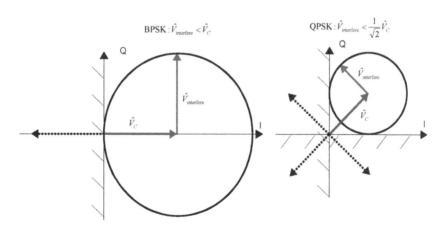

Figure 1.53   Maximum interference voltages for BPSK and QPSK, where $\hat{V}_C$ is carrier voltage of the desired signal and $\hat{V}_{interfere}$ is the carrier voltage of the interfering signal.

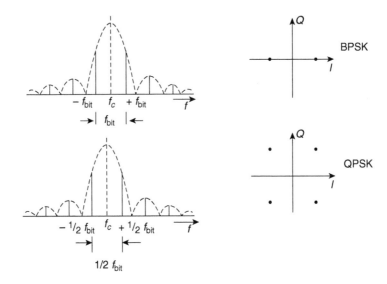

Figure 1.54 Bandwidth requirements (left) and constellation diagrams (right) for BPSK and QPSK. For each emission, all symbols are transmitted at the same amplitude, as is indicated by their equidistance from the constellation origin.

results in 16-PSK. Increasing the data rate by increasing the number of bits per symbol does not give us something for nothing, however; the smaller the phase difference between adjacent symbols, the likelihood that phase shifts resulting from phase noise, multipath reception, and other sources of phase perturbation will cause demodulation errors.

Modulation schemes that differentiate their data symbols through a combination of phase and amplitude shifts are susceptible to disturbances in amplitude in addition to phase. Figure 1.55 shows the modulator and constellation diagram for 16-state quadrature

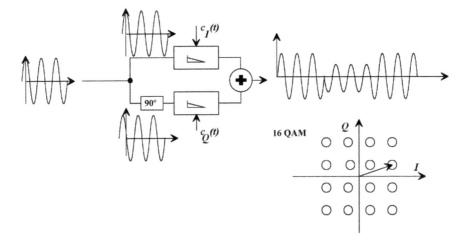

Figure 1.55 Quadrature amplitude modulation (QAM), with constellation diagram for 16-state QAM (16 QAM). Because transitions between QAM symbols involve shifts in phase and amplitude; even unfiltered QAM is a combination of amplitude and angle modulation.

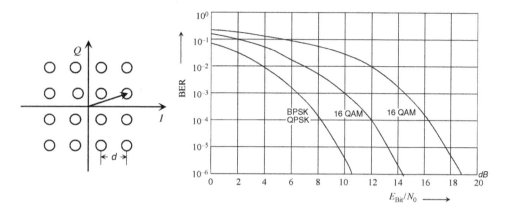

Figure 1.56 Bit error rate versus E_{bit}/N_0 for BPSK/QPSK, 16-QAM and 64-QAM, showing the significantly greater SNR necessary for a given BER as the number of signal states is increased. In 64-QAM, each symbol represents six bits.

amplitude modulation (16-QAM), in which each symbol represents four bits. Figure 1.56 shows bit error rate versus SNR for BPSK/QPSK, and 16-QAM and 64-QAM. QAM systems can support very high bit rates—if sufficient SNR can be maintained and disturbances in signal amplitude and phase can be kept under control [5].

The infinite number of sidebands produced by angle modulation must be reduced by filtering to minimize the spectrum occupied by an emission to just that necessary for communication (Figure 1.57). Considerations of filter realizability and transmitter frequency agility dictate that this filtering must be done at baseband (Figure 1.58). The filter characteristics are optimized in accordance with the modulation scheme used. The filtering is done in using DSP, minimizing component count, and resulting in characteristics that are essentially temperature independent and identical from unit to unit.

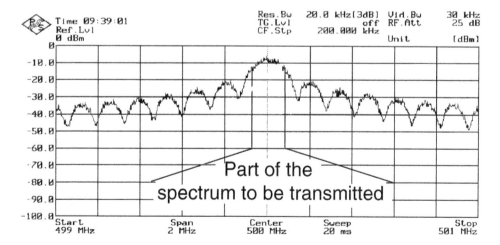

Figure 1.57 QPSK spectrum resulting from pseudorandom binary sequence (PRBS) data. Most of an angle-modulated emission's infinite sidebands are unnecessary for communication and can be removed by filtering—at the cost of introducing amplitude variations that must be sufficiently preserved to keep the bit error rate low.

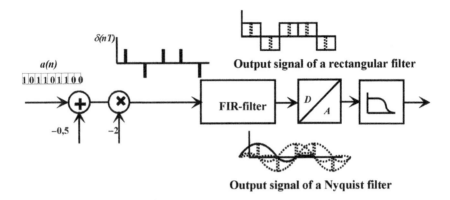

Output signal of a rectangular filter

Output signal of a Nyquist filter

Figure 1.58 Baseband filtering. Figure 1.59 shows the result for QPSK.

Fundamental angle-modulation theory tells us that the envelope of an angle-modulated signal does not vary in amplitude. Such an emission could be amplified in highly nonlinear stages without distortion. But because angle-modulated signals must be band-limited for practical use, and because band-limiting an angle-modulated signal strips away sideband energy that would contribute to its envelope constancy, the emissions used in modern wireless systems, even those that do not involve intentional amplitude modulation, vary in amplitude with modulation. How much an wireless emission's amplitude varies depends on the degree of filtering and the particular modulation scheme used. Various schemes, many of which rely on particular synergies of coding, modulation, and filtering, have been devised for minimizing amplitude variation in digitally modulated signals. Figure 1.60 shows one (offset DQPSK [OQPSK]); Figure 1.61 compares the spectra of three (QPSK, minimum shift keying [MSK], and Gaussian MSK [GMSK]). Detailed discussion of such techniques is more concerned with coding, logic circuitry, and software than with radio hardware, and is therefore beyond the scope of this book. What is important to the circuit designer is that stages handling the emission(s) on which a system depends must be sufficiently amplitude

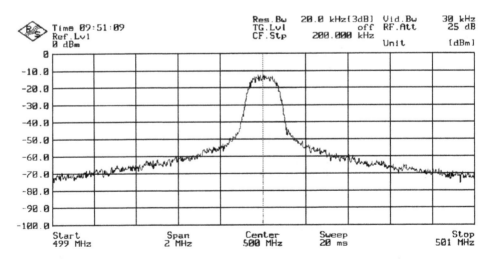

Figure 1.59 Spectrum of a band-limited QPSK signal.

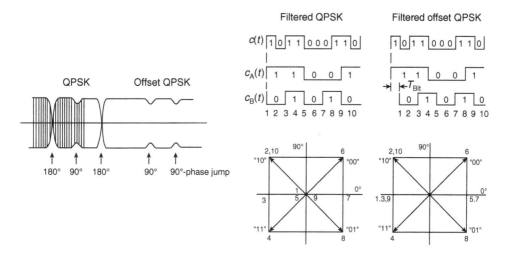

Figure 1.60 Offset QPSK (OQPSK) exhibits reduced envelope variations compared to standard QPSK by time-offsetting the *I* and *Q* channels.

linear to maintain the modulation integrity of the signal and, in the case of transmitters, to keep adjacent-channel interference and other spurious emissions within acceptable levels.

1.4.2.1 Spread-Spectrum and CDMA Modulation Techniques

A spread-spectrum techniques is, by definition, a modulation technique that spreads a baseband signal over a broader bandwidth for transmission—without improving the signal-to-noise ratio (SNR) compared to AM. Common FM or PM therefore are not considered spread spectrum, since increasing the bandwidth yields a reduction in SNR. The two techniques most commonly used for spread spectrum in commercial standards are frequency hopping, and direct-sequence spread spectrum. Let us have a look into these schemes in

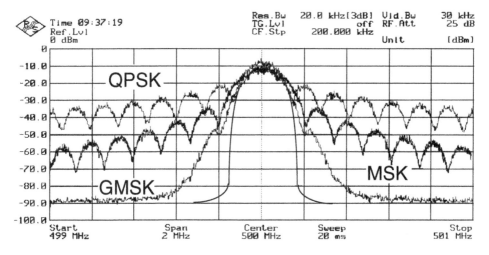

Figure 1.61 Comparison of QPSK, MSK, and GMSK.

order to motivate why it can be advantageous to spread the signal instead of keeping it as narrowband as possible [19, 20].

Frequency Hopping   The name already implies how it works: the carrier frequency is not fixed. During transmission, the signal hops between different frequency channels in a pseudorandom sequence that is known only to transmitter and receiver. This approach is advantageous in a hostile environment. The technique is dating back to the 1920s, and was first used as a measure to establish secure military radio links. Without knowledge of the hopping sequence, it is next to impossible in an analog world to track the transmitted signal. The transmission might even be unnoticed by the enemy since the average power for the full band is low and might even be below the total noise power. For one narrow band, on the other hand, the signal only appears for a very short time until it hops away. Another advantage is that it is not easy to disrupt the transmission by jamming. A jammer could practically just block one narrowband channel, while the transmit signal hops around it without experiencing any problem.

Today, frequency hopping is commonly used in standards like Bluetooth that use an unlicensed ISM band. An unlicensed band can be regarded as a quite hostile environment, since many other devices might be active, for example, microwave ovens. Bluetooth uses an adaptive frequency-hopping spread-spectrum technique to ensure good transmission properties. In fact, the hopping sequence is adaptively updated, mainly by removal of channels that provide low-quality transmission, for example, due to fading or due to another device that is sending there.

Direct Sequence Spread-Spectrum   DSSS is used in code-division multiple access scenarios (CDMA). In DSSS, the digital baseband signal is multiplied with a fast pseudo-random bit sequence. One bit of information is now expressed by a long series of so-called chips. Since the chip length is a small fraction of the bit length, the respective spectrum is spread accordingly. On the receiver side, the original signal can only be reconstructed, if the received signal is multiplied with the original pseudorandom bit sequence.

This signal is also hard to detect. Without knowledge of the spreading code, it looks like broadband noise. The power level might only be slightly higher or even below the natural noise level. This technique therefore also has its roots in the military sector, and the spreading code was also used to encrypt the transmission.

An important feature of the DSSS is that the received signal is again multiplied with the pseudorandom bit sequence. Figure 1.62 shows a sketch of a spread-spectrum transmission. The original signal is multiplied with the spreading sequence, which can be done through an XOR operation. The figure shows the respective signals in the baseband. Spreading results in a sequence of shorter symbols, thereby the original spectrum is spread. The channel will add white noise, and possibly a narrowband interfering signal.

The received signal is the superposition of these signals. Despreading is performed by applying an XOR operation of the received signal with the original spreading code. The transmitted signal, therefore gets reconstructed. In the time domain, the double XOR operation cancels out, and in the frequency domain, it is despread to its original bandwidth. Any other received signal, on the contrary, is spread, and its power is distributed over a wide band. However, spreading will not affect white noise, which is still white at the same power level after spreading. Therefore, the noise power that defines the SNR of the DSSS transmission equals the noise power within the original bandwidth, no change compared to AM transmission.

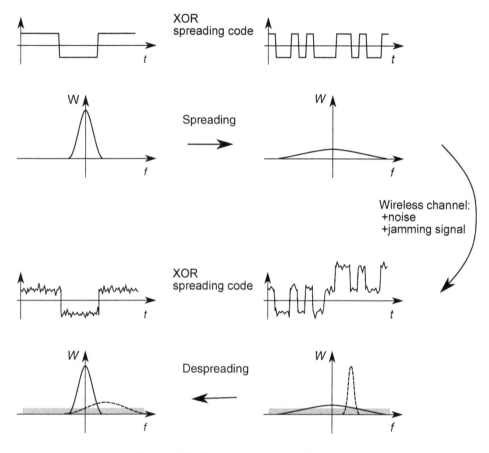

Figure 1.62 Direct-sequence spread spectrum.

Regarding a narrowband jamming signal, a clear advantage over AM is seen. The despreading procedure is identical to the spreading procedure for any signal except the desired transmitted signal. A jammer is therefore spread, and its power is distributed over a wide bandwidth. Most of the power is then filtered.

In CDMA, or wideband-CDMA (W-CDMA) systems like UMTS, the different users share the same bandwidth by using different orthogonal spreading codes. Ideally, the cross-correlation between the different bit sequences would be zero, which means that the average of one bit sequence multiplied with another one is equal to zero. In UMTS, these bit sequences are realized through feedback shift registers providing so-called Gold Codes. The codes of this family, however, are not fully orthogonal. The nonzero cross-correlation results in a higher noise floor, depending on the number of parallel users.

W-CDMA can handle multipath propagation that causes reception of delayed echos of the signal. If the spreading code is orthogonal to a delayed copy of itself, all echos will just be discarded when despreading the main signal coherently. However, a rake receiver that basically despreads each of the echos individually, can even take advantage of the multipath propagation. It is, however, required to know in advance when the echos will arrive, which is accomplished through channel estimation techniques.

UMTS mobiles operate in full duplex, they transmit their broadband signal at a quite low-power spectral density. The W-CDMA scheme enables flexible usage of the available bandwidth; more users at a time increase the noise floor for all, which in turn reduces data throughput. But the number of users is not a priori fixed as in pure TDMA or DFMA schemes.

1.4.2.2 Orthogonal Frequency Division Modulation (OFDM) and Single-Carrier Frequency-Division Multiple Access (SC-FDMA)

Orthogonal frequency division modulation (OFDM) starts from a quite simple thought: if very long symbols are transmitted, multipath transmission and echos are not an issue, as long as the difference in delay for the different paths is short compared to the duration of the symbols. This requires a very long duration for a symbol that requires only narrow bandwidth. Thus, for a high-bitrate data transmission, a high number of subchannels is defined, and a high number of bits is transmitted in parallel.

Regarding the different subchannels, it is required that they are independent or orthogonal. OFDM assumes at first a rectangular shape of the bits to be transmitted, which results in a $\sin(x)/x$-shaped spectrum $S_v(\omega)$ that has periodic zero crossings as follows:

$$S_v(\omega) = \left| \frac{\sin([\omega - \omega_v]T_S/2)}{(\omega - \omega_v)T_S/2} \right|^2 \tag{1.16}$$

with the bitlength T_S and the subcarrier frequency ω_v. The distance of the zeros in the spectrum defines the channel spacing f_s as follows:

$$f_s = \frac{1}{T_S} \tag{1.17}$$

The subcarrier frequencies therefore are given as $\omega_v = v \cdot 2 \cdot \pi f_s$. Figure 1.63 shows the channel definition for five channels. It is obvious that the channels are only independent at the discrete frequencies f_s; for all other frequencies, it is not possible to distinguish between the different channels. This channel definition is, however, only processed in the digital domain. Prior to transmission, the signal is inverse Fourier transformed into time domain, bandpass filtered, and converted into analog. The underlying principle is basically the same as the well-known Nyquist criterion for digital data transmission in the time domain. This criterion defines that the bits in a bitstream need to be independent from each other at the sampling time. And assuming an ideal low-pass channel, the bit symbols become $\sin(x)/x$-shaped, and the time difference between the zeros of one bit signal defines the minimum data rate. OFDM, as stated before, uses basically the same principle, but with time and frequency domain interchanged. OFDM, therefore, requires perfect frequency

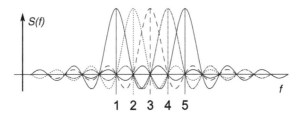

Figure 1.63 OFDM spectrum for five channels.

synchronization between transmitter and receiver, since otherwise the subchannels are no longer orthogonal.

OFDM is inherently robust against multipath transmission due to the long duration of the single bits. In order to completely cancel the impact of multipath effects, a guard period is inserted before each of the symbols, the so-called cyclic prefix. The end of the symbol is copied into this guard period, and all the echo signals are expected to arrive within this time. Before reconstructing the bitstream through FFT, this guard period is discarded. Therefore, in OFDM the signals arriving through the different paths superimpose for each symbol. If they add, it is even able to take advantage of the multipath propagation. And narrowband fading only affects a few subbands, therefore only a few bits are lost without any impact on the other bits transmitted in parallel.

From the point of view of a circuit designer, however, OFDM has a major drawback, that lies in the high number of independent subchannels. The resulting time-domain waveform is almost chaotic and has a high variation in amplitude—the minimum in case that all subchannels transmit the symbol of lowest amplitude, and the maximum being the case, where all channels transmit at maximum amplitude. Even if a clever symbol mapping algorithm is able to reduce the peak-to-average ratio, the time-domain signal's amplitude variation is still the peak-to-average ratio of the single-carrier signal multiplied by the number of subchannels.

Therefore, in order to simplify circuit design, we would prefer a single-carrier modulation due to its lower peak-to-average power. But regarding robustness against multipath propagation, OFDM would be the preferred. Fortunately, it is to a certain extent possible to unite the two approaches. Regarding OFDM, the system looks like this: original signal → IFFT → add cyclic prefix → radio channel → discard cyclic prefix → FFT → equalize → received signal.

However, in a linear system, it is possible to exchange the building blocks without changing the final result. A single-carrier scenario could look like this: original signal → add cyclic prefix → radio channel → discard cyclic prefix → FFT → equalize → IFFT → received signal. The only change is that the IFFT block was moved to after the equalizer. This scheme is called single-carrier frequency-domain equalization (SC-FDE). On the transmitter side, the single-carrier signal is only changed by adding a cyclic prefix to a certain block of data, but no mapping to subcarriers is done. On the receive side, the guard interval is again discarded, which removes the impact of echos on the respective block of data, as in OFDM. Remains frequency-selective fading, which is addressed by equalization in frequency domain. SC-FDE also reduces the requirements regarding frequency synchronization between transmitter and receiver compared to OFDM.

The SC-FDE scheme therefore reduces the requirements on the side of the transmitter power amplifier, since the peak-to-average ratio is the same as for the single-carrier modulation. On the other hand, it requires increased processing power on the receiver side, since an adaptive frequency-domain equalization is required. This scheme is therefore very attractive for the upstream in mobile communication systems, where the mobile unit is transmitting and the base station receives.

In case of LTE, however, the SC-FDE is implemented in a way that additionally allows for multiple access by frequency-domain multiplexing. While in SC-FDE, the full channel bandwidth is dedicated to one transmission, in single-carrier frequency-domain multiple access (SC-FDMA), the different channels defined for equalization are dedicated to different subscribers. It can either be a contiguous block of subchannels or the channels that are interleaved. On the mobile unit's side, it is now required that the single-carrier signal is Fourier

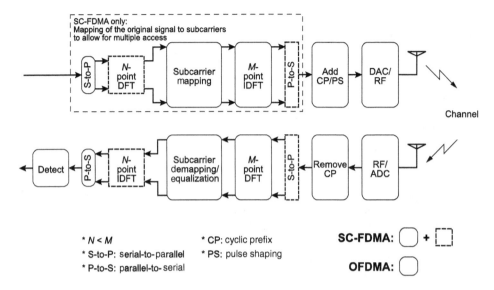

Figure 1.64 SC-FDMA signal flow. Neglecting the dashed box yields a SC-FDE system that transmits a single-carrier signal enhanced by the cyclic prefix and performs the equalization on the receive side in the frequency domain. SC-FDMA in addition first converts the single-carrier signal to frequency domain and maps the discrete spectrum to the available subcarriers (adapted from the public-domain graphic published in the SC-FDMA Wikipedia article).

transformed and shifted to the allowed subchannels. Figure 1.64 shows a block diagram. An N-point FFT is performed first, and the N subchannels are then mapped to a certain subset of M subchannels, unused subchannels are filled with zeros. The IFFT transforms the whole signal back into the time domain for transmission, the guard interval/cyclic prefix is inserted, and the signal finally is converted to analog and transmitted. The receive path is almost unchanged, except that the equalizer needs to be aware of the channel mapping in order to reconstruct the signal.

Further details on the SC-FDMA coding scheme and how it is implemented in LTE can be found in the literature [21, 22].

1.5 ANALYSIS OF WIRELESS SYSTEMS

1.5.1 Analog and Digital Receiver Designs

For the purpose of showing the capability of modern CAD, we now show first an analog receiver and following this an analog/digital receiver with its functionality and performance.

1.5.1.1 Receiver Design Examples

Analog Receiver Design   A transmitter modulates information signals onto an RF carrier for the purpose of efficient transmission over a noise filled air channel. The RF receiver's job is to demodulate that information signal while maintaining a sufficient signal-to-noise ratio (SNR). This must be done for widely varying input RF power level and with the presence of noise and interferers.

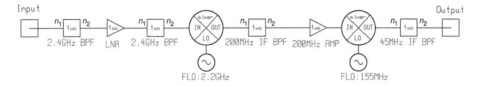

Figure 1.65 Dual-downconversion receiver schematic.

Modern communication standards place requirements on key system specifications such as RF sensitivity and spurious response rejection. These system specifications must then be separated into individual circuit specifications via an accurate overall system model. Symphony can play a major role in modeling systems and in determining the individual component requirements. A 2.4-GHz dual down-conversion system (see Figure 1.65) will be used as an example. The first IF is 200 MHz and the second IF is 45 MHz.

As a signal propagates from the transmitter to the receiver, it is subject to path loss and multipath resulting in extremely low signal levels at the receive antenna. RF sensitivity is a measure of how well a receiver can respond to these weak signals. It is specified differently for analog and digital receivers. For analog receivers, there are several sensitivity measures including minimum discernible signal (MDS), signal-to-noise plus distortion ratio (SINAD), and noise figure. For digital receivers, the typical sensitivity measure is maximum bit error rate (BER) at a given RF level. Typically, a required SINAD at the baseband demodulator output is specified over a given RF input power range. For example, audio measurements may require 12 dB SINAD at the audio output over RF input powers ranging from –110 dBm to –35 dBm. This can then be translated to a minimum carrier-to-noise (C/N) ratio at the demodulator input to achieve a 12 dB SINAD at the demodulator output (see Figure 1.66).

Determining receiver sensitivity requires accurate determination of each component's noise contribution. Modern CAD software, such as Symphony by Ansoft, can model noise from each stage in the system including oscillator phase noise. The C/N ratio can then be plotted as a function of input power (see Figure 1.67).

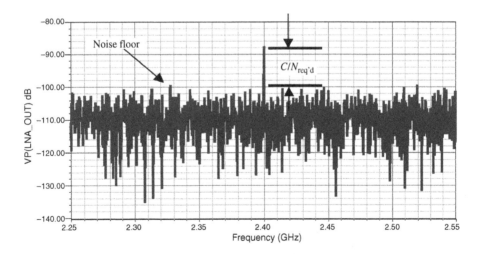

Figure 1.66 Receiver sensitivity measure showing required C/N.

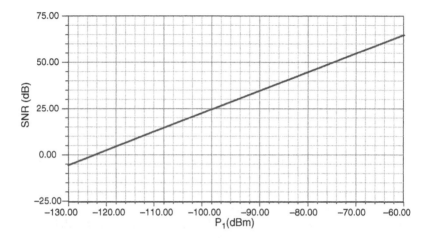

Figure 1.67 C/N ratio versus input RF power level.

This plot enables straightforward determination of the minimum input power level to achieve a certain *C/N* ratio. Once the sensitivity has been specified, the necessary gain or loss of each component can be determined. A budget calculation (see Table 1.5) can be used to examine the effects of each component on a particular system response.

Another key system parameter is receiver spurious response. The receiver's mixers typically cause the spurious responses. The RF and LO harmonics mix and create spurious responses at the desired IF frequency. These spurious responses can be characterized by the equation [23]

$$\pm m f_{RF} \pm n f_{LO} = \pm f_{IF} \tag{1.18}$$

Some spurious responses, or spurs, can be especially problematic because they may be too close to the intended IF to filter thus masking the actual information bearing signal. These spurious responses and their prediction can be especially troublesome if the receiver is operated near saturation. The analysis can then be refined by including a spur table that predicts the spur level relative to the IF signal. The spur's power at the output of a mixer is calculated using the spur table provided for this mixer. Once generated, each spur is carried

Table 1.5 Budget Calculation for the ΔS_{21} Across Each System Component Accounting for Impedance Mismatches

Names	ΔS_{21} (dB)
2.4 GHz BPF	−0.82
LNA	22.04
2.4 GHz BPF	−0.93
MIXER_1	−5.34
200 MHz IF BPF	−3.85
200 MHz AMP	22.21
MIXER_2	−8.43
45 MHz IF BPF	−0.42

Table 1.6 Spur Output Powers and the Corresponding RF and LO Harmonic Indices

Frequency (MHz)	P_{out} (dBm)
45.00	−26
20.00	−106
65.00	−86
90.00	−76
110.00	−66
135.00	−96
155.00	−44
175.00	−96
180.00	−106
200.00	−56
220.00	−96
245.00	−76
245.00	−46
265.00	−76

through the remainder of the system with all mismatches accounted for. The output power level of each spur is shown in Table 1.6. Another useful output is the RF and LO indices that indicate the origin of each spur.

In addition to sensitivity and spurious response calculations, an analog system can be analyzed in a CAD tool such as Symphony for gain, output power, noise figure, third-order intercept point, IMD (due to multiple carriers), system budget, and dynamic range.

Mixed-Mode MFSK Communication System   Next, a mixed-mode, digital and RF, communication system will be described and simulated. In this example, digital symbols will be used to modulate an 8 GHz carrier. The system uses multiple frequency shift keying (MFSK) bandpass modulation with a data rate of 40 Mbps. Convolutional coding is employed as a means of forward error correction. The system includes digital signal processing sections as well as RF sections and channel modeling. Several critical system parameters will be examined including bit error rate (BER). FSK modulation can be described by the equation [24]

$$s_i(t) = \sqrt{\frac{2E}{T}} \cos(\omega_i t + \phi) \qquad (1.19)$$

where $i = 1 \ldots M$. So the frequency term will have M discrete values with almost instantaneous jumps between each frequency value (see Figure 1.68). These rapid jumps between frequencies in an FSK system lead to increased spectral content.

Figure 1.69 shows a block diagram of the complete system. The system can be split into several major functional subsystems, the baseband modulator, the RF transmitter section, channel model, RF receiver, clock recovery circuitry, and baseband demodulator. Looking specifically at the baseband modulator circuitry (Figure 1.70), a pseudorandom bit source is used and the bit rate is set to 40 MHz.

A convolutional encoder then produces two coded bits per data bit and increases the bit rate to 80 MHz. The purpose of the convolutional encoder is to add redundancy to improve the received BER. A binary-to-M-ary encoder then assigns one symbol to every two bits

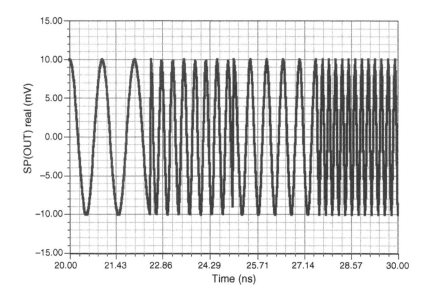

Figure 1.68 4FSK signal generated in Symphony.

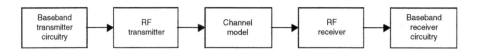

Figure 1.69 Mixed RF/DSP MFSK communication system simulated in Symphony.

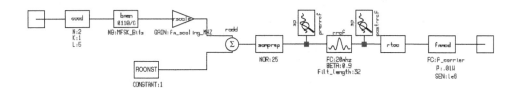

Figure 1.70 MFSK baseband circuitry including frequency modulator.

creating the four levels for the 4FSK effectively halving the bit rate down to 40 MHz again. The signal is then scaled and upsampled. To decrease the bandwidth, a root-raised cosine filter is used to shape the pulses. The filtered signal then serves as the input to the frequency modulator. The RF section (see Figure 1.71) includes the transmitter that modulates the baseband signal onto an 8-GHz carrier.

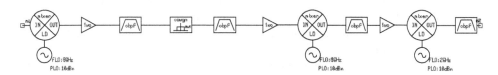

Figure 1.71 RF section including Gaussian noise channel model.

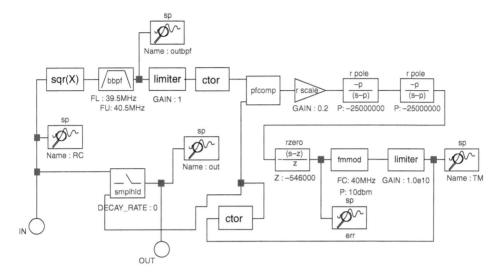

Figure 1.72 Clock-recovery circuitry.

That signal is amplified and then filtered to remove any harmonics. The signal is then passed through an additive white Gaussian noise model. The received signal is filtered, amplified, and downconverted twice to baseband. After carrier demodulation, the signal is then sent through the clock recovery circuitry. Clock recovery is employed in order to ensure that sampling of the received signal is executed at the correct instances. This recovered timing information is then used as a clock signal for sample and hold circuitry. Clock recovery in this system was achieved using a PLL configuration. The schematic for the clock-recovery circuit is shown in Figure 1.72.

At the heart of the clock recovery circuit is the phase comparator element and the frequency modulator. The inputs to the phase comparator consist of a sample of the received data signal and the output of the feedback path that contains the frequency modulator. The frequency modulator reacts to phase differences in its own carrier and the received data signal. The output of the frequency modulator is fed back into the phase comparator whose output is dependent on the phase differential at its inputs. The phase of the frequency modulator continually reacts to the output of the phase comparator and eventually lock is achieved. Once the timing of the received signal is locked onto, the clock that feeds the sample and hold will be properly aligned and correct signal sampling will be assured.

Figure 1.73 shows the received signal (filtered and unfiltered) as well as the recovered clock and the final data. Equalizer circuitry is then used to compensate for channel effects that degrade the transmitted signal. The equalizer acts to undo or adapt the receiver to the effects of the channel. The equalizer employed in this MFSK system is the recursive least square equalizer. The equalizer consists of a filter of N taps that undergoes an optimization in order to compensate for the channel effects. The equalizer depends on a known training sequence in order to adapt itself to the channel. The equalizer model updates the filter coefficients based on the input signal and the error signal (i.e., the difference between the output of the equalizer and the actual desired output). The update (optimization) is based on the recursive least square algorithm. Several equalizers are available in Symphony, including complex least mean square equalizer, complex recursive least square equalizer, least mean square

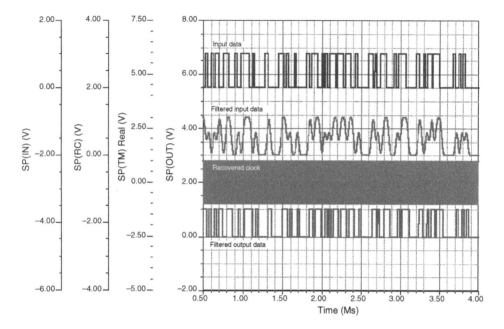

Figure 1.73 Input data, filtered input data, recovered clock, and final output data for the MFSK communication system.

equalizer, recursive least square equalizer, and the Viterbi equalizer. After equalization, the BER of the system is analyzed versus SNR (see Figure 1.74).

1.5.1.2 PLL CAD Simulation

From the clock recovery circuit, it is logical to ask the question about the various frequency sources and their performance. Now to answer this question, we are resorting to Symphony 8.0, where the system simulator allows us to evaluate phase-locked loops. Figure 1.75 shows

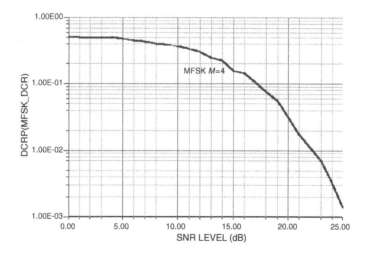

Figure 1.74 BER versus SNR for the MFSK system.

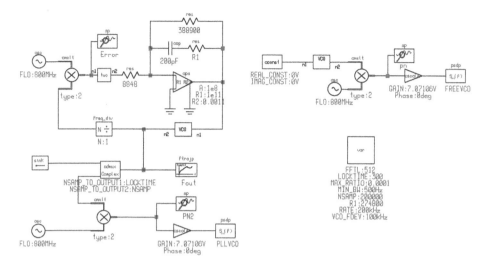

Figure 1.75   Block diagram of a CAD-based PLL system.

the block diagram of a PLL in which the VCO is synchronized against a reference. For the purpose of demonstrating the capability, we have selected a 1:1 loop with a crossover point of about 100 kHz, meaning that the loop gain is 1 at this frequency.

Its noise performance is best seen by using a CAD tool to show both the open- and closed-loop phase-noise performance. At the crossover point, the loop is running out of gain. The reason why the closed-loop noise increases above 2 kHz has to do with the noise contribution of various components of the loop system. The highest improvement occurs at 100 Hz, and as the loop gain decreases, the improvement goes away. Therefore, it is desirable to make the loop bandwidth as wide as possible as this also improves the switching speed. On the bad side, as the phase noise of the free-running oscillator crosses over the phase noise of the reference noise, divider noise, and other noise contributors, one can make the noise actually worst than that of the free-running state. In a single loop, there is always a compromise necessary between phase noise, switching speed, and bandwidth. A first-order approximation for switching speed is $2/f_L$, where f_L is the loop bandwidth. Assuming a loop bandwidth of 100 kHz, the switching speed will be 20 μs. Looking at Figure 1.76, we can clearly see the trend from 200 Hz to 100 kHz. Because of the resolution of the sampling time (computation time), the open-loop phase noise below 150 Hz is too low and could be corrected by a straight line extrapolation from 500 Hz toward 100 Hz. We did not correct this drawing so we could show the reader the effect of not-quite-adequate resolution.

Having said this, Table 1.7 shows cellular and cordless standards, everything referring to 2G. Figure 1.77 shows the multiplexing schemes used. Table 1.8 shows the parameters of cellular and cordless systems. Figure 1.78 shows the cellular structure.

1.5.2 Transmitters

Since it is possible to generate any type of modulation using DSP, including those described in this chapter, a DSP-based transmitter can also be built. A good example of this is the Philips SA900, which is truly universal.

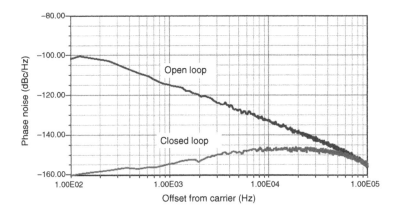

Figure 1.76 Open- and closed-loop phase noise for a CAD-based test phase-locked loop.

Table 1.7 Cellular and Cordless Standards

Cellular Telephone Networks	
Analog	Digital
AMPS	**IS-54**
Advanced Mobile Phone System	North American Digital Cellular
TACS	**IS-95**
Total Access Communication System	North American Digital Cellular
NMT	**GSM**
Nordic Mobile Telephone	Global System for Mobile Communications
	PDC
	Personal Digital Cellular

Cordless Telephone Networks	
Analog	Digital/PCN
CTO	**CT2/CT2+**
Cordless Telecom 0	Cordless Telecom 2
JCT	**DECT**
Japanese Cordless Telecom	Digital European Cordless Telecom
CT1/CT1+	**PHS**
Cordless Telecom 1	Personal Handy Phone System
	DCS-18OO

Introduction to the SA900 [1]

The SA900 (Figure 1.79) is a truly universal in-phase and quadrature (*I/Q*) radio transmitter that can perform many types of analog and digital modulation including AM, FM, SSB, QAM, BPSK, QPSK, FSK, and so on. It is a highly integrated system that saves space

[1] Based on portions of the Philips Semiconductors/Signetics RF Communications Products Application Note AN1892, "SA900 I/Q transmit modulator for 1 GHz applications," August 20, 1997. Used with permission.

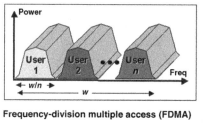

Frequency-division multiple access (FDMA)

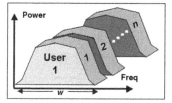

Time-division multiple access (TDMA)

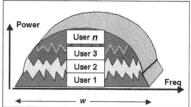

Code-division multiple access (CDMA)

Figure 1.77 Multiplexing schemes.

and cost for the manufacturers producing cellular and wireless products. The device allows baseband signals to directly modulate the I/Q carriers, which are generated by internal phase shift network, in the 1-GHz range, and to maintain good linearity required for linear modulation scheme (e.g., $\pi/4$-DQPSK). It contains an on-chip frequency divider, phase detector, and VCO, which can be built into a PLL frequency synthesizer to create a transmit offset frequency. Its unique internal design allows frequency conversion without having an external image rejection filter for eliminating the sum term after mixing. The SA900 meets

Table 1.8 Parameters of Cellular and Cordless Systems

Characteristics	Cellular Telephones			Cordless Telephones		
	IS–54	IS–95	GSM	CT2	DCS1800	DECT
Band (downlink)	869–894	869–894	935–960	864–868	1805–1880	1880–1900
Band (uplink)	824–849	824–849	890–915	864–868	1710–1785	1880–1900
Bandwidth	50 MHz	50 MHz	50 MHz	2 MHz	150 MHz	20 MHz
Channelization	TDMA/FDMA	CDMA/FDMA	TDMA/FDMA	FDMA	TDMA/FDM	TDMA
Channel spacing	30 KHz	1250 kHz	200 KHz	100 kHz	200 kHz	1728 kHz
Channels/carrier	3	55–62?	8	1	16	12
Number of channels	832	20	124	40	750	10
	(3 users/ch.)	(798 users/ch.)	(8 users/ch.)		(16 users/ch.)	(12 users/ch.)
Duplex method	FDD	FDD	FDD	TDD	FDD	TDD
Channel bit rate	48.6 kbps	1.2288 Mb/s	271 kbps	72 kbps	271 kbps	1152 kbps
Speech codec	VSELP	CELP	RPE-LTP	ADPCM	RPE-LTP	ADPCM
Bit rate (voice)	8 kbps	1.2–9.6 kbps	13 kbps	32 kbps	13 kbps	32 kbps
Modulation	$\pi/4$DQPSK	BPSK/OQPSK	GMSK	FSK	GMSK	GMSK
Mobile peak power	0.6–3 W	0.2–2 W	2–20 W	10 mW	0.25–2 W	250 mW
Mobile average power	0.6–3 W	0.2–2 W	0.25–2.5 W	5 mW	0.03–0.25 W	10 mW
Cell radius	30 miles	30 miles	1–5 miles	?	?	40–140 m
Cluster size (min)	7	1	3	N/A	3	N/A

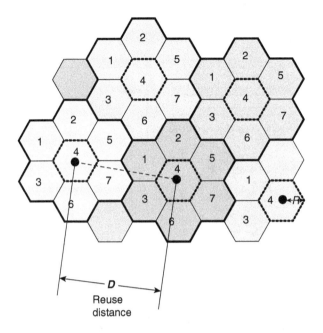

Figure 1.78 Cellular structure. $W = N \times B_C$, where W is the full bandwidth and B_C is the bandwidth per cell. The number of channels/cell $= B_C/B_U$, where B_U is bandwidth per user (FDMA) or N users (TDMA). $D = R\sqrt{3N}$.

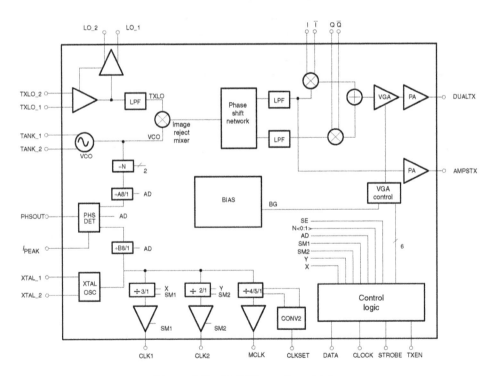

Figure 1.79 SA900 transmit modulator.

the specifications required by the IS-54, the industry standard for North America Digital Cellular (NADC) system. This application note reviews the basic concept of *I/Q* modulation and discusses the key points when designing the SA900 for an RF transmitter.

I/Q Modulation   Any bandpass RF signals can be represented in polar form by

$$s(t) = A(t)\cos[\omega_c t + \phi(t)] \tag{1.20}$$

where $A(t)$ is the signal envelope and $\phi(t)$ is the phase. By using the trigonometric identities, we can represent (1.20) in rectangular form by

$$s(t) = I(t)\cos[\omega_c t] - Q(t)\sin[\omega_c t],$$
$$I(t) = A(t)\cos[\phi(t)]$$
$$Q(t) = A(t)\sin[\phi(t)] \tag{1.21}$$

Since the baseband signals $I(t)$ and $Q(t)$ modulate two exactly 90° out-of-phase carriers $\cos(\omega_c t)$ and $-\sin(\omega_c t)$, respectively, we call the system implementing (1.21) an in-phase and quadrature (*I/Q*) modulator. Figure 1.80 shows the mathematics and hardware implementation of an *I/Q* modulator.

The local oscillator, usually a VCO within a PLL, generates the carrier and is split into two equal signals. One goes directly into a double-balanced mixer to form the *I*-channel and the other one goes into the other mixer via a 90° phase shifter (realized by passive elements) to provide the *Q*-channel. The baseband signals $I(t)$ and $Q(t)$, either analog or digital in nature, modulate the carrier to produce the *I* and *Q* components, which are finally combined to form the desired RF transmitting signal. Since any RF signal can be represented in the *I/Q* form, any modulation scheme can be implemented by an *I/Q* modulator.

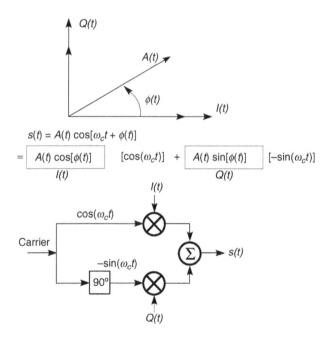

Figure 1.80 Mathematical representation and hardware implementation of *I/Q* modulator.

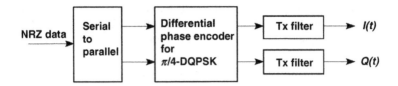

Figure 1.81 QPSK and π/4-DQPSK baseband generator.

1.5.2.1 Linear Digital Modulation

Linear digital modulation techniques depend on varying the phase and/or magnitude of an analog carrier according to some digital information: ones and zeros. This digital information can be the output of an analog-to-digital converter (e.g., voice codec) or it can be digital data in some standard formats (e.g., ASCII). The most popular digital signaling format is non-return-to-zero (NRZ), where 1s and 0s are converted into signal with amplitude of 1 and -1, respectively, in a symbol duration. Since NRZ signal has infinite bandwidth, transmit filters have to be used to limit the spectral spreading. To ensure each NRZ symbol does not smear into its neighbors due to low-pass filtering and channel distortion causing intersymbol interference (ISI), the frequency response of the low-pass filter has to satisfy Nyquist criteria. One example of this type of filter is the linear phase square-root-raised cosine filter. Together with the same type of filter for receive low-pass filtering, the signal is guaranteed ISI-free in a Gaussian environment. One straightforward technique of transmitting these band-limited signals through communication channels would be applying it directly to the mixer of the I channel to generate the RF signal. This is known as binary phase shift keying (BPSK), where the phase of the carrier is shifted 180° to transmit a data change from 0 to 1 or 1 to 0.

Quadrature or quaternary phase shift keying (QPSK) is a much more common type of modulation scheme used in mobile and satellite communications. It has four possible states (90° apart) and each of them represents two bits of data. Figure 1.81 shows the baseband generator for QPSK (without the differential phase encoder). NRZ data bits go through the serial-to-parallel converter (see Figure 1.82) and are mapped in accordance to some rules to generate I and Q values. The generic rule will be the values of I and Q components are 1 and 1 for the data bits "11" (45°) and -1 and -1 for the data bits "00" ($-135°$). These discrete signals have to be band limited by Nyquist low-pass filters to be ISI free.

A more sophisticated way of mapping results in π/4-DQPSK (D for differential encoding), which is chosen for North America Digital Cellular (IS-54), Personal Digital Cellular (PDC) in Japan, and Personal Handy Phone System (PHS) in Japan. In this scheme, consecutive pairs of bits are encoded into one of the four possible phases: π/4 for "11," 3π/4 for "01," -3π/4 for "00," and $-\pi$/4 for "10." However, unlike the previous case that "11" is always π/4 and "00" is always -3π/4, the encoded phases are the degrees that the carrier has to shift at each sampling instances. Thus, the information is contained in the phase difference (differential) instead of absolute phase for π/4-DQPSK.

A better way to tell the difference between QPSK and π/4-DQPSK is by looking at the signal constellation diagram, shown in Figure 1.83, which displays the possible values of I and Q vectors and change of states. Constellation diagram is also known as phase diagram because it shows the phase of the carrier at the sampling point. Note that the phases of QPSK are assigned for every two bits of data; therefore, it can transmit twice as much information as BPSK in a given bandwidth, that is, more bandwidth efficient. 8-PSK is another type of

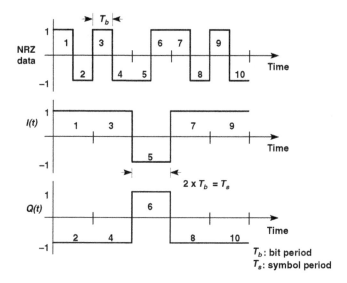

Figure 1.82 Serial-to-parallel conversion.

modulation used for high-efficiency requirements. It maps three bits into eight phases, 45° apart, in the constellation. More spectrally efficient modulation can be created by mapping more bits into one phase at each sampling point. However, as you put more dots in the signal constellation, the signal susceptibility to noise is lower because the decision distance is shorter (dots are closer). Then, it requires higher carrier-to-noise (C/N) ratio to maintain the same bit error rate (BER).

One common misconception is that since $\pi/4$-DQPSK has eight states in the constellation, it is just another type of 8-PSK. Note that at every sampling instant, the carrier of $\pi/4$-DQPSK is only allowed to switch to one of the four possible states (see Figure 1.83). So, we still have two data bits that get encoded into four phases. Thus, it has the same spectral efficiency as QPSK for the same carrier power. The reason for using this modulation scheme is twofold. First, the envelope fluctuation, which causes spectral spreading due to nonlinearity of transmitter and amplifier, is reduced because the maximum phase

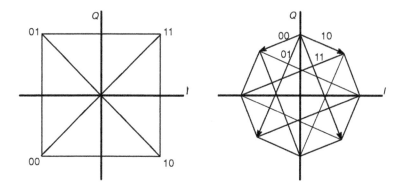

Figure 1.83 Signal constellation of QPSK and $\pi/4$-DQPSK.

shift is 135° instead of 180°. Second, the signal can be demodulated noncoherently, which simplifies the receiver circuitry by eliminating the need for carrier recovery.

1.5.2.2 Digital and Analog FM

Another family of digital modulation is categorized by frequency change of the carrier instead of phase and/or amplitude change. One of them is frequency shift keying (FSK), where the carrier switches between two frequencies. FSK is also known as digital FM because it can be generated by feeding the NRZ data stream into an analog VCO. FSK appears as a unit circle in the signal constellation because the RF signal envelope is constant and the phase is continuous. Baseband filtering is usually applied for FSK to limit the RF bandwidth of the signal so that more channels can fit into a given frequency band.

One common modulation of this type is known as Gaussian minimum shift keying (GMSK), which is used for GSM and some other wireless applications. GMSK can be generated by following its definition: band-limit the NRZ data stream by a Gaussian low-pass filter, then modulate a VCO with modulation index (2 × frequency deviation/bit rate) set to 0.5. In other words, the single-sided frequency deviation is one fourth of the bit rate ($\Delta f = R/4$).

Another way of generating GMSK is by I/Q modulator. Referring back to (1.21), any RF signal can be split into I and Q components. Unlike the QPSK mentioned before, baseband $I(t)$ and $Q(t)$ are not discrete points for FM signals; rather, they are continuous functions of time. The way to produce FM is shown in Figure 1.84. We first store all the possible values of $\cos[\phi(t)]$ and $\sin[\phi(t)]$ in a ROM lookup table, which will be addressed by the incoming data to generate the I and Q samples. The output data from the ROM is then applied to D/A converters, after low-pass filtering for signal smoothing, to produce the analog baseband I and Q signals. This method guarantees the modulation index to be exactly 0.5, which is required for coherent detection of GMSK (e.g., the GSM system). The same I/Q principle can also be applied to generating analog FM signals.

1.5.2.3 Single Sideband AM (SSB-AM)

AM signals can be divided into three types: the conventional AM, double sideband suppressed carrier AM (DSB-AM), and single sideband suppressed carrier AM (SSB-AM). The first type is not attractive because for 100% modulation, two thirds of the transmit signal power appears in the carrier, which itself conveys none of the information added by modulation. By using a balanced mixer (e.g., a Gilbert cell), one can generate DSB-AM, where the carrier is totally suppressed and only the upper and lower sidebands are present. However, this is still not the best because the information is transmitted twice, once in each sideband. To further increase the efficiency of transmission, only one sideband is needed to deliver the information. The SSB-AM can be generated by an I/Q modulator with the baseband information feeding the modulator (by quadrature), as shown in Figure 1.85. This modulation technique can greatly reduce the bandwidth of the signal and allows more signals to be transmitted in a given frequency band. This topic is discussed in detail in Ref. [26].

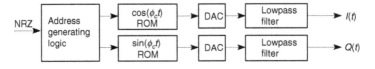

Figure 1.84 Digital FM (e.g., GMSK) baseband I/Q generator.

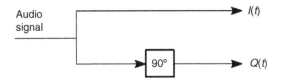

Figure 1.85 Baseband processing for SSB-AM.

System Architecture There are usually two schemes, the dual conversion and direct conversion, used for implementing transmit modulators. Dual conversion is simpler to implement by modulating an oscillator at lower frequency and then upconverting to the carrier frequency. This scheme, however, is more expensive due to the need for additional filtering and more PC board space. By using only one mixer, direct conversion requires fewer components but is harder to implement.

The problems that direct conversion suffers are carrier leakage and modulated signal coupling. Poor RF isolation of the surface mount packages will allow the carrier to be present at the transmitter output thus making it difficult to have −40 dBc carrier suppression. In addition to that, modulated RF signal would couple back to the oscillator (usually a VCO in a PLL synthesizer loop) and cause modulation distortion.

Based on the concept of dual conversion, the SA900 uses an image-reject mixer to eliminate the need for IF filtering and allow monolithic integration. The transmit carrier (LO) is downconverted by the frequency synthesized by the on-chip VCO, which operates from 90 to 140 MHz. This LO is then modulated by the baseband *I/Q* signals to obtain a complex modulation scheme. The image (sum term) after mixing and LO is sufficiently suppressed by the image rejection mixer. Any residual amounts can be further suppressed by an external duplex filter.

Figures 1.86 and 1.87, respectively, show how the SA900 can be used in frequency division duplex (FDD) and time division duplex (TDD) transceivers. Note that the LO for both systems is running at a frequency that is higher than the transmit frequency, thus minimizing carrier leakage. In the FDD system, only one external VCO is required for generating both transmit and receive LO when using the SA900.

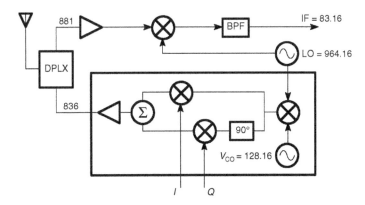

Figure 1.86 FDD system using SA900.

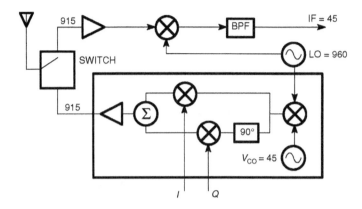

Figure 1.87   TDD system using SA900.

Figure 1.88 shows the IS-54 front-end chip set that consists of the SA601, SA7025, SA900, and SA637. This receiver architecture (SA637) supports a digital magnitude/phase baseband demodulator. An alternate configuration will be using the SA606 FM/IF receiver in conjunction with an external *I/Q* demodulator IC. Table 1.9 shows the possible configurations for the IS-54 handsets using the SA900 as transmitters.

1.5.2.4 Designing with the SA900

Baseband I/Q Inputs   The baseband modulation inputs are designed to be driven differentially for the SA900 to operate at its best. The *I* and *Q* inputs should have a dc offset of $V_{CC}/2$, which is externally provided by common DSP chips. If all four inputs are biased from the same source, the device can tolerate ± 0.5 V dc error; however, inaccuracy of dc

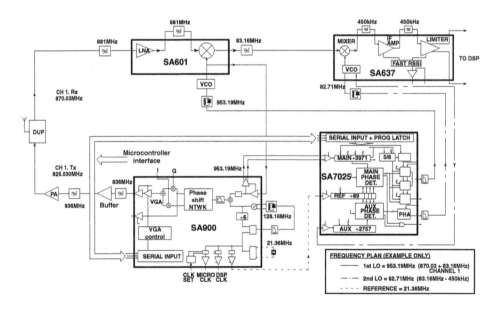

Figure 1.88   IS-54 front-end chipset from Philips.

Table 1.9 Possible Configurations for IS-54 Handsets

Rx 1st IF (MHz)	On-Chip VCO Frequency (MHz)	On-Chip + N Value (MHz)	Crystal Frequency (MHz)
83.16	128.16	6	21.36
71.64	116.64	6	19.44
45	90	6	15
84.6	129.6	9	14.4

bias between I_1/I_2 or Q_1/Q_2 causes reduced suppression of the carrier. Thus, it is important to have a well-regulated dc supply for I and Q signal biasing. The bandwidth of the inputs is much higher than the specified 2 MHz. Approximately 2 dB of power loss will be experienced if the I and Q inputs are 50 MHz.

The SA900 generates a minimum of 0 dBm of power to a 50-Ω load when the amplitude of the I and Q signals are 400 mV P-P. The output power will decrease by 6 dB for every 50% decrease in I/Q amplitude. Single-ended I and Q sources can be used but are not recommended due to the degradation in carrier suppression (more than 10 dB compared to differential). In addition, the entire noise performance of the device will suffer. $V_{CC}/2$ should be applied to I_2 and Q_2 pins if the part is driven single endedly.

Transmit Local Oscillator The transmit local oscillator path consists of a TXLO input buffer, LO output buffer, VCO, image rejection mixer, and phase shift network. Together with a few external components, this section provides the I and Q carrier for modulation.

The TXLO inputs and LO outputs are designed to be used in an external PLL that synthesizes different frequencies for channel selection. The RF signal being generated is fed into TXLO inputs and then comes out of LO outputs to complete the system synthesizer loop. The TXLO inputs are differential in nature and have a VSWR of 2:1 with input impedance of 50 Ω. Single-ended sources can be used by ac grounding the TXLO_2, as done on the demo board. This signal should also be ac coupled into the TXLO_1. The frequency range for these inputs is from 900 to 1040 MHz while the input power should be between -10 and -13 dBm. The output level will be changed significantly if the input level is below -25 dBm.

The output power of the LO-buffered signal changes by about 2 dB when the SA900 is in a different mode of operation. Typical values are -13.5 dBm and -15.5 dBm for DUAL mode and STANDBY mode, respectively.

The 90° phase shift network, realized by RC networks, is capable of operating over a wide frequency range. Even though their frequency characteristics are optimized for cellular band, the part can also be used in other applications in a different band. In such cases, designers have to test the part experimentally to find out the performance, such as sideband suppression, carrier suppression, and image rejection.

Crystal Oscillator The crystal oscillator (XTAL_1 and XTAL_2 pins) is used to provide reference frequency between 10 and 45 MHz for the phase detector and the three on-chip clocks. It can be configured as a crystal oscillator using external crystal and capacitors or it can be driven by an external source. In the latter case, pin XTAL_2 can be left floating. Information regarding crystal oscillator design can be found in Ref. [27].

VCO The VCO, together with the phase detector, the divider, and external low-pass filter, can form a PLL for the transmit offset frequency. The image-rejected mixer down converts

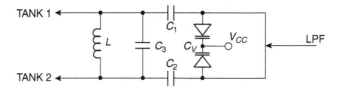

Figure 1.89 VCO tank circuit.

the TXLO signal to the RF carrier by the amount of VCO frequency. Thus, the TXLO frequency should be the desired channel frequency plus the IF offset generated by the VCO. Note that the part will not function if the VCO section is not used.

The VCO is designed for generating IF frequency between 90 and 140 MHz. Together with an external varactor diode and resonator, it can be configured as an oscillator shown in Figure 1.89.

The resonant frequency of such a circuit is

$$f_{VCO} = \frac{1}{2\pi\sqrt{LC_T}} \tag{1.22}$$

where $C_T = (C_1 \| C_2 \| C_V) + C_3$. C_V is a varactor diode, which changes capacitance as the voltage across it varies.

Calculation:

$C_1 = C_2 = 33\,\text{pF}$
$C_3 = 5.6\,\text{pF}$
$C_V = 33.5\,\text{pF at 2.5V}$
$L = 100\,\text{nH}$
$C_T = 5.6 + (1/33 + 1/33 + 1/33.5) - 1 = 16.7\,\text{pF}$

$$f_{vco} = \frac{1}{2\pi}\left(100\ 10^{-9} \cdot 16.7\ 10^{-12}\right)^{0.5} = 123\,\text{MHz}$$

On the demo board, a 1:1 ratio RF transformer is also included to allow single-ended external source driving differential inputs when the VCO is not used.

When designing the VCO, careful PCB layout has to be made. Traces have to be short to avoid the parasitic capacitance and inductance that may cause unwanted oscillation. Referring to (1.22), there is a large combination of L and C_T values that will give the same resonant frequency. If undesired spurs are found in the design due to PCB layout, experimenting with a different set of LC values may sometimes solve the problem.

Output Impedance Matching   The equivalent output impedance at the DUALTX pin is approximately equal to 600 Ω in parallel with 2 pF at 830 MHz. It has to be matched properly to generate maximum power into a 50-Ω load (e.g., a SAW filter). Figure 1.90 shows the recommended matching network. The shunt inductor (L_1) is used to provide maximum swing at the output (short at dc) and also provide reactance to make the real impedance 50-Ω looking into the matching network. The remaining negative reactance is canceled by the series inductor (L_2). The values used on the demo board can be used as

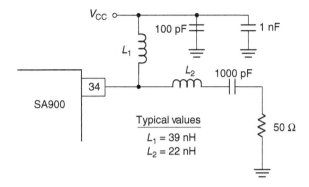

Figure 1.90 DUALTX output matching network.

a reference but may not be suitable if a different layout is implemented. The two shunt capacitors are included to bypass the high-frequency RF signal, avoiding direct coupling into V_{CC}. The series ac coupling capacitor is used to maintain the proper bias for the output stage. Their values are big enough to be left out in impedance matching calculation.

Using a network analyzer to measure the S characteristic is necessary for obtaining optimum matching, which generates maximum output power. Figure 1.91a, b shows how to match the output impedance to a 50-Ω load at 915 MHz. First, calibrate the network analyzer

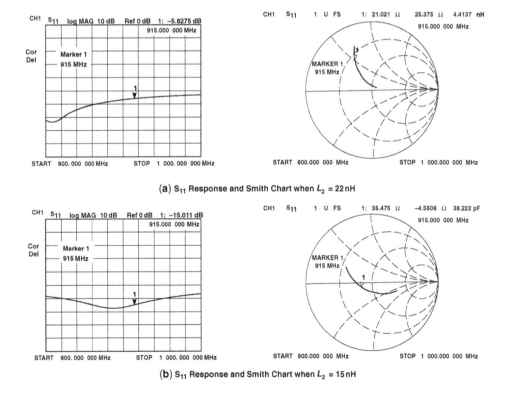

Figure 1.91 SA900 output matching using S parameters.

Table 1.10 Possible SA900 Clock Outputs

CLK1	Divide by 3	X (bit 18) = 1
	Divide by 1	X (bit 18) = 0
CLK2	Divide by 2	Y (bit 19) = 1
	Divide by 1	Y (bit 19) = 0
MCLK	Divide by 4	CLKSET pin = V_{CC}
	Divide by 5	CLKSET pin = $V_{CC}/2$
	Divide by 1	CLKSET pin grounded

to the DUALTX SMA connector on the demo board. Then, short the point where the series inductor is located and use the DELAY feature of the network analyzer to move the point of reference in the Smith Chart to the leftmost point. Now the network analyzer is calibrated to the beginning of the matching network, not just the SMA connector. The frequency response (Figure 1.91a) shows that the "dip" is around 830 MHz, the frequency where the board was originally matched. The Smith Chart shows that it requires less inductance to bring the marker to the center of the chart (50 Ω). By using a 15-nH series inductor, the "dip" was moved closer to 915 MHz (−15 dB) and a better matching is achieved (Figure 1.91b).

On-Chip Clocks   The crystal oscillator is buffered to provide three external clock signals: CLK1, CLK2, and MCLK. Table 1.10 shows the divide ratio and the controlling mechanism.

   CLK1 is usually used for the system synthesizer (e.g., SA7025) reference. Since MCLK is active all the time, it is ideal for providing the master clock for the microcontroller. When the device is in STANDBY mode, CLK1 and MCLK provide the clock signals necessary for receiving RF signals. CLK2 can also be used as a clock for digital signal processing (DSP) chip.

Modes of Operation   The SA900 is intended for either AMPS mode (analog cellular) or DUAL mode (digital cellular, IS-54) operation. When the device is running in AMPS mode, the *I/Q* modulator, variable gain amplifier (VGA), and phase shifter are disabled. The fixed-gain amplifier is powered up during AMPS mode operation. However, since the divide ratio is too low (6, 7, or 8), the comparison frequency of the on-board PLL is too high, making it very difficult for the loop bandwidth to be less than 300 Hz for analog frequency modulation. The device includes two power saving modes of operation that disable partial circuitry to reduce the power consumption of the overall chip. The SLEEP mode disables all the circuitry except the master clock (MCLK pin) of the SA900. The STANDBY mode shuts down everything except the TXLO buffer, MCLK, and CLK1, which allows the system synthesizer (e.g., SA7025) to continue running. These two power saving modes are common to both AMPS and DUAL mode operation. The SA900 draws 60 mA in DUAL mode, reduced to 3 mA and 8 mA in SLEEP and STANDBY modes, respectively.

   TXEN pin is for hardware powering down the modulator and synthesizer. The falling edge of the signal disables the modulator and synthesizer while the rising edge enables the modulator. To power down the synthesizer using software, send a data word with SE bit set to "0" ("1" for enable). The synthesizer will be disabled right after the strobe signal is transmitted. Either SE or TXEN going low will turn off the synthesizer. This operation is common to both AMPS and DUAL mode.

Performance of the SA900

Performance Criteria   Since the *I/Q* modulator is a universal transmitter, measuring only the frequency stability and modulation index of a generated FM signal would not be useful for other modulation schemes. Measurement parameters should be general enough so that they can represent the performance of modulators when applying different types of modulation and allow fair comparisons among different *I/Q* modulators. Based on this idea, two measurement techniques, in-phase modulation and quadrature modulation, are used for evaluating *I/Q* modulators.

The in-phase modulation relies on injecting two equal frequencies and phase signals at f_{mod} into the *I* and *Q* inputs. The result of this modulation is two sidebands appearing at f_{mod} offset from the carrier, with the carrier totally suppressed. This is also known as double-sideband (DSB) conversion. The quadrature modulation requires two equal frequencies (but 90° out-of-phase signals) being injected into the *I* and *Q* inputs. The result is a single-sideband suppressed carrier (SSB-SC) signal with either the upper or the lower sideband at f_{mod} carrier offset being suppressed. This is also known as single-sideband (SSB) upconversion. Figure 1.92 summarizes these two tests.

In a practical system, imperfection of an *I/Q* modulator is directly related to these two measurements. Sideband and carrier suppression from the quadrature modulation test will show the amount of gain imbalance, phase imbalance, and dc offset. On the other hand, intermodulation product suppression from the in-phase modulation test will show the linearity of an *I/Q* modulator. When making measurements, it is important to have well-balanced *I* and *Q* baseband modulating signals for measurement since the signal imperfection will translate into degradation in sideband and carrier suppression.

Performance Graphs   In making those measurements for the demo board, the following parameters were used.

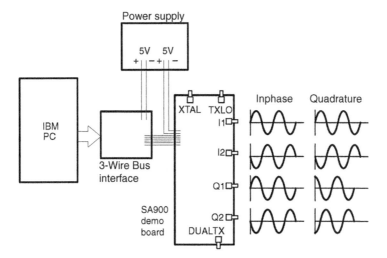

Figure 1.92 In-phase and quadrature modulation test.

In-phase modulation.

PIN 43 $I_1 = 400$ mV P-P, dc $= V_{CC}/2$ at 200 kHz, Phase $= 0°$
PIN 42 $I_2 = 400$ mV P-P, dc $= V_{CC}/2$ at 200 kHz, Phase $= 180°$
PIN 41 $Q_1 = 400$ mV P-P, dc $= V_{CC}/2$ at 200 kHz, Phase $= 0°$
PIN 40 $Q_2 = 400$ mV P-P, dc $= V_{CC}/2$ at 200 kHz, Phase $= 180°$

Quadrature modulation.

PIN 43 $I_1 = 400$ mV P-P, dc $= V_{CC}/2$ at 200 kHz, Phase $= 0°$
PIN 42 $I_2 = 400$ mV P-P, dc $= V_{CC}/2$ at 200 kHz, Phase $= 180°$
PIN 41 $Q_1 = 400$ mV P-P, dc $= V_{CC}/2$ at 200 kHz, Phase $= 90°$
PIN 40 $Q_2 = 400$ mV P-P, dc $= V_{CC}/2$ at 200 kHz, Phase $= 270°$

Figure 1.96a and b illustrates what the typical output spectrum would be if in-phase and quadrature modulation were applied to an *I/Q* modulator. Quadrature modulation will produce lower sideband (LSB) or upper sideband (USB) signal, depending on the phase angle between the *I* and *Q* signals. The SA900 was designed to have USB suppressed when the *I* signal is leading the *Q* signal. The undesired signals are carrier breakthrough and the harmonic products of the baseband modulating signals sitting at $f_c \pm n \cdot f_{mod}$, where n is an integer ≥ 2.

Referring to Figure 1.93a, the output power is 1.3 dBm (cable loss $= 0.7$ dB) for the LSB while better than -38 dBc of carrier, sideband, and harmonics suppression is measured. The USB better than -26 dBc implies that the residual AM of the transmit signal is better than 5%, a requirement of the IS-54 specification.

In-phase modulation test will generate both LSB and USB. Beside these two tones, the carrier breakthrough and the harmonics, intermodulation (IM) products will all appear at the output. The odd IM products are dominant, and they satisfy the following rules.

Let $f_1 = f_c - f_{mod}$, $f_2 = f_c + f_{mod}$.

Third-order IM: $2f_1 - f_2 = f_c - 3f_{mod}$, $2f_2 - f_1 = f_c + 3f_{mod}$
Fifth-order IM: $3f_1 - 2f_2 = f_c - 5f_{mod}$, $3f_2 - 2f_1 = f_c + 5f_{mod}$
Seventh order IM: $4f_1 - 3f_2 = f_c - 7f_{mod}$, $4f_2 - 3f_1 = f_c + 7f_{mod}$

Referring to Figure 1.93b, both LSB and USB are -1.6 dBm (cable loss $= 0.7$ dB) in power, which is 3 dB less than the measured power for the quadrature modulation test. The IM_3 is better than -35 dBc. IM products of much higher orders are totally suppressed.

Amplitude and Phase Imbalance   Both amplitude and phase imbalance (error) of an *I/Q* modulator can be calculated directly from the SSB performance plots. Assume phase error equals ϕ radian and amplitude error equals K, the sideband suppression, X, in dBc can be expressed as follows (see the *I/Q* Modulator Equations section for derivation):

$$\text{SSB suppression, } X(\text{dBc}) = 10\log\left(\frac{K^2 + 2 \times K \times \cos(\phi) + 1}{K^2 - 2 \times K \times \cos(\phi) + 1}\right) \qquad (1.23)$$

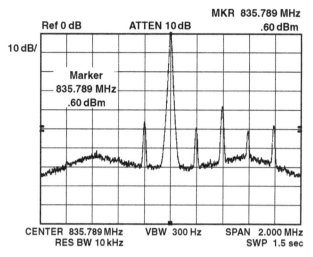

(**a**) Quadrature test (SSB upconversion)

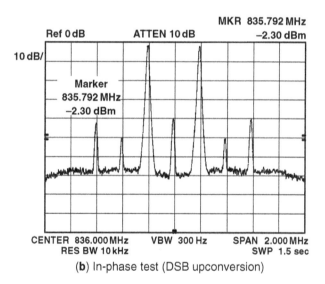

(**b**) In-phase test (DSB upconversion)

Figure 1.93   Typical SA900 output spectra.

Collecting the like terms and expressing ϕ in terms of K and X, it becomes

$$\phi = \cos^{-1}\left(\frac{10^{X/10} \times K^2 + 10^{X/10} - 1 - K^2}{2 \times K + 2 \times K \times 10^{X/10}}\right) \qquad (1.24)$$

For a given X, there will be a set of ϕ and K that satisfies (1.24). We can represent this relationship graphically, as shown in Figure 1.94. The contours show the phase and amplitude errors for SSB suppression, X, from –44 to –26 dBc. When X equals –40 dBc, phase error is less than 1.2° with an amplitude error of 0 dB. By the same token, the amplitude error is less than 0.2 dB with a 0° phase error.

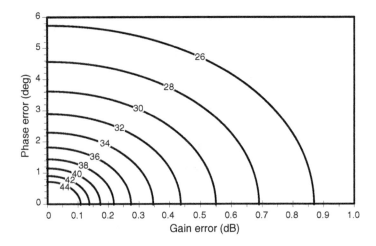

Figure 1.94 SSB suppression contours.

Spectral Mask To fully characterize the performance of an *I/Q* modulator, measurements of the power spectral density of various digital modulation schemes have to be made. Figure 1.95a and b shows the measured spectral masks of IS-54 and PDC standards, which designate $\pi/4$-DQPSK as the modulation format.

GMSK is a digital modulation scheme widely used for wireless and mobile communications. Figure 1.95c shows the spectral mask of the modulation format required by GSM, the digital cellular standard in Europe. At 200 kHz and 300 kHz carrier offset, the power of the signal is suppressed by 46 dB and 58 dB, respectively, which is well within the GSM specification.

Power ON Time The power ON time for the SA900 is mainly determined by the loop bandwidth of the on-board PLL frequency synthesizer. It can be measured by using the HP 53310A modulation domain analyzer set to the EXTERNAL TRIGGERED mode. The STROBE signal from three-wire bus is used to trigger the equipment. Figure 1.96 shows that the part can be powered up and locked in about 62 μs.

1.5.2.5 ISM Band Application
The FCC has assigned three bands for ISM type of application. The one below 1 GHz is from 902 to 928 MHz. This band becomes very attractive because users are allowed, without having a license, to transmit up to 1 W of power when frequency hopping or direct sequence CDMA is used. The wide bandwidth nature of the SA900 fits well into this application. Figures 1.97a and b are the output spectrum of the SA900 showing how well the image reject mixer works. A common IF (45 MHz) was chosen to be the offset frequency, and then injected externally into the VCO pins. The closest images are sitting at 45 MHz apart and are better than –36 dBc.

I/Q Modulator Equations Assume an imperfect *I/Q* modulator with gain error, K, and phase error, ϕ, modulated by quadrature *I/Q* signals (SSB upconversion) ω_m. Then the

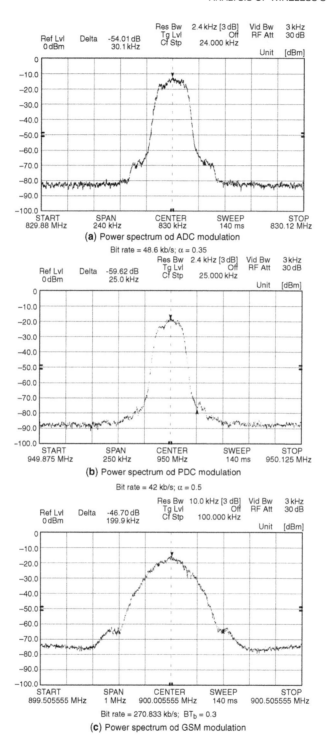

(a) Power spectrum od ADC modulation

Bit rate = 48.6 kb/s; α = 0.35

(b) Power spectrum od PDC modulation

Bit rate = 42 kb/s; α = 0.5

(c) Power spectrum od GSM modulation

Bit rate = 270.833 kb/s; BT_b = 0.3

Figure 1.95 SA900 ADC, PDC, and GSM outputs.

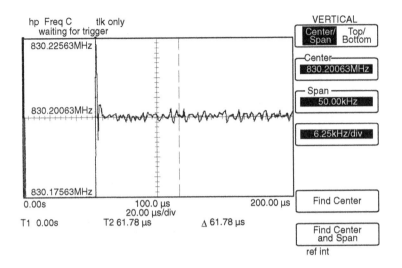

Figure 1.96 Power ON time measurement.

signal, $s(t)$, at the output of the *I/Q* modulator becomes

$$s(t) = K \cos(\omega_c t + \phi) \cos(\omega_m t) - \sin(\omega_c t) \cos(\omega_m t + 90°) \qquad (1.25)$$

Using trigonometric identity and letting $\omega_c - \omega_m = A$ and $\omega_c + \omega_m = B$, we obtain,

$$s(t) = \frac{K}{2} \cos[At + \phi] + \frac{K}{2} \cos[Bt + \phi] + \frac{1}{2} \cos[At] - \frac{1}{2} \cos[Bt] \qquad (1.26)$$

Assume that the information is in LSB—that is, A—and the spur is the USB, that is, B, we have

$$\text{Signal} = \frac{K}{2} \cos A \cos \phi + \frac{1}{2} \cos A - \frac{K}{2} \sin A \sin \phi \qquad (1.27)$$

$$\text{Noise} = \frac{K}{2} \cos B \cos \phi + \frac{1}{2} \cos B - \frac{K}{2} \sin B \sin \phi \qquad (1.28)$$

To find the power, we have to evaluate the envelope (amplitude) of these two signals. Recall that for any given bandpass signal in rectangular form

$$\text{Bandpass signal} = X \cos \omega t - Y \sin \omega t$$

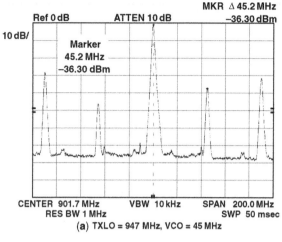

(a) TXLO = 947 MHz, VCO = 45 MHz

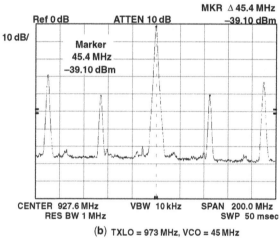

(b) TXLO = 973 MHz, VCO = 45 MHz

Figure 1.97 Output spectrum of the SA900 in the ISM band.

and the envelope is

$$\text{Envelope} = \sqrt{X^2 + Y^2}$$

Therefore, from (1.27) and (1.28)

$$\text{Signal} = \left[\left(\frac{K}{2} \cos \phi + \frac{1}{2} \right)^2 + \left(\frac{K}{2} \sin \phi \right)^2 \right]^{0.5} \qquad (1.29)$$

$$\text{Noise} = \left[\left(\frac{K}{2} \cos \phi - \frac{1}{2} \right)^2 + \left(\frac{K}{2} \sin \phi \right)^2 \right]^{0.5} \qquad (1.30)$$

Finally, the *S/N* ratio can be found by taking 20 log the ratio of (1.29) and (1.30):

$$S/N = 10 \log \left(\frac{K^2 + 2K \cos \phi + 1}{K^2 - 2K \cos \phi + 1} \right) \tag{1.31}$$

1.6 BUILDING BLOCKS

Our earlier block diagram already referred to a variety of integrated circuits that can be used to build a wireless system—in particular, a portable telephone. If we open any of the earlier-mentioned magazines and go through the advertising section, we will find many suppliers of subsystems/components that fit this requirement.

They can be split into various technologies. The following table is a product overview of next-generation advances in wireless technology taken from an ad of Stanford Microdevices, one of the very aggressive component/subsystem companies.

- New SX amplifiers features 2.4-GHz MMICs
- GaAs heterojunction bipolar transistor (HBT) discrete transistors (100 mW to 10 W)
- InP/GaAs (indium phosphate/gallium arsenide) high-linearity gain blocks
- Millimeter-wave product line featuring ICs for LMDS
- Silicon-germanium product line (gain blocks and low-noise amplifiers)
- SXQ power modules for cellular and PCS applications
- Power modules featuring 4- to 25-W power amplifiers for land mobile amplifiers
- 30- to 120-W silicon LDMOS power transistors for PCS, CDMA, W-CDMA
- SAW files featuring integrated transceiver modules
- Electro-optic and datacomm products.

In the beginning of this chapter, we looked at block diagrams and now into building blocks. Most manufacturers will try to move to the highest possible integration while avoiding external components. Figure 1.98, a test circuit for Motorola's MC13109 cordless telephone subsystem IC, shows the level of complexity needed to test such an IC with all its functionality.

Another key issue at those frequencies is to really know a component's frequency dependencies. Figure 1.99 shows a good example. Typical components are capacitors and inductors as reactive elements, and resistors as passive elements. Depending on the type of integration, one may also have to look into the noise performance of integrated resistors. To look into this feature is best discussed with the foundries. Along these lines, the dielectric material on which these surface-mounted components are mounted plays a big role and must be addressed individually.

Besser Associates of Los Altos, California, among other firms, offers a very nice one-day short course titled "Wireless Circuit Components: Measurements, Models and Data Extraction."

Table 1.11 presents a sampling of RFICs for wireless applications. These types of ICs, which were taken from Motorola are made in similar form by many companies. The selection guide is useful in starting off with a design that is less integrated, therefore allowing the designer to have access to portions of the circuit that later "disappear" inside the IC.

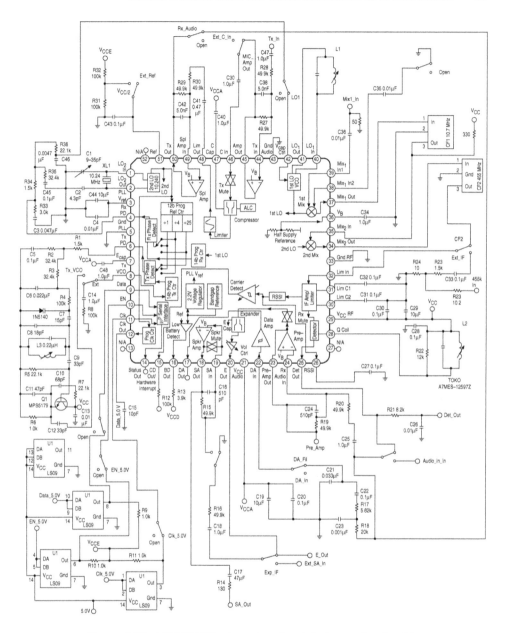

Figure 1.98   MC13109FB universal cordless telephone subsystem IC.

1.7   SYSTEM SPECIFICATIONS AND THEIR RELATIONSHIP TO CIRCUIT DESIGN

Wireless communication involves a large range of signal powers—from levels on the order of 10–18 W (at the receiver input) to 102 W (at a base-station transmitter output). A receiver must be able to demodulate signals that have been attenuated billions of time through propagation; a transmitter must be able to produce a properly modulated signal at a frequency

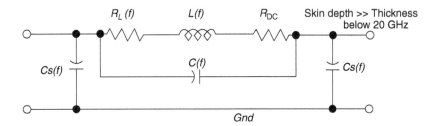

Figure 1.99   Equivalent model of a film resistor above a ground plane.

suitable for propagation, at a level high enough to overcome worst-case propagation losses and provide a useful signal at the receiver. Gain is therefore an essential attribute of wireless systems. Because no single active device can provide all the gain required for transmission or reception, we must distribute the gain among multiple stages, designing each for optimum performance across the power span it bridges.

Two inescapable realities impose limits on the gain and absolute power output, we may achieve with a given circuit. All real electrical and electronic networks generate noise to some degree, and all real electrical and electronic networks distort the signals applied to them to some degree. A signal weaker than a circuit's inherent noise cannot be amplified by that circuit because it remains indistinguishable from the noise. A signal that exceeds the power-handling capability of the circuit to which it is applied may be degraded, even rendered useless, by the resulting distortion. The following sections examine issues particular to system noise and linearity performance.

1.7.1   System Noise and Noise Floor

Assuming that a system's gain is sufficient, the weakest signal that may be processed satisfactorily, a figure of merit referred to as noise floor or (in receivers) minimum detectable signal (MDS), is limited by thermal noise, assumed to be equal to the noise power available from a resistor at 290 K (about 17°C or 62°F), an arbitrary reference value near standard room temperature. The noise power is equal to

$$P_n = kTB \tag{1.32}$$

where P_n is the noise power, k is Boltzmann's constant ($1.38 \cdot 10^{-23}$ W/K), T is the temperature in kelvins, and B is the bandwidth (in Hertz) in which the noise appears. For $T = 290$ K, P_n is, therefore, $4.00 \cdot 10^{-18}$ W, or -174 dBm in a 1-Hz bandwidth. Increasing B to a value suitable for digital communications, such as 160 kHz for a GSM system, admits more noise to the network, raising the minimum noise power against which an incoming signal must compete to -122 dBm.

If the noise figure and bandwidth are known, the system noise floor can be calculated using the equation

$$\text{Noise floor} = -174\,\text{dBm} + NF + 10\log B \tag{1.33}$$

The trouble with this equation is that the "integrated" bandwidth depends so much on the selectivity shape factor, which is not always known. Figure 1.100 shows the translation of the bandwidth of a single tuned circuit with its Gaussian shape into its rectangular equivalent. The transformation is done by sizing the rectangle such that $A' = A$ and $B' = B$; when this is true, the area of the rectangular equals the area under the curve.

Table 1.11 Sample RF Component/IC Selector Guide

RFICs

Downconverters

Device	RF Freq. Range (MHz)	Supply Volt. Range (V/dc)	Supply Current (mA) (Typ)	LNA Gain (dB) (Typ)	LNA NF (dB) (Typ)	Mixer Conv. Gain (dB) (Typ)	Mixer NF(dB) (Typ)	Package	System Applicability
MC13142	dc to 1800	2.7–6.5	13.5	17	1.8	–3.0	12	SO-8	ISM, Cellular, PCS
MRFIC1502	1575	4.5–5.5	52	20	–	45	9.5	LQFP-48	GPS
MRFIC1814	1800 to 2000	2.7–4.5	10	17	2.5	8.0	10	TSSOP-16	DCS1800, PCS, PHS

Upconverters/Exciters

Device	RF Freq. Range (MHz)	Supply Volt. Range (V/dc)	Supply Current (mA) (Typ)	Standby Current (Typ)	Conv. Gain (dB) (mA) (Typ)	Output IP3 (dBm) (Typ)	Package	System Applicability
MRFIC0954	800 to 1000	2.7–5.0	65	5.0	31	28	TSSOP-20EP	CDMA, TDMA, ISM
MRFIC1813	1700 to 2000	2.7–4.5	25	0.1	15	11	TSSOP-16	DCS1800, PCS
MRFIC1854	1700 to 2000	2.7–5.0	70	5.0	31	23	TSSOP-20EP	CDMA, TDMA, PCS

Power Amplifiers

Device	Freq. Range (MHz)	Supply Volt. Range (V/dc)	Saturated P_{out} (dBm) (Typ)	PAE% (Typ)	Gain P_{out}/P_{in} (dB) (Typ)	Package	System Applicability
MRFIC0913	800 to 1000	2.7–7.5	35	50	25	PFP-16	GSM
MRFIC0917	800 to 1000	2.7–5.5	34.5	45	22.5	PFP-16	GSM
MRFIC0919	800 to 1000	3.0–5.5	35.3	48	32.3	TSSOP-16EP	GSM
MRFIC1805	1500 to 2200	2.7–5.0	25	28	20	TSSOP-16	PHS, DECT, PCS

(continued)

Table 1.11 *(Continued)*

Power Amplifiers

Device	Freq. Range (MHz)	Supply Volt. Range (V/dc)	Saturated P_{out} dBm (Typ)	PAE% (Typ)	Gain P_{out}/P_{in} dB (Typ)	Package	System Applicability
MRFIC1807	1500 to 2200	3.0–5.0	26.8	35	8.0	SO-16	DECT, PCS
MRFIC1817	1700 to 2000	2.7–5.0	33.5	42	30.5	PFP-16	DCS1800, PCS
MRFIC1818	1700 to 2000	2.7–6.0	34.5	42	31.5	PFP-16	DCS1800, PCS
MRFIC1819	1700 to 2000	3.0–5.0	33	40	27	TSSOP-16EP	DCS1800, PCS
MRFIC1856	800 to 1000	3.0–5.6	32	50	32	TSSOP-20EP	TDMA, CDMA, AMPS
	1700 to 2000		30	35	30		TDMA, CDMA, PCS
MRFIC2006	500 to 1000	1.8–4.0	15.5	25	23	SO-8	Cellular, ISM, CT2

RF Building Blocks

Amplifiers

Device	RF Freq. Range (MHz)	Supply Volt. Range V/dc	Supply Current mA (Typ)	Standby Current μA(Typ)	Small Signal Gain dB (Typ)	Output IP3 dBm (Typ)	NF dB (Typ)	Package	System Applicability
MC13144	100 to 2000	1.8–6.0	8.5	1	17	−5.0	1.4	SO-8	ISM, PCS, Cellular
MRFIC0915	100 to 2500	2.7–5.0	2.0	–	16.2	4.0	1.9	SOT-143	ISM, PCS, Cellular
MRFIC0916	100 to 2500	2.7–5.0	4.7	–	18.5	11	1.9	SOT-143	ISM, PCS, Cellular
MRFIC0930	800 to 1000	2.7–4.5	8.5	20	19	10	1.7	SO-8	GSM, AMPS, ISM
MRFIC1808DM	1700 to 2100	2.7–4.5	5.0	8.0	18	13	1.6	Micro-8	DCS1800, PCS
MRFIC1830	1700 to 2100	2.7–4.5	9.0	20	18.5	8.5	2.1	Micro-8	DCS1800, PCS
MRFIC1501	1000 to 2000	3.0–5.0	5.9	–	18	10	1.1	SO-8	GPS

Mixers

Device	RF Freq. Range (MHz)	Supply Volt. Range (V/dc)	Supply Current mA (Typ)	Standby Current μA (Typ)	Conv. Gain dB (Typ)	Input IP3 dBm (Typ)	Package	System Applicability
MC13143	DC to 2400	1.8–6.0	4.1	–	−2.6	16	SO-8	ISM, PCS, Cellular

PLL Synthesizers

Frequency (MHz)	Supply Voltage (V)	Nominal Supply Current (mA)	Phase Detector	Standby	Interface	Device	Suffix/Case
20 @ 5.0 V	3.0–9.0	7.5 @ 5 V	Single-ended 3-state, double-ended	No	Parallel	MC145151-2	DW1751F
			Double-ended			MC145152-2	DW/751F
			Single-ended 3-state, double-ended		Serial	MC145157-2	DW/751G
						MC145158-2	DW/751G
60 @ 3.0 V	2.5–5.5	3 @ 3 V	Two single-ended 3-state	Yes		MC145162[a]	P/648, D/751B
85 @ 3.0 V	2.5–5.5	3 @ 3 V			Serial	MC145162-1[a]	D/751B
100 @ 3.0 V 185 @ 5.0 V	2.7–5.5	2 @ 3 V 6 @ 5 V	Single-ended 3-state, double-ended	No		MC145170-2	P/648, D/751B, DT/948C
550,60	1.8–3.6	3	Loop 1 = Current source/sink/float Loop 2 = Three-state	Yes		MC145181[a](46a)	FTA873C
1000	2.7–5.5	4.25	Current source/sink/float	No	Parallel	MC12181	D/751B
1100	4.5–5.5	7 @ 5 V	Current source/sink/float, double-ended	Yes	Serial	MC145191	F/751J, DT/948D
1100	2.7–5	6 @ 2.7 V				MC145192	F/751J, DT/948D

(continued)

Table 1.11 (Continued)

Frequency (MHz)	Supply Voltage (V)	Nominal Supply Current (mA)	Phase Detector	Standby	Interface	Device	Suffix/ Case
1100	2.7–5.5	12	Two current source/sink/float, double-ended			MC145220[a]	F/803C, DT/9480
1200, 550	1.8–3.6	4	Loop 1 = current source/sink/float Loop 2 = three-state			MC145225[a](46a)	FTA/873C
2000	4.5–5.5	12 @ 5 V	Current source/sink, double-ended			MC145201	F/751J, DT/948D
2000	2.7–5.5	4 @ 3 V				MC145202	F/751J, DT/9480
2200, 550	1.8–3.6	5	Loop 1 = current source/sink/float Loop 2 = Three-state			MC145230[a](46a)	FTA/873C
2500	2.7–5.5	9.5	Current source/sink/float with dual outputs	No		MC12210	D/751B, DT/948E
2800	4.5–5.5	3.5	Current source/sink/float		None	MC12179	D/751

[a]Dual PLL.

84

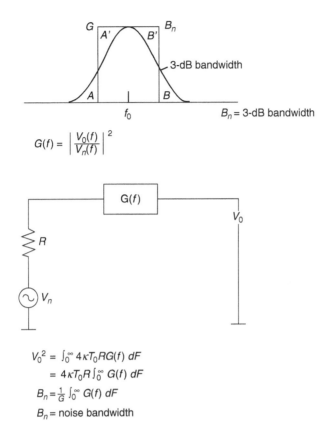

$$G(f) = \left| \frac{V_0(f)}{V_n(f)} \right|^2$$

$$V_0^2 = \int_0^\infty 4\kappa T_0 R G(f)\, dF$$
$$= 4\kappa T_0 R \int_0^\infty G(f)\, dF$$
$$B_n = \frac{1}{G} \int_0^\infty G(f)\, dF$$
$$B_n = \text{noise bandwidth}$$

Figure 1.100 Graphical and mathematical explanation of the noise bandwidth from a comparison of the Gaussian-shaped bandwidth to the rectangular filter response.

Signal-to-Noise Ratio (S/N, SNR) and Sensitivity Successful radiocommunication depends on the achievement of a specified minimum ratio of signal power to noise power, expressed in decibels, at the output of the receiver. The input voltage, expressed in absolute units or decibels relative to a microvolt ($dB\mu V$), necessary to achieve a particular signal-to-noise ratio in a particular bandwidth may be specified as a figure of merit called sensitivity. Because techniques used to measure S/N actually measure the ratio of signal-plus-noise to noise, specifications may refer to (or imply) $S + N/N$ or $(S + N)/N$ rather than S/N. The difference between $(S + N)/N$ and S/N becomes negligible at high ratios of signal to noise; even at an SNR of 10 dB—a common value—the difference is only 0.46 dB [28].

Most receivers are designed to operate optimally when connected to an antenna system of a specified impedance (commonly 50 Ω), but relatively few receivers exhibit this design load impedance at their input terminals; that is, they are not designed for a conjugate input match. It is therefore customary to specify sensitivity in terms of "open circuit" voltage— the signal voltage that, with the receiver's antenna input terminated in its design antenna impedance, results in the desired ratio of signal to noise. If, as is commonly the case, the input voltage for a given SNR is determined using instrumentation calibrated in terms of closed-circuit voltage—that is, in terms of voltage across a load resistance equal to the instrument's source resistance—the voltage indicated will be 1/2 the open-circuit value for the SNR specified [29]. By convention, the open-circuit measurement condition is indicated by a sensitivity specification in volts of electromotive force (EMF). Specifying sensitivity in terms of available signal power (usually decibels relative to 1 mW or dBm) eliminates this open/closed-circuit confusion.

SINAD Ratio   Extending the measurement of signal-plus-noise to noise to include distortion results in a figure of merit called SINAD (signal-plus-noise-and-distortion), commonly applied to FM receivers

$$\text{SINAD} = 10\log_{10}\frac{S+N+D}{N+D} \tag{1.34}$$

where SINAD is in decibels, S is signal power, N is noise power, and D is distortion power. At a SINAD ratio of 12 dB—a common specification—the noise-and-distortion power is 25% that of the desired signal. As is true of $(S+N)/N$, SINAD closely approximates S/N at high ratios of signal to noise.

Bit Error Rate and Noise   For digital systems, signal-to-noise ratio and bit error rate are related. As introduced earlier, depending on the waveform, coding, and filtering, different BERs are related to particular SNRs (Figure 1.101). The SNR also depends on the interference from the synthesizer, as will be seen in the section on adjacent channel power ratio (ACPR). Since a digital transmitter is always involved, in the on-the-air testing, the laboratory measurement of SINAD cannot quite be correlated. At the IF level, we find the handover point between the transceiver's analog front end and its digital portion, and a standard noise-figure test setup, such as the one by Agilent, is still a valuable instrument for evaluating noise figure and signal-to-noise ratio.

Here is an example for a digital measurement. Assume that we use a test generator like the SMIQ to evaluate a front end having an 8.8-dB noise figure. A 384-kbps $\pi/4$ DQPSK digital modulation signal (–5 dBm) is applied to the transceiver front end. Testing at 2.47 GHz,

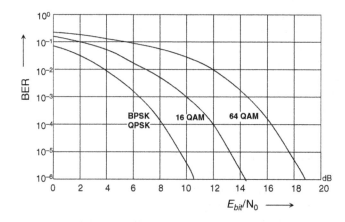

Figure 1.101   Bit error rate versus Ebit/N0 for BPSK/QPSK, 16-QAM and 64-QAM, showing the significantly greater SNR necessary for a given BER as the number of signal states is increased.

we find that the error vector magnitude (EVM; defined later in this section) is about 3.89% in transmit and about 1.37% in receive. To determine the receive sensitivity of the RF front end, the signal from the digitally modulated generator (at –95 dBm) is applied to the RF front end in receive mode. The measured EVM at 2.47 GHz is now 20%. By using the equation

$$S/N = -20 \log(\text{EVM}) \tag{1.35}$$

we see that the ratio that corresponds to an EVM of 20% is 14 dB.

Since

$$S/N = \frac{(E_S R_S)}{N_0 B} \tag{1.36}$$

where E_S is the energy of a symbol, R_S is the symbol rate, N_0 is the noise power density, and B is the bandwidth. Assuming a $\pi/4$ DQPSK signal and ratio of 14 dB, it can be determined that

$$E_S/N_0 = S/N(B/R_S) = 17 \, \text{dB} \tag{1.37}$$

Assuming that the BER is 10^{-5}, the required E_S/N_0 is 10 dB. Hence, from the definition of the sensitivity

$$\begin{aligned}
\text{Sensitivity} &= -144 + NF + R_S + E_S/N_0 \\
&= -144 + 8.8 + 10 \log(384/2) + 10 \\
&= -102 \, \text{dBm}
\end{aligned} \tag{1.38}$$

The value –144 is derived from kT_0 by taking the bandwidth (1 kHz) into consideration.

The result obtained at (1.38) is consistent with the obtained E_S/N_0 of 17 dB from the measurement at –95 dBm input power (since $-102 + [17 - 10] = -95$). Hence, it was determined that the receive sensitivity of the RF front end is approximately –102 dBm. We can then determine the dynamic range (DR) from

$$DR = P_{-1\text{dB}} - \text{Sensitivity} = -21 - 102 = 81 \, \text{dB} \tag{1.39}$$

Noise Factor and Noise Figure   The noise floor predicted by (1.32) cannot be achieved and maintained in any real network or system of networks because all real networks generate noise. Determining how closely the SNR achieved at a given input level approaches the SNR achievable at that input level in a noiseless network is therefore of high interest to the circuit and system designer. The degree to which a network's noise contribution degrades the noise floor predicted by (1.32) is evaluated by its noise factor (F), which is expressed as the ratio

$$F = \frac{N_{\text{in}} + N_{\text{added}}}{N_{\text{in}}} \tag{1.40}$$

where F is noise factor, N_{in} is the noise power available from the source and N_{added} is the noise power added by the network, with both powers determined in the same bandwidth. Expressing this ratio in decibels ($10 \log_{10} F$) returns noise figure (NF), a bandwidth-independent figure of merit of great value in evaluating the noise performance of networks and communication systems. We can also express NF as the ratio of the network's input

SNR to its output SNR:

$$\text{NF} = 10 \log_{10}\left[\frac{(S_{\text{in}}/N_{\text{in}})}{(S_{\text{out}}/N_{\text{out}})}\right] \tag{1.41}$$

where NF is noise figure in decibels, S is signal power, and N is noise power, with the input and output values of these quantities signified by the subscripts and all powers determined in the same bandwidth. The noise figure of an ideal noiseless network is 0 dB; for all real, noisy networks, NF is positive. The NF of a lossy passive device is equal to its insertion loss.

Noise figure can also be defined for antennas and antenna systems:

$$\text{NF}_{\text{ant}} = 10 \log_{10} \frac{N_t + N_{\text{ant}}}{N_t} \tag{1.42}$$

where NF_{ant} is the antenna system noise figure in decibels, N_t is the antenna's system's thermal noise power, as defined in (1.32), and N_{ant} is the total noise power picked up by the antenna system. From the lower end of the radio spectrum, and decreasingly up to approximately 400 MHz, noise intercepted by an antenna system from atmospheric, man-made, and galactic sources will dominate NF_{ant} (Figure 1.102), and N_t can be considered as equivalent to the noise power of a resistor at 290 K. Atmospheric noise subsides above 40 MHz; from this region to perhaps 700 MHz, N_t is still largely negligible, with noise from man-made and/or sky sources largely determining an antenna's noise figure. At these frequencies, an antenna's directivity and orientation relative to noise sources play a critical role in determining its noise figure; a strongly directive antenna located in a city suburb, for instance, exhibits a significantly higher noise figure when pointed toward the city center than it exhibits when pointed toward a more sparsely populated area. At frequencies above about 700 MHz, the RF environment is generally quieter, allowing receiver NF, not antenna NF, to more routinely limit a wireless receiver's sensitivity. At these frequencies, the concept

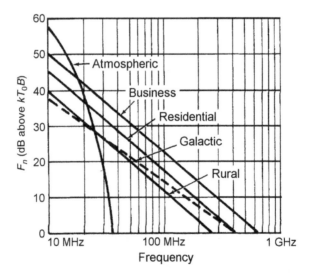

Figure 1.102 The higher the frequency, the more quiet the RF environment becomes, although the noise profile of specific sources may contradict this general rule.

of noise temperature is commonly used to evaluate the quietness of a receiver, its antenna system, and its RF environment. Noise temperature is particularly useful in designing systems for space communication, an antenna for which may be cooled, through radiation of a portion its own noise into the RF-cold sky, to a noise temperature an order of magnitude below 290 K.

Noise Figure of Cascaded Networks   The noise figure of two networks in cascade may be determined from

$$\text{NF}_{\text{total}} = 10 \, \log_{10} \left(F_1 + \frac{F_2 - 1}{G_1} \right) \tag{1.43}$$

where NF is noise figure in dB, F_1 is the noise factor of the first network, F_2 is the noise factor of the second network, and G_1 is the gain (as a numerical ratio, not in dB) [30]. The noise figure of a system with more than two stages can be evaluated through repeated iterations of (1.43). Note that (1.43) assumes two conditions: (1) that F_1 and F_2 are determined in the same bandwidth, and (2) that the networks' input and output terminations are resistive—a condition that is commonly not true of RF amplifiers optimized for lowest noise. Equation (1.43) can be expanded to account for bandwidth, but accounting for complex terminations requires the use of noise correlation matrix techniques as described in Ref. [28].

1.7.2   System Amplitude and Phase Behavior

If we could build electronic systems that were absolutely amplitude- and phase-linear, radiocommunication system design would be greatly simplified. An amplifier designed for a power gain of 10 dB, for example, would merely increase the magnitude of all signals at its input by a factor of 10, regardless of their frequencies, while perfectly maintaining their relative phases. But all real electrical and electronic networks, even those designed (or supposed) to be amplitude- and phase-linear, exhibit amplitude and phase nonlinearity to some degree, just as they all generate noise to some degree.

The effects of amplitude nonlinearity, generically referred to as nonlinear distortion, include the generation, through harmonic distortion and/or intermodulation distortion (IMD), of output signals at frequencies not present at a system's input. Nonlinear distortion also results in gain compression—changes in system gain with changes in input-signal level. By convention, when workers in electronics refer to or consider a network's "linearity," they usually mean its amplitude linearity; likewise, by "distortion" they usually mean nonlinear distortion.

The effects of phase or frequency nonlinearity are generically referred to as linear distortion because they occur independently of signal amplitude and polarity. We often intentionally apply linear distortion through filtering, which modifies the amplitude relationships among existing spectral components of a signal without creating any new frequencies. Another linear-distortion effect, phase or delay distortion, results in the delay of signals of differing frequencies by differing amounts of time. In a system where signal phase conveys information, as is true of most wireless links, phase distortion can seriously degrade communication.

The fact that all real networks are amplitude- and angle-nonlinear to some degree means that all real networks modify the amplitude and angle characteristics of the signals they handle. What is perhaps less obvious is that subjecting a signal to amplitude and angle nonlinearities causes "crosstalk" between its amplitude and angle characteristics.

For example, through a nonlinear distortion effect called AM-to-PM conversion, changes in a signal's amplitude result in changes in its phase. Filtering an angle-modulated signal produces the reverse effect; deprived by the filter's selectivity of spectral components necessary to maintain its envelope constancy, it emerges from the filter as a combination of angle and amplitude modulation.

Spectral Considerations of Analog and Digitally Modulated Signals   Many wireless modulation schemes exist and even more are proposed; all use angle (usually phase) modulation or a combination of phase and amplitude modulation using an emission format engineered to achieve multiple goals of information throughput, robustness, spectral and hardware efficiency, and reproducibility. Digital modulation is standard in mainstream wireless applications because it allows increased channel capacity, and immunity to noise and distortion, compared to analog systems. At the circuit-design level, the particulars of whether, or to what degree, a modulation scheme is AM, PM, analog, or digital, matters less than the actual angle and amplitude characteristics of the signal(s) involved, and the tolerances within which these characteristics can be expected and allowed to vary.

A system's amplitude linearity is of major concern because of its relation to energy efficiency, receiver dynamic range, and transmitter spectral purity. Energy efficiency, important because many wireless applications use battery power, generally decreases inversely with amplitude linearity. Yet, sufficient amplitude linearity must be guaranteed in the front-end circuitry of wireless receivers subjected to multiple strong signals, and in all wireless receiver and transmitter stages handling variable-envelope signals.

Some digital modulation schemes, designed to be distortion-tolerant, produce constant-envelope signals; Gaussian minimum shift keying (GMSK) and Feher's quadrature phase-shift keying (FQPSK) are examples of these. Such signals can be processed in highly nonlinear circuitry—for example, an RF power amplifier operated in saturation to maximize efficiency—without degrading their spectral composition. Other digital modulation schemes result in the emission of, ideally, of a single PM sideband, with the carrier and all other sideband components suppressed. Such a signal exhibits, of necessity, phase and amplitude variations; the IS-54 cellular standard's $\pi/4$ differential quadrature phase-shift keying ($\pi/4$ DQPSK) scheme, which results in envelope fluctuations of 3–6 dB, exemplifies this [31]. Intermodulation distortion among such a signal's spectral components can generate products that fall outside the bandwidth painstakingly controlled in the modulation process. This spectral regrowth must be minimized to prevent adjacent-channel interference (Figure 1.103).

For a better understanding, Figure 1.104 shows the various channels and the energy levels associated with them.

Amplitude Linearity Issues and Figures of Merit   A network's amplitude nonlinearity can be characterized by the expansion

$$y = k_1 f(x) + k_2 [f(x)]^2 + k_3 [f(x)]^3 + \text{higher-order terms} \qquad (1.44)$$

where y represents the output, the coefficients k_n represent complex quantities whose values can be determined by an analysis of the output waveforms, and $f(x)$ represents the input. Even though all practical networks exhibit amplitude nonlinearity, we can (and often do) refer to many networks as "linear." We say this of networks that are sufficiently amplitude-linear for our purposes—for example, weakly nonlinear networks in which small-signal

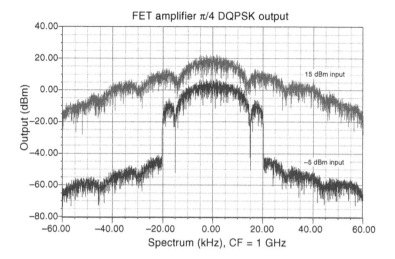

Figure 1.103 Spectral regrowth results when a variable-envelope emission, π/4 DQPSK in this case, is subjected to significant nonlinear distortion. This graph shows the simulated performance of a MESFET power amplifier operating at 1 GHz; the amplifier is 6 dB into compression when driven at 15 dBm.

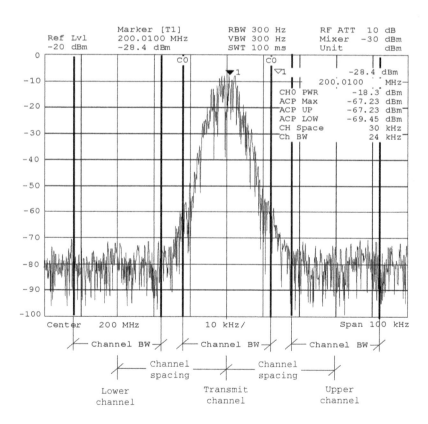

Figure 1.104 NADC signal and parameters, including channel spacing and channel bandwidth.

operation is assumed even though the signal levels involved are sufficient to cause slight distortion. For many practical purposes, the first three terms of (1.44) adequately describe such a network's nonlinearity:

$$y = k_1 f(x) + k_2 \left[f(x) \right]^2 + k_3 \left[f(x) \right]^3 \qquad (1.45)$$

In adopting this simplification, we also assume that the nonlinearity is frequency independent—that is, the network has sufficient bandwidth to allow all the products predicted by (1.45) to appear at its output terminals unperturbed [32].

When multiple signals are present in a network, even weak nonlinearity can result in profound consequences. To illustrate this, we will let $f(x)$ consist of two sinusoidal signals:

$$f(x) = A_1 \cos \omega_1 t + A_2 \cos \omega_2 t \qquad (1.46)$$

We will assume that ω_1 and ω_2 are close enough so that the coefficients k_i can be considered equal for both signals. We will also assume for simplicity that all the k_i are real. If (1.45) describes the network's response to an input $f(x)$, the response will be

$$
\begin{aligned}
y = &\, k_1(A_1 \cos \omega_1 t + A_2 \cos \omega_2 t) + k_2(A_1 \cos \omega_1 t + A_2 \cos \omega_2 t)^2 \\
&+ k_3(A_1 \cos \omega_1 t + A_2 \cos \omega_2 t)^3 \\
= &\, k_1(A_1 \cos \omega_1 t + A_2 \cos \omega_2 t) \\
&+ k_2 \left[A_1^2 \frac{1+\cos 2\omega_1 t}{2} + A_2^2 \frac{1+\cos 2\omega_2 t}{2} + A_1 A_2 \frac{\cos(\omega_1+\omega_2)t + \cos(\omega_1-\omega_2)t}{2} \right] \\
&+ k_3 \left\{ \left[A_1^3 \left(\frac{\cos \omega_1 t}{2} + \frac{\cos \omega_1 t}{4} + \frac{\cos 3\omega_1 t}{4} \right) + A_2^3 \left(\frac{3\cos \omega_2 t}{4} + \frac{\cos 3\omega_2 t}{4} \right) \right] \right. \\
&+ A_1^2 A_2 \left[\frac{3}{2} \cos \omega_2 t + \frac{3}{4} \cos(2\omega_1 - \omega_2)t + \frac{3}{4} \cos(2\omega_1 - \omega_2)t \right] \\
&\left. + A_2^2 A_1 \left[\frac{3}{2} \cos \omega_1 t + \frac{3}{4} \cos(2\omega_2 + \omega_1)t + \frac{3}{4} \cos(2\omega_2 - \omega_1)t \right] \right\} \qquad (1.47)
\end{aligned}
$$

The k_1 term of equation (1.47) represents the results of amplitude-linear behavior. No new frequency components have appeared; the two sine waves have merely been "rescaled" by k_1.

The second- and third-order terms of (1.47) represent the effects of harmonic distortion and intermodulation distortion. Second-order effects include second-harmonic distortion (the production of new signals at $2\omega_1$ and $2\omega_2$) and IMD (the production of new signals at $\omega_1 + \omega_2$ and $\omega_1 - \omega_2$). Third-order effects include gain compression, third-harmonic distortion (the production of new signals at $3\omega_1$ and $3\omega_2$), and IMD (the production of new signals at $2\omega_1 \pm \omega_2$ and $2\omega_2 \pm \omega_1$).

Gain Compression Gain compression occurs when a network cannot increase its output amplitude in linear proportion to an amplitude increase at its input; gain saturation occurs when a network's output amplitude stops increasing (in practice, it may actually decrease) with increases in input amplitude. We can deduce from (1.47) that the amplitude of the $\cos \omega_1 t$ signal has become

$$A_1' = k_1 A_1 + k_3 \left(\frac{3}{4} A_1^3 + \frac{3}{2} A_1 A_2^2 \right) \qquad (1.48)$$

Because k_3 will normally be negative, a large signal $A_2 \cos \omega_2 t$ can effectively mask a smaller signal $A_1 \cos \omega_1 t$ by reducing the network's gain. This third-order effect, known as blocking or desensitization when it occurs in a receiver, is a special case of gain compression. The presence of additional signals results a greater reduction in gain; the gain reduction for each signal is a function of the relative levels of all signals present. A receiver's blocking behavior may be characterized in terms of the level of off-channel signal necessary to reduce the strength of an in-passband signal by a specified value, typically 1 dB; alternatively, the decibel ratio of the off-channel signal's power to the receiver's noise-floor power may be cited as blocking dynamic range. Desensitization may be also characterized in terms of the off-channel-signal power necessary to degrade a system's SNR by a specified value.

Multiple signals need not be present for gain compression to occur. If only one signal is present, the ratio of gain with distortion to the network's idealized (linear) gain is

$$A_1' = \frac{k_1 + k_3 \left(\frac{3}{4} A_1^2\right)}{k_1} \tag{1.49}$$

This is referred to as the single-tone gain-compression factor. Figure 1.105 shows how the k_3 term causes a network's gain to deviate from the ideal. The point at which a network's power gain is down 1 dB from the ideal for a single signal is a figure of merit known as the 1-dB compression point (P_{-1dB}). Many networks (including many receiving and low-level transmitting circuits, such as low-noise amplifiers, mixers, and IF amplifiers) are usually operated under small-signal conditions—at levels sufficiently below P_{-1dB} to maintain high linearity. As we will see, however, some networks (including power amplifiers for wireless systems) may be operated under large-signal conditions—near or in compression— to achieve optimum efficiency at some specified level of linearity. Figure 1.106 shows what happens when a digital emission that uses amplitude to convey information is subjected to amplitude compression.

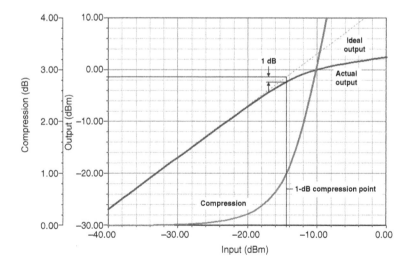

Figure 1.105   The power level at which a network's power output is down 1 dB relative to that of its ideally linear equivalent is a figure of merit known as the 1-dB compression point (P_{-1dB}). The 1-dB compression point can be expressed relative to input power ($P_{-1dB,in}$) or output power ($P_{-1dB,out}$). For the amplifier simulated here, $P_{-1dB,in} \approx -14.5\,dBm$ and $P_{-1dB,out} \approx -1.3\,dBm$.

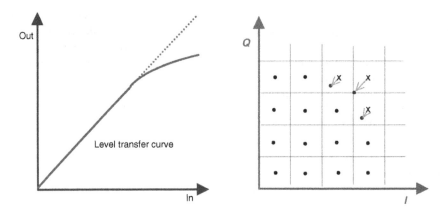

Figure 1.106 Influence of differential amplitude error (compression) on a QAM constellation.

Intermodulation The new signals produced through intermodulation distortion (IMD) can profoundly affect the performance even of systems operated far below gain compression (Figure 1.107). IMD products of significant power can appear at frequencies remote from, in and/or near the system passband, resulting in demodulation errors (in reception) and interference to other communications (in transmission). Where an IMD product appears relative to the passband depends on the passband width and center frequency, the frequencies of the signals present at the system input, and the order of the nonlinearity involved. These factors also determine the strength of an IMD product relative to the desired signal.

Second-order IMD (IM_2) results, for an input consisting of two signals ω_1 and ω_2, in the production of new signals at $\omega_1 + \omega_2$ and $\omega_1 - \omega_2$; third-order IMD (IM_3) results, for an input consisting of two signals ω_1 and ω_2, in the production of new signals at $2\omega_1 \pm \omega_2$ and $2\omega_2 \pm \omega_1$.

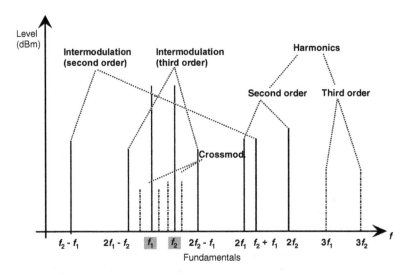

Figure 1.107 Relationships between fundamental and spurious signals, including harmonics and products of intermodulation.

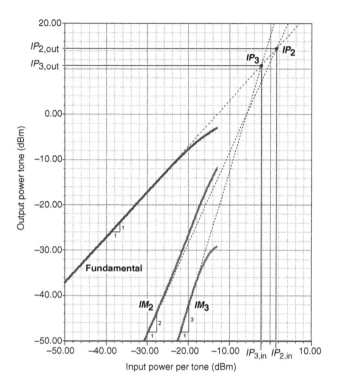

Figure 1.108 The level at which the power of one of a network's IM products equals that of the network's linear output is a figure of merit known as the intermodulation intercept point (IP). The intercept point for a given IM order n can be expressed, and should always be characterized, relative to input power ($IP_{n,\text{in}}$) or output power ($IP_{n,\text{out}}$); the IP_{in} and IP_{out} values differ by the network's linear gain. For the amplifier simulated here, $IP_{2,\text{in}} \approx 1.5\,\text{dBm}$, $IP_{2,\text{out}} \approx 14.5\,\text{dBm}$, $IP_{3,\text{in}} \approx -2.3\,\text{dBm}$, and $IP_{3,\text{out}} \approx 10.7\,\text{dBm}$. Each curve depicts the power in one tone of the response evaluated.

Under small-signal conditions—that is, at levels well below compression—the power of an IM_2 product varies by 2 dB, and the power of an IM_3 product varies by 3 dB, per decibel change in input power level. This allows us to derive a network figure of merit, the intermodulation intercept point (IP), for a given IM order by extrapolating a network's linear and IM responses to their point of intersection (Figure 1.108)—the point at which their powers would be equal if compression did not occur. Because of the system noise and/or intermodulation distortion products, there is a minimum discernible signal (MDS) that limits the dynamic range at the lower end. Theoretically, Figure 1.108 should show a noise floor or IMD-spur floor for a given input signal that represents a lower limit below which signals cannot be detected. The intercept point for a given IM order n can be expressed, and should always be characterized, relative to input power ($IP_{n,\text{in}}$) or output power ($IP_{n,\text{out}}$); the IP_{in} and IP_{out} values differ by the network's linear gain. For equal-level test tones, IP_n, in can be determined by

$$IP_{n,\text{in}} = \frac{n P_A - P_{\text{IM}n}}{n - 1} \tag{1.50}$$

where n is the order, P_A is the input power (of one tone), $P_{\text{IM}n}$ is the power of the IM product, and IP is the intercept point. The intercept point for cascaded networks can be determined from

$$IP_{2,\text{in}} = \frac{1}{\left(\frac{1}{\sqrt{IP_1}} + \frac{G}{\sqrt{IP_2}}\right)^2} \tag{1.51}$$

for IP_2 and from

$$IP_{3,\text{in}} = \frac{1}{\frac{1}{IP_1} + \frac{G}{IP_2}} \tag{1.52}$$

for IP_3. In both equations, IP_1 is the input intercept of Stage 1 in watts, IP_2 is the input intercept of Stage 2 in watts, and G is the gain of Stage 1 (as a numerical ratio, not in decibels). Both equation assume the worst-case condition, in which the distortion products of both stages add in-phase.

The ratio of the signal power to the IM-product power, the distortion ratio, can be expressed as

$$R_{dn} = (n-1)\left[IP_{n(\text{in})} - P_{(\text{in})}\right] \tag{1.53}$$

where n is the order, R_{dn} is the distortion ratio, $IP_{n(\text{in})}$ is the input intercept point, and $P_{(in)}$ is the input power of one tone.

Discussions of IMD have traditionally downplayed the importance of IM_2 because the incidental distributed filtering contributed by the tuned circuitry once common in radiocommunication systems was usually enough to render out-of-passband IM_2 products caused by in-passband signals, and in-passband IM_2 products caused by out-of-passband signals, vanishingly weak compared to fundamental and IM_3 signals. In broadband systems that operate at bandwidths of an octave or more, however, in-passband signals may produce significantly strong in-passband IM_2 and second-harmonic products. In such applications, balanced circuit structures (such as push–pull amplifiers and balanced mixers) can be used to minimize IM_2 and other even-order nonlinear products.

As with IM_2, which IM_3 products are important depends on the spacing of the signals involved and the relative width of the system passband. If ω_1 and ω_2 are of approximately the same frequency, the additive products $2\omega_1 + \omega_2$ and $2\omega_2 + \omega_1$ will be outside the passband of a narrowband system. The subtractive products $2\omega_1 - \omega_2$ and $2\omega_2 - \omega_1$, however, will likely appear near or within the system passband. The IM_3 performance of any network subjected to multiple signals is therefore of critical importance, and an array of IM_3-related, sometimes application-specific, figures of merit has evolved as a result.

Dynamic Range   As we have seen, thermal noise sets the lower limit of the power span over which a network can operate. Distortion—that is, degradation by distortion of the signal's ability to convey information—sets the upper limit of a network's power span. Because the power level at which distortion becomes intolerable varies with signal type and application, a generic definition has evolved; the upper limit of a network's power span is the level at which the power of one IM product of a specified order is equal in power to the network's noise floor. The ratio of the noise-floor power to the upper-limit signal power is referred to as the network's dynamic range (DR), often more carefully characterized as two-tone IMD dynamic range, which, when evaluated with equal-power test tones, is a figure of merit commonly used to characterize receivers. The MDS relative to the input, as already defined, is

$$\text{MDS}_{\text{in}} = kTB + 3 \text{ dB} + \text{NF}$$

When $IP(n)$ in and MDS are known, IMD DR can be determined from

$$\text{DR}_n = \frac{(n-1)[IP_{n(\text{in})} - \text{MDS}_{\text{in}}]}{n} \tag{1.54}$$

where DR is the dynamic range in decibels, n is the order, $IP_{(\text{in})}$ is the input intercept power in dBm, and MDS is the minimum detectable signal power in dBm. The so-called spurious-free dynamic range (SFDR or DRSF) is calculated from

$$\text{DR}_{\text{SF}} = \frac{2}{3} (IP_3 - 174 \text{ dBm} + \text{NF} + 3 \text{ dB})$$

This equation allows us to determine how to measure the spurious-free dynamic range. This is done by applying the two-tone signals (in the case of IP_3) and increasing the two signals to the point where the signal-to-noise ratio deteriorates by 3 dB or, if the measurement is done relative to MDS, where the noise floor rises by 3 dB. The factor 2/3 is derived from the fact that the levels of IM_3 outputs increase 3 dB for 1 dB of input increase. This definition of dynamic range now is referenced to a noise figure rather than a minimum level in dBm, and is therefore independent of bandwidth. (By choosing smaller bandwidths (1 kHz instead of 10 kHz), a dynamic range measurement can be made to look better. Biasing the specification on noise figure directly avoids this problem.)

Triple-Beat Distortion and Cross-Modulation $P_{-1\text{dB}}$ is a single-tone figure of merit; blocking, intercept point and dynamic range evaluate two-tone behavior. For networks that must handle AM and composite (AM and angle modulation) signals very linearly, such as television transmitters and cable TV distribution systems, a three-tone figure of merit called triple-beat distortion has gained acceptance. Signals at ω_1 and ω_2 (closely spaced) and ω_3 (positioned far away from ω_1 and ω_2) are applied to the network under test, at levels, frequencies, and spacings that vary with the application. One triple-beat distortion figure of merit is the ratio, expressed in decibels, of the IM product at $\omega_3 + (\omega_2 - \omega_1)$ to one of the network's linear outputs at a specified output level. Alternatively, the triple-beat figure of merit may express the network output level at which a specified triple-beat ratio occurs.

Triple-beat distortion is the mechanism underlying cross-modulation, a form of intermodulation in which one or more AM signals present in a network amplitude-modulate all signals present in the network [33]. Angle-modulation-based wireless systems are largely immune to such effects.

Figures 1.108 and 1.109 graph the results of gain compression, two-tone intermodulation, cross-modulation, and triple-beat testing on a wideband (5–1000 MHz) amplifier.

Noise Power Ratio   Triple-beat testing is one way of improving on two-tone testing as a means of evaluating a network's intermodulation behavior in the presence of multiple signals. Another figure of merit, noise power ratio (NPR), uses thermal noise as a test signal. The test measures the introduction, by IM, of noise into a quiet slot created by the insertion of a band-stop filter, equal in stopband width to the width of the measurement channel, between the noise generator and the network under test (Figure 1.110).

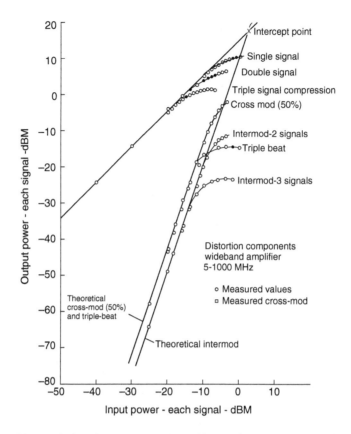

Figure 1.109 Measured distortion components in a wideband (5–1000 MHz) amplifier. Figure 1.108 shows a magnified view of the gain-compression region [34].

Large-Signal Effects Except for P_{-1dB}, the figures of merit discussed so far evaluate amplitude nonlinearity under small-signal conditions. At input powers that drive a network into gain compression and saturation, IM products of odd orders higher than 3 become significant, and curves, dips, and nulls appear in their characteristics (Figure 1.111). Phase shifts related to nonlinear (primarily voltage-dependent) capacitances in solid-state devices are one cause of these effects. Under such conditions, a network may exhibit hysteresis, with its behavior at any given instant depending not only on the voltage or current applied to it but also on its recent history [35].

AM-to-PM Conversion The nonlinear distortion effects we have discussed so far can be termed AM-to-AM distortion—distortion that, to a degree that depends on the amplitude of the signal(s) applied to the network, results in changes in the network's gain, and/or production of signals at new frequencies. AM-to-PM distortion can also occur. As a network nears saturation, part of its driving signal goes into shifting the bias point(s) of its active device(s), changing their drive-dependent reactances and shifting the phase of the output signal relative to its value at input levels below compression (Figures 1.112 and 1.113). This effect, AM-to-PM conversion, can cause incidental phase modulation that degrades the performance of digital communication systems.

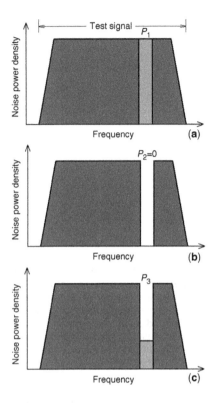

Figure 1.110 Determining a network's noise power ratio (NPR) involves the application of a test signal consisting of thermal noise [34]. The reference measurement-channel noise power, P_1, is then measured (a). Next, a stop-band filter is placed between the noise generator and network under test to keep the test signal out of the measurement channel (b). Assuming sufficient filter attenuation, if the network were absolutely noiseless and linear, the ideal noise power in the measurement channel, P_2, would then be zero. In practice, the network's own thermal noise and intermodulation between noise components outside the measurement channel result in an actual measurement-channel noise power (P_3) greater than zero. The noise power ratio equals P_1/P_3.

Spectral Regrowth and Adjacent-Channel Power Ratio   Spectral regrowth occurs largely as a result of third, fifth, and seventh-order IMD in power amplifiers operated near or in compression—at power levels where hysteretic IM effects result in poor agreement between measured behavior and predictions based on small-signal IM figures of merit [36]. We therefore evaluate the impact of spectral regrowth more directly, using a figure of merit called adjacent channel power ratio (ACPR). ACPR measurement techniques that incorporate memory can be used to increase ACPR predictions for networks with that exhibit saturation hysteresis [37]. Figure 1.114 shows the critical relationship between compression, power-added efficiency, and ACPR in a MESFET power amplifier.

Phase Response Issues and Figures of Merit   We have already seen how large-signal nonlinear distortion can result in amplitude-dependent phase shifts through AM-to-PM conversion. Because phase linearity is critical at all signal levels in PM systems, especially those using digital modulation, we must also consider linear distortion in evaluating networks used in wireless systems.

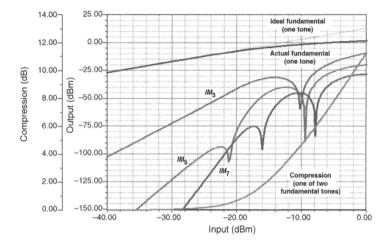

Figure 1.111 As a network is driven into compression, IM products at odd orders higher than 3 become significant, and phase shifts in power-dependent device capacitances cause curves and dips in the IM characteristics. The onset of these departures from IM-response linearity occurs at generally lower input power levels for higher IM orders; their severity, and their position on the IM curves, differs among the various products of a given order and varies with network topology and tone spacing. Figures of merit based on straight-line IM responses fail to usefully predict nonlinear network behavior under these conditions. This graph shows the simulated performance of a single-BJT broadband amplifier driven by two equal-amplitude tones at 10 and 11 MHz.

Differential Group Delay Very frequency-selective network subjects signals passing through it to some degree of time delay. Ideally, this delay, also known as group or envelope delay, does not vary with frequency; that is, the network's phase-shift versus frequency response is monotonic and linear. In practice, a network's time delay varies across its

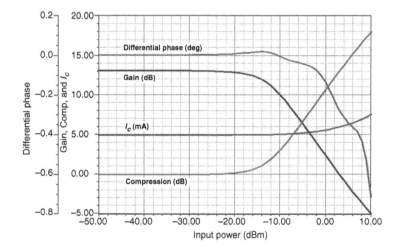

Figure 1.112 Driving a network into compression and saturation shifts the bias point(s) of its active device(s), changing their drive-dependent reactances and shifting the phase of the output signal relative to its value at input levels below compression. This graph shows the simulated performance of a single-BJT broadband amplifier driven by a single tone at 10 MHz.

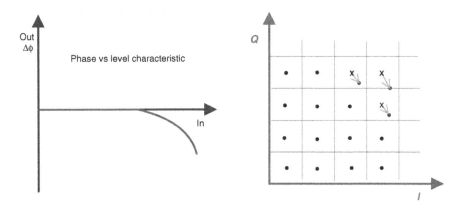

Figure 1.113 Influence of differential phase error (AM-to-PM conversion) on a QAM constellation.

passband, transition bands, and stopbands, exhibiting curvature, ripple, and transition-band peaks (Figure 1.115). The network's differential group delay—its group-delay spread—is therefore of considerable importance. This is especially so in digitally modulated systems, where the resulting phase distortion can cause errors in modulation and demodulation.

Effects of Phase Noise   As Chapter 5 will discuss in detail, the phase of an oscillator's output signal is subject to random phase variations (Figures 1.116 and 1.117). Called phase noise, this effect is often quantified as the decibel ratio of the phase noise power in a single (the upper or lower) phase-noise sideband, in a 1-Hz bandwidth centered at a specified frequency offset from the oscillator carrier, to the carrier power (Figure 1.118); alternatively, it may be specified in degrees rms. A microwave voltage-controlled oscilla-

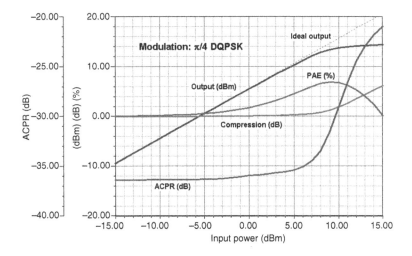

Figure 1.114 Keeping adjacent-channel interference under control can involve a critical trade-off between a wireless transmitter's power-added efficiency (PAE) and ACPR, as shown in this graph of the simulated behavior of the 1-GHz MESFET power amplifier introduced in Figure 1.103. As the amplifier is driven into compression, a peak occurs in the PAE response—in this case, at P_{-1dB}—and ACPR rises sharply.

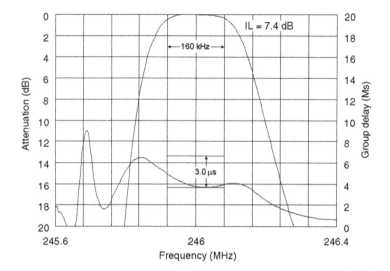

Figure 1.115 Close-in amplitude and group delay responses for a 246-MHz SAW filter designed for GSM applications [38]. This filter is well within its 3.0 μs differential group delay specification across its passband (160 kHz at −3 dB); the peaks just outside the passband limits are characteristic of a network's transition-band phase response.

tor, for instance, might exhibit an SSB phase noise of −95 dBc/Hz at 10 kHz. Oscillator phase noise may manifest itself, through a mechanism known as reciprocal mixing, as the emission of unacceptably strong noise outside a transmitter's occupied bandwidth or as an increase in receiver noise floor. Phase noise may also directly introduce phase errors that result in modulation and demodulation errors.

Because the oscillators used for frequency translation in wireless systems are usually embedded in phase-locked loops, their phase-noise characteristics differ from those of "bare" oscillators as shown in Figures 1.119 and 1.120. Figures 1.121 and 1.122 show the measured phase noise of the Rohde & Schwarz SMY and SMIQ signal generators.

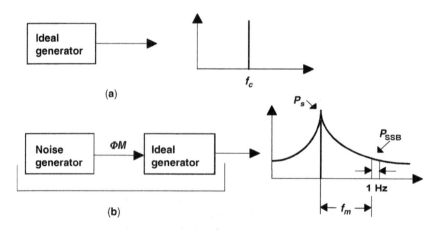

Figure 1.116 SSB phase noise. An ideal signal generator (a) would produce an absolutely pure carrier. A real signal generator (b) acts like an ideal generator driven by a noise generator, producing a noise-modulated carrier.

SYSTEM SPECIFICATIONS AND THEIR RELATIONSHIPTO CIRCUIT DESIGN

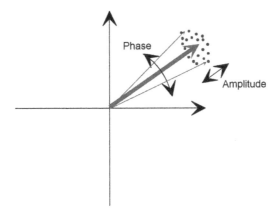

Figure 1.117   Oscillator noise can be split into amplitude and phase components.

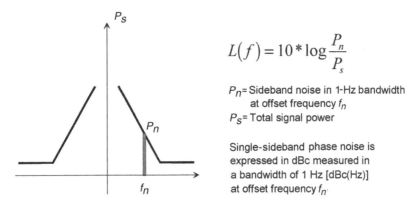

$$L(f) = 10 * \log \frac{P_n}{P_s}$$

P_n= Sideband noise in 1-Hz bandwidth
at offset frequency f_n

P_s= Total signal power

Single-sideband phase noise is
expressed in dBc measured in
a bandwidth of 1 Hz [dBc(Hz)]
at offset frequency f_n.

Figure 1.118   Phase-noise calculation.

The SMY is a low-cost signal source, while the SMIQ is a very high-performance signal generator capable of being programmed for all digital modulations; therefore, their PLL systems exhibit different phase noise versus frequency responses as the measured results show.

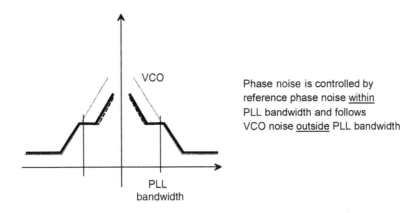

Phase noise is controlled by
reference phase noise <u>within</u>
PLL bandwidth and follows
VCO noise <u>outside</u> PLL bandwidth

Figure 1.119   Phase noise of an oscillator controlled by a phase-locked loop.

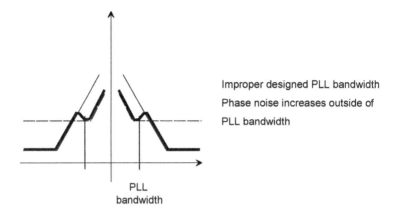

Improper designed PLL bandwidth
Phase noise increases outside of
PLL bandwidth

PLL
bandwidth

Figure 1.120   Effect of improper loop-filter design.

Reciprocal Mixing   In reciprocal mixing, incoming signals mix with LO-sideband energy to produce IF output (Figure 1.123). Because one of the two signals is usually noise, the resulting IF output is usually noise. (Reciprocal mixing effects are not limited to noise; discrete-frequency oscillator sideband components, such as those resulting from crosstalk to or reference energy on a VCO's control line, or the discrete-frequency spurious signals endemic to direct digital synthesis, can also mix incoming signals to IF.) In practice, the resulting noise-floor increase can compromise the receiver's ability to detect weak signals and achieve a high IMD dynamic range; on the test bench, noise from reciprocal mixing may invalidate desensitization, cross-modulation, and IM testing by obscuring the weak signals that must be measured in making these tests.

Figure 1.124 shows a typical arrangement of a dual-conversion receiver with local oscillators. The signal coming from the antenna is filtered by an arrangement of tuned circuits referred to providing as input selectivity. For a minimum attenuation in the passband, an operating bandwidth of

$$B = \frac{f}{\sqrt{2} \cdot Q_L}$$

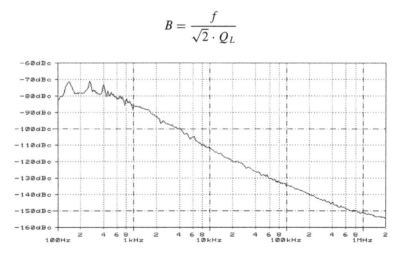

Figure 1.121   Measured phase noise of the Rohde & Schwarz SMY signal generator at 1 GHz. This signal generator has no provision for digital modulation, and therefore shows the best possible phase noise in its class.

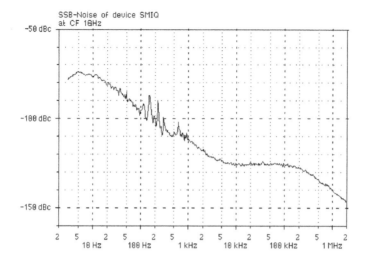

Figure 1.122 Measured phase noise of the Rohde & Schwarz SMIQ signal generator at 1 GHz. This signal generator is optimized for all digital modulation capabilities and can be configured via appropriate programming. Above 10 kHz, the influence of the wideband loop becomes noticeable; above 200 kHz, the resonator Q takes over.

This approximation formula is valid for the insertion loss of about 1 dB due to loaded Q.

The filter in the first IF is typically either a SAW filter (in the frequency range from 500 MHz to 1 GHz) or a crystal filter (45–120 MHz). Typical insertion loss is 6 dB. Since these resonators have significantly higher Q than LC circuits, the bandwidth for the first

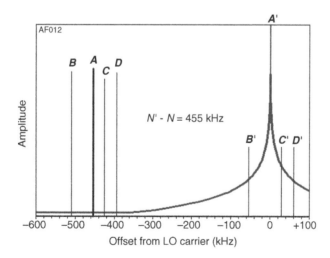

Figure 1.123 Reciprocal mixing occurs when incoming signals mix energy from an oscillator's sidebands to the IF. In this example, the oscillator is tuned so that its carrier, at A', heterodynes the desired signal, A, to the 455 kHz as intended; at the same time, the undesired signals B, C, and D mix the oscillator noise-sideband energy at B', C', and D', respectively, to the IF. Depending on the levels of the interfering signals and the noise-sideband energy, the result may be a significant rise in the receiver noise floor.

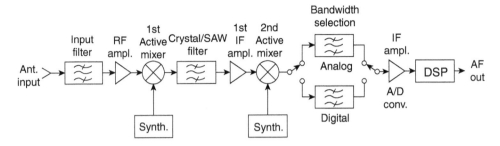

Figure 1.124 Block diagram of an analog/digital receiver showing the signal path from antenna to audio output. No AGC or other auxiliary circuits are shown. This receiver principle can be used for all types of modulation, since the demodulation is done in the DSP block.

IF will vary from ±5 kHz to ±500 kHz. It is now obvious that the first RF filter does not protect the first IF because of its wider bandwidth. For typical communication receivers, IF bandwidths from 150 Hz to 1 MHz are found; for digital modulation, the bandwidth varies roughly from 30 kHz to 1 MHz. Therefore, the IF filter in the second IF has to accommodate these bandwidths, otherwise, the second mixer easily gets into trouble from overloading. This also means that both synthesizer paths must be of low-noise and low-spurious design. The second IF of this arrangement (Figure 1.124) can be either analog or digital, or even zero-IF. In practical terms, there are good reasons for using IFs like 50/3 kHz, as in hi-fi audio equipment, with DSP processing at this frequency (50/3 kHz) using the low-cost modules found in mass-market consumer products.

The following two pictures (Figures 1.125 and 1.126) show the principle of selectivity measurement both for analog and digital signals. The main difference is that the occupied bandwidth for the digital system can be significantly wider, and yet both signals can be interfered with by either a noise synthesizer/first LO or a synthesizer that has unwanted spurious frequencies. Such a spurious signal is shown in Figure 1.125. In the case of Figure 1.125, the

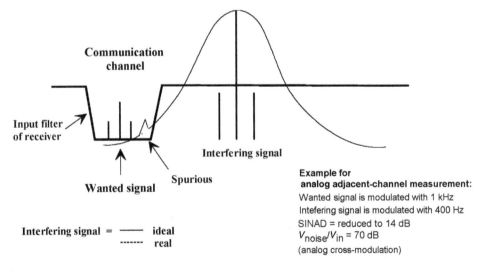

Figure 1.125 Principle of selectivity measurement for analog receivers.

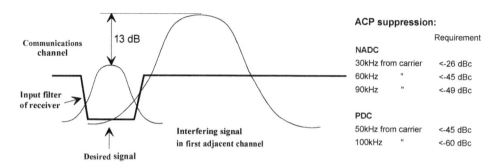

Figure 1.126   Principle of selectivity measurement for digital receivers.

analog adjacent-channel measurement has some of the characteristics of cross-modulation and intermodulation, while in the digital system, the problem with the adjacent-channel power suppression in modern terms is more obvious. Rarely has the concept of adjacent-channel power (ACP) been used with analog systems. Also, to meet the standards, signal generators have to be used that are better, with some headroom, that the required dynamic measurement. We have therefore included in Figure 1.126 the achievable performance for a practical signal generator—in this case, the Rohde & Schwarz SMHU58.

Because reciprocal mixing produces the effect of noise leakage around IF filtering, it plays a role in determining a receiver's dynamic selectivity (Figure 1.127). There is little value in using IF filtering that exhibits more stopband rejection than a value 3–10 dB greater than that results in an acceptable reciprocal mixing level.

Although additional RF selectivity can reduce the number of signals that contribute to the noise, improving the LO's spectral purity is the only effective way to reduce reciprocal mixing noise from all signals present at a mixer's RF port.

Factoring in the effect of discrete spurious signals with that of oscillator phase noise can give us the useful dynamic range of which an instrument or receiver is capable (Figure 1.128).

In evaluating the performance of digital wireless systems, we are interested in determining a receiver's resistance to adjacent-channel signals.

Phase Errors   In PM systems, especially those employing digital modulation, oscillator phase noise contributes directly to modulation and demodulation errors by introducing random phase variations that cannot be corrected by phase-locking techniques (Figure 1.129). The greater the number of phase states a modulation scheme entails, the greater its sensitivity to phase noise.

When an output signal is produced by mixing two signals, the resulting phase noise depends on whether the input signals are correlated—all referred to the same system clock—or uncorrelated, as shown in Figure 1.130.

Error Vector Magnitude   Several earlier figures (Figures 1.106, 1.113, and 1.129) have shown how particular sources of amplitude and/or phase error can shift the values of a digital emission's data states toward decision boundaries, resulting in increased BER due to intersymbol interference. Figures 1.131–1.133 show three additional sources of such errors.

A figure of merit known as error vector magnitude (EVM) has been developed as sensitive indicator of the presence and severity of such errors. An emission's error vector magnitude

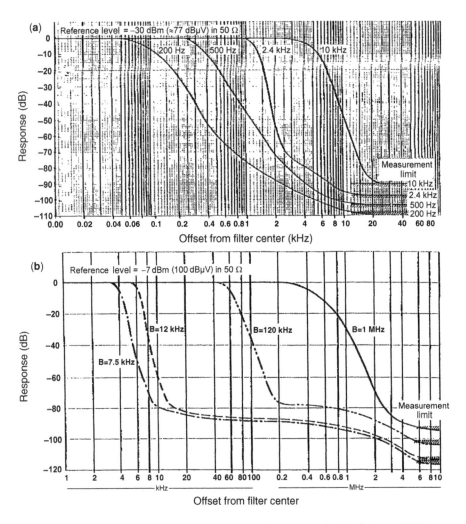

Figure 1.127 Dynamic selectivity versus IF bandwidth for (a) the Rohde & Schwarz ESH-2 test receiver (9 kHz to 30 MHz) and (b) the Rohde and Schwarz ESV test receiver (10 MHz to 1 GHz). Reciprocal mixing widens the ESH-2's 2.4 kHz response below −70 dB (−100 dBm) at (a) and the ESV's 7.5, 12, and 120 kHz responses below approximately −80 dB (−87 dBm) at (b).

is the magnitude of the phasor difference as a function of time between an ideal reference signal and the measured transmitted signal after its timing, amplitude, frequency, phase, and dc offset have been modified by circuitry and/or propagation. Figure 1.134 illustrates the EVM concept.

1.8 TESTING

1.8.1 Introduction

Testing of digital circuits deviates from the typical analog measurements, and yet the analog measurements are still necessary and related. As an example, the second-generation cellular

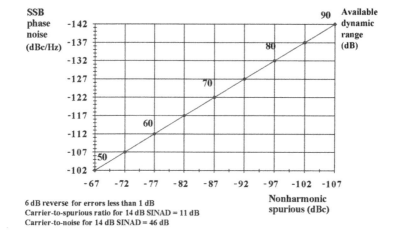

Figure 1.128 This graph shows the available dynamic range, which is determined either by the masking of the unwanted signal by phase noise or by discrete spurii. As far as the culprit synthesizer is concerned, it can be either the local oscillator or the effect of a strong adjacent-channel signal that takes over the function of the local oscillator.

telephones standards really have addressed only the transfer of voice and were just beginning to look into the transfer of data, mostly in the form of SMS, text messages limited to 140 characters. Current 3G standards with adaptive bandwidth do address high-speed data and video. Since the major use of 2G phones is voice, information such as signal-to-noise ratio as a function of many things is important. In particular, because of the Doppler effect and the use of digital rather than analog signals, where the phase information is significant, the designer ends up using coding schemes for error-correction—specifically, forward error correction (FEC). The signal-to-noise ratio, as we know it from analog circuits, now determines the bit

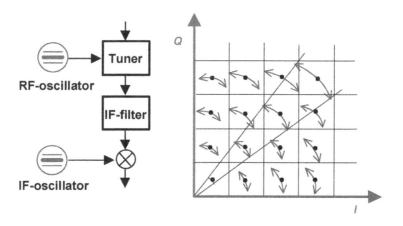

Figure 1.129 Phase noise is critical to digitally modulated communication systems because of the modulation errors it can introduce. Intersymbol interference (ISI), accompanied by a rise in BER, results when state values become so badly error blurred that they fall into the regions of adjacent states. This drawing depicts only the results of phase errors introduced by phase noise; in actual systems, thermal noise, AM-to-PM conversion, differential group delay, propagation, and other factors may also contribute to the spreading of state amplitude and phase values.

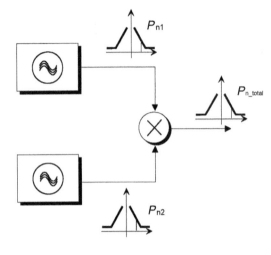

Uncorrelated signals:
Noise sideband power is added

$$P_{n\_total} = P_{n1} + P_{n2}$$

$$\Delta\mathcal{L}(f) = 10 * \log \frac{P_{n1} + P_{n2}}{P_s}$$

Correlated signals:
Noise sideband voltage is added
or substracted

$$V_{n\_total} = V_{n1} \pm V_{n2}$$

$$\Delta\mathcal{L}(f) = 20 * \log \frac{V_{n1} \pm V_{n2}}{V_s}$$

Figure 1.130   The noise sideband power of a signal that results from mixing two signals depends on whether the signals are correlated—referenced to the same system clock—or uncorrelated.

error rate (BER), and its tolerable values depend on the type of modulation used. Because of correlation, it is possible to "rescue" a voice with a bit error rate of 10^{-4}, but for higher data rates with little, if any, correlation, we are looking for BERs down to 10^{-9}. The actual bit error rate depends on the type of filtering, coding, modulation, and demodulation. In several earlier figures (Figures 1.52 and 1.56), we related BER to signal-to-noise ratio; this is a key issue in receiver design.

Another problem in receivers that has to do with the transmitter of a second station is adjacent-channel power ratio (ACPR). Given the fact that a transmitter handling digital modulation delivers its power in pulses, its transmission may affect adjacent channels by producing transient spurious signals similar to what we call splatter in analog SSB systems. This is a function of the linearity of the transmitter system all the way out to the antenna, and forces most designers to resort to less-efficient Class A operation. As possible alternatives,

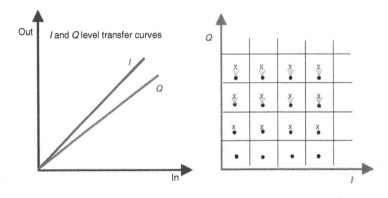

Figure 1.131   Effect of gain imbalance between *I* and *Q* channels on a data signal's phase constellation.

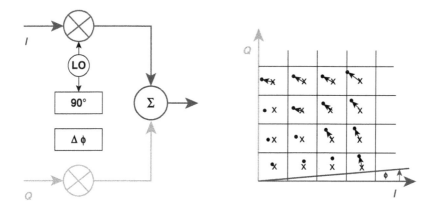

Figure 1.132   Effect of quadrature offset on a data signal's phase constellation.

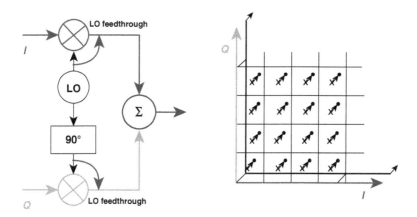

Figure 1.133   Effect of LO-feedthrough-based *IQ* offset on a data signal's phase constellation.

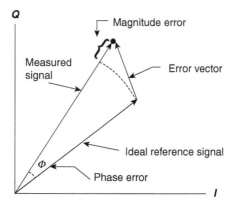

Figure 1.134   Representing errors in a digitally modulated signal's in-phase (*I*) and quadrature (*Q*) values relative to the ideal as a single error vector allows us to derive the resulting error vector magnitude (EVM), a sensitive indicator of transmission quality [39].

some researchers are looking into Classes D and E modulation, more about which is found in the power amplifier section of Chapter 3.

It is actually not uncommon to do many linear measurements, and then by using correlation equations translate these measured results into their digital equivalents.

Therefore, the robustness of the signal as a function of antenna signal at the receiver site, constant or known phase relationships, and high adjacent power ratios will provide good transfer characteristics.

1.8.2 Transmission and Reception Quality

The acoustic transmission and reproduction quality of a mobile phone is a highly important characteristic in everyday use. Accurate and reproducible measurements of a cell phone's frequency response cannot be achieved with static sinewave tones (SINAD measurements) because of coder and decoder algorithms. In this case, test signals simulating the characteristic of the human voice—that is, tones that are harmonic multiples of the fundamental, are required. A so-called vocoder is used to produce the lowest possible data rate. Instead of the actual voice, only the filter and fundamental parameters required for signal reconstruction are transmitted. Particularly, in the medium and higher audio frequency ranges, the static sinusoidal tones become a more or less stochastic output signal. For example, if a tone of approximately 2.5 kHz is applied to the telephone at a constant sound pressure, the amplitude of the signal obtained at the vocoder output varies by approximately 20 dB. In type-approval tests, where highly accurate measurements are required, the coder and docoder are excluded from the measurement.

Whether the results obtained for the fundamental are favorable depends on how far the values coincide with the clock of the coding algorithm. Through a skillful choice of fundamental frequencies, test signals with overlapping spectral distribution can be generated, giving a sufficient number of test points in subsequent measurements at different fundamental frequencies so that a practically continuous frequency response curve is obtained. Evaluation can be done by means of FFT analysis with s special window function and selection of result bins. The results are sorted and smoothed by the software and display in the form of a frequency response curve. Depending on the measurement, a program calculates the sending or receiving loudness rating in line with CCITT P.79 and shows the result.

Acoustic measurements relevant to GSM 11.10 include the following.

- Sending frequency response
- Sending loudness rating
- Receiving frequency response
- Receiving loudness rating
- Sidetone masking rating
- Listener sidetone rating
- Echo loss
- Stability margin
- Distortion sending
- Distortion receiving
- Idle channel noise receiving
- Idle channel noise sending.

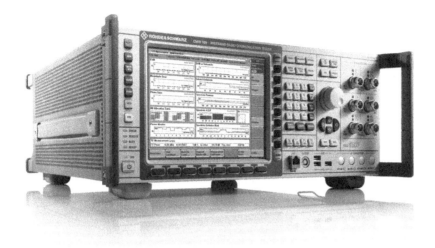

Figure 1.135   The Rohde & Schwarz CMW 500 Wideband Radio Communication Tester can be used to measure all the relevant specifications of cellular telephones.

There are two categories of testing. One is a full-compliance test for all channels and all combinations of capabilities, and the other is a production tester with evaluation of the typical characteristic data. Both Agilent and Rohde & Schwarz are the leading companies in this area, and as the technology develops further, different equipment will be made available. Figure 1.134 shows a Rohde & Schwarz radiocommunication tester capable of evaluating cellular systems. Table 1.12 shows the typical specifications required to make the appropriate cellular telephone measurements.

The test setup must be capable of measuring key IS-95 parameters, using the following.

- Frame errors
- Waveform quality
- Error vector magnitude
- Phase error
- Magnitude error
- Carrier feedthrough
- Frequency accuracy
- Power measurements
- Base station signaling for mobile testing.

1.8.3   Base Station Simulation

Simulating a CDMA (IS-95) base station involves the following.

- Synchronization of mobile
- Registration of mobile
- Incoming/outgoing call origination
- Handoff.

Table 1.12 Technical Specification of the Rohde & Schwarz CMW 500

General Technical Specifications

RF Generator

Frequency range		70 MHz to 3300 MHz Up to 6000 MHz with the R&S®CMW-KB036 option
Frequency resolution		0.1 Hz
Frequency uncertainty		Sema as timebase + frequency resolution

Output level range		
RF1 COM, RF2 COM	70 MHz to 100 MHz	
	Continuous wave (CW)	−130 dBm to −15 dBm
	Peak envelope power (PEP)	Up to −15 dBm
	Overranging (PEP)	Up to −10 dBm
	100 MHz to 3300 MHz	
	Continuous wave (CW)	−130 dBm to −5 dBm
	Peak envelope power (PEP)	Up to −5 dBm
	Overranging (PEP)	Up to 0 dBm
	3300 MHz to 6000 MHz	
	Continuous wave (CW)	−120 dBm to −15 dBm
	Peak envelope power (PEP)	Up to −15 dBm
	Overranging (PEP)	Up to −10 dBm
	Maximum input DC level	0 V DC
RF1 OUT	70 MHz to 100 MHz	
	Continuous wave (CW)	−120 dBm to −2 dBm
	Peak envelope power (PEP)	Up to −2 dBm
	Overranging (PEP)	Up to +3 dBm
	100 MHz to 3300 MHz	
	Continuous wave (CW)	−120 dBm to +8 dBm
	Peak envelope power (PEP)	Up to +8 dBm
	Overranging (PEP)	Up to +13 dBm
	3300 MHz to 6000 MHz	
	Continuous wave (CW)	−110 dBm to −2 dBm
	Peak envelope power (PEP)	Up to −2 dBm
	Overranging (PEP)	Up to +3 dBm
	Maximum input DC level	0 V DC

Output level uncertainty	In temperature range +20 °C to +35 °C, no overranging	
RF1 COM, RF2 COM	Output level > −120 dBm	
	70 MHz to 100 MHz	< 1.2 dB[a]
	100 MHz to 3300 MHz	< 0.6 dB[a]
	3300 MHz to 6000 MHz	< 1.2 dB[a]
RF1 OUT	Output level > −110 dBm	
	70 MHz to 100 MHz	< 1.6 dB[a]
	100 MHz to 3300 MHz	< 0.8 dB[a]
	3300 MHz to 6000 MHz	< 1.6 dB[a]

Output level uncertainty	In temperature range +5 °C to +45 °C, no overranging	
RF1 COM, RF2 COM	Output level > −120 dBm	
	70 MHz to 100 MHz	< 2.0 dB[a]
	100 MHz to 3300 MHz	< 1.0 dB[a]
	3300 MHz to 6000 MHz	< 2.0 dB[a]
RF1 OUT	Output level > −110 dBm	
	70 MHz to 100 MHz	< 2.0 dB[a]
	100 MHz to 3300 MHz	< 1.0 dB[a]
	3300 MHz to 6000 MHz	< 2.0 dB[a]

Table 1.12 (*Continued*)

Output level linearity with fixed RF output attenuator setting	In temperature range +20 °C to +35 °C, GPRF generator list mode, Level range 0 dB to –30 dB	
RF1 COM, RF2 COM	No overranging	< 0.2 dB, typ. < 0.1 dB

Output level resolution		0.01 dB
Output level repeatability	Typical values after 1 h warm-up time, always returning to same level and frequency, no temperature change, insignificant time change	
	Output level ≥ –80 dBm	< 0.01 dB
	Output level < –80 dBm	< 0.05 dB

VSWR		
RF1 COM, RF2 COM	70 MHz to 3300 MHz	< 1.2
	3300 MHz to 5000 MHz	< 1.5
	5000 MHz to 6000 MHz	< 1.6
RF1 OUT	70 MHz to 3300 MHz	< 1.5
	3300 MHz to 5000 MHz	< 1.5
	5000 MHz to 6000 MHz	< 1.6

Attenuation of 2nd harmonic		
RF1 COM, RF2 COM	70 MHz to 6000 MHz, P < –10 dBm	> 30 dB
RF1 OUT	70 MHz to 6000 MHz, P < 0 dBm	> 30 dB

Attenuation of 3rd harmonic		
RF1 COM, RF2 COM	70 MHz to 6000 MHz, P < –10 dBm	> 40 dB
RF1 OUT	70 MHz to 6000 MHz, P < 0 dBm	> 40 dB

Attenuation of nonharmonics	> 5 kHz offset from carrier, For output level > –40 dBm, For full scale CW signal	
	400 MHz to 3300 MHz, Except $f_{nonharmonic} = 3900\ MHz - f_{carrier}$, Except $f_{nonharmonic} = 3900\ MHz$ Except $f_{carrier} = (899\ to\ 901)\ MHz + n \times 800\ MHz$ with n = 1, 2, 3	> 60 dB
	3300 MHz to 3600 MHz	> 25 dB
	3600 MHz to 6000 MHz, Except $f_{nonharmonic} = 2 \times f_{carrier} - 6400\ MHz$	> 40 dB

Phase noise	Single sideband, 70 MHz to 3300 MHz	
Carrier offset	≥ 1 MHz	< –120 dBc, 1 Hz

Phase noise	Single sideband, 3300 MHz to 6000 MHz	
Carrier offset	≥ 1 MHz	< –117 dBc, 1 Hz

Signal-to-noise ratio	70 MHz to 3300 MHz	
RF1 COM, RF2 COM	5 MHz offset from carrier, For output level > –30 dBm	> 95 dB, typ. > 101 dB, 1 kHz (> 125 dB, typ. > 131 dB, 1 Hz)

Signal-to-noise ratio	3300 MHz to 6000 MHz	
RF1 COM, RF2 COM	5 MHz offset from carrier, For output level > –30 dBm	> 92 dB, 1 kHz

(*continued*)

Table 1.12 (*Continued*)

Modulation source: arbitrary waveform generator (ARB) (R&S®CMW-B110A option)

Memory size		1.024 Gbyte
Word length	I	16 bit
	Q	16 bit
	Marker	4 bit to 16 bit
Sample length	With 4-bit marker	Up to 227.55 Msample
Sample rate	Minimum	400 Hz
	Maximum	100 MHz
Maximum possible RF bandwidth	Depending on arbitrary waveform file	80 MHz

Trigger		
Trigger sources		BASE: external TRIG A, BASE: external TRIG B

RF analyzer

VSWR		
RF1 COM, RF2 COM	70 MHz to 3300 MHz	< 1.2
	3300 MHz to 5000 MHz	< 1.5
	5000 MHz to 6000 MHz	< 1.6

Inherent spurious response	Without input signal, 70 MHz to 6000 MHz, Except 4000 MHz, 4800 MHz, 5162.5 MHz, 5600 MHz, 6000 MHz	
	Expected nominal power setting ≤ -10 dBm	< −100 dBm
	Expected nominal power setting > −10 dBm	< −90 dB below expected nominal power setting

Spurious response	For full scale single tone input signal	
	70 MHz to 3300 MHz Except f_{in} = 1962.5 MHz and 3925 MHz Except f_{in} = 1962.5 MHz + $f_{selected}$	< −55 dB
	3300 MHz to 3700 MHz, Except f_{in} = 6400 MHz − $f_{selected}$, Except f_{in} = 6400 MHz − 0.5 × $f_{selected}$	< −40 dB
	3700 MHz to 6000 MHz, Except f_{in} = 6400 MHz − 0.5 × $f_{selected}$	< −40 dB

Harmonic response	2nd harmonic	
RF1 COM, RF2 COM	f_{in} = 70 MHz to 1650 MHz, $f_{selected}$ = 140 MHz to 3300 MHz	< −30 dB
	f_{in} = 1650 MHz to 3000 MHz, $f_{selected}$ = 3300 MHz to 6000 MHz	< −30 dB

Harmonic response	3rd harmonic	
RF1 COM, RF2 COM	f_{in} = 70 MHz to 900 MHz, $f_{selected}$ = 210 MHz to 2700 MHz	< −50 dB
	f_{in} = 900 MHz to 1100 MHz, $f_{selected}$ = 2700 MHz to 3300 MHz	< −45 dB
	f_{in} = 1100 MHz to 2000 MHz, $f_{selected}$ = 3300 MHz to 6000 MHz	< −50 dB

Phase noise	Single sideband, 70 MHz to 3300 MHz	
Carrier offset	≥ 1 MHz	< −120 dBc, 1 Hz

Phase noise	Single sideband, 3300 MHz to 6000 MHz	
Carrier offset	≥ 1 MHz	< −117 dBc, 1 Hz

Table 1.12 (*Continued*)

Trigger		
Trigger sources		BASE: external TRIG A, BASE: external TRIG B, GPRF: free run, GPRF: IF power, BB generators, BB signaling

Power meter

Frequency range		70 MHz to 3300 MHz Up to 6000 MHz with the R&S®CMW-KB036 option
Frequency resolution		0.1 Hz
Resolution bandwidths		Gaussian, 1 kHz to 10 MHz, in 1/3/5 steps, Bandpass, 1 kHz to 30 MHz, in 1/3/5 Steps, RRC, α = 0.1, 3.84 MHz, RRC, α = 0.22, WCDMA filter, 1.2288 MHz, CDMA filter
Expected nominal power setting range	For ADC full scale	
RF1 COM, RF2 COM	70 MHz to 100 MHz	−37 dBm to +42 dBm[b]
	100 MHz to 3300 MHz	−47 dBm to +42 dBm[b]
	3300 MHz to 6000 MHz	−37 dBm to +42 dBm[b]

Level range		
RF1 COM, RF2 COM	70 MHz to 100 MHz	
	Continuous power (CW)	−74 dBm[c] to +34 dBm
	Peak envelope power (PEP)	up to +42 dBm[b]
	100 MHz to 3300 MHz	
	Continuous power (CW)	−84 dBm[c] to +34 dBm
	Peak envelope power (PEP)	up to +42 dBm[b]
	3300 MHz to 6000 MHz	
	Continuous power (CW)	−74 dBm[c] to +34 dBm
	Peak envelope power (PEP)	up to +42 dBm[b]
	Maximum input DC level	0 V DC

Level uncertainty	In temperature range +20 °C to +35 °C	
RF1 COM, RF2 COM	70 MHz to 100 MHz	< 1.0 dB[d]
	100 MHz to 3300 MHz	< 0.5 dB[d]
	3300 MHz to 6000 MHz	< 1.0 dB[d]

Level uncertainty	In temperature range +5 °C to +45 °C	
RF1 COM, RF2 COM	70 MHz to 100 MHz	< 1.2 dB[d]
	100 MHz to 3300 MHz	< 0.7 dB[d]
	3300 MHz to 6000 MHz	< 1.2 dB[d]

Level linearity with fixed expected nominal power setting	In temperature range +20 °C to +35 °C	
RF1 COM, RF2 COM	Level range 0 dB to −40 dB	< 0.15 dB, typ. < 0.1 dB

Level resolution		0.01 dB

Level repeatability	Typical values after 1 h warm-up time, always returning to same level and frequency, no temperature change, insignificant time change	
	Input level ≥ −40 dBm	< 0.01 dB
	Input level < −40 dBm	< 0.03 dB

(*continued*)

Table 1.12 (*Continued*)

Dynamic range	70 MHz to 3300 MHz, *RBW → 1 kHz*, With fixed expected nominal power setting	> 100 dB
Expected nominal power setting for full dynamic range		
RF1 COM, RF2 COM		−8 dBm to +42 dBm[e]

Dynamic range	3300 MHz to 6000 MHz, *RBW → 1 kHz*, With fixed expected nominal power setting	> 97 dB
Expected nominal power setting for full dynamic range		
RF1 COM, RF2 COM		+2 dBm to +42 dBm[e]

Spectrum measurements

FFT spectrum analyzer (R&S®CMW-KM010 option)		
Frequency range		70 MHz to 3300 MHz
		Up to 6000 MHz with the R&S®CMW-KB036 option
Frequency span		1.25 MHz, 2.5 MHz, 5 MHz, 10 MHz, 20 MHz, 40 MHz
FFT length		1k, 2k, 4k, 8k, 16k
Detector		peak, RMS
Level range		See general technical specifications
Level uncertainty	For center frequency and *detector → peak*	See general technical specifications

Dynamic range	70 MHz to 3300 MHz, *For FFT length → 16k and span → 5 MHz* (equivalent to RBW → 781 Hz)	> 100 dB
Expected nominal power setting for full dynamic range		
RF1 COM, RF2 COM		−8 dBm to +42 dBm[e]

Dynamic range	3300 MHz to 6000 MHz, *For FFT length → 16k and span → 5 MHz* (equivalent to RBW → 781 Hz)	> 97 dB
Expected nominal power setting for full dynamic range		
RF1 COM, RF2 COM		+2 dBm to +42 dBm Fehler! Textmarke nicht definiert.

Inherent spurious response	Without input signal	See general technical specifications

Table 1.12 (*Continued*)

Possible configurations with two RF paths[f]

Necessary hardware (H570, H590X):
Selections: R&S®CMW-S590A RF frontend (BASIC) or R&S®CMW-S590D RF frontend (ADV.) and R&S®CMW-S570 RF TRX.
Options: R&S®CMW-B590A RF frontend (BASIC) or R&S®CMW-B590D RF frontend (ADV.) and R&S®CMW-B570 RF TRX.

Configuration with two H570 (RF TRX) and two H590A (RF frontend (BASIC))

The R&S®CMW-B570 and R&S®CMW-B590A options make the second RF path (RF path 2) available on the front of the instrument with three additional RF connectors, i.e. RF3 COM, RF4 COM and RF3 OUT.

RF3 COM	Equivalent to RF1 COM	See general technical specifications
RF4 COM	Equivalent to RF2 COM	See general technical specifications
RF3 OUT	Equivalent to RF1 OUT	See general technical specifications

Configuration with two H570 (RF TRX) and one H590D (RF frontend (ADV.))

The R&S®CMW-B570 option and R&S®CMW-S590D selection make the second RF path (RF path 2) available on the front of the instrument at connectors RF1 COM, RF2 COM and RF1 OUT.

RF path 1 and RF path 2 routed to separate connectors

RF generator 1 and RF generator 2	Switchable to RF1 COM, RF2 COM, RF1 OUT	See general technical specifications
RF analyzer 1 and RF analyzer 2	Switchable to RF1 COM, RF2 COM	See general technical specifications
Expected nominal power setting for full	70 MHz to 3300 MHz	−5 dBm to +42 dBm[g]
dynamic range	3300 MHz to 6000 MHz	+5 dBm to +42 dBm[g]

RF path 1 and RF path 2 routed to common connector

RF generator 1 and RF generator 2	Switchable to RF1 COM, RF2 COM, RF1 OUT	See general technical specifications
Output level range	Peak envelope power (PEP)	The specified value is valid for the total power of the two RF generators, see general technical specifications
Output level uncertainty	For each carrier	see general technical specifications + 0.2 dB
Signal-to-noise ratio	For the carrier with the highest output level (at least 3 dB higher than the other carrier)	see general technical specifications
RF analyzer 1 and RF analyzer 2	Switchable to RF1 COM, RF2 COM	See general technical specifications
Level uncertainty	70 MHz to 3300 MHz	See general technical specifications + 0.2 dB
	3300 MHz to 6000 MHz	See general technical specifications + 0.3 dB
Expected nominal power setting for full	70 MHz to 3300 MHz	−5 dBm to +42 dBm[g]
dynamic range	3300 MHz to 6000 MHz	+5 dBm to +42 dBm[g]

(continued)

Table 1.12 (*Continued*)

Possible configurations with four RF paths[h]

Necessary hardware (H570, H571B, H590D):
Selections: R&S®CMW-S590D RF frontend (ADV.) and R&S®CMW-S570 RF TRX.
Options: R&S®CMW-B590D RF frontend (ADV.) and R&S®CMW-B570 RF TRX and two R&S®CMW-B571B RF TX.

Configuration with two H570 (RF TRX), two H571   B (RF TX) and two H590D (RF frontend (ADV.))

The R&S®CMW-B570 option, two R&S®CMW-B571 options and R&S®CMW-B590D option make the four RF paths (RF path 1 RX and TX, RF path 2 RX and TX, RF path 3 TX only, RF path 4 TX only) available on the front of the instrument at connectors RF1 COM, RF2 COM, RF1 OUT and RF3 COM, RF4 COM, RF3 OUT.

RF path 1, 2, 3 and 4 routed to separate connectors

RF generator 1 and RF generator 3	Switchable to RF1 COM, RF2 COM, RF1 OUT	See general technical specifications
RF generator 2 and RF generator 4	Switchable to RF3 COM, RF4 COM, RF3 OUT	See general technical specifications
RF analyzer 1	Switchable to RF1 COM, RF2 COM	See general technical specifications
Expected nominal power setting for full dynamic range	70 MHz to 3300 MHz	–5 dBm to +42 dBm[i]
	3300 MHz to 6000 MHz	+5 dBm to +42 dBm[i]
RF analyzer 2	Switchable to RF3 COM, RF4 COM	see general technical specifications
Expected nominal power setting for full dynamic range	70 MHz to 3300 MHz	–5 dBm to +42 dBm[i]
	3300 MHz to 6000 MHz	+5 dBm to +42 dBm[i]

RF path 1, RF path 3 routed to common connector and RF path 2, RF path 4 routed to common connector

RF generator 1 and RF generator 3	Switchable to RF1 COM, RF2 COM, RF1 OUT	See general technical specifications
Output level range	Peak envelope power (PEP)	The specified value is valid for the total power of the two RF generators, see general technical specifications
Output level uncertainty	For each carrier	See general technical specifications + 0.2 dB
Signal-to-noise ratio	For the carrier with the highest output level (at least 3 dB higher than the other carrier)	See general technical specifications
RF generator 2 and RF generator 4	Switchable to RF3 COM, RF4 COM, RF3 OUT	See general technical specifications
Output level range	Peak envelope power (PEP)	The specified value is valid for the total power of the two RF generators, see general technical specifications
Output level uncertainty	For each carrier	See general technical specifications + 0.2 dB
Signal-to-noise ratio	For the carrier with the highest output level (at least 3 dB higher than the other carrier)	See general technical specifications
RF analyzer 1	Switchable to RF1 COM, RF2 COM	See general technical specifications
Level uncertainty	70 MHz to 3300 MHz	See general technical specifications + 0.2 dB
	3300 MHz to 6000 MHz	See general technical specifications + 0.3 dB
Expected nominal power setting for full dynamic range	70 MHz to 3300 MHz	–5 dBm to +42 dBm[i]
	3300 MHz to 6000 MHz	+5 dBm to +42 dBm[i]
RF analyzer 2	Switchable to RF3 COM, RF4 COM	See general technical specifications
Level uncertainty	70 MHz to 3300 MHz	See general technical specifications + 0.2 dB
	3300 MHz to 6000 MHz	See general technical specifications + 0.3 dB
Expected nominal power setting for full dynamic range	70 MHz to 3300 MHz	–5 dBm to +42 dBm[i]
	3300 MHz to 6000 MHz	+5 dBm to +42 dBm[i]

Table 1.12 (*Continued*)

Timebase

Timebase TCXO

Max. frequency drift	In temperature range +5 °C to +45 °C	$\pm 1 \times 10^{-6}$
Max. aging	At +25 °C, after 14 days of continuous operation	$\pm 1 \times 10^{-6}$/year

Timebase basic OCXO (R&S®CMW-B690A option)

Max. frequency drift	In temperature range +5 °C to +45 °C	$\pm 5 \times 10^{-8}$
Retrace	At +25 °C, after 24 hours power ON / 2 hours power OFF / 1 hour power ON	$\pm 2 \times 10^{-8}$
Max. aging	At +25 °C, after 10 days of continuous operation	$\pm 1 \times 10^{-7}$/year $\pm 1 \times 10^{-9}$/day
Warm-up time	At +25 °C, the frequency is in the range that is 10 times the frequency drift ($\pm 5 \times 10^{-7}$)	approx. 10 min

Timebase highly stable OCXO (R&S®CMW-B690B option)

Max. frequency drift	In temperature range +5 °C to +45 °C, referenced to +25 °C	$\pm 5 \times 10^{-9}$
	With instrument orientation	$\pm 1 \times 10^{-9}$
Retrace	At +25 °C, after 24 hours power ON / 2 hours power OFF / 1 hour power ON	$\pm 5 \times 10^{-9}$
Max. aging	At +25 °C, after 10 days of continuous operation	$\pm 3 \times 10^{-8}$/year $\pm 5 \times 10^{-10}$/day
Warm-up time	At +25 °C, the frequency is in the range that is 10 times the frequency drift ($\pm 5 \times 10^{-8}$)	approx. 10 min

Reference frequency inputs/outputs

Synchronization input		BNC connector REF IN, rear panel
Frequency	Sine wave	10 MHz to 80 MHz, step: 1 Hz
	square wave (TTL level)	1 MHz to 80 MHz, step: 1 Hz
Max. frequency variation		$\pm 10 \times 10^{-6}$
Input voltage range		0.5 V to 2 V, RMS
Impedance		50 Ω

Synchronization output 1		BNC connector REF OUT 1, rear panel
Frequency		10 MHz from internal reference or frequency at synchronization input
Output voltage		> 1.4 V, peak-to-peak
Impedance		50 Ω

[a] Valid for a 12-month calibration interval.

[b] The maximum permissible continuous power is +34 dBm due to thermal limits.

[c] RBW → 1 kHz.

[d] Valid for a 12-month calibration interval.

[e] The maximum permissible continuous power is +34 dBm due to thermal limits.

[f] R&S®CMW500 only.

[g] The maximum permissible continuous power is +34 dBm due to thermal limits.

[h] R&S®CMW500 only.

[i] The maximum permissible continuous power is +34 dBm due to thermal limits.

During the call, the tester must verify the handset's RF performance and checks the correct operation of the basic signaling features. The best testers do not rely on any special test modes in the mobile station, performing their measurements under conditions almost identical to those in a real network. A voice loop-back allows quick verification of the performance of a mobile as it is perceived by the user.

1.8.4 GSM

Measurement, test, and adjustment capabilities for GSM should include the following.

- Synchronization of mobile phone with base station (which is simulated by CTS)
- Location update
- Call setup (incoming/outgoing)
- Call release (incoming/outgoing)
- Control and measurement of transmitter power
- Handover (channel change)
- Sensitivity, including bit error rate (BER) and raw bit error rate (RBER), limit sensitivity via search routine, RxLev and RxQual
- Phase and frequency error
- Power ramp versus time
- Timing error
- AFC (automatic frequency correction) and RSSI (radio signal strength indication)
- *I/Q* modulator adjustment
- Echo test (voice test, includes also testing of loudspeaker and microphone)
- Functional test of mobile's keypad through display of dialed number
- Display of IMSI (international mobile subscriber identity), IMEI (international mobile equipment identity), power class, and revision level
- Short message service (SMS)

1.8.5 DECT

Measurement, test, and adjustment capabilities for DECT should include the following.

- Synchronization of DUT with the CTS
- Call setup
- Call release
- Echo test
- Detection and display of RFPI (FP)
- Normal transmit power (NTP)
- Power ramp versus time
- Modulation characteristics versus time
- Frequency offset
- Maximum modulation deviation
- Frequency drift

- Timing (jitter, packet delay)
- Bit error rate (BER), frame error rate (FER)

1.9 CONVERTING C/N OR SNR TO E_B/N_0

Figure 1.136 shows an application note for converting carrier-to-noise ratio (C/N) or SNR to energy per bit / normalized noise power (E_b/N_0).

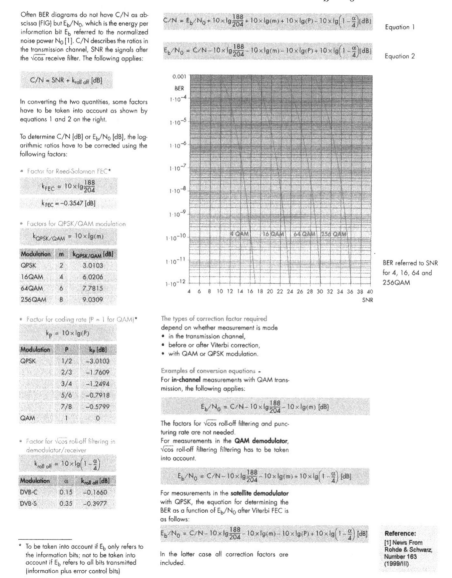

Figure 1.136 Conversion of C/N or BER to E_b/N_0.

REFERENCES

1. J. C. Lin, "Risk of human exposure to cellular mobile communication radiation ," *IEEE Microwave Magazine*, Vol. 2, No. 2, June 2001, pp. 36–40.
2. J. C. Lin, "The multinational study of brain tumors in cell phone users' heads [health effects]," *IEEE Microwave Magazine*, Vol. 10, No. 4, June 2009, pp. 138, 140–140, 144.
3. Z. Kostic, "Digital signal processors in cellular radio communications," *IEEE Communications Magazine*, pp. 22–35, December 1997.
4. B. Razavi, "Recent advanced in RF integrated circuits," *IEEE Communications Magazine*, pp. 36–43, December 1997.
5. W. Reuter, "Source and synthesizer phase noise requirements for QAM radio applications," Part 1, *Microwave Product Digest*, pp. 36, 38, 50–51, 54–56, January 1999.
6. A. Springer and R. Weigel, *UMTS: The Physical Layer of the Universal Mobile Telecommunications System (Signals and Communication Technology)*. Berlin & Heidelberg, Germany: Springer Verlag, 2002.
7. H. Holma and A. Toskala (eds.), *WCDMA for UMTS: HSPA Evolution and LTE*. New York: Wiley 2007.
8. C. Gessner, "UMTS long term evolution (LTE) technology introduction," Rohde & Schwarz Application Note 1MA111, 09.2008.
9. M. Kottkamp, "HSPA+ technology introduction," Rohde & Schwarz Application Note, 05.2009-2E.
10. M. Kottkamp, "LTE-advanced technology introduction," Rohde & Schwarz White Paper 07.2010-1MA169_2e.
11. M. Kottkamp, A. Rössler, J. Schlienz, and J. Schütz "LTE release 9 technology introduction," Rohde & Schwarz White Paper, December 2011.
12. G. Jue, "3GPP W-CDMA systems: Design and testing," *IEEE Microwave Magazine*, pp. 56–64, June 2002.
13. S. Chia, T. Gill, L. Ibbetson, D. Lister, A. Pollard, R. Irmer, D. Almodovar, N. Holmes, and S. Pike, "3GPP evolution," *IEEE Microwave Magazine*, pp. 52–63, August 2008.
14. M. Steer, "Beyond 3GPP," *IEEE Microwave Magazine*, pp. 76–82, February 2007.
15. T. Zemen, *OFDMA/SC-FDMA Basics for 3GPP LTE*, Forschungszentrum Telekommunikation Wien, 2008.
16. Agilent Technologies, *3GPP Long Term Evolution, System Overview, Product Development, and Test Challenges*, Application Note 5989–8139EN, 2009.
17. M. Rumney, (ed.), *LTE and the Evolution to 4G Wireless, Design and Measurement Challenges*, Agilent, 2009.
18. J. Zyren and W. McCoy, "Overview of the 3GPP long term evolution physical layer," Freescale Semiconductor White Paper, 3GPPEVOLUTIONWP, July 2007.
19. G. R. Cooper and C. D. McGillem, *Modern Communications and Spread Spectrum*. New York: McGraw–Hill, 1986.
20. M. K. Simon, J. K. Omura, R. A. Scholtz, and B. K. Levitt, *Spread Spectrum Communications Handbook, Electronic Edition*. New York: McGraw–Hill, 2002.
21. H. G. Myung, J. Lim, and D. J. Goodman, "Single carrier FDMA for uplink wireless transmission," *IEEE Vehicular Technology Magazine*, pp. 30–38, September 2006.
22. H. G. Myung and D. Goodman, *Single Carrier FDMA: A New Air Interface for Long Term Evolution*. New York: Wiley 2008.
23. P. Vizmuller, RF Design Guide: Systems, Circuits, and Equations. Boston, MA: Artech House, 1995.

24. B. Sklar, *Digital Communications: Fundamentals and Applications*. New York: Prentice-Hall, 1995.

25. L. N. Lee and J. Mullen, "Third-generation cellular advances toward voice and data," *Wireless Systems Design*, pp. 32–38, December 1997.

26. New Low-Power Single-Sideband Circuits. Philips RF Application Note AN1981.

27. *Apply the Oscillator of the SA602 in Low-Power Mixer Applications.* Philips RF Application Note AN1982.

28. U. L. Rohde, J. C. Whitaker, and T. T. N. Bucher, *Communications Receivers: Principles and Design*. New York: McGraw-Hill, 1997, p. 58.

29. W. E. Sabin and E. O. Schoenike (eds.), *Single Sideband Systems and Circuits*, 2nd ed. New York: McGraw-Hill, 1995, pp. 123–124.

30. H. T. Friis, "Noise figures of radio receivers," *Proceedings of IRE*, Vol. 32, No. 7, pp. 419–422, July 1944.

31. M. Williams, F. Bonn, C. Gong, and T. Quach, "GaAs RF ICs target 2.4-GHz frequency band," *Microwaves & RF*, July 1994.

32. J. Smith, *Modern Communication Circuits*, 2nd ed. New York: McGraw-Hill, 1997, pp. 93–99.

33. S. M. Perlow, "Third-order distortion in amplifiers and mixers," *RCA Review*, pp. 234–265, June 1976.

34. E. C. Jordan (ed.), *Reference Data for Engineers*, 7th ed. Indianapolis, IN: Howard W. Sams & Co, Inc, 1985, pp. 12-33–12-35.

35. S. A. Maas, *Nonlinear Microwave Circuit*. Norwood, MA: Artech House, 1988, pp. 18, 169.

36. J. Staudinger, "Specifying power amplifier linearity via intermodulation distortion and channel spectral regrowth," *Applied Microwaves and Wireless*, pp. 62, 64, 66, 68, 70, July/August 1997.

37. M. S. Heutmaker, E. Wu, and J. R. Welch, "Envelope distortion models with memory improve some amplifier spectral regrowth predictions," *Applied Microwave & Wireless*, pp. 72, 74, 76, 78, July/August 1997.

38. Toko SF246ZA-019 filter specifications. Available at http://tokoam.com/saw/applications.html

39. *Using Vector Modulation Analysis in the Integration, Troubleshooting and Design of Digital RF Communications Systems*, Product Note HP89400-8, Hewlett-Packard Co., 1994. Available at http://www.tmo.hp.com/tmo/Notes/English/5091-8687E.html.

40. G. D. Vendelin, A. M. Pavio, and U. L. Rohde, *Microwave Circuit Design Using Linear and Nonlinear Techniques*. New York: Wiley, 1990, pp. 275–278.

41. W. G. Scanlon, "Health aspects of low-level exposure to RF electromagnetic waves," *RF Design*, pp. 40, 44, 46, 48, 50, 52, 54, 56, 64, 66, 68, July 1999.

42. Van Valkenburg (ed.), *Reference Data for Engineers: Radio, Electronics, Computer, and Communications*, 8th ed. Carmel: Prentice-Hall, 1993, pp. 25-19–25-20.

FURTHER READING

W. Y. Ali-Ahmad, "Improving the receiver intercept point using selectivity," *RF Design*, pp. 22, 24, 26, 28, 30, December 1997.

AM Signal Analysis in a Dual Down-Conversion Receiver, Symphony Examples Volume, Ansoft Corporation, 1999.

Cellular System Dual-Mode Mobile Station-Base Station Compatibility Standard, IS-54-B, EIA/TIA, April 1992.

P. Danzer (ed.), *The ARRL Handbook for Radio Amateurs*, 75th ed. Newington: American Radio Relay League, 1997, p. 26.44.

Double Down Conversion Communications Receiver, Symphony Examples Volume, Ansoft Corporation, 1999.

W. Gosling (ed.), *Radio Receivers*. London: Peter Peregrinus Ltd., 1986, p. 29.

W. H. Hayward, *Introduction to Radio Frequency Design*. Newington: American Radio Relay League, 1994.

M. C. Jeruchim, P. Balaban, and K. S. Shanmugan, *Simulation of Communication Systems*. New York: Plenum Press, 1992.

L. E. Larson (ed.), *RF and Microwave Circuit Design for Wireless Communications*. Boston, MA: Artech House, 1996.

B. J. Minnis, *Designing Microwave Circuits by Exact Synthesis*. Boston, MA: Artech House, 1996.

M. K. Nezami, "Evaluate the impact of phase noise on receiver performance," *Microwaves & RF*, pp. 165–172, May 1998.

Physical Layer on the Radio-Path, GSM Standard, July 1988.

U. L. Rohde, *Microwave and Wireless Synthesizers: Theory and Design*. New York: Wiley 1997.

R. Steele, *Mobile Radio Communications*. New York: IEEE Press, 1992.

T. Stroud, "Tutorial on all-digital modulation techniques," *RF Design*, pp. 52–58, June 1997.

J. Uddenfeldt, "Digital cellular—its roots and its future," *Proceedings IEEE*, Vol. 86, No. 7, pp. 1319–1324, July 1998.

P. Vizmuller, *RF Design Guide: Systems, Circuits and Equations*. Norwood: Artech House, 1995.

D. Vondrian, "Noise figure measurement: Corrections related to match and gain," *Microwave Journal*, pp. 22–38, March 1999.

2

MODELS FOR ACTIVE DEVICES

Because we are dealing with amplifiers, both small-signal and power, oscillators and active mixers, active devices will play a major role in circuit design. At our frequency range of interest (about 1000 MHz and above), distributed elements predominate, and even in device modeling, we must consider some distributed effects such as packaging.

As we dive into providing functional explanations of the active devices used in practical circuits, there is some temptation not to provide enough information on semiconductor behavior. Because manufacturers of microwave products will use the services of a foundry or specify semiconductors from their suppliers, we will cover the device operation in depth. That said, for the beginner looking into designing circuits from high frequencies to low microwave frequencies, we will provide examples of a variety of devices unique to this frequency range. Some treatment of device physics will therefore be useful in providing a basic understanding of the nonlinear behavior of active devices and modeling in preparing for practical circuits using computer-aided-design (CAD) tool. For CAD applications, we assume the user will have access to nonlinear CAD software, such as a high-performance SPICE program or a harmonic-balance program, such as Ansoft Designer or its derivatives. The general trend is that models for CAD are provided, in addition to the datasheets. For some devices, S-parameter sets, general-purpose SPICE model parameters, or highly sophisticated design kits can be directly downloaded from the company web site, while models for others are available in design kits from third-party companies specialized in device modeling, for example, from Modelithics, Inc. What CAD does for us—and this applies to SPICE and harmonic-balance programs alike—is predict the dc bias point of a device in the context of its surrounding circuitry. Configuring the simulation to drive the device into large-signal operation provides us with insight into the various forms of distortion and compression.

However, it is necessary to be careful when relying on a model. Without validation under realistic conditions, one cannot be sure it is performing as required.

RF/Microwave Circuit Design for Wireless Applications, Second Edition. Ulrich L. Rohde and Matthias Rudolph.
© 2013 John Wiley & Sons, Inc. Published 2013 by John Wiley & Sons, Inc.

First, semiconductor manufacturers tend to provide models that can be used in all simulators, which restricts the choice of models to basically the classic models that are widespread but also sometimes grossly outdated. Such models usually can only give a first guess of how the device might behave, not more.

Some foundries, however, spend quite some effort in developing good models for their devices. One example is Freescale who provide a design kit for their packaged LDMOS power devices that even is based on a large-signal model specifically developed for these devices. The drawback is, however, that these design kits are usually only available for a few simulators, in this case it is Agilent's ADS.

But even if a sophisticated design kit is provided, careful validation is imperative.

The issue is that a model cannot be expected to be highly accurate for all possible types of operation. For example, for LNA design, one needs a model with good prediction of white noise and of the weak nonlinearity. When an oscillator is to be designed, $1/f$ noise needs to be described well, together with the nonlinearities due to the large-signal oscillation. But this nonlinearity is again probably different from the case of a power amplifier. What we consider to be good accuracy thus depends on the application; and a generally good model might fail for certain applications, since this case was not considered during parameter extraction.

In evaluating active devices for RF circuit design, we have a few different technologies to consider. First, we will examine the semiconductor junction diode—a basic *pn* junction— and venture from there to the bipolar transistor and heterobipolar transistor. An investigation of metal-oxide semiconductor (MOS) devices, another important set of semiconductors for medium frequencies, will follow. We will also examine the metal-gate Schottky field-effect transistor (MESFET) and heterojunction field-effect transistor (HEMT) on III–V materials such as gallium arsenide and gallium nitride. Since critical large-signal parameters for some devices may not be available from their manufacturers, we will examine parameter-extraction techniques and problems. For device type, we will provide insight into the noise properties of the device.

2.1 DIODES[1]

The diode model contains a nonlinear current source that follows the Shockley equation:

$$\text{Current} = IS \left(e^{\frac{V_j}{NV_t}} - 1 \right) \tag{2.1}$$

where

V_j = voltage across the junction
V_t = thermal voltage $(= kT/q)$

These values, with the model parameters IS and N, are used to model the current-voltage effects of the semiconductor junction. This does not include the nonideal operation of real

---

[1] Portions of this chapter's diode coverage are based on material in Alpha Industries Semiconductor Application Reports 80800, 80200, and 80500. Used with permission.

diodes. For example, at low currents (less than 1 nA), other semiconductor processes that increase the flow of currents become noticeable.

By setting IS to different values, we can obtain the characteristics of other devices, such as a Schottky-barrier diode or a silicon diffused-junction diode. High-current effects are modeled, grossly, by including a series resistance that is intended to combine the effects of bulk resistance (the material on each side of the junction) and high-level injection. At high currents, the observed diode current stops following the Shockley form:

$$I_{\text{forward}} = IS \times e^{\frac{V_j}{NV_t}} \tag{2.2}$$

and approaches a modified form

$$I_{\text{forward}} = IS \times e^{\frac{V_j}{2NV_t}} \tag{2.3}$$

2.1.1 Large-Signal Diode Model

Three diode models are used in the industry. These are as follows.

- Microwave diode model
- PIN diode model
- Enhanced SPICE diode model

Figure 2.1 shows the large-signal microwave diode model. Its key words appear in Table 2.1. This model can also be used to simulate varactor diodes.

Table 2.2 lists SPICE parameters for a selection of Schottky mixer diodes by Alpha.

In most cases, the diode capacitance is modeled by a voltage-dependent capacitor, which is connected in parallel with the nonlinear current generator described previously, to represent the charge-storage effects of the junction. There are two components to this charge:

- the reverse-voltage capacitive effect of the depletion region, and
- the forward-voltage charge represented by mobile carriers in the diode junction.

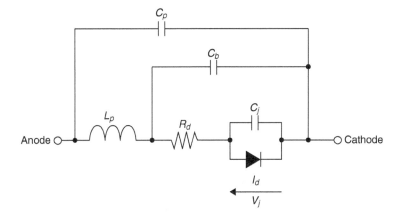

Figure 2.1 The large-signal microwave diode model. This model is temperature dependent.

Table 2.1 Large-Signal Microwave Diode Model Key Words

Key Word	Description	Unit	Default
	Intrinsic Model		
JS	Saturation current	amp	0
ALFA	Slope factor of conduction current	/volt	38.696
JB	Breakdown saturation curren	amp	10 mA
VB	Breakdown voltage	volt	$-\infty$
E	Power-law parameter of breakdown current	–	10.0
CT0	Zero-bias depletion capacitance	farad	0
FI	Built-in barrier potential	volt	0.8
GAMA	Capacitance power-law parameter		0.5
GC1	Varactor capacitance polynomial coefficient 1	/volt	0.0
GC2	Varactor capacitance polynomial coefficient 2	/volt^2	0.0
GC3	Varactor capacitance polynomial coefficient 3	/volt^3	0.0
CD0	Zero-bias diffusion capacitance ($p - n$ diodes)	farad	0
AFAC	Slope factor of diffusion capacitance	/volt	38.696
R0	Bias-dependent part of series resistance in forward-bias condition	ohm	0
T	Intrinsic time constant of depletion layer for abrupt-junction diodes	s	0
KF	Flicker noise coefficient	–	0.0
AF	Flicker noise exponent	–	1.0
FCP	Flicker noise frequency shape factor	–	1.0
AREA	Area multiplier	–	1.0
	Extrinsic Model		
CP	Package parasitic capacitance	farad	0.0
CB	Beam-lead parasitic capacitance	farad	0.0
LP	Package parasitic inductance	henry	0.0

Reverse-voltage capacitance follows the simple approximation that the depletion region (the area of the junction that is depleted of carriers) serves as the gap between the "plates" of a capacitor. This region varies in thickness, and therefore the capacitance varies with applied voltage. For a step (abrupt) junction, or linearly graded junction, the capacitance approximation is

$$\text{Capacitance} = \frac{C_{J0}}{\left(1 - \frac{V_j}{\phi}\right)^M} \tag{2.4}$$

where C_{J0} is the zero-bias value, ϕ (phi) is the junction barrier potential, and M is the grading coefficient that varies (1/2 is used for step junctions and 1/3 is used for linearly graded junctions, and most junctions are somewhere in between).

There is often confusion about the barrier potential, which appears in the capacitance equation. From capacitance measurements, ϕ (model parameter VJ, and not to be confused with V_j in the equations) takes on a value of nearly 0.7 V for regular (silicon) junction diodes and a range of 0.58–0.85 V for various Schottky-barrier diodes. This value is sometimes

Table 2.2 Diode SPICE Parameters

Parameter	Unit	SMS1546	SMS3922	SMS3923	SMS3924	SMS3926	SMS3927	SMS3928	SMS7621	SMS7630
I_S	A	3E-7	3E-8	5E-9	2E-11	2.5E-07	1.3E-09	9E-13	4E-8	5E-06
R_S	Ω	4	9	11	11	4	4	4	12	30
n	–	1.04	1.08	1.05	1.08	1.04	1.04	1.04	1.05	1.05
Td	S	1E-11	8E-11	8E-11	8E-11	1E-11	1E-11	1E-11	1E-11	1E-11
C_{J0}	PF	0.38	0.9	0.93	1.6	0.42	0.39	0.39	0.1	0.14
m	–	0.36	0.26	0.24	0.4	0.32	0.37	0.42	0.35	0.4
E_G	EV	0.69	0.69	0.69	0.69	0.69	0.69	0.69	0.69	0.69
Xti	–	2	2	2	2	2	2	2	2	2
F_C	–	0.5	0.5	0.5	0.5	0.5	0.5	0.5	0.5	0.5
B_V	V	3	20	46	100	2	3	4	3	2
I_{BV}	A	1E-5	1E-5	1E-5	1E-5	1.00E-05	1.00E-05	1.00E-05	1E-5	0.0001
V_J	–	0.51	0.65	0.15	0.84	0.495	0.595	0.800	0.51	0.34

confused with the forward current voltage drop of the diode or the energy gap of the material; it is *neither* of these.

Varying M generates a variety of reverse-bias capacitance characteristics. Inspection of the capacitance formula reveals that it predicts infinite capacitance for a forward bias, which is not the case for a real junction. Several depletion–capacitance formulas have been proposed that more correctly fit observed operation; however, SPICE uses a simple approach. For forward biases beyond some fraction (set by the parameter FC) of the value for ϕ, the capacitance is calculated as the linear extrapolation of the capacitance at the departure. This provides a continuous numerical result, and does not affect circuit operation significantly because, for forward bias, the device capacitance is normally dominated by diffusion capacitance.

The diffusion charge (and therefore the capacitance) varies with forward current and is simply modeled as a transit time (model parameter TT) for the carriers to cross the diffusion region of the junction. The total charge is

$$\text{Diffusion charge} = \text{Device current} \times \text{Transit time} \qquad (2.5)$$

and capacitance is the derivative, with respect to bias, of this

$$\text{Diffusion capacitance} = \text{TT}\frac{I_S}{NV_t}e^{\frac{V_t}{NV_t}} \qquad (2.6)$$

Diffusion charge manifests itself as the *storage time* of a switching diode, which is the time required to discharge the diffusion charge in the junction, which must happen before the junction can be reverse biased (switched off). Storage time is normally specified as the time to discharge the junction so that it is supporting only a fraction (typically 10%) of the initial reverse current. First, a forward current is supplied to the device to charge the junction. Then, as quickly as possible, a reverse current is supplied to the device. Internally, the junction is still forward biased to a voltage nearly the same as before the switch in current; the junction is still conducting at the forward current rate. This internal current adds to the external current as the total current discharging the junction. As the junction voltage decreases, the internal current falls off exponentially (according to the Shockley equation). This system is a relatively simple differential equation that can be solved to an explicit equation for the TT parameter (assuming complete discharge) as follows:

$$\text{Transit time} = \frac{\text{Storage time}}{\ln\left(\frac{I_F - I_R}{-I_R}\right)} \qquad (2.7)$$

The diffusion charge dominates the reverse recovery characteristic of the diode. During the last part of the recovery, as the junction becomes reverse biased, the depletion capacitance dominates. This causes the small tail at the end of the discharge cycle. Total capacitance is taken to be the sum of these capacitances: the depletion approximation dominates for reverse bias as the device current is small, and the diffusion proximity dominates for forward bias as the device current is large.

A special case for diode application is the switching diode, and its description and application will be part of a later chapter. Figure 2.2 shows the dc I–V curves, which indicate the different voltage potential, that are a result of the different doping profiles.

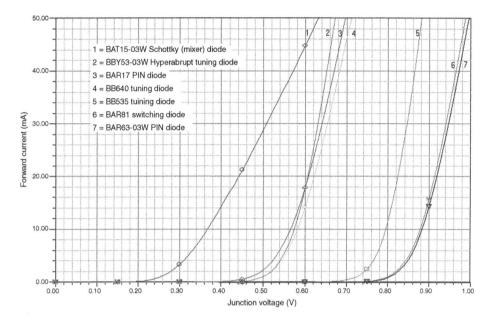

Figure 2.2 DC *I–V* curves for seven diodes, showing the various barrier voltages that result from different doping profiles.

2.1.2 Mixer and Detector Diodes

Electrical Characteristics and Physics of Schottky Barriers Schottky barrier diodes differ from junction diodes in that current flow involves only one type of carrier instead of both types. That is, in *n*-type Schottkys, the forward current consists of holes flowing from the *n*-type material into the metal.

Diode action results from a contact potential set up between the metal and the semiconductor, similar to the voltage between the two metals in a thermocouple. When metal is brought into contact with an *n*-type semiconductor (during fabrication of the chip), electronics diffuse out of the semiconductor, into the metal, leaving a region under the contact that has no free electrons ("depletion layer"). This region contains donor atoms that are positively charged (because each lost its excess electron), and this charge makes the semiconductor positive with respect to the metal. Diffusion continues until the semiconductor is so positive with respect to the metal that no more electrons can go into the metal. The internal voltage difference between the metal and the semiconductor is called the contact potential, and is usually in the range of 0.3–0.8 V for typical Schottky diodes. A cross-section is shown in Figure 2.3.

When a positive voltage is applied to the metal, the internal voltage is reduced, and electrons can flow into the metal. The process is similar to thermionic emission of electrons from the hot cathode of a vacuum tube, except that the electrons are "escaping" into a metal instead of into a vacuum. Unlike the vacuum tube case, room temperature is "hot" enough for this to happen if enough voltage is applied. However, only those electronics whose thermal energy happens to be many times the average can escape, and these "hot electrons" account for all the forward current from the semiconductor into the metal.

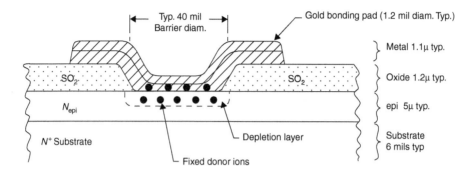

Figure 2.3 Schottky diode chip cross-section.

One important thing to note is that there is no flow of minority carriers from the metal into the semiconductor and thus no neutral plasma of holes and electrons is formed. Therefore, if the forward voltage is removed, current stops "instantly," and reverse voltage can be established in a few picoseconds. There is no delay effect to charge storage as in junction diodes. This accounts for the exclusive use of barrier diodes in microwave mixers, where the diode must switch conductance states at microwave oscillator rates.

The current–voltage relationship for a barrier diode is described by the Richardson equation (which also applies to thermionic emission from a cathode). The derivation is given in many textbooks (e.g., Sze).

$$I = A A_{RC} T^2 \exp\left(-\frac{q\phi_B}{kT}\right)\left[\exp\left(\frac{qV_J}{kT}\right) - M\right] \tag{2.8}$$

where A = area [cm^2], A_{RC} = modified Richardson constant [A/(K$^2 \cdot$ cm^2)], k = Boltzmann's constant, T = absolute temperature [K], ϕ_B = barrier height [V], V_J = external voltage across the depletion layer (positive for external voltage) = $V - IR_S$, R_S = series resistance, M = avalanche multiplication factor, and I = diode current in amperes (positive forward current).

The barrier height, ϕ_B, is typically a few tenths of a volt higher than the contact potential, ϕ_C (about 0.15 V higher than ϕ_C for silicon). This equation agrees well with experimental data for diodes without surface leakage, but is difficult to use because A_{RC}, ϕ_B, and M are all dependent on applied voltage.

The major cause for variation in ϕ_B with voltage is the "image effect," in which the barrier height is lowered as the electric field near the metal is increased, especially at the edges.

A better equation for circuit designers to use is one in which all parameters are independent of voltage and current. The simplest one that agrees reasonably well with Richardson's equation is

$$I = I_S \left[\exp\left(\frac{V_t}{0.028\ V}\right) - 1 + \frac{K}{\left(1 - \frac{V_B}{V}\right)}\right] \tag{2.9}$$

where I_S = saturation current (a temperature-dependent quantity), $0.028\ V = nkT/q$ at room temperature ($n = 1.08$), n = "forward slope factor" (derived from the variation of ϕ_B

with forward voltage), K = reverse slope factor (expressing the variation of ϕ_B with reverse voltage), and V_B = breakdown voltage (the voltage at which $M = 1$).

As before, V and I are considered positive for forward bias and negative for reverse bias.

Typical ranges for these parameters for microwave Schottky and point-contact mixer diodes are

$$I_S = 10^{-12} \text{ to } 10^{-5}\,\text{A}$$
$$n = 1.04 - 1.10$$
$$R_S = 2 - 20\,\Omega$$
$$K = 8 - 100$$
$$V_B = 2 - 20\,\text{V}$$

The quantities I_S and $0.028\ V$ are strongly temperature dependent, while both R_S and V_B increase with temperature to a slight degree. R_S increases with current at high current levels (due to carrier velocity saturation) but is essentially independent of current at $10\,\text{mA}$ and below for mixer diodes. Thus, for normal mixer and detector operation, R_S can be considered constant.

Agreement between equations (2.8) and (2.9) is not perfect but equation (2.9) is much easier to use and is preferred by most circuit designers. A comparison of the two equations near zero bias gives the following relationship between zero-bias barrier height, ϕ_0, and saturation current

$$I_S = AA_{RC}T^2\exp\left(-\frac{q\phi_0}{kT}\right)$$
$$\left(\frac{10^7 A}{\text{cm}^2}\right) A\exp\left(-\frac{\phi_0}{0.026}\right) \quad (\text{for } n \text{ silicon at room temperature}) \qquad (2.10)$$

Small-Signal Parameters By combining equations (2.9) and (2.10), the values of the parameters in equation (2.9) can be derived from a few simple measurements. Many specific equations can be derived, but the following are commonly used for production measurements:

$$R_S = \frac{V_{F10} - V_{F1} - 0.065}{0.009} \quad (\text{for } n = 1.08) \qquad (2.11)$$

$$\phi_0 = \frac{V_{F1} - 0.001R_S + 0.28 + 0.12\log_{10}D}{1.08} \qquad (2.12)$$

$$n = \frac{V_{F1} - V_{F0.1} - 0.0009R_S}{0.060} \qquad (2.13)$$

$$K = \left(\frac{I_{R1}}{I_S} - 1\right)V_B \qquad (2.14)$$

$$I_S = \exp\left(-\frac{V_{F1}}{0.028} + \frac{R_S}{28}\right) \quad (\text{in mA}) \qquad (2.15)$$

where $V_{F0.1}$, V_{F1}, and V_{F10} are the forward voltages at 0.1, 1, and 10 mA, respectively, and I_{R1} is the reverse current at 1 V. (The derivation of these equations requires that I_S be small compared to 0.1 mA.) The quantity D is the diameter of the metal-silicon contact in mils.

Measuring V_{F1} at 1 mA and 10 mA instead of some other current levels leads to the best accuracy for typical mixer diodes.

The total dynamic resistance for a forward-biased diode is given by

$$R_t = \frac{dV}{dI} = R_S + \frac{nkT}{q(I + I_S)} = R_S + R_B \tag{2.16}$$

and

$$R_B = \frac{28}{I + I_S} \text{ at room temperature (with } I \text{ and } I_S \text{ in mA, } n = 1.08) \tag{2.17}$$

This equation is also good at zero bias (unless K is very large or there is significant surface leakage). That is,

$$R_0 = R_S + \frac{28}{I_S} \tag{2.18}$$

For reverse voltages of a few volts, the dynamic resistance is dominated by the K term:

$$R_R = \text{Reverse resistance} = \frac{dV}{dI} \simeq \frac{V_B}{KI_S} \tag{2.19}$$

For typical values of I_S, R_O is larger than $5000\,\Omega$ and R_R is larger than $100\,\mathrm{k\Omega}$. For some zero-bias Schottky applications, it is desirable for R_O to be made smaller than this.

The factors that determine R_S are: (1) the thickness of the epitaxial layer, (2) the epi doping level (N_D), (3) the barrier diameter, (4) the substrate resistivity (spreading resistance), (5) the contact resistances of the metals used for the barrier and the substrate contact, and (6) the resistance associated with the bonding wire or whisker. The barrier height is about 0.15 V higher than the contact potential between the barrier metal and the semiconductor, and is influenced by the method used to apply the metal, conditions at the edge of the junction, and the doping level. Saturation current depends on barrier height, junction area, and temperature, and the slope factors, n and K, depend on doping level, punch-through voltage, and edge conditions.

2.1.2.1 Junction Capacitance

The capacitance of a Schottky barrier chip results mainly from two sources: the depletion layer under the metal-semiconductor contact and the capacitance of the oxide layer under the bonding pad (the so-called overlay capacitance). The bonding pad is required because the typical Schottky barrier diameter is so small that it is impractical to bond directly to the metal on the junction. If the semiconductor epitaxial layer is uniformly doped, the capacitance-voltage characteristic is similar to that of a textbook "abrupt-junction" diode.

$$C_t = \frac{\varepsilon_S \varepsilon_0 A'}{X_D} + C_0 \tag{2.20}$$

$$X_D = \sqrt{\frac{2\varepsilon_S \varepsilon_0 (\phi_c - V)}{qN}} \tag{2.21}$$

where ϕ = contact potential, C_0 = overlay (bonding pad) capacitance, ε_S = dielectric constant of the semiconductor ($\cong 12$ for Si or GaAs), N = doping level for the epitaxial layer, and A' = effective contact area, including fringing corrections.

In practical terms, the capacitance can be related to the 0-V barrier capacitance defined by

$$C_{BO} = \frac{\varepsilon_S \varepsilon_0 A'}{X_{DO}} \qquad (2.22)$$

where

$$X_{DO} = \sqrt{\left(\frac{1.3 \times 10^{15}}{N_D}\right)\phi_c} \quad \text{(in } \mu\text{m)} \qquad (2.23)$$

The resulting $C - V$ relationship can be written as

$$C_j = \frac{C_{BO}}{\sqrt{1 - \frac{V}{\phi_c}}} + C_0 \qquad (2.24)$$

The contact potential, ϕ_c, is related to the barrier height as follows:

$$\phi_c = \phi_B - 0.026\left[1 + L_n\left(\frac{N_c}{N}\right)\right] \qquad (2.25)$$

The theoretical meaning of these terms can be clarified by looking at Figure 2.4.

2.1.2.2 Parameter Trade-Offs

Barrier Height The barrier height of a Schottky diode is important because it directly determines the forward voltage. In order to get good noise figure, the LO drive voltage, V_L, must be large compared to V_T, which is essentially V_{F1}. Normally, it is best to have a low forward voltage (low V_{F1}), or low drive diode, to reduce the amount of LO power needed. However, if high dynamic range is important, high LO power is needed, and the diode can have a higher V_F and should also have a high V_B (see Table 2.3).

Noise Figure Versus LO Power At low LO drive levels, noise figure is poor because of poor conversion loss, due to a too-low conduction angle. At very high LO drive levels, noise figure again increases due to diode heating, excess noise, and reverse conduction.

If high LO drive level is needed, for example, to get higher dynamic range, high $V_B(> 5V)$ should be specified. However, nature requires that you play for this with higher R_S (lower f_C), so the noise figure will be degraded compared to what could be obtained with diodes designed for lower LO drive. Forward voltage and breakdown are basically independent parameters, but high breakdown is not needed or desirable unless high LO power is used.

Table 2.3 LO Power and V_{F1} for Various Applications

Type	Typical V_{F1}	LO Power (mW)	Application
"Zero bias"	$0.10 - 0.25$	<0.1	Mainly for detectors
Low barrier	$0.25 - 0.35$	$0.2 - 2$	Low-drive mixers
Medium barrier	$0.35 - 0.50$	$0.5 - 10$	General purpose
High barrier	$0.50 - 0.80$	>10	High dynamic range

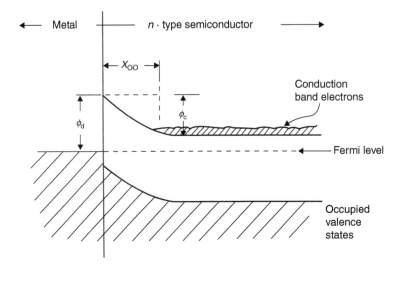

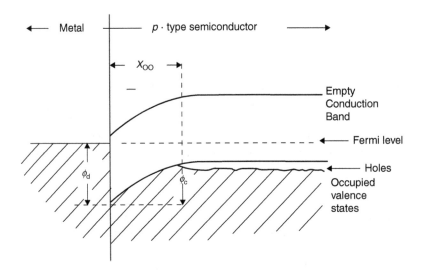

Figure 2.4   Schottky diode band diagrams.

Such a high-breakdown diode will have low reverse current (which is important only if the diode has to run hot).

Silicon Versus GaAs   Typical silicon Schottky diodes have cutoff frequencies up to the lower GHz range. In order to get higher in frequency, GaAs-based diodes are required. Dedicated technologies are even available for THz applications.

However, if your IF frequency is low, be careful; GaAs diodes have high $1/f$ noise. They also have high V_{F1}, so more LO power is required.

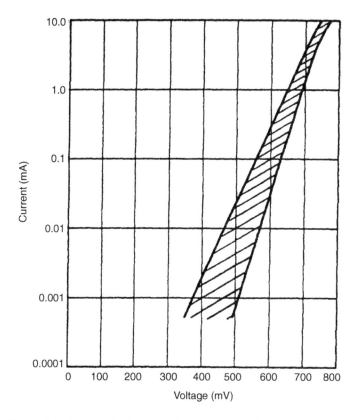

Figure 2.5 Forward dc characteristic curve range-voltage versus current.

C_J Versus Frequency   There is quite a lot of latitude in choosing C_J. However, in general, the capacitive reactance should be a little lower than the transformed line impedance (Z_0). If Z_0 is not known, a good way to start is to use $X_C = 100\,\Omega$. Experience has shown that most practical mixers use an X_C near this value (a little higher in waveguide and lower in 50-Ω systems). This translates to follow the rule of thumb for choosing the junction capacitance of a diode for operation at frequency f (in GHz):

$$C_{J0} \approx \frac{100}{\omega}$$
$$\approx \frac{1.6}{f} \quad \text{(in pF)} \tag{2.26}$$

In order to evaluate possible tolerances, we show the range of forward current currents as a function of diode voltage (Figure 2.5), the junction capacitance as a function of the bias voltage (Figure 2.6), and finally some important RF parameters, such as noise figure and IF impedance, as a function of the local-oscillator drive (Figure 2.7).

2.1.2.3 Mixer Diodes
As an example of some of the parameters for mixer diodes, Table 2.4 gives data on some of the X-band mixer diodes. *NF* is measured at 9.375 GHz.

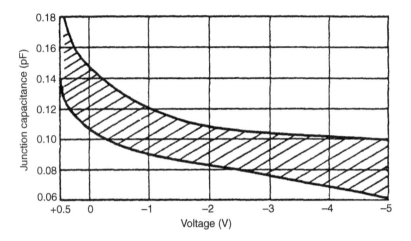

Figure 2.6   Junction capacitance range versus voltage.

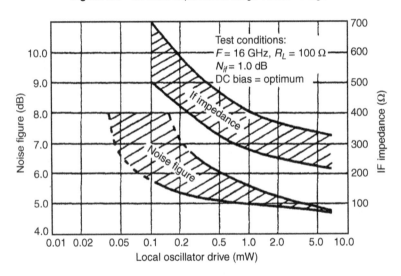

Figure 2.7   RF parameters versus local-oscillator drive level.

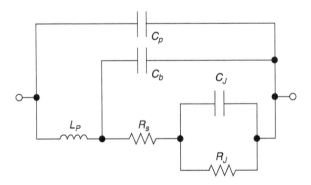

Figure 2.8   The linear diode model. This model is temperature-dependent.

Table 2.4 X-Band Mixer Diode Data

Material	Barrier	Typ. V_F (@1 mA)	Typ. F_{C0} (GHz)	Typ. R_S (ohms)	Typ. C_{J0} (pF)	Max. NF (dB)
n GaAs	High	0.70	1000	–	0.15	5.0[a]
n GaAs (BL)	High	0.70	500	–	0.15	6.0[a]
n GaAs (chip)	High	0.70	1000	–	0.15	5.3[a]
n Silicon (BL)	Low	0.28	150	6	0.20	6.5
n Silicon (quad)	Low	0.28	150	6	0.20	6.5
p Silicon (BL)	Low	0.28	150	12	0.20	6.5
n Silicon (BL)	High	0.60	100	8	0.20	6.5
n Silicon (quad)	High	0.60	100	8	0.20	6.5
n Silicon	Low	0.28	200	6	0.15	5.5
p Silicon	Low	0.28	200	18	0.14	6.0
p Silicon	Medium	0.40	150	12	0.12	6.5
n Silicon	Low	0.28	150	8	0.18	6.5
p Silicon	Low	0.28	150	12	0.18	6.5

[a]Specified for $N_{1F} = 1.0$ dB

2.1.2.4 Linear Diode Model

Figure 2.8 shows the linear diode model. Its key words appear in Table 2.5.

The circuit simulators, such as those supplied by Ansoft, Agilent, or AWR, provide a model library that has SPICE-type parameters for diodes (regular diodes, varactor diodes, and PIN diodes) as well as bipolar transistors and FETs, which will be discussed later.

2.1.3 PIN Diodes

2.1.3.1 Introduction

The PIN diode, in comparison with other microwave semiconductor devices, is fairly easy to understand. This makes it possible to reduce complex behavior to simple terms and enables the microwave engineer to grasp the operating principles and design details of this family of devices.

We do not attempt to describe the many possible microwave circuits in which PIN diodes are used. Rather, we attempt to explain the behavior of the diode in all aspects, giving the facts and some of the theory behind the facts. We offer the circuit designer the opportunity

Table 2.5 Linear Diode Model Key Words

Key Word	Description	Unit	Default
LP	Package inductance	henry	0.0
CB	Beam lead capacitance	farad	0.0
CP	Package capacitance	farad	0.0
RS	Contact resistance	ohm	0.0
RJ	Junction resistance	ohm	
CJ	Junction capacitance	farad	

to understand the PIN, so that he can understand its behavior in his circuits. We assume the reader knows the circuit equations; to that knowledge we hope to add diode equations.

Most of the material presented consists of generalized data and explanations of the behavior of PIN diodes; we conclude with a brief description of circuit performance, test methods, and some hints on proper PIN specification writing.

The user can then evaluate the trade-offs involved in diode design and performance and be able to select the most nearly optimum diode from the wide range of diodes offered.

2.1.3.2 Large-Signal PIN Diode Model

Figure 2.9 shows the large-signal model for a PIN diode. Table 2.6 lists its key words.

Notes on the PIN Diode Model

1. The PIN diode model is used to model a bias-dependent RF resistance for use in PIN diode circuits such as attenuators and switches. The resistance varies from R_{max} to R_S using the R function above. A typical R versus I characteristic is shown in Figure 2.10 with parameters ($IS = 5.96\,\text{nA}, R_S = 2.016, R_{max} = 6500, K1 = 0.1272, K2 = 1.0, N = 2.077$).

2. The transit-time parameter, TT, can also be used to approximately model a switching PIN diode's reverse-recovery time—a value often provided by diode manufacturers.

3. Diode breakdown can be modeled by specifying IBV and BV parameters.

4. The reverse-bias capacitance characteristics can be more accurately modeled than the common expression derived from *pn* junction theory. The capacitance grading coefficient exponent can be expressed as a polynomial function of voltage by specifying values for GC1, GC2, and GC3.

5. The PIN diode model was derived from J. Walston, "SPICE circuit yields recipe for PIN diode," *Microwaves & RF*, November 1992.

6. Following Sze, *Physics of Semiconductor Devices*, the variable resistance may be modeled by setting

$$K1 = \frac{3}{8} V_T \frac{W^2}{D_a \tau_a} \qquad K2 = 1.0$$

where W is the width of the intrinsic region, D_a is the ambipolar diffusion coefficient, τ_a is the ambipolar lifetime, and V_T is the thermal voltage.

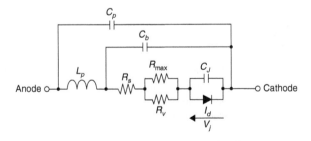

Figure 2.9   The large-signal PIN diode model. This model is temperature dependent.

Table 2.6 PIN Diode Model Key Words

Key Word	Description	Unit	Default
	Intrinsic Model		
IS	Saturation current	amp	1.0E−14
N	Emission coefficient		1.0
IBV	Magnitude of current at the reverse breakdown voltage	amp	1.0E−10
BV	Magnitude of the reverse breakdown voltage	volt	∞
FC	Coefficient for forward-bias depletion capacitance		0.5
CJ0	Zero-bias *pn* junction capacitance	farad	0.0
VJ	Built-in junction potential	volt	1.0
M	*pn* junction grading coefficient		0.5
GC1	Varactor capacitance polynomial coefficient 1	/volt	0.0
GC2	Varactor capacitance polynomial coefficient 2	/volt^2	0.0
GC3	Varactor capacitance polynomial coefficient 3	/volt^3	0.0
TT	Transit time	s	0.0
K1	Variable resistance coefficient	volt	0.0
K2	Variable resistance current exponent		1.0
RMAX	Maximum resistance of PIN intrinsic region	ohm	0.0
KF	Flicker noise coefficient		0.0
AF	Flicker noise exponent		1.0
FCP	Flicker noise frequency shape factor		1.0
AREA	Area multiplier		1.0
	Extrinsic Model		
RS	Series resistance (min. resistance of PIN diode)	ohm	0.0
CP	Package parasitic capacitance	farad	0.0
CB	Beam-lead parasitic capacitance	farad	0.0
LP	Package parasitic inductance	henry	0.0

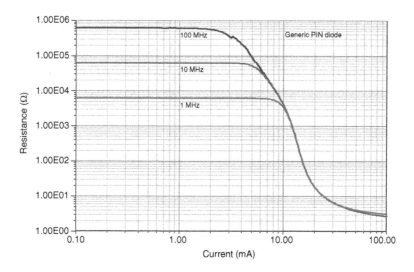

Figure 2.10 Simulated PIN diode resistance as a function of dc at 1, 10, and 100 MHz.

2.1.3.3 Basic Theory: Variable Resistance

Intrinsic or "pure" silicon as it can be grown in a laboratory is an almost lossless dielectric. Some of is physical properties include:

Dielectric constant (relative)	12
Dielectric strength	400 V/mil (approximate)
Specific density	2.3
Specific heat	0.72 J/g/°C
Thermal conductivity	1.5 W/cm/°C
Resistivity	300,000 Ω/cm

Since a PIN diode is valuable essentially because it is a variable resistor, let us concentrate initially on the resistivity. Consider a volume comparable to a typical PIN diode chip, say 20 mils in diameter and 2 mils thick. This chip has a dc resistance of about 0.75 mΩ. High resistivity in any material indicates that most of the likely carriers of electric charge, electrons and holes, are tightly held in the crystal lattice and cannot "conduct."

In real life there are impurities typically like boron, that cannot be segregated out of the crystal. Such impurities contribute carriers, holes or electrons, that are not very tightly bound to the lattice and therefore lower the resistivity of the silicon.

Through various techniques we can adjust the level of impurities, called *dopants*, to produce resistivities ranging from 10 kΩ/cm (for good PIN diodes) to 0.001 Ω/cm (for substrates).

If the impurity adds "electrons" to the crystal, it is called a *donor*; if it adds a hole it is called an *acceptor*. Boron adds holes, hence it is an acceptor, and the silicon plus boron combination is called *p-type*, or *positive*, because it has an excess of positive carriers. Phosphorus, on the other hand, is a *donor*, adding electrons, and the corresponding mix is *n*-type, or *negative*.

There are many concepts important to the physicist but not to the diode user, that elaborate upon the impact of impurities on the behavior of silicon. These can be studied in Ref. [1]. The more carriers added, the lower the resistivity.

If one wished to vary the resistance of a given diode, in principle he could bring it into a semiconductor laboratory, add or subtract carriers as desired, and perhaps even make the process reversible. However, this is a slow, expensive, and impractical way to make a variable resistor; one would be better advised to take a wrench and a soldering iron and replace a component.

The PIN diode derives its value from the fact that the free charge carrier concentration in silicon, and hence its resistance, can be varied electronically by means of current from a simple bias supply. This can be done rapidly (in nanoseconds in some cases), reversibly, repeatably, and accurately. The thing that makes this possible is called a *junction*, the interface between the relatively pure silicon in the middle of the PIN diode (the I stands for *intrinsic*) and the heavily doped layers on either end, p^+ and n^+. The p^+ region is rich in holes; the n^+ region is rich in electrons. Both of these regions have low resistance. The I region is the variable resistive element in the diode (see Figure 2.11). In the absence of any external bias, internal effects within the crystal keep the charges fixed; the resistance of the I region is high.

When the p^+ region (anode) is biased positively with respect to the n^+ region (cathode), the interface potential "barrier" is overcome. and direct current flows in the form of holes

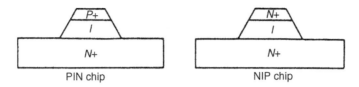

PIN chip   NIP chip

Note: Chips and diodes of either PIN of NIP polarity are
generally available for any application. The only basic
difference is the polarity of the heat sink and of the
diode.

Figure 2.11   General outline of PIN diode construction.

streaming from p^+ toward n^+, with electrons moving in the opposite direction; we say that free carriers have been injected into the I region. The resistance of the I region becomes low.

The number of free carriers within the I region determines the resistivity of the region and thus the resistance of the diode.

Consider "one hole" and "one electron" drifting in opposite directions in the I region under the impetus of the applied field. Under certain conditions, imperfections in the silicon may cause these carriers to recombine. They are no longer available to constitute current or to lower the resistivity of the I region.

It can be shown that the amount of "recombination" between holes and electrons that continuously takes place in a semiconductor is governed by a property of the lattice called *lifetime*. In fact, lifetime is defined as the reciprocal of recombination rate.

Thus, $Q_S = Q_0\exp(-t/T_L)$, where Q_S is the total amount of free charge "stored" in the I region, and T_L is the lifetime or the mean time between recombination events. In steady-state condition, the bias supply must deliver current to maintain constant Q_S. The required current is

$$I_{dc} = \frac{dQ_S}{dt} = \frac{-Q_S}{T_L} \tag{2.27}$$

or $Q_S = I_{dc}T_L$, dropping the minus sign.

Ignoring some details that are not crucial to this section, we can now calculate the resistance of a given diode, of area A and thickness W. (W stands for *base width*, the width or thickness of the intrinsic layer). The p^+ and n^+ regions have essentially zero resistance, as they are very heavily doped.

The resistivity of a given material is inversely proportional to the number of free carriers, N, and the mobility (not quite the same as velocity) of the carriers. Thus

$$\rho = \frac{1}{q\left(\mu_n N + \mu_p P\right)} \tag{2.28}$$

both holes and electrons, μ_n, μ_p are mobilities of electrons and holes, and N, P are numbers of electrons and holes. Simplifying,

$$\rho = \frac{C}{dQ_S} \tag{2.29}$$

where C is a collection of constants, and dQ_S is the stored charge density (numbers per unit volume). For our piece of silicon, volume is WA, and

$$dQ_S = \frac{Q_S}{WA} \tag{2.30}$$

The resistivity is

$$\rho = \frac{CWA}{Q_S} \tag{2.31}$$

and the resistance is

$$R_S = \rho \frac{W}{A} = \frac{CW^2}{I_{dc}T_L} \tag{2.32}$$

This is a fundamental equation in PIN diode theory and design.
Rigorous analysis shows

$$R = \frac{2kt/q}{I_f} \sinh\left(\frac{W}{2\sqrt{DT_L}} \tan^{-1}\left[\sinh\frac{W}{2\sqrt{DT_L}}\right]\right) \tag{2.33}$$

where k = Boltzmann's constant, T = temperature in Kelvins, and D = diffusion coefficient = $\mu kT/q$.

For most PIN diodes, W/DT_L is less than unity, and the equation simplifies to the simple equation above.

Typical data on R_S as a function of bias current are shown in Figure 2.12. A wide range of design choices is available, as the data indicate. Many combinations of W and T_L have been developed to satisfy the full range of applications.

2.1.3.4 *Breakdown Voltage, Capacitance, Q Factor*

The previous section on R_S explained how a PIN can become a low resistance, or a "short." This paragraph will describe the other state: a high impedance, or an "open." Clearly, the better PIN diode is the one that has the better on–off ratio at the frequency and power level of interest.

If we return to the undoped or intrinsic I region, we note that it is an almost lossless dielectric. As such, it has a dielectric strength of about 400 V/mil, and all PIN diodes have a parameter called V_b, breakdown voltage, which is a direct measure of the width of the I region. Voltage in excess of this parameter results in a rapid increase in current flow (called avalanche current) shown in Figure 2.13. When the negative bias voltage is below the bulk breakdown of the I region, a few nanoamperes will be drawn. As V_b is approached, the leakage current increases often gradually, as is exaggerated in the curve. This current is primarily caused by less-than-perfect diode fabrication, although there is some contribution from temperature. Typically, the leakage current occurs at the periphery of the I region. For this reason, various *passivation* materials (silicon dioxide, silicon nitride, hard glass) are grown or deposited to protect and stabilize this surface and minimize leakage. These techniques have been well advanced over the years, and PIN diode reliability has improved as a result.

Most diodes are specified in terms of minimum V_b for a nominal leakage, usually 10 μA.

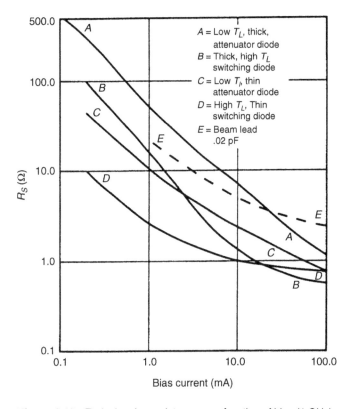

Figure 2.12 Typical series resistance as a function of bias (1 GHz).

It will be noted later that RF voltage swings in excess of the rated V_b are permitted, for the mechanisms causing leakage current do not always respond at radio frequencies. However, bulk breakdown is effectively instantaneous, and that voltage should never be exceeded.

The next characteristic of our "open" circuit is the capacitance. In simplest form, the capacitance of a PIN diode is determined by the area and width of the I region and the dielectric constant of silicon; however, we have discussed the fact that intrinsic material does contain some carriers and therefore has some conductivity. An E field could not exist unless all these carriers were swept out, or depleted.

Application of a reverse bias accomplishes this. At zero bias, the excess carriers on either side of the junction are separated, held apart, by "built in" fields. This is the contact potential (about 0.5 V for silicon). If there are only a few excess carriers in the I region, this "potential" can separate the charges more easily. The junction "widens" in the sense that, start at the p^+ and I interface, there is a region of no free carriers called the *depletion zone*. Beyond this depletion zone the I region still contains the free charges started with. With the application of reverse bias, the depletion zone widens. Eventually, at a bias equal to a so-called *punchthrough* voltage (V_{PT}), the depletion zone fills the entire I region. At this voltage, the 1-MHz capacitance bottoms out and the diode Q reaches its maximum. Figure 2.14 illustrates the equivalent circuit of the I region before punchthrough.

Some very interesting facts can be derived from this model. Consider the undepleted region; this is a lossy dielectric consisting of a volume (area A, length ℓ). of silicon of

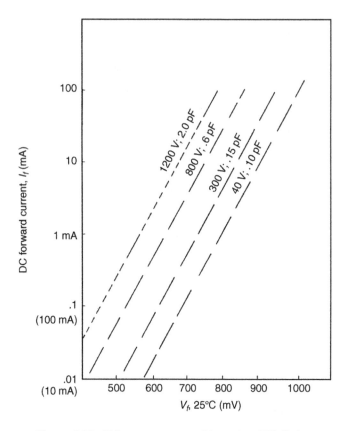

Figure 2.13   Voltage versus current for various PIN diodes.

permittivity 12 and resistivity ε. The capacitance is

$$12\frac{\varepsilon_0 A}{\ell} \tag{2.34}$$

and the admittance is

$$2\pi\frac{(12\varepsilon_0 A)}{\ell} \tag{2.35}$$

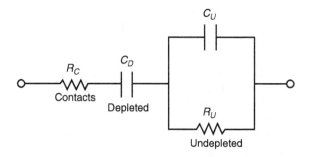

Figure 2.14   Equivalent circuit of I region before punchthrough.

The resistance is $\varepsilon\ell/A$ and the conductance is $A/\varepsilon\ell$.

At very low frequencies, the undepleted zone looks like a pure resistor. At very high frequencies it looks like a lossy capacitor. The "crossover" frequency depends on the resistivity of the I region material. For ε of 160 Ω/cm, the frequency is 1 GHz. Higher resistivity is generally used for PINs—say, 1000 Ω/cm, and the crossover frequency is 160 MHz.

Diode manufacturers measure junction capacitance at 1 MHz; clearly, what is measured is the depletion-zone capacitance.

If the I region thickness is W and the depletion with X_d, the undepleted region is $(W - X_d)$.

The capacitance of the depleted zone is, proportionally,

$$\frac{1}{X_d} \tag{2.36}$$

of the undepleted zone,

$$\frac{1}{W - X_d} \tag{2.37}$$

The 1-MHz capacitance as a function of reverse bias is seen in Figure 2.15.

The 1-MHz capacitance decreases with bias until punchthrough, we $X_d = W$. However, at microwave frequencies well above the crossover, the junction looks like two capacitors in series.

$$C_T = \frac{C_d C_u}{C_d + C_u} \propto \frac{1}{W} \tag{2.38}$$

That is, the microwave capacitance tends to be constant, independent of X_d and bias voltage.

However, since the undepleted zone is lossy, an increase in bias up to the punchthrough voltage reduces the loss.

At any given frequency, the equivalent network can now be drawn as Figure 2.16.

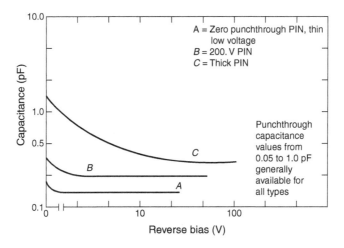

Figure 2.15 Typical 1-MHz capacitance.

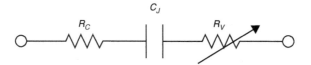

Figure 2.16 Simplified equivalent circuit, series.

R_v is now the equivalent series resistance of the undepleted region. Typical R_v data is shown in Figure 2.17.

An alternate equivalent network is shown in Figure 2.18, and typical R shunt data is shown in Figure 2.19.

A good way to understand the effects of series resistance is to observe the insertion loss of a PIN chip shunt mounted in a 50-Ω line, as shown in Figure 2.20.

An accepted way to include reverse loss in the figure of merit of a PIN diode is to write

$$Q = \frac{1}{2\pi + \sqrt{R_S R_v C}} \tag{2.39}$$

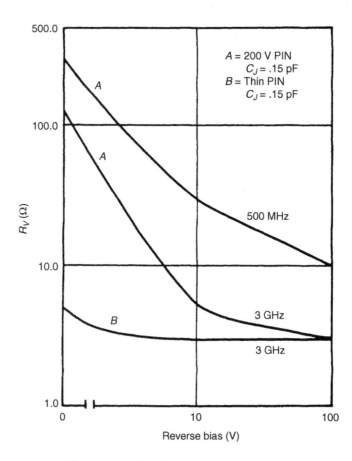

Figure 2.17 Simplified equivalent circuit, series.

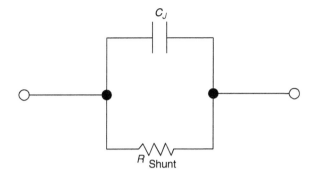

Figure 2.18 Simplified equivalent circuit, shunt.

where R_s and R_v are measured under the expected forward and reverse bias conditions at the frequency of interest.

The punchthrough voltage is a function of the resistivity and thickness of the I region. It is advisable to measure loss as a function of bias voltage and RF voltage to determine if the correct diode has been selected for your application. (*Note*: At frequencies below crossover, and diodes with thin I regions, the effective junction capacitance can increase substantially at low forward bias, on the order of 1–200 μA.)

Incidentally, if you are working with PIN or NIP chips that do not have an opaque covering, note that PIN diodes are photosensitive. Incident light causes photogeneration of carriers in the I region, increasing the chip's insertion loss.

2.1.3.5 PIN Diode Applications

If the intrinsic zone is thick (10–100 μm), we then have a high-reverse-voltage rectifier with a low forward voltage drop at high current or, in other words, a highly efficient rectifier. The low forward voltage results from the fact that the conductivity of the I zone can be modulated by large amounts of charge carriers injected from the *p* and the *n* zones.

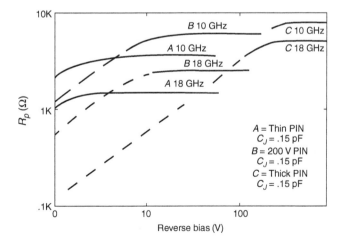

Figure 2.19 Reverse shunt resistance, R_p.

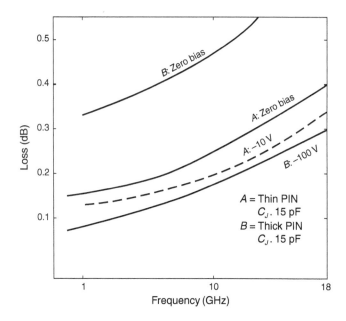

Figure 2.20 Insertion loss versus frequency.

Another application of PIN diodes is the high-frequency (HF) field. Here, the fact is exploited that, due to the long carrier lifetime at frequencies beyond approximately 10 MHz, a rectifying effect will no longer occur and that the PIN diode rather behaves like a real resistance, the magnitude of which depends on the forward direct current passed by the device and produces an equal effect on both half waves of the HF signal. In view of this behavior, the PIN diode can be used as a switch or a variable resistor for HF signals. Thus, it becomes possible, for example, to subject an HF signal to amplitude modulation by means of an AF-controlled PIN diode.

An important application of PIN diodes that has found favor in recent times is their application to dc-operated attenuators in TV tuners and antenna distribution amplifiers. Figure 2.21 shows the real HF forward resistance r_f as a function of the forward current I_f, measured at 100 MHz.

Figures 2.22 and 2.23 show second-order IMD and cross-modulation for PIN diodes.

Applying the PIN in Amplitude Control of High-Frequency Signals In conventional transistorized TV tuners, automatic gain control (AGC) is usually achieved by varying the emitter current of the input transistor. This method exploits the fact that the input transistor exhibits maximum gain at a certain level of emitter current and that this gain decreases when the emitter current is either increased or decreased relative to this point. Since modulation capacity and cross-modulation resistance grow with the emitter current, "upward" gain control has lately become the preferred solution, and according to it, gain reduction is achieved by an increase of the input transistor's emitter current. An input stage based on this design, therefore, is least resistant to cross-modulation when receiving a signal from a weak transmitter because it is operating at or below the point at which AGC action begins. This effect is particularly disadvantageous when the signal of a weak transmitter is to be

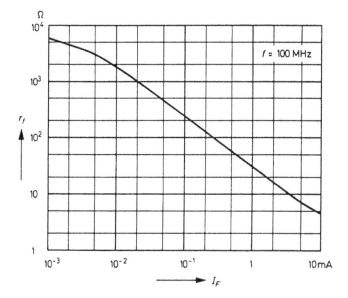

Figure 2.21   Forward resistance versus forward current.

received in the presence of a strong local transmitter. The unsatisfactory cross-modulation properties of such an input stage manifest themselves more and more frequently in the form of image perturbations as the TV reception band is crowded by an increasing number of transmitters. Another disadvantage is the variation of the transistor parameters that results

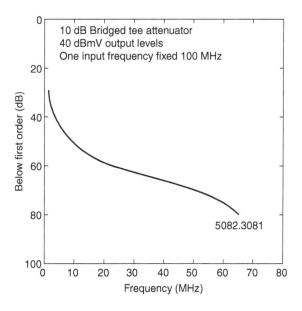

Figure 2.22   Second-order IMD in a PIN diode. Equation (1.47) can be used to determine the diode's IP2 from the absolute values of the its fundamental and IM2 outputs.

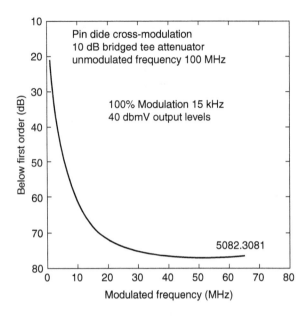

Figure 2.23 Cross-modulation in a PIN diode. Equation (2.62) relates cross-modulation level to IMD.

from the control action. The variations affect antenna matching at the input and the response of the RF filter at the output.

To obviate these problems, methods being adopted today provide for the input stage to be equipped with cross-modulation-resistant transistors with fixed bias, gain control being provided by a variable attenuator preceding the transistor(s). The input-filter termination provided by this attenuator must be independent of the degree of attenuation to preserve the filter's response characteristics.

The first two requirements are met by the properties of the PIN diode and the latter requirement can be met by designing a PIN diode control in the form of a π network, a schematic of which is shown in Figure 2.24. The attenuation of a matched π network is

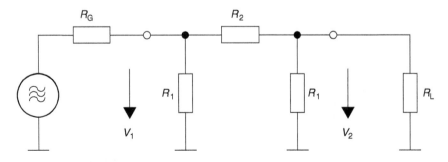

Figure 2.24 Basic circuit of a π-mode filter attenuator.

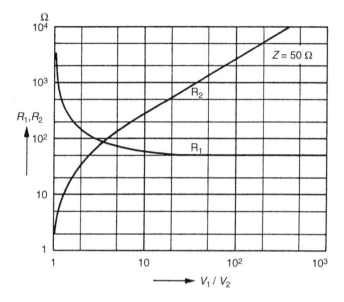

Figure 2.25 Values of R_1 and R_2 in Figure 2.24 versus relation of input and output voltage.

expressed by the following equations:

$$\frac{\alpha}{dB} = 20 \log \frac{Z + R_2}{R_2 - Z} \tag{2.40}$$

$$\frac{\alpha}{dB} = 20 \log \left(\frac{R_1}{Z} + \sqrt{\frac{R_1^2}{Z^2} + 1} \right) \tag{2.41}$$

Figure 2.25 plots the relationship between the resistance levels of R_1 and R_2 and of the relationship between the input and output voltages of the π network. Figure 2.26 shows

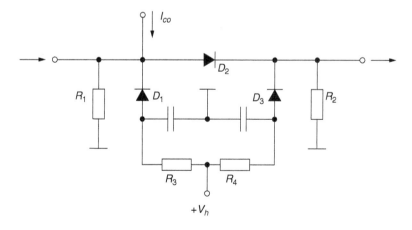

Figure 2.26 PIN diode π-mode attenuator.

the circuit diagram of an implementation of the π network composed of PIN diodes. The resistance characteristic required according to Figure 2.25 is achieved approximately by varying the control current I_{CO}.

If the control current I_{CO} in Figure 2.26 equals zero, then—due to the fact that $R_1 = R_2$ and $R_3 = R_4$—the auxiliary voltage V_h causes forward currents of equal magnitude to flow through diodes D_1 and D_3. The voltage drops across R_1 and R_2 are of equal magnitude, so that no current passes through diode D_2. Therefore, the latter presents a high resistance, whereas D_1 and D_3 present a low resistance. Under these conditions, the network produces maximum attenuation. A control current I_{CO} reduces the forward currents through D_1 and D_3 and allows a forward current to be passed by diode D_2. When it has attained its maximum value, diodes D_1 and D_3 are blocked—that is, high-ohmic—and D_2 low-ohmic. Under these conditions, the network produces its minimum attenuation.

If the maximum attenuation of the π network is to be increased, it is possible to make the two shunt diodes D_1 and D_3 more low-ohmic than the impedance Z, although this would reduce the reflection coefficient of attenuation. For example, if a reflection coefficient of about 7 dB is accepted over the entire control range, then a maximum attenuation of 25–30 dB could be achieved in the VHF range. The series inductances of the transverse diodes and the shunt capacitance of the series diode are responsible for this relatively unsatisfactory value. Improved attenuation is brought about the integrated π network described in the next section.

2.1.3.6 Example: A PIN Diode π Network for TV Tuners

The since-discontinued TDA1053 is an example of an integrated PIN π network attenuator intended for TV-tuner use. It comprises three silicon planar PIN diodes connected to form a π network (Figure 2.27) and serves for the electronic amplitude control of the input signals of TV tuners and antenna distribution amplifiers in the 40–1000 MHz range. Its input and output impedances remain constant over the entire control range. This can also be achieved with discrete diodes.

The TDA1053 was normally supplied with vertical leads. The characteristics stated below apply to devices of this configuration.

These data reveal that the compact design of the three PIN diodes in a common 50B4 plastic package guarantees favorable values for minimum and maximum attenuation, as

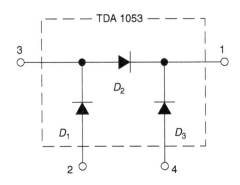

Figure 2.27 Internal circuitry of the TDA1053.

Maximum Ratings of Individual Diodes			
Reverse voltage	V_R	30	V
Forward current at $T_{amb} = 25°C$	I_F	50	mA
Junction temperature	T_j	125	°C
Storage temperature range	T_s	-55 to $+125$	°C
Maximum Ratings of the π Network			
Ambient operating temperature range when operating according to Figure 2.28	T_{amb}	100	°C
Characteristics of Individual Diodes at $T_{amb} = 25°C$			
Forward voltage at $I_F = 50$ mA	V_F	<1.2	V
Forward current at $V_R = 15$ V	I_R	<500	nA
Differential forward resistance			
At $I_F = 10$ mA, $f = 100$ MHz	r_f	5	Ω
At $I_F = 10$ μA, $f = 100$ MHz	r_f	1.4	kΩ
Characteristics in the Test Circuit (Figure 2.28) at $T_{amb} = 25°C$			
Voltage for 1% cross-modulation	V_{cr}	1	V
Attenuation in the 40–1000 MHz range			
At $V_{co} = 1.5$ V (1–2 V)	a_{max}	45(>36)	dB
At $V_{co} = 5$ V (4–5 V)	a_{min}	1.5(<2)	dB
Reflection coefficient in the 40–1000 MHz range over the entire control range, depending on circuit design	a_{refl}	20(>16)	dB

well as reflection attenuation. The test and application circuit shown in Figure 2.28 also comprises the transistor control-signal amplifier. The typical characteristic of the attenuation and the reflection attenuation for this circuit are shown in Figure 2.29, as a function of the control voltage V_{co}. Figure 2.30 shows the attenuation at different control voltages, as a function of frequency.

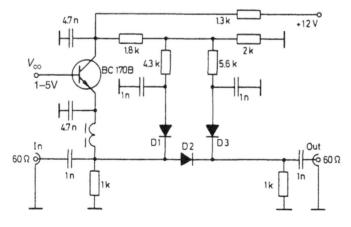

Figure 2.28 Test and application circuit.

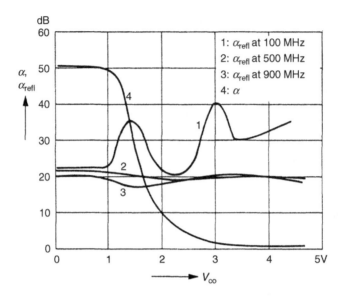

Figure 2.29   Attenuation and reflection attenuation versus control voltage.

2.1.4 Tuning Diodes

2.1.4.1 Introduction

In recent years, continuous development of tuning diodes—also known as *varactors or varicaps*—together with increased commercial and military use has led to substantial improvement in Q, reproducibility and reliability. Concurrently, new techniques for producing and controlling a hyperabrupt dopant profile in the semiconductor permit the capacitance–voltage law to be much faster than the classical square root or cube root behavior.

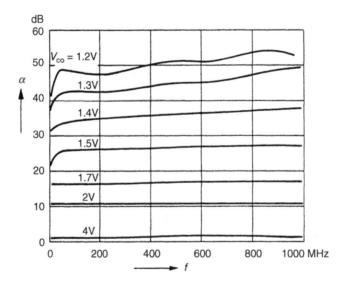

Figure 2.30   Attenuation and reflection attenuation versus control voltage.

Current tuning diode materials include silicon and gallium arsenide; silicon is favored for low cost and lower Q applications from HF through microwave frequencies. Hyperabrupt varactors, also of silicon, are finding many applications in commercial television tuner applications, where their high tuning ratios, linear tuning, and low cost are needed. New developments include low-capacitance hyperabrupts for microwave and wireless applications.

Gallium arsenide used with high operating frequency dictates the highest Q possible, as in parametric amplifiers and millimeter multipliers.

This section will acquaint the reader with tuning diodes: how they work, and what they can or cannot be expected to do in an electronic circuit. The basic properties of a tuning diode will be described in terms of the parameters that manufacturers use in characterizing them. The following topics will also be addressed.

- Capacitance ratio with respect to voltage and voltage breakdown
- Q as a function of design and operating conditions
- Stability–leakage current, temperature coefficient, and post-tuning drift
- Distortion products
- Packaging parasitics
- Applications–suggestions on how to specify a varactor

2.1.4.2 Tuning Diode Physics

Introduction   All junction diodes are made up of the same physical parts: a *pn* junction, a carefully controlled epitaxial layer, and a very-low-resistance substrate. These parts are shown in Figure 2.31.

No matter what type of junction device we are discussing—a tuning diode, a step-recovery diode, or a PIN diode—these parts are all present; the main difference between these devices is the resistivity and thickness of the epitaxial layer. Tuning diodes and multiplier diodes need epitaxial layers where both the resistivity and thickness are carefully controlled.

Abrupt Junction   An abrupt junction diode is one in which the p^+, diffused, region of the diode is much more highly doped that the epitaxial layer. Also, the high doping drops to the doping level of the epitaxial layer in a distance that is short compared to the epitaxial layer thickness, and the doping level of the epitaxial layer is constant over its thickness. This is shown in Figure 2.32, with the corresponding *C–V* curve shown in Figure 2.33. When these requirements are satisfied, the diode capacity, diode area, epitaxial layer doping level, and diode voltage are related by

$$\frac{C(V)}{A} = K \left(\frac{N}{V + \phi} \right)^n \tag{2.42}$$

where $C(V)$ = capacitance of the diode at voltage V, A = area of the diode, N = doping level of the epitaxial layer, V = voltage applied to the diode, ϕ = built-in potential of the diode (0.6–0.8 V), n = slope of diode *C–V* curve; $n \approx 0.5$ for an abrupt-junction diode, and K = constant.

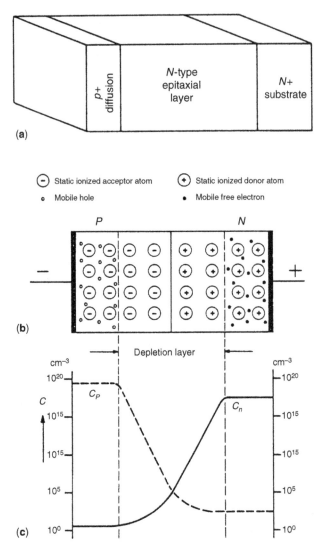

Figure 2.31 (a) Basic PIN structure. (b) Cross-section of a reverse-biased *pn* junction. (c) Density distribution of free charge carriers.

As a consequence of the physical properties of a *pn* junction, a depletion layer is formed between the *p* and *n* regions whose width depends on the voltage applied to the diode. The capacitance of the diode is inversely proportional to the width of the depletion layer, that is, $C \propto 1/\ell_0$. In addition, the series resistance of the diode is proportional to the width of the undepleted epitaxial layer. Thus, as diode reverse bias is increased, the depletion layer increases, causing a decrease in capacitance and a decrease in series resistance. As the diode reverse bias is increased further, a point is reached where the electric field caused by the reverse bias reaches a critical level, and current through the diode increases rapidly; this is the breakdown voltage of the diode. If, at the breakdown voltage, the epitaxial layer is not completely depleted, the diode will have excessive series resistance. Conversely, if the

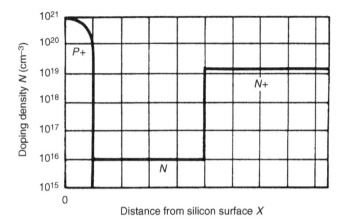

Figure 2.32   *N–X* abrupt-junction diode.

epitaxial layer is depleted before the breakdown voltage is reached, no further capacitance decrease occurs after the total depletion, and a condition called punchthrough occurs.

While, in the ideal case, voltage breakdown will occur just as the epitaxial layer is totally depleted, this seldom occurs in practice, and we generally have a condition of either punchthrough or excess series resistance.

Linearly Graded Junction   If, instead of the junction profile shown in Figure 2.32, we have a p^+ region and an n region whose doping levels increase linearly with distance from

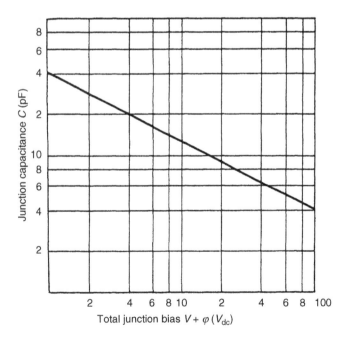

Figure 2.33   Capacitance versus total junction bias for abrupt-junction diode.

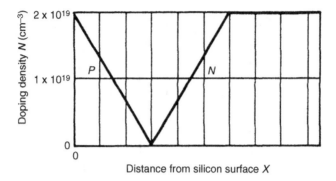

Figure 2.34 $N-X$ linearly graded junction.

the *pn* junction as shown in Figure 2.34, with its corresponding $C-V$ curve in Figure 2.35, we then have what is called a linearly graded junction diode. This diode follows equation (2.42) with the exception that the exponent n is equal to 1/3. This means that, for a given voltage change, the linearly graded junction will have a smaller capacitance change than an abrupt junction diode. Since, in most cases, the designer is looking for the maximum capacitance change obtainable, the linearly graded junction is not used as a tuning diode. This structure found its greatest use several years ago as a "cube law" multiplier, but even this use has decreased as new structures have been developed.

Hyperabrupt Junction The hyperabrupt diode provides a greater capacitance change than the abrupt junction diode for a given voltage change, as well as a linear frequency versus

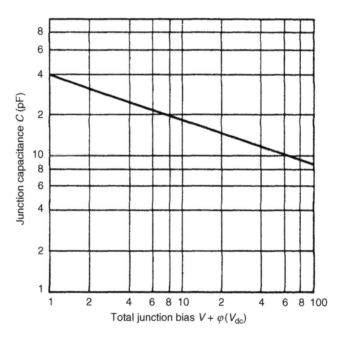

Figure 2.35 Capacitance versus total junction bias for linearly graded junction.

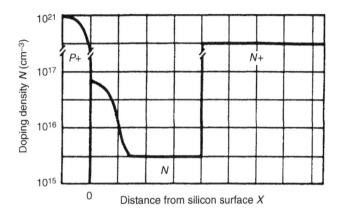

Figure 2.36 *N–X* hyperabrupt junction diode.

voltage characteristic over a limited voltage range. The structure of the hyperabrupt diode is shown in Figure 2.36 and can be seen to be an abrupt junction diode with an additional, increased doping level at the *pn* junction. This diode also follows equation (2.42) with the exception that *n* is not a function of voltage and is generally in the range of 0.5–2. A typical curve of *n* versus voltage is shown in Figure 2.37.

The *C–V* curve in a hyperabrupt diode is shown in Figure 2.38 and is seen to start at a high value of capacitance per unit area at low bias (high epitaxial doping) and change to a lower value of capacitance per unit area (low epitaxial doping) at high bias. The details of the curve depend on details of the shape of the more highly doped region near the *pn* junction.

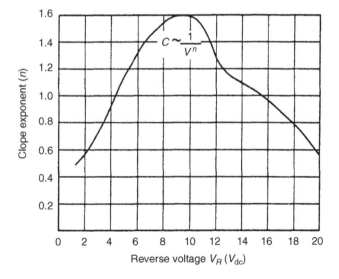

Figure 2.37 Typical *n* versus reverse voltage for hyperabrupt junction diode.

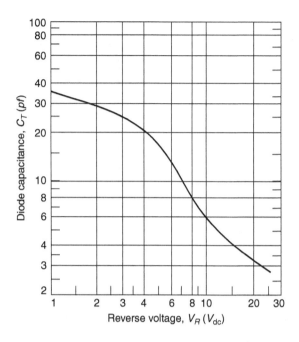

Figure 2.38   Capacitance versus junction bias for hyperabrupt junction diode.

Unfortunately, with a hyperabrupt diode, you must settle for a lower Q than an abrupt junction diode with the same breakdown voltage and same capacitance at 4 V.

It should be noted that any diode that has an n value that exceeds 0.5 at any bias voltage is, by definition, a hyperabrupt diode. Thus, the hyperabrupt diode family can have an infinite number of different $C-V$ curves. Since the abrupt junction diode has a well-defined $C-V$ curve, the capacitance value at one voltage is sufficient to define the $C-V$ capacitance at any other voltage. This is not the case for the hyperabrupt diode. In order to adequately define the $C-V$ characteristics of a hyperabrupt diode, two and sometimes three points on the curve must be specified.

Silicon Versus Gallium Arsenide   Everything mentioned so far applies to both silicon and gallium arsenide (GaAs) diodes. The main difference between silicon and GaAs from a user's point of view is that higher Q can be obtained from GaAs devices. This is due to the lower resistivity of GaAs from a given doping level N. The resistivity of the epitaxial layer, or substrate, of a diode is given by

$$\rho = \frac{1}{Ne\mu} \tag{2.43}$$

where ϱ is the resistivity, N is the doping level of the layer, e is the charge on an electron, and μ is the mobility of the charge carriers in the layer.

Gallium arsenide has a mobility about four times that of silicon and, thus, a lower resistivity and higher Q for a given doping level N. Since diode capacitance is proportional to $\sqrt{N}$, independent of resistivity, a silicon diode and a GaAs diode of equal area and doping will have a capacitance difference proportional to the square root of the dielectric constant ratio. This gives the GaAs diode a 5% higher capacitance and is thus of little practical significance. The penalty paid for using GaAs is an unpassivated diode and a more expensive diode due to higher material and processing costs. If the higher Q of the GaAs device is not really needed, a substantial price saving will be obtained by using a silicon device.

Planar Versus Mesa Construction   The two basic construction techniques used to manufacture tuning diodes are planar and mesa; a cross section of each of these devices is shown in Figure 2.39. The planar process, which is the backbone of the integrated circuit industry, lends itself to large volume production techniques and is the one use for the 1 N series of tuning diodes. Mesa processing, on the other hand, requires more processing steps and is generally done on a wafer-by-wafer basis. This results in a more costly process and thus a more expensive diode. All microwave tuning diodes are of mesa design because of greatly higher Q. Due to the relatively small radius of curvature at the junction edge of a planar diode, the electric field in this area is greater than the electric field in the center (flat) portions of the junction. As a result, the breakdown voltage of the diode is determined by both the epitaxial resistivity and the radius of curvature of the junction edge. Thus, for a given breakdown voltage, a planar diode must use higher resistivity epitaxial material than a mesa diode, which has a completely flat junction. The end result is that the planar diode has a greater series resistance than a mesa diode for the same capacitance and breakdown voltage, and thus lower Q.

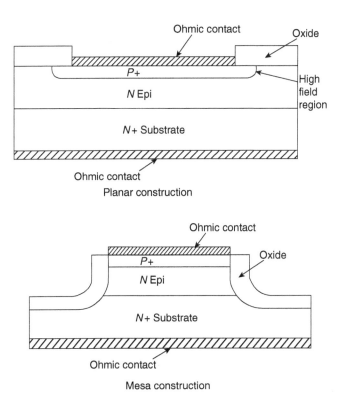

Figure 2.39   Cross sections of planar and mesa devices.

2.1.4.3 Capacitance

Capacitance Ratio From the user's point of view, ratio is simply the capacitance available in the circuit. Thus, a user tuning from, say, -4 to -45 V defines ratio as

$$R = \frac{C_T(-4)}{C_T(-45)} \tag{2.44}$$

where C_T includes C_J plus C_P plus C_F. The manufacturer, however, defines C_T as $C_J + C_P$.

To explore the significance of this difference, let us take two examples, a large C_J and a small C_J, in chip, package and "typical" fringe situations. Both are 45-V tuning varactors.

Device	C_{J0}	C_{J45}	Ratio
A	0.6 pF	0.1 pF	6.0
B	15.0 pF	2.5 pF	6.0

Put both devices in a standard 023 package with C_P (strap and ceramic) of 0.18 pF:

Device	C_{T0}	C_{T45}	Ratio
C	0.78 pF	0.28 pF	2.75
D	15.18 pF	2.68 pF	5.67

Note the drop in ratio, especially for the low-C_J diode. If we now add a typical 0.04 pF for external fringe capacitance, we get

Device	C_{T0}	C_{T45}	Ratio
C	0.82 pF	0.32 pF	2.56
D	15.22 pF	2.72 pF	5.6

The reduction in ratio, and thus circuit tuning capability, by the fringing fields is quite obvious and amounts to 7% in this example.

Because of the often stringent specifications on tuning ratio, it is mandatory that the manufacturer and customer clearly agree on the exact design of the holder used to measure the varactor in question.

Having described how to measure capacitance, it is relatively easy to describe the results. The section on diode physics described the various types of "laws," or C–V curves, and we will not repeat them here. Nonetheless, several important points must be covered.

The first is "available capacitance swing." The laws indicate a steadily decreasing capacitance with voltage, which indicates that the epi region is widening and the electric field is increasing. (For an abrupt junction, since $C \propto 1/\sqrt{V}$, the depletion zone with W is increasing as $\sqrt{V}$, and the electric field V/W increases as $\sqrt{V}$.)

Two things can happen.

1. The junction width widens so that the entire intrinsic region is depleted. The capacitance bottoms out, resulting in voltage punchthrough.

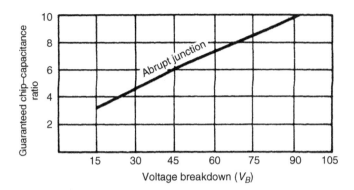

Figure 2.40   Capacitance ratio versus breakdown voltage.

2. The electric field exceeds the dielectric strength of silicon (or GaAs), and "solid-state discharge" or "avalanche" current is drawn.

The diode impedance drops, the varactor no longer "varacts," and circuit operation ceases. Moreover, if more than a few milliamperes of current are drawn, localized overheating may destroy the diode. All varactors are characterized for breakdown voltage—for example, 45 V minimum.

The theoretical tuning varactor is designed so that the punchthrough occurs at a voltage equal to the voltage breakdown of the diode. Logically, then this means that in order to obtain greater tuning ratios, it is necessary to be able to increase the depletion-layer width without reaching punchthrough or breakdown. You must have a thicker epi region to make this possible.

Figure 2.40 shows catalog ratio values, from zero bias to breakdown, as a function of breakdown voltage necessary. In the next section, on Q, we will discuss other elements in your choice of V_B.

To complete this section, we should mention that semiconductor processing control has been refined so well that capacitance tracking to within $\pm 1\%$ over the full range from zero to breakdown is now readily obtainable in production quantities.

Temperature Coefficient of Capacitance (T_{CC})   Unfortunately, since most datasheets give the value of T_{CC} at 4 V, it is sometimes assumed that this value applies at all bias voltages. This is not the case. Consider equation (2.44), a rewritten form of equation (2.42).

$$C(V) = \frac{C(O)}{(V + \phi)^n} \tag{2.45}$$

Taking the derivative of this with respect to temperature T we have

$$\frac{dC(V)}{dT} = \frac{+nC(O)}{(v + \phi)(v + \phi)^n}\frac{d\phi}{dT} \tag{2.46}$$

or, after substituting equation (2.44)

$$T_{CC} = \frac{1}{C(V)} \cdot \frac{dC(V)}{dT} = \frac{-n}{V + \phi} \cdot \frac{d\phi}{dT} \tag{2.47}$$

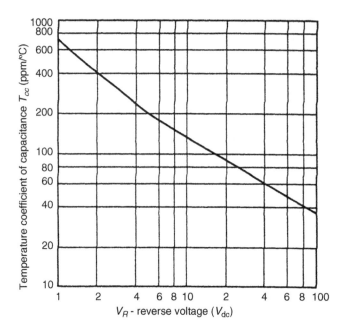

Figure 2.41   Temperature coefficient of capacitance versus tuning voltage–abrupt junction diode.

As a first approximation, we can say that $d\phi/dT = -2.3\,\text{mV/°C}$ over the temperature range of interest.

From equation (2.47), we can draw the following conclusions:

1. the temperature coefficient is inversely proportional to the applied voltage, and
2. the temperature coefficient is directly proportional to the diode slope, n.

For an abrupt junction diode that has a constant value of n (0.5), the temperature coefficient has a smooth curve in the form $K/(v + \phi)$. However, in the case of hyperabrupt diodes, n is a function of voltage, and the shape of the T_{CC} curve depends on the details of the $n(V)$ curve. A typical T_{CC} curve for an abrupt junction diode is shown in Figure 2.41 and for an Alpha DKV6520 series hyperabrupt diode in Figure 2.42. The inflection in the hyperabrupt T_{CC} is due to the fact that in this voltage range $n(V)$ is increasing faster than $1/V$, giving an increase in T_{CC}. It should also be noted, however, that over the range of the T_{CC} minimum the temperature coefficient is relatively constant, and operation in this area may be advantageous in some applications where a restricted tuning range can be used.

2.1.4.4 Q Factor or Diode Loss

Definitions   The classical definition of the Q of any device or circuit is

$$Q = \frac{2\pi \times \text{Energy stored}}{\text{Energy dissipated per cycle}} \tag{2.48}$$

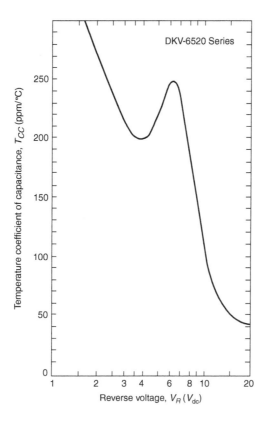

Figure 2.42 Temperature coefficient of capacitance versus tuning voltage ($T_A = 25°C$) hyperabrupt junction diode.

For a capacitor, two formulations are possible.

(a) Series C $Q = \dfrac{1}{2\pi \, f \, R_S \, C}$

R_S

(b) Parallel C R_P $Q = 2\pi \, f \, R_P \, C$

Clearly, the two definitions must be equal at any frequency, which establishes

$$R_p = \frac{1}{(2\pi f)^2 \, C^2 \, R_S} \tag{2.49}$$

In the case of a high-Q tuning diode, the proper physical model is the series configuration, for the depleted region is almost perfectly pure capacitance, and the undepleted region, due to its relatively low resistivity, is almost a pure resistor in series with the capacitance.

Furthermore, the contact resistances are also clearly in series. Q, then, for a tuning diode, is given by

$$Q_{(-v)} = \frac{1}{2\pi f_o R_{(-v)} C_{(-v)}} \qquad (2.50)$$

where f_o is the operating frequency, $C_{(-v)}$ is the junction capacitance, and $R_{(-v)} = R(\text{epi}) + R'_C$, the sum of the resistance of the undepleted epi and the fixed contact resistance.

Cutoff frequency, f_c is defined as that frequency at which Q equals unity. Thus,

$$f_{c(-v)} = \frac{1}{2\pi R_{(-v)} C_{(-v)}} \qquad (2.51)$$

Historically, the tuning diode business a habit of specifying Q at 50 MHz, despite the fact that Q values of microwave diodes are so high that it is almost impossible to measure them at 50 MHz. Instead, as discussed below, Q is measured at microwave frequencies (e.g., 1–3 GHz) and related to 50 MHz by the relationship

$$Q_{(f1)} = Q_{(f2)} \frac{f_2}{f_1} \qquad (2.52)$$

which derives quickly from the assumption that f_c is independent of the measuring frequency.

Since both junction capacitance and epi resistance are functions of the applied bias, it is not possible to calculated Q as a function of bias from a measurement of capacitance alone. Catalog specifications typically show Q at -4 V, together with the capacitance at two or more voltages.

Relative to $Q_{(-4)}$, Q increases faster than the reduction in capacitance for bias greater than 4 V and, conversely, decreases faster for bias less than 4 V.

In the following section, we will discuss the diode design parameters that determine Q. Following this, we will describe some elementary Q measurement techniques.

Causes   In the discussion of device physics, the resistivity of the epi region was discussed, together with its impact on punchthrough and breakdown. For example, Table 2.7 supplies typical resistivity and relative parameters of 0.6 pF ($C_{J_{(4)}}$) diodes of different breakdowns.

The entry R_{sp} ($R_{\text{spreading}}$) is the series resistance between the epi region and the low resistivity substrate. The calculations are for idealized cylindrical epi regions of uniform resistivity, low resistivity contact on the anode (top) and low resistivity substrate on the

Table 2.7 Parameters for 0.6-pF Diode

Breakdown Voltage V_B	Resistivity ($\Omega \cdot$cm)	Junction Diameter (mil.)	Epi Region (μm)	Depelted Epi $V = -4$V (μm)	Undepleted Epi $V = -4$V (μm)	R Undepleted (Ω)	R_{sp} (Ω)	Q_4
30	0.31	2.4	1.37	0.54	0.83	0.81	0.18	5200
45	0.52	2.8	2.25	0.73	1.52	1.86	0.15	2600
60	0.74	3.2	3.20	0.90	2.30	3.24	0.13	1500
90	1.25	3.7	5.27	1.21	1.06	7.17	0.10	700

cathode. This resistance is constant, independent of bias; also shown are epi thickness and width of the depletion zone at -4 V bias.

Note the substantial reduction in Q for high-voltage diodes caused by the increased epi resistance; this is true for any value of capacitance or any type of junction. For greater voltage breakdown, the epi thickness must be increased, which requires an increase in epi layer resistivity; the higher the resistivity of the undepleted zone, multiplied by the fact that it is much wider for high-voltage diodes, means the resistance increases substantially.

Consequently, a rule of thumb emerges: For maximum Q, never choose a tuning diode with a voltage breakdown in excess of what is needed for the necessary tuning range. If the required tuning range is an octave, requiring a 4 to 1 ratio, the selection of a 30-V diode will result in diode losses half those of a 60-V diode.

Table 2.8 lists capacitance and Q for each of these chips as a function of bias. Please remember that Q is calculated at 50 MHz.

Table 2.9 rewrites the data of Table 2.8 to show available capacitance ratios between zero bias and breakdown. The first column is the theoretical optimum, as tabulated. The second column is the typical catalog specification. The reduction in tuning ratio below theoretical optima is caused by nonideal junction fabrication. The junctions are never perfectly abrupt.

Although the tables and numbers given refer to abrupt-junction silicon devices, the principle applies without exception to all types of tuning diodes. For comparison, Table 2.10 lists available ratios and Q values for a number of different varactors. The high Q values for GaAs and the low values for hyperabrupts are apparent.

One last point: The Q values and series resistance refer to chips only. The effects of package parasitics will be discussed later, but it is important to consider circuit contact losses here. For low-capacitance diodes—for example, $C_J(4) = 0.6$ pF—the epi region contributes a high value of resistance and dominates Q except at punchthrough. Diode contact losses are less significant.

Post-Tuning Drift Post-tuning drift (PTD) is the change in oscillator frequency with time after the tuning voltage has stabilized. The minimization of PTD has assumed greater importance with the design of more sophisticated electronic countermeasure systems, where rapid, accurate frequency changes are required.

Post-tuning drift can characterized as short-term and long-term. Short-term PTD occurs in the time range of tens of nanoseconds to a few seconds, while long-term PTD is in the time range of seconds to minutes, hours or days.

Short-term PTD is mainly dependent on the thermal properties of the diode and is improved by high Q (low power loss) and flip chip construction. Long-term PTD depends on oxide stability and freedom of mobile charge in the oxide. It should be noted that actual oscillation frequency change may occur even with a perfect tuning diode because of variation with frequency in the power dissipated by the diode, changes in the diode heat-sink temperature, and frequency changes due to other circuit elements. Less than 0.01% short- and long-term PTD can be obtained.

2.1.4.5 Distortion Products

Inasmuch as nonlinear components generate harmonics and other distortion products, an understanding of this mechanism is of prime interest to the circuit designer. In some instances, the distortion products are the desired end result of the circuit design, as in frequency multipliers, where harmonics of the input signal frequency are the required output signal. For other applications, such tuning-diode-tuned linear circuits, distortion products are

Table 2.8 Q Versus Bias for 0.6-pF Diode

Breakdown Voltage V_B	C_{J0}	$Q(0)$	$C_J(-4)$	$Q(-4)$	$C_J(-10)$	$Q(-10)$	$C_J(-30)$	$Q(-30)$	$C_J(-45)$	$Q(-45)$	$C_J(-60)$	$Q(-60)$	$C_J(-90)$	$Q(-90)$
30	1.43	1700	0.6	5200	0.4	10k	0.23	64k						
45	1.43	850	0.6	2600	0.4	5k	0.23	20k	0.19	90k				
60	1.43	550	0.6	1500	0.4	3k	0.23	9k	0.19	23k	0.17	170k		
90	1.43	270	0.6	700	0.4	1.3k	0.23	3.5k	0.19	6k	0.17	10k	0.14	220k

Table 2.9 Capacitance Ratios ($C_{J0}/C_J V_B$)

Breakdown Voltage V_B	Optimum Ratio	Minimum Guaranteed Ratio	Q_{-4} Typical
30	6.2	4.5	3000
45	7.5	6.0	2500
60	8.4	7.5	1400
90	10.2	8.7	650

extremely undesirable, and in some instances the end product specification may set a maximum limit to the distortion products allowed.

Cross-Modulation   Cross-modulation is the transfer of the modulation on one signal to another signal and is caused by third-order and higher odd-order nonlinearities in the behavior of the device. Rewriting equation (2.42), we have

$$C(V) = \frac{C_0}{\left(1 + \frac{V}{\phi}\right)^n} \qquad (2.53)$$

where C_0 = capacitance, V = applied voltage = $V_0 + v$, V_0 = dc applied voltage, and v = ac applied voltage.

Then, for a desired signal of

$$S_1 = v_1 \sin \omega_1 t \qquad (2.54)$$

and a second, amplitude-modulates signal of

$$S_2 = v_2 (1 + m \cos \omega_m t) \sin \omega_2 t \qquad (2.55)$$

it can be shown that the cross-modulation, γ, defined by

$$\text{Ouput signal } v_1 \sim \sin \omega_1 t + \gamma \sin (\omega_1 \pm \omega_m) t \qquad (2.56)$$

Table 2.10 Comparative Tuning Diodes

Type	C_{J0}	Breakdown Voltage V_B	Q^a_{-4}	Ratio $C_{J0}/C_J VB^a$
Silicon abrupt	1.0	30	5,000	4.5
Silicon abrupt	2.5	30	4,600	4.5
Silicon abrupt	5.0	30	3,800	4.5
Gallium arsenide abrupt	1.0	25	10,000	3.6
Gallium arsenide abrupt	0.5	10	17,000	2.5
Silicon hyperabrupt	50.0	22	300	17.0
Silicon hyperabrupt	2.5	22	500	14

[a]Minimum guaranteed.

is found to be

$$\gamma = \frac{n(n+1)mv_2^2}{4(V_0 + \phi)^2} \tag{2.57}$$

From this equation, it can be seen that cross-modulation is

- proportional to the square of the interfering signal;
- directly proportional to the interfering signal's modulation index, m;
- independent of the strength of the desired signal;
- independent of the frequencies of the desired and interfering signals (assuming that the nonlinearity to which both signals are subjected is sufficiently frequency-indiscriminate so this is the case); and
- present for all values of n. That is, no value of n gives zero cross-modulation.

Solving equation (2.57) for the signal level v_2 required to produce cross-modulation of value γ we have

$$v_2 = \frac{2(V_0 + \phi)\gamma}{n(n+1)^m} \tag{2.58}$$

The interfering signal levels required to produce 1% cross-modulation from a 30%-modulated interfering signal applied to an abrupt junction diode and a hyperabrupt junction diode are shown in Figure 2.43.

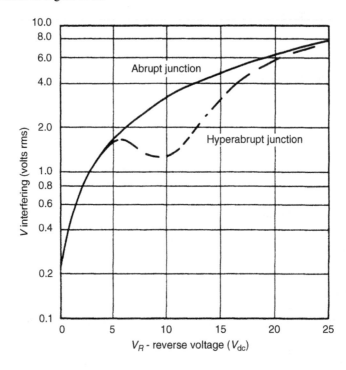

Figure 2.43 Interfering signal (30% amplitude modulated) level versus bias for 1% cross-modulation—abrupt and hyperabrupt junction diodes.

From this figure, it can be seen that the hyperabrupt diode is more susceptible to cross-modulation than the abrupt-junction diode in the region of maximum slope of the hyperabrupt diode. For many applications, however, distortion products will be generated by other devices, such as transistors, at signal levels considerably below those given in Figure 2.43.

Intermodulation   Intermodulation is the production of undesired frequencies in the form

$$\sin(2\omega_1 t - \omega_2 t) \qquad \text{and} \qquad \sin(\omega_1 t - 2\omega_2 t) \tag{2.59}$$

from an input signal in the form of

$$v(\cos \omega_1 t + \cos \omega_2 t) \tag{2.60}$$

From an analysis similar to that done for cross-modulation, it can be shown that

$$\text{Intermodulation} = \frac{n(n+1)v^2}{8(V_0 + \phi)^2} \tag{2.61}$$

or

$$\text{Cross-modulation} = (2m) \times \text{Intermodulation} \tag{2.62}$$

Harmonic Distortion   Harmonic distortion products are integral multiples of the signal frequencies and decrease in amplitude as the harmonic number decreases. Due to passband considerations and amplitude decrease with harmonic number, the second harmonic is the one of prime concern. Again, it can be shown that the second harmonic, v_2, of a signal of amplitude v_1 is

$$v_2 = \frac{n}{3(V_0 + \phi)}v_1^2 \tag{2.63}$$

Figure 2.44 shows the signal level required to produce 10% second-harmonic distortion in an abrupt-junction and a hyperabrupt-junction diode. Again, as in the case of cross-modulation, the hyperabrupt diode is slight worse than the abrupt-junction diode in the region of maximum slope of the hyperabrupt diode.

Reduction of Distortion Products   In some cases, the signal levels applied to the diode generate distortion products larger than desirable for the circuit application. In this case, significant reduction in the distortion products can be achieved by using two diodes in a back-to-back configuration, as shown in Figure 2.45. Analysis shows that the fundamental signal components through the diodes are in phase and add, while some distortion products are out of phase and cancel, thus improving distortion performance.

Since the gradient of the electrical field produced in the depletion layer by a reverse bias applied to the device is proportional to the space charge density, the following equations can be written for the junction width W as a function of the reverse bias V_R:

For an abrupt *pn* junction (alloyed diodes)

$$W = \sqrt[2]{2 \cdot \frac{\varepsilon_R \cdot \varepsilon_0}{q}\left(\frac{1}{C_p} + \frac{1}{C_n}\right) \cdot (V_R + V_D)} \tag{2.64}$$

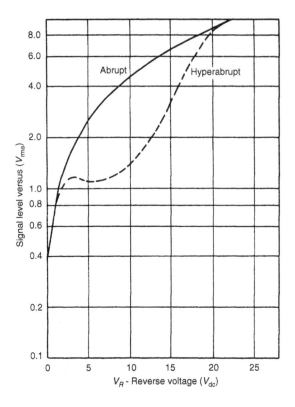

Figure 2.44 Signal level versus reverse voltage for 10% harmonic distortion.

For linear *pn* junctions (single-diffused diodes, such as BA110–BA112)

$$W = \sqrt[3]{12\frac{\varepsilon_R \cdot \varepsilon_0}{a \cdot q}(V_R + V_D)} \qquad (2.65)$$

wherein a is the impurity gradient within the depletion layer, ε_0 is the absolute dielectric constant (8.85×10^{-14} As/V cm), and $\varepsilon_r \approx 12$, the relative dielectric constant of silicon.

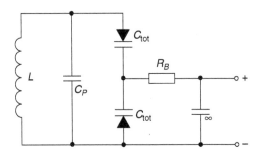

Figure 2.45 Back-to-back diodes. CP is a fixed parallel capacitance, RB is a bias decoupling resistor, and the capacitor marked ∞ is a low-impedance bypass.

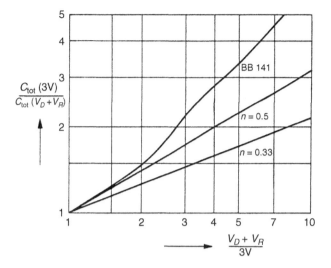

Figure 2.46 Capacitance/voltage characteristic for (a) an alloyed capacitance diode; (b) a diffused capacitance diode; and (c) a wide-range tuner diode (BB141).

The junction capacitance, which is inversely proportional to the junction width, therefore varies in alloyed diodes with the square root, and in single-diffused diodes with the cube root of the externally applied reverse bias, and can be calculated from the general equation

$$C = \frac{\varepsilon_R \varepsilon_0 S}{W} \tag{2.66}$$

wherein S is the surface area of the pn junction. By way of approximation, we can also use the equation

$$C = \frac{K}{(V_R + V_D)^n} \tag{2.67}$$

wherein all constants and all parameters determined by the manufacturing process are contained in K. The exponent n is a measure of the slope of the capacitance/voltage characteristics and is 0.5 for alloyed diodes, 0.33 for single-diffused diodes, and (on average) 0.75 for tuner diodes with a hyperabrupt pn junction. Figure 2.46 shows the capacitance/voltage characteristics of an alloyed, a diffused, and a tuner diode.

Recently, an equation is indicated which, although purely formal, describes the practical characteristics better than equation (2.67):

$$C = C_0 \left(\frac{A}{A + V_R} \right)^m \tag{2.68}$$

where C_0 is the capacitance at $V_R = 0$, and A is a constant whose dimension is a voltage. The exponent m is much less dependent on voltage that the exponent n in equation (2.67).

Equations (2.64)–(2.68) express the pure junction capacitance of the capacitance diode, but to this must still be added a constant capacitance, determined by structure parameters, in order to obtain the diode capacitance C_{tot}, which interests the user. With high inverse

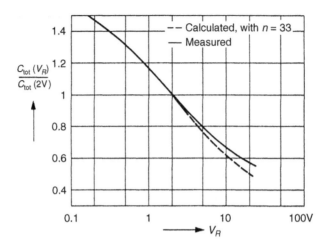

Figure 2.47   Capacitance/voltage characteristic of the BA110 diode.

voltages—that is, low junction capacitance—a difference will therefore arise between the theoretical capacitance/voltage characteristic according to equation (2.67) and the practical characteristic, as shown in Figure 2.47.

The operating range of a capacitance diode or its useful capacitance ratio

$$\frac{C_{max}}{C_{min}} = \frac{C_{tot}(V_{Rmin})}{C_{tot}(V_{Rmax})} \tag{2.69}$$

is limited by the fact that the diode must not be driven by the alternating voltage superimposed on the tuning voltage either into the forward mode or the breakdown mode. Otherwise, rectification, which would shift the diode's bias and considerably affect its figure of merit, would take place. Figure 2.48 plots the capacitance/voltage characteristics of a capacitance diode to clarify the relationship. The useful operating range lies between the voltages

$$V_{min} > \hat{V} - V_F \qquad \text{and} \qquad V_{max} < V_{(BR)R} - \hat{V} \tag{2.70}$$

As has already been indicated, the exponent n for large capacitance ratio or tuning diodes used nowadays for TV tuners is not constant, but voltage dependent and subject to manufacturing tolerances. This means that the capacitance/voltage characteristic of these diodes is likewise subject to manufacturing tolerances. Since, in a TV tuner, it is necessary for two or three circuits to be tuned uniformly, tuner diodes must be selected empirically for identical characteristics and supplied in equipment lots.

2.1.4.6 Electrical Properties of Tuning Diodes

In this section, the electrical properties of capacitance diodes with reference to data published on the tuner diodes BB141 and BB142 is explained.

Equivalent Circuit   Since a capacitance diode is not an ideal capacitor, it is useful to introduce an equivalent circuit that can serve as a basis for discussing the diode's electrical properties. Figure 2.49 shows various types of equivalent circuits.

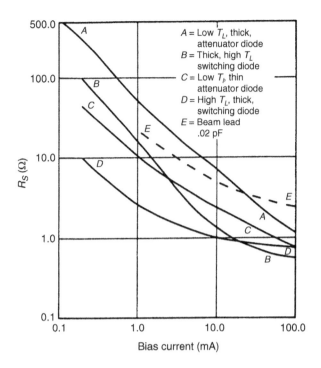

Figure 2.48 Basic current/voltage and capacitance/voltage characteristics.

The complete equivalent circuit (Figure 2.49a), which conforms closely to physical conditions, comprises, in addition to the diode capacitance C_{tot}, the series resistance r_S and the series inductance L_S, and also the reverse resistance $R = dV_R/dI_R$. At higher frequencies, this resistance can usually be disregarded, so that the equivalent circuit is simplified in terms of Figure 2.49b, the configuration usually employed. In many cases, also the series inductance can be disregarded, in which case we obtain Figure 2.49c.

Capacitance Figure 2.50 shows the capacitance/voltage characteristic borrowed from the data sheets of diodes BB141 and BB142. As this characteristic, like most characteristics published in data sheets, represents merely a typical curve, it would not convey sufficient

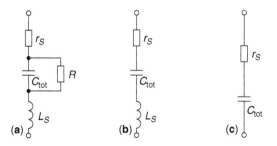

Figure 2.49 Equivalent circuits for capacitance diodes: (a) complete circuit, (b) simplified circuit, and (c) further simplification for low frequencies.

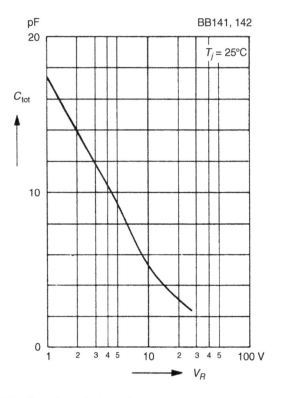

Figure 2.50   Capacitance/voltage characteristic of diodes BB141 and BB142.

information for the user to dimension a tuner. Therefore, the data sheets contain additional data on capacitance and useful capacitance ratio:

Capacitance		BB141A (pF)	BB141B (pF)	BB142 (pF)
At $V_R = 1$ V	C_{tot}	16	19	17
At $V_R = 3$ V	C_{tot}	11	13	12
At $V_R = 25$ V	C_{tot}	2–2.35	2.25–2.65	2–3

$$\text{Useful capacitance ratio} = \frac{C_{tot}(3V)}{C_{tot}(25V)} = 4 \text{ to } 6$$

On the basis of the spread of the diode capacitance C_{tot} guaranteed for $V_R = 25$ V, it is possible to calculate the tuner with the aid of the spread guaranteed for the capacitance ratio.

Three further capacitance graphs are shown in Figures 2.51–2.53. Figure 2.51 shows a (typical) curve representing the normalized slope as a function of the reverse voltage, Figure 2.52 the normalized capacitance as a function of the junction temperature, and Figure 2.53 the temperature coefficient of the capacitance as a function of the reverse voltage.

The variation of diode capacitance with ambient temperature, as shown in Figures 2.52 and 2.53, is really only a function of the temperature dependence of the diffusion voltage (see equation 2.66). The diffusion voltage drops with rising temperatures by about 2 mV/°C, which means that the diode capacitance rises with temperature. The influence of

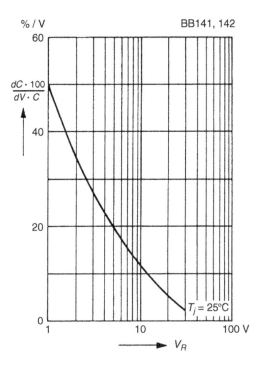

Figure 2.51   Slope (normalized) as a function of the reverse voltage.

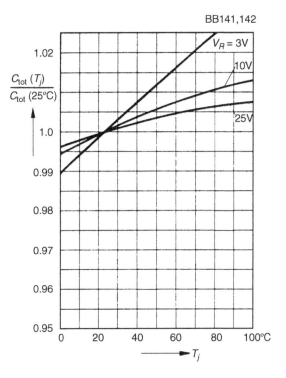

Figure 2.52   Capacitance (normalized) as a function of the junction temperature.

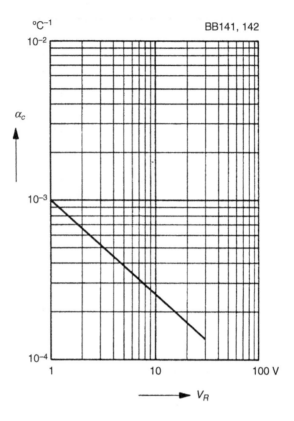

Figure 2.53 Temperature coefficient of capacitance as a function of the reverse voltage.

the diffusion voltage, and thus the temperature coefficient of the capacitance, decreases as the reverse voltage rises. It is therefore advisable to run a capacitance diode with as high a reverse voltage as the required capacitance ratio permits. Compensation for the dependence of capacitance on temperature will be discussed later.

Series Resistance, Figure of Merit (Q) Since capacitance diodes are intended for employment in resonant circuits, indications are required as to the circuit attenuation that they produce. Usually, the series resistance or the figure of merit that can be calculated therefrom, is indicated. With reference to the equivalent circuit shown in Figure 2.49a, the figure of merit (or Q factor) is calculated.

$$Q = \frac{1}{\omega C_{tot} r_S + \frac{1}{\omega C_{tot} R}} \tag{2.71}$$

Figure 2.54 shows the theoretical (normalized) Q as a function of frequency. With the frequency range of 10–1000 MHz, which is essential for tuner diodes, the parallel resistance R caused by the reverse current (see Figure 2.49b) can be disregarded, and equation (2.71) can be simplified to read as follows (see Figure 2.55):

$$Q = \frac{1}{\omega C_{tot} r_S} \tag{2.72}$$

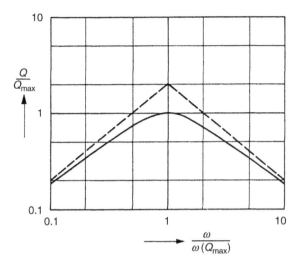

Figure 2.54 *Q* (normalized) as a function of frequency.

In Figure 2.52, Q is plotted as a function of frequency for the tuner diodes BB141 and BB142. The data sheet indicates, in addition to this curve, the value of the series resistance at $f = 470\,\text{MHz}$ and $C_{\text{tot}} = 9\,\text{pF}$:

	BB141A, BB141B	BB142
Series resistance (r_s)	0.6 (<0.8)	0.9 (<1.2)

Since the diode capacitance and the series resistance decrease as the reverse voltage rises, Q is likewise dependent on the reverse voltage. It is smallest at a low reverse voltage, that is, at a low frequency. The series and parallel capacitances inevitable in a diode-tuned resonant circuit influence the effective Q of the tuning capacitance, as will be described later.

Series Inductance, Series Resonant Frequency, Cutoff Frequency for $Q = 1$ These three parameters define the behavior of the diode at high frequencies when the influence of the series inductance can no longer be neglected. The parameter on which the calculation is based is the series inductance L_S, which is determined by the connection leads. As indicated in Figure 2.49a, it is in series with the capacitance and, given a suitable frequency, forms a series-resonant circuit therewith. In that case

$$\omega L_S = \frac{1}{\omega C_{\text{tot}}} \tag{2.73}$$

and the series-resonant frequency is

$$f_0 = \frac{1}{2\pi \sqrt{C_{\text{tot}} L_S}} \tag{2.74}$$

Above this frequency, the impedance of the diode is inductive, but it depends on the reverse voltage. Figure 2.56 shows for the tuner diodes BB141 and BB142 how the (normalized) series-resonant frequency depends upon the reverse voltage.

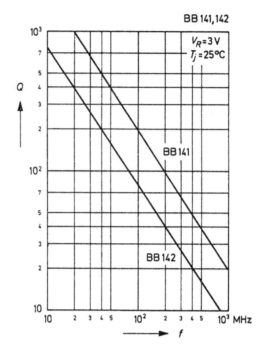

Figure 2.55 *Q* as a function of frequency.

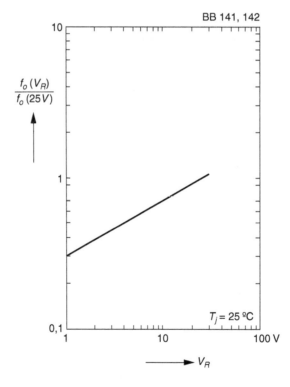

Figure 2.56 Series-resonant frequency (normalized) as a function of the reverse voltage.

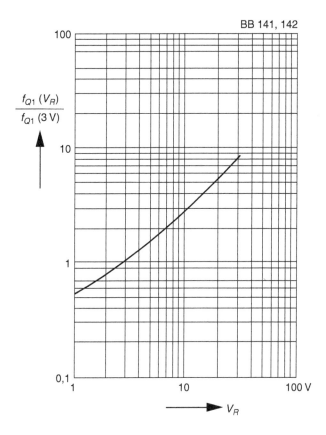

Figure 2.57 Cutoff frequency for $Q = 1$ (normalized) as a function of the reverse voltage.

The cutoff frequency for $Q = 1$ is defined as the frequency at which Q assumes the value of unity. In a rough calculation, the reactive impedance of the diode capacitance can be neglected and the frequency f_{Q1} obtained for the condition in which the inductive reactance equals the series resistance

$$\omega L_S = r_S \tag{2.75}$$

$$f_{Q1} = \frac{r_S}{2\pi L_S} \tag{2.76}$$

For the tuner diodes BB141 and BB142, Figure 2.57 shows how the (normalized) cutoff frequency depends on the reverse voltage. The data sheets contain the following numerical information.

		BB141	BB142	
Series inductance measured 1.5 mm from the case	L_S	2.5	2.5	nH
Series-resonant frequency at $V_R = 25$ V	f_0	2	1.8	GHz
Cutoff frequency for $Q = 1$ at $V_R = 3$ V	f_{Q1}	24	16	GHz

Depending on the intended application, either the series-resonant frequency or the cutoff frequency for $Q = 1$ determines the maximum useful frequency range.

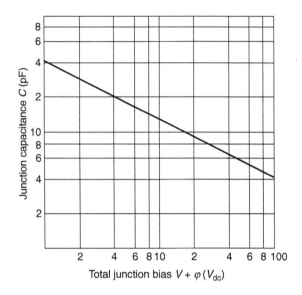

Figure 2.58 Leakage current as a function of reverse voltage.

Leakage Current, Breakdown Voltage These two parameters mark the dc performance of the diode when a reverse bias is applied to it. In the data book for the tuner diodes BB141 and BB142, the breakdown voltage is quoted as follows:

$$\text{Reverse breakdown voltage at } I_R = 100 \ \mu A: \qquad V_{(BR)R} > 30V$$

This means that the maximum reverse voltage that may be applied to the diode as a tuning voltage is 30 V.

The leakage current for the above-mentioned tuner diodes is guaranteed as follows:

$$\text{Leakage current at } V_R = 28 \ V: \qquad I_R < 50 \, nA$$

Moreover, Figure 2.58 shows a graph that illustrates the leakage current as a function of the reverse voltage. Since this current is temperatures dependent (as with every silicon diode)—it doubles with each temperature rise of about 10°C—care should be taken when dimensioning the tuner circuit that the leakage current does not cause inadmissible voltage variations at increased ambient temperatures.

Matching of Tuner Diodes, Uniform Parameters As has already been mentioned, the capacitance versus voltage characteristics of modern tuner diodes is subject to a certain scatter, so that it becomes necessary to test these diodes to obtain equipment lots empirically.

2.1.4.7 Diode-Tuned Resonant Circuits

The Tuner Diode in the Parallel Resonant Circuit Figures 2.59–2.61 illustrate three basic circuits for the tuning of parallel-resonant circuits by means of capacitance diodes.

In the circuit diagram of Figure 2.59, the tuning voltage is applied to the tuner diode via the input coil and the bias resistor R_B. Series connected to the tuner diode is the series capacitor C_S, which completes the circuit for ac but isolates the cathode of the tuner diode

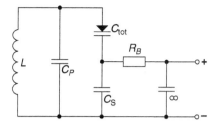

Figure 2.59   Parallel-resonant circuit with tuner diode, and bias resistor parallel to the series capacitor.

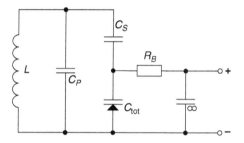

Figure 2.60   Parallel-resonant circuit with tuner diode, and bias resistor parallel to the diode.

from the coil and thus from the negative terminal of the tuning voltage. Moreover, a fixed parallel capacitance C_P is provided. The decoupling capacitor preceding the bias resistor is large enough for its value to be disregarded in the following discussion. Since for high-frequency purposes the biasing resistor is connected in parallel with the series capacitor, it is transformed into the circuit as an additional equivalent shunt resistance R_C. We have the equation

$$R_c = R_B \left(1 + \frac{C_S}{C_{\text{tot}}} \right)^2 \tag{2.77}$$

If in this equation the diode capacitance is substituted by the resonant circuit frequency ω, we obtain

$$R_c = R_B \left(\frac{\omega^2 L C_S}{1 - \omega^2 L C_p} \right)^2 \tag{2.78}$$

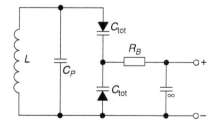

Figure 2.61   Parallel-resonant circuit with two tuner diodes.

The resistive loss R_C caused by the bias resistor R_B is seen to be highly frequency-dependent, and this may result in the bandwidth of the tuned circuit being independent of frequency if the capacitance of the series capacitor C_S is not chosen sufficiently high.

Figure 2.60 shows that the tuning voltage can also be applied directly and in parallel to the tuner diode. For the parallel loss resistance transformed in to the circuit, we have the expression

$$R_c = R_B \left(\frac{1 + C_{\text{tot}}}{C_S} \right)^2 \tag{2.79}$$

and

$$R_c = R_B \left(\frac{\omega^2 L C_S}{\omega^2 L \left(C_S + C_p \right) - 1} \right)^2 \tag{2.80}$$

The influence of the bias resistor R_B in this case is larger than in the circuit of Figure 2.60, provided that

$$C_S^2 > C_S \left(C_{\text{tot}} + C_p \right) + C_{\text{tot}} C_p \tag{2.81}$$

This is usually the case because the largest possible capacitance will be preferred for the series capacitor C_S and the smallest for the shunt capacitance C_P. The circuit of Figure 2.59 is therefore normally preferred to that of Figure 2.60. An exception would be the case in which the resonant circuit is meant to be additionally damped by means of the bias resistor at higher frequencies.

In the circuit of Figure 2.61, the resonant circuit is tuned by two tuner diodes that are connected in parallel via the coil for tuning purpose, but series connected in opposition for high-frequency signals. This arrangement has the advantage that the capacitance shift caused by the ac modulation (see "Modifying the Diode Capacitance by the Applied ac Voltage") takes effect in opposite directions in these diodes and therefore cancels itself. The bias resistor R_B, which applies the tuning voltage to the tuner diodes, is transformed into the circuit at a constant ratio throughout the whole tuning range. Given two identical, loss-free tuner diodes, we obtain the expression

$$R_c = 4R_B \tag{2.82}$$

Capacitances Connected in Parallel or Series with the Tuner Diode Figures 2.59 and 2.60 show that a capacitor is usually in series with the tuner diode, in order to close the circuit for alternating current and, at the same time, to isolate one terminal of the tuner diode from the rest of the circuit with respect to direct current, so as to enable the tuning voltage to be applied to the diode. As far as possible, the value of the series capacitor C_S will be chosen such that the effective capacitance variation is not restricted. However, in some cases, as, for example, in the oscillator circuit of receivers whose intermediate frequency is of the order of magnitude of the reception frequency, this is not possible and the influence of the series capacitance will then have to be taken into account. By connecting the capacitor C_S, assumed to be lossless, in series with the diode capacitance C_{tot}, the tuning capacitance is reduced to the value

$$C^* = C_{\text{tot}} \frac{1}{1 + \frac{C_{\text{tot}}}{C_S}} \tag{2.83}$$

The Q of the effective tuning capacitance, taking into account the Q of the tuner diode, increases to

$$Q^* = Q\left(1 + \frac{C_{tot}}{C_S}\right) \tag{2.84}$$

The useful capacitance ratio is reduced to the value

$$\frac{C^*_{max}}{C^*_{min}} = \frac{C_{max}}{C_{min}} \frac{1 + \frac{C_{min}}{C_S}}{1 + \frac{C_{max}}{C_S}} \tag{2.85}$$

wherein C_{max} and C_{min} are the maximum and minimum capacitance of the tuner diode.

On the other hand, the advantage is gained that, due to capacitive potential division, the amplitude of the alternating voltage applied to the tuning diode is reduced to

$$\hat{v}^* = \hat{v} \frac{1}{1 + \frac{C_{tot}}{C}} \tag{2.86}$$

so that the lower value of the tuning voltage can be smaller, and this results in a higher maximum capacitance C_{max} of the tuner diode and a higher useful capacitance ratio. The influence exerted by the series capacitor, then, can actually be kept lower than equation (2.84) would suggest.

The parallel capacitance C_P that appears in Figures 2.59–2.61 is always present, since wiring capacitances are inevitable and every coil has its self-capacitance. By treating the capacitance C_P, assumed to be lossless, as a shunt capacitance, the total tuning capacitance rises its value and, if C_S is assumed to be large enough to be disregarded, we obtain

$$C^* = C_{tot}\left(1 + \frac{C_p}{C_{tot}}\right) \tag{2.87}$$

The Q of the effective tuning capacitance, as derived from the Q of the tuner diode, is

$$Q^* = Q\left(1 + \frac{C_p}{C_{tot}}\right) \tag{2.88}$$

or, in other words, it rises with the magnitude of the parallel capacitance. The useful capacitance ratio is reduced as follows:

$$\frac{C^*_{max}}{C^*_{min}} = \frac{C_{max}}{C_{min}} \frac{1 + \frac{C_p}{C_{max}}}{1 + \frac{C_p}{C_{min}}} \tag{2.89}$$

In view of the fact that even a comparatively small shunt capacitance reduces the capacitance ratio considerably, it is necessary to ensure low wiring and coil capacitances in the circuit-design stage.

Tuning Range   The frequency range over which a parallel-resonant circuit according to Figure 2.59 can be tuned by means of the tuner diode depends upon the useful capacitance ratio of the diode and on the parallel and series capacitances present in the circuit.

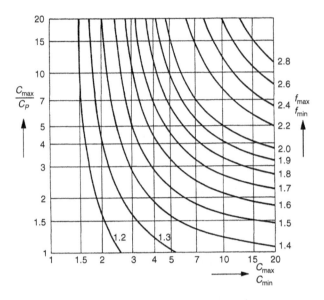

Figure 2.62   Diagram for determining the capacitance ratio and minimum capacitance.

The ratio can be found from

$$\frac{f_{\text{max}}}{f_{\text{min}}} = \sqrt{\frac{1 + \dfrac{C_{\text{max}}}{C_p\left(1 + \frac{C_{\text{max}}}{C_S}\right)}}{1 + \dfrac{C_{\text{max}}}{C_p\left(\frac{C_{\text{max}}}{C_{\text{min}}} + \frac{C_{\text{max}}}{C_S}\right)}}} \tag{2.90}$$

In many cases, the series capacitor can be chosen large enough for its effect to be negligible. In that case, equation (2.90) is simplified as follows:

$$\frac{f_{\text{max}}}{f_{\text{min}}} = \sqrt{\frac{1 + \frac{C_{\text{max}}}{C_p}}{1 + \frac{C_{\text{min}}}{C_p}}} \tag{2.91}$$

From this equation, the diagram shown in Figure 2.62 is computed. With the aid of this diagram, the tuner diode parameters required for tuning a resonant circuit over a stipulated frequency range—that is, the maximum capacitance and the capacitance ratio—can be determined. Whenever the series capacitance C_S cannot be disregarded, the effective capacitance ratio is reduced according to equation (2.85).

Tracking   Some applications require the maintenance of a fixed frequency relationship between two or more tuned circuits as their tuning is simultaneously adjusted. Referred to as tracking, this technique requires narrow tolerances of capacitance versus tuning voltage. Minimizing tracking error requires special care if the tracking circuits must cover the same frequency span beginning a different start and end frequencies, as is necessary when simultaneously tuning oscillator and mixer/RF circuitry in a superheterodyne receiver. Then, tracking error must be minimized by means of series and shunt capacitances in accordance with methods known from variable capacitors. The frequency deviations that must be anticipated are summarized as

$$\frac{df}{f} = -\frac{1}{2}\frac{dC_0}{C_0} - \frac{1}{2}\frac{d(L-L_0)}{L} - \frac{1}{2}\frac{dL_0}{L_0} + \frac{n}{2}\frac{dV_R}{V_R+V_D} \tag{2.92}$$

The spread of parameters dC_0/C_0 and dL_0/L_0 can only be compensated by varying the circuit inductances $d(L-L_0)/L$ or by varying the bias $dV_R/(V_R+V_D)$.

Modulating the Diode Capacitance by the Applied ac Voltage In normal operation, the sum of the tuning voltage and the alternating signal voltage of the resonant circuits are applied to the tuner diode. The bias, and thus the capacitance, of the tuner diode therefore varies at the rhythm of the alternating voltage. Due the nonlinear character of the capacitance versus voltage curve, voltage distortions and capacitance shifts are inevitable, and these must be kept within adequate limits. This is done by maintaining the ac applied to the diode(s) at sufficiently low ac amplitude and by choosing an adequate minimum value for the tuning voltage. In the resonant circuit, a tuner diode is modulated predominantly by a current free from harmonics, according to

$$i = \hat{i}\cos\omega t \tag{2.93}$$

The alternating voltage across the diode is

$$v = (V_R + V_D)\left[\left(1 + \frac{\hat{i}(1-n)}{\omega\, C_{\text{tot}} V_R}\sin\omega t\right)^{\frac{1}{1-n}} - 1\right] \tag{2.94}$$

An evaluation of this equation shows that especially the first harmonic makes its appearance. The capacitance shift caused by the alternating voltage superimposed on the tuning voltage is shown in Figure 2.63. However, the voltage distortion, and thus the capacitance shift, can be largely avoided if two tuner diodes are used, as in Figure 2.61.

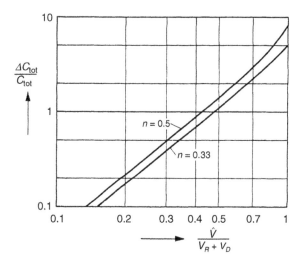

Figure 2.63 Capacitance increase as a function of the ac voltage drop across the tuner diode.

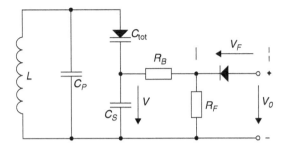

Figure 2.64   Temperature-compensation circuit with diode.

Compensating Temperature Dependence   As has already been mentioned, variations are caused in the capacitance of the diode mainly by the dependence of the diffusion voltage on temperature. In a diode-tuned resonant circuit, the temperatures coefficient of the resonant frequency, therefore, depends on the tuning voltage and thus on the resonant frequency. It is therefore impossible to compensate the temperature dependence of the resonant frequency by means of temperature-dependent capacitors—the method usually adopted for mechanically tuned circuits.

To achieve satisfactory compensation, the tuning voltage should be increased by an amount equal to that by which the diffusion voltage of the diode is reduced with rising temperature, that is, by approximately 2 mV/°C. This can be done easily by connecting a forward-based silicon diode in series with the source of the tuning voltage, as shown in Figure 2.64. Since the forward voltage of the diode varies by $-2\,\text{mV}/°\text{C}$ with ambient temperature, the voltage

$$V = V_0 + V_D - V_F$$

which determines the bias of the tuner diode, and therefore also the diode capacitance, is almost temperature-independent. In practice, it is advisable to feed this diode with additional current via the resistor R_F to keep its differential forward resistance sufficiently low and thus to prevent capacitive noise voltages from leaking into the tuning circuit.

To prevent the load placed by the resistor R_F on the source of the tuning voltage from fluctuating in dependence on the amount of the tuning voltage, a circuit such as shown in Figure 2.65 may be used in which the emitter diode of the transistor effects the compensation, and the only load placed on the source of the tuning voltage is the base current of this transistor.

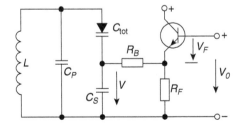

Figure 2.65   Temperature-compensation circuit with transistor.

Dynamic Stability The resonant frequency of a diode-tuned RF circuit follows the equation

$$f = f_0 \sqrt{\left(\frac{V_R + V_D}{V_{R0} + V_D}\right)^n} \tag{2.95}$$

The frequency f_0 is set with the aid of the circuit inductance when the tuning voltage equals V_{R0}. Since the signal amplitude in the circuit depends upon the field strength of the received transmitter and since the resonant frequency of the circuit depends on the amplitude, undesirable feedback between the tuning voltage and the received alternating voltage may occur at the tuner diode. A disturbance that has a similar effect is the microphonic effect, which can sometimes be observed with mechanically variable capacitors. In an investigation of the feedback in diode-tuned resonant circuits, the following assumptions were made.

- The impedance of the antenna (or of the signal generator) is negligible.
- The full alternating voltage is applied to the diode (whereas, in practice, the series capacitor brings about a potential division).
- The capacitance versus voltage characteristic is an exponential function.
- The Q of this circuit is independent on the amplitude of the alternating voltage.

As a result of a high alternating voltage at the resonant circuit, three undesirable effects—harmonic generation, frequency shift, and cross-modulation—are encountered.

Harmonic Generation The nonlinearity of the capacitance versus voltage characteristic produces harmonics. However, the selectivity of the receiver reduces the effects of these harmonics to noise level.

Frequency Shift When a sinusoidal voltage v is applied to the tuner diode of the RF circuit, the variation of the diode capacitance does not follow the sine law, but the equation

$$C_\omega/C = \left(1 + \frac{v}{V_R + V_D} \sin \omega t\right)^{-n} \tag{2.96}$$

This results in a change of the resonant frequency of the circuit. The inductive slope of the resonant curve steepens and may even turn back. This could lead to bistable behavior of the resonant circuit. The change of frequency takes place as if the tuning voltage, and thus the resonant frequency, has decreased. The order of magnitude of this change of resonant frequency was determined with the aid of a computer from Fourier analysis of

$$\frac{\Delta f}{f} = 1 + \frac{1}{2\pi} \int_0^{2\pi} \left(1 + \frac{v}{V_R + V_D} \sin \omega t\right)^{\frac{n}{2}} d\omega t \tag{2.97}$$

Figure 2.66 shows the admissible alternating voltage at the diode, computed for a 2% frequency shift, as a function of the tuning voltage. The measured values differ by not more than 10% from the computed values. Since the Q of the circuits is about 50 for the VHF as well as the UHF range, a 2% frequency shift is permissible.

Cross-Modulation Cross-modulation is a disturbance caused by nonlinearities of the characteristics of the components concerned. The modulation of an undesired

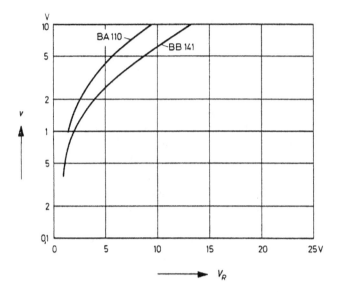

Figure 2.66   Admissible alternating voltage for a 2% frequency shift as a function of tuning voltage.

amplitude-modulated signal is here transferred to the carrier of the amplitude-modulated intelligence signal. Cross-modulation is virtually independent of the amplitude of the intelligence signal.

The cross-modulation factor for tuner diodes is

$$k = \frac{v^2}{(V_R + V_D)^2} \, m_v \frac{n(n+1)}{2} \tag{2.98}$$

Resolving this equation in terms of v, we obtain

$$v = \frac{V_R + V_D}{n} \sqrt{\frac{2kn}{m_v(n+1)}} \tag{2.99}$$

The usual definition of the cross-modulation percentage for measurement purposes is as follows. A 100% modulated noise signal ($m_v = 1$) of amplitude v is received simultaneously with the intelligence signal in the diode-tuned RF circuit. If a 1% modulation of the signal carrier by the noise signal takes place, this amounts to a cross-modulation factor $k = 0.01$. We thus obtain for tuner diodes

$$V = 0.14 \frac{V_R + V_D}{n} \sqrt{\frac{n}{n+1}} \tag{2.100}$$

Figure 2.67 gives the graphs for this equation. In practical operating, FM and VHF/UHF tuners equipped with the tuner diodes BB121 or BB141 exhibit merely the cross-modulation caused by the BJTs or FETs employed. For bipolar transistors, $v \approx 15$–$100\,\text{mV}$; for FETs, 100–$300\,\text{mV}$.

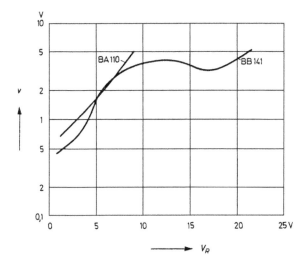

Figure 2.67 Graphical representation of equation (2.100).

Static Stability To ensure trouble-free operation, the static parameters of the tuner diodes also, especially the leakage current, must be taken into account. When differentiating the expression for the tuning voltage

$$V_R = V_B - I_R R_B \qquad (2.101)$$

we obtain for a known leakage current IR the tuning voltage variation

$$dV_R = dV_B - R_B I_R \left(\frac{dR_B}{R_B} + \frac{dI_R}{I_R} + dV_B \right) \qquad (2.102)$$

This leads to a change in the resonant frequency

$$\frac{df}{f} = \frac{n}{2\,(V_R + V_D)} \left[dV_D - R_B I_R \left(\frac{dR_B}{R_B} + \frac{dI_R}{I_R} \right) + dV_B \right] \qquad (2.103)$$

The resonant frequency is more stable the smaller the exponent n and the higher the tuning voltage V_R. The influence of the temperature dependence of the diffusion voltage V_D and its compensation has already been discussed. If the series bias resistance R_B is not too high, the effect of its variation (dR_B/R_B) and the effect of the temperature dependence of the leakage current—doubling for every 10°C temperature rise—may be neglected.

Generating the Tuning Voltage The supply voltage V_B in equation (2.103) exerts an influence in many respects. The temperature coefficient of the voltage source—usually, for economic reasons, only a Zener diode is provided for stabilization—affects conditions not only with ambient temperature variations but also when the Zener diode heats up after the set is switched on, or when its operating current varies. Because external temperature compensation of the operating voltage of the Zener diode cannot cover all detrimental effects and, moreover, increases the series bias resistance R_B, as well as any variations the latter may undergo, the use of a temperature-compensated Zener diode, such as the 1N4065A (nominal Zener voltage, 33 V; temperature coefficient, 0.002°C/W), is recommended.

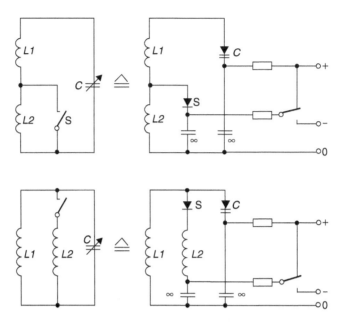

Figure 2.68 Comparison of mechanically and electronically tuned and switched resonant circuits.

Diode Switches The diode switches described here differ somewhat from the switching diodes used in computer and pulse technology. While in the case of switching diodes the signal itself triggers the switching operation—current does or does not pass through the switching diode in dependence on the signal level—diode switches allow an alternating current to be switched on or off by means of a direct voltage or a direct current. Figure 2.68 shows an example of the use of diode switches.

Diode Switch Technology Special manufacturing techniques are needed to attain the low junction capacitance, low differential forward resistance r_f and the low-inductance structure required for the special applications envisaged. Two conflicting requirements have to be met: on the one hand, a high-resistivity basic material should be used and the diode area should be kept as small as possible to minimize the junction capacitance for a given reverse voltage; on the other hand, the bulk resistance should be kept low. The size of the junction area, therefore, can represent only a compromise between these two requirements. To obtain a low bulk resistance R_B despite the use of high-resistivity silicon around the junction, the diodes are manufactured by the epitaxial process, characterized by a thin high-resistivity layer on a low-resistivity substrate. If the forward current I_F is sufficiently high, then the specific resistance of the epitaxial layer does not affect the bulk resistance, because under these conditions, the resistivity of this layer is considerably reduced by being flooded with carriers.

The junction impedance Z_j is somewhat more critical (Figure 2.69). It is composed of the junction capacitance C_S, which does not depend on current, the current-dependent diffusion capacitance

$$C_D = \frac{\iota_p I_F}{V_t} \tag{2.104}$$

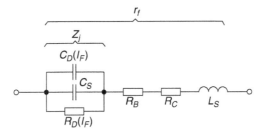

Figure 2.69 Equivalent circuit of diode switches in the frequency range from 1 MHz to 1 GHz.

and the diffusion resistance

$$R_D = \frac{V_t}{I_F} \qquad (2.105)$$

where in ι_p is the lifetime of the minority charge carriers and

$$V_t = \frac{kT}{q} = 26 \text{ mV} \qquad (2.106)$$

is the voltage equivalent of thermal energy. The resistive component of the impedance Z_j is calculated as follows:

$$\text{Re}(Z_t) = \frac{R_D}{1 + \omega^2 R_D^2 (C_S + C_D)^2} \qquad (2.107)$$

As can be seen from the following approximations, this resistance becomes rather dependent on current at higher frequencies:

$$\text{for } I_F \to 0 \qquad \text{Re}(Z_j) \approx \frac{I_F}{\omega^2 C_S^2 V_t} \qquad (2.108)$$

$$\text{for } I_F \to \infty \qquad \text{Re}(Z_j) \approx \frac{I_F}{\omega t_p I_F} \qquad (2.109)$$

In Figure 2.70, the differential forward resistance is plotted as a function of the forward current. The only parameter by which the differential forward resistance can be reduced for a given current and a given frequency according to equation (2.109) is the carrier lifetime. Steps will have to be taken, therefore, to make the carrier lifetime as long as possible.

Because diode switches are designed for applications up to the gigahertz range, a geometry resulting in a low inherent capacitance and inductance had to be used. Furthermore, the device had to be assembled in such a way that virtually no additional contact resistance R_C was introduced. For this reason, a double-plug construction (hard-glass pressure contacts) was chosen to avoid the S- or U-bends customary with other encapsulations. The system is pressure clamped between two stud-shaped enlargements of the connecting leads, the necessary high pressure being produced as the glass shrinks during cooling. This construction results in an encapsulation of short length (only 4 mm) and low series inductance, the advantages of which can be fully exploited because soldered connections to the leads can be made directly at the glass body.

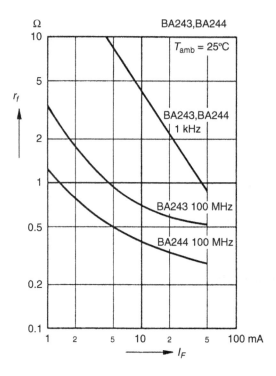

Figure 2.70   Differential forward resistance as a function of forward current.

Diode Switch Data   The silicon planar diode switches BA243 and BA244 are designed for electronic band selection in tuners in the frequency range from 10 to 1000 MHz. The differential forward resistance remains constant and is very low over a wide frequency and current range. The diode capacitance is likewise small and independent of voltage through a wide range. The devices are characterized by low-inductance geometry. The BA243 is intended for the VHF range and the BA244 for the UHF range. See Table 2.11.

Resonant Circuits Incorporating Diode Switches   Figure 2.71 illustrates the function of the diode switch with reference to a parallel-resonant circuit whose resonant frequency can be switched by short-circuiting part of the circuit inductance. Such a switching procedure may be necessary for a VHF tuner, for example, to switch from band I to band III, because the frequencies of these bands are too far apart.

Capacitor C_K represents the capacitance of the resonant circuit. The symbol ∞ at capacitor C is meant to indicate that the value of this capacitor should be chose to be much higher than the circuit capacitance. C completes the ac circuit when the inductor L_2 is shorted and enables the switching or reverse voltage to be applied at a point of virtually zero RF potential. The resistor R_1 limits the diode current in the forward direction.

At a reverse voltage of $V_R = 15$ V, for example, the diode switch represents a 1-pF capacitor, as shown in Figure 2.71b. This represents parasitic capacitance C' in the circuit, which reduces the effective capacitance variation of the resonant circuit in the case of capacitive tuning. The capacitance C' is small enough to be negligible in most cases relative to the wiring and winding capacitances. Figure 2.72 shows the relative capacitance of the blocked-diode switches, as a function of the reverse voltage.

Table 2.11 BA243/244 Specifications

Maximum ratings			
Reverse voltage	V_R	20	V
Forward current at $T_{amb} = 25°C$	I_F	100	mA
Junction temperature	T_j	150	°C
Storage temperature range	T_s	-55 to $+150$	°C
Characteristics at $T_{amb} = 25°C$			
Foward voltage at $I_F = 100$ mA	V_F	<1	V
Leakage current at $V_R = 15$ V	I_R	<100	nA
Dynamic forward resistance at $f = 50 - 1000$ MHz, $I_F = 10$ mA			
BA243	r_f	0.7(<1)	Ω
BA244	r_f	0.4(<0.5)	Ω
Relative variation of dynamic forward resistance with the variation of foward current in the range of $I_F = 2 - 40$ mA	$\dfrac{\Delta r_F 100}{r_f \Delta I_f}$	5	%/mA
Capacitance at $V_R = 15$ V, $f = 1$ MHz	C_{tot}	1.3(<2)	pF
Relative variation of capacitance with the variation of reverse voltage in the range of $V_R = 7 - 20$ V, $f = 100$ MHz	$\dfrac{\Delta C_{tot} 100}{C_{tot} \Delta V_R}$	1	%/V
Series inductance across case	L_S	2.5	nH

At a forward current of, for example, 10 mA, the diode has a differential resistance of approximately 0.5 Ω (see Figure 2.70). The effect of this resistance r_F on the resonant circuit is illustrated in Figure 2.71c. Conversion of a series loss resistance r_F into a parallel loss resistance r'_F is based on the following considerations: the equivalent parallel resistance r'_F affects the tuned circuit in the same way as the series resistance r_F, that is, the Q of the circuit should be the same in both cases. If one assumes that $\omega L \gg r'_F$, the equivalent of parallel resistance can be expressed as follows:

$$Q = \frac{\omega L_1}{r_F} = \frac{r'_F}{\omega L_1} \tag{2.110}$$

$$r'_F = \frac{(\omega L_1)^2}{r_F} \tag{2.111}$$

An evaluation of this equation shows that a differential resistance $r_F \approx 0.5$ Ω can be tolerated in most applications as regards amplification and selectivity.

Instead of altering the resonant frequency of a tuned circuit by short-circuiting part of its inductance, the same result can be obtained by connecting a second inductance in parallel. The use of the diode switch for this purposes is illustrated in Figure 2.73. This differs from the one shown in Figure 2.71 in that the circuit inductance consists of either L_1 or of L_1 and L_2 in parallel. By applying a permanent negative bias to the diode via a high-value resistor R_2, it is possible to dispense with the use of one switch contact. Figure 2.73 shows the equivalent circuit for the diode under reverse-bias conditions (corresponding to a low resonant frequency) and Figure 2.73c shows the equivalent circuit under forward-bias conditions (corresponding to a higher resonant frequency).

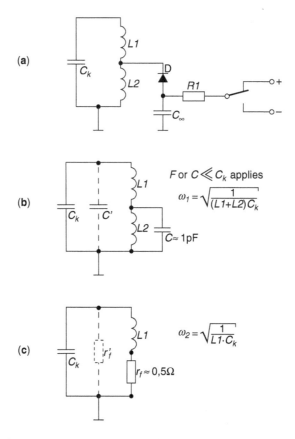

Figure 2.71 (a) Diode switch as bandswitch, shorting a partial inductance. (b) Equivalent circuit with blocked diode switch. (c) Equivalent circuit with conducting diode switch.

In radio and television receivers, it is usually possible to derive the necessary reverse bias for the diode by rectification of the local oscillator signal, since the power required for blocking the diodes is extremely small. Note that under reverse-bias conditions the inductance L_2 and the capacitance C form a series-tuned acceptor circuit. However, this capacitance being extremely small, any undesirable effects due to its presence can normally be avoided. The same diode switching arrangements described for single-tuned circuits can also be employed in multisection bandpass and broadband filters.

In the following paragraph, a practical example for the use of diode switches will be given, and it will be seen that it is possible in principle to replace all the mechanical RF switch contacts normally employed in receivers by diode switches. In cases where application of the diode dc potentials to a "cold" point in the circuit is not feasible, it may be necessary to apply them via LC or RC isolating networks.

Use of the Diode Switch in a Television Receiver Figure 2.74 shows the schematic of an electronically tuned and switched band I (48–62 MHz)/band III (175–224 MHz) TV tuner that requires switching of the input filter, bandpass filter, and oscillator circuit. These switching functions are performed by BA243 diode switches. As is customary in most European countries, in this example a broadband input filter between the antenna and the first RF transistor is used rather than a selective front-end circuit. In view of the need to

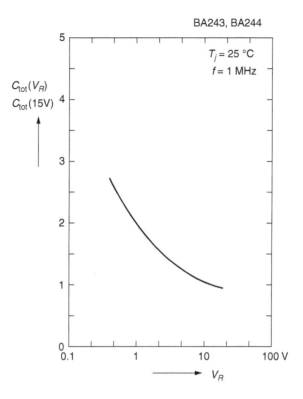

BA243, BA244

$T_j = 25\ °C$

$f = 1\ MHz$

$\dfrac{C_{\mathrm{tot}}(V_R)}{C_{\mathrm{tot}}(15V)}$

Figure 2.72 Capacitance as a function of the reverse voltage.

correct noise and power matching over a relatively large frequency range, a switchable filter is employed in most tuners, switching being effected in this case by means of diodes D_1 and D_2.

Diodes D_3–D_6 are used for band-filter switching, for switching the coupling to the mixer transistor, and for switching the oscillator frequency. To minimize the bias-current requirements (each diode requires a bias current of approximately 10 mA), all diodes that are to be activated at the same time are connected in series. This arrangement requires only a few additional RF chokes, which are easily provided because of the high frequencies involved). The direct current required for the diode switches could be obtained by simple rectification, for example, from the heater circuit of the television receiver CRT, and the diode reverse bias could be obtained by rectification of the local oscillator signal, in which case the minute additional power requirement would have to be allowed for in the oscillator design.

A technically sophisticated design for a television tuner takes the form of a combined VHF/UHF unit incorporating electronically switched circuits in which several elements would be common. Such an all-band tuner can be assembled from only three transistors and a combined VHF/UHF band filter in λ/2 or λ/4 technology for UHF reception. The VHF circuits are short circuited when the UHF position is selected.

Multistandard TV receivers require a large number of switch contacts in the tuner, the IF amplifier, the video amplifier, and the time base. The use of diode switches is particularly advantageous in this situation because it renders the physical layout of the equipment virtually independent of the location of the controls, and obviates the need for long RF cable runs, which may be prone to interference.

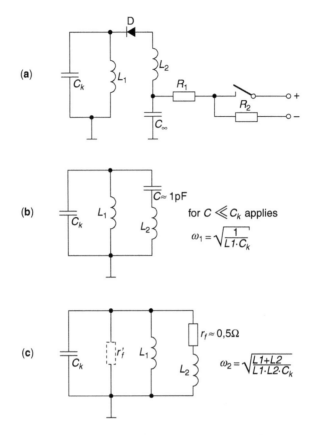

Figure 2.73 (a) Diode switch as a bandswitch, with an inductance in parallel. (b) Equivalent circuit with blocked-diode switch. (c) Equivalent circuit with conducting diode switch.

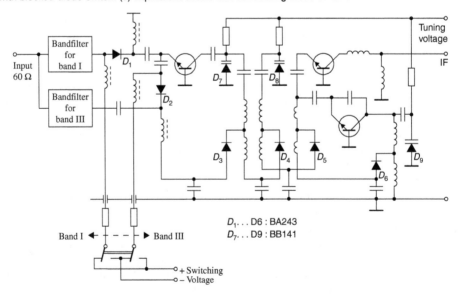

Figure 2.74 Schematic of a VHF television tuner with electronic band selection.

2.2 BIPOLAR TRANSISTORS

2.2.1 Transistor Structure Types

Bipolar transistors come in different flavors.

- Silicon-based bipolar junction transistors (BJTs) are the classical type; they basically consist of three semiconductor layers, either of *npn* or of *pnp*-type (see Figure 2.75). In case of *npn*, the current is based on electrons, while the *pnp*-type current relies on holes. Due to the higher mobility of electrons, *npn*-type HBTs are commonly faster than their *pnp* counterparts from the same technology. The different semiconductor *n* and *p* regions can be formed by means of ion implantation.

- Advanced silicon-based transistors rely on a certain germanium content in the base region. These are called SiGe heterojunction bipolar transistors (HBTs), because there is not only an ordinary *pn* junction between base and emitter (or collector) but also the semiconductor crystal changes from SiGe to Si. The technology required to produce these transistors is more involved compared to the BJT case, as it requires to grow the SiGe and emitter Si crystals during the processing using selective epitaxy. An advantage of SiGe technology is that it is in principle compatible with CMOS technology. Processes providing both types of transistors on a single wafer are called BiCMOS. These processes allow for an integration of logic circuits and for baseband signal processing together with a wireless front-end providing reasonable power levels.

- HBTs are also available on GaAs substrates. These transistors use InGaP emitters, while base and collector are GaAs. InGaP emitters have now replaced AlGaAs emitters in commercial processes, due to the better electrical properties and due to the fact that the aluminum-free semiconductors showed higher lifetimes. The first steps in GaAs-based transistor technologies is commonly an epitaxial growth of the required transistor layers on the whole wafer. The transistors commonly are realized by etching active layers away where not required—these HBTs are therefore called mesa transistors, as they resemble the table mountains under the microscope.

- HBT on InP are beyond the scope of this book—these devices target applications in the mm-wave region and beyond. The InP-based HBT is similar to the GaAs HBT, but commonly not used in the wireless sector for cost reasons.

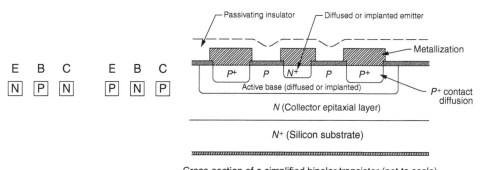

Cross-section of a simplified bipolar transistor (not to scale)

Figure 2.75 Cross-section of a Si BJT.

The main advantage of bipolar transistors is that the current flows through the bulk material, not just through a thin channel along a surface, as in case of FETs. This results in the following properties of BJTs and HBTs.

- Bipolars can carry higher current densities compared to the respective FET of the same size. Smaller transistor size means less capacitance, thus easier matching without reducing output power. On the negative side, it means higher dissipated power densities resulting in self-heating.
- Bipolars show less flicker $(1/f)$ noise than FETs, for the same reason that the current does not flow along an interface or crystal surface, where the density of defects and traps is much higher than in the bulk material.
- Transit times in bipolars are mainly determined by the thickness of epitaxial layers, not by the length of a metal gate like in FETs. In epitaxial growth, it is not a real problem to control thicknesses of a few nanometer, at least it is much easier achieved than similar dimensions of metal strips defined by lithographical techniques. Limiting factors are rather breakdown of the collector–base junction, and base sheet resistance.
- Bipolar transistors require only one positive supply voltage, in contrast, for example, to GaAs HEMTs that additionally need negative gate bias. This reduces complexity of the whole system.

The main applications for bipolar transistors are therefore power amplifiers and low phase-noise oscillators. For power amplifiers (PAs) providing about 1 W at 2 GHz, for example, the technology of choice could be GaAs-based HBT. This power level is so far unreachable for CMOS, calling for the PA to be designed in another technology. The GaAs HBT, on the other hand, still has enough headroom to reach these specification; its breakdown is in excess of 15 V, and its cutoff frequencies are around 30 GHz and beyond. Silicon BJTs and HBTs offer the lowest flicker noise, thus they are the candidates of choice for oscillator applications.

2.2.2 Large-Signal Behavior of Bipolar Transistors[2]

In this section, the large-signal or dc behavior of bipolar transistors is considered. Large-signal models are available in virtually all circuit simulators, and foundries commonly provide highly sophisticated design kits to their customers. This section introduces the physical effects that these models attempt to describe. Second-order effects, such as current-gain and cutoff-frequency variation with collector current, and self-heating can be important in many circuits and are treated in detail.

2.2.2.1 Electrical Characteristics and Specifications

Electrical characteristics may be conveniently classified into two main types: dc and ac.

DC Characteristics The importance of dc characteristics of high-frequency transistors lies primarily in biasing and reliability considerations. However, certain dc characteristics are also directly related to high-frequency performance. For example, high-frequency noise figure is affected by dc current gain. The dc characteristics discussed here are those usually found on high-frequency transistor datasheets.

---

[2]Portions of this section, Sections 2.2.3–2.2.7, and Sections 2.4.2–2.4.5, are based on Paul R. Gray and Robert G. Meyer, *Analysis and Design of Analog Integrated Circuits*, 3rd ed., ©1977, 1984, 1993 by John Wiley & Sons, Inc. Reprinted by permission of John Wiley & Sons, Inc.

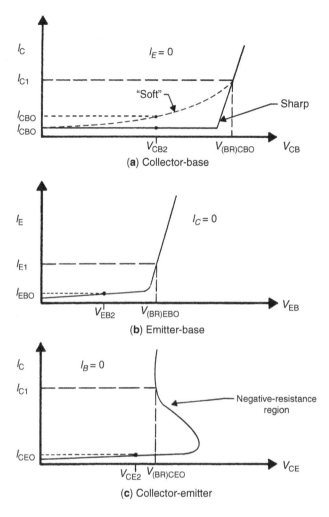

Figure 2.76 Transistor reverse V–I characteristics.

V_{CBO}, I_{CBO} These two parameters serve to characterize the reverse-biased collector–base *pn* junction and are defined as follows (with the aid of Figure 2.76a). The collector-base breakdown voltage, V_{CBO}, identifies the voltage at which collector current tends to increases without limit, usually due to the high electric field developed across the junction. This voltage sets a limit on the maximum transistor operating voltage and, as mentioned before under maximum ratings, usually is the basis for the collector-base maximum voltage rating. V_{CBO} should be specified at a value of $I_C = I_{C1}$ in the figure, which is within the avalanche (or high slope) region of the device's reverse characteristic. Typical values of I_{C1} are in the 1–10-μA region for high-frequency transistors.

To further define the quality of the reverse V–I characteristic, a specification is usually placed on the collector cutoff current, I_{CBO}, which is measured at some value of collector-base voltage less than V_{CBO}. For a good-quality silicon junction ("sharp" instead of soft; see Figure 2.76a), I_{CBO} is in the nanoampere range.

V_{EBO}, I_{EBO} These two parameters characterize the reverse-biased emitter–base *pn* junction in an analogous manner to the collector–base junction parameters V_{CBO} and I_{CBO}, given above, and are shown in Figure 2.76b. No further discussion of these parameters will be given here.

V_{CEO}, I_{CEO} The collector–emitter breakdown voltage and cutoff current are somewhat more complex in nature than the collector–base and emitter–base parameters. In the latter two, only a single *pn* diode is involved. In the collector–emitter case, two diodes are involved. Moreover, each is influenced by the other through transistor action, since the reverse current of the collector–base diode flows through the emitter–base junction as forward current. Thus, the collector–base reverse current is amplified by the dc current gain of the transistor, resulting in

1. I_{CEO} being greater than I_{CBO} (for a given voltage);
2. typically, the familiar negative-resistance region in the I–V characteristic as shown in Figure 2.76c.

Consequently, V_{CEO} is typically specified at collector currents one to three orders of magnitude higher than in the case of V_{CBO} and V_{EBO} to establish the minimum value of this characteristic.

Practical Relevance of the Three Breakdown Voltages   Now that we have defined three breakdown voltages for one device, the question arises which of them is actually setting a limit in a common-emitter circuit?

The first guess would be that only the collector–emitter breakdown voltage V_{CEO} is relevant. If we deal with InGaP/GaAs HBTs in a mobile application, we might not run into trouble, since V_{CEO} is easily four times higher than the collector–emitter dc voltage V_{CE}. However, in SiGe HBTs, V_{CEO} might be around or below 5 V, and restricting the voltage swing to well below this maximum would not yield acceptable output power and efficiency. Given the fact that $V_{CBO} \gg V_{CEO}$, it also seems that we will not exploit the full capabilities of the device if we restrict the RF swing to the lower of the two.

To identify V_{CEO} with the collector–emitter breakdown voltage in a practical circuit application, however, would be a misinterpretation. By definition, it requires that the base current is zero, which means that the base is terminated by an open circuit. V_{CBO}, on the other hand, is detemined with zero base–emitter voltage, which means that the base terminal is terminated by a short circuit. In a practical case, the RF termination will be somewhere between these extreme values, and choosing a low-impedance RF base termination allows for operation well beyond V_{CEO}. Figure 2.77 shows measurement of the collector–emitter breakdown behavior of an Infineon BFP450 transistor. In this measurement, the base is connected to ground through resistances of different values [2].

Providing a high impedance at the base also potentially has a negative effect on base–emitter breakdown, for example, in class-AB power amplifiers [3]. When increasing input power, one would expect the dc current to rise, as it is equal to the average of the collector current. In deep class-AB, the current can be approximated by positive sinusoidal half-waves, thus the dc current depends on the RF amplitude. Fixing the base current by providing a high-impedance termination prevents this effect from happening—instead, the RF signal is reflected and reverse voltage peaks rise with input power, until, finally, the a base–emitter breakdown occurs.

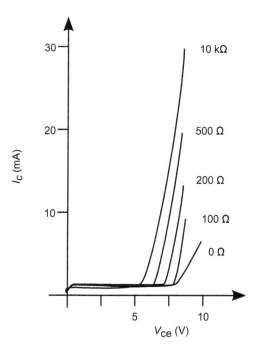

Figure 2.77   Dependence of collector–emitter breakdown on the internal resistance of the base voltage source (data from Ref. [2]).

Thus, regarding breakdown, it can be concluded that providing a low impedance to the base in common emitter is beneficial. In order to account for this fact, datasheets often provide two collector–emitter breakdown parameters, V_{CEO} (base open) and $V_{CES} \approx V_{CBO}$ (base shortened to ground).

h_{FE} This parameter is simply the dc common-emitter current gain, that is, the ratio of collector current to base current at some specified collector voltage and current.

AC Characteristics   Of the numerous ac characteristics defined for transistors, only relatively few are commonly used in characterizing high-frequency transistors. Some of the more pertinent parameters are briefly covered here (see also Figure 2.78).

Transition Frequency   One of the better known, but perhaps least understood, figures of merit for high-frequency transistors is the so-called transition frequency, f_T. Part of the misunderstanding that appears to exist is due to the use of the misleading (but common, for historical reasons) term short-circuit gain-bandwidth product for this parameter. By definition, f_T is that characteristic frequency described by the equation

$$f_T = h_{fe} \times f_{\text{meas}} \qquad (2.112)$$

where h_{fe} is the magnitude of small-signal common-emitter short-circuit current gain, and f_{meas} is the frequency of measurement, chosen such that

$$2 \leq h_{fe} \leq \frac{h_{fe0}}{2} \qquad (2.113)$$

where h_{fe0} denotes the low-frequency value of h_{fe}.

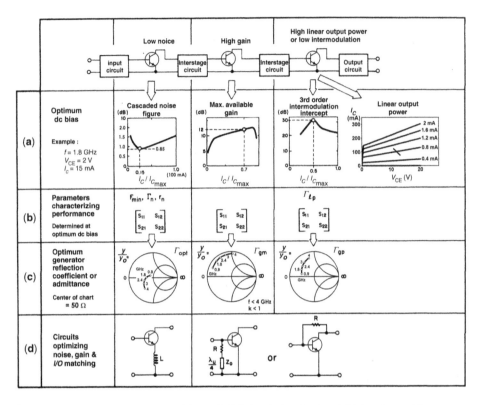

Figure 2.78 Key parameters in applying a BJT in low-noise front-end, high-gain, and linear-power stages. At (a), the BJT characteristics that lead to optimum dc bias for a low-noise front end; at (b), parameters characterizing device performance; at (c), Γ_{opt} for the first stage, and S_{11} and S_{22} for the output stage as simulated by CAD software. The example is based on the Infineon BFP420 (low-noise stage), BFP450 (high-gain stage), and BFG235 (output stage). Circuits to optimize stage noise, gain, and I/O matching are shown at (d). (Presentation based on Figure 14 in Charles A. Liechti, "Microwave field-effect transistors—1976," *IEEE Trans. Microwave Theory Tech.*, vol. MTT-24, pp. 279–300, June 1976.)

To varying degrees of approximation, depending on the transistor type, f_T is the frequency at which h_{fe} approximates unity. It is not, in general, the frequency at which h_{fe} is precisely equal to unity. To clarify these points further, consider the plot of h_{fe} against frequency sketched in Figure 2.79.

At low frequencies, $f \ll f_B$, h_{fe} is constant and equal to h_{fe0}.

At f_B, h_{fe} has decreased to $0.707h_{fe0}$; that is, f_B is the 3-dB cutoff frequency for common-emitter short-circuit current gain, h_{fe}.

For frequencies such that

$$2f_B < f < f_T \qquad (2.114)$$

h_{fe} varies inversely proportional to frequency. That is, the f_T defining relationship

$$h_{fe} \times f_{meas} = \text{constant} = f_T \qquad (2.115)$$

holds.

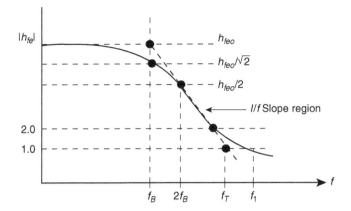

Figure 2.79 $|h_{fe}|$ frequency characteristics.

At frequencies approaching f_T, other parameters, especially package parasitics, can cause $|h_{fe}|$ to depart significantly from this $1/f$ variation. Therefore, the frequency, f_1, at which $|h_{fe}|$ actually equals unity can be somewhat different from f_T.

Applying this frequency-gain characteristics to the common-emitter wideband, low-pass amplifiers give rise to the terminology of f_T being a "gain-bandwidth product." However, this is an optimistic approximation at best, since the product of low-frequency circuit gain and the 3-dB cutoff frequency is reduced from f_T by an amount that depends on circuit impedances.

The real significance of f_T lies in the fact that it is a measure of certain internal transistor parameters that do, in fact, affect high-frequency performance, for example, gain (though not in the convenient quantitative manner implied by the term "gain-band product"). In particular, good high-frequency noise performance requires that f_T be high. Thus, f_T is included on transistor datasheets primarily as a figure of merit, not as a parameter to be used directly in design.

Collector–Base Time Constant, $r_b' C_c$ This is an internal device parameter that relates only indirectly to high-frequency performance. It is primarily a measure of internal feedback within the transistor. It also relates to transistor high-frequency impedance. As the name says (in symbols), it is a measure of transistor base resistance and collector capacitance in combination; however, except for certain low-frequency transistors, it cannot be considered the simple two-element lumped RC time constant implied by the terminology. (In high-frequency transistors, base resistance and collector impedance must be considered to be distributed when considered in detail). As a figure of merit, it is included on transistor datasheets to indicate how well base resistance and collector impedance have been minimized. It also allows the estimation of certain gain properties of the transistor (see the fmax parameter, which we will describe shortly).

Collector–Base Capacitance, C_{cb}   This parameter is simply the total collector–base *pn* junction capacitance measured at a low frequency (typically, 1 MHz) where it can be considered a single lumped element. For high-frequency transistors it is, of course, desirable that C_{cb} be small from bandwidth and stability considerations as well as from gain considerations alone.

Maximum Frequency of Oscillation, f$_{max}$ This is another figure-of-merit parameter, as opposed to measurable parameters directly usable in the application of transistors. Its importance stems from the following approximate relationships (which will not be derived here):

$$f_{max} \approx \left(\frac{f_t}{8\pi r_b' C_c} \right)^{1/2} \tag{2.116}$$

$$G_{max} \approx \left(\frac{f_{max}}{f_{oper}} \right)^2 \tag{2.117}$$

These expressions illustrate in a quantitative way the importance and the interrelationship between high f_T and low $r_B' C_C$ insofar as high-frequency gain is concerned. However, since the approximations and their derivation involves several assumptions that are not always valid, they must be interpreted with caution. For example, the expression for G_{max} is obviously not applicable at low frequencies, since as $f \to 0$, $G_{max} \to \infty$, according to this expression. As a rule of thumb, the G_{max} expression is a reasonable approximation for frequencies such that

$$5 > \frac{f_{max}}{f_{oper}} > 1 \tag{2.118}$$

For accurate analysis of transistor gain and stability, a complete set of two-port parameters must be employed in exact expressions, such as those from which the approximations shown above were derived.

Datasheet Transistor Specifications   The following is a reprint of a Infineon microwave transistor data sheet. Key parameters describing the transistor are

1. dc parameters, such as maximum current and voltage and dissipation;
2. noise figure as a function of generator impedance, bias point, and frequency.

Detailled numerical data that can be used to actually design a circuit, such as

1. frequency-dependent behavior, typically expressed in S parameters, and also as a function of bias point and frequency;
2. large-signal performance to be calculated based upon a nonlinear model, such as the Gummel-Poon;

are today usually provided by for download by the manufacturers, in compatible format for major circuit simulators.

The following figures, reproduced with permission from the datasheet for the BFR380F transistor by Infineon Technologies are a typical representation of the dc–IV curves, S parameters, and noise parameters of a normal, packaged microwave transistor. Besides its dc parameters and capacitance, there are other critical parameters. A BJT's noise figure is a strong function of its dc current gain, and one can determine a 3-dB cutoff frequency referred to as f_T, which shows the gain roll-off as a function of collector current and frequency.

BFR380F

NPN Silicon RF Transistor

- High linearity low noise driver amplifier
- Output compression point 19.5 dBm @ 1.8 GHz
- Ideal for oscillators up to 3.5 GHz
- Low noise figure 1.1 dB at 1.8 GHz
- Collector design supports 5V supply voltage
- Pb-free (RoHS compliant) package
- Qualified according AEC Q101

  RoHS

ESD (**E**lectro**s**tatic **d**ischarge) sensitive device, observe handling precaution!

Type	Marking	Pin Configuration			Package
BFR380F	FCs	1 = B	2 = E	3 = C	TSFP-3

Maximum Ratings

Parameter	Symbol	Value	Unit
Collector-emitter voltage	V_{CEO}	6	V
Collector-emitter voltage	V_{CES}	15	
Collector-base voltage	V_{CBO}	15	
Emitter-base voltage	V_{EBO}	2	
Collector current	I_C	80	mA
Base current	I_B	14	
Total power dissipation[1] $T_S \leq 95°C$	P_{tot}	380	mW
Junction temperature	T_J	150	°C
Ambient temperature	T_A	-65 ... 150	
Storage temperature	T_{Stg}	-65 ... 150	

Thermal Resistance

Parameter	Symbol	Value	Unit
Junction - soldering point[2]	R_{thJS}	≤ 145	K/W

[1] T_S is measured on the collector lead at the soldering point to the pcb

[2] For calculation of R_{thJA} please refer to Application Note AN077 Thermal Resistance

2010-09-13

BFR380F

Electrical Characteristics at T_A = 25°C, unless otherwise specified

Parameter	Symbol	min.	typ.	max.	Unit
DC Characteristics					
Collector-emitter breakdown voltage I_C = 1 mA, I_B = 0	$V_{(BR)CEO}$	6	9	-	V
Collector-emitter cutoff current	I_{CES}				nA
V_{CE} = 5 V, V_{BE} = 0		-	1	30	
V_{CE} = 15 V, V_{BE} = 0		-	-	1000	
Collector-base cutoff current V_{CB} = 5 V, I_E = 0	I_{CBO}	-	-	30	
Emitter-base cutoff current V_{EB} = 1 V, I_C = 0	I_{EBO}	-	1	500	
DC current gain I_C = 40 mA, V_{CE} = 3 V, pulse measured	h_{FE}	90	120	160	-

BFR380F

Electrical Characteristics at T_A = 25°C, unless otherwise specified

Parameter	Symbol	Values			Unit
		min.	typ.	max.	
AC Characteristics (verified by random sampling)					
Transition frequency I_C = 40 mA, V_{CE} = 3 V, f = 1 GHz	f_T	11	14	-	GHz
Collector-base capacitance V_{CB} = 5 V, f = 1 MHz, V_{BE} = 0, emitter grounded	C_{cb}	-	0.5	0.7	pF
Collector emitter capacitance V_{CE} = 5 V, f = 1 MHz, V_{BE} = 0, base grounded	C_{ce}	-	0.2	-	
Emitter-base capacitance V_{EB} = 0.5 V, f = 1 MHz, V_{CB} = 0, collector grounded	C_{eb}	-	1	-	
Minimum noise figure I_C = 8 mA, V_{CE} = 3 V, Z_S = Z_{Sopt}, f = 1.8 GHz I_C = 8 mA, V_{CE} = 3 V, Z_S = Z_{Sopt}, f = 3 GHz	NF_{min}	 - -	 1.1 1.6	 - -	dB
Power gain, maximum available[1] I_C = 40 mA, V_{CE} = 3 V, Z_S = Z_{Sopt}, Z_L = Z_{Lopt}, f = 1.8 GHz I_C = 40 mA, V_{CE} = 3 V, Z_S = Z_{Sopt}, Z_L = Z_{Lopt}, f = 3 GHz	G_{ma}	 - -	 13.5 9.5	 - -	dB
Transducer gain I_C = 40 mA, V_{CE} = 3 V, Z_S = Z_L = 50 Ω, f = 1.8 GHz f = 3 GHz	$|S_{21e}|^2$	 - -	 11 7	 - -	dB
Third order intercept point at output[2] V_{CE} = 3 V, I_C = 40 mA, Z_S=Z_L=50 Ω, f = 1.8 GHz	IP_3	-	29	-	dBm
1dB compression point at output I_C = 40 mA, V_{CE} = 3V, f = 1.8 GHz Z_S=Z_L=50 Ω Z_S = Z_{Sopt}, Z_L = Z_{Lopt}	P_{-1dB}	 - -	 17 19.5	 - -	

[1]$G_{ma} = |S_{21e} / S_{12e}| (k-(k^2-1)^{1/2})$
[2]IP3 value depends on termination of all intermodulation frequency components.
Termination used for this measurement is 50 Ω from 0.1 MHz to 6 GHz

Source: Courtesy Infineon Technologies.

Total power dissipation $P_{tot} = f(T_S)$

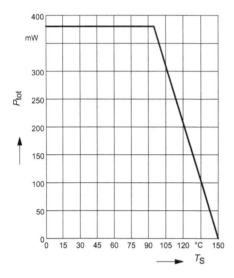

Permissible Pulse Load $R_{thJS} = f(t_p)$

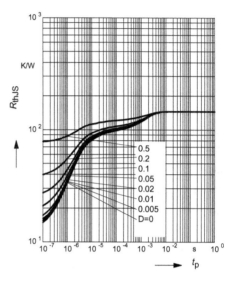

Permissible Pulse Load

$P_{totmax}/P_{totDC} = f(t_p)$

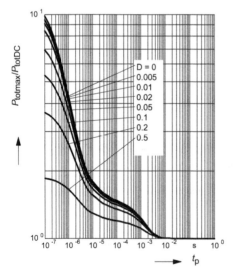

Collector-base capacitance $C_{cb} = f(V_{CB})$

$f = 1MHz$

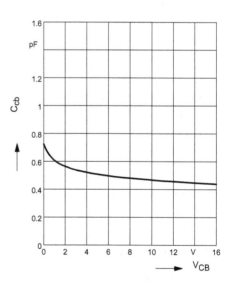

2010-09-13

BFR380F

Third order Intercept Point $IP_3 = f(I_C)$
(Output, $Z_S = Z_L = 50\ \Omega$)
V_{CE} = parameter, f = 1.8 GHz

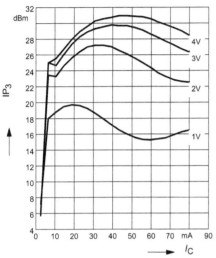

Third order Intercept Point $IP_3 = f(I_C)$
(Output, $Z_S = Z_L = 50\ \Omega$)
V_{CE} = parameter, f = 900 MHz

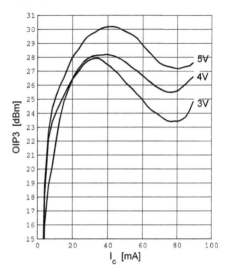

Transition frequency $f_T = f(I_C)$
f = 1 GHz
V_{CE} = parameter

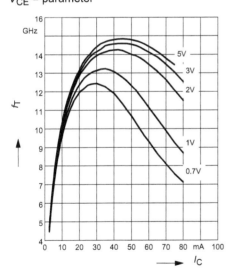

Power gain G_{ma}, $G_{ms} = f(I_C)$
f = 1.8 GHz
V_{CE} = parameter

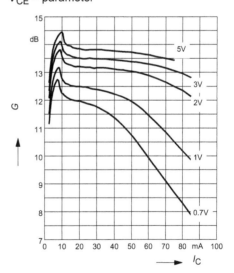

2010-09-13

Source: Courtesy Infineon Technologies.

BFR380F

Power Gain G_{ma}, G_{ms} = $f(f)$
V_{CE} = parameter

Power Gain $|S_{21}|^2$ = $f(f)$
V_{CE} = parameter

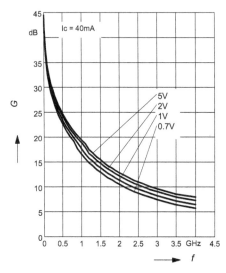

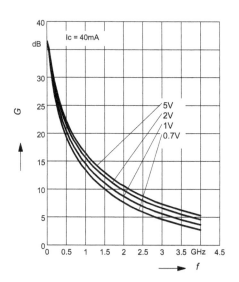

Power Gain G_{ma}, G_{ms} = $f(V_{CE})$: ——
$\quad\quad\quad |S_{21}|^2$ = $f(V_{CE})$: - - - -
f = parameter

Power gain G_{ma}, G_{ms} = $f(I_C)$
V_{CE} = 3 V
f = parameter

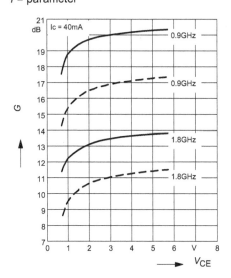

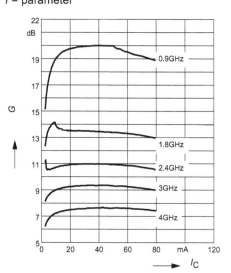

2010-09-13

BFR380F

Minimum noise figure $NF_{min} = f(I_C)$
V_{CE} = 3V, $Z_S = Z_{Sopt}$

Noise figure $F = f(I_C)$
V_{CE} = 3 V, f = 1.8 GHz

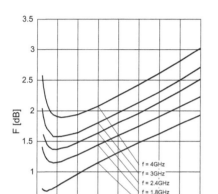

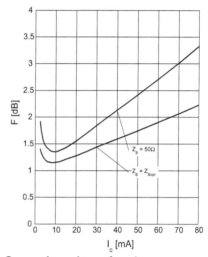

Minimum noise figure $NF_{min} = f(f)$
V_{CE} = 3V, $Z_S = Z_{Sopt}$

Source impedance for min.
noise figure vs. frequency
V_{CE} = 3 V, I_C = 8.0 mA/40.0 mA

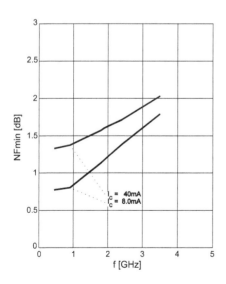

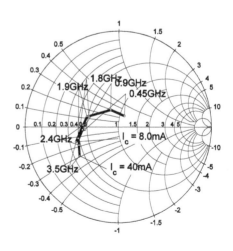

2010-09-13

Figure 2.80 Photograph of the BFP420 die. The bonding pad for the base is at the upper right and for the collector is at the upper left. The two lower pads connect to the emitter. Courtesy Infineon Technologies.

Figure 2.80 shows a photograph of the die for the BFP420.

2.2.3 Large-Signal Transistors in the Forward-Active Region

We saw in the diode case that its description is physics based and it is, therefore, logical to apply the same modeling for the transistors. The first complete modeling for transistors was described in the Gummel–Poon model [4, 5]. By combining two diodes, the parasitic elements, and the base separating resistor, we can show the "intrinsic" nonlinear model for the bipolar transistor.

The advantages of the 40-year-old Gummel–Poon model is its simplicity, and that it is definitely available in every circuit simulator. The way it explains the whole transistor operation through the charge-control theory, connecting the electrical behavior with device physics, namely, the charge stored in the base Q_b, is simply beautiful. This is the reason why this model still has its place in textbooks like ours: it is well suited to explain the basics.

Of course, this model does not cover all the effects observed in todays SiGe HBTs. Modern models are expected to account also for the following effects, which today can be considered to be of first-order inportance.

> *Dynamical Self-Heating.* The temperature of the transistor is not static, given by the ambient temperature. It depends on the power dissipated during operation and needs to be determined during circuit simulation as a function of dc and RF fed to the device, and RF power delivered to the load. Especially regarding GaAs-based power amplifier chips, the power transistor might heat up surrounding components that will impact circuit performance. Self-heating impacts everything from dc operating point to cutoff frequency.

Dependence of Transit-Time on Collector Current. The presence of many electrons leads to interactions with the doping—and changes transit times and capacitances as a function of bias. These effects are only expected under high-current conditions, that is, when the mobile carrier density cannot be neglected compared to the doping density. In modern bipolar transistors, however, this condition is reached almost immediately. GaAs and Si-based devices, however, show a different behavior. Unfortunately, it is not a solution to simulate using an average transit time; the current depencency is the dominant source of distortion products in HBTs [8, 9].

Heterojunctions. These are generally improving transistor performance. But they might not be invisible. A modern transistor model needs to account for these variations, especially if the high-current low-voltage region is affected.

Parasitic Transistors. They can exist if a *pn*-junction is used to isolate a silicon transistor from the rest of the chip.

Comprehensive models for silicon technologies are the MEXTRAM [7] and HICUM [6] models. In the III–V world, the AgilentHBT [10] model is widely used due to the fact that it is versatile and the only model for these devices that is available by default in Agilent's ADS simulation software. However, some companies developed their own derivatives, such as RFMD. Another alternative is the FBH-HBT [10] model that is of reduced complexity, but is published in the Verilog-A language, which requires that the circuit simulator supports this interface.

The drawback of modern transistor models is their complexity. HICUM, for example, comes with roughly 120 parameters, which is not an uncommon number. Models for GaAs HBTs are less complicated, which seems to keep them alive, although MEXTRAM and HICUM claim to be suited for this material system as well.

Explaining modern models, and their physical derivation, in detail is a book in itself, for example, Refs [10, 11]. Model details can also be found in the respective documentations.

For complex models, parameter extraction is involved, and requires a lot of time, experience, and measurement effort. That is why foundries are today expected to provide external designers with design kits—basically plug-ins for the major circuit simulators—instead of just data sheets.

Consequently, we will review the basic bipolar physics in the following, with some additions for the today's transistors.

A typical *npn* planar bipolar transistor structure is shown in Figure 2.81a, where collector, base, and emitter are labeled C, B, and E, respectively. The impurity doping density in the base and the emitter of such a transistor is not constant but varies with distance from the top surface. However, many of the characteristics of such a device can be predicted by analyzing the idealized transistor structure shown in Figure 2.81b. In this structure, the base and emitter doping densities are assumed constant, and this is sometimes called a "uniform-base" transistor. Wherever possible in the following analyses, the equations for the uniform-base analysis are expressed in a form that applies also to nonuniform-base transistors.

A cross section AA' is taken through the device of Figure 2.81b and carrier concentrations along this section are plotted in Figure 2.81c. Hole concentrations are denoted by p and electron concentrations by n with subscripts p or n representing p-type or n-type regions. The n-type emitter and collector regions are distinguished by subscripts E and C, respectively. The carrier concentrations shown in Figure 2.81c apply to a device in the *forward-active*

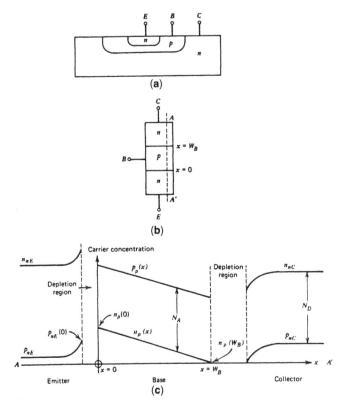

Figure 2.81   (a) Cross section of a typical *npn* planar bipolar transistor structure. (b) Idealized transistor structure. (c) Carrier concentrations along the cross section *AA'* of the transistor in (b). Uniform doping densities are assumed. (Not to scale.)

region. That is, the base–emitter junction is forward biased and the base–collector junction is reverse biased. The minority-carrier concentrations in the base at the edges of the depletion regions can be calculated from a Boltzmann approximation to the Fermi–Dirac distribution function to give [17]

$$n_p(0) = n_{po}\exp\frac{V_{\text{BE}}}{V_T} \tag{2.119}$$

$$n_p(W_B) = n_{po}\exp\frac{V_{\text{BC}}}{V_T} \simeq 0 \tag{2.120}$$

where W_B is the width of the base from the base–emitter junction depletion layer edge of the base–collector depletion layer edge and n_{po} is the equilibrium concentration of electrons in the base. Note that V_{BC} is negative for an *npn* transistor in the forward-active region and thus $n_p(W_B)$ is very small. Low-level injection conditions are assumed in the derivation of (2.119) and (2.120). This means that the minority-carrier concentrations are always assumed much smaller than the majority-carrier concentration.

If *recombination* of holes and electrons in the base is small, it can be shown that the minority-carrier concentration $n_p(x)$ in the base varies linearly with distance [18]. Thus, a straight line can be drawn joining the concentrations at $x = 0$ and $x = W_B$ in Figure 2.81c.

For charge neutrality in the base, it is necessary that

$$N_A + n_p(x) = p_p(x) \tag{2.121}$$

and thus

$$p_p(x) - n_p(x) = N_A \tag{2.122}$$

where $p_p(x)$ is the hole concentration in the base and N_A is the base doping density that is assumed constant. Equation (2.122) indicates that the hole and electron concentrations are separated by a constant amount and thus $p_p(x)$ also varies linearly with distance.

Collector current is produced by minority-carrier electrons in the base diffusing in the direction of the concentration gradient and being swept across the collector–base depletion region by the field existing there. The diffusion current density due to electrons in the base is

$$J_n = q D_n \frac{dn_p(x)}{dx} \tag{2.123}$$

where D_n is the diffusion constant for electrons. From Figure 2.81c,

$$J_n = -q D_n \frac{n_p(0)}{W_B} \tag{2.124}$$

If I_C is the collector current and is taken as positive flowing into the collector, it follows from (2.124) that

$$I_c = q A D_n \frac{n_p(0)}{W_B} \tag{2.125}$$

where A is the cross-sectional area of the emitter. Substitution of (2.119) into (2.125) gives

$$I_c = \frac{q A D_n n_{po}}{W_B} \exp \frac{V_{BE}}{V_T} \tag{2.126}$$

$$I_S \exp \frac{V_{BE}}{V_T} \tag{2.127}$$

where

$$I_S = \frac{q A D_n n_{po}}{W_B} \tag{2.128}$$

and I_S is a constant to describe the transfer characteristic of the transistor in the forward-active region. Equation (2.128) can be expressed in terms of the base doping density by noting that [19]

$$n_{po} = \frac{n_i^2}{N_A} \tag{2.129}$$

and substitution of (2.129) in (2.128) gives

$$I_S = \frac{q A D_n n_i^2}{W_B N_A} = \frac{q A \bar{D}_n n_i^2}{Q_B} \tag{2.130}$$

where $Q_B = W_B N_A$ is the number of doping atoms, n the base per unit area of the emitter, and n_i is the intrinsic carrier concentration in silicon. In this form (2.130) applies to both uniform and nonuniform base transistors and D_n has been replaced by $\bar{D}_n$, which is an average effective value of the electron diffusion constant in the base. This is necessary for nonuniform-base devices because the diffusion constant is a function of impurity concentration. Typical values of I_S as given by equation (2.130) are 10–14 to 10–16 A.

Equation (2.127) gives the collector current as a function of base–emitter voltage. The base current I_B is also an important parameter and, at moderate current levels, consists of two major components. One of these (I_{B1}) represents recombination of holes and electrons in the base and is proportional to the minority-carrier charge Q_e in the base. From Figure 2.81c, the minority-carrier charge in the base is

$$Q_e = \frac{1}{2}n_p(0)W_B q A \tag{2.131}$$

and we have

$$I_{B1} = \frac{Q_e}{\tau_b} = \frac{1}{2}\frac{n_p(0)W_B q A}{\tau_b} \tag{2.132}$$

where τ_b is the minority-carrier lifetime in the base. I_{B1} represents a flow of majority holes from the base lead into the base region. Substitution of (2.119) in (2.132) gives

$$I_{B1} = \frac{1}{2}\frac{n_{po}W_B q A}{\tau_B}\exp\frac{V_{BE}}{V_t} \tag{2.133}$$

The second major component of base current is due to injection of holes from the base into the emitter. This current component depends on the gradient of minority carrier holes into the emitter and is [20]

$$I_{B2} = \frac{qAD_p}{L_p}p_{nE}(0) \tag{2.134}$$

where D_p is the diffusion constant for holes and L_p is the diffusion length (assumed small) for holes in the emitter. $p_{nE}(0)$ is the concentration of holes in the emitter at the edge of the depletion region and is

$$p_{nE}(0) = p_{nE0}\exp\frac{V_{BE}}{V_T} \tag{2.135}$$

If N_D is the donor atom concentration in the emitter (assumed constant) then

$$p_{nE0} \simeq \frac{n_i^2}{N_D} \tag{2.136}$$

The emitter is deliberated doped much more heavily than the base, making N_D large and p_{nEo} small, so that the base-current component, I_{B2}, is minimized.

Substitution of (2.135) and (2.136) in (2.134) gives

$$I_{B2} = \frac{qAD_p}{L_p}\frac{n_i^2}{N_D}\exp\frac{V_{BE}}{V_T} \tag{2.137}$$

The total base current, I_B, is the sum of I_{B1} and I_{B2}

$$I_B = I_{B1} + I_{B2} = \left(\frac{1}{2} \frac{n_{po} W_B q A}{\tau_B} + \frac{qAD_p}{L_p} \frac{n_i^2}{N_D} \right) \exp \frac{V_{BE}}{V_T} \qquad (2.138)$$

Although this equation was derived assuming uniform base and emitter doping, it gives the correct functional dependence of I_B on device parameters for practical double-diffused nonuniform-base devices. Second-order components of I_B, which are important at low current levels, are considered later.

Since I_C in (2.127) and I_B in (2.128) are both proportional to $\exp(V_{BE}/V_T)$ in this analysis, the base current can be expressed in terms of collector current as

$$I_B = \frac{I_c}{\beta_F} \qquad (2.139)$$

where β_F is the forward current gain. An expression for β_F can be calculated by substituting (2.126) and (2.138) in (2.139) to give

$$\beta_F = \frac{\frac{qAD_n n_{po}}{W_B}}{\frac{1}{2} \frac{n_{po} W_B q A}{\tau_B} + \frac{qAD_p n_i^2}{L_p N_D}} = \frac{1}{\frac{W_B^2}{2\tau_B D_n} + \frac{D_p}{D_n} \frac{W_B}{L_p} \frac{N_A}{N_D}} \qquad (2.140)$$

where (2.129) has been substituted for n_{p0}. Equation (2.140) shows that β_F is maximized by minimizing the base width W_B and maximizing the ratio of emitter to base doping densities N_D/N_A. Typical values of β_F for *npn* transistors in integrated circuits are 50–500, whereas lateral *pnp* transistors have values 10–100. Finally, the emitter current is

$$I_E = -(I_c + I_B) = -\left(I_c + \frac{I_c}{\beta_F} \right) = -\frac{I_c}{\alpha_F} \qquad (2.141)$$

where

$$\alpha_F = \frac{\beta_F}{1 + \beta_F} \qquad (2.142)$$

The value of α_F can be expressed in terms of device parameters by substituting (2.140) in (2.142) to obtain

$$\alpha_F = \frac{1}{1 + \frac{1}{\beta_F}} = \frac{1}{1 + \frac{W_B^2}{2\tau_B D_n} + \frac{D_p}{D_n} \frac{W_B}{L_p} \frac{N_A}{N_D}} \simeq \alpha_t \gamma \qquad (2.143)$$

where

$$\alpha_t = \frac{1}{1 + \frac{W_B^2}{2\tau_B D_n}} \qquad (2.144)$$

$$\gamma = \frac{1}{1 + \frac{D_p}{D_n} \frac{W_B}{L_p} \frac{N_A}{N_D}} \qquad (2.145)$$

The validity of (2.143) depends on $W_B^2/(2\tau_B D_n) \ll 1$ and $\left(D_p/D_n \right) \left(W_B/L_p \right) (N_A/N_D) \ll 1$, and this is always true if β_F is large (see 2.140). The term γ in (2.143) is called the *emitter injection efficiency* and is equal to the ratio of the electron current

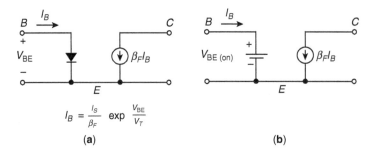

$$I_B = \frac{I_S}{\beta_F} \exp \frac{V_{BE}}{V_T}$$

(a)   (b)

Figure 2.82   Large-signal models of *npn* transistors for use in bias calculations. (a) Circuit incorporating an input diode. (b) Simplified circuit with an input voltage source.

(*npn* transistor) injected into the base from the emitter to the total hole and electron current crossing the base–emitter junction. Ideally, $\gamma \rightarrow 1$, and this is achieved by making N_D/N_A large and W_B small. In that case, very little reverse injection occurs from base to emitter.

The terms α_T in (2.143) is called the *base transport factor* and represents the fraction of carriers injected into the base (from the emitter) that reach the collector. Ideally, $\alpha_T \rightarrow 1$ and this is achieved by making W_B small. It is evident from the above development that fabrication changes that cause α_T and γ to approach unity also maximize the value of β_F of the transistor.

The results derived from the above allow formulation of a large-signal model of the transistor suitable for bias-circuit calculations with devices in the forward-active region. One such circuit is shown in Figure 2.82 and consists of a base–emitter diode to model (2.138) and a controlled collector-current generator to model (2.139). Note that the collector voltage ideally has no influence on the collector current and the collector node acts as a high-impedance current source. A simpler version of this equivalent circuit, which is often useful, is shown in Figure 2.82b, where the input diode has been replaced by a battery with a value $V_{BE(on)}$, which is usually 0.6–0.7 V. This represents the fact that in the forward-active region, the base–emitter voltage varies very little because of the steep slope of the exponential characteristic. In some circuits, the temperature coefficient of $V_{BE(on)}$ is important and a typical value for this is -2 mV/°C. The equivalent circuits of Figure 2.82 apply for *npn* transistors. For *pnp* devices, the corresponding equivalent circuits are shown in Figure 2.83.

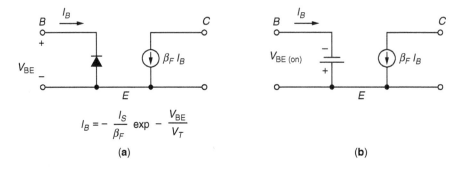

$$I_B = -\frac{I_S}{\beta_F} \exp -\frac{V_{BE}}{V_T}$$

(a)   (b)

Figure 2.83   Large-signal models of *pnp* transistors corresponding to the circuits of Figure 2.82.

2.2.4 Improving RF Performance by Means of Heterostructures

Let us recall a few equations that were defined in the previous sections. It was shown in (2.143) that the common-base current gain α_F consists of the emitter efficiency γ multiplied by the base transport factor α_T.

$$\alpha_t = \frac{1}{1 + \frac{W_B^2}{2\tau_B D_n}} \tag{2.144}$$

$$\gamma = \frac{1}{1 + \frac{D_p}{D_n} \frac{W_B}{L_p} \frac{N_A}{N_D}} \tag{2.145}$$

The RF performance of bipolar transistors is characterized by the cutoff frequencies f_T and $f_{\max}$:

$$f_{\max} \approx \sqrt{\frac{f_t}{8\pi r_b' C_c}} \tag{2.116}$$

where f_t is basically the inverse of the emitter, base, and collector transit times. The base transit time is given by

$$\tau_b = \frac{W_B^2}{2D_B} \tag{2.146}$$

These four formulas point to a dilemma, as it is impossible to increase all figures of merit at a time. To begin with, high emitter efficiency is imperative, which means $\gamma \to 1$. But the only degree of freedom, if we assume that the crystal quality is perfect, is the ratio of base doping to emitter doping $\frac{N_A}{N_D}$. We need the emitter doping to be much higher than base doping. Since there is a practical limit to doping the emitter, base doping will not be too high.

On the other hand, we need a thin base for high speed (low τ_b), and a low base resistance r_b' for high $f_{\max}$. However, the resistance decreases with base doping and base thickness. A thin base needs to be highly doped in order to allow for low r_b'.

These interdependence of parameters sets a natural limit for downscaling of BJTs and required to develop HBTs.

In order to overcome this dilemma, an additional degree of freedom is required. The theory was derived already in 1957 by H. Kroemer [12], but it was not before the mid-1990 that the epitaxial layers could be realized in good quality.

Consider an HBT where the emitter is made of a material of wider bandwidth than the base, for example, InGaP emitter versus GaAs base in Figure 2.84a. If the material system is well chosen, there will be a step in the valence band at the pn junction due to the hetero interface. This step prevents holes from being emitted from the base to the emitter, while the desired electron current is not affected. The valence band offset ΔE_v alters the formula for the emitter efficiency:

$$\gamma = \frac{1}{1 + \left(\frac{D_p}{D_n} \frac{W_B}{L_p} \frac{N_A}{N_D}\right) e^{-\Delta E_v}} \tag{2.147}$$

For a reasonably high ΔE_v, $\gamma \approx 1$ is achieved independently of the doping levels. Now, the base can be thinned and still provide a low sheet resistance due to high doping.

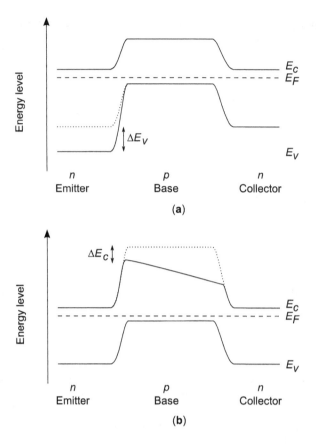

Figure 2.84 Band diagrams of HBTs. The dotted lines denote the respective bands without het-erostructure (a) InGaP emitter, GaAs base, and collector. The valence-band offset at the BE interface suppresses injections of holes into the emitter. (b) Si emitter and collector, SiGe base with Ge content increased toward the collector. The Ge content lowers the conduction band compared to Si and supports electron injection from the emitter. The slope of the conduction band accelerates the electrons toward the collector.

The same effect can be achieved with Si-based transistors, when SiGe is used in the base. Adding Ge to the base lowers the bandgap, and supports electron injection into the base, see Figure 2.84b. Another positive effect is used in some of the SiGe technologies. By varying the germanium content along the base, the conduction band gets a negative slope toward the collector. This results in an acceleration of electrons through a static built-in field. Since drift is much faster than diffusion, the base transit time is significantly reduced. For these transistors, the factor 2 in equation (2.146) needs to be replaced by a factor η that accounts for reduction of the τ_b due to drift current.

A further benefit of a highly doped base is that base width modulation due to variations of V_{CE}—the Early effect that is explained in the following section—is in general negligible.

2.2.5 Effects of Collector Voltage on Large-Signal Characteristics in the Forward-Active Region of BJTs

In the analysis of the previous section, the collector–base junction was assumed reverse biased and ideally had no effect on the collector currents. This is a useful approximation

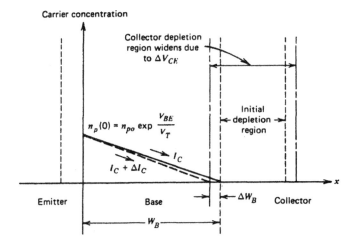

Carrier concentration

Collector depletion
region widens due
to ΔV_{CE}

$n_p(0) = n_{po} \exp \dfrac{V_{BE}}{V_T}$

Initial
depletion
region

I_C

$I_C + \Delta I_C$

ΔW_B

Emitter

Base

Collector

W_B

Figure 2.85   Effect of increases in V_{CE} on the collector depletion region and base width of a bipolar transistor.

for first-order calculations, but it is not strictly true in practice. There are occasions where the influence of collector voltage on collector current is important, and this will now be investigated.

The collector voltage has a dramatic effect on the collector current in two regions of device operation. These are the saturation (V_{CE} approaches zero) and breakdown (V_{CE} very large) regions that will be considered later. For values of collector–emitter voltage V_{CE} between these extremes, the collector current increases slowly as V_{CE} increases. The reason for this can be seen from Figure 2.85, which is a sketch of the minority-carrier concentration in the base of the transistor. Consider the effect of changes in V_{CE} on the carrier concentration for constant V_{BE}. Since V_{BE} is constant, the change in V_{CB} equals the change in V_{CE}, and this causes an increase in the collector-base depletion-layer width as shown. The change in the base width of the transistor, ΔW_B, equals the change in depletion-layer width and causes an increase ΔI_C in the collector current.

From (2.127) and (2.130), we have

$$I_c = \frac{qA\bar{D}_n n_i^2}{Q_B}\exp\frac{V_{BE}}{V_T} \tag{2.148}$$

Differentiation of (2.148) yields

$$\frac{\partial I_c}{\partial V_{CE}} = \frac{-qA\bar{D}_n n_i^2}{Q_B^2}\left(\exp\frac{V_{BE}}{V_T}\right)\frac{dQ_B}{dV_{CE}} \tag{2.149}$$

and substitution of (2.148) in (2.149) gives

$$\frac{\partial I_c}{\partial V_{CE}} = -\frac{I_c}{Q_B}\frac{dQ_B}{dV_{CE}} \tag{2.150}$$

For a uniform-base transistor, $Q_B = W_B N_A$ and (2.150) becomes

$$\frac{\partial I_c}{\partial V_{CE}} = -\frac{I_c}{W_B}\frac{dW_B}{dV_{CE}} \tag{2.151}$$

Note that since the base width decreases as V_{CE} increases, dW_B/dV_{CE} in (2.151) is negative and thus $\partial I_C/\partial V_{CE}$ is positive. dW_B/dV_{CE} is a function of the bias value of V_{CE},

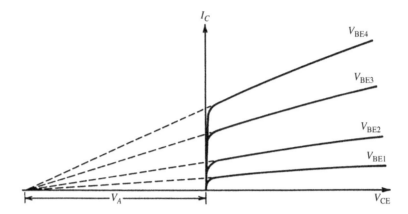

Figure 2.86   Bipolar transistor output characteristics showing the Early voltage, V_A.

but the variation is typically small for a reverse-biased junction and dW_B/dV_{CE} is often assumed constant. The resulting predictions agree adequately with experimental results.

Equation (2.151) shows that $\partial I_C/\partial V_{CE}$ is proportional to the collector-bias current and inversely proportional to the transistor base width. Thus, narrow-base transistors show a greater dependence of I_C on V_{CE} in the forward-active region. The dependence of $\partial I_C/\partial V_{CE}$ on I_C results in typical transistor output characteristics as shown in Figure 2.86. In accordance with the assumptions made in the foregoing analysis, these characteristics are shown for constant values of V_{BE}. Extrapolation of the characteristics of Figure 2.86 back to the V_{CE} axis gives an intercept V_A called the Early voltage, where

$$V_A = \frac{I_c}{\frac{\partial I_c}{\partial V_{CE}}} \tag{2.152}$$

Substitution of (2.151) in (2.152) gives

$$V_A = -W_B \frac{dV_{CE}}{dW_B} \tag{2.153}$$

which is a constant, independent of I_C. Thus, all the characteristics extrapolate to the same point on the V_{CE} axis. The variation of I_C with V_{CE} is called the Early effect and V_A is a common model parameter for circuit-analysis computer programs. Typical values of V_A for integrated-circuit transistors are 15–100 V. The inclusion of Early effect in dc bias calculations is usually limited to computer analysis because of the complexity introduced into the calculation. However, the influence of the Early effect is often dominant in small-signal calculations for high-gain circuits and this point will be considered later.

Finally, the influence of the Early effect on the transistor large-signal characteristics in the forward-active region can be represented approximately by modifying (2.127) to

$$I_c = I_S \left(1 + \frac{V_{BE}}{V_A} \right) e^{\frac{V_{BE}}{V_T}} \tag{2.154}$$

This is a common means of representing the device output characteristics for computer simulation.

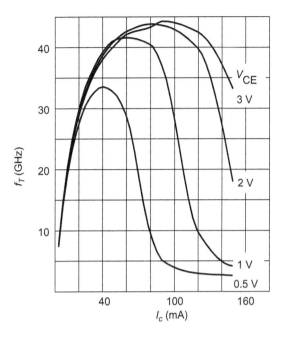

Figure 2.87 Transit frequency f_T of an Infineon BFP650 SiGe HBT as a function of bias point (from the datasheet, reprinted with permission). Courtesy Infineon Technologies.

2.2.6 Effects of Collector Current and Voltage on Large-Signal Characteristics in the Forward-Active Region of HBTs

An important issue is the bias dependence of the RF performance in HBTs. As a figure of merit, we will look at the transit frequency f_T, in order to discuss the various physical effects. f_T is a measure of the overall transistor behavior. It always corresponds to the total time delay of the transistor, and to some variation in HBT capacitances.

Even at first look it becomes obvious that f_T is no constant in SiGe transistors, see Figure 2.87. It shows a maximum at an optimum collector current I_C and degrades toward lower and higher currents. In order to explain this behavior, we will consider the main contributors to f_T, which are the base-emitter time constant τ_{be}, the base transit time τ_b, and the collector transit time τ_c:

$$f_T \approx \frac{1}{2\pi(\tau_{be} + \tau_b + \tau_c)} \tag{2.155}$$

Some of the time constants depend on the emitter current I_e, others on collector current I_c. However, a good bipolar transistor provides high current gain $\alpha \approx 1$. Therefore, we consider $I_c \approx I_e$ in this section.

> τ_{be} to be regarded here is the RC time constant of the base-emitter junction's resistance R_{be} and depletion capacitance C_{beD}. The differential resistance of a pn junction is inverse proportional to the diode current, and therefore R_{be} decreases with $1/I_e$. On the other hand, the base-emitter voltage is close to the diffusion voltage of the junction. The depletion capacitance C_{beD} will be approximately constant. Thus, $\tau_{be} \propto 1/I_e$ is

observed. Since R_{be} is significant at low currents, τ_{be} dominates f_T for $I_e \rightarrow 0$. In Figure 2.87, it is shown that f_T starts at quite low frequencies at low current, and then increases rapidly. There is no dependence on V_{ce} in the low-current regime, since τ_{be} is dominant.

τ_b used to be the dominant time constant in old-fashioned BJTs, but HBTs have very thin base layers, and often even built-in drift fields. In absence of the two pn junctions, this time constant would define the absolute minimum time constant.

τ_c is dominant at higher current densities. First of all, it is observed in Figure 2.87, that the device obviously becomes faster if higher voltages V_{ce} are applied. Second, there is a distinct maximum of f_T increases toward higher currents and voltages. Responsible for this behavior is the so-called Kirk effect, or base push-out [13, 14].

Base push-out occurs if the collector is only weakly doped and collector current is high. Under these conditions, it is no longer possible to neglect the electrons travelling through the space-charge region compared to the doping. It is rather necessary to define an effective collector doping N_{Ceff}:

$$N_{Ceff} = N_D - \frac{J_c}{q v_s} \qquad (2.156)$$

with the collector doping N_D, the collector current density J_c, electron charge q, and the average electron velocity v_s. Thus, even if the collector is n-doped, it can effectively change its doping to p-type if the current is high enough. As a consequence, the pn junction is shifted from the base–collector interface to the collector–subcollector interface. The subcollector is a highly n-doped semiconductor layer used to form the contact to the collector. This might not be an issue if the space-charge region fills the whole collector. But with increasing p-type N_{Ceff}, the space-charge region is reduced, and a neutral p-type region is formed at the base side of the collector. Concerning transistor operation, this effect looks like an increased base width, and it is therefore called base push-out.

Base push-out reduces the base transport factor, which might be of minor importance for high-quality HBTs. But in any case, it significantly increases τ_b, even though the effective base extends far into the collector.

Base push-out takes place at a certain collector current, but it can be delayed by applying higher collector voltages, as is clearly seen in Figure 2.87.

The behavior of GaAs-based HBTs is at first sight similar, but differs in some important points, see Figure 2.88.

Obviously, in contrast to the Si HBT, the device becomes slower with increasing V_{ce}. Second, there seems to be a region at moderate currents, where f_T rises more or less linearly with current. But before discussing the differences, let us point out what is similar to the SiGe case. First of all, it is the behavior toward low currents. It is dominated by τ_{be} also for GaAs-based HBTs, even though the interesting part is not measured in our figure here.

The second similarity is the region of base push-out. f_T decreases beyond a critical current, and the effect is delayed toward higher currents at higher voltages.

This leaves two major differences to be explained: Why is the GaAs-based HBT becoming slower at higher V_{ce}, and why is it getting faster with increasing I_c?

III–V semiconductors do not just have a constant electron mobility and an electron saturation velocity. Instead, for very high fileds, the electrons are scattered into the Γ valley.

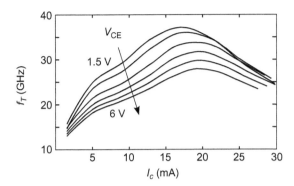

Figure 2.88 Transit frequency f_T of an InGaP/GaAs HBT, emitter size $3 \times 15 \, \mu m^2$ as a function of bias point (courtesy Ferdinand-Braun-Institut). Parameter is $V_{CE} = 1.5, \ 2, \ 3, \ 4, \ 5, \ 6$ V.

In the Γ valley, the effective mass is higher, thus the velocity is lower. The typical velocity-field characteristic of GaAs therefore starts with a constant mobility; at an electric field of about $E_c \approx 4 \, kV/cm$, a maximum velocity of almost $2 \times 10^7 \, /cm/s$ is reached. The velocity decreases toward higher currents to about a bit more than half of the maximum velocity [11, 15].

However, collector layers are quite short, in the region of $1 \, \mu m$ or less. Therefore, the electrical fields in the collector will always be beyond E_c. Therfore, increasing V_{ce} means lowering electron velocity, and therefore reducing f_T.

Altough it is not as pronounced, self-heating also reduces f_T. It supports the effect observed, since the dissipated power is higher at higher supply voltages, at least when measuring S-parameters.

What remains to be explained is that f_T seems to increase more or less proportional to I_c at moderate currents.

This region is dominated by the effect that the collector doping is compensated by the current according to equation (2.156), but we are still below the critical current. As a consequence, the space-charge region extends wider into the collector, and also the field profile changes. Remember that the electrons do not travel with constant velocity, thus the detailed investigation gets a bit involved.

However, it turns out that the compensation of the collector charge improves the transistor operation by lowering τ_c, but only until base push-out takes place. This effect is called velocity modulation and was derived in detail by Camnitz [10, 16].

Velocity modulation and base push-out also impact the base-collector capacitance C_{bc}. While it is reduced in the velocity modulation conditions in GaAs HBTs, C_{bc} increases rapidly at base push-out for both SiGe and GaAs HBTs.

2.2.7 Saturation and Inverse Active Regions

Saturation is a region of device operation that is usually avoided in analog circuits because the transistor gain is very low in this region. Saturation is much more commonly encountered in digital circuits, where it provides a well-specified output voltage that represents a logic state.

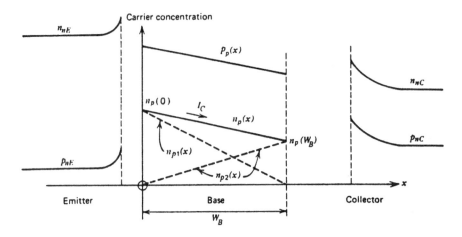

Figure 2.89   Carrier concentrations in a saturated *npn* transistor. (Not to scale.)

In saturation, both emitter–base and collector–base junctions are forward biased. Consequently, the collector–emitter voltage V_{CE} is quite small and is usually in the range of 0.05–0.3 V. The carrier concentrations in a saturated *npn* transistor with uniform base doping are shown in Figure 2.89. The minority-carrier concentration in the base of the edge of the depletion region is again given by (2.120) as

$$n_p(W_B) = n_{po}\exp\frac{V_{BC}}{V_t} \tag{2.157}$$

but since V_{BC} is now positive, the value of $n_p(W_B)$ is no longer negligible. Consequently, changes in V_{CE} with V_{BE} held constant (which cause equal changes in V_{BC}) directly affect $n_p(W_B)$. Since the collector current is proportional to the slope of the minority-carrier concentration in the base [see (2.123)], it is also proportional to $[n_p(0) - n_p(W_B)]$ from Figure 2.89. Thus changes in $n_p(W_B)$ directly affect the collector current, and the collector node of the transistor appears to have a *low impedance*. As V_{CE} is decreased in saturation with V_{BE} held constant, V_{BC} increases as does $n_p(W_B)$ from (2.157). Thus from Figure 2.89 the collector current decreases because the slope of the carrier concentration decreases. This gives rise to the saturation region of the I_C–V_{CE} characteristic shown in Figure 2.90. The slope of the I_C–V_{CE} characteristic in this region is largely determined by the resistance in series with the collector lead due to the finite resistivity of the *n*-type collector material. A useful model for the transistor in this region is shown in Figure 2.91 and consists of a fixed voltage source to represent $V_{BE(on)}$, and a fixed voltage source to represent the collector-emitter voltage. $V_{CE(sat)}$. A more accurate but more complex model includes a resistor in series with the collector. This resistor can have a value ranging from 20 to 500 Ω, depending on the device structure.

An additional aspect of transistor behavior in the saturation region is apparent from Figure 2.89. For a given collector current, there is now a much larger amount of stored charge in the base than there is in the forward-active region. Thus the base-current contribution represented by (2.133) will be larger in saturation. In addition, since the collector-base junction is now forward biased, there is a new base current component due to injection of carriers from the base to the collector. These two effects result in a base current I_B in

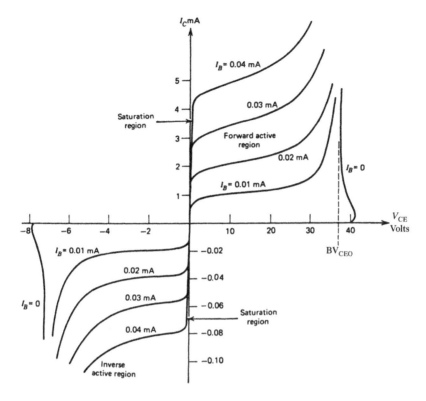

Figure 2.90 Typical I_C-V_{CE} characteristics for an *npn* bipolar transistor. Note the different scales for positive and negative currents and voltages.

saturation, which is larger than in the forward-active region for a given collector current I_C. Ratio I_C/I_B in saturation is often referred to as the *forced β* and is always less than $β_F$. As the forced β is made lower with respect to $β_F$, the device is said to be more *heavily saturated*.

The minority-carrier concentration in saturation shown in Figure 2.89 is a straight line joining the two end points assuming that recombination is small. This can be represented by a linear superposition of the two dotted distributions as shown. The justification for this is that the terminal currents depend *linearly* on the concentrations $n_p(0)$ and $n_p(W_B)$. This picture

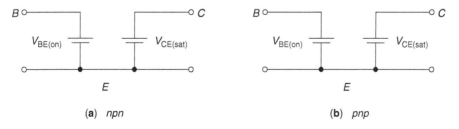

Figure 2.91 Large-signal models for bipolar transistors in the saturation region.

of device-carrier concentrations can be used to derive some general equations describing transistor behavior. Each of the distributions in Figure 2.89 is considered separately and the two contributions are combined. The *emitter* current that would result from $n_{p1}(x)$ above is given by the classical diode equation

$$I_{EF} = -I_{ES}\left(e^{\frac{V_{BE}}{V_t}} - 1\right) \tag{2.158}$$

where I_{ES} is a constant that is often referred to as the "saturation current" of the junction (no connection with the transistor saturation described above). Equation (2.158) predicts that the junction current is given by $I_{EF} \cong I_{ES}$ with a reverse-bias voltage applied. However, in practice, (2.158) is applicable only in the forward-bias region, since second-order effects dominate under reverse-bias conditions and typically result in a junction current several orders of magnitude larger than I_{ES}. The junction current that flows under reverse-bias conditions is often called the "leakage current" of the junction.

Returning to Figure 2.89, we can describe the *collector* current resulting from $n_{p2}(x)$ alone as

$$I_{CR} = -I_{CS}\left(\exp\frac{V_{BC}}{V_T} - 1\right) \tag{2.159}$$

where I_{CS} is a constant. The total collector current I_C is given by I_{CR} plus the fraction of I_{EF} that reaches the collector (allowing for recombination and reverse emitter injection). Thus

$$I_c = \alpha_F I_{ES}\left(\exp\frac{V_{BE}}{V_t} - 1\right) - I_{CS}\left(\exp\frac{V_{BC}}{V_T} - 1\right) \tag{2.160}$$

where α_F has been defined previously by (2.143). Similarly, the total emitter current is composed of I_{EF} plus the fraction of I_{CR} that reaches the emitter with the transistor acting in an inverted mode. Thus

$$I_E = -I_{ES}\left(\exp\frac{V_{BE}}{V_t} - 1\right) + \alpha_R I_{CS}\left(\exp\frac{V_{BC}}{V_T} - 1\right) \tag{2.161}$$

where α_R is the ratio of emitter to collector current with the transistor operating inverted (i.e., with the collector–base junction forward biased and emitting carriers into the base and the emitter–base junction reverse biased and collecting carriers). Typical values of α_R are 0.5–0.8. An inverse current gain β_R is also defined as

$$\beta_R = \frac{\alpha_R}{1 - \alpha_R} \tag{2.162}$$

and has typical values 1–5. This is the current gain of the transistor when operated inverted and is much lower than β_F because the device geometry and doping densities are designed to maximize β_F. The inverse-active region of device operation occurs for V_{CE} negative in an *npn* transistor and is shown in Figure 2.90. In order to display these characteristics adequately in the same figure as the forward-active region, the negative voltage and current scales have been expanded. The inverse-active mode of operation is rarely encountered in analog circuits.

Equations (2.160) and (2.161) describe *npn* transistor operation in the saturation region when V_{BE} and V_{BC} are both positive, and also in the forward-active and inverse-active regions. These equations are from the *Ebers–Moll* equations. In the forward-active region, they degenerate into a form similar to that of (2.127), (2.139), and (2.141) derived earlier. This can be shown by putting V_{BE} positive and V_{BC} negative in (2.160) and (2.161) to obtain

$$I_c = \alpha_F I_{ES} \left(\exp\frac{V_{BE}}{V_T} - 1 \right) + I_{CS} \tag{2.163}$$

$$I_E = -I_{ES} \left(\exp\frac{V_{BE}}{V_T} - 1 \right) - \alpha_R I_{CS} \tag{2.164}$$

Equation 2.163 is similar in form to (2.127) except that leakage currents that were previously neglected have now been included. This minor difference is significant only at high temperatures or very low operating currents. Comparison of (2.163) with (2.127) allows us to identify $I_S = \alpha_F I_{ES}$ and it can be shown [21] in general that

$$\alpha_F I_{ES} = \alpha_R I_{CS} = I_S \tag{2.165}$$

where this expression represents a reciprocity condition. Using (2.165) in (2.160) and (2.161) allows the Ebers–Moll equations to be expressed in the general form

$$I_c = I_S \left(\exp\frac{V_{BE}}{V_T} - 1 \right) - \frac{I_S}{\alpha_R} \left(\exp\frac{V_{BC}}{V_T} - 1 \right) \tag{2.166}$$

$$I_E = -\frac{I_S}{\alpha_F} \left(\exp\frac{V_{BE}}{V_T} - 1 \right) + I_S \left(\exp\frac{V_{BC}}{V_T} - 1 \right) \tag{2.167}$$

This form is often used for computer representation of transistor large-signal behavior.

The effect of leakage currents mentioned above can be further illustrated as follows. In the forward-active region, we have, from (2.164),

$$I_{ES} \left(\exp\frac{V_{BE}}{V_T} - 1 \right) = -I_E - \alpha_R I_{CS} \tag{2.168}$$

Substitution of (2.168) in (2.163) gives

$$I_c = -\alpha_F I_E + I_{CO} \tag{2.169}$$

where

$$I_{CO} = I_{CS} (1 - \alpha_R \alpha_F) \tag{2.170}$$

and I_{CO} is the collector–base leakage current with the emitter open. Although I_{CO} is given theoretically by (2.170), in practice, surface leakage effects dominate when the collector–base junction is reverse biased and I_{CO} is typically several orders of magnitude larger than the value given by (2.170). However, (2.169) is still valid if the appropriate measured value for I_{CO} is used. Typical values of I_{CO} are 10^{-10} to 10^{-12} A at 25°C, and the magnitude doubles about every 8°C. As a consequence, these leakage terms can become very significant at high temperatures. For example, consider the base current I_B. This is

$$I_B = -(I_c + I_E) \tag{2.171}$$

If I_E is calculated from (2.169) and substituted in (2.171), the result is

$$I_B = \frac{1 - \alpha_F}{\alpha_F} I_c - \frac{I_{CO}}{\alpha_F} \tag{2.172}$$

But from (2.142)

$$\beta_F = \frac{\alpha_F}{1 - \alpha_F} \tag{2.173}$$

and use of (2.173) in (2.172) gives

$$I_B = \frac{I_c}{\beta_F} - \frac{I_{CO}}{\alpha_F} \tag{2.174}$$

Since the two terms in (2.174) have opposite signs, the effect of I_{CO} is to *decrease* the magnitude of the external base current at a given value of collector current.

2.2.8 Self-Heating

Temperature is a dynamic parameter in any circuit, and it impacts the electrical performance especially for semiconductor devices. Three temperatures are important for circuit design:

- T_0, the ambient temperature during for parameter extraction;
- T_{amb}, the ambient temperature in simulation. Could be, for example, swept between −40 and 120°C;
- T_j, the dynamical junction temperature during operation.

The actual temperature during operation therefore is $T_j = T_{amb} + \Delta T_j$, where ΔT_j denotes the increase in temperature due to self-heating.

For accurate simulation, it therefore is necessary that the transistor mathematics do not just calculate currents and charges as functions of voltages, but also as a function of temperature. The models therefore do have quite a number of parameters that describe how transistor performance changes with temperature. But that is not all, since the actual temperature depends on the device dissipated power and on the heatsinking measures applied.

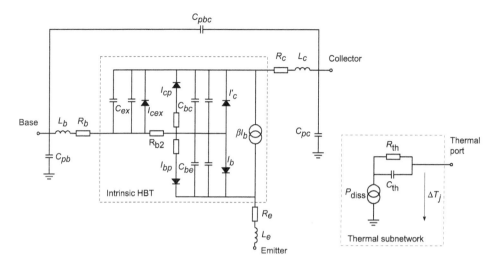

Figure 2.92 Equivalent circuit of the large signal FBH–HBT model with thermal subcircuit.

Fortunately, the flow of heat is described through differential equations like the flow of current. This allows for calculation of temperature within a circuit simulator without the need for a dedicated solver. The relation between dissipated power and temperature is given by a thermal resistance R_{th} (in K/W), and a thermal capacitance C_{th} (in J/W). The dissipated power is hence treated like a current, and the temperature is calculated through an equivalent voltage.

Transistor models commonly use thermal subcircuits like the one seen in Figure 2.92. It consists of the equivalent current source that forces the current equivalent to the total dissipated power:

$$P_{\text{diss}} = \sum_{\text{all branches}} |i_i(t) \cdot v_i(t)|^2 \qquad (2.175)$$

The thermal time constant accounts for the fact that the temperature does not follow the electrical power with infinite speed. In an integrated circuit, the $R_{\text{th}} C_{\text{th}}$ time constant is typically around a microsecond.

The thermal subcircuit provides a thermal node in this model. In the basic application, this node would just be shortened to ground, and the increase of device's junction temperature compared to the ambient temperature, ΔT_j would equal the equivalent voltage across R_{th} and C_{th}. It is clearly seen that this subcircuit is fully separated from the rest of the circuit.

Commonly, the $R_{\text{th}} C_{\text{th}}$ time constant is only used to separate dc from RF, since dc power heats the device, while RF power does not. Only for wideband applications, or for baseband applications, the real thermal frequency response can be interesting. In this case, detailed analysis of the thermal response is required, and usually multiple time constants are to be taken into account, for example, one for the die, another for the package, a third one for the heatsink.

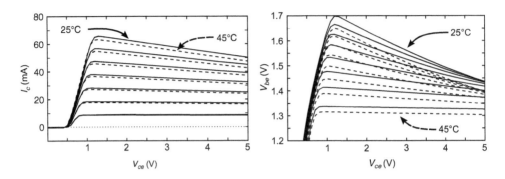

Figure 2.93 Output IV curves of a 3 × 30 μm² HBT. (Courtesy Ferdinand-Braun-Institut.)

The thermal port also enables the exchange of thermal energy between different devices and thus can be used to determine the impact of mutual heating between different devices.

Impact of Self-Heating on HBT DC Characteristics The output IV curves of GaAs-based HBTs look a bit different from the curves previously discussed for BJTs, see Figure 2.93. Instead of increasing with V_{ce} due to the Early effect, these curves decrease due to self-heating. Early effect is not pronounced in these devices due to the high base doping, anyway.

The figure shows two measurements, one at 25°C ambient temperature and one at 45°C. When looking at the IV curve first, it is observed that increasing the ambient temperature basically reduces current gain. The negative slope of the curves, on the other hand, is due to self-heating, it is most pronounced in the region of high dissipated powers $P_{\text{diss}} = I_C \cdot V_{\text{CE}}$. Comparing how the two types of heating impact device performance points to the fact that the junction temperature can easily reach about 100°C in power amplifiers.

The temperature dependence of transistor model parameters are usually linearly approximated, since the span for which the model needs to be accurate is rather narrow, say, for example, −40 to 120°C ambient temperature. For diode currents, however, an exponential dependence of the saturation current I_s needs to be considered:

$$I(V, T_j) = I_S(T_j) \cdot \left(e^{V/(nV_{T_j})} - 1\right) = I_{S,T_0} \cdot \left(e^{E_g/(kT_0)-E_g/(kT_j)}\right) \cdot \left(e^{V/(nV_{T_j})} - 1\right) \quad (2.176)$$

with the bandgap energy E_g, the Boltzmann constant k, and the junction temperature at time of parameter extraction (the reference temperature) T_0, and the actual junction temperature T_j. At fixed base voltage, the current increases with temperature. This positive feedback allows GaAs HBTs become thermally unstable with low-impedance base matching. In the worst case, the devices melts down or the current concentrates in one part of the device. The latter effect, hot-spot formation, is sometimes hard to be detected and suppressed, but it is of course a threat regarding reliability and lifetime.

When forcing current, on the other hand, base voltage is reduced with self-heating in case of HBTs. This effect is shown in Figure 2.93b. During dc characterization of HBTs, this $V_{\text{BE}} - V_{\text{CE}}$ plot should always be taken together with the $I_C - V_{\text{CE}}$ plot, as it shows the temperature dependence of the base–emitter diode much more pronounced than the collector current alone.

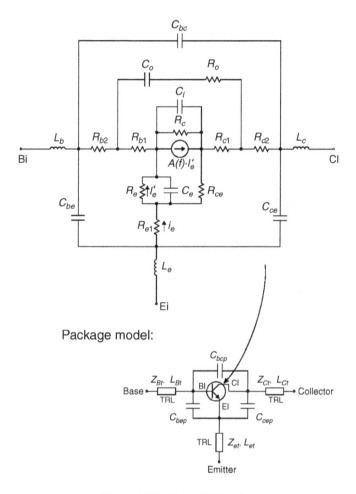

Figure 2.94 Linear BJT model.

In the case of SiGe HBTs, the situation is similar but a bit relaxed. First, Si conducts the heat about three times better than GaAs. Second, the material systems offers the possibility to compensate the temperature behavior, at least to a certain extent.

2.2.9 Small-Signal Models of Bipolar Transistors

Figure 2.94 shows the small-signal equivalent BJT model. It consists of an intrinsic model and a package model. The model key words are listed in Table 2.12.

2.3 FIELD-EFFECT TRANSISTORS

Both diodes and *pnp*/*npn* transistors work on the injection principle, which means the base–emitter junction generates either free electrons or other carriers that are being "collected" by the collector connection and the base connection controls it.

Table 2.12 Small-Signal BJT Model Key Words

Key Word	Description	Unit	Default
	Intrinsic Model		
A	Ratio of I_C to I_E at dc		0.0
RE	Emitter resistance	ohm	0.0
F	Current generator roll-off frequency	Hz	∞
T	Time delay	s	0.0
CE	Emitter capacitance	farad	0.0
CI	Collector capacitance	farad	0.0
RCE	Collector emitter resistance	ohm	∞
RC	Collector resistance	ohm	∞
RO	Extrinsic base collector resistance	ohm	0.0
CO	Extrinsic base collector capacitance	farad	0.0
RB1	Intrinsic base resistance (R_{bb})	ohm	0.0
RC1	Parasitic collector resistance	ohm	0.0
RE1	Parasitic emitter resistance	ohm	0.0
RB2	Parasitic base resistance	ohm	0.0
RC2	Parasitic collector resistance	ohm	0.0
CBE	Base-to-emitter package capacitance	farad	0.0
CBC	Base-to-collector package capacitance	farad	0.0
CCE	Collector-to-emitter package capacitance	farad	0.0
LB	Base lead inductance	henry	0.0
LC	Collector lead inductance	henry	0.0
LE	Emitter lead inductance	henry	0.0
TJ	Chip temperature	K	298
NFAC	Noise factor proportional to drive		1.0
FC	Flicker noise ($1/f$ noise) corner frequency	Hz	50
	Package Model		
CBCP	Base-to-collector package capacitance	farad	0.0
CBEP	Base-to-emitter package capacitance	farad	0.0
CCEP	Collector-to-emitter package capacitance	farad	0.0
ZBT	Base transmission line impedance	ohm	50
ZCT	Collector transmission line impedance	ohm	50
ZET	Emitter transmission line impedance	ohm	50
LBT	Base transmission line length at $\varepsilon_r = 1$	meter	0.0
LCT	Collector transmission line length at $\varepsilon_r = 1$	meter	0.0
LET	Emitter transmission line length at $\varepsilon_r = 1$	meter	0.0

Notes:
1. $A \equiv \alpha = I_c/I_e$; $\beta = $ dc current gain $= \alpha/(1 - \alpha)$.
2. The bipolar current gain in this model is described by

$$A = A(0) = \frac{e^{-j\omega T}}{1 + jf/F}$$

where $\omega = 2\pi f$ and $f = $ frequency.

3. The current source is controlled by the current through R_e. The current generator has a cutoff frequency with respect to the total emitter current, I_E:

$$F = g_m/2\pi C_e$$

where $g_m = 1/R_e$. This frequency becomes infinity for the default value for $C_e(0.0)$. The parameter F specifies the frequency roll-off for the current generator with respect to the current through R_e. Effectively, this frequency parameter may be used to model additional delays in the device.

In opposition to this, the FET controls the electronic conduction in a solid by an electric field, and this concept actually predates the invention of the bipolar transistor. J. E. Lilienfeld filed for a patent on such a device in 1925 (U.S. Patent No. 1,745,175), and W. Shockley presented a comprehensive theory of the FET in 1952. It took until 1960 before the first commercially available FETs came to the market.

Several types of semiconductor materials have been used for making FETs: silicon, germanium, gallium arsenide, gallium nitride, and other have been used. By far, the most widely used is silicon, followed by gallium arsenide.

The field-effect transistor (FET) is a class of electronic semiconductor device in which the conduction of a "channel" between source (*S*) and drain (*D*) terminals is controlled by an electric field impressed upon the channel via a gate (*G*) terminal. The conduction channel may utilize *n*-type carriers (electrons) or *p*-type carriers (holes). The electric field that controls the channel conduction may be introduced via a *pn* junction (for a "junction" FET [JFET]), a metal plate separated from the semiconductor channel by an oxide dielectric (for a metal-oxide semiconductor [MOS] FET), or a combination of the two methods. The channel of heterostructure FETs (HFETs or HEMTs) consists of a quantum well of high conductivity and the gate voltage controls the number of available electrons in the channel. The polarity of the controlling electrical field is a function of the type of carriers in the channel. A FET "family tree" is shown in Figure 2.95.

For today's microwave and RF integrated circuits, the JFET with its high-frequency problems (low cutoff frequency) and large variation in parameters, such as transconductance and pinchoff voltage, has fallen out of favor. It has, however, still its place in frequency generation up to 500 MHz due to its unparalleled low flicker noise corner frequency. The most-used FET types are MOS for small-signal applications, LDMOS (lateral diffused MOS), and as an emerging technology, GaN HEMTs for power applications, GaAs HEMT for lower-power applications (low-voltage applications) as well as high-power applications.

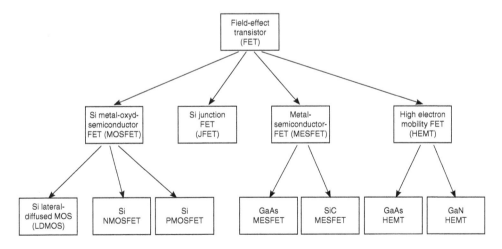

Figure 2.95   A RF-FET family tree.

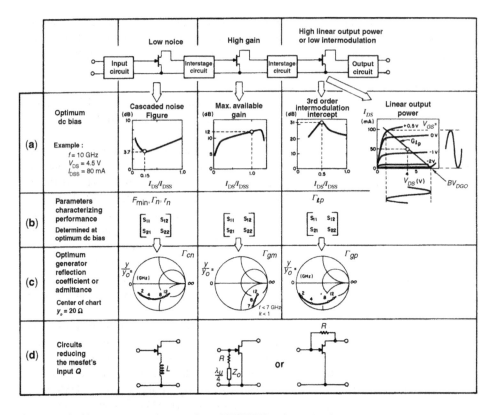

Figure 2.96 Key parameters in operating a MESFET in a low-noise front-end stage, a high-gain intermediate stage, and a linear power stage, showing at (a), the device characteristics that lead to optimum dc bias for each service; at (b), the parameters that characterize the device performance; at (c), the optimum generator reflection coefficient that must be synthesized by each circuit; and at (d), circuit arrangements that reduce the transistor's input Q or stabilize it at low frequencies. Although this example is based on an HP MESFET with a gate length of 1 μm and a gate width of 500 μm, this principle applies to all members of the FET family [22]. ©1976 IEEE.

All members of the FET family are high-impedance-input devices. Figure 2.96 shows the key parameters in operating for MOSFETs, MESFETs, and HEMTs in low-noise stages, high-gain stages, and linearized power stages. The linearization also improves the matching. This figure is similar to its counterpart in Section 2.2.2.

The following several pages present a typical example of a datasheet[3] for a JFET (U310) and the chip (NZA) on which it is based, followed by an example of a device with a much higher dynamic range (the CP640 family). The CP640 family and its similars are no longer made and are being replaced by low-power LDMOS, which are not yet well documented. At least one company, InterFET, is devoted entirely to producing discrete JFETs and JFET-related hybrid and small-scale-integration IC products.

---

[3] Reproduced with permission.

n-channel JFETs
designed for . . .

- ■ **VHF Amplifiers**
- ■ **Front End High Sensitivity Amplifiers**
- ■ **Oscillators**
- ■ **Mixers**

Performance Curves NZA
See Section 4

BENEFITS
- ● Industry Standard
- ● High Power Gain
  16 dB at 105 MHz, Common-Gate
  11 dB at 450 MHz, Common-Gate
- ● Low Noise
  2.7 dB Noise Figure at 450 MHz
- ● Wide Dynamic Range
  Greater than 100 dB
- ● 75 Ω Input Match Common Gate

Siliconix

ABSOLUTE MAXIMUM RATINGS (25°C)

Gate-Drain or Gate-Source Voltage . . . . –25 V
Gate Current . . . . 20 mA
Total Power Dissipation at T_A = 25°C . . . . 500 mW
Power Derating to 150°C . . . . 4.0 mW/°C
Storage Temperature Range . . . . –65 to +200°C
Lead Temperature
  (1/16″ from case for 10 seconds) . . . . 300°C

TO-52
See Section 6

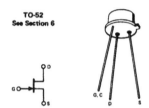

ELECTRICAL CHARACTERISTICS (25°C unless otherwise noted)

	Characteristic		U308 Min	Typ	Max	U309 Min	Typ	Max	U310 Min	Typ	Max	Unit	Test Conditions	
1	I_{GSS}	Gate Reverse Current			–150			–150			–150	pA	V_{GS} = –15 V,	
2					–150			–150			–150	nA	V_{GS} = 0	T_A = 125°C
3	BV_{GSS}	Gate-Source Breakdown Voltage	–25			–25			–25			V	I_G = –1 μA, V_{DS} = 0	
4	$V_{GS(off)}$	Gate-Source Cutoff Voltage	–1.0		–6.0	–1.0		–4.0	–2.5		–6.0		V_{DS} = 10 V, I_D = 1 nA	
5	I_{DSS}	Saturation Drain Current (Note 1)	12		60	12		30	24		60	mA	V_{DS} = 10 V, V_{GS} = 0	
6	$V_{GS(f)}$	Gate-Source Forward Voltage			1.0			1.0			1.0	V	I_G = 10 mA, V_{DS} = 0	
7	g_{fg}	Common-Gate Forward Transconductance (Note 1)	10	17		10	17		10	17		mmho	V_{DS} = 10 V, I_D = 10 mA	f = 1 kHz
8	g_{og}	Common-Gate Output Conductance			250			250			250	μmho		
9	C_{gd}	Drain-Gate Capacitance			2.5			2.5			2.5	pF	V_{GS} = –10 V, V_{DS} = 10 V	f = 1 MHz
10	C_{gs}	Gate-Source Capacitance			5.0			5.0			5.0			
11	$\overline{e_n}$	Equivalent Short Circuit Input Noise Voltage		10			10			10		$\frac{nV}{\sqrt{Hz}}$	V_{DS} = 10 V, I_D = 10 mA	f = 100 Hz
12	g_{fg}	Common-Gate Forward Transconductance		15			15			15				f = 105 MHz
13				14			14			14				f = 450 MHz
14	g_{og}	Common-Gate Output Conductance		0.18			0.18			0.18		mmho		f = 105 MHz
15				0.32			0.32			0.32			V_{DS} = 10 V, I_D = 10 mA	f = 450 MHz
16	G_{pg}	Common-Gate Power Gain (Note 2)	14	16		14	16		14	16				f = 105 MHz
17			10	11		10	11		10	11				f = 450 MHz
18	NF	Noise Figure		1.5	2.0		1.5	2.0		1.5	2.0	dB		f = 105 MHz
19				2.7	3.5		2.7	3.5		2.7	3.5			f = 450 MHz

NOTES:
1. Pulse test duration = 2 ms.
2. Gain (G_{pg}) measured at optimum input noise match.

NZA

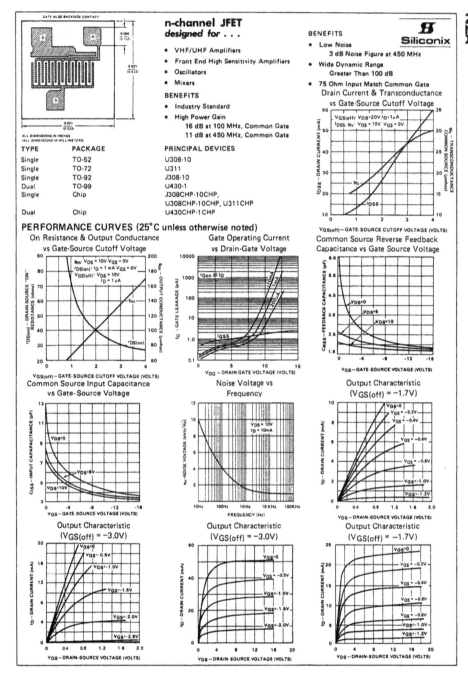

GATE ALSO BACKSIDE CONTACT

ALL DIMENSIONS IN INCHES
(ALL DIMENSIONS IN MILLIMETERS)

n-channel JFET
designed for . . .

- VHF/UHF Amplifiers
- Front End High Sensitivity Amplifiers
- Oscillators
- Mixers

BENEFITS
- Industry Standard
- High Power Gain
  16 dB at 100 MHz, Common Gate
  11 dB at 450 MHz, Common Gate

BENEFITS
- Low Noise
  3 dB Noise Figure at 450 MHz
- Wide Dynamic Range
  Greater Than 100 dB
- 75 Ohm Input Match Common Gate

Siliconix

NZA

TYPE	PACKAGE
Single	TO-52
Single	TO-72
Single	TO-92
Dual	TO-99
Single	Chip
Dual	Chip

PRINCIPAL DEVICES

U308-10
U311
J308-10
U430-1
J308CHP-10CHP,
U308CHP-10CHP, U311CHP
U430CHP-1CHP

Drain Current & Transconductance
vs Gate-Source Cutoff Voltage

PERFORMANCE CURVES (25°C unless otherwise noted)

On Resistance & Output Conductance
vs Gate-Source Cutoff Voltage

Gate Operating Current
vs Drain-Gate Voltage

Common Source Reverse Feedback
Capacitance vs Gate Source Voltage

Common Source Input Capacitance
vs Gate-Source Voltage

Noise Voltage vs
Frequency

Output Characteristic
($V_{GS(off)} = -1.7V$)

Output Characteristic
($V_{GS(off)} = -3.0V$)

Output Characteristic
($V_{GS(off)} = -3.0V$)

Output Characteristic
($V_{GS(off)} = -1.7V$)

NZA

PERFORMANCE CURVES (Con't) (25°C unless otherwise noted)

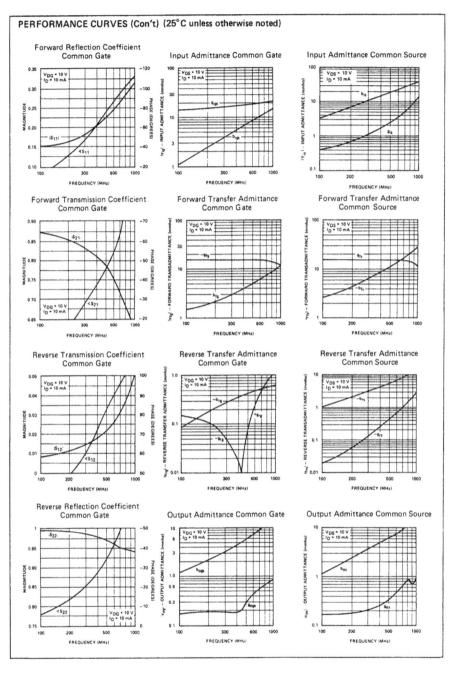

BROADBAND RF FET
SILICON EPITAXIAL JUNCTION
N-CHANNEL FIELD EFFECT TRANSISTOR

HIGH DYNAMIC RANGE HF AND VHF AMPLIFIER
FOR USE IN COMMON GATE CONFIGURATION
- USABLE TO OVER 300 MHz
- 50 Ohm VSWR < 1.5:1 0.5-50 MHz (FIG. 1)
- LOW NOISE FIGURE — 2.2 dB TYPICAL @ 50 MHz
- INPUT Z CONSTANT 0.5-50 MHz
- HIGH IM INTERCEPT POINT — > + 40 dBm
- HIGH TRANSCONDUCTANCE — 100,000 μmhos (TYP.)
- 1 dB COMPRESSION POINT > + 20 dBm
- DYNAMIC RANGE > 140 dB (TO 1 dB COMPRESSION)
- HIGH VOLTAGE—TO 50 V.

CP640
CP664
CP665
CP666

ELECTRICAL DATA   ABSOLUTE MAXIMUM RATINGS

PARAMETER	SYMBOL	CP 640	CP 664	CP 665	CP 666	UNITS
Drain to Source Voltage	BV$_{DSO}$	20	30	40	50	Volts
Drain to Gate Voltage	BV$_{DGO}$	20	30	40	50	Volts
Gate to Source Voltage	BV$_{GSO}$	– 15	– 20	– 20	– 20	Volts
Peak Drain Current	I$_D$	1.2	1.2	1.2	1.2	Amps
Power Dissipation 25 °C CASE	P$_D$	8.0	8.0	8.0	8.0	Watts
Derating Factor (slope)	DF	22	22	22	22	°C/W
Junction Temp.(Oper. & Store)	T$_J$	– 55 °C to + 200 °C				

TYPICAL TWO TONE 3rd ORDER IM
PRODUCTS — CIRCUIT FIGURE 1

Tones at 3MHz/5MHz

Signal Level EMF	3rd Order Product
1 Volt	-44 dB
0.3 Volt	-75 dB
0.25 Volt (OdBM)	-80 dB

ELECTRICAL CHARACTERISTICS: T$_{CASE}$ = 25 °C (UNLESS OTHERWISE STATED)

PARAMETERS	CONDITIONS		SYMBOL	Min.	Typ.	Max.	UNITS
Gate Leakage Current	V$_{GS}$ = 15V, V$_{DS}$ = 0	25 °C	I$_{GSS}$		5	100	nA
		150 °C	I$_{GSS}$			10	μA
Operating Transconductance	V$_{DS}$ = 15V, I$_{DS}$ = 40 mA		g$_{fg}$	40	60	80	mmho
Zero Bias Transconductance	V$_{DS}$ = 15V, V$_{GS}$ = 0(1)		g$_{fg}$	75	100	200	mmho
Gate-Source Cut-Off Voltage	V$_{DS}$ = 5V, I$_{DS}$ = 1.0 mA		V$_{GS}$ (off)	2	5	10	Volts
Zero Bias Drain Current	V$_{DS}$ = 15V, V$_{GS}$ = 0 (1)		I$_{DSS}$	100	400	800	mA
Gate to Source Cap.	V$_{GS}$ = – 20V		C$_{GS}$		15	20	pf
Gate to Drain Cap.	V$_{GD}$ = – 20V		C$_{GD}$		15	20	pf
Power Gain	I$_{DS}$ = 40mA, f = 50MHz, Fig. 1		Gpg	8	8.5	9.5	dB
Noise Figure	I$_{DS}$ = 40mA, f = 30MHz, Fig. 1		N.F.		2.2	3.0	dB
Voltage Standing Wave Ratio	f = 0.5-50MHz, 50 Ω Source, Fig. 1		VSWR			1.5:1	
Common Gate Input Conductance	f = 0.5-50MHz, V$_{DS}$ = 15, I$_D$ = 40mA		g$_{igs}$		60		mmho
Common Gate Output Conductance	f = – 50MHz, V$_{DS}$ = 15, I$_D$ = 40mA		g$_{ogs}$		0.4		mmho

'Pulse Measurement 1% Duty Cycle 10 mS Max.

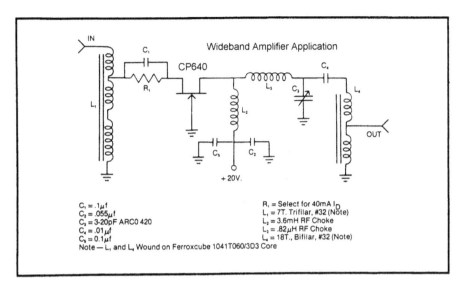

Wideband Amplifier Application

$C_1 = .1\mu f$
$C_2 = .055\mu f$
$C_3 = 3\text{-}20pF \text{ ARCO } 420$
$C_4 = .01\mu f$
$C_5 = 0.1\mu f$
Note — L_1 and L_4 Wound on Ferroxcube 1041T060/3D3 Core

$R_1 = \text{Select for 40mA } I_D$
$L_1 = 7T. \text{ Trifilar, } \#32 \text{ (Note)}$
$L_2 = 3.6mH \text{ RF Choke}$
$L_3 = .82\mu H \text{ RF Choke}$
$L_4 = 18T., \text{ Bifilar, } \#32 \text{ (Note)}$

TYPICAL INTERCEPT AND COMPRESSION POINT

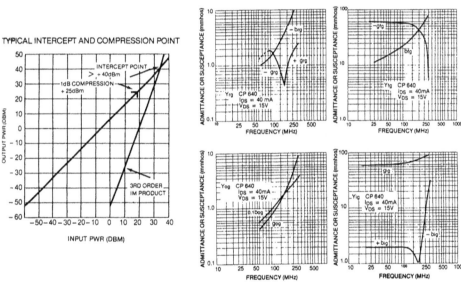

2.4   LARGE-SIGNAL BEHAVIOR OF JFETs

A section view of a junction field-effect transistor (JFET) is shown in Figure 2.97. This structure contains, between the source and the drain contacts, an n-type "channel" embedded in a p-type silicon substrate. If it is assumed that the pn junction forms a barrier to current flow, then it can be seen that channel conduction is a function of the channel width, length, and thickness and of the density and mobility of the carriers. In this structure, current can flow equally well in either direction through the channel; that is, the drain can be positive or negative with respect to the source.

Since we have found out that the SPICE models for JFETs are really incomplete and give poor answers, we will only briefly touch here on the JFET's performance. The modified Materka model, one of the best GaAsFET models, turns out to be best suited for junction FETs because both transistors act similarly in the dc area. Both have a gate-to-source diode that becomes conductive above 0.7 V at the input and they are also quite similar in other respects. Their equivalent circuits are also quite similar, as parameter extractions for both approaches have shown.

In general, in a three-terminal device, the drain current I_D is a function of two variables: V_{DS} and V_{GS}. This function is best represented by families of characteristic curves, as shown in Figure 2.98. These curves are for the "common source" configuration, with the drain as the output and the gate as the input. They reveal that for this device, if V_{DS} is greater than about 2 V (but less than the drain breakdown), I_D is primarily determined by the gate voltage V_{GS}. Under these conditions, it is valid for small signals to characterize the FET by a single transfer characteristic curve, commonly called the "forward transconductance curve," such as the upper curve shown in Figure 2.98b. Some of the relationships between the forward transfer curve and the output characteristic curves of Figure 2.98a will be examined. The value of V_{GS} that reduces I_D to approximately zero is the gate-source cutoff voltage, $V_{GS(off)}$. With reference to the output curve, Figure 2.98a, note that the drain current at $V_{GS} = 0$ tends to become saturated at a drain voltage approximately equal in magnitude to $-V_{GS(off)}$. This drain voltage is often referred to as the pinchoff voltage V_P; however, in this section, pinchoff voltage is used interchangeably with gate-source cutoff voltage. V_P will have the same meaning as $V_{GS(off)}$. The symbol I_{DSS} is commonly used to indicate the value of saturated drain current at $V_{GS} = 0$.

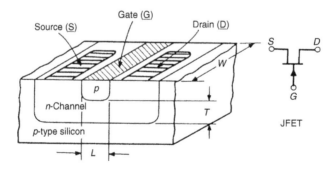

Figure 2.97   Junction field-effect transistor [24].

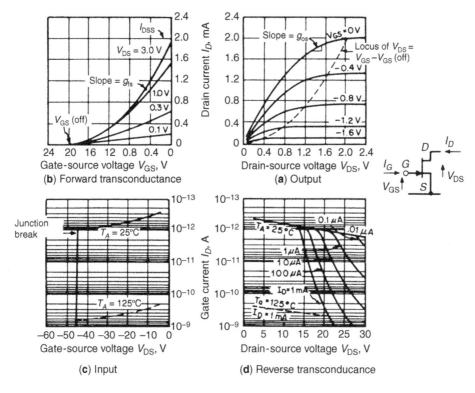

Figure 2.98   Static characteristics of an n-channel JFET.

The forward transconductance characteristic of Figure 2.98b can be approximated by a power law relation expressed as

$$I_D = I_{DSS}\left(1 - \frac{V_{GS}}{V_{GS(off)}}\right)^n \tag{2.177}$$

if $V_{DS} \geq -V_{GS(off)}$. By differentiation the small-signal transconductance, g_{fs} is given by

$$g_{fs} = \frac{dI_D}{dV_{GS}} = -n\frac{I_{DSS}}{V_{GS(off)}}\left(1 - \frac{V_{GS}}{V_{GS(off)}}\right)^{n-1} \tag{2.178}$$

Some texts indicate a value of 3/2 for n; however, experimental measurements on a number of n-channel JFET geometries indicate that the exponent n is close to 2, which is the value derived in an approximate treatment by R. D. Middlebrook [25].

A useful relationship between g_{fs0}, I_{DSS}, and $V_{GS(off)}$ is derived from the ratio of equations (2.177) and (2.178).

$$\frac{g_{gs}}{I_D} = n\left(V_{GS} - V_{GS(off)}\right)^{-1} \tag{2.179}$$

At $V_{GS} = 0$, $I_D = I_{DSS}$ and $g_{fs} = g_{fso}$. Using 2 as the value of the constant n leads to

$$g_{fso} = -2\frac{I_{DSS}}{V_{GS(off)}} \tag{2.180}$$

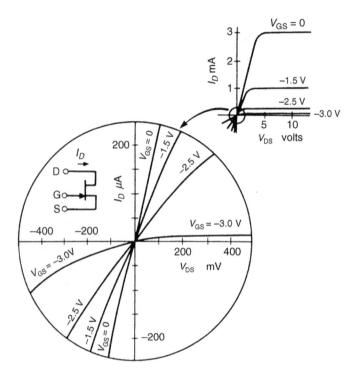

Figure 2.99 Enlargement of *n*-channel output characteristic around $V_{DS} = 0$.

For *n*-channel FETs, I_{DSS} is positive and $V_{GS(off)}$ is negative; for *p*-channel FETs I_{DSS} is negative and $V_{GS(off)}$ is positive; thus g_{fs} is a positive quantity for both *p*- and *n*-channel FETs.

Equation (2.177) indicates that for $V_{GS} = V_{GS(off)}$, $I_D = 0$. In a real device, this does not happen. Starting from zero, as the gate voltage is made more negative, the drain current decreases until it reaches a very low value equal to the drain-leakage current. At this value, the source current will consist of source-gate leakage. Any further increase in the magnitude of the negative gate voltage will result in an increase in I_D leakage. For the small-signal device illustrated, this minimum I_D is on the order of 2×10^{-13} A.

For some types of applications of the FET, it is helpful to understand the characteristics at very low values of V_{DS}, such as those shown in Figure 2.99. A "very low value" is one that is small compared to the magnitude of $V_{GS} - V_{GS(off)}$. In this region, V_{DS} is small enough to have little effect upon channel thickness, so that the I_D/V_{DS} slope is nearly linear. Since the slope is a function of V_{GS}, the FET can be utilized as a voltage-controlled resistor. The conductance slope ($\Delta I_D/V_{DS}$) at $V_{DS} = 0$ is approximately a linear function of $V_{GS} - V_{GS(off)}$.

If g_{ds} at $V_{GS} = 0$ is given the term g_{ds0}, then

$$g_{ds} = g_{ds0}\left(1 - \frac{V_{GS}}{V_{GS(off)}}\right) \tag{2.181}$$

with $V_{DS} = 0$. A plot of this characteristic is shown in Figure 2.100, along with g_{fs} and I_D characteristics.

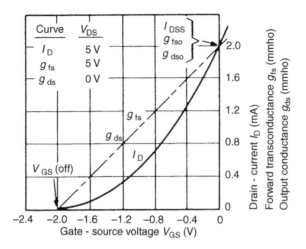

Figure 2.100   JFET I_D, g_{fs}, and g_{ds}.

The relationship between g_{ds0}, $V_{GS(off)}$, and I_{DSS} is given by the equation

$$g_{ds0} = -2\frac{I_{DSS}}{V_{GS(off)}} \tag{2.182}$$

where I_{DSS} and $V_{GS(off)}$ are as indicated in Figure 2.98. It is important to note that equations (2.181) and (2.182) and the g_{ds} curve of Figure 2.100 are valid only for the case where V_{DS} is very small compared to V_P. Drain-source conductance at higher values of V_{DS} will be discussed later. FETs designed to be used as voltage-controlled resistors typically have a high $V_{GS(off)}$ because the $V_{DS}/(V_{GS} - V_{GS(off)})$ ratio should be low to keep distortion low. The actual large-signal JFET model, as used in most simulators, is shown in Figure 2.101, including a list of its intrinsic key words (Table 2.13). Again, we had very little luck using this model as published by several SPICE CAD software manufacturers as offered. Although we cannot judge on the quality of the parameters that came with the software, it was not possible to closely match even optimized sets of parameters to this large-signal model and obtain acceptable results. However, since the modified Materka model worked very well for this, we recommend not to use this popular JFET model at frequencies above 10 MHz.

2.4.1   Small-Signal Behavior of JFETs

Figure 2.102 shows the small-signal equivalent transistor model we would recommend. It consists of an empirical FET model and a package model. Table 2.14 lists its key words. Its advantage is that its built-in noise model is very accurate, even for JFETs and metal semiconductor FETs.

Note 1: The transconductance of this model may be approximately described by

$$g_m = G\frac{e^{-j\omega T}}{1 + jf/F}$$

where $\omega = 2\pi f$ and f = frequency.

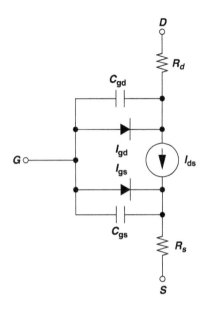

Figure 2.101   The nonlinear JFET model. Table 2.13 lists its intrinsic and extrinsic parameters. In our opinion, this model is limited to frequencies below 10 MHz; above this frequency, the modified Materka model should replace it.

Table 2.13   Large-Signal JFET Model Key Words

Key Word	Description	Unit	Default
	Intrinsic Model		
NJF/PJF	Channel-type selection (NJF=N-channel, PJF=P-channel)		NJF
VT0	Threshold voltage	volt	−2.0
BETA	Transconductance coefficient	amp/volt2	1.0e-4
LAMB	Channel-length modulation	/volt	0.0
IS	Gate-junction saturation current	amp	1.0e-14
PB	Gate-junction potential	volt	1.0
FC	Forward-bias depletion capacitance coefficient		0.5
CGS	Zero-bias gate-source junction capacitance	farad	0.0
CGD	Zero-bias gate-drain junction capacitance	farad	0.0
T	Channel transit-time delay		0.0
KF	Flicker noise coefficient		0.0
AF	Flicker noise exponent		1.0
FCP	Flicker noise frequency shape factor		1.0
SN	Switch to turn device shot noise on (1) or off (0)		1
AREA	Area multiplier		1.0
NOIS	Reference label to a set of noise data		
NAME	Required user-specified name up to 8 characters		
	Extrinsic Model		
RD	Drain ohmic resistance	ohm	0.0
RS	Source ohmic resistance	ohm	0.0

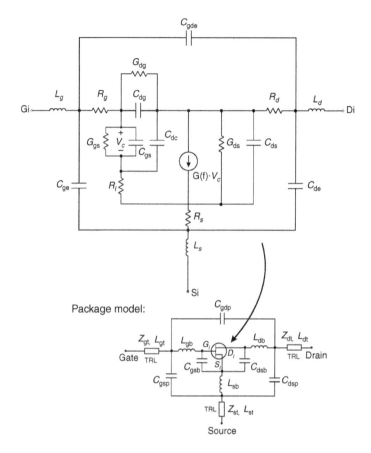

Figure 2.102   Linear FET model.

Note 2: The flicker noise frequency dependence is given by

$$\frac{1}{(f/F_c)^{\text{FCP}}}$$

Electrical noise generated within a junction FET is usually represented by equivalent noise sources, $\bar{E}_n$ and $\bar{i}_n$. Both noise voltage $\bar{E}_n$ and noise current $\bar{i}_n$ are frequency dependent and have the characteristics shown in Figure 2.103.

An equivalent noise circuit is shown in Figure 2.104. Above the frequency f_1, $\bar{E}_n$ is approximately given by

$$\bar{e}_n \simeq \left(4KTB\frac{0.67}{g_{fs}}\right)^{1/2} \tag{2.183}$$

where $K = 1.374 \times 10^{-23}$ J/K, $T = $ absolute temperature in Kelvins (273 K $= 0°$C), $B = $ frequency range in Hertz, and $g_{fs} = $ transconductance of FET.

With the input short circuited, the noise voltage across the load R_L resulting from the FET is

$$\text{Output noise voltage} = \bar{e}_n A_V \tag{2.184}$$

Table 2.14 Small-Signal JFET Model Key Words

Key Word	Description	Unit	Default
	Intrinsic Model		
G	Transconductance at dc, G_0 (see Note 1)	/ohm	
CGS	Gate-source capacitance	farad	
F	3-dB roll-off frequency	Hz	∞
T	Time delay	s	0.0
TDS	Drain source time delay	s	0.0
GGS	Gate-source conductance	/ohm	0.0
CDG	Drain-gate capacitance	farad	0.0
CDC	Dipole layer capacitance	farad	0.0
CDS	Drain-source capacitance	farad	0.0
GDS	Drain-source conductance	/ohm	0.0
RI	Channel resistance	ohm	0.0
RG	Gate resistance	ohm	0.0
RD	Drain resistance	ohm	0.0
RS	Source resistance	ohm	0.0
CGE	External gate capacitance	farad	0.0
CDE	External drain capacitance	farad	0.0
LG	Gate lead inductance	henry	0.0
LD	Drain lead inductance henry	0.0	
LS	Source lead inductance henry	0.0	
CGDE	External gate-drain capacitance	farad	0.0
GDG	Gate drain conductance	/ohm	0.0
TJ	Chip temperature	K	298
	Package Parasitics		
LGB	Gate wirebond inductance	henry	0.0
LDB	Drain wirebond inductance	henry	0.0
LSB	Source wirebond inductance	henry	0.0
CGSB	Gate bondpad to source capacitance	farad	0.0
CDSB	Drain bondpad to source capacitance	farad	0.0
CGSP	Gate to source package capacitance	farad	0.0
CDSP	Drain to source package capacitance	farad	0.0
CGDP	Gate to drain package capacitance	farad	0.0
ZGT	Gate transmission line impedance	ohm	50
ZDT	Drain transmission line impedance	ohm	50
ZST	Source transmission line impedance	ohm	50
LGT	Gate transmission line length for at $\varepsilon_r = 1$	meter	0.0
LDT	Drain transmission line length for at $\varepsilon_r = 1$	meter	0.0
LST	Source transmission line length for at $\varepsilon_r = 1$	meter	0.0
FC	Corner frequency of flicker $(1/f)$ noise (See Note 2)	hertz	10 MHz
FCP	Shape factor of the $1/f$ noise response		1.0
Label	User-defined term that refers to temperature coefficient		

Notes:

1. The transconductance of this model may be approximately described by $g_m = G \dfrac{e^{-j\omega T}}{1 + j(f/F)}$ where $\omega = 2\pi f$ and f is frequency.

2. The flicker noise frequency dependence is given by $1/(f/F_c)^{\text{FCP}}$.

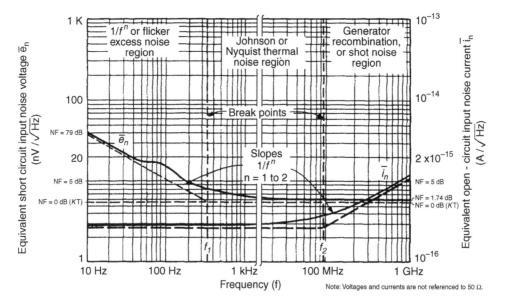

Figure 2.103 JFET noise characteristics.

Below the frequency f_1, $\bar{e}_n$ increases proportional to $1/f^n$ and is expressed as

$$\bar{e}_n = \left[4KTB \left(\frac{0.67}{g_{fs}} \right) \left(1 + \frac{f_1}{f^n} \right)^{1/2} \right] \qquad (2.185)$$

The low-frequency corner frequency f_1 for JFETs is typically in the 100 Hz to 1 kHz range, and the exponent n is usually between 1 and 2. As indicated by the equations, $\bar{E}_n$ is inversely proportional to the square root of g_{fs}.

The equivalent input noise current $\bar{i}_n$ is caused by the current in the gate-to-channel junction. Its approximate value below f_2 is

$$\bar{i}_n = \left(2qI_G B \right)^{1/2} \qquad (2.186)$$

where $q = 1.602 \times 10^{-19}$, $B = $ frequency range in hertz, and $I_G = $ dc gate current.

This expression is fairly accurate when I_G is the result of the active device conductance. Typically, $\bar{i}_n$ will be lower than the calculated value because part of I_G is due to conductance across the device package.

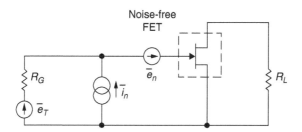

Figure 2.104 Equivalent noise circuit of the FET.

At higher frequencies (above f_2)

$$\bar{i}_n = \left(\frac{4KTB}{R_p}\right)^{1/2} \tag{2.187}$$

where R_P = real part of gate input impedance. R_P, in terms of Y_{11}, can range from several tens of mΩ (at audio frequencies) to 1 kΩ or less (at VHF/UHF frequencies). The high-frequency corner f_2 is typically in the range of 5–50 kHz and more than 100 MHz for high-frequency devices.

Another form of noise is known as "popcorn" or burst noise, the causes of which have not been completely identified. It shows up as a random short-duration step-function change in drain current, equivalent to an input gate-source voltage change of a few tenths of a microvolt. It is not unlike big bubbles on the surface of boiling water.

As far as the noise is concerned, we can also state that the Materka built-in proprietary noise model is significantly more accurate than the one mentioned above [26–29]. This model is implemented, with minimal documentation of its inner workings because of its proprietary nature, in the linear portion of the circuit simulator in Ansoft Designer. In many cases, the manufacturer supplies only limited data, so it is most convenient that after SPICE-type parameter extraction is done by using the Scout program, one even gets good noise prediction for JFETs using the modified Materka model.

2.4.2 Large-Signal Behavior of MOSFETs

Metal-oxide-semiconductor field-effect transistors (MOSFETs) are important components in contemporary analog integrated circuits. While initial applications were centered on all-MOS processes, combined bipolar and MOS processes give the designer the best of both worlds. A major advantage of MOS processes for realizing analog functions is that complex, dense digital functions can be realized on the same chip.

2.4.2.1 Transfer Characteristics of MOS Devices

A cross section of a typical enhancement-mode *n*-channel MOS transistor (NMOS) is shown in Figure 2.105a. Heavily doped *n*-type source and drain regions are fabricated in a *p*-type substrate (often called the body). A thin layer of silicon dioxide is grown over the substrate material and a conductive gate material (metal or polycrystalline silicon) covers this oxide between source and drain. The operation of the device is very similar to that of a JFET in that the gate-source voltage is used to modify the conductance of the region under the gate. This allows the gate voltage to control the current flowing between source and drain, giving gain in analog circuits and switching characteristics in digital circuits.

The enhancement-mode NMOS device of Figure 2.105a shows significant conduction between source and drain only when an *n*-type channel exists under the gate, and this is the origin of the " *n*-channel" designation. The term "enhancement mode" refers to the fact that no conduction occurs for $V_{GS} = 0$, and thus the channel must be "enhanced" to cause conduction. MOS device can be made equally well by using an *n*-type substrate with a *p*-type conducting channel. Such devices are called enhancement-mode *p*-channel MOS transistors (PMOS). In technologies employing one or the other of these device types, the circuit symbol of Figure 2.105b is commonly used for either one. In complementary MOS technology (CMOS), both device types are present and the circuit symbols of Figure 2.105c is used to distinguish them. In PMOS or NMOS technologies, the substrate is common to

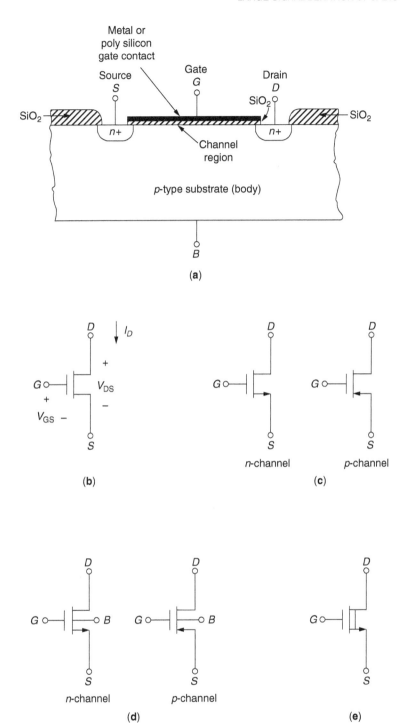

Figure 2.105 (a) Typical enhancement-mode NMOS structure. (b) Enhancement-mode NMOS or PMOS circuit symbol when one device type only is present. (c) NMOS and PMOS symbols used in CMOS circuits. (d) NMOS and PMOS symbols used when the substrate connection is nonstandard. (e) Depletion MOS device symbol.

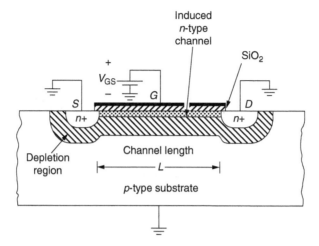

Figure 2.106 Idealized NMOS device cross section with positive V_{GS} applied, showing depletion regions and the induced channel.

all devices, invariably connected to a dc power supply voltage, and usually not shown on the circuit diagram. In CMOS technology, however, devices or one type or another are fabricated in individual, separate isolation regions, which may or may not be connected to a power supply voltage. If these isolation regions *are* connected to the appropriate power supply, the symbols of Figure 2.105c will be used and the substrate connection will not be shown. If the individual isolation regions are connected elsewhere, however, the devices will be represented by the symbols of Figure 2.105d, where the substrate is labeled *B*. Finally, in NMOS technology, an additional device type called a "depletion-mode" device is usually available. This is a conducting channel implanted between the source and the drain so that conduction occurs for $V_{GS} = 0$. This device has characteristics that are almost identical to that of a JFET and will be represented by the symbol of Figure 2.105e.

The derivation of the transfer characteristics of the enhancement-mode NMOS device of Figure 2.105a begins by noting that with $V_{GS} = 0$, the source and drain regions are separated by back-to-back *pn* junctions. These junction are formed between the *n*-type source and drain regions and the *p*-type substrate, resulting in an extremely high resistance (about 1012 Ω) between drain and source when the device is off.

Now consider substrate, source, and grain grounded and a positive voltage V_{GS} applied to the gate as shown in Figure 2.106. The gate and substrate then form the plates of a capacitor with the SiO$_2$ as a dielectric. Positive charge accumulates on the gate and the negative charge on the substrate. Initially, the negative charge in the *p*-type substrate is manifested by creation of a *depletion region* and resulting exclusion of holes under the gate. This is shown in Figure 2.106. The depletion-layer width X under the oxide is

$$X = \left(\frac{2\epsilon\phi}{qN_A} \right)^{1/2} \tag{2.188}$$

where ϕ is the potential in the depletion layer at the oxide–silicon interface, N_A atoms/cm^3 is the doping density (assumed constant) of the *p*-type substrate, and ε is the permittivity of the silicon. The charge per unit area in this depletion region is

$$Q = qN_A X = \sqrt{2qN_A\epsilon\phi} \tag{2.189}$$

When the potential in the silicon reaches a critical value equal to twice the Fermi level $\phi_f \approx 0.3$ V, a phenomenon known as "inversion" occurs [30]. Further increases in gate voltage produce no further changes in the depletion-layer width but instead a thin layer of electronics is induced in the depletion layer directly under the oxide. This produces a continuous n-type region with the source and drain regions and is the conducting channel between source and drain. This channel can then be modulated by increases or decreases in the gate voltage. In the presence of an inversion layer, and with no substrate bias, the depletion region contains a fixed charge

$$Q_{b0} = \sqrt{2qN_A\epsilon 2\phi_f} \tag{2.190}$$

If a substrate-bias voltage V_{SB} (source positive for n-channel devices) is applied between source and substrate, the potential required to produce inversion becomes $(2\phi_f + V_{SB})$ and the charge stored in the depletion region, in general, is

$$Q_b = \sqrt{2qN_A\epsilon(2\phi_f + V_{SB})} \tag{2.191}$$

The gate voltage V_{GS}, required to produce an inversion layer, is called the threshold voltage V_t and can now be calculated. This voltage consists of several components. First, a voltage $[2\phi_f + (Q_b/C_{ox})]$ is required to sustain the depletion-layer charge Q_b, where C_{ox} is the gate oxide capacitance per unit area. Second, a work-function difference ϕ_{ms} exists between the gate metal and the silicon. Third, there is always charge density Q_{ss} (positive) in the oxide at the silicon interface. This is caused by crystal discontinuities at the Si–SiO$_2$ interface and must be compensated by a gate voltage contribution of $-Q_{ss}/C_{ox}$. Thus, we have a threshold voltage

$$
\begin{aligned}
V_t &= \phi_{ms} + 2\phi_f + \frac{Q_b}{C_{ox}} - \frac{Q_{ss}}{C_{ox}} \\
&= \phi_{ms} + 2\phi_f + \frac{Q_{b0}}{C_{ox}} - \frac{Q_{ss}}{C_{ox}} + \frac{Q_b - Q_{b0}}{C_{ox}} \tag{2.192} \\
&= V_{t0} + \gamma\left(\sqrt{2\phi_f + V_{SB}} - \sqrt{2\phi_f}\right) \tag{2.193}
\end{aligned}
$$

where (2.190) and (2.191) have been used and V_{t0} is the threshold voltage with $V_{SB} = 0$. The parameter γ is defined as

$$\gamma = \frac{1}{C_{ox}}\sqrt{2q\epsilon N_A} \tag{2.194}$$

and

$$C_{ox} = \frac{\epsilon_{ox}}{t_{ox}} \tag{2.195}$$

where ε_{ox} and t_{ox} are the permittivity and thickness of the oxide, respectively. A typical value of γ is 0.5 V$^{1/2}$ and $C_{ox} = 3.5 \times 10^{-4}$ pF/μm^2 for $t_{ox} = 0.1$ μm.

In practice, the value of V_{t0} is usually adjusted in processing by implanting additional impurities into the channel region. Extra p-type impurities are implanted in the channel to

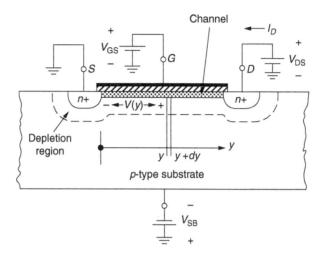

Figure 2.107   NMOS device with bias voltages applied.

make $V_{t0} \approx 0.5$ to 1.5 V for n-channel enhancement devices. By implanting n-type impurities in the channel region, a conducting channel can be formed, even for $V_{GS} = 0$. This MOS transistor is called a depletion device, with typical values of V_{t0} in the range -1 to -4 V. If Q_i is the charge density per unit area due to the implant, then the threshold voltage given by (2.192) is shifted by approximately Q_i/C_{ox}.

The preceding equations can now be used to calculate the large-signal characteristics of an NMOS transistor. For purposes of analysis, the source is assumed grounded and bias voltages V_{GS}, V_{DS}, and V_{SB} are applied as shown in Figure 2.107. If V_{GS} is greater than V_t, a conducting channel exists and V_{DS} causes a larger reverse bias from drain to substrate than exists from source to substrate, and thus a wider depletion region exists at the drain. However, for simplicity, we assume that the voltage drop along the channel itself is small so that the depletion layer is constant along the channel.

At a distance y along the channel, the voltage with respect to the source is $V(y)$ and the gate-to-channel voltage at that point is $V_{GS} - V(y)$. We assume that this voltage exceeds the threshold voltage V_t, and thus the induced electron charge per unit area in the channel is

$$Q_I(y) = C_{ox}[V_{GS} - V(y) - V_t] \qquad (2.196)$$

The resistance dR of a length dy of the channel is

$$dR = \frac{dy}{W\mu_n Q_I(y)} \qquad (2.197)$$

where W is the width of the device, perpendicular to the plane of Figure 2.107, and μ_n is the average electron mobility of the channel.

The voltage drop dV along the length of the channel dy is

$$dV = I_D dR = \frac{I_D}{W\mu_n Q_I(y)} dy \qquad (2.198)$$

If L is the total channel length, then substitution of (2.196) in (2.198) and integration gives

$$\int_0^L I_D dy = \int_0^{V_{DS}} W\mu_n C_{ox}(V_{GS} - V - V_t)dV \qquad (2.199)$$

This results in

$$I_D = \frac{k'}{2}\frac{W}{L}\left[2(V_{GS} - V_t)V_{DS} - V_{DS}^2\right] \qquad (2.200)$$

where

$$k' = \mu_n C_{ox} = \frac{\mu_n \epsilon_{ox}}{t_{ox}} \qquad (2.201)$$

Equation (2.200) is important and describes the I–V characteristics of an MOS transistor, assuming a continuous induced channel. A typical value of k' for $t_{ox} = 0.1\ \mu$m is about $20\ \mu$A/V^2 for an n-channel device.

As the value of V_{DS} is increased, the induced conducting channel narrows at the drain end and (2.196) indicates that Q_I at the drain end approaches zero as V_{DS} approaches $(V_{GS} - V_t)$. This results in the same pinch-off phenomenon as occurs in a JFET, and further increases in V_{DS} produce little change in I_D. Equation (2.200) is thus no longer valid if V_{DS} is greater than $(V_{GS} - V_t)$. The value of I_D in this region is obtained by substituting $V_{DS} = (V_{GS} - V_t)$ in (2.200), giving

$$I_D = \frac{k'}{2}\frac{W}{L}(V_{GS} - V_t)^2 \qquad (2.202)$$

for an MOS transistor in the pinch-off region. As in the case of a JFET, the drain current in the pinch-off region varies slightly as the drain voltage is varied. This is due to the presence of a depletion region between the physical pinch-off point in the channel at the drain end and the drain region itself. If this depletion-layer width is X_d, then the *effective* channel length is given by

$$L_{eff} = L - X_d \qquad (2.203)$$

If L_{eff} is used in place of L in (2.202), we obtain a more accurate formula for the pinch-off region

$$I_D = \frac{k'}{2}\frac{W}{L_{eff}}(V_{GS} - V_t)^2 \qquad (2.204)$$

The fact that X_d (and thus L_{eff}) are functions of the drain, source voltage results in a variation of I_D and V_{DS} in the pinch-off region. Using (2.203) and (2.204), we obtain

$$\frac{\partial I_D}{\partial V_{DS}} = -\frac{k'}{2}\frac{W}{L_{eff}^2}(V_{GS} - V_t)^2\frac{dL_{eff}}{dV_{DS}} \qquad (2.205)$$

and thus

$$\frac{\partial I_D}{\partial V_{DS}} = \frac{I_D}{L_{eff}}\frac{dX_d}{dV_{DS}} \qquad (2.206)$$

This equation is analogous to (2.151) for bipolar transistors. Following a similar procedure, a voltage analogous to the early voltage can be defined as

$$V_A = \frac{I_D}{\partial I_D / \partial V_{DS}} \tag{2.207}$$

and thus

$$V_A = L_{\text{eff}} \left(\frac{dX_d}{dV_{DS}} \right)^{-1} \tag{2.208}$$

For MOS transistors, the most widely used parameter for the characterization of output resistance is

$$\lambda = \frac{1}{V_A} \tag{2.209}$$

As in the bipolar case, the large-signal properties of the transistor can be approximated by assuming that λ and V_A are constants, independent of bias conditions. Thus, we can formulate a better approximation to the I–V characteristics as

$$I_D = \frac{k'}{2} \frac{W}{L} (V_{GS} - V_t)^2 (1 + \lambda V_{DS}) \tag{2.210}$$

In practical, MOS transistors variation of X_d with voltage is complicated by the fact that the field distribution in the drain depletion region is not one dimensional. A vertical component in the field distribution is introduced by the potential difference between the gate and channel and the gate and drain. As a result, the calculation of λ from the device structure is quite different [31], and it is usually necessary to develop effective values of λ from experimental data. The parameter λ is a linear function of effect channel length and is an increasing function of the doping level in the channel. Typical values of λ are in the range 0.05–0.005 V^{-1}. Note that, in the case of a JFET, the pinch-off region for MOS devices is often called the *saturation* region.

A plot of I_D versus V_{DS} for an NMOS transistor is shown in Figure 2.108. Below pinch-off, the device behaves as a nonlinear voltage-controlled resistor, which is often called the *ohmic* or *triode* region. Above pinch-off, the device approximates a voltage-controlled current source. Note that for depletion MOS devices, V_t is negative and I_D is finite, even for $V_{GS} = 0$. For PMOS devices, all polarities of voltage and current are reversed.

The results as previously derived can be used to form a large-signal model of an MOS transistor. The model topology in the pinch-off region is the same as that for the JFET, but using (2.202) for the controlled-current generator.

2.4.2.2 MOS Device Voltage Limitations

The voltage limitations of MOS transistors depend on the gate length L. For small values of L (less than about 10 μm), the drain-depletion region exerts an appreciable influence on the channel. This causes I_D to rise with increasing V_{DS} in a similar fashion to the bipolar transistor curves of Figure 2.90.

For long channel lengths, the drain-depletion region has little effect on the channel and the I_D-versus-V_{DS} curves follow closely the ideal curves of Figure 2.108. Eventually, the drain-substrate pn-junction breakdown voltage is exceeded and a sharp breakdown characteristic is obtained, which is similar to the JFET characteristic.

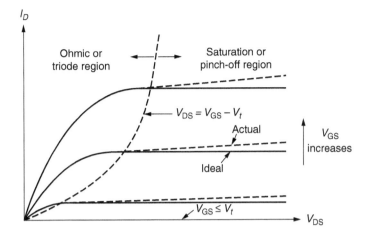

Figure 2.108   NMOS device characteristics.

In addition to V_{DS} limitations, MOS devices must also be protected against excessive gate voltages. Typical gate oxides break down with about 25–50 V applied from gate to channel, and this process is destructive to the transistor.

2.4.3   Small-Signal Model of the MOS Transistor in Saturation

The preceding large-signal equations can now be used to derive the small-signal model of the MOS transistor in the saturation or pinch-off region. The important equations are (2.207) and (2.193). Note from (2.193) that the source-substrate voltage V_{BS} affects V_t, and thus I_D. This is due to the influence of the substrate acting as a second gate and is called *body effect*. As a consequence, I_D is a function of both V_{GS} and V_{BS}, and we require *two* transconductance generators in the small-signal model as shown in Figure 2.17. Variations in the voltage V_{bs} from source to body cause current gm_{bvbs} to flow from drain to source. Note that the body (or substrate) of an NMOS integrated circuit is usually connected to the most negative supply voltage and is thus an ac ground. However, the source connection can have a significant ac voltage impressed upon it. Parasitic resistances due to the channel contact regions should be included in series with the source and drain of the model, but are usually neglected in hand calculations. These resistances have an inverse dependence on channel width W and have typical values of 50–100 Ω for devices with W of about 1 μm.

The parameters of Figure 2.109 can be determined from (2.210) by differentiating.

$$g_m = \frac{\partial I_D}{\partial V_{GS}} = k'\frac{W}{L}(V_{GS} - V_t)(1 + \lambda V_{DS}) \tag{2.211}$$

If $\lambda V_{DS} \ll 1$, this is often approximated as

$$g_m = k'\frac{W}{L}(V_{GS} - V_t) = \sqrt{2k'\frac{W}{L}I_D} \tag{2.212}$$

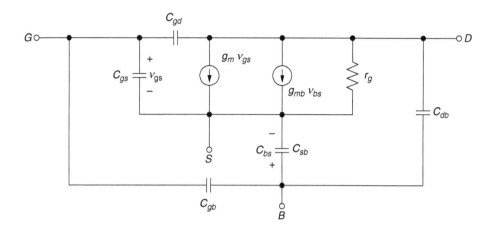

Figure 2.109   Small-signal MOS transistor equivalent circuit.

Like the JFET and unlike the bipolar transistor, the transconductance of the MOSFET depends on both bias current and the W/L ratio (also on the oxide thickness via k'). Similarly

$$g_{mb} = \frac{\partial I_D}{\partial V_{BS}} = -k' \frac{W}{L}(V_{GS} - V_t)(1 + \lambda V_{DS})\frac{\partial V_t}{\partial V_{BS}} \qquad (2.213)$$

From (2.193)

$$\frac{\partial V_t}{\partial V_{BS}} = -\frac{\gamma}{2\sqrt{2\phi_f + V_{SB}}} = -\chi \qquad (2.214)$$

where this equation defines a factor χ. This parameter is equal to the rate of change of threshold voltage with body bias voltage. One obtains

$$\chi = \frac{C_{js}}{c_{ox}} \qquad (2.215)$$

where C_{js} is the capacitance per unit area of the depletion region under the channel, assuming a one-sided step junction with a built-in potential $\psi_0 = 2\phi_f$.

Substitution of (2.214) in (2.213) gives

$$g_{mb} = \frac{\gamma k'(W/L)(V_{GS} - V_t)(1 + \lambda V_{DS})}{2\sqrt{2\phi_f + V_{SB}}} \qquad (2.216)$$

Again, if $\lambda V_{DS} \ll 1$, we have

$$g_{mb} = \frac{\gamma\sqrt{k'(W/L)I_D}}{\sqrt{2(2\phi_f + V_{SB})}} \qquad (2.217)$$

An important quantity is the ratio g_{mb}/g_m, and from (2.211) and (2.216) we find

$$\frac{g_{mb}}{g_m} = \frac{\gamma}{2\sqrt{2\phi_f + V_{SB}}} = \chi \qquad (2.218)$$

The factor χ is typically in the range 0.1–0.3.

Finally, the small-signal output resistance can be obtained directly from (2.205)

$$r_o = \left(\frac{\partial I_D}{\partial V_{DS}}\right)^{-1} = \frac{L_{eff}}{I_D}\left(\frac{dX_d}{dV_{DS}}\right)^{-1} \qquad (2.219)$$

and using (2.208) and (2.209), we find

$$r_o = \frac{1}{\lambda I_D} = \frac{V_A}{I_D} \qquad (2.220)$$

Small-signal model capacitances are also shown in Figure 2.109. Of these, only the gate-source capacitance C_{gs} is intrinsic to the device operation in the saturation region. Capacitances C_{sb} and C_{db} are parasitic depletion-region capacitances between the substrate and the source and drain regions, respectively. These can be expressed as follows:

$$C_{sb} = \frac{C_{sb0}}{\left(1 + \dfrac{V_{SB}}{\Psi_0}\right)^{1/2}} \qquad (2.221)$$

$$C_{db} = \frac{C_{db0}}{\left(1 + \dfrac{V_{DB}}{\Psi_0}\right)^{1/2}} \qquad (2.222)$$

These capacitances are proportional to the gate and source region areas (including sidewalls), and C_{sb} also includes depletion-region capacitance from the induced channel in the body.

Capacitance C_{gb} between gate and substrate models parasitic oxide capacitance between the gate contact material and the substrate outside the active-device area. This is a constant capacitance, and models coupling between polysilicon and metal interconnects and the underlying substrate. In fact, parasitic capacitance of this type underlies all polysilicon and metal traces on the chip and should be taken into account when simulating and calculating high-frequency circuit and device performance. Typical values depend on oxide thicknesses and range from about 0.04 to 0.15 fF per square micrometer of interconnect, with fringing effects becoming important for narrow lines (several micrometers or less in width).

Capacitances C_{gs} and C_{gd} exist from gate to source and drain, respectively. If C_{ox} is the oxide capacitance per unit area from gate to channel then the total capacitance under the gate is $C_{ox}WL$. This capacitance is intrinsic to the device operation and models the gate control of the channel conductance. In the ohmic region of device operation, this capacitance is split equally between source and drain so that $C_{gs} = C_{gd} = 1/2C_{ox}WL$. However, in the saturation region, the channel is very narrow at the drain end and the drain voltage exerts little influence on either the channel or the gate charge. As a consequence, the intrinsic portion of C_{gd} is essentially zero in the saturation region and C_{gd} then consists of a constant parasitic oxide-capacitance contribution due to gate overlap of the drain region. This is on the order of 1^{-10} fF for small devices.

In order to calculate the corresponding value of C_{gs} in the saturation region, we must calculate the total charge Q_T stored in the channel. This can be obtained by integrating (2.196) to obtain

$$Q_T = WC_{ox} \int_0^L [V_{GS} - V(y) - V_t]dy \tag{2.223}$$

Substituting for dy/dV from (2.198) in (2.223), we find

$$Q_T = \frac{W^2 C_{ox}^2 \mu_n}{I_D} \int_0^{V_{GS}-V_t} (V_{GS} - V - V_t)^2 dV \tag{2.224}$$

where the limit $y = L$ corresponds to $V = (V_{GS} - V_t)$ in saturation. Solution of (2.224) and use of (2.201) and (2.202) gives

$$Q_T = \frac{2}{3} WLC_{ox}(V_{GS} - V_t) \tag{2.225}$$

and thus

$$C_{gs} = \frac{\partial Q_T}{\partial V_{GS}} = \frac{2}{3} WLC_{ox} \tag{2.226}$$

In addition, there is a contribution to C_{gs} from the constant parasitic oxide capacitance due to gate overlap of the source region.

The f_T of the MOSFET is given by

$$f_T = \frac{1}{2\pi} \frac{g_m}{C_{gs} + C_{gd} + C_{gb}} \tag{2.227}$$

The dependence of MOSFET f_T on device and process parameters can be seen by assuming that the intrinsic device capacitance C_{gs} dominates. Thus, from (2.227) we have

$$f_T = \frac{1}{2\pi} \frac{g_m}{C_{gs}} \tag{2.228}$$

Substituting in (2.228) for g_m from (2.212) and C_{gs} from (2.226), we find for a MOSFET

$$f_T = 1.5 \frac{\mu_n}{2\pi L^2} (V_{GS} - V_t) \tag{2.229}$$

It is interesting to compare this with the intrinsic f_T of a bipolar transistor when parasitic depletion-layer capacitance is neglected. From

$$f_T = \frac{1}{2\pi \tau_F} \tag{2.230}$$

and substituting for τ_f and using the Einstein relationship $D_n/\mu_n = kT/q = V_t$, we find for a bipolar transistor

$$f_T = 2\frac{\mu_n}{2\pi W_B^2}V_T \tag{2.231}$$

The similarity in form between (2.229) and (2.231) is striking. In both cases, the intrinsic device f_T increases as the inverse square of the critical device dimension across which carriers are in transit. The voltage $V_T = 26$ mV is fixed for a bipolar transistor, but MOSFET f_T can be increased by operating at high values of $(V_{GS} - V_t)$. Note that the base width W_B in a bipolar transistor is a vertical dimension determined by diffusions or implants and can typically be made much smaller than the channel length L of a MOSFET, which depends on surface geometry and photolithographic processes. Thus, bipolar transistors generally have higher f_T than MOSFETs made with comparable processing. Finally, (2.229) was derived assuming that the MOSFET square law is valid. However, as discussed in Section 2.4.4, submicrometer MOSFETs depart significantly from square-law characteristics, and we find that for such devices f_T increases as L^{-1} rather than L^{-2}.

2.4.4 Short-Channel Effects in FETs

The evolution of integrated-circuit processing techniques has led to continuing reductions in both the horizontal and vertical dimensions of the active devices (the minimum allowed dimension of passive devices has also increased). This trend is driven primarily by economics in that more devices and circuits can be processed at one time on a given wafer. A second benefit has been that the frequency capability of the active devices continues to increase, as intrinsic f_T values increase with smaller dimensions while parasitic capacitances decrease.

Vertical dimensions such as the base width of a bipolar transistor in production processes may now be on the order of 0.05 μm or less, whereas horizontal dimensions such as bipolar emitter width or FET gate length may be in the order of some ten nm. Even at these very small dimensions, the large-signal and small-signal models of bipolar transistors given in previous sections remain valid. However, significant short-channel effects become important in FETs of all types at channel lengths on the order of 1 μm or less and require modifications to the FET models given previously. The primary effect is to modify the classical FET square-law transfer characteristic in the saturation region to make the device more closely approach an ideal linear transfer function. Note, however, that even in processes with submicrometer capability, many of the FETs in a given analog circuit may be deliberately chosen to be larger than the minimum size and may be well approximated by the square-law model.

The most important short-channel effect in FETs is due to velocity saturation of carriers in the channel [32]. At low electric field values, the linear relation between carrier velocity and field implied by (2.197) and (2.198) is valid. At high fields, however, the carrier velocities approach the thermal velocities and subsequently the carrier velocities increase more slowly with increasing field. This is illustrated in Figure 2.110, which shows typical measured electron drift velocity v_d versus tangential electric field strength magnitude $\mathcal{E}$ in an NMOS surface channel. Note that at low field values, the velocity is proportional to the field, while at high fields the velocity approaches a constant value called the scattering-limited velocity v_{scl}. A first-order analytical approximation to this curve is

$$v_d = \frac{\mu_n \mathcal{E}}{1 + \mathcal{E}/\mathcal{E}_c} \tag{2.232}$$

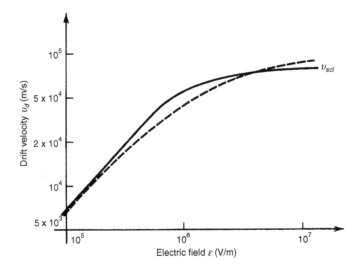

Figure 2.110 Typical measured electron drift velocity v_d versus tangent electric field $\mathcal{E}$ in an MOS surface channel (solid line). Also shown (dotted line) is the analytical approximation of equation (1.204) with $\mathcal{E}_c = 1.5 \times 10^6$ V/m and $\mu_n = 0.07$ m²/Vs.

where $\mathcal{E} \cong 1.5 \times 10^6$ V/m and $\mu_n \cong 0.07$ m²/Vs in the low-field mobility. Equation (2.232) is also plotted in Figure 2.110. From (2.232), as $\mathcal{E} \to \infty$, we have $v_{scl} = \mu_n \mathcal{E}_c$. At the critical field value $\mathcal{E}_c$, the carrier velocity is a factor of 2 less than the low-field formula would predict. In a device with a channel length $L = 1$ μm, we need a voltage drop of only 1.5 V along the channel to have an average field equal to $\mathcal{E}_c$, and this condition is readily achieved in small MOSFETs. Similar results are found for PMOS devices.

As an example of the effects of velocity saturation on FET characteristics, we consider the example of the MOSFET. In the analysis of "Transfer Characteristics of MOS Devices" in Section 2.4.2, we now use the more general expression

$$I_D = W Q_I(y) v_d(y) \tag{2.233}$$

Substituting (2.232) in (2.233) and using

$$I_D = W Q_I(y) v_d(y) \tag{2.234}$$

for the magnitude of the field, we find that

$$I_D \left(1 + \frac{1}{\mathcal{E}_c} \frac{dV}{dy} \right) = W Q_I(y) \mu_n \frac{dV}{dy} \tag{2.235}$$

Note that as $\mathcal{E}_c \to \infty$ and velocity saturation becomes negligible, (2.235) approaches the original equation (2.198).

Integrating (2.235) along the channel, we obtain

$$\int_0^L I_D \left(1 + \frac{1}{\mathcal{E}_c}\frac{dV}{dy}\right) dy = \int_0^{V_{DS}} WQ_I(y)\mu_n dV \tag{2.236}$$

and thus

$$I_D = \frac{\dot{\mu}_n C_{ox}}{2\left(1 + \frac{1}{\mathcal{E}_c}\frac{V_{DS}}{L}\right)}\frac{W}{L}\left[2(V_{GS} - V_t)V_{DS} - V_{DS}^2\right] \tag{2.237}$$

The quantity V_{DS}/L in (2.237) can be interpreted as the average field in the channel. If this is comparable to $\mathcal{E}_c$ the drain current for a given V_{DS} is less than the simple expression (2.200) would predict.

Equation (2.237) is valid in the triode region. The MOSFET transfer function in saturation can be obtained by using $V_{DS} = (V_{GS} - V_t)$ in (2.237) to obtain

$$I_D = \frac{k'}{2[1 + \theta(V_{GS} - V_t)]}\frac{W}{L}(V_{GS} - V_t)^2 \tag{2.238}$$

where $\theta = 1/L\mathcal{E}_c$ and has the dimension V^{-1}. For $L = 1 \ \mu m$, a typical value is $\theta \approx 0.7 \ V^{-1}$. Note that in the presence of velocity saturation effects, the device enters the saturation region for $V_{DS} < (V_{GS} - V_t)$. However, (2.238) still gives a good estimate of the saturation current.

Thus far, we have only considered the effects of the tangential field due to the V_{DS} along the channel when considering velocity saturation effects. However, there exists a normal field originating from the gate voltage that also inhibits channel carrier mobility. Since the normal field depends on the value of V_{GS}, we find that an empirical modification to θ in (2.238) can adequately model this effect. In practice, θ is determined by a best fit to measured device characteristics.

Returning to (2.238), we note that for very short channel lengths, θ becomes large and (2.238) reduces to

$$I_D \propto (V_{GS} - V_t) \tag{2.239}$$

Thus, the FET characteristics tend toward a *linear* transfer function as the channel length becomes very small (less than 1 μm).

Equation (2.238) has a simple circuit representation. Consider the circuit of Figure 2.111, where an ideal square-law MOSFET has a resistance R_{SX} in series with the source of the FET. Assume

$$I_D = \frac{\mu C_{ox}}{2}\frac{W}{L}(V'_{GS} - V_t)^2 \tag{2.240}$$

Now

$$V_{GS} = V'_{GS} + I_D R_{SX} \tag{2.241}$$

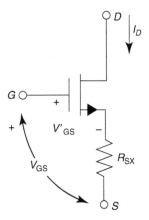

Figure 2.111 Model of velocity saturation of a MOSFET by addition of a series source resistance to an ideal square-law device.

and substituting (2.241) into (2.240) we find that

$$I_D = \frac{\mu C_{\text{ox}}}{2} \frac{W}{L} (V_{\text{GS}} - I_D R_{\text{SX}} - V_t)^2 \tag{2.242}$$

Rearranging (2.242), we find

$$I_D = \frac{\mu C_{\text{ox}}}{2 \left[1 + \mu C_{\text{ox}} \dfrac{W}{L} R_{\text{SX}} (V_{\text{GS}} - V_t) \right]} \frac{W}{L} (V_{\text{GS}} - V_t)^2 \tag{2.243}$$

This has exactly the same form as (2.238) if we identify

$$\theta = \mu C_{\text{ox}} \frac{W}{L} R_{\text{SX}} \tag{2.244}$$

Substituting $\theta = 1/L\mathcal{E}_c$ into (2.244), we have

$$R_{\text{SX}} = \frac{1}{\mathcal{E}_c} \frac{1}{\mu C_{\text{ox}}} \frac{1}{W} \tag{2.245}$$

Thus, the influence of velocity saturation on the large-signal characteristics of a FET can be modeled to first order by a resistor R_{SX} in series with the source of an ideal square-law device. Note that R_{SX} varies inversely with W, as does the intrinsic physical series resistance due to source and drain contact regions. Typically, R_{SX} is larger than the physical series resistance. For $W = 2\,\mu\text{m}$, $\mu C_{\text{ox}} = 40\,\mu\text{A/V}^2$, and $\mathcal{E}_c = 1.5 \times 10^6\,\text{V/m}$, we find $R_{\text{SX}} = 8\,\text{k}\Omega$.

The foregoing analysis has developed a modified large-signal mode for the MOSFET including velocity saturation effects. Small-signal MOSFET modeling for small devices can still be done using the equivalent circuit of Figure 2.109 if the values of g_m and g_{mb} are modified to account for the effects of velocity saturation using (2.238).

Figure 2.112 shows a cross-section of VMOS (vertical MOS) and DMOS (double-diffused MOS) transistors. The probably most popular device of this family is the LDMOS (laterally diffused MOS) power transistor (Figures 2.113 and 2.114). LDMOS

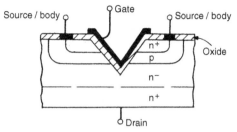

(a) VMOS (vertical channel MOS)

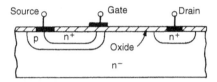

(b) DMOS (double diffused MOS)

Figure 2.112   Cross-section of (a) VMOS and (b) DMOS FETs.

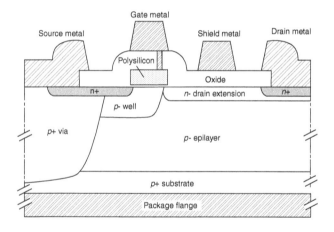

Figure 2.113   Cross-section of a Philips (now NXP) LDMOS FET [33].

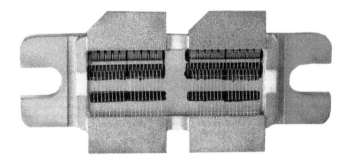

Figure 2.114   Photo of the underside of an LDMOS FET.

FETs operate from a single supply (approximately 8 V and up) and are available for output powers surpassing 100 W at 2 GHz. The feedback capacitance is much lower than VMOS, TMOS, and earlier models, significantly improving circuit stability.

2.4.5 Small-Signal Models of MOSFETs

2.4.5.1 Subthreshold Conduction in MOSFETs

The MOSFET analysis of Section 2.4.2 focused on the normal region of operation where this is a well-defined conducting channel under the gate. Changes in the gate voltage are assumed to cause changes in the channel charge only, and not in the depletion region below. However, for gate voltages less than the extrapolated threshold voltage V_t, the applied gate potential affects both the depletion region charge and the channel charge, which is then very small but not zero. The device can thus conduct finite (but small) current for $V_{GS} < V_t$, so that (2.202) is not valid in this region. The electrons in the n^+ source region of an NMOS transistor can overcome the potential barrier to the p-type substrate and enter the channel region. This process is very similar to the turn-on of a bipolar transistor, and in fact the MOSFET characteristics in this subthreshold region (also called weak inversion) are very similar to those of a bipolar transistor. Analysis [34] shows that in the subthreshold region, MOSFET characteristics can be defined by the equation

$$I_D = k_x \frac{W}{L} e^{v_{GS}/nV_T}(1 - e^{-V_{DS}/V_T}) \tag{2.246}$$

where k_x depends on process parameters and $n \cong 1.5$. The value of n is not equal to unity in this case (as it is for the bipolar transistor) because the applied voltage V_{GS} appears partially at the silicon surface and partially across the depletion layer.

To illustrate this effect, we show measured NMOS characteristics plotted on three different scales in Figures 2.115–2.117. In Figure 2.115, we show the transfer characteristic in the

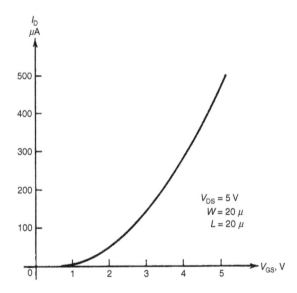

Figure 2.115 Measured NMOS transfer characteristic in the forward-active region.

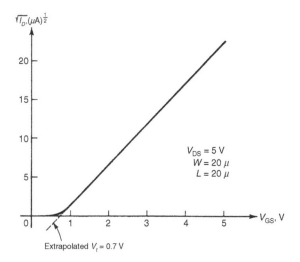

Figure 2.116 Data from Figure 2.115 plotted as $\sqrt{I_D}$ versus VGS showing the square-law characteristic.

forward-active region plotted on linear scales. For this device, $W = 20\,\mu\text{m}$ and $L = 20\,\mu\text{m}$ and short-channel effects are negligible. The same data is plotted in Figure 2.116 as $\sqrt{I_D}$ versus V_{GS}. The resulting straight line shows that the device characteristic is close to an ideal square law. Plots like this are commonly used to obtain V_t by extrapolation (0.7 V in this case) and also k' from the slope of the curve (54 $\mu\text{A/V}^2$ in this case). Note that near the threshold voltage, the curve deviates from the straight line representing the square law. This is the subthreshold region. The data are plotted a third time in Figure 2.117 on log-linear

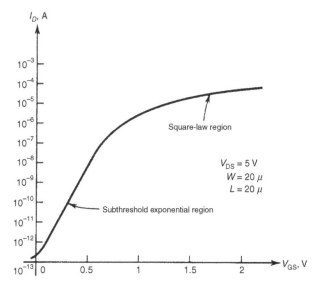

Figure 2.117 Data from Figure 2.115 plotted on log-linear scales showing the exponential characteristic of the subthreshold region.

scales. The straight line obtained for $V_{GS} < V_t$ fits (2.246) with $n = 1.5$. At currents below 10^{-12} A, the effect of leakage currents becomes evident.

The major application of subthreshold operation is in very low power applications at relatively low signal frequencies. The limitation to low signal frequencies occurs because the MOSFET f_T becomes very small. Since the device capacitances at very low bias currents are essentially fixed, and the small-signal g_m calculated from (2.246) becomes proportional to I_D, we see that the value of f_T becomes very small at very low values of I_D.

2.4.5.2 Substrate Flow in MOSFETs

As the reverse-bias voltages on the device are increased, carriers traversing the depletion regions gain sufficient energy to create new hole-electron pairs in lattice collisions by a process known as *impact ionization*. Eventually, at high enough bias voltages, the process results in large avalanche currents. For collector–base bias voltages well below the breakdown value, a small enhanced current flow may occur across the collector–base junction due to this process with little apparent effect on the device characteristics.

Impact ionization also occurs in MOSFETs but has a significantly different effect on the device characteristics. This is because the channel electrons (for the case of NMOS) create hole-electron pairs in lattice collisions in the drain depletion region, and some of the resulting holes then flow to the substrate, creating a substrate current. (The electrons created in the process flow out the drain terminal.) The carriers created by impact ionization are thus not confined within the device as they are in a bipolar transistor. The effect of this phenomenon can be modeled by inclusion of a controlled current generator I_{DB} from drain to substrate, as shown in Figure 2.118 for an NMOS device. The magnitude of this substrate current depends on the voltage across the drain depletion region (which determines the energy of the ionizing channel electrons) and also on the drain current (which is the rate at which channel electrons enter the depletion region). It has been found empirically [35] that the current I_{DB} can be expressed as

$$I_{DB} = K_1(V_{DS} - V_{DS\ sat})I_D e^{-[K_2/(V_{DS} - V_{DS\ sat})]} \qquad (2.247)$$

where K_1 and K_2 are process-dependent parameters and $V_{DS\ sat}$ is the value of V_{DS} where the drain characteristics enter the saturation region. Typical values for NMOS devices are $K_1 = 5V^{-1}$ and $K_2 = 30$ V. The effect is generally much less significant in PMOS devices

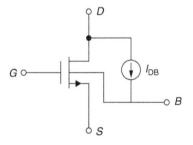

Figure 2.118 Representation of impact ionization in a MOSFET by a drain-substrate current generator.

because the holes carrying the charge in the channel are much less efficient in creating hole-electron pairs than are energetic electrons.

The major impact of this phenomenon on circuit performance is that a parasitic resistance from drain to substrate now exists. Since the substrate of an NMOS device in a p-substrate process is an ac ground (the common substrate terminal must always be connected to the most negatives supply voltage in the circuit), the parasitic resistance shunts the drain to ground and can be a limiting factor in many circuit designs. Differentiating (2.247), we find for the drain-substrate small-signal conductance

$$g_{db} = \frac{\partial I_{DB}}{\partial V_D}$$

$$= K_2 \frac{I_{DB}}{(V_{DS} - V_{DS\ sat})^2} \qquad (2.248)$$

Its main advantage is the BiCMOS process, which allows the mixing of analog and digital circuits nicely. Heterojunction transistors, such as SiGe, are inherently much more linear and, for the same performance, require less current than MOS technology.

In the following pages, we reproduce the datasheet for the Infineon BF999 single-gate MOSFET.[4] Above 500 MHz, a dual-gate (also referred to as tetrode) MOSFET, such as the BF998, is preferred over a single-gate device. The main reasons for this are the following.

1. Internally, it is equivalent to two single-gate devices connected in a cascode arrangement, with the output device operating in the grounded-gate configuration, which entirely avoids the Miller effect and therefore remains stable at higher frequencies, operating up to 1.2 GHz.
2. The second gate can be used for AGC. As the Gate 2 voltage changes, the gain varies heavily.
3. By applying a local oscillator signal to Gate 2, this configuration can be used as a mixer.

The major drawbacks of these devices are their wide tolerances and temperature sensitivity. They follow to some degree the CMOS design, but seem to have better performance.

To demonstrate the capabilities of LDMOS power transistors, we show the PTFA211801E datasheet. This device is intended for CDMA base-station power amplification and delivers up to 180 W. These transistors are key components for high-power amplification for mobile communications. Besides its low feedback capacitance, the LDMOS features higher breakdown. However, these devices have an extremely large total gate width in order to be able to provide the required currents. This results in large output capacitances and low output impedances (around 1 Ω) at microwave frequencies. Such transistors are only useful if the first matching stage is already at the chip inside the package, a technique that is called prematching. Due to prematching, power LDMOS transistors are commonly designed for specificly defined frequency bands, for example, 2110–2170 MHz in case of the PTFA211801E.

---

[4] Reproduced with permission.

BF999

Silicon N-Channel MOSFET Triode

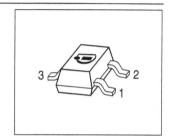

- For high-frequency stages up to 300 MHz preferably in FM applications
- Pb-free (RoHS compliant) package[1]
- Qualified according AEC Q101

ESD (Electrostatic discharge) sensitive device, observe handling precaution!

Type	Marking	Pin Configuration					Package
BF999	LBs	1=G	2=D	3=S	-	-	-	SOT23

Maximum Ratings

Parameter	Symbol	Value	Unit
Drain-source voltage	V_{DS}	20	V
Continuous drain current	I_D	30	mA
Gate-source peak current	$\pm I_{GSM}$	10	mA
Total power dissipation $T_S \leq 76\ °C$	P_{tot}	200	mW
Storage temperature	T_{stg}	-55 ... 150	°C
Channel temperature	T_{ch}	150	

Thermal Resistance

Parameter	Symbol	Value	Unit
Channel - soldering point[2]	R_{thchs}	≤ 370	K/W

[1]Pb-containing package may be available upon special request
[2]For calculation of R_{thJA} please refer to Application Note Thermal Resistance

2007-04-20

Source: Courtesy Infineon Technologies.

BF999

Electrical Characteristics at T_A = 25°C, unless otherwise specified

Parameter	Symbol	Values			Unit
		min.	typ.	max.	
DC Characteristics					
Drain-source breakdown voltage I_D = 10 µA, $-V_{GS}$ = 4 V	$V_{(BR)DS}$	20	-	-	V
Gate-source breakdown voltage $\pm I_{GS}$ = 10 mA, V_{DS} = 0	$\pm V_{(BR)GSS}$	6.5	-	12	
Gate-source leakage current $\pm V_{GS}$ = 5 V, V_{DS} = 0	$\pm I_{GSS}$	-	-	50	nA
Drain current V_{DS} = 10 V, V_{GS} = 0	I_{DSS}	5	10	16	mA
Gate-source pinch-off voltage V_{DS} = 10 V, I_D = 20 µA	$-V_{GS(p)}$	-	0.8	1.5	V

Electrical Characteristics at T_A = 25°C, unless otherwise specified

Parameter	Symbol	Values			Unit
		min.	typ.	max.	
AC Characteristics					
Forward transconductance V_{DS} = 10 V, I_D = 10 mA	g_{fs}	14	20	-	mS
Gate input capacitance V_{DS} = 10 V, I_D = 10 mA, f = 10 MHz	C_{gss}	-	2.5	-	pF
Output capacitance V_{DS} = 10 V, I_D = 10 mA, f = 10 MHz	C_{dss}	-	0.9	-	pF
Power gain V_{DS} = 10 V, I_D = 10 mA, f = 45 MHz	G_p	-	27	-	dB
Noise figure V_{DS} = 10 V, I_D = 10 mA, f = 45 MHz	F	-	2.1	-	dB

2

2007-04-20

Source: Courtesy Infineon Technologies.

Total power dissipation $P_{tot} = f(T_S)$

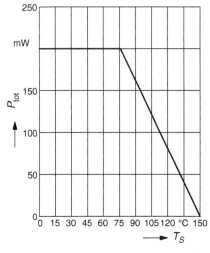

Output characteristics $I_D = f(V_{DS})$

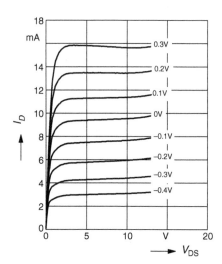

Gate transconductance $g_{fs} = f(V_{GS})$

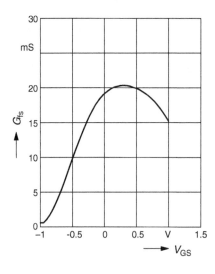

Drain current $I_D = (V_{GS})$

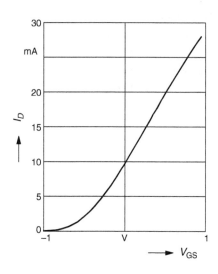

2007-04-20

Source: Courtesy Infineon Technologies.

BF999

Gate input capacitance $C_{gss} = f(V_{GS})$

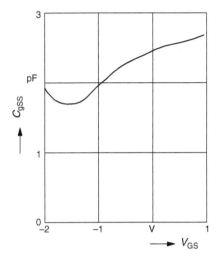

Output capacitance $C_{dss} = f(V_{DS})$

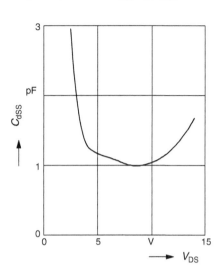

2007-04-20

PTFA211801E

Thermally-Enhanced High Power RF LDMOS FET
180 W, 2110 – 2170 MHz

Description

The PTFA211801E is a thermally-enhanced, 180-watt, internally matched LDMOS FET intended for WCDMA applications. It is characacterized for single- and two-carrier WCDMA operation from 2110 to 2170 MHz. Manufactured with Infineon's advanced LDMOS process, this device provides excellent thermal performance and superior reliability.

PTFA211801E
Package H-36260-2

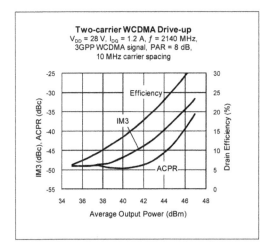

Features

- Broadband internal matching
- Typical two-carrier WCDMA performance at 2140 MHz, 28 V
  - Average output power = 45.5 dBm
  - Linear Gain = 15.5 dB
  - Efficiency = 27.5%
  - Intermodulation distortion = −36 dBc
  - Adjacent channel power = −41 dBc
- Typical CW performance, 2170 MHz, 30 V
  - Output power at P_{1dB} = 180 W
  - Efficiency = 52%
- Integrated ESD protection
- Excellent thermal stability, low HCI drift
- Capable of handling 10:1 VSWR @ 28 V, 150 W (CW) output power
- Pb-free and RoHS-compliant

RF Characteristics

WCDMA Measurements (tested in Infineon test fixture)
V_{DD} = 28 V, I_{DQ} = 1.2 A, P_{OUT} = 35 W average, f_1 = 2135 MHz, f_2 = 2145 MHz, 3GPP signal, channel bandwidth = 3.84 MHz, peak/average = 8 dB @ 0.01% CCDF

Characteristic	Symbol	Min	Typ	Max	Unit
Gain	G_{ps}	14.5	15.5	–	dB
Drain Efficiency	ηD	26	27.5	–	%
Intermodulation Distortion	IMD	–	−36	−34	dBc

All published data at T_{CASE} = 25°C unless otherwise indicated

ESD: Electrostatic discharge sensitive device—observe handling precautions!

RF Characteristics (cont.)

CW Measurements (tested in Infineon test fixture)
V_{DD} = 28 V, I_{DQ} = 1.2 A, P_{OUT} = 150 W average, f = 2170 MHz

Characteristic	Symbol	Min	Typ	Max	Unit
Gain Compression	G_{comp}	—	0.5	1.0	dB

Two-tone Measurements (not subject to production test—verified by design/characterization in Infineon test fixture)
V_{DD} = 28 V, I_{DQ} = 1.2 A, P_{OUT} = 140 W PEP, f = 2140 MHz, tone spacing = 1 MHz

Characteristic	Symbol	Min	Typ	Max	Unit
Gain	G_{ps}	—	15.5	—	dB
Drain Efficiency	ηD	—	38.5	—	%
Intermodulation Distortion	IMD	—	−28	—	dBc

DC Characteristics

Characteristic	Conditions	Symbol	Min	Typ	Max	Unit
Drain-Source Breakdown Voltage	V_{GS} = 0 V, I_{DS} = 10 mA	$V_{(BR)DSS}$	65	—	—	V
Drain Leakage Current	V_{DS} = 28 V, V_{GS} = 0 V	I_{DSS}	—	—	1.0	μA
Drain Leakage Current	V_{DS} = 63 V, V_{GS} = 0 V	I_{DSS}	—	—	10.0	μA
On-State Resistance	V_{GS} = 10 V, V_{DS} = 0.1 V	$R_{DS(on)}$	—	0.05	—	Ω
Operating Gate Voltage	V_{DS} = 28 V, I_{DQ} = 1.2 A	V_{GS}	2.0	2.5	3.0	V
Gate Leakage Current	V_{GS} = 10 V, V_{DS} = 0 V	I_{GSS}	—	—	1.0	μA

Maximum Ratings

Parameter	Symbol	Value	Unit
Drain-Source Voltage	V_{DSS}	65	V
Gate-Source Voltage	V_{GS}	−0.5 to +12	V
Junction Temperature	T_J	200	°C
Storage Temperature Range	T_{STG}	−40 to +150	°C
Thermal Resistance (T_{CASE} = 70°C, 150 W CW)	$R_{\theta JC}$	0.31	°C/W

Ordering Information

Type and Version	Package Outline	Package Description	Shipping
PTFA211801E V5	H-36260-2	Thermally-enhanced slotted flange, single-ended	Tray
PTFA211801E V5 R250	H-36260-2	Thermally-enhanced slotted flange, single-ended	Tape & Reel

Typical Performance (data taken in a production test fixture)

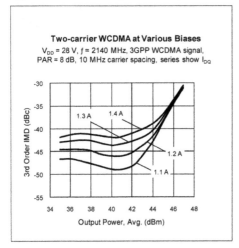

Two-carrier WCDMA at Various Biases
V_{DD} = 28 V, f = 2140 MHz, 3GPP WCDMA signal, PAR = 8 dB, 10 MHz carrier spacing, series show I_{DQ}

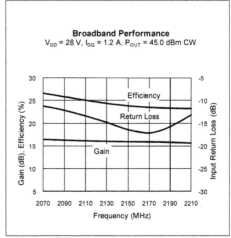

Broadband Performance
V_{DD} = 28 V, I_{DQ} = 1.2 A, P_{OUT} = 45.0 dBm CW

PTFA211801E

Typical Performance (cont.)

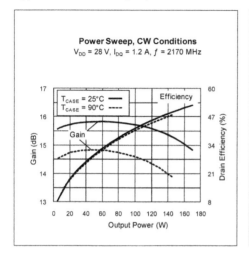

Power Sweep, CW Conditions
V_{DD} = 28 V, I_{DQ} = 1.2 A, f = 2170 MHz

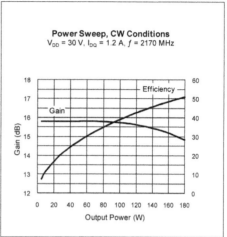

Power Sweep, CW Conditions
V_{DD} = 30 V, I_{DQ} = 1.2 A, f = 2170 MHz

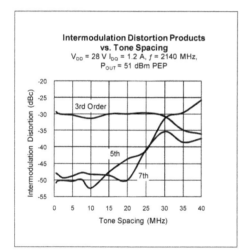

**Intermodulation Distortion Products
vs. Tone Spacing**
V_{DD} = 28 V I_{DQ} = 1.2 A, f = 2140 MHz,
P_{OUT} = 51 dBm PEP

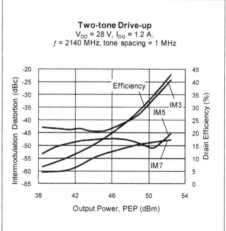

Two-tone Drive-up
V_{DD} = 28 V, I_{DQ} = 1.2 A,
f = 2140 MHz, tone spacing = 1 MHz

PTFA211801E

Typical Performance (cont.)

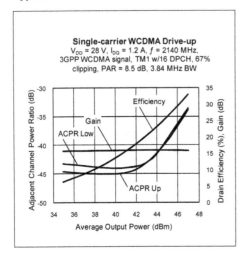

Single-carrier WCDMA Drive-up
V_{DD} = 28 V, I_{DQ} = 1.2 A, f = 2140 MHz,
3GPP WCDMA signal, TM1 w/16 DPCH, 67%
clipping, PAR = 8.5 dB, 3.84 MHz BW

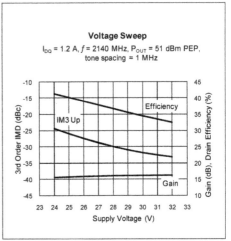

Voltage Sweep
I_{DQ} = 1.2 A, f = 2140 MHz, P_{OUT} = 51 dBm PEP,
tone spacing = 1 MHz

Broadband Circuit Impedance

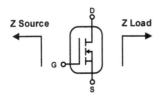

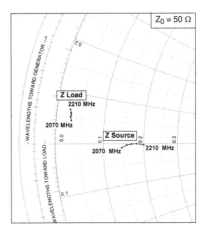

Frequency	Z Source Ω		Z Load Ω	
MHz	R	jX	R	jX
2070	7.2	−0.5	1.5	2.3
2110	7.8	−0.2	1.4	2.6
2140	8.4	−0.0	1.4	2.8
2170	9.1	0.0	1.4	3.0
2210	10.0	−0.2	1.3	3.4

Reference Circuit

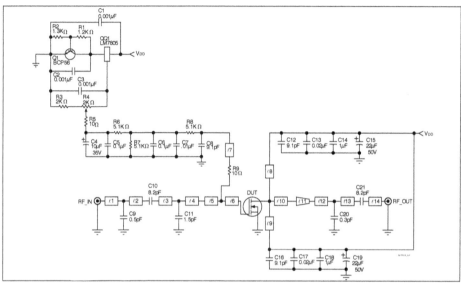

Reference circuit schematic for f = 2140 MHz

Electrical Characteristics at 2140 MHz

Transmission Line	Electrical Characteristics	Dimensions: L x W (mm)	Dimensions: L x W (in.)
$\ell 1$	$0.097\ \lambda$, $50.0\ \Omega$	7.37 x 1.40	0.290 x 0.055
$\ell 2$	$0.267\ \lambda$, $50.0\ \Omega$	19.86 x 1.40	0.782 x 0.055
$\ell 3$	$0.136\ \lambda$, $42.0\ \Omega$	10.24 x 1.85	0.403 x 0.073
$\ell 4$	$0.087\ \lambda$, $42.0\ \Omega$	6.50 x 1.85	0.256 x 0.073
$\ell 5$	$0.018\ \lambda$, $11.4\ \Omega$	1.24 x 10.24	0.049 x 0.403
$\ell 6$	$0.077\ \lambda$, $6.9\ \Omega$	5.23 x 17.78	0.206 x 0.700
$\ell 7$	$0.207\ \lambda$, $48.0\ \Omega$	15.70 x 1.50	0.618 x 0.059
$\ell 8, \ell 9$	$0.256\ \lambda$, $45.0\ \Omega$	19.30 x 1.65	0.760 x 0.065
$\ell 10$	$0.087\ \lambda$, $5.0\ \Omega$	5.84 x 25.40	0.230 x 1.000
$\ell 11$ (taper)	$0.073\ \lambda$, $5.0\ \Omega$ / $40.0\ \Omega$	5.59 x 25.40 / 1.98	0.220 x 1.000 / 0.078
$\ell 12$	$0.019\ \lambda$, $40.0\ \Omega$	1.45 x 1.98	0.057 x 0.078
$\ell 13$	$0.087\ \lambda$, $50.0\ \Omega$	6.65 x 1.40	0.262 x 0.055
$\ell 14$	$0.403\ \lambda$, $50.0\ \Omega$	30.73 x 1.40	1.210 x 0.055

PTFA211801E

Reference Circuit (cont.)

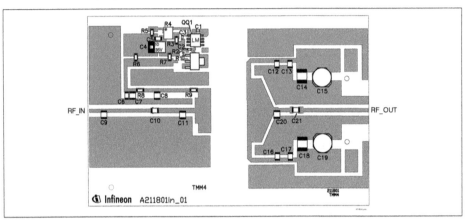

Reference circuit assembly diagram (not to scale)*

Circuit Assembly Information

DUT	PTFA211801E	LDMOS Transistor	
PCB	0.76 mm [.030"] thick, εr = 4.5	Rogers TMM4	2 oz. copper

Component	Description	Suggested Manufacturer	P/N
C1, C2, C3	Capacitor, 0.001 µF	Digi-Key	PCC1772CT-ND
C4	Tantalum capacitor, 10 µF, 35 V	Digi-Key	PCS6106TR-ND
C5, C6	Capacitor, 0.1 µF	Digi-Key	PCC104BCT
C7	Capacitor, 0.01 µF	ATC	200B103
C8, C12, C16	Ceramic capacitor, 9.1 pF	ATC	100B 9R1
C9	Ceramic capacitor, 0.5 pF	ATC	100B 0R5
C10, C21	Ceramic capacitor, 8.2 pF	ATC	100B 8R2
C11	Ceramic capacitor, 1.5 pF	ATC	100B 1R5
C13, C17	Ceramic capacitor, 0.02 µF	ATC	200B 203
C14, C18	Ceramic capacitor, 1 µF	ATC	920C105
C15, C19	Electrolytic capacitor, 22 µF, 50 V	Digi-Key	PCE3374CT-ND
C20	Ceramic capacitor, 0.3 pF	ATC	100B 0R3
Q1	Transistor	Infineon Technologies	BCP56
QQ1	Voltage regulator	National Semiconductor	LM7805
R1	Chip resistor, 1.2 k Ω	Digi-Key	P1.2KGCT-ND
R2	Chip resistor, 1.3 k Ω	Digi-Key	P1.3KGCT-ND
R3	Chip resistor, 2 k Ω	Digi-Key	P2KECT-ND
R4	Potentiometer, 2 k Ω	Digi-Key	3224W-202ETR-ND
R5, R9	Chip resistor, 10 Ω	Digi-Key	P10ECT-ND
R6, R7, R8	Chip resistor, 5.1 k Ω	Digi-Key	P5.1KECT-ND

* Gerber Files for this circuit available on request

PTFA211801E

Package Outline Specifications

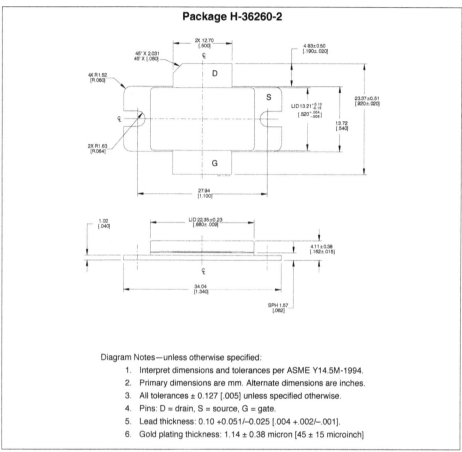

Package H-36260-2

Diagram Notes—unless otherwise specified:

1. Interpret dimensions and tolerances per ASME Y14.5M-1994.
2. Primary dimensions are mm. Alternate dimensions are inches.
3. All tolerances ± 0.127 [.005] unless specified otherwise.
4. Pins: D = drain, S = source, G = gate.
5. Lead thickness: 0.10 +0.051/–0.025 [.004 +.002/–.001].
6. Gold plating thickness: 1.14 ± 0.38 micron [45 ± 15 microinch]

Find the latest and most complete information about products and packaging at the Infineon Internet page
http://www.infineon.com/rfpower

PTFA211801 E V5

Revision History:	2011-01-11	Data Sheet
Previous Version:	2010-08-04, Data Sheet	

Page	Subjects (major changes since last revision)
1	Updated ESD protection feature
All	Removed earless package

We Listen to Your Comments

Any information within this document that you feel is wrong, unclear or missing at all?
Your feedback will help us to continuously improve the quality of this document.
Please send your proposal (including a reference to this document) to:

highpowerRF@infineon.com

To request other information, contact us at:
+1 877 465 3667 (1-877-GO-LDMOS) USA
or +1 408 776 0600 International

Edition 2011-01-11
Published by
Infineon Technologies AG
81726 Munich, Germany
© 2005 Infineon Technologies AG
All Rights Reserved.

Legal Disclaimer

The information given in this document shall in no event be regarded as a guarantee of conditions or characteristics. With respect to any examples or hints given herein, any typical values stated herein and/or any information regarding the application of the device, Infineon Technologies hereby disclaims any and all warranties and liabilities of any kind, including without limitation, warranties of non-infringement of intellectual property rights of any third party.

Information

For further information on technology, delivery terms and conditions and prices, please contact the nearest Infineon Technologies Office (**www.infineon.com/rfpower**).

Warnings

Due to technical requirements, components may contain dangerous substances. For information on the types in question, please contact the nearest Infineon Technologies Office.

Infineon Technologies components may be used in life-support devices or systems only with the express written approval of Infineon Technologies, if a failure of such components can reasonably be expected to cause the failure of that life-support device or system or to affect the safety or effectiveness of that device or system. Life support devices or systems are intended to be implanted in the human body or to support and/or maintain and sustain and/or protect human life. If they fail, it is reasonable to assume that the health of the user or other persons may be endangered.

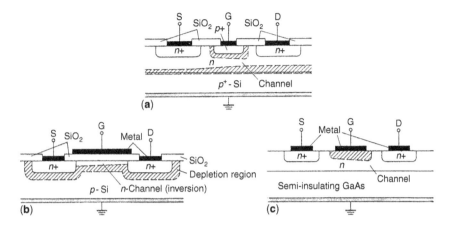

Figure 2.119 This picture is provided to differentiate between the structure of the three technologies. (a) (JFET) and (b) (MOSFET) are based on silicon, and (c), the GaAsFET, on semiinsulating GaAs. While the MOS transistor has silicon dioxide (SiO_2) as gate insulation, both the JFET and GaAs MESFET have a diode from gate to source, with a barrier voltage of 0.65 V for silicon and 0.8 V for GaAs.

2.4.6 III–V MESFETs and HEMTs

2.4.6.1 Introduction

For RF applications, more than 20 years ago, the JFET was the dominant FET. As mentioned earlier, its major disadvantage is the fact that the f_T cutoff frequency, defined as

$$f_T = g_m/2\pi C_{GS} \tag{2.249}$$

limits the operating range of this semiconductor device to 1 GHz at most. Also, the JFET's temperature sensitivity and loose tolerances in transconductance and other RF-related values made it difficult to use. Finally, no RFICs made use of this technology. While at this moment silicon-based MOS technology probably is the most cost effective and produced in largest quantities, it does not rival the metal-semiconductor FET designed on GaAs technology. Figure 2.119 shows the structure of the silicon MOSFET, silicon JFET, and the GaAs MESFET.

2.4.6.2 HEMTs

The best performance in terms of cutoff frequencies and white noise is achieved today by heterojuction FETs. The channels of these transistors is a two-dimensional electron gas at a hetero interface, which allows much migher electron velocities. The devices are therefore called high electron mobility transistors (HEMTs).

The basic principle of an HEMT can be explained regarding the band diagram shown in Figure 2.120. A material with a wider bandgap (e.g., AlGaAs) is grown on a III–V compound semiconductor (e.g., GaAs). A step in the conduction and valence bands will be present at the heterointerface. If the wider bandgap material now is n-doped, it will not take long until the electrons fall down the step—but are confined there directly at the interface due to the electric field from the donors. As a result, the bands bow as shown in the figure;

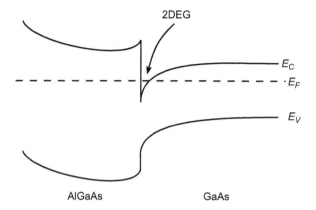

Figure 2.120 Band diagram of a HEMT structure.

and directly at the interface, a spike in the conduction band reaches below the fermi level. Inside this narrow spike, we find the two-dimensional electron gas.

Transistor operation is achieved by applying a gate voltage vertically to this structure, which controls how deep the spike is. In the figure, the gate contact would be on the left-hand side.

Commonly, HEMTs are normally-on and negative gate voltages are required to pinch them off. There are, however, special HEMT structures that are normally-off.

The performance of the HEMT is therefore mainly determined by the properties of the two-dimensional electron gas. In order to prevent the electrons from being scattered at the doping atoms, the first layer of wider-bandgap material will be undoped. It serves as a spacer. For better performance, it is desirable to replace the GaAs channel by InP, but without switching to the more expensive wafers. Simply growing InP crystals on GaAs substrates, on the other hand, is not easily possible since the lattice constants are too different. The solution are today's pseudomorphic HEMTs (pHEMTs) and metamorphic HEMTs (mHEMTs) that rely on extremely thin layers of strained InP.

The most popular HEMTs are the GaAs ones; InP-based HEMTs are rather used for highest performance at highest frequencies. For highest power, GaN devices are the devices of choice. Since GaN crystals are not available, it is required to grow the whole transistor layer stack on a suitable wafer. GaN on Si promises to be cheaper, but it is also technologically even more challenging due to the difference in lattice constants. Thermal conduction is also not too high. GaN on SiC offers best performance and thermal management.

Luckily, HEMTs outperform MESFETs, but they generally do not require a completely new model for circuit design. The typical small-signal and large-signal behavior of HEMTs and MESFETs is quite similar, and the same equivalent circuits and parameter extraction approaches are used. Although today, dedicated HEMT models are available, these should also fit MESFET devices with reasonable accuracy.

2.4.6.3 Large-Signal Behavior of MESFETs and HEMTs

The mathematics for the large-signal behavior of the GaAsFET is basically quite similar to that for the JFET; however, the computation of JFET channel current, diode currents, and capacitance are much simpler than necessary for the GaAsFET. Temperature effects are embedded in all the equations for all the transistors mentioned so far. A rather high number of models is available for MESFET and HEMT devices; these were developed over the past almost 40 years. Over the years, complexity and accuracy increased, of course. Early models such as the famous Curtice–Ettenberg cubic model required to anticipate the limited computer ressures available at these times. Just to give one example, IV curves needed to be approximated by a power series up to the cubic term. Today, a tangens hyperbolicus function is used since, besides other advantages, it is defined and behaves well for all arguments. The additional load for the computation is today not even noticed. Modern transistor models also incorporate sophisticated models for the bias-dependent capacitances where the conservation of energy and charge is an issue. Last but not least, it is to mention that the dynamical self-heating is extremely important in many applications.

The III–V FET models that are mostly in use, among the latest generation models, are the EEHEMT and Chalmers (Angelov) models. As mentioned, these might be much more complex than their predecessors, and it is left to the designer or device modeling engineer to choose the optimum complexity-to-accuracy ratio. For example, we cose to provide the description of the Materka model since its mathematics is much less involved and therefore easier to understand than for the most advanced model. In many cases, it can still be the model of choice.

The following models are usually available in commercial circuit simulators.

- Curtice–Ettenberg cubic
- Tajima
- Modified Materka–Kacprzak
- Raytheon (Statz)
- TOM (TriQuint's Own Model, in different versions)
- Root
- EEHEMT
- Chalmers (Angelov)

However, when switching between different circuit simulators, it is necessary to take into consideration that there is no standardizing process in the GaAs world as it happens for silicon transistors. Models for silicon transistors need to be approved by the Compact Modeling Council, which requires publication of the model *code* before any major company agrees to provide model parameters or to incorporate the model into their simulation software. The GaAs world is smaller and less regulated. The *philosophy* and some of the *math* behind these models are usually published in scientific papers, but it is more or less left to the software companies how they turn a scientific publication into code. This results in slight different implementations and can lead to different model behavior for different tools.

The main advantages of GaAsFET technology derive from the fact that it uses a metal-semiconductor junction with a barrier voltage of 0.8 V, its input capacitance is typically less than 0.2 pF, and the reverse feedback capacitance is less than 0.02 pF, or roughly 10% of the input capacitance. As a result of this, f_{max} is approximately five times higher than f_T.

Table 2.15 Comparison of Si BJT, SiGe HBT, and GaAsFET Technologies

	Si Bipolar	SiGe HBT	GaAs FET
f_T	25 GHz	100 GHz	22 GHz
f_{max}	40 GHz	200 GHz	110 GHz
Features	Low cost, low $1/f$ noise (5 kHz)	Low $1/f$ noise, very low distortion, high cost, IC only	Highest flexibility, lowest NF_0, well established.

Table 2.15 shows a comparison of silicon BJT and GaAsFET technologies. While f_{max} for the bipolar transistor is

$$f_{max} \simeq \sqrt{\frac{f_t}{8\pi r'_{bb}C_c}} \tag{2.250}$$

the f_{max} determination for the GaAsFET is given by

$$f_{max} = \frac{f_t}{2}\sqrt{\frac{R_0}{R_i + R_S + R_g}} \tag{2.251}$$

As a sample calculation

$$f_{max} = \frac{21.9\,\text{GHz}}{2}\sqrt{\frac{450}{1 + 1.5 + 2}} = 110\,\text{GHz} \tag{2.252}$$

The three major drawbacks of the GaAsFET are the following.

1. Much higher flicker corner frequency (somewhere between 10 and 100 MHz), probably due to a lack of a surface passivation.

2. Much higher output conductance. This tends to load down any circuit connected to the drain. On the other hand, since the transconductance is quite high for even low currents, these devices have very high gains at low frequencies, which can make them quite unstable. In the saturated mode, it is not uncommon to find a drain-source resistance of 100–500 Ω, while BJTs and JFETs offer values of several kilohms and higher.

3. Because of the very high flicker corner frequency (from 10 to 100 MHz), MESFETs are really not useful for low-noise mixers and oscillators, and unless there are no devices available in the frequency range above 30 GHz, they should be avoided for these applications.

As to the MESFET's construction and dc properties, Figure 2.121 shows a MESFET's cross-section and dc *I–V* characteristics.

It was outlined in the beginning that while the GaAsFET is a close relative to the JFET and MOSFET, its actual behavior was found to be best described by a set of analytic equations. The first such model was the one by Curtice in the form of quadratic and cubic models, but it does not have enough derivatives to give enough insight into subtle things such as third and higher-order intermodulation distortion and accurate harmonic generation. Other researchers have addressed various areas, but we still find that the Materka model has the

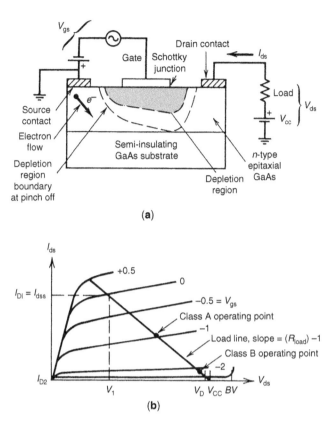

(a)

(b)

Figure 2.121 (a) Cross-section and bias circuit and (b) dc *I–V* curve, including ac load line, for a MESFET.

best approximation, especially using the two-transistor approach outlined later [35, 36]. The large-signal topology for all FETs consists of an intrinsic model with some extrinsic parameters, further complicated by the package, as shown by Figures 2.122 and 2.123. Table 2.16 lists their key words. The actual intrinsic model and its parameter definition

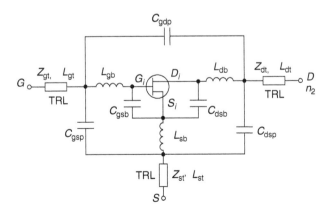

Figure 2.122 MESFET extrinsic model.

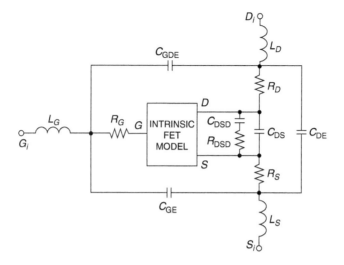

Figure 2.123   MESFET package model.

Table 2.16 The Modified Materka–Kacprzak Model: Extrinsic Key Words

Key Word	Description	Unit	Default
RG	Gate bulk and ohmic resistance	ohm	0.0
RD	Drain bulk and ohmic resistance	ohm	0.0
RS	Source bulk and ohmic resistance	ohm	0.0
LG	Gate lead inductance (metallization)	henry	0.0
LD	Drain lead inductance (metallization)	henry	0.0
LS	Source lead inductance (via)	henry	0.0
CDS	Drain-source capacitance	farad	0.0
CDSD	Low frequency trapping capacitor	farad	0.0
RDSD	Channel trapping resistance	ohm	∞
CGE	Gate-source electrode capacitance	farad	0.0
CDE	Drain-source electrode capacitance	farad	0.0
CGDE	Gate-drain electrode capacitance	farad	0.0
LGB	Gate wirebond inductance	henry	0.0
LDB	Drain wirebond inductance	henry	0.0
LSB	Source wirebond inductance	henry	0.0
CGSB	Gate bondpad to source capacitance	farad	0.0
CDSB	Drain bondpad to source capacitance	farad	0.0
CGSP	Gate to source package capacitance	farad	0.0
CDSP	Drain to source package capacitance	farad	0.0
CGDP	Gate to drain package capacitance	farad	0.0
ZGT	Gate transmission line impedance	ohm	50
ZDT	Drain transmission line impedance	ohm	50
ZST	Source transmission line impedance	ohm	50
LGT	Gate transmission line length for $\varepsilon_r = 1$	meter	0.0
LDT	Drain transmission line length for $\varepsilon_r = 1$	meter	0.0
LST	Source transmission line length for $\varepsilon_r = 1$	meter	0.0

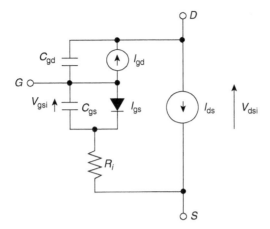

Figure 2.124 Intrinsic model of the modified Materka–Kacprzak MESFET.

depends on the particular model and since designs using GaAsFET will always be done using CAD tools, we will not go into any detail of the equations but will list them. They are not dissimilar from the JFET and MOSFET equations.

2.4.6.4 The Modified Materka–Kacprzak Model

Figure 2.124 shows the intrinsic model of the Materka FET. Table 2.17 lists its key words.

Large-Signal Equations   Device equations

V_{gsi} = Intrinsic gate-source voltage
V_{dsi} = Intrinsic drain-source voltage
V_1 = Voltage across C_{GS} and R_i
V_{gdi} = Intrinsic gate-drain voltage
$V_T = kTJ/q$ (thermal voltage)
k = Boltzmann's constant
q = Electron charge
T_J = Analysis temperature (Kelvins)
Channel Current

$$I_{ds} = \text{IDSS}\left(1 + \text{SS}\frac{V_{dsi}}{\text{IDSS}}\right)\left(1 - \frac{V_{gsi}(t - T)}{\text{VP0} + \text{GAMA } V_{dsi}}\right)^{(E + \text{KE}V_{gsi}(t-T))}$$

$$\times \tan h\left(\frac{\text{SL } V_{dsi}}{\text{IDSS}(1 - \text{KG}V_{gsi}(t - T))}\right)$$

Diode

$$I_{gd} = I_{gdc} - \begin{cases} \text{IB0exp}\left(-\text{AFAB}\left(V_{gdi} + \text{VBC}\right)\right) \\ \frac{\text{GMAX}}{4}\left(\tan h\left(\text{K1D}\left(V_{gsi} - \text{K2D}\right)\right) - 1\right) \\ \times \left(V_{gdi} + \text{VBC} - \sqrt{\left(V_{gdi} + \text{VBC}\right)^2 + \text{K3D}}\right) \end{cases}$$

where

$$I_{gdc} = \text{IG0}\left(\exp\left(\text{AFAG}V_{gdi}\right) - 1\right)$$

Table 2.17 The Modified Materka–Kacprzak Model: Key Words

Key Word	Description	Unit	Default
	Area, Noise, and Name Key Words		
AREA	Area multiplier		1.0
KFN	Flicker noise coefficient (Materka model only)[a]		0
AF	Flicker noise exponent		1.0
FCP	Flicker noise frequency shape factor		1.0
	Channel Current Model Key Words		
IDSS	Drain saturation current for $V_{GS} = 0$	ampere	0.1
VP0	Pinch-off voltage for $V_{DS} = 0$	volt	−2.0
GAMA	Voltage slope parameter of pinch-off voltage	/volt	0.0
E	Constant part of power law parameter		2.0
KE	Dependence of power law on V_{GS}	/volt	0.0
SL	Slope of the $V_{GS} = 0$ drain characteristic in the linear region	amp/volt	0.15
KG	Drain dependence on V_{GS} in the linear region	/volt	0.0
SS	Slope of the drain characteristic in the saturated region	amp/volt	0.0
T	Channel transit-time delay	s	0.0
IG0	Diode saturation current	ampere	0
AFAG	Slope factor of forward diode current	/vol	38.696
IB0	Breakdown saturation current	amp	0
AFAB	Slope factor of breakdown current	/volt	0
VBC	Breakdown voltage	volt	∞
GMAX	Breakdown conductance	amp/volt	0
K1D	Fitting parameter	/volt	0
K2D	Fitting parameter	volt	0
K3D	Fitting parameter	volt^2	0
R10	Intrinsic channel resistance for $V_{GS} = 0$	ohm	0.0
KR	Slope factor of intrinsic channel resistance	/volt	0.0
	Materka Capacitance Model Key Words		
C10	Gate–source Schottky barrier capacitance for $V_{GS} = 0$	farad	0.0
K1	Slope parameter of gate-source capacitance	/volt	1.25
MGS	Gate–source grading coefficient		0.5
C1S	Constant parasitic component of gate-source capacitance	farad	0.0
CF0	Gate-drain feedback capacitance for $V_{GD} = 0$	farad	0.0
KF	Slope parameter of gate-drain feedback capacitance	/volt	1.25
MGD	Gate-drain grading coefficient		0.5
FCC	Forward-bias depletion capacitance coefficient		0.8

[a]The flicker noise parameter of the Materka model is KFN so as not to conflict with the KF parameter in the capacitance model.

Channel Resistance

$$R_i = \begin{cases} \text{R10}\left(1 - \text{KR}\,V_{\text{gsi}}\right) & \text{KR}\,V_{\text{gsi}} < 1.0 \\ 0 & \text{KR}\,V_{\text{gsi}} \geq 1.0 \end{cases}$$

Capacitance Model

$$C_{\text{gs}} = \text{CGS0}\,\frac{F1F2}{\sqrt{1 - \dfrac{V_{\text{new}}}{\text{VBI}}}} + \text{CGD0}F3$$

$$C_{\text{gd}} = \text{CGS0}\,\frac{F1F3}{\sqrt{1 - \dfrac{V_{\text{new}}}{\text{VBI}}}} + \text{CGD0}F2$$

where
$$F1 = \frac{1}{2}\left(1 + \frac{V_{\text{eff}} - VT}{\sqrt{(V_{\text{eff}} - VT)^2 + \delta^2}}\right)$$

$$F2 = \frac{1}{2}\left(1 + \frac{V_{\text{gsi}} - V_{\text{gdi}}}{\sqrt{\left(V_{\text{gsi}} - V_{\text{gdi}}\right)^2 + (1/\text{ALFA})^2}}\right)$$

$$F3 = \frac{1}{2}\left(1 - \frac{V_{\text{gsi}} - V_{\text{gdi}}}{\sqrt{\left(V_{\text{gsi}} - V_{\text{gdi}}\right)^2 + (1/\text{ALFA})^2}}\right)$$

$$V_{\text{new}} = \begin{cases} A1 & A1 < VMAX \\ \text{VMAX} & A1 \geq VMAX \end{cases}$$

$$A1 = \frac{1}{2}\left(V_{\text{eff}} + VT + \sqrt{(V_{\text{eff}} - VT)^2 + \delta^2}\right)$$

$$V_{\text{eff}} = \frac{1}{2}\left(V_{\text{gsi}} + V_{\text{gdi}} + \sqrt{\left(V_{\text{gsi}} - V_{\text{gdi}}\right)^2 + (1/\text{ALFA})^2}\right)$$

$$VT = \text{VP0} + \text{GAMA}\,V_{\text{dsi}}$$
$$\delta = 0.2$$

Some of the modifications to the Materka model have been done by Ansoft under various Department of Defense contracts, and by Raytheon and Texas Instruments under similar contracts, with Ansoft being a subcontractor.

The most relevant equation is really the channel current. Its derivatives are largely responsible for the accuracy of the intermodulation distortion, power-added efficiency, and, of course, its dc I–V curves.

2.4.6.5 Enhancement/Depletion FETs

To make the designer's life more difficult, it turns out that there are two types of GaAsFETs.

1. *Depletion FETs (DFETs).* Most similar to the JFET; here V_G must be negative to control the device. They are the most commonly produced and are the FET type most referred to in this book.

   On one hand

   - they require a negative gate voltage with respect to the source;
   - self-bias allows operation from a single supply voltage.

   On the other hand

   - for low-voltage operation, a negative voltage generator may be required;
   - supply voltage must be doubled to accommodate full-swing operation.

2. *Enhancement FETs (EFETs).* Most similar to the MOSFET; here V_G must be positive to bring life to the device. Practically speaking, EFETs are used mostly in integrated circuits; they are typically not available in discrete, packaged form.

   On one hand

   - they need only positive supply for biasing;
   - they provide higher gm/mA (for same device width)—5.1 mS versus 3.9 mS at 8 mA;
   - they are good for low-power LNAs, giving slightly better NF than DFETs, an NF of 1 dB at 1 GHz, and $I_{dd} < 10$ mA.

   On the other hand

   - they have a very limited gate bias range (V_{GS} between 0.15 and 0.7 V);
   - the gate conduction degrades NF and input impedance;
   - the gate capacitance is higher than that of DFETs;
   - the linearity is not as good as that of DFETs.

With today's technologies, all GaAs devices are n-channel; we have not seen any p-channels yet. More information about biasing will be given in the next chapter.

Figure 2.125 is a lumped element, two-part equivalent circuit of a MESFET showing the location of lumped-element components.

2.4.7 Small-Signal GaAs MESFET and HEMT Model

Figure 2.126 shows the applicable linear equivalent circuit for a MESFET and Table 2.18 lists its key words. As with the MOS transistors, there is a gallium-arsenide dual-gate MOSFET available that is mostly used in special circuits, such as preamplifiers and mixers, whose IF frequency has to be significantly higher than the flicker corner frequency, for example, higher than 200 MHz.

The following two datasheets[5] show a typical low-noise enhancement GaAs pHEMT and a high-power GaN HEMT device.

---

[5] Copyright © 2009 by Avago Technologies. All rights reserved, and copyright © 2009–2011 by Cree Inc., respectively. Reprinted with permission.

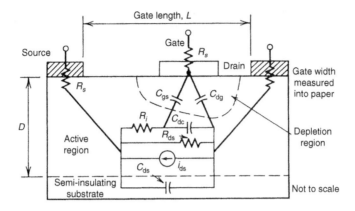

Figure 2.125 Location of lumped-element components for a MESFET.

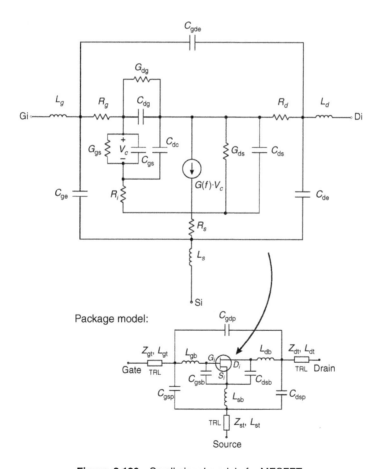

Figure 2.126 Small-signal model of a MESFET.

Table 2.18 Small-Signal MESFET Model Key Words

Key Word	Description	Unit	Default
G	Transconductance at dc, G_0 (see Note 1)	/ohm	
CGS	Gate-source capacitance	farad	
F	3-dB Roll-off frequency	Hz	∞
T	Time delay	s	0.0
TDS	Drain source time delay	s	0.0
GGS	Gate-source conductance	/ohm	0.0
CDG	Drain-gate capacitance	farad	0.0
CDC	Dipole layer capacitance	farad	0.0
CDS	Drain-source capacitance	farad	0.0
GDS	Drain-source conductance	/ohm	0.0
RI	Channel resistance	ohm	0.0
RG	Gate resistance	ohm	0.0
RD	Drain resistance	ohm	0.0
RS	Source resistance	ohm	0.0
CGE	External gate capacitance	farad	0.0
CDE	External drain capacitance	farad	0.0
LG	Gate lead inductance	henry	0.0
LD	Drain lead inductance henry	0.0	
LS	Source lead inductance henry	0.0	
CGDE	External gate-drain capacitance	farad	0.0
GDG	Gate drain conductance	/ohm	0.0
TJ	Chip temperature	K	298
	Package Parasitics		
LGB	Gate wirebond inductance	henry	0.0
LDB	Drain wirebond inductance	henry	0.0
LSB	Source wirebond inductance	henry	0.0
CGSB	Gate bondpad to source capacitance	farad	0.0
CDSB	Drain bondpad to source capacitance	farad	0.0
CGSP	Gate to source package capacitance	farad	0.0
CDSP	Drain to source package capacitance	farad	0.0
CGDP	Gate to drain package capacitance	farad	0.0
ZGT	Gate transmission line impedance	ohm	50
ZDT	Drain transmission line impedance	ohm	50
ZST	Source transmission line impedance	ohm	50
LGT	Gate transmission line length for at $\varepsilon_r = 1$	meter	0.0
LDT	Drain transmission line length for at $\varepsilon_r = 1$	meter	0.0
LST	Source transmission line length for at $\varepsilon_r = 1$	meter	0.0
FC	Corner frequency of flicker ($1/f$) noise (see Note 2)	Hz	10 MHz
FCP	Shape factor of the $1/f$ noise response		1.0
label	User-defined term that refers to temperature coefficient		

Notes:
1. The transconductance of this model may be approximately described by

$$g_m = G \frac{e^{-j\omega T}}{1 + j(f/F)}$$

where $\omega = 2\pi f$ and f = frequency.
2. The flicker noise frequency dependence is given by

$$\frac{1}{(f/F_c)^{\text{FCP}}}$$

ATF-53189

Enhancement Mode[1] Pseudomorphic HEMT in SOT 89 Package

Data Sheet

Description

Avago Technologies's ATF-53189 is a single-voltage high linearity, low noise E-pHEMT FET packaged in a low cost surface mount SOT89 package. The device is ideal as a high-linearity, low noise, medium-power amplifier. Its operating frequency range is from 50 MHz to 6 GHz.

ATF-53189 is ideally suited for Cellular/PCS and WCDMA wireless infrastructure, WLAN, WLL and MMDS application, and general purpose discrete E-pHEMT amplifiers which require medium power and high linearity. All devices are 100% RF and DC tested.

Pin Connections and Package Marking

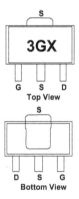

Top View

Bottom View

Notes:
Package marking provides orientation and identification:

"3G" = Device Code
"x" = Month code indicates the month of manufacture.
D = Drain
S = Source
G = Gate

Features

- Single voltage operation
- High Linearity and Gain
- Low Noise Figure
- Excellent uniformity in product specifications
- SOT 89 standard package
- Point MTTF > 300 years[2]
- MSL-1 and lead-free
- Tape-and-Reel packaging option available

Specifications

2 GHz, 4.0V, 135 mA (Typ.)

- 40.0 dBm Output IP3
- 23.0 dBm Output Power at 1dB gain compression
- 0.85 dB Noise Figure
- 15.5 dB Gain
- 46% PAE at P1dB
- LFOM[3] 12.7 dB

Applications

- Front-end LNA Q1 and Q2, Driver or Pre-driver Amplifier for Cellular/PCS and WCDMA wireless infrastructure
- Driver Amplifier for WLAN, WLL/RLL and MMDS applications
- General purpose discrete E-pHEMT for other high linearity applications

Notes:
1. Enhancement mode technology employs a single positive V_{gs}, eliminating the need of negative gate voltage associated with conventional depletion mode devices.
2. Refer to reliability datasheet for detailed MTTF data.
3. Linearity Figure of Merit (LFOM) is OIP3 divided by DC bias power.

ATF-53189 Absolute Maximum Ratings[1]

Symbol	Parameter	Units	Absolute Maximum
V_{ds}	Drain–Source Voltage[2]	V	7
V_{gs}	Gate–Source Voltage[2]	V	-5 to 1.0
V_{gd}	Gate Drain Voltage[2]	V	-5 to 1.0
I_{ds}	Drain Current[2]	mA	300
I_{gs}	Gate Current	mA	20
P_{diss}	Total Power Dissipation[3]	W	1.0
$P_{in\,max.}$	RF Input Power	dBm	+24
T_{ch}	Channel Temperature	°C	150
T_{stg}	Storage Temperature	°C	-65 to 150

Thermal Resistance[2,4]

$\theta_{ch\text{-}b} = 70°C/W$

Notes:
1. Operation of this device above any one of these parameters may cause permanent damage.
2. Assuming DC quiescent conditions.
3. Board (package belly) temperature T_B is 25°C. Derate 14.30 mW/°C for $T_B > 80°C$.
4. Channel-to-board thermal resistance measured using 150°C Liquid Crystal Measurement method.

ATF-53189 Electrical Specifications

$T_A = 25°C$, DC bias for RF parameters is Vds = 4.0V and Ids = 135 mA unless otherwise specified.

Symbol	Parameters and Test Conditions		Units	Min.	Typ.	Max.
Vgs	Operational Gate Voltage	Vds = 4.0V, Ids = 135 mA	V	—	0.65	—
Vth	Threshold Voltage	Vds = 4.0V, Ids = 8 mA	V	—	0.30	—
Ids	Drain to Source Current	Vds = 4.0V, Vgs = 0V	µA	—	3.70	—
Gm	Transconductance	Vds = 4.0V, Gm = ΔIds/ΔVgs; ΔVgs = Vgs1 – Vgs2 Vgs1 = 0.6V, Vgs2 = 0.55V	mmho	—	650	—
Igss	Gate Leakage Current	Vds = 0V, Vgs = -4V	µA	-10.0	-0.34	—
NF	Noise Figure	f=900 MHz	dB	—	0.80	—
		f=2.0 GHz	dB	—	0.85	1.3
		f=2.4 GHz	dB	—	1.00	—
G	Gain[1]	f=900 MHz	dB	—	17.2	—
		f=2.0 GHz	dB	14.0	15.5	17.0
		f=2.4 GHz	dB	—	15.0	—
OIP3	Output 3rd Order Intercept Point[1]	f=900 MHz	dBm	—	42.0	—
		f=2.0 GHz	dBm	36.0	40.0	—
		f=2.4 GHz	dBm	—	38.6	—
P1dB	Output 1dB Compressed[1]	f=900 MHz	dBm	—	21.7	—
		f=2.0 GHz	dBm	—	23.0	—
		f=2.4 GHz	dBm	—	23.2	—
PAE	Power Added Efficiency	f=900 MHz	%	—	33.8	—
		f=2.0 GHz	%	—	46.0	—
		f=2.4 GHz	%	—	49.0	—
ACLR	Adjacent Channel Leakage Power Ratio[1,2]	Offset BW = 5 MHz	dBc	—	-54.0	—
		Offset BW = 10 MHz	dBc	—	-64.0	—

Notes:
1. Measurements at 2 GHz obtained using production test board described in Figure 1.
2. ACLR test spec is based on 3GPP TS 25.141 V5.3.1 (2002-06)
   - Test Model 1
   - Active Channels: PCCPCH + SCH + CPICH + PICH + SCCPCH + 64 DPCH (SF=128)
   - Freq = 2140 MHz
   - Pin = -8 dBm
   - Channel Integrate Bandwidth = 3.84 MHz

2

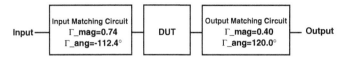

Figure 1. Block diagram of the 2 GHz production test board used for NF, Gain, OIP3 , P1dB, PAE and ACLR measurements. This circuit achieves a trade-off between optimal OIP3, P1dB and VSWR. Circuit losses have been de-embedded from actual measurements.

Product Consistency Distribution Charts[1,2]

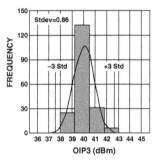

Figure 2. OIP3 @ 2 GHz, 4V, 135 mA.
LSL = 36 dBm, Nominal = 40 dBm.

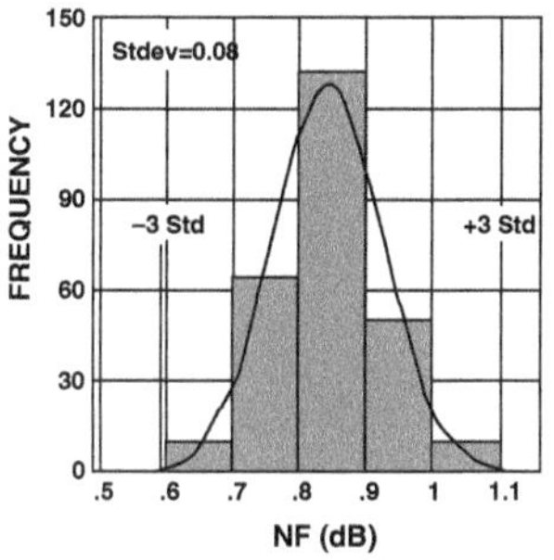

Figure 3. NF @ 2 GHz, 4V, 135 mA.
USL = 1.30 dBm, Nominal = 0.84 dBm.

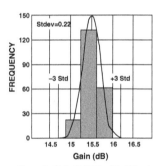

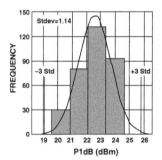

Figure 4. Gain @ 2 GHz, 4V, 135 mA.
LSL = 14 dBm, Nominal = 15.5 dBm,
USL = 17 dBm.

Figure 5. P1dB @ 2 GHz, 4V, 135 mA.
Nominal = 23 dBm.

Notes:
1. Distribution data sample size is 500 samples taken from 3 different wafers. Future wafers allocated to this product may have nominal values anywhere between the upper and lower limits.
2. Measurements are made on production test board, which represents a trade-off between optimal OIP3, P1dB and VSWR. Circuit losses have been de-embedded from actual measurements.

3

Gamma Load and Source at Optimum OIP3 Tuning Conditions

The device's optimum OIP3 measurements were determined using a Maury Load Pull System at 4.0V, 135 mA quiesent bias.

Typical Gammas at Optimum OIP3[1]

Freq (GHz)	Gamma Source		Gamma Load		OIP3 (dBm)	Gain (dB)	P1dB (dBm)	PAE (%)
	Mag	Ang (deg)	Mag	Ang (deg)				
0.9	0.8179	−143.28	0.0721	124.08	42.0	17.2	21.7	33.8
2.0	0.7411	−112.36	0.4080	119.91	41.6	15.6	23.4	44.2
3.9	0.6875	−94.23	0.4478	174.74	41.3	11.2	23.1	41.4
5.8	0.5204	−75.91	0.3525	−120.13	36.9	5.6	22.4	25.7

Note:
1. Typical describes additional product performance information that is not covered by the product warranty.

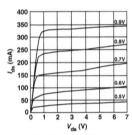

Figure 6. Typical IV Curve.

4

ATF-53189 Typical Performance Curves (at 25°C unless specified otherwise)
Tuned for Optimal OIP3 at Vd = 4.0V, Ids = 135 mA.

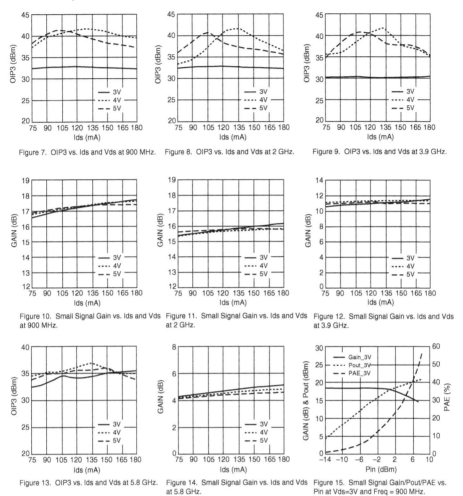

Figure 7. OIP3 vs. Ids and Vds at 900 MHz.   Figure 8. OIP3 vs. Ids and Vds at 2 GHz.   Figure 9. OIP3 vs. Ids and Vds at 3.9 GHz.

Figure 10. Small Signal Gain vs. Ids and Vds at 900 MHz.   Figure 11. Small Signal Gain vs. Ids and Vds at 2 GHz.   Figure 12. Small Signal Gain vs. Ids and Vds at 3.9 GHz.

Figure 13. OIP3 vs. Ids and Vds at 5.8 GHz.   Figure 14. Small Signal Gain vs. Ids and Vds at 5.8 GHz.   Figure 15. Small Signal Gain/Pout/PAE vs. Pin at Vds=3V and Freq = 900 MHz.

Note:
Bias current for the above charts are quiescent conditions. Actual level may increase depending on amount of RF drive.

5

ATF-53189 Typical Performance Curves (at 25°C unless specified otherwise), continued
Tuned for Optimal OIP3 at Vd = 4.0V, Ids = 135 mA.

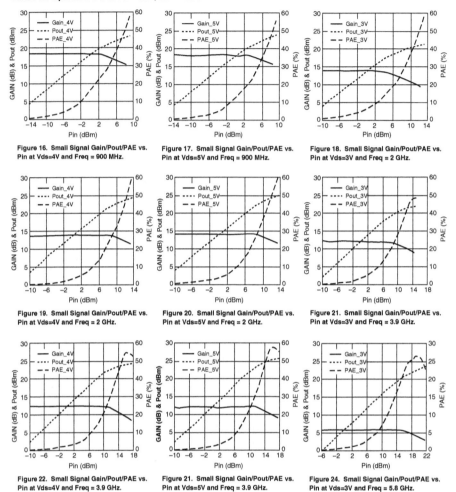

Figure 16. Small Signal Gain/Pout/PAE vs. Pin at Vds=4V and Freq = 900 MHz.

Figure 17. Small Signal Gain/Pout/PAE vs. Pin at Vds=5V and Freq = 900 MHz.

Figure 18. Small Signal Gain/Pout/PAE vs. Pin at Vds=3V and Freq = 2 GHz.

Figure 19. Small Signal Gain/Pout/PAE vs. Pin at Vds=4V and Freq = 2 GHz.

Figure 20. Small Signal Gain/Pout/PAE vs. Pin at Vds=5V and Freq = 2 GHz.

Figure 21. Small Signal Gain/Pout/PAE vs. Pin at Vds=3V and Freq = 3.9 GHz.

Figure 22. Small Signal Gain/Pout/PAE vs. Pin at Vds=4V and Freq = 3.9 GHz.

Figure 21. Small Signal Gain/Pout/PAE vs. Pin at Vds=5V and Freq = 3.9 GHz.

Figure 24. Small Signal Gain/Pout/PAE vs. Pin at Vds=3V and Freq = 5.8 GHz.

Note:
Bias current for the above charts are quiescent conditions. Actual level may increase depending on amount of RF drive.

6

ATF-53189 Typical Performance Curves (at 25°C unless specified otherwise), continued
Tuned for Optimal OIP3 at Vd = 4.0V, Ids = 135 mA.

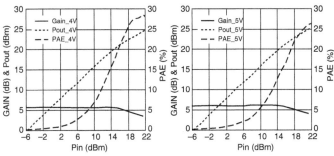

Figure 25. Small Signal Gain/Pout/PAE vs.
Pin at Vds = 4V and Freq = 5.8 GHz.

Figure 26. Small Signal Gain/Pout/PAE vs.
Pin at Vds = 5V and Freq = 5.8 GHz.

ATF-53189 Typical Performance Curves, continued
Tuned for Optimal OIP3 at Vd = 4.0V, Ids = 135 mA, Over Temperature and Frequency

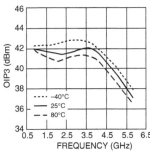

Figure 27. OIP3 vs. Temperature and
Frequency at optimum OIP3.

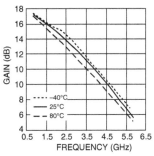

Figure 28. Gain vs. Temperature and
Frequency at optimum OIP3.

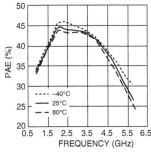

Figure 29. PAE vs. Temperature and
Frequency at optimum OIP3.

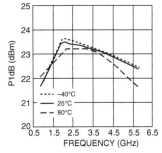

Figure 30. P1dB vs. Temperature and
Frequency at optimum OIP3.

Note:
Bias current for the above charts are quiescent conditions. Actual level may increase depending on amount of RF drive.

7

ATF-53189 Typical Performance Curves (at 25°C unless specified otherwie), continued
Tuned for Optimal OIP3 at Vd = 4.0V, Ids = 135 mA

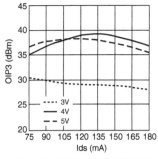

Figure 31. OIP3 vs. Ids and Vds at 2.4 GHz.

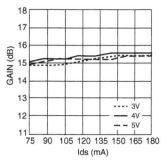

Figure 32. Small Signal Gain vs. Ids and Vds at 2.4 GHz.

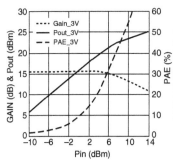

Figure 33. Small Signal Gain/Pout/PAE vs. Pin at Vds 3V and Freq = 2.4 GHz.

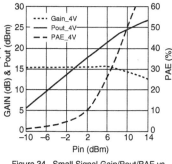

Figure 34. Small Signal Gain/Pout/PAE vs. Pin at Vds 4V and Freq = 2.4 GHz.

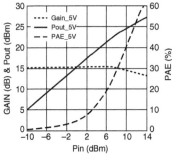

Figure 35. Small Signal Gain/Pout/PAE vs. Pin at Vds 5V and Freq = 2.4 GHz.

Note:
Bias current for the above charts are quiescent conditions. Actual level may increase depending on amount of RF drive.

8

ATF-53189 Typical Scattering and Noise Parameters at 25°C, V_{DS} = 4.0V, I_{DS} = 180 mA

Freq. GHz	S_{11} Mag.	Ang.	dB	S_{21} Mag.	Ang.	dB	S_{12} Mag.	Ang.	S_{22} Mag.	Ang.	MSG/MAG dB
0.1	0.544	-133.2	31.0	35.531	110.9	-37.7	0.013	31.7	0.692	-163.7	34.4
0.2	0.704	-158.7	25.6	19.023	97.1	-37.1	0.014	25.2	0.738	-173.2	31.3
0.3	0.777	-169.4	22.2	12.872	90.4	-36.5	0.015	24.9	0.749	-177.6	29.3
0.4	0.813	-176.1	19.7	9.705	85.7	-35.9	0.016	26.3	0.752	179.3	27.8
0.5	0.856	178.5	17.7	7.687	84.4	-35.4	0.017	30.4	0.756	175.7	26.6
0.6	0.866	174.5	16.2	6.438	81.7	-34.9	0.018	32.6	0.755	173.5	25.5
0.7	0.872	170.9	14.9	5.582	79.2	-34.4	0.019	34.5	0.755	171.4	24.7
0.8	0.874	167.5	13.9	4.939	76.5	-33.6	0.021	35.9	0.753	169.4	23.7
0.9	0.876	164.1	12.9	4.433	73.8	-33.2	0.022	36.8	0.755	167.5	23.0
1.0	0.880	161.0	12.1	4.026	70.9	-32.4	0.024	37.1	0.753	165.6	22.2
1.5	0.881	150.2	9.3	2.910	59.6	-30.5	0.030	35.8	0.753	158.4	19.2
2.0	0.882	137.1	6.5	2.123	45.9	-28.6	0.037	31.0	0.752	150.1	16.0
2.5	0.879	124.9	4.3	1.647	33.4	-27.3	0.043	25.0	0.768	142.3	13.4
3.0	0.874	112.7	2.3	1.304	21.1	-26.6	0.047	18.3	0.766	135.5	11.5
3.5	0.882	99.5	0.5	1.062	11.3	-26.0	0.050	12.6	0.773	131.8	10.0
4.0	0.889	92.6	-0.7	0.921	1.5	-25.8	0.051	7.1	0.779	123.3	9.4
5.0	0.903	78.2	-3.5	0.669	-19.8	-25.2	0.055	-5.3	0.793	102.9	7.0
6.0	0.918	61.3	-5.8	0.515	-41.5	-25.7	0.052	-22.4	0.806	84.7	5.2
7.0	0.948	41.2	-8.2	0.389	-59.6	-26.0	0.050	-39.5	0.809	69.9	3.2
8.0	0.960	24.3	-10.2	0.308	-79.9	-26.7	0.046	-55.9	0.844	54.6	2.1
9.0	0.941	11.8	-12.4	0.239	-100.5	-28.4	0.038	-73.5	0.882	37.0	1.4
10.0	0.946	10.8	-14.6	0.187	-109.4	-31.1	0.028	-81.6	0.896	27.1	0.1
11.0	0.937	0.3	-16.0	0.158	-124.9	-34.4	0.019	-108.3	0.872	20.3	-1.8
12.0	0.914	-8.0	-17.7	0.131	-138.0	-46.0	0.005	-147.3	0.916	7.0	-1.3
13.0	0.951	-12.1	-19.2	0.110	-153.4	-40.0	0.010	71.0	0.877	-1.1	-4.4
14.0	0.948	-20.6	-21.0	0.089	-168.9	-37.1	0.014	30.2	0.882	-7.5	-6.3
15.0	0.939	-23.6	-21.4	0.085	177.8	-39.2	0.011	-4.9	0.865	-19.2	-7.2
16.0	0.948	-23.1	-21.1	0.088	165.9	-37.7	0.013	-8.8	0.864	-26.2	-6.9
17.0	0.947	-24.3	-18.9	0.114	155.2	-41.9	0.008	-173.5	0.856	-33.6	-4.7
18.0	0.903	-32.5	-17.1	0.140	133.4	-35.4	0.017	161.7	0.835	-42.5	-3.2

Freq GHz	Fmin dB	Gamma Opt Mag	Ang	Rn/50	Ga dB
0.5	0.65	0.394	163.6	0.11	25.82
0.9	0.76	0.417	172.4	0.09	21.83
1.0	0.79	0.423	175.3	0.08	21.71
1.5	0.86	0.465	-165.4	0.08	18.70
2.0	0.94	0.509	-147.7	0.06	17.63
2.4	1.00	0.545	-134.6	0.08	16.45
3.0	1.10	0.600	-116.7	0.16	14.90
3.5	1.17	0.645	-103.3	0.28	13.53
5.0	1.41	0.777	-70.0	0.35	11.35
5.8	1.53	0.840	-56.1	0.41	10.31
6.0	1.56	0.855	-52.9	0.42	10.38
7.0	1.72	0.920	-39.0	0.51	9.79
8.0	1.87	0.970	-27.5	0.97	7.91
9.0	2.03	0.993	-19.1	1.88	6.11
10.0	2.18	0.997	-7.5	2.54	4.56

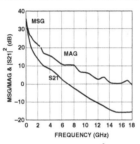

Figure 36. MSG/MAG & $|S21|^2$ vs. and Frequency at 4.0V/180 mA.

Notes:

1. F_{min} values at 2 GHz and higher are based on measurements while the F_{min} below 2 GHz have been extrapolated. The F_{min} values are based on a set of 16 noise figure measurements made at 16 different impedances using an ATN NP5 test system. From these measurements a true F_{min} is calculated. Refer to the noise parameter application section for more information.

2. S and noise parameters are measured on a microstrip line made on 0.025 inch thick alumina carrier. The input reference plane is at the end of the gate lead. The output reference plane is at the end of the drain lead.

9

ATF-53189 Typical Scattering and Noise Parameters at 25°C, V_{DS} = 4.0V, I_{DS} = 135 mA

Freq. GHz	S_{11} Mag.	Ang.	dB	S_{21} Mag.	Ang.	dB	S_{12} Mag.	Ang.	S_{22} Mag.	Ang.	MSG/MAG dB
0.1	0.544	-133.2	31.0	35.531	110.9	-37.7	0.013	31.7	0.692	-163.7	34.4
0.2	0.704	-158.7	25.6	19.023	97.1	-37.1	0.014	25.2	0.738	-173.2	31.3
0.3	0.777	-169.4	22.2	12.872	90.4	-36.5	0.015	24.9	0.749	-177.6	29.3
0.4	0.813	-176.1	19.7	9.705	85.7	-35.9	0.016	26.3	0.752	179.3	27.8
0.5	0.856	178.5	17.7	7.687	84.4	-35.4	0.017	30.4	0.756	175.7	26.6
0.6	0.866	174.5	16.2	6.438	81.7	-34.9	0.018	32.6	0.755	173.5	25.5
0.7	0.872	170.9	14.9	5.582	79.2	-34.4	0.019	34.5	0.755	171.4	24.7
0.8	0.874	167.5	13.9	4.939	76.5	-33.6	0.021	35.9	0.753	169.4	23.7
0.9	0.876	164.1	12.9	4.433	73.8	-33.2	0.022	36.8	0.755	167.5	23.0
1.0	0.880	161.0	12.1	4.026	70.9	-32.4	0.024	37.1	0.753	165.6	22.2
1.5	0.881	150.2	9.3	2.910	59.6	-30.5	0.030	35.8	0.753	158.4	19.2
2.0	0.882	137.1	6.5	2.123	45.9	-28.6	0.037	31.0	0.752	150.1	16.0
2.5	0.879	124.9	4.3	1.647	33.4	-27.3	0.043	25.0	0.768	142.3	13.4
3.0	0.874	112.7	2.3	1.304	21.1	-26.6	0.047	18.3	0.766	135.5	11.5
3.5	0.882	99.5	0.5	1.062	11.3	-26.0	0.050	12.6	0.773	131.8	10.0
4.0	0.889	92.6	-0.7	0.921	1.5	-25.8	0.051	7.1	0.779	123.3	9.4
5.0	0.903	78.2	-3.5	0.669	-19.8	-25.2	0.055	-5.3	0.793	102.9	7.0
6.0	0.918	61.3	-5.8	0.515	-41.5	-25.7	0.052	-22.4	0.806	84.7	5.2
7.0	0.948	41.2	-8.2	0.389	-59.6	-26.0	0.050	-39.5	0.809	69.9	3.2
8.0	0.960	24.3	-10.2	0.308	-79.9	-26.7	0.046	-55.9	0.844	54.6	2.1
9.0	0.941	11.8	-12.4	0.239	-100.5	-28.4	0.038	-73.5	0.882	37.0	1.4
10.0	0.946	10.8	-14.6	0.187	-109.4	-31.1	0.028	-81.6	0.896	27.1	0.1
11.0	0.937	0.3	-16.0	0.158	-124.9	-34.4	0.019	-108.3	0.872	20.3	-1.8
12.0	0.914	-8.0	-17.7	0.131	-138.0	-46.0	0.005	-147.3	0.916	7.0	-1.3
13.0	0.951	-12.1	-19.2	0.110	-153.4	-40.0	0.010	71.0	0.877	-1.1	-4.4
14.0	0.948	-20.6	-21.0	0.089	-168.9	-37.1	0.014	30.2	0.882	-7.5	-6.3
15.0	0.939	-23.6	-21.4	0.085	177.8	-39.2	0.011	-4.9	0.865	-19.2	-7.2
16.0	0.948	-23.1	-21.1	0.088	165.9	-37.7	0.013	-8.8	0.864	-26.2	-6.9
17.0	0.947	-24.3	-18.9	0.114	155.2	-41.9	0.008	-173.5	0.856	-33.6	-4.7
18.0	0.903	-32.5	-17.1	0.140	133.4	-35.4	0.017	161.7	0.835	-42.5	-3.2

Freq GHz	Fmin dB	Gamma Opt Mag	Ang	Rn/50	Ga dB
0.5	0.30	0.162	150.8	0.05	26.27
0.9	0.41	0.291	161.3	0.05	22.12
1.0	0.44	0.302	164.2	0.05	22.02
1.5	0.53	0.369	-174.2	0.04	18.95
2.0	0.62	0.433	-154.6	0.04	17.05
2.4	0.69	0.484	-140.2	0.05	15.87
3.0	0.80	0.556	-120.6	0.10	14.63
3.5	0.89	0.613	-106.1	0.19	13.21
5.0	1.16	0.764	-71.0	0.26	11.19
5.8	1.31	0.832	-56.6	0.30	10.26
6.0	1.34	0.848	-53.4	0.30	10.04
7.0	1.52	0.914	-39.3	0.39	9.64
8.0	1.71	0.963	-27.9	0.77	8.68
9.0	1.89	0.991	-18.2	0.96	6.57
10.0	2.07	0.998	-9.2	1.58	4.51

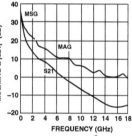

Figure 37. MSG/MAG & $|S21|^2$ vs. and Frequency at 4.0V/135 mA.

Notes:

1. F_{min} values at 2 GHz and higher are based on measurements while the F_{min} below 2 GHz have been extrapolated. The F_{min} values are based on a set of 16 noise figure measurements made at 16 different impedances using an ATN NP5 test system. From these measurements a true Fmin is calculated. Refer to the noise parameter application section for more information.
2. S and noise parameters are measured on a microstrip line made on 0.025 inch thick alumina carrier. The input reference plane is at the end of the gate lead. The output reference plane is at the end of the drain lead.

10

ATF-53189 Typical Scattering and Noise Parameters at 25°C, V_{DS} = 4.0V, I_{DS} = 75 mA

Freq. GHz	S_{11} Mag.	S_{11} Ang.	dB	S_{21} Mag.	S_{21} Ang.	dB	S_{12} Mag.	S_{12} Ang.	S_{22} Mag.	S_{22} Ang.	MSG/MAG dB
0.1	0.544	-133.2	31.0	35.531	110.9	-37.7	0.013	31.7	0.692	-163.7	34.4
0.2	0.704	-158.7	25.6	19.023	97.1	-37.1	0.014	25.2	0.738	-173.2	31.3
0.3	0.777	-169.4	22.2	12.872	90.4	-36.5	0.015	24.9	0.749	-177.6	29.3
0.4	0.813	-176.1	19.7	9.705	85.7	-35.9	0.016	26.3	0.752	179.3	27.8
0.5	0.856	178.5	17.7	7.687	84.4	-35.4	0.017	30.4	0.756	175.7	26.6
0.6	0.866	174.5	16.2	6.438	81.7	-34.9	0.018	32.6	0.755	173.5	25.5
0.7	0.872	170.9	14.9	5.582	79.2	-34.4	0.019	34.5	0.755	171.4	24.7
0.8	0.874	167.5	13.9	4.939	76.5	-33.6	0.021	35.9	0.753	169.4	23.7
0.9	0.876	164.1	12.9	4.433	73.8	-33.2	0.022	36.8	0.755	167.5	23.0
1.0	0.880	161.0	12.1	4.026	70.9	-32.4	0.024	37.1	0.753	165.6	22.2
1.5	0.881	150.2	9.3	2.910	59.6	-30.5	0.030	35.8	0.753	158.4	19.2
2.0	0.882	137.1	6.5	2.123	45.9	-28.6	0.037	31.0	0.752	150.1	16.0
2.5	0.879	124.9	4.3	1.647	33.4	-27.3	0.043	25.0	0.768	142.3	13.4
3.0	0.874	112.7	2.3	1.304	21.1	-26.6	0.047	18.3	0.766	135.5	11.5
3.5	0.882	99.5	0.5	1.062	11.3	-26.0	0.050	12.6	0.773	131.8	10.0
4.0	0.889	92.6	-0.7	0.921	1.5	-25.8	0.051	7.1	0.779	123.3	9.4
5.0	0.903	78.2	-3.5	0.669	-19.8	-25.2	0.055	-5.3	0.793	102.9	7.0
6.0	0.918	61.3	-5.8	0.515	-41.5	-25.7	0.052	-22.4	0.806	84.7	5.2
7.0	0.948	41.2	-8.2	0.389	-59.6	-26.0	0.050	-39.5	0.809	69.9	3.2
8.0	0.960	24.3	-10.2	0.308	-79.9	-26.7	0.046	-55.9	0.844	54.6	2.1
9.0	0.941	11.8	-12.4	0.239	-100.5	-28.4	0.038	-73.5	0.882	37.0	1.4
10.0	0.946	10.8	-14.6	0.187	-109.4	-31.1	0.028	-81.6	0.896	27.1	0.1
11.0	0.937	0.3	-16.0	0.158	-124.9	-34.4	0.019	-108.3	0.872	20.3	-1.8
12.0	0.914	-8.0	-17.7	0.131	-138.0	-46.0	0.005	-147.3	0.916	7.0	-1.3
13.0	0.951	-12.1	-19.2	0.110	-153.4	-40.0	0.010	71.0	0.877	-1.1	-4.4
14.0	0.948	-20.6	-21.0	0.089	-168.9	-37.1	0.014	30.2	0.882	-7.5	-6.3
15.0	0.939	-23.6	-21.4	0.085	177.8	-39.2	0.011	-4.9	0.865	-19.2	-7.2
16.0	0.948	-23.1	-21.1	0.088	165.9	-37.7	0.013	-8.8	0.864	-26.2	-6.9
17.0	0.947	-24.3	-18.9	0.114	155.2	-41.9	0.008	-173.5	0.856	-33.6	-4.7
18.0	0.903	-32.5	-17.1	0.140	133.4	-35.4	0.017	161.7	0.835	-42.5	-3.2

Freq GHz	Fmin dB	Gamma Opt Mag	Gamma Opt Ang	Rn/50	Ga dB
0.5	0.32	0.175	127.6	0.05	26.45
0.9	0.41	0.224	143.8	0.04	21.98
1.0	0.43	0.235	148.3	0.03	21.50
1.5	0.49	0.306	173.6	0.03	18.55
2.0	0.56	0.375	-163.6	0.03	16.33
2.4	0.61	0.428	-147.2	0.04	15.18
3.0	0.69	0.507	-125.3	0.08	13.86
3.5	0.75	0.569	-109.3	0.14	12.68
5.0	0.95	0.738	-72.0	0.20	10.81
5.8	1.05	0.814	-57.4	0.24	10.64
6.0	1.08	0.831	-54.2	0.24	9.97
7.0	1.21	0.907	-40.5	0.30	9.25
8.0	1.34	0.961	-29.3	0.60	7.78
9.0	1.47	0.992	-19.3	0.71	6.96
10.0	1.60	0.996	-8.9	1.01	4.46

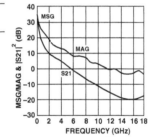

Figure 38. MSG/MAG & $|S21|^2$ vs. and Frequency at 4.0V/75 mA.

Notes:

1. F_{min} values at 2 GHz and higher are based on measurements while the F_{min} below 2 GHz have been extrapolated. The F_{min} values are based on a set of 16 noise figure measurements made at 16 different impedances using an ATN NP5 test system. From these measurements a true Fmin is calculated. Refer to the noise parameter application section for more information.
2. S and noise parameters are measured on a microstrip line made on 0.025 inch thick alumina carrier. The input reference plane is at the end of the gate lead. The output reference plane is at the end of the drain lead.

11

ATF-53189 Typical Scattering and Noise Parameters at 25°C, V_{DS} = 5.0V, I_{DS} = 135 mA

Freq. GHz	S_{11} Mag.	Ang.	dB	S_{21} Mag.	Ang.	dB	S_{12} Mag.	Ang.	S_{22} Mag.	Ang.	MSG/MAG dB
0.1	0.544	-133.2	31.0	35.531	110.9	-37.7	0.013	31.7	0.692	-163.7	34.4
0.2	0.704	-158.7	25.6	19.023	97.1	-37.1	0.014	25.2	0.738	-173.2	31.3
0.3	0.777	-169.4	22.2	12.872	90.4	-36.5	0.015	24.9	0.749	-177.6	29.3
0.4	0.813	-176.1	19.7	9.705	85.7	-35.9	0.016	26.3	0.752	179.3	27.8
0.5	0.856	178.5	17.7	7.687	84.4	-35.4	0.017	30.4	0.756	175.7	26.6
0.6	0.866	174.5	16.2	6.438	81.7	-34.9	0.018	32.6	0.755	173.5	25.5
0.7	0.872	170.9	14.9	5.582	79.2	-34.4	0.019	34.5	0.755	171.4	24.7
0.8	0.874	167.5	13.9	4.939	76.5	-33.6	0.021	35.9	0.753	169.4	23.7
0.9	0.876	164.1	12.9	4.433	73.8	-33.2	0.022	36.8	0.755	167.5	23.0
1.0	0.880	161.0	12.1	4.026	70.9	-32.4	0.024	37.1	0.753	165.6	22.2
1.5	0.881	150.2	9.3	2.910	59.6	-30.5	0.030	35.8	0.753	158.4	19.2
2.0	0.882	137.1	6.5	2.123	45.9	-28.6	0.037	31.0	0.752	150.1	16.0
2.5	0.879	124.9	4.3	1.647	33.4	-27.3	0.043	25.0	0.768	142.3	13.4
3.0	0.874	112.7	2.3	1.304	21.1	-26.6	0.047	18.3	0.766	135.5	11.5
3.5	0.882	99.5	0.5	1.062	11.3	-26.0	0.050	12.6	0.773	131.8	10.0
4.0	0.889	92.6	-0.7	0.921	1.5	-25.8	0.051	7.1	0.779	123.3	9.4
5.0	0.903	78.2	-3.5	0.669	-19.8	-25.2	0.055	-5.3	0.793	102.9	7.0
6.0	0.918	61.3	-5.8	0.515	-41.5	-25.7	0.052	-22.4	0.806	84.7	5.2
7.0	0.948	41.2	-8.2	0.389	-59.6	-26.0	0.050	-39.5	0.809	69.9	3.2
8.0	0.960	24.3	-10.2	0.308	-79.9	-26.7	0.046	-55.9	0.844	54.6	2.1
9.0	0.941	11.8	-12.4	0.239	-100.5	-28.4	0.038	-73.5	0.882	37.0	1.4
10.0	0.946	10.8	-14.6	0.187	-109.4	-31.1	0.028	-81.6	0.896	27.1	0.1
11.0	0.937	0.3	-16.0	0.158	-124.9	-34.4	0.019	-108.3	0.872	20.3	-1.8
12.0	0.914	-8.0	-17.7	0.131	-138.0	-46.0	0.005	-147.3	0.916	7.0	-1.3
13.0	0.951	-12.1	-19.2	0.110	-153.4	-40.0	0.010	71.0	0.877	-1.1	-4.4
14.0	0.948	-20.6	-21.0	0.089	-168.9	-37.1	0.014	30.2	0.882	-7.5	-6.3
15.0	0.939	-23.6	-21.4	0.085	177.8	-39.2	0.011	-4.9	0.865	-19.2	-7.2
16.0	0.948	-23.1	-21.1	0.088	165.9	-37.7	0.013	-8.8	0.864	-26.2	-6.9
17.0	0.947	-24.3	-18.9	0.114	155.2	-41.9	0.008	-173.5	0.856	-33.6	-4.7
18.0	0.903	-32.5	-17.1	0.140	133.4	-35.4	0.017	161.7	0.835	-42.5	-3.2

Freq GHz	Fmin dB	Gamma Opt Mag	Ang	Rn/50	Ga dB
0.5	0.36	0.266	149.9	0.05	26.51
0.9	0.46	0.315	162.4	0.04	22.79
1.0	0.49	0.327	165.6	0.04	22.09
1.5	0.59	0.388	-172.7	0.04	18.92
2.0	0.69	0.448	-153.0	0.04	17.04
2.4	0.77	0.495	-138.6	0.06	15.87
3.0	0.88	0.563	-116.3	0.12	14.50
3.5	0.98	0.617	-104.9	0.21	13.11
5.0	1.28	0.764	-70.5	0.31	11.19
5.8	1.44	0.830	-56.5	0.37	10.10
6.0	1.48	0.845	-53.4	0.38	10.08
7.0	1.68	0.912	-39.7	0.42	9.39
8.0	1.88	0.960	-28.3	0.84	8.78
9.0	2.08	0.988	-18.3	1.24	8.05
10.0	2.28	0.994	-8.5	1.78	4.74

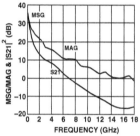

Figure 39. MSG/MAG & $|S21|^2$ vs. and Frequency at 5.0V/135 mA.

Notes:
1. F_{min} values at 2 GHz and higher are based on measurements while the F_{min} below 2 GHz have been extrapolated. The F_{min} values are based on a set of 16 noise figure measurements made at 16 different impedances using an ATN NP5 test system. From these measurements a true Fmin is calculated. Refer to the noise parameter application section for more information.
2. S and noise parameters are measured on a microstrip line made on 0.025 inch thick alumina carrier. The input reference plane is at the end of the gate lead. The output reference plane is at the end of the drain lead.

ATF-53189 Typical Scattering and Noise Parameters at 25°C, V_{DS} = 3.0V, I_{DS} = 135 mA

Freq. GHz	S_{11} Mag.	Ang.	S_{21} dB	Mag.	Ang.	S_{12} dB	Mag.	Ang.	S_{22} Mag.	Ang.	MSG/MAG dB
0.1	0.544	-133.2	31.0	35.531	110.9	-37.7	0.013	31.7	0.692	-163.7	34.4
0.2	0.704	-158.7	25.6	19.023	97.1	-37.1	0.014	25.2	0.738	-173.2	31.3
0.3	0.777	-169.4	22.2	12.872	90.4	-36.5	0.015	24.9	0.749	-177.6	29.3
0.4	0.813	-176.1	19.7	9.705	85.7	-35.9	0.016	26.3	0.752	179.3	27.8
0.5	0.856	178.5	17.7	7.687	84.4	-35.4	0.017	30.4	0.756	175.7	26.6
0.6	0.866	174.5	16.2	6.438	81.7	-34.9	0.018	32.6	0.755	173.5	25.5
0.7	0.872	170.9	14.9	5.582	79.2	-34.4	0.019	34.5	0.755	171.4	24.7
0.8	0.874	167.5	13.9	4.939	76.5	-33.6	0.021	35.9	0.753	169.4	23.7
0.9	0.876	164.1	12.9	4.433	73.8	-33.2	0.022	36.8	0.755	167.5	23.0
1.0	0.880	161.0	12.1	4.026	70.9	-32.4	0.024	37.1	0.753	165.6	22.2
1.5	0.881	150.2	9.3	2.910	59.6	-30.5	0.030	35.8	0.753	158.4	19.2
2.0	0.882	137.1	6.5	2.123	45.9	-28.6	0.037	31.0	0.752	150.1	16.0
2.5	0.879	124.9	4.3	1.647	33.4	-27.3	0.043	25.0	0.768	142.3	13.4
3.0	0.874	112.7	2.3	1.304	21.1	-26.6	0.047	18.3	0.766	135.5	11.5
3.5	0.882	99.5	0.5	1.062	11.3	-26.0	0.050	12.6	0.773	131.8	10.0
4.0	0.889	92.6	-0.7	0.921	1.5	-25.8	0.051	7.1	0.779	123.3	9.4
5.0	0.903	78.2	-3.5	0.669	-19.8	-25.2	0.055	-5.3	0.793	102.9	7.0
6.0	0.918	61.3	-5.8	0.515	-41.5	-25.7	0.052	-22.4	0.806	84.7	5.2
7.0	0.948	41.2	-8.2	0.389	-59.6	-26.0	0.050	-39.5	0.809	69.9	3.2
8.0	0.960	24.3	-10.2	0.308	-79.9	-26.7	0.046	-55.9	0.844	54.6	2.1
9.0	0.941	11.8	-12.4	0.239	-100.5	-28.4	0.038	-73.5	0.882	37.0	1.4
10.0	0.946	10.8	-14.6	0.187	-109.4	-31.1	0.028	-81.6	0.896	27.1	0.1
11.0	0.937	0.3	-16.0	0.158	-124.9	-34.4	0.019	-108.3	0.872	20.3	-1.8
12.0	0.914	-8.0	-17.7	0.131	-138.0	-46.0	0.005	-147.3	0.916	7.0	-1.3
13.0	0.951	-12.1	-19.2	0.110	-153.4	-40.0	0.010	71.0	0.877	-1.1	-4.4
14.0	0.948	-20.6	-21.0	0.089	-168.9	-37.1	0.014	30.2	0.882	-7.5	-6.3
15.0	0.939	-23.6	-21.4	0.085	177.8	-39.2	0.011	-4.9	0.865	-19.2	-7.2
16.0	0.948	-23.1	-21.1	0.088	165.9	-37.7	0.013	-8.8	0.864	-26.2	-6.9
17.0	0.947	-24.3	-18.9	0.114	155.2	-41.9	0.008	-173.5	0.856	-33.6	-4.7
18.0	0.903	-32.5	-17.1	0.140	133.4	-35.4	0.017	161.7	0.835	-42.5	-3.2

Freq GHz	Fmin dB	Gamma Opt Mag	Ang	Rn/50	Ga dB
0.5	0.34	0.225	146.2	0.05	26.30
0.9	0.43	0.282	157.0	0.04	22.19
1.0	0.45	0.296	160.2	0.04	22.07
1.5	0.53	0.362	-177.0	0.03	19.00
2.0	0.61	0.427	-156.3	0.03	17.13
2.4	0.68	0.478	-141.3	0.05	15.89
3.0	0.78	0.551	-121.1	0.09	14.59
3.5	0.86	0.608	-106.2	0.17	13.17
5.0	1.10	0.763	-70.8	0.24	11.22
5.8	1.24	0.832	-56.6	0.28	10.16
6.0	1.27	0.848	-53.5	0.30	9.93
7.0	1.43	0.915	-39.7	0.38	9.57
8.0	1.60	0.964	-28.4	0.74	8.78
9.0	1.76	0.991	-18.5	0.95	7.27
10.0	1.93	0.995	-8.6	1.55	3.39

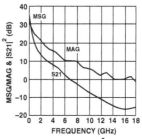

Figure 40. MSG/MAG & $|S21|^2$ vs. and Frequency at 3.0V/135 mA.

Notes:
1. F_{min} values at 2 GHz and higher are based on measurements while the F_{min} below 2 GHz have been extrapolated. The F_{min} values are based on a set of 16 noise figure measurements made at 16 different impedances using an ATN NP5 test system. From these measurements a true Fmin is calculated. Refer to the noise parameter application section for more information.
2. S and noise parameters are measured on a microstrip line made on 0.025 inch thick alumina carrier. The input reference plane is at the end of the gate lead. The output reference plane is at the end of the drain lead.

13

CGH40180PP

180 W, RF Power GaN HEMT

Cree's CGH40180PP is an unmatched, gallium nitride (GaN) high electron mobility transistor (HEMT). The CGH40180PP, operating from a 28 volt rail, offers a general purpose, broadband solution to a variety of RF and microwave applications. GaN HEMTs offer high efficiency, high gain and wide bandwidth capabilities making the CGH40180PP ideal for linear and compressed amplifier circuits. The transistor is available in a 4-lead flange package.

Package Types: 440199
PN: CGH40180PP

FEATURES

- Up to 2.5 GHz Operation
- 20 dB Small Signal Gain at 1.0 GHz
- 15 dB Small Signal Gain at 2.0 GHz
- 220 W typical P_{SAT}
- 70 % Efficiency at P_{SAT}
- 28 V Operation

APPLICATIONS

- 2-Way Private Radio
- Broadband Amplifiers
- Cellular Infrastructure
- Test Instrumentation
- Class A, AB, Linear amplifiers suitable for OFDM, W-CDMA, EDGE, CDMA waveforms

Large Signal Models Available for SiC & GaN

Rev 1.3 – September 2011

Subject to change without notice.
www.cree.com/wireless

1

Absolute Maximum Ratings (not simultaneous) at 25 °C Case Temperature

Parameter	Symbol	Rating	Units
Drain-Source Voltage	V_{DSS}	84	Volts
Gate-to-Source Voltage	V_{GS}	-10, +2	Volts
Storage Temperature	T_{STG}	-65, +150	°C
Operating Junction Temperature	T_J	225	°C
Maximum Forward Gate Current	I_{GMAX}	60	mA
Soldering Temperature[1]	T_S	245	°C
Screw Torque	τ	80	in-oz
Thermal Resistance, Junction to Case[2]	$R_{\theta JC}$	0.9	°C/W
Case Operating Temperature[2,3]	T_C	-40, +150	°C

Note:
[1] Refer to the Application Note on soldering at www.cree.com/products/wireless_appnotes.asp
[2] CGH40180PP at P_{DISS} = 224 W.
[3] See also, the Power Dissipation De-rating Curve on Page 6.

Electrical Characteristics (T_C = 25°C)

Characteristics	Symbol	Min.	Typ.	Max.	Units	Conditions
DC Characteristics						
Gate Threshold Voltage	$V_{GR(th)}$	-3.8	-3.3	-2.3	V_{DC}	V_{DS} = 10 V, I_D = 57.6 mA
Gate Quiescent Voltage	$V_{GS(Q)}$	–	-3.0	–	V_{DC}	V_{DS} = 28 V, I_D = 2.0 A
Saturated Drain Current[2]	I_{DS}	46.4	56.0	–	A	V_{GS} = 6.0 V, V_{DS} = 2.0 V
Drain-Source Breakdown Voltage	V_{BR}	120	–	–	V_{DC}	V_{GS} = -8 V, I_D = 57.6 mA
RF Characteristics[3,4] (T_C = 25 °C, F = 1.3 GHz unless otherwise noted)						
Small Signal Gain	G_{SS}	17.5	19	–	dB	V_{DD} = 28 V, I_{DQ} = 2.0 A
Power Output at Saturation[5]	P_{SAT}	180	220	–	W	V_{DD} = 28 V, I_{DQ} = 2.0 A
Drain Efficiency[6]	η	50	65	–	%	V_{DD} = 28 V, I_{DQ} = 2.0 A, P_{OUT} = P_{SAT}
Output Mismatch Stress	VSWR	–	–	10 : 1	Ψ	No damage at all phase angles, V_{DD} = 28 V, I_{DQ} = 2.0 A, P_{OUT} = 180 W CW
Dynamic Characteristics[7]						
Input Capacitance	C_{GS}	–	35.7	–	pF	V_{DS} = 28 V, V_{gs} = -8 V, f = 1 MHz
Output Capacitance	C_{DS}	–	9.6	–	pF	V_{DS} = 28 V, V_{gs} = -8 V, f = 1 MHz
Feedback Capacitance	C_{GD}	–	1.6	–	pF	V_{DS} = 28 V, V_{gs} = -8 V, f = 1 MHz

Notes:
[1] Measured on wafer prior to packaging.
[2] Scaled from PCM data.
[3] Measured in CGH40180PP-TB, including all coupler losses.
[4] I_{DQ} of 2.0 A is by biasing each device at 1.0 A.
[5] P_{SAT} is defined as: Q1 or Q2 = I_G = 2.8 mA.
[6] Drain Efficiency = P_{OUT} / P_{DC}
[7] Capacitance values are for each side of the device.

Source: Copyright © 2009-2011 Cree Inc.

Typical Performance

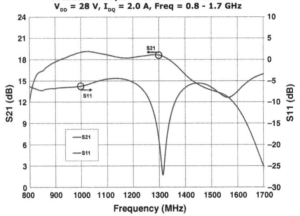

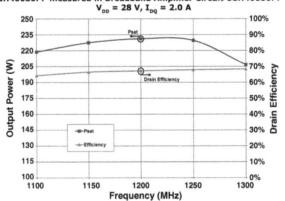

Source: Copyright © 2009-2011 Cree Inc.

CREE®

Typical Performance

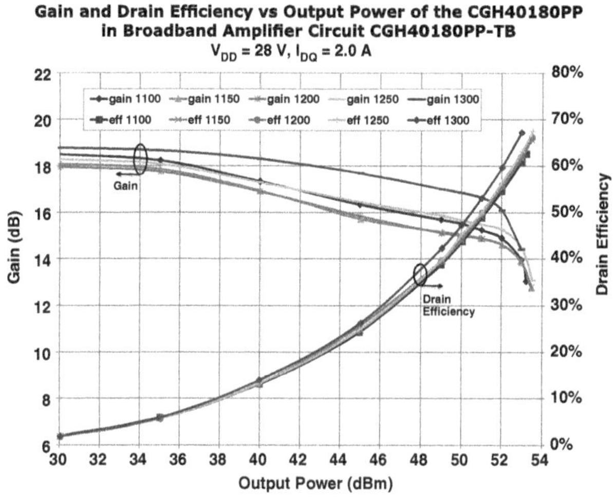

Cree, Inc.
4600 Silicon Drive
Durham, NC 27703
USA Tel: +1.919.313.5300
Fax: +1.919.869.CREE
Fax: +1.919.869.2733
www.cree.com/wireless

Source: Copyright © 2009-2011 Cree Inc.

Typical Performance

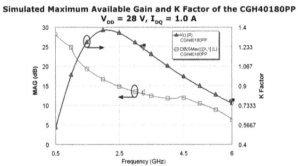

Simulated Maximum Available Gain and K Factor of the CGH40180PP
V_{DD} = 28 V, I_{DQ} = 1.0 A

Typical Noise Performance

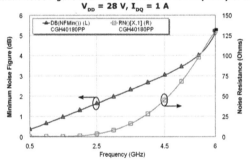

Simulated Minimum Noise Figure and Noise Resistance vs Frequency of the CGH40180PP
V_{DD} = 28 V, I_{DQ} = 1 A

Electrostatic Discharge (ESD) Classifications

Parameter	Symbol	Class	Test Methodology
Human Body Model	HBM	1A > 250 V	JEDEC JESD22 A114-D
Charge Device Model	CDM	1 < 200 V	JEDEC JESD22 C101-C

5 CGH40180PP Rev 1.3 Preliminary

Source: Copyright © 2009-2011 Cree Inc.

CGH40180PP Power Dissipation De-rating Curve

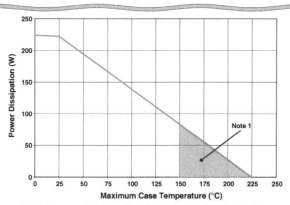

Note 1. Area exceeds Maximum Case Operating Temperature (See Page 2).

CGH40180PP Transient Power Dissipation De-rating Curve

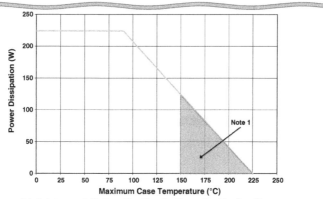

Note 1. Area exceeds Maximum Case Operating Temperature (See Page 2).
Note 2. This transient de-rating curve assumes a 1msec pulse with a 20%
duty cycle with no power dissipated during the "off-cycle."

Source: Copyright © 2009-2011 Cree Inc.

Thermal Resistance as a Function of Pulse Width

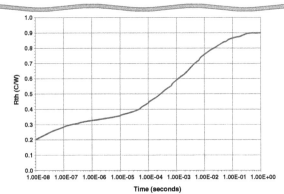

Note 1: This heating curve assumes zero power dissipation during the "off" portion of the duty cycle.

Note 2: This data is for transient power dissipation at 224 W, Duty Cycle = 20 %.

Simulated Source and Load Impedances

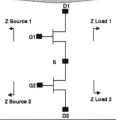

Frequency (MHz)	Z Source	Z Load
500	2.85 + j1.99	5.27 + j0.68
1000	0.8 + j0.42	4.91 + j0.36
1500	0.84 - j1.69	4.65 - j0.24
2000	0.88 - j3.05	2.8 - j1.05
2500	1.08 - j4.5	3.1 - j2.47
3000	1.25 - j6.06	3.1 - j4.01

Note 1. V_{DD} = 28V, I_{DQ} = 2.0 A in the 440199 package.

Note 2. Optimized for power gain, P_{SAT} and PAE.

Note 3. When using this device at low frequency, series resistors should be used to maintain amplifier stability.

Source: Copyright © 2009-2011 Cree Inc.

CGH40180PP-TB Demonstration Amplifier Circuit Schematic

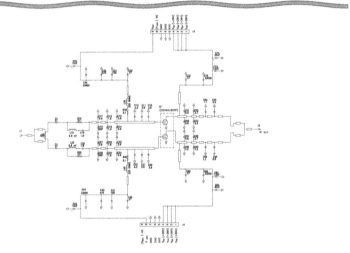

CGH40180PP-TB Demonstration Amplifier Circuit Outline

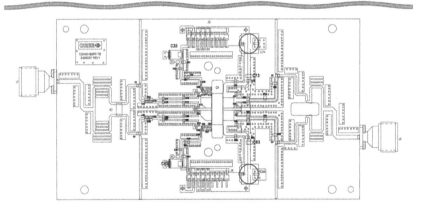

Cree, Inc.
4600 Silicon Drive
Durham, NC 27703
USA Tel: +1.919.313.5300
Fax: +1.919.869.CREE
Fax: +1.919.869.2733
www.cree.com/wireless

Source: Copyright © 2009-2011 Cree Inc.

CGH40180PP-TB Demonstration Amplifier Circuit Bill of Materials

Designator	Description	Qty
R1	RES, 100 Ohm, +/-1%, 1 W, 2512	1
R10,R20	RES, 511 Ohm, +/- 5%, 1/16W, 0603	2
R30,R40	RES, 1/16W, 0603, 1%, 5.1 OHMS	2
C1,C2,C3,C4,C30,C40,C70,C80	CAP, 27 pF, +/-5% 0805,ATC600F	8
C10,C11,C13,C14,C20,C21,C23,C24	CAP, 3.9PF, +/-0.1 pF, 0603, ATC600S	8
C12,C22	CAP, 3.3PF, +/-0.1 pF, 0603, ATC600S	2
C15,C19,C25,C29	CAP, 1.8PF, +/-0.1 pF, 0603, ATC600S	4
C16,C26	CAP, 1.0PF, +/-0.1 pF, 0603, ATC600S	2
C17,C27	CAP, 0.9PF, +/-0.1 pF, 0603, ATC600S	2
C31,C41	CAP, 100 pF,+/-5%, 0603,ATC600S	2
C32,C42	CAP, 470 pF, 5%, 100V, 0603, X7R	2
C34,C44,C72,C82	CAP, 33000 pF, 0805, 100V, X7R	4
C35,C45	CAP, 10 uF, 16V, TANTALUM	2
C50,C51,C60,C61	CAP, 5.6 pF, +/-0.1 pF, 0805, ATC600F	4
C52,C62	CAP, 2.7 pF, +/-0.1 pF, 0805, ATC600F	2
C53,C63	CAP, 2.2 pF, +/-0.1 pF, 0805, ATC600F	2
C54,C64	CAP, 1.1 pF, +/-0.05 pF, 0805, ATC600F	2
C55,C65	CAP, 0.5 pF, +/-0.05 pF, 0805, ATC600F	2
C73,C83	CAP, 1.0 uF, +/-10%, 1210, 100V, X7R	2
C74,C84	CAP, 33 uF, 100V, ELECT, FK, SMD	2
L10,L20	IND, 6.8 nH, 0603, L-14C6N8ST	2
L30,L40	FERRITE, 220 OHM, 0603, BLM21PG221SN1	2
J1,J2	CONN, N-Type, Female, 0.500 SMA Flange	2
J3,J4	CONN, Header, RT> PLZ, 0.1 CEN, LK, 9 POS	2
	PCB, RO4350, Er = 3.48, h = 20 mil	1
Q1	CGH40180PP	1

CGH40180PP-TB Demonstration Amplifier Circuit

Source: Copyright © 2009-2011 Cree Inc.

Typical Package S-Parameters for CGH40180PP, Single Side
(Small Signal, V_{DS} = 28 V, I_{DQ} = 1000 mA, angle in degrees)

Frequency	Mag S11	Ang S11	Mag S21	Ang S21	Mag S12	Ang S12	Mag S22	Ang S22
500 MHz	0.957	-177.48	4.22	79.26	0.007	10.74	0.798	-179.16
600 MHz	0.957	-178.74	3.51	76.30	0.007	12.14	0.800	-179.41
700 MHz	0.957	-179.78	3.00	73.47	0.007	13.71	0.802	-179.63
800 MHz	0.957	179.32	2.62	70.74	0.007	15.38	0.804	-179.84
900 MHz	0.957	178.51	2.33	68.08	0.007	17.15	0.807	179.96
1.0 GHz	0.957	177.76	2.09	65.49	0.007	18.99	0.809	179.74
1.1 GHz	0.957	177.06	1.90	62.95	0.007	20.87	0.812	179.52
1.2 GHz	0.957	176.38	1.73	60.46	0.007	22.80	0.814	179.28
1.3 GHz	0.957	175.72	1.60	58.02	0.008	24.73	0.817	179.03
1.4 GHz	0.956	175.08	1.48	55.63	0.008	26.66	0.820	178.76
1.5 GHz	0.956	174.44	1.38	53.29	0.008	28.57	0.823	178.46
1.6 GHz	0.956	173.81	1.29	50.98	0.008	30.44	0.825	178.15
1.7 GHz	0.956	173.18	1.22	48.72	0.008	32.25	0.828	177.82
1.8 GHz	0.955	172.55	1.15	46.50	0.009	33.98	0.831	177.47
1.9 GHz	0.955	171.91	1.09	44.32	0.009	35.62	0.833	177.10
2.0 GHz	0.955	171.27	1.04	42.17	0.009	37.17	0.835	176.71
2.1 GHz	0.954	170.62	0.99	40.06	0.010	38.61	0.838	176.30
2.2 GHz	0.954	169.96	0.95	37.98	0.010	39.93	0.840	175.87
2.3 GHz	0.953	169.29	0.91	35.93	0.011	41.14	0.842	175.42
2.4 GHz	0.952	168.60	0.87	33.91	0.011	42.22	0.844	174.95
2.5 GHz	0.952	167.90	0.84	31.92	0.012	43.18	0.845	174.47
2.6 GHz	0.951	167.18	0.82	29.95	0.013	44.01	0.847	173.96
2.7 GHz	0.950	166.45	0.79	28.00	0.013	44.73	0.848	173.44
2.8 GHz	0.949	165.69	0.77	26.07	0.014	45.32	0.849	172.89
2.9 GHz	0.948	164.91	0.75	24.15	0.015	45.79	0.850	172.33
3.0 GHz	0.946	164.10	0.73	22.24	0.016	46.15	0.850	171.74
3.2 GHz	0.943	162.39	0.71	18.45	0.018	46.53	0.851	170.51
3.4 GHz	0.939	160.55	0.69	14.64	0.020	46.47	0.850	169.19
3.6 GHz	0.935	158.53	0.67	10.80	0.023	45.97	0.848	167.76
3.8 GHz	0.929	156.31	0.67	6.86	0.027	45.03	0.845	166.21
4.0 GHz	0.922	153.83	0.67	2.78	0.031	43.63	0.841	164.53
4.2 GHz	0.913	151.03	0.68	-1.51	0.036	41.72	0.834	162.69
4.4 GHz	0.901	147.82	0.69	-6.12	0.042	39.23	0.825	160.65
4.6 GHz	0.886	144.10	0.72	-11.16	0.049	36.07	0.813	158.39
4.8 GHz	0.866	139.68	0.76	-16.81	0.059	32.05	0.797	155.86
5.0 GHz	0.838	134.36	0.81	-23.30	0.073	26.92	0.775	153.00
5.2 GHz	0.799	127.78	0.88	-30.99	0.091	20.30	0.747	149.76
5.4 GHz	0.742	119.49	0.97	-40.41	0.117	11.55	0.708	146.16
5.6 GHz	0.658	108.92	1.08	-52.33	0.157	-0.34	0.657	142.31
5.8 GHz	0.534	95.85	1.21	-67.76	0.219	-16.90	0.594	138.62
6.0 GHz	0.373	82.93	1.34	-87.69	0.321	-40.38	0.534	134.70

Download this s-parameter file in ".s2p" format at http://www.cree.com/products/wireless_s-parameters.asp

Source: Copyright © 2009-2011 Cree Inc.

2.5 PARAMETER EXTRACTION OF ACTIVE DEVICES

2.5.1 Introduction

We have already seen that all linear models that are used on the market are really derived from large-signal models, and it is our point that the quality of the model depends mostly on the quality of the parameters used to describe the model.

As far as modeling is concerned, there are two options. The first option is a physics-based model, which can be applied to items such as diodes and bipolar transistors, including its most advanced versions, such as HBTs, SiGe versions, and other future derivatives. The bipolar transistor is essentially a combination of two diodes whereby the base–emitter junction generates the electrons or holes (depending on how the reader is accustomed to viewing the process) and then transfers those to the collector. The emitter, as its name states, emits the charged particles and the collector collects them, minus some current losses, which are expressed in the current gain of the device. A current gain (β) of 100 means that the collector receives 1% less than the emitter emits.

Earlier in this chapter, we went through a detailed physics-based derivation about the inner workings of these devices, explaining how, for a given dc bias point, one can simplify them to a small-signal equivalent circuit. This is possible since we assume that the RF current and voltages are less than 1% of the dc equivalent currents and voltages. Reducing a device's complex nonlinear equivalent circuit to a quasistatic linear equivalent circuit allows its manufacturer to generate and publish sets of S parameters for it. These parameters are bias-, voltage-, and temperature dependent. At higher frequencies, parasitics play an enormous role in determining a device's RF performance.

A device operating at signal levels that are more than 1% of its dc equivalent currents and voltages must be evaluated in terms of large-signal performance. Most of the large-signal-equivalent circuits for nonbipolar devices, such as FETs (MOS, silicon JFETs, and members of the GaAsFET family) the large-signal equivalent, is an analytic approximation that essentially bears no physical insight into the inner working of the transistor. This fact is particularly painful because to look at high-order intermodulation distortion, one needs to have third and fourth derivatives of mathematically continuous equations. The SPICE approach cannot easily be translated into modern harmonic-balance simulators because such cop-outs as the use of IF THEN ELSE statements in programming provide everything else but a continuous model. In the case of our own work, we used complicated curve-fitting equations to obtain analytic equations and its derivatives to accurately provide a large-signal model that is valid over a wide range of dc and RF.

2.5.2 Typical SPICE Parameters

Since we are about to evaluate bipolar microwave transistors, junction FETs, MOSFETs (model level 3), and GaAsFETs. Tables 2.19–22 list typical parameters for the devices we have used. These parameters can be obtained by the Scout program with the appropriate measurements. The BSIM model for MOSFETs, applicable for submicrometer technology transistors, requires an enormous level of parameter extraction and has not fully been validated for the LDMOS-type transistors currently favored for RF and microwave applications.

The meaning and significance of the various parameters is best explored in a book on SPICE or semiconductor physics [37–39].

Table 2.19 BFR193W Bipolar Junction Transistor

IS	0.2738	fA	BF	125		NF	0.95341	
VAF	24	V	IKF	0.26949	A	ISE	10.627	fA
NE	1.935		BR	14.267		NR	1.4289	
VAR	3.8742	V	IKR	0.037925	A	ISC	0.037409	fA
NC	0.94371		RB	1.8368	Ω	IRB	0.91763	mA
RBM	1	Ω	RE	0.76534		RC	0.11938	Ω
CJE	1.1824	fF	VJE	0.70276	V	MJE	0.48654	
TF	18.828	ps	XTF	0.69477		VTF	0.8	V
ITF	0.96893	mA	PTF	0	deg	CJC	935.03	fF
VJC	1.1828	V	MJC	0.30002		XCJC	0.053563	
TR	1.0037	ns	CJS	0	fF	VJS	0.75	V
MJS	0		NK	0		EG	1.11	eV
XTI	3		FC	0.72063		TNOM	300	K

Table 2.20 2SK125 JFET

IDSS	0.0525	VP0	−3.111	GAMA	−0.1867 E-01	E	15.20
KE	−0.3856 E-03	SL	0.2818 E-01	KG	−0.2398	T	0
SS	0.7448 E-04	IG0	0.2 E-14	AFAG	38.46	IB0	0.1 E-04
AFAB	38	VBC	30	R10	17.11	KR	0
C10	0.6609 E-11	K1	1.675	C1S	0.6818 E-33	CF0	0.7261 E-11
KF	1.156	RG	0.5	RD	1.542	RS	1.333
LG	0.6098 E-09	LD	0.5159 E-08	LS	0.1482 E-08	CDS	0.4813 E-16
CGE	0.1590 E-11	CDE	0.3394 E-26	CGSP	0.8282 E-13	CDSP	0.4832 E-12
ZGT	50	LGT	0.4712 E-01	ZDT	50	LDT	0.3998 E-01
CGDP	0.3653 E-12	ZST	50	LST	0.1495 E-01	CGDE	0.3831 E-12
CGSB	0.3120 E-13	CDSB	0.5896 E-12	VDMX	10		

Note for junction FETs: The currently implemented model for junction FET is too primitive for serious RF applications. We have therefore taken the approach (liberty) to use the Materka parameter extraction approach for silicon junction FETs. This has resulted in unparalleled high-quality parameters; in particular, the knee voltage behavior has significantly improved, as well as the overall frequency response.

Table 2.21 GaAs MESFET

IDSS	0.1077	VP0	−1.8	GAMA	−0.5741 E-01	E	1.29
KE	−0.1155 E-01	SL	0.1652	KG	−0.1782	T	0
SS	−0.1208 E-02	IG0	0.213 E-11	AFAG	27.4	IB0	0.5680 E-09
AFAB	1.826	VBC	9	R10	8.382	KR	0.6359
C10	0.5964 E-12	K1	1.296	C1S	0	CF0	0.611 E-13
KF	0.9775	RG	1.996	RD	1.296	RS	1.234
CDS	0.7852 E-13	CDSD	1. E-08	RDSD	158.1	CGE	0.1609 E-12
CDE	0.8674 E-13	VDMX	8				

Table 2.22 1 μm × 750 μm Level 3 LDMOS FET

CBD	0.863 E-12	CGD0	166. E-12	CGS0	246. E-12	GAMA	0.211		
IS	6.53 E-16	KAPA	0.809	MJ	0.536	NSUB	1.E 15		
PB	0.71	PBSW	0.71	PHI	0.579	RD	39		
RS	0.1	THET	0.588	TOX	4. E-8	U0	835		
VMAX	3.38 E5	VT0	2.78	XQC	0.41				

Table 2.23 Diode Noise Model Key Words

Key Word	Description	Unit	Default
ID	Required bias current for the data point	Ampere	
KF	Flicker noise coefficient		0.0
AF	Bias exponent of the flicker noise model		1.0
FCP	Frequency exponent of the flicker noise model		1.0
FC	Flicker noise corner frequency	Hz	

2.5.3 Noise Modeling

2.5.3.1 Diode Noise Model

The noise model for the diodes (Figure 2.127) consists of two contributions: the shot noise and the flicker noise. The shot noise is computed automatically and does not require any parameters. The flicker noise can be specified in two ways:

1. using the enhanced SPICE noise model by specifying KF, AF, and FCP in the model/parameter list (this option is usually sufficient for most applications);
2. using bias-dependent flicker noise coefficients (specifying KF and AF at multiple bias points).

The noise generators in the diode noise model are the series parasitic resistance, R_S, and the intrinsic junction. The figure below illustrates the intrinsic junction noise generator. Let Δf be the bandwidth (usually normalized to 1 Hz). The intrinsic noise generator has a mean-square value of

$$\langle i_{Dn}^2 \rangle = 2qI_D\Delta f + KF\frac{I_D^{\text{AF}}}{f^{\text{FCP}}}\Delta f \tag{2.253}$$

Notes on the diode noise model.

1. Shot noise is always present unless the SN parameter is set to zero. Turning noise off is useful for comparing the total circuit noise that is generated by the nonlinear devices and that generated by the linear circuit components.

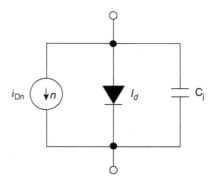

Figure 2.127 Equivalent noise circuit for a diode chip.

2. If the value of KF is specified as zero, then the flicker noise will not be contributed by the device and only shot noise is considered in the intrinsic model.

3. The corner frequency noise model uses the system noise floor to internally compute the flicker noise coefficient, KF. The system noise floor is computed by the program using the diode parameters and kT.

4. This noise model of course considers the actual operating temperature, which must be supplied to the model.

2.5.3.2 BJT Noise Model

The noise model for the Gummel–Poon BJT model consists of two contributions: shot noise and the flicker noise. The shot noise is computed automatically and does not require any parameters. The flicker noise can be specified in two ways:

1. using the enhanced SPICE noise model by specifying KF, AF, and FCP in the model parameter list (this option is usually sufficient for most applications);

2. using bias-dependent flicker noise coefficients (specifying KF and AF at multiple bias points).

Option 1: *Specifying the Bias-Independent Flicker Noise Coefficient*

This option involves the straightforward specification of KF, AF, and FCP that are constant with bias, as in the SPICE noise model. Notes on Option 1.

1. Shot noise is always present unless it is turned off. Turning noise off is useful for comparing the total circuit noise that is generated by the nonlinear devices and that generated by the linear circuit components.

2. If the value of KF is specified as zero, flicker noise will not be contributed by the device and only shot noise is considered in the intrinsic model.

Option 2: *Specifying the Bias-Dependent Flicker Noise Coefficient or Flicker Corner Frequency*

Option 2 allows a bias-dependent flicker noise coefficient (i.e., KF and AF vary with the bias point).

Notes on the BJT noise model.

1. KF, AF, and FC can be specified as bias dependent. If only one set of noise data is specified, the corresponding bias point is not meaningful because all parameters are considered constant over all bias values. However, the bias point is needed for the program to identify the data as bipolar noise data.

2. The corner frequency noise model option uses the system noise floor to compute the flicker noise coefficient, KF. The system noise floor is computed by the program using the transistor parameters and k_T.

3. This noise model of course considers the actual operating temperature, which must be supplied to the model.

Figure 2.128 shows the BJT noise model. Let Δf be the bandwidth (usually normalized to a 1-Hz bandwidth). The noise generators introduced in the intrinsic device are shown

Table 2.24 BJT Noise Model Key Words

Key Word	Description	Unit	Default
IB	Required base bias current for the data point	ampere	
VCE	Required collector-emitter voltage for the data point	volt	
VBS	Base-substrate voltage required for LPNP type when four nodes are used.	volt	
VCS	Collector-substrate voltage required for NPN or PNP type when four nodes are used.	volt	
KF	Flicker noise coefficient		0.0
AF	Bias exponent of the flicker noise model		1.0
FCP	Frequency exponent of the flicker noise model		1.0
FC	Flicker noise corner frequency	Hz	

below, and have mean-square values of

$$\langle i_{bn}^2 \rangle = 2qI_B\Delta f + KF\frac{I_B^{AF}}{f^{FCP}}\Delta f \tag{2.254}$$

$$\langle i_{cn}^2 \rangle = 2qI_c\Delta f \tag{2.255}$$

$$\langle i_{R_{bb}}^2 \rangle = \frac{4kT}{R_{bb}}\Delta f \tag{2.256}$$

$$\langle i_{R_{el}}^2 \rangle = \frac{4kT}{R_{el}}\Delta f \tag{2.257}$$

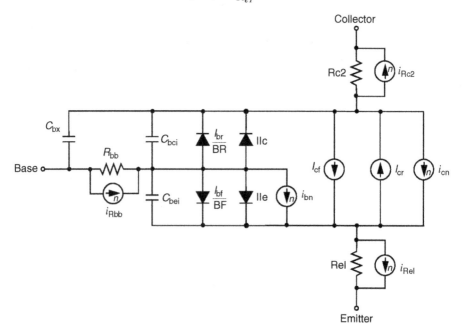

Figure 2.128 BJT noise model (not showing extrinsic parasitics). Current sources with n are noise sources.

$$\langle i^2_{R_{c2}}\rangle = \frac{4kT}{R_{c2}}\Delta f \qquad (2.258)$$

$$I_B = \frac{I_{\mathrm{bf}}}{BF} + I_{le} \qquad (2.259)$$

$$I_c = I_{cf} - I_{cr} \qquad (2.260)$$

2.5.3.3 JFET and MESFET Noise Model

The noise model for the FETs consists of two contributions: the shot noise and the flicker noise. There are two options to specify noise in the FET model.

1. Using the enhanced SPICE noise model by specifying KF, AF, and FCP in the model parameter list to determine the flicker noise (this option is usually sufficient for most applications). The shot noise will be automatically computed using the SPICE equation.
2. Using bias-dependent flicker noise coefficients through a reference in the DATA block (specifying KF and AF at multiple bias points) and specifying the four noise parameters ($F_{\min}$, MG_{opt}, PG_{opt}, and R_n) at multiple bias points.

Option 1: *Specifying the Enhanced SPICE Noise Model*

Option 1 is the straightforward specification of KF, AF, and FCP that are constant with bias, as in the SPICE noise model.

The drain noise model has the form

$$\langle |I_{\mathrm{dn}}|^2\rangle = 4K_BT\frac{2g_m}{3}\Delta f + KF\frac{|I_D|^{\mathrm{AF}}}{f^{\mathrm{FCP}}}\Delta f \qquad (2.261)$$

where the shot noise is derived from g_m and the flicker noise is proportional to KF and the drain channel current, ID, and inversly proportional to frequency. The AF and FCP parameters tailor the flicker noise dependence on bias and frequency, respectively.

Notes on Option 1.

1. Shot noise is always present unless it is turned off. Turning noise off is useful for comparing the total circuit noise that is generated by the nonlinear devices and that generated by the linear circuit components.
2. If the value of KF is specified as zero, then flicker noise will not be contributed by the device and only shot noise is considered in the intrinsic model.

Option 2: *Specifying the Bias-Dependent Flicker Noise Coefficient or Flicker Corner Frequency*

Option 2 allows the specification of the complex bias-dependent nature of the shot noise and flicker noise. At high frequencies, the equivalent noise sources are correlated (the SPICE noise model does not account for this correlation). The complete evaluation of the shot noise sources can be determined from the four noise parameters. Since these are functions of bias, they can be specified over the (V_{GS}, V_{DS}) bias plane.

Additionally, a bias-dependent flicker noise coefficient (i.e., KF and AF vary with current) can be specified.

Table 2.25 FET Noise Model Key Words

Key Word	Description	Unit	Default
FN	Noise data measurement frequency	Hz	1.0 GHz
VGS	Required gate-source voltage for the data point	volt	
VDS	Required drain-source voltage for the data point	volt	
FMIN	Required minimum noise figure in dB at FN		
MGO	Required magnitude of optimum noise reflectioncoefficient at FN		
PGO	Required phase of optimum noise reflection coefficient at FN		
RN	Required normalized noise resistance at FN		
KF	Flicker noise coefficient		0.0
AF	Bias exponent of the flicker noise model		1.0
FCP	Frequency exponent of the flicker noise model		1.0
FC	Flicker noise corner frequency	Hz	

The MESFET noise model uses the four measured noise data (F_{min}, Γ_{opt}, and R_n) at one frequency and multiple arbitrary bias points. The program uses this data and the FET model parameters to de-embed the noise data to an intrinsic noise model. The intrinsic model is accurate at all frequencies, and therefore can predict the noise performance at all frequencies given data at just one frequency point. Built-in bias-dependent characteristics are used if multibias noise data is not provided.

Notes on the FET noise model.

1. The corner frequency noise model option uses the system noise floor to compute the flicker noise coefficient, KF. The system noise floor is computed by the program using the transistor parameters and k_T.

2. This noise model of course considers the actual operating temperature, which must be supplied to the model.

Noise in a MESFET is produced by sources intrinsic to the device. The same approach, but with different flicker corner frequencies, is highly applicable to JFETs and MOSFETs. For more detail as to simulation, see the Element library book for the active device portion of Ansoft's Designer. The equivalent noisy circuit of an intrinsic FET is represented in Figure 2.129.

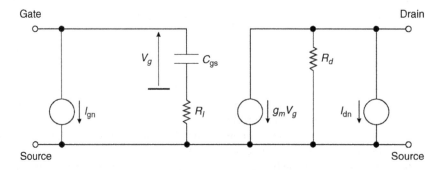

Figure 2.129 Equivalent noise circuit of an intrinsic FET device.

The intrinsic FET is internally represented as a noiseless nonlinear two-port with one equivalent noise current connected across the gate-source terminal and one across the drain-source terminal. The correlations of the gate and drain noise current sources are

$$\langle |I_{\text{gn}}|^2 \rangle = 4K_B T \Delta f \frac{\omega^2 C_{\text{gs}}^2}{g_m} R \tag{2.262}$$

$$\langle |I_{\text{gn}}|^2 \rangle = 4K_B T \Delta f g_m P \tag{2.263}$$

$$\langle I_{\text{gn}} I_{\text{dn}}^* \rangle = 4K_B T \Delta f j\omega C_{\text{gs}} \sqrt{PR}\, C \tag{2.264}$$

The correlation matrix of the noise current sources is

$$C_{\text{dc}}(\omega) = \frac{2}{\pi} K_B T\, d\omega \begin{bmatrix} \frac{\omega^2 C_{\text{gs}}^2}{g_m} R & -j\omega C_{\text{gs}} \sqrt{PR}\, C \\ j\omega C_{\text{gs}} \sqrt{PR}\, C & g_m P \end{bmatrix} \tag{2.265}$$

The gate and drain noise parameters R and P and the correlation coefficient C are related to the physical noise sources acting in the channel and are functions of the device structure and bias noise parameters. By defining measured noise parameters, F_{min}, R_n and Γ_{opt}, and using a noise-de-embedding procedure, the parameters R, P, and C and the intrinsic noise correlation matrix of a FET device as functions of device bias are determined by the program.

In addition to the noise sources shown above, the flicker ($1/f$) noise can also be modeled by means of a noise current source connected in parallel with the intrinsic drain port. The flicker noise component in a narrow band, Δf, is expressed in the form

$$\langle |I_F|^2 \rangle = Q \Delta f \frac{|I_D|^{\text{AF}}}{f^{\text{FCP}}} \tag{2.266}$$

where I_D is the instantaneous value of the channel current, and Q, AF, and FCP are empirical parameters. In most practical cases, AF and FCP are directly obtained from measurements (typically, AF = 2 and FCP = 1), while Q is not. In Ansoft's Serenade Design Environment, Q is either provided directly using KF or is computed by providing the flicker corner frequency (FC). FC is the frequency at which the flicker noise equals the shot/diffusion noise. The corner frequency is defined by the equation

$$Q \frac{|I_D|^{\text{AF}}}{f_c^{\text{FCP}}} = g_m P \tag{2.267}$$

Given the corner frequency FC and the measurement bias point V_{gs} and V_{ds}, the program automatically computes I_D, g_m, and P, and finally Q.

More information on FET noise modeling can be found in [40–46].

2.5.3.4 MOSFET Noise Model

The MOSFET noise model (Figure 2.130) consists of two contributions: the shot noise and the flicker noise. The shot noise is computed automatically and does not require any

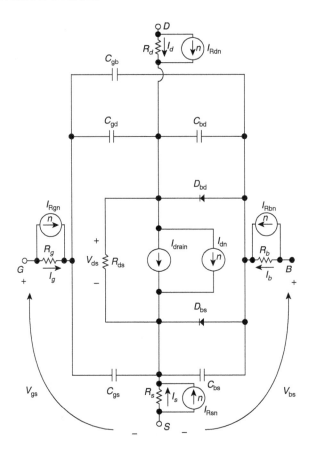

Figure 2.130   Equivalent noise circuit of an intrinsic MOSFET device.

parameters. It can be turned off by specifying SN = 0. The flicker noise can be specified in two ways:

1. using the enhanced SPICE noise model by specifying KF, AF, and FCP in the model parameter list (this option is usually sufficient for most applications);
2. using bias-dependent flicker noise coefficients through a reference (specifying KF and AF at multiple bias points).

Option 1: *Specifying the Enhanced SPICE Noise Model*

This option is the straightforward specification of KF, AF, and FCP that are constant with bias, as in the SPICE noise model (the flicker noise is considered bias dependent). Notes on the MOSFET noise model.

1. Shot noise is always present unless the SN parameter is set to zero. Turning noise off is useful for comparing the total circuit noise that is generated by the nonlinear devices and that generated by the linear circuit components.

Table 2.26 Noise Model Key Words

Key Word	Description	Unit	Default
VGS	Required gate-source bias for the data point	volt	
VDS	Required drain-source for the data point	volt	
VBS	Required drain-bulk for the data point	volt	
KF	Flicker noise coefficient		1.0E−13
AF	Bias exponent of the flicker noise model		2.0
FCP	Frequency exponent of the flicker noise model		1.0
FC	Flicker noise corner frequency	Hz	

2. If the value of KF is specified as zero, then the flicker noise will not be contributed by the device and only shot noise is considered in the intrinsic model.

Option 2: *Specifying the Bias-Dependent Flicker Noise Coefficient or Flicker Corner Frequency*

This option allows a bias-dependent flicker noise coefficient (i.e., KF and AF varies with drain current). The MOSFET noise model data is given and referenced by a model parameter.

Notes on the MOSFET noise model.

1. KF, AF, and FC can be specified as bias dependent. If only one set of noise data is specified, the corresponding bias point is not meaningful because all parameters are considered constant over all bias values. However, the bias point is needed for the program to identify the data as MOSFET noise data.
2. The corner frequency noise model option uses the system noise floor to compute the flicker noise coefficient, KF. The system noise floor is computed by the program using the transistor parameters and k_T.
3. This noise model of course considers the actual operating temperature, which must be supplied to the model.

Let Δf be the bandwidth (normalized to 1 Hz). The noise generators introduced in the intrinsic device are shown below, and have mean-square values of

$$\langle i_{dn}^2 \rangle = \frac{8kTg_m}{3}\Delta f + KF\frac{I_D^{AF}}{f^{FCP}}\Delta f$$

$$\langle i_{Rgn}^2 \rangle = 4\frac{kT}{R_g}\Delta f$$

$$\langle i_{Rdn}^2 \rangle = 4\frac{kT}{R_d}\Delta f$$

$$\langle i_{Rsn}^2 \rangle = 4\frac{kT}{R_s}\Delta f$$

$$\langle i_{Rbn}^2 \rangle = 4\frac{kT}{R_b}\Delta f$$

We include this MOSFET noise model (used for quite awhile) for completeness. At the moment, we do not know which MOSFET noise model the industry will settle on in the future.

Modern CAD tools, such as Ansoft's Serenade product, use these models allowing to generate quite accurate noise data based on a good linear equivalent model. Internally, it uses the noise-correlation-matrix method (first introduced by Russer). The same noise-correlation techniques apply to BJTs, all FETs except MOS, and HBTs (such as SiGe devices).

2.5.4 Scalable Device Models

Since diodes and transistors are scalable, here are guidelines for how to use and scale them:

Microwave Diode
$ID = \text{area} \times ID$
$C_j = \text{area} \times C_j$
$RD = RD/\text{area}$

PIN Diode
$ID = \text{area} \times ID$
$C_j = \text{area} \times C_j$
$RS = RS/\text{area}$
$R_{\max} = R_{\max}/\text{area}$

Bipolar

$I_{bf} = \text{area} \times I_{bf}$	$I_{br} = \text{area} \times I_{br}$
$I_{le} = \text{area} \times I_{le}$	$I_{lc} = \text{area} \times I_{lc}$
$I_{cf} = \text{area} \times I_{cf}$	$I_{cr} = \text{area} \times I_{cr}$
$C_{bc} = \text{area} \times C_{bc}$	$C_{be} = \text{area} \times C_{be}$
$C_{bx} = \text{area} \times C_{bx}$	$R_{bb} = R_{bb}/\text{area}$
$RB_2 = RB_2/\text{area}$	$RC_2 = RC_2/\text{area}$
$RE_1 = RE_1/\text{area}$	$I_{jss} = I_{jss} \times \text{area}$
	$C_{jss} = C_{jss} \times \text{area}$

Materka FET
$IGSS = IGSS \times \text{area}$
$CGS0 = CGS0 \times \text{area}$
$CGS1 = CGS1 \times \text{area}$
$CDVC = CDVC \times \text{area}$
$CDVS = CDVS \times \text{area}$
$RG = RG \times \text{area} / (\text{number of fingers} \times 2)$
$RD = RD/\text{area}$
$RS = RS/\text{area}$

2.5.5 Generating a Databank for Parameter Extraction

Given the fact that one can measure with an automated system (e.g., by Rohde & Schwarz, Agilent) the bias-dependent S parameters of any active device, bipolar and FET, we can generate a huge databank that allows us to have corresponding values for both RF and dc operating points. To validate this, one needs to run a large-signal simulator, such as

Ansoft's Serenade Design Environment, or its equivalent from other manufacturers, enter the large-signal parameters and generate a set of S parameters that correspond to the dc values for which we have measurements. The agreement between measured and predicted points for both RF and dc values is a good measure to evaluate the accuracy of the parameter extraction and the simulator's performance. Model accuracy, however, is a relative measure with respect to the actual circuit to be designed. For example, a model that describes noise and weakly nonlinear behavior of an LNA with high accuracy might easily fail to predict the power-added efficiency of a switch-mode amplifier. In any case, a model must always be validated against relevant nonlinear measurement. A very helpful tool in this respect are modern nonlinear vector network analyzers or large-signal S-parameter measurement systems that allow for measurement of dynamic $I–V$ trajectories as a function of frequency, bias point, driving power, and load impedance. These measurements give a wealth of information for model validation. We cannot stress often enough, however, the fact that while most simulator models are quite good, the weakest point of the link is the parameter extraction program.

Currently, the best parameter extraction for bipolar transistors and FETs, excluding the BSIM model, is implemented in the Scout program by Ansoft, based on modified techniques found in the literature. These techniques are based on the fact that present-day nonlinear modeling of microwave devices, especially FETs, is not exactly adequate to describe all effects found within these devices. In the case of FETs, the most troublesome parameter is R_{DS}, which is really more bias dependent than practically all existing models take into consideration. Techniques used in published extraction methods extract nonlinear models independently for each section of a device model, and then assume that the proper response will be generated when all parts are put together. For instance, many extraction techniques independently fit the $I–V$ data of the section describing the $I–V$ curves and then fit the $C–V$ data to the equivalent small-signal models. This results in an approximation to the bias-dependent response of the complete model.

In the technique used in the Scout program, model parameters are extracted simultaneously so the effects between model sections are accounted for and the device is treated as a whole. The most critical area happens to be in the linear region, typically around the pinchoff area, and for very small supply voltages. The Scout techniques use measured data taken at a wide variety of bias ranges and frequencies to define the model response. The possible solution region for the complete model becomes well identified and uniqueness of solution is vastly improved. The Scout program is interactive in nature and displays simultaneous measured versus predicted data. This allows for fine-tuning of the parameters.

Over the years, the industry has adapted a large number of models. Those supported by the Scout parameter–extraction program include:

- *npn* Gummel-Poon BJT
- *pnp* Gummel-Poon BJT
- *npn* heterojunction BJT
- *pnp* heterojunction BJT
- general-purpose *N*-channel JFET
- general-purpose *P*-channel JFET
- Chalmers (Angelov) MESFET/HEMT
- Curtice-Ettenberg cubic MESFET/HEMT
- Curtice quadratic MESFET/HEMT
- IAF (Berroth) MESFET/HEMT
- modified Materka-Kacprzak MESFET/HEMT

- Raytheon (Statz) MESFET/HEMT
- TriQuint (TOM-1 and TOM-2) MESFET/HEMT
- Ansoft physics-based MESFET

The Ansoft physics-based MESFET model is unique and politically sensitive because its users are essentially required to enter all device fabrication parameters to get the appropriate answer, and most, if not all, companies will stay away from handing out such company proprietary recipes. The validation of this model was done using manufacturing data for transistors that were developed under government contract and therefore accessible for the team members. Compact Software/Ansoft over the last 10 years has been involved in a large number of government-funded research and development programs, and practically all modern modeling implementations and their validation were done in conjunction with the largest semiconductor manufacturers.

As far as further validation is concerned, besides our looking at the matching of dc $I-V$ curves and S parameters, important parameters such as 1-dB compression point, power-added efficiency, and harmonic content are important criteria to evaluate the quality of such an undertaking. As far as hardware is concerned, one needs to have a network analyzer, such as the ZVA (Rohde & Schwarz) or PNA (Agilent). Its selection is determined by the highest cutoff frequency where measurements are necessary. A dc feed capability, up to several amperes, through the measuring ports and appropriate power supplies, are essential, as is a dc $I-V$ curve tracer. The built-in various optimizers of the program establish the best possible match between the selected parameters and the measured results.

The area of proper modeling has always been fascinating because the active device model can make or break first-pass success. Figure 2.131 shows the doubler gain comparison of a model supplied by NEC (the device manufacturer, parameter extraction done with Scout,

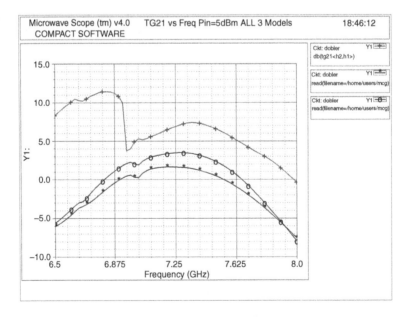

Figure 2.131   Gain comparison of a MESFET doubler based on the three device models used: NEC (+), Ansoft (O), and Microwave Engineering Europe (*).

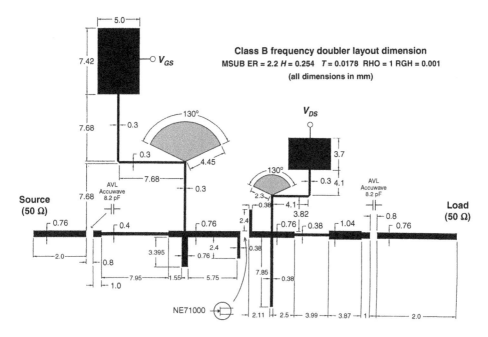

Figure 2.132   Layout of the MESFET doubler.

and parameter extraction supplied from the *Microwave Engineering Europe* magazine in its CAD review of May 1994. It is somewhat unclear where the *MEE* model came from. The actual task given by *MEE* was to simulate a frequency doubler against measured data. This type of simulation required two areas of high precision, one being the model and other being the electromagnetic modeling of the discontinuities of the circuit—in this case, a stub element. The circuit itself was very simple. The manufacturer-supplied data gave the poorest results (we believe the parameter extraction came from a third party not quite up to speed in this area). The results that came from the Scout parameter extraction and the third one that *MEE* supplied were fairly close, yet the Scout solution, including the electromagnetic simulation part, gave the best answer. More simulators have since appeared in the market, and it would be interesting to revisit this topic. Figure 2.132 shows physical layout of the doubler circuit.

Test Setup   The measuring equipment required for the actual extraction includes the following.

- Semiconductor parameter analyzer (HP4145)
- Vector network analyzer (ZVR, HP8510, W360)
- Bias supplies and meters
- Power RF source, power meters, spectrum analyzer, and so on

The required software tools include:

- GPIB controller: Compact NETCOM (network analyzer control program)
- Parameter extraction performed by Scout

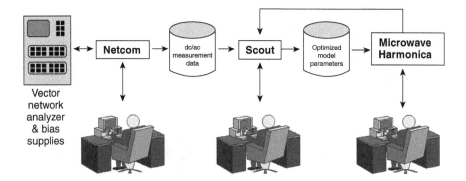

Figure 2.133 Flow chart of the parameter extraction process.

- Small-signal and harmonic-balance validation: Serenade nonlinear analysis (Microwave Harmonica) or its equivalent.

After we have assembled all these tools, the device extraction process can begin. It will follow the flow as shown in Figure 2.133. Depending on the numbers for the dc/ac measurement, this process can take up to 2 h. The Scout program, with its built-in optimizer, can take another 2 h. It is therefore reasonable to assume a thorough parameter extraction of a device will take about 1 day.

Parameter Extraction As we have outlined, the parameter extraction program Scout, which is optimized for high-frequency and millimeter-wave application, is a Windows-based parameter extraction and optimization utility. It requires a network-communication program to acquire dc and S-parameter measurement data. It supports interactive device fitting and optimization, and supports the various models listed earlier. For ease of use, it has an attractive graphical user interface with multiple windows, as can be seen from Figure 2.134. Agilent offers more general RF parameter extraction programs, but we did not have access to those. One of the unfortunate deficiencies of most BJT parameter extraction programs known to us is the fact that (as one example) the sensitivity to determine the actual base-spreading resistor for bipolar transistors is poor. The only useful workaround is considering the noise figure of the device, given the fact that the midrange noise factor of any bipolar transistor can be calculated from

$$F = 1 + \frac{R_{bb'}}{R_g} + \frac{0.5\left(\dfrac{26\,\mathrm{mV}}{I_C}\right)}{R_g} \quad \text{(approximate equation at medium frequencies)} \quad (2.268)$$

By solving the equation for R'_{bb}, since we know that R_g, the generator impedance, is 50 Ω, and the emitter diffusion resistance (26 mV/I_c) can be determined from the bias point, we can already eliminate the need to extract R'_{bb} from the test equipment. By using a generic algorithm based on S-parameters alone, errors of a factor up to 10 are not uncommon because of the low sensitivity for this parameter. Another headache in the case of the BJT is the determination of the excess phase (PTF) and forward and reverse transit time (TF and TR, respectively). The lack of quality modeling can be best found by examining the

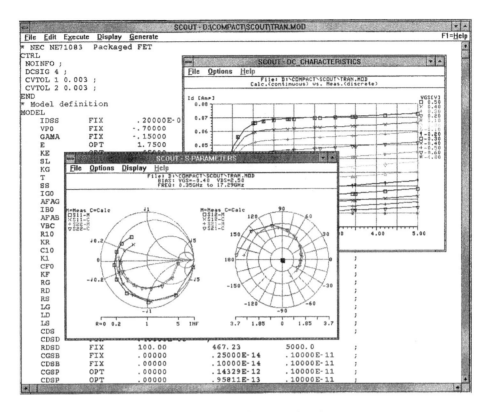

Figure 2.134 The Scout user interface.

measured versus predicted values of S_{22}. Again, practically all the parameter extraction programs we have seen so far suffer from this deficiency.

2.5.5.1 MESFETs

In the case of a member of the MESFET family, a model as shown in Figure 2.135 is used. It consists of an intrinsic model and a complete chip/package parasitic model. In the case of the MESFET, to obtain first-class results, the following measurements are necessary:

- dc I–V measurements
- dc diode measurements
- cold-FET and forward-biased-gate S-parameters
- S-parameter measurements over bias
- harmonic power measurements

Z-parameter equations for Figure 2.136:

$$Z_{11} = R_S + R_g + R_{gs} + \alpha_g R_{ch} + j\omega\left(L_g + L_S\right) \tag{2.269}$$

$$Z_{12} = Z_{21} = R_S + \alpha R_{ch} + j\omega L_S \tag{2.270}$$

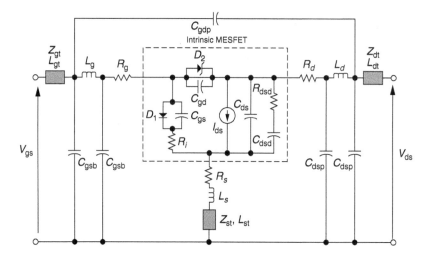

Figure 2.135   Intrinsic model and complete chip/package model.

$$Z_{22} = R_S + R_D + 2\alpha R_{ch} + j\omega (L_S + L_D) \tag{2.271}$$

$$C_B = -\frac{\text{Im}(Y_{12})}{\omega} \tag{2.272}$$

$$C_{pg} = \frac{\text{Im}(Y_{11})}{\omega} - 2C_B \tag{2.273}$$

$$C_{pd} + C_{ds} = \frac{\text{Im}(Y_{22})}{\omega} - C_B \tag{2.274}$$

In order to simplify the modeling, it is useful to do cold-FET (Figure 2.135 and equations 2.269–2.271) and pinched-FET (Figure 2.136 and equations 2.272–2.274)

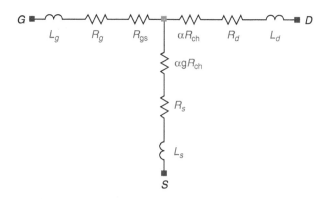

Figure 2.136   Forward-biased gate FET model for cold-FET ($V_{DS} = 0$, $V_{GS} > 0$) measurements. This approach allows direct extraction of parasitic resistances and inductances.

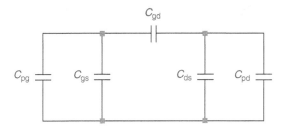

Figure 2.137 Pinched-FET ($V_{DS} = 0$, $V_{GS} <$ VP) model. This approach allows direct extraction of parasitic gate and drain capacitances.

measurements to derive some of the intrinsic or extrinsic values. For the highest accuracy, the following extraction strategies are recommended.

- Preextraction using basic dc and S-parameters—for example, C_{GS} from S_{11}

- Six extraction steps to determine parameters:
  - basic dc model
  - advanced dc model
  - basic ac model
  - advanced ac model
  - package determination
  - diode characterization
- Large-signal validation
  - power compression and bias shift
  - harmonic power comparison
  - intermodulation characterization

As an example, we are going to use the popular and well-characterized NE71000, which is not a particularly modern device, but has been used for various research projects. Figure 2.138 shows measured versus simulated dc I–V curves for the NE71000. The agreement between measured and predicted dc IV curves is quite good. In a similar fashion, we obtain obtain the S parameter fits over bias. They are shown in Figures 2.116 and 2.117.

As mentioned above, 1-dB compression, power saturation, and drain current as a function of drive. This particular model point is not particularly impressive. We believe this has to do with the original Materka model.

Having brought up the subject of models, Figure 2.143 compares measured data with that obtained with the Angelov, cubic, quadratic, TriQuint, Raytheon, and Materka models. The figure shows that there is no perfect model yet. To put this in perspective, in Figure 2.144 we show the dc I–V error bars for these six models. This graph shows that the least errors for the dc measurement versus modeling are obtained with the Curtice cubic model, followed by the modified Materka model (implemented in Serenade) and TriQuint models. The difference in S parameter matching for RF application is significantly more relevant for the designer.

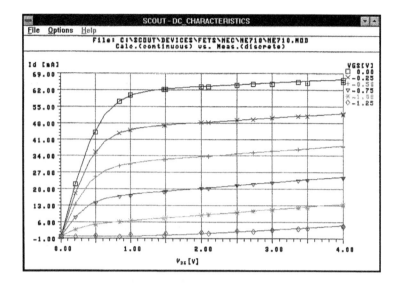

Figure 2.138 Measured NE71000 dc *I–V* curves versus those simulated using the Ansoft Materka model.

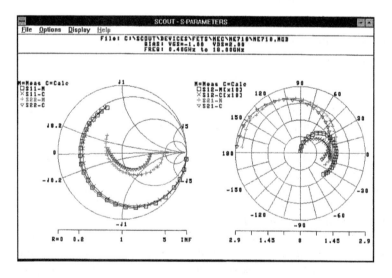

Figure 2.139 Measured versus calculated NE71000 *S* parameters for $V_{GS} = -1\,\text{V}$ and $V_{DS} = 2\,\text{V}$.

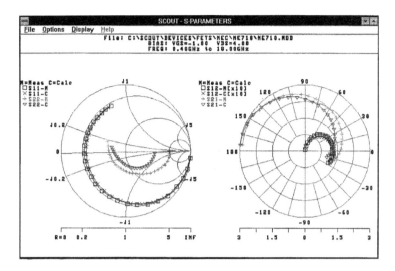

Figure 2.140 Measured versus calculated NE71000 S parameters for $V_{GS} = -1\,V$ and $V_{DS} = 4\,V$.

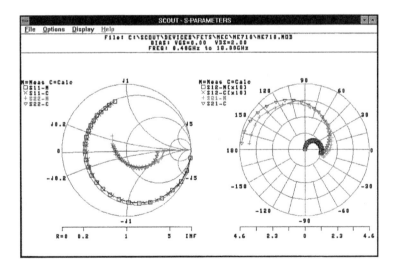

Figure 2.141 Measured versus calculated NE71000 S parameters for $V_{GS} = 0\,V$ and $V_{DS} = 2\,V$.

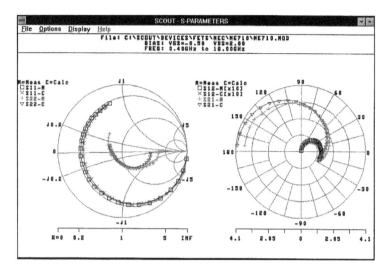

Figure 2.142 Measured versus calculated NE71000 S parameters for $V_{GS} = -0.5\,\text{V}$ and $V_{DS} = 2\,\text{V}$.

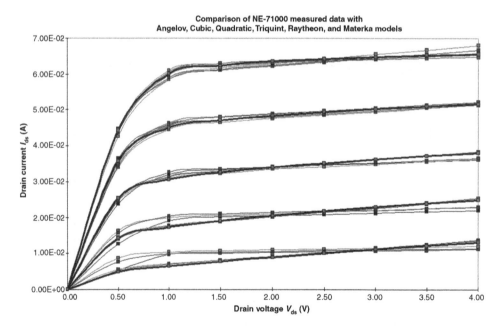

Figure 2.143 Comparison of dc models: Chalmers (Angelov), Curtice cubic, Curtice quadratic, Materka, Raytheon, and TOM.

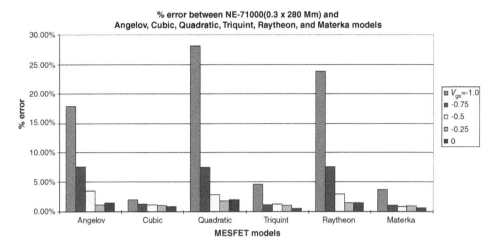

Figure 2.144 DC *I–V* error bars: Chalmers (Angelov), Curtice cubic, Curtice quadratic, Materka, Raytheon, and TOM.

Things become much more interesting if we compare the S parameter errors between the Materka shown in Figure 2.145, the Curtice cubic model (Figure 2.146), and the TOM model (Figure 2.147).

Since the trend is to go to low-voltage power RF operation, several issues must be battled.

- Models are most inaccurate at low V_{DS} and near pinchoff
- Modeling of low-voltage operation is adversely impacted
- Thermal effects must be considered, but older models do not account for this

2.5.5.2 A Case Study

As a case study, we will consider the Oki KGF1608 power MESFET (0.5 μm $\times$ 28 mm). Figures 2.148 and 2.149 show the dc *I–V* curves and S parameters, respectively. The resolution of the S parameters, based on the scale used, is not particularly impressive.

Figure 2.150 shows dc IV curve, with load line, for the KGF1608; the self-heating effect of the transistor can be seen in the droop of the top two curves. Figures 2.151 and 2.152 show drain current and output power versus drive, respectively. The test conditions for these three figures were $f = 850$ MHz, $V_{DS} = 3.4$ V, and $V_{DS} = 2.025$ V.

2.5.6 Conclusions

From all plots of dc and ac error contours, dc *I–V* curves, and dc error bar chart, we can draw the following conclusions.

DC Errors

- *In the Saturation Region*: All the models considered work very well, with a relative error of less than 15%. This means that all models are suitable for typical applications where the device operates in the saturation region.

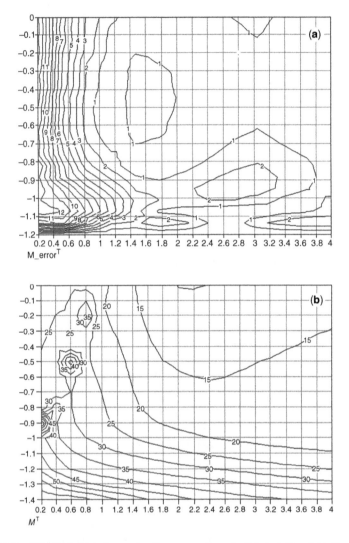

Figure 2.145   NE71000, Materka model, % error contours for (a) dc *I–V* and (b) *S* parameters.

- *In the Linear Region*: Although the errors in the saturation region are similar for different models, the errors in the linear region are quite different. The Materka, Curtice cubic, and TriQuint models behave better than other models in the linear region. The relative errors of these three models are less than 10%, while other models have more than 60% relative errors at some points.

AC Errors   From the ac error contour plots, we can see that the Chalmers (Angelov) model emerges as the most accurate for ac small-signal operation. The comparison also shows that all mores considered are good for ac operations biased in the saturation region with a typical relative error of less than 15%. However, the ac relative errors are much higher in the linear region, particularly for the Curtice quadratic model, which exhibits more 80% error at some points.

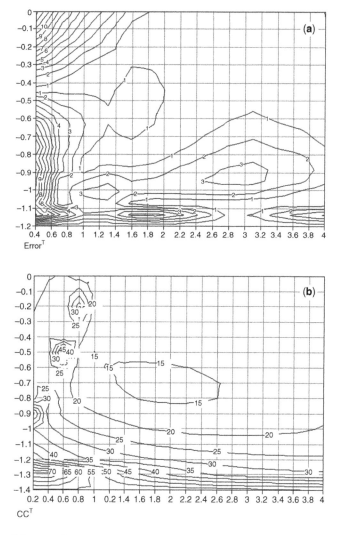

Figure 2.146   NE71000, Curtice cubic model, % error contours for (a) dc I–V and (b) S parameters.

Notes   The above conclusions are based on the matching between the simulated responses of different models and the measured data from a typical NE71000 device; the models may behave differently for a different device. It is not possible to draw any broad-based conclusions as to what model is the best. In order to select the most suitable model for your application, you must consider all the factors, such as bias range, frequency range, small-signal, or large-signal operation, according to your experiences in device modeling.

2.5.7   Device Libraries

All these measurements lead to a nonlinear device library. Figures 2.153 and 2.154 show a typical datasheet for a device in the library with measured and modeled data. The previous high-power example does not always lead to acceptable results as the models are far from

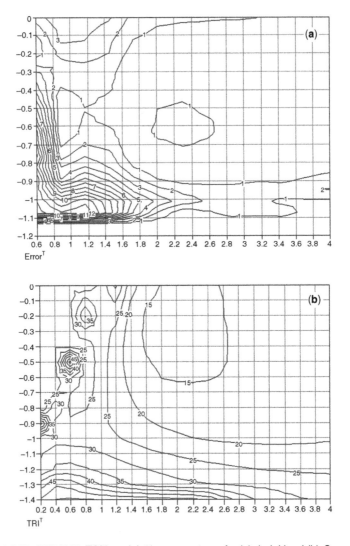

Figure 2.147   NE71000, TOM model, % error contours for (a) dc $I-V$ and (b) S parameters.

being fully developed. We also show the simulation results using the physics-based model for FETs.

2.5.8 Physics-Based MESFET Modeling

We mentioned a physics-based model earlier. Figure 2.155 shows the measured versus predicted dc IV, I/O power, and S parameters for an Alpha Industries MBE MESFET developed under one of the government contracts. The simulation does not handle the trapped energy, which explains the negative slope on the third trace of the output dc $I-V$ curve.

The process for bipolar transistors is quite similar and less tricky than the ones just described. Good examples can be found in the User Manual for Compact Scout.

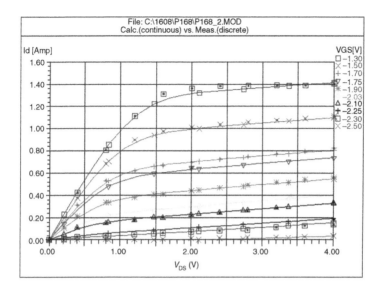

Figure 2.148   Calculated versus measured dc *I–V* curves for the Oki KGF1608 package MESFET.

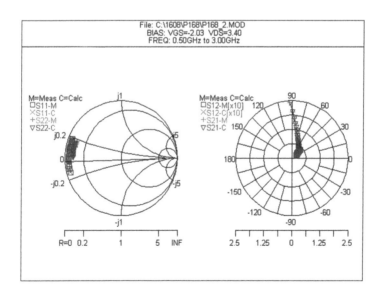

Figure 2.149   Calculated versus measured *S* parameters for the Oki KGF1608 package MESFET.

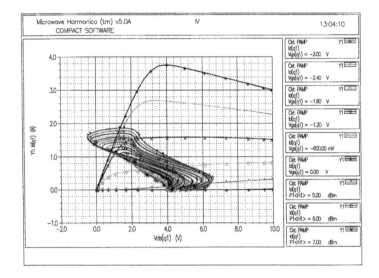

Figure 2.150 Load line and dc *I–V* curves for the KGF1608 MESFET.

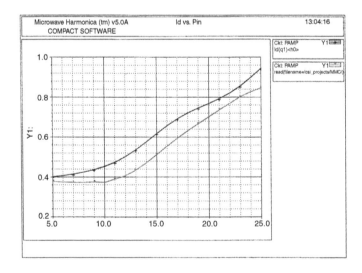

Figure 2.151 Measured versus calculated drain current for the KGF1608 MESFET.

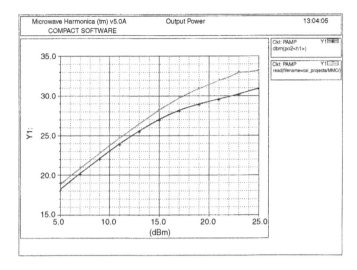

Figure 2.152 Measured versus calculated output power for the KGF1608 MESFET.

2.5.9 Example: Improving the BFR193W Model

We have already mentioned the difficulty encountering the proper base spreading resistor values and others by using standard parameter–extraction programs. Table 2.19 lists the SPICE Parameters for the Gummel-Poon Model, Berkley-SPICE 2G.6 syntax, from the Infineon datasheet.

A certain issue arises when measuring noise performance of the same device at an operating point of 10 mA. Using above-mentioned simplified equation

$$F = 1 + \frac{R_{bb'}}{R_g} + \frac{0.5 \left(\dfrac{26\,\text{mV}}{I_C} \right)}{R_g} \quad \text{(approximate equation at medium frequencies)} \quad (2.275)$$

we obtain the following results.

A. Inserting 1.8 Ω (value for R_B provided by the data sheet) and $R_B = 15\,\Omega$ (measured by the authors), here are the interesting results:

$$F = 1 + \frac{1.8}{50} + \frac{2.6}{100} = 1.062 \quad (2.276)$$

as noise factor or, as noise figure, $NF = 0.26\,\text{dB}$.

B. Or using the measured 15 Ω, one obtains

$$F = 1 + \frac{15}{50} + \frac{2.6}{100} = 1.326 \quad (2.277)$$

as noise factor or, as noise figure, $NF = 1.22\,\text{dB}$.

The Infineon datasheet claims a noise figure of 1 dB and 1.3 dB at 900 MHz and 1.8 GHz, respectively. Therefore, the measured versus calculated data differs by 0.2 dB or less,

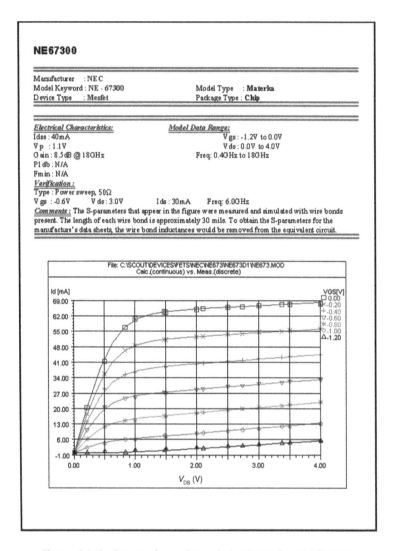

Figure 2.153   Page 1 of a nonlinear device library for the NE67300.

compared to 0.74 dB or more. This simple example shows how easy it is to provide measurement data that are inconsistent with the laws of physics. It is therefore highly recommended to add the noise figure as an additional parameter in the parameter extraction to reduce the number of variables—specifically, those for which the standard extraction has little sensitivity. The collector–emitter junction capacitance is another important parameter that frequently gets measured incorrectly. The way around this is to remember the definition

$$f_{\max} = \sqrt{\frac{f_t}{8\pi R'_{bb} C_c}} \tag{2.278}$$

f_T can be measured from the 3-dB point of the emitter current gain (β) and R'_{bb} was just computed above, so the equation can be solved for C_c, as there is a relationship between

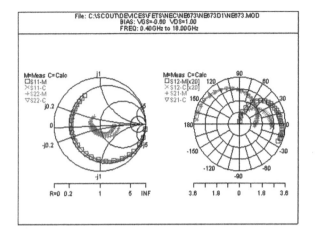

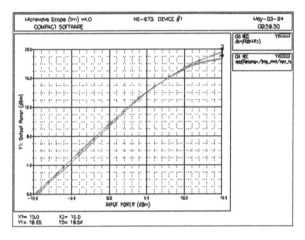

Figure 2.154 Page 2 of a nonlinear device library for the NE67300.

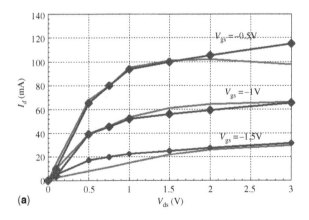

(a)

Figure 2.155 Comparison of measured data with PHYSFET and TOM models for an Alpha Industries MBE MESFET (0.25 μm × 400 μm) at $V_{GS} = -1$, $V_{DS} = 2$.

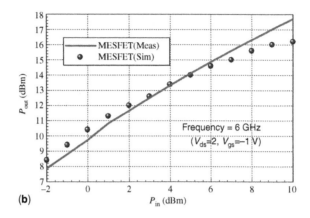

(b)

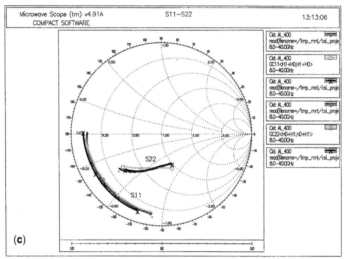

(c)

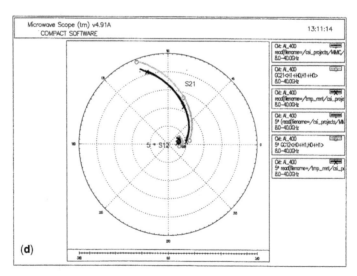

(d)

Figure 2.155   (*Continued*)

measured g_{max} and f_{max}. There are several correlations between the parameters of the equivalent circuit as published in different places, and therefore one can restrict the degree of freedom within the optimization process and obtain much better results. This approach is applicable to all types of transistors, bipolar and FET, but can not only be used to improve the accuracy of parameters that affect a transistor's RF performance but also its noise performance.

REFERENCES

1. H. A. Watson, *Microwave Semiconductor Devices and Their Circuit Applications*. New York: McGraw-Hill, 1969, Chapter 2.

2. Infineon Technologies, "The VCEO-mystery or how to use low-VCE0-transistors with high operatin voltages," Application Note No. 003, rev. 2.0, November 2006.

3. M. Rudolph, F. Schnieder, and W. Heinrich, "Modeling emitter breakdown in GaAs-based HBTs," in *IEEE MTT-S International Microwave Symposium Digest*, pp. 651–654, 2003.

4. H. K. Gummel and H. C. Poon, "An integral charge control model of bipolar transistors," *Bell System Technical Journal*, pp. 827 – 852, May–June 1970.

5. I. Getreu, *Modeling the Bipolar Transistor*. Beaverton, OR: Tektronix Inc., 1976.

6. *HICUM documentation and Verilog-A Code*. Available at http://www.iee.et.tu-dresden.de/iee/eb/hic_new/hic_intro.html

7. *MEXTRAM Documentation and Verilog-A and C Code*. Available at http://mextram.ewi.tudelft.nl

8. M. Iwamoto, P. M. Asbeck, T. S. Low, C. P. Hutchinson, J. B. Scott, A. Cognata, X. Qin, L. H. Camnitz, and D. C. D'Avanzo, "Linearity characteristics of GaAs HBTs and the influence of collector design," *IEEE Trans. Microwave Theory Tech.*, Vol. 48, pp. 2377–2388, December 2000.

9. M. Rudolph and R. Doerner, "Large-signal HBT model requirements to predict nonlinear behaviour" in *IEEE MTT-S International Microwave Symposium Digest*, pp. 43–46, 2004.

10. M. Rudolph, *Introduction to Modeling HBTs*. Boston, MA: Artech House, 2006.

11. F. Schwierz and J. J. Liou, *Modern Microwave Transistors*. Hoboken, NJ: Wiley, 2003.

12. H. Kroemer, "Theory of a wide-gap emitter for transistors," *Proceedings of IRE*, vol. 45, pp. 1535–1537, November 1957.

13. C. T. Kirk, Jr., "A theory of transistor cutoff frequency (f_t) falloff at high current densities," *IRE Transactions Electron Device*, No. 3, pp. 164–174, March 1962.

14. R. J. Whittier and D. A. Tremere, "Current gain and cutoff frequency falloff at high currents," *IEEE Transactions Electron Devices*, Vol. ED-16, No. 1, pp. 39–57, January 1969.

15. S. M. Sze, *Physics of Semiconductor Devices*. Hoboken, NJ: Wiley, 1981.

16. L. H. Camnitz and N. Moll, "An analysis of the behaviour of microwave heterostructure bipolar transistors," *Compound Semiconductor Transistors—Physics and Technology*, S. Tiwari (ed). New York: IEEE Press, 1993, pp. 21–45.

17. A. S. Grove, Op. cit., Ch. 4.

18. A. S. Grove, Op. cit., Ch. 7.

19. P. E. Gray, D. DeWitt, A. R. Boothroyd, and J. G. Gibbons, *Physical Electronics and Circuit Models of Transistors*. New York: Wiley, 1964, p 10.

20. P. E. Gray et al. Op. cit., p. 129.

21. P. E. Gray, et al. Op. cit., p. 180.

22. C. A. Liechti, "Microwave field-effect transistors—1976," *IEEE Transactions Microwave Theory and Technology*, Vol. MTT-24, pp. 279–300, June 1976.

23. E. A. Wolff and R. Kaul, *Microwave Engineering and Systems Applications*. New York: Wiley, 1988, p. 395.

24. A. D. Evans (ed.), *Designing with Field-Effect Transistors*. New York: McGraw-Hill, 1981.

25. R. D. Middlebrook, "A simple derivation of field-effect transistor characteristics," Proceedings of IEEE, Vol. 50, pp. 1146–1147, 1963.

26. T.-Hwa and C. P. Snapp, "Low noise microwave bipolar transistor with sub-half micrometer emitter width," *IEEE Transactions Electron Devices*, Vol. ED-25, pp. 723–730, June 1978.

27. U. L. Rohde, "Designing a match low noise amplifier using CAD tools," *Microwave Journal*, October 1986.

28. R. A. Pucel, W. Strubel, R. Hallgren, and U. L. Rohde, "A general noise de-embedding procedure for packaged two-port linear active devices," *IEEE Transactions Microwave Theory Technology*, November 1992.

29. R. A. Pucel and U. L. Rohde, "An accurate expression for noise resistance Rn of a bipolar transistor for use with the Hawkins noise model," *Letter in MTT Transactions*, 1993.

30. R. S. Muller and T. I. Kamins, *Device Electronics for Integrated Circuits*. New York: Wiley, 1986, p. 386.

31. D. Frohman-Bentchkowsky and A. S. Grove, "Conductance of MOS transistors in saturation," *IEEE Transactions Electron Devices*, Vol. ED-16, pp. 108–113, January 1969.

32. R. S. Muller and T. I. Kamins, *Device Electronics for Integrated Circuits*. New York: Wiley, 1986, p. 480.

33. "New LDMOST technology superb for RF applications," *Philips Semiconductors World News*, Vol. 7, No. 3, Article 29.

34. Y. P. Tsividis, *Operation and Modeling of the MOS Transistor*. New York: McGraw-Hill, 1987, p. 136.

35. A. Materka and T. Kacprzak, "Computer calculation of large-signal GaAs FET amplifier characteristics," *IEEE Transactions Microwave Theory Technology*, Vol. MTT-33, No. 2, pp. 129–135, February 1985.

36. R. E. Anholt and S. E. Swirhun, "Experimental investigation of the temperature dependence of GaAs FET equivalent circuits," *IEEE Transactions Electron Devices*, Vol. 39, No. 9, pp. 2029–2036, September 1992.

37. P. R. Gray and R. G. Meyer, *Analysis and Design of Analog Integrated Circuits*. New York: Wiley, 1986.

38. P. W. Tuinenga, *SPICE: A Guide to Circuit Simulation and Analysis Using PSpice®*. Englewood Cliffs NJ: Prentice-Hall, 1992.

39. A. Vladimirescu, *The SPICE Book*. New York: Wiley, 1994.

40. H. Statz, H. Haus, and R. A. Pucel, "Noise characteristics of gallium arsenide field-effect transistors," *IEEE Transactions Electron Devices*, Vol. ED-21, pp. 549–562, September 1974.

41. V. Rizzoli, F. Mastri, and C. Cecchetti, "Computer-aided noise analysis of MESFET and HEMT mixers," *IEEE Transactions Microwave Theory Technology*, Vol. MTT-37, pp. 1401–1410, September 1989.

42. V. Rizzoli, F. Mastri, and D. Masotti, "General-purpose noise analysis of forced nonlinear microwave circuits," *Military Microwave*, 1992.

43. U. L. Rohde, and C.-R. Chang, "Parameter extraction for large signal noise models and simulation of noise in large signal circuits like mixers," *Microwave Journal*, pp. 24–28, May 1993.

44. U. L. Rohde, "Parameters extraction for large signal noise models and simulation of noise in large signal circuits like mixers and oscillators," *23rd European Microwave Conference*, Madrid, Spain, September 6–9, 1993.

45. U. L. Rohde, C.-R. Chang, and J. Gerber, *Design and optimization of low-noise oscillators using non-linear CAD tools," presented at the Frequency Control Symposium*, Boston, MA, June 1–2, 1994.

46. U. L. Rohde, *Microwave and Wireless Synthesizers: Theory and Design*. New York: Wiley, 1997, Appendix B-3, "Bias Independent Noise Model."

FURTHER READING

ITT Capacitance Diodes, Tuner Diodes, Diode Switch, PIN Diodes, Basics and Applications, 1976.

X. McDade and X. Schiavone, "Switching Time Performance of Microwave PIN Diodes," *Microwave Journal*, September 1974.

M. Rudolph, Ch. Fager, and D. E. Root (eds.), *Nonlinear Transistor Model Parameter Extraction Techniques*. Cambridge, U.K. Cambridge University Press, 2012.

3

AMPLIFIER DESIGN
WITH BJTs AND FETs

3.1 PROPERTIES OF AMPLIFIERS

3.1.1 Introduction

The goal of this chapter is to move from the details on the semiconductor devices themselves to our first practical application. One of the more interesting features of amplifiers is that they can easily become what will be covered in Chapter 5: *oscillators*. This is due to the high gain of the devices, unaccounted for parasitic elements, and other design flaws. The amplifiers we need will fall into three categories:

- low-noise amplifiers,
- high-gain amplifiers, and
- medium- to high-power amplifiers.

See Figure 3.1. The low-noise amplifier always operates in Class A, typically at 15%–20% of its maximum useful current. The high-gain amplifier can operate in Class A, as well as B (mostly push–pull). The higher dc current for the same device in a higher noise figure and more gain, and ultimately more output power. Class C operating mode is really reserve to either FM transmissions or constant-carrier modes like CW. Some of the modern digital modulation types are sensitive to phase distortion rather than amplitude changes, and because of the resulting output spectrum, designers have stayed away from Class C operation.

Some of these amplifiers, such as bipolar versions, can be dc coupled with very few difficulties; others, like those in the FET families, will cause more headaches. An interesting example, although much too high in frequency for the purpose of this book, is shown in Figures 3.2–3.5. It shows a three-stage amplifier that has noise feedback for the first two

RF/Microwave Circuit Design for Wireless Applications, Second Edition. Ulrich L. Rohde and Matthias Rudolph.
© 2013 John Wiley & Sons, Inc. Published 2013 by John Wiley & Sons, Inc.

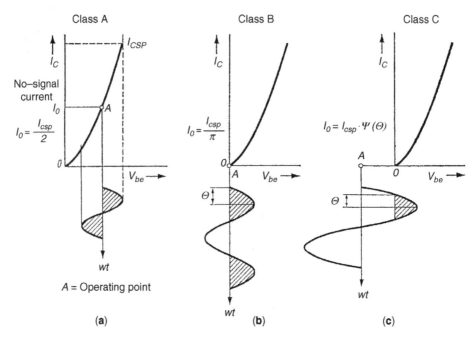

Figure 3.1 Definition of Classes A–C operation and resulting bias, including conduction angle (Θ) to be discussed in Section 3.2.2. The transfer characteristic can be either quadratic or exponential. This results in different distortion, but does not change the basic operating mode.

stages and resistive feedback for the output stage. These are the three types of amplifiers, regardless of technology, we will be evaluating. The achievement of accurate noise analysis of this circuit, reported in Ref. [1], marked the first time a complete linear noise model with essentially no frequency limitation had been developed. Its accuracy depends solely on the accuracy of the element values of the linear equivalent circuit, which can be obtained from measured S parameters by an optimization process. Needless to say, this technique is also

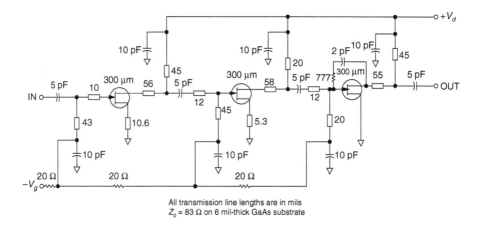

Figure 3.2 Schematic of the X-band GaAs monolithic low-noise amplifier (Texas Instruments EG8021). For reasons of linearity, all three stages operate in Class A.

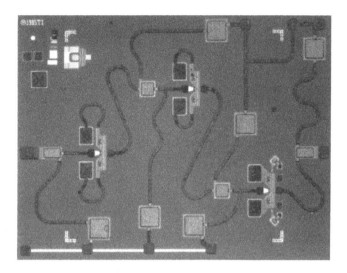

Figure 3.3 Photograph of the EG8021 monolithic amplifier chip. The area pictured is 0.09 in by 0.12 in in size.

applicable at lower frequencies, such as 1.5–3 GHz, and its deviation from measured values has never been worse than 0.2 dB or, relative to the noise figure, 10% expressed in dB.

The results of Figure 3.4 were obtained by using the linear FET model as supplied by Texas Instruments. It becomes somewhat obvious that a match for S_{11} of −60 dB is not likely. We then replaced the linear transistor with the SPICE-type nonlinear model, resulting in the responses shown in Figure 3.5. S_{11} now looks much more realistic in the center of

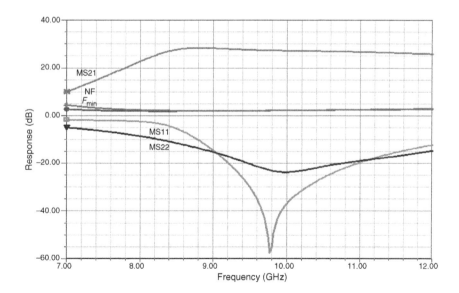

Figure 3.4 Simulated F_{min}, noise figure, S_{11}, S_{21}, and S_{22} responses for the three-stage GaAsFET amplifier using the TI linear FET model. The values at 10 GHz are: F_{min}, 2.20 dB; NF, 2.21 dB; S_{11}, −36.8 dB; S_{21}, 27.2 dB; and S_{22}, −23.7 dB.

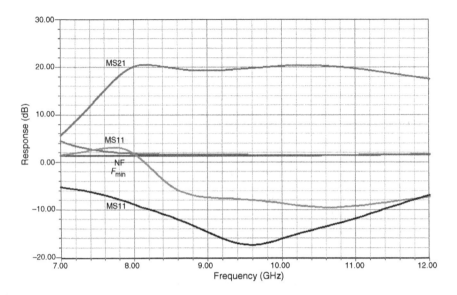

Figure 3.5 Simulated F_{min}, noise figure, S_{11}, S_{21}, and S_{22} responses for the three-stage GaAsFET amplifier using the nonlinear FET model. The values at 10 GHz are: F_{min}, 1.53 dB; NF, 1.65 dB; S_{11}, −8.5 dB; S_{21} 20.3 dB; and S_{22}, −15.9 dB.

the range, but at the same time exhibits a tendency toward instability at frequencies around 8 GHz. The actual circuit, when measured, did not exhibit this potential instability; on the other hand, the noise figure and gain agreed quite well. The discrepancies between the results with the linear and nonlinear models emphasize the importance of accurate device modeling.

Figure 3.6 shows a wideband amplifier, specifically a distributed amplifier, that covers 1–20 GHz. Figure 3.7 shows its simulated frequency-dependent gain, matching, and noise performance. This example is given here as an outlook to circuit concepts applyed at frequencies well beyond the wireless range. The other reason is to show that the modern CAD tools can handle such high frequencies accurately, which means one can trust them at the significantly lower wireless frequencies of 500 MHz to 3 GHz. Above 1500 MHz, one really should model using distributed elements. We will come to this later in this chapter.

Input/output and interstage matching will become another challenging task, specifically if one requires sufficient bandwidth. As circuits become more complex and as the operating frequency increases, passive components play an increasingly important role; the use of

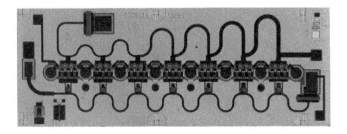

Figure 3.6 Photo of the 1–20-GHz distributed amplifier.

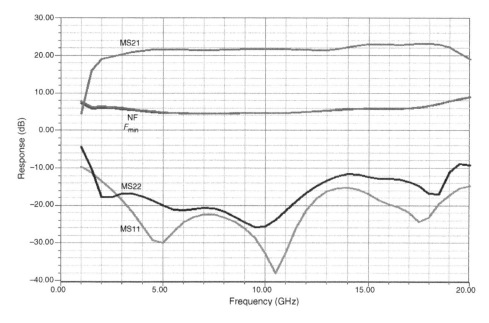

Figure 3.7 Simulated F_{min}, NF, S_{11}, S_{21}, and S_{22} responses. At the center of the range, the following numbers were found: F_{min}, 4.80 dB; NF, 4.80 dB; S_{11}, −32.8 dB; S_{21}, 21.8 dB; and S_{11}, −25.2 dB.

distributed elements is practically a necessity above 1500 MHz. For this reason, we begin this chapter by covering the application of transistors as amplifiers, following this with a discussion of the surrounding external circuitry, such as transmission lines, inductors, capacitors, and resistors.

The goal of the applications presented here is to address, where possible, high-performance wireless stages that are less subject to constraints of space, power consumption, and cost, than portable/mobile applications. This is consistent with our preface, in which we have clearly stated our goal. The designers of handsets typically cannot consider the full variety of circuits we will present here. These types of stages (low noise, high gain, and power) frequently will be used with filtering, and we will give some examples of input filters, including a tracking high-performance design.

A key topic besides the frequently mentioned gain is the whole noise issue. The following subchapter gives the reader a complete introduction into this important area, followed by a similar introduction into the gain and matching of transistors. A *caution*, the majority of applications published go through the exercise of using S parameters manually to design low-noise and small-signal amplifiers. While we have been tempted to repeat the information spread around in several textbooks, we feel that the engineers today will have access to CAD tools one way or another. It is highly unlikely that readers will undertake the inefficient process of going through all the design steps manually (a procedure that can take days); hopefully, engineers today will resort to a linear/nonlinear CAD tool to do so. Only for the purpose of summarizing the necessary steps, we will provide the relevant information regarding gain, stability, and related topics, demonstrating their usefulness or possible misuse with a modern CAD tool—in our case, the Serenade Design Environment product from Ansoft. Although our examples are primarily simulated with Serenade, designers can basically choose from three comprehensive microwave-CAD tools: Ansoft's

Designer (the successor of Ansoft's Serenade), AWR's Microwave Office, and Agilent's Advanced Design System (ADS). These tools are comprehensive in a sense that the full range of simulation and layout software is integrated, for example, circuit simulation in time and frequency domain, electromagnetic simulation of the transmission lines, and lay-out generation. Besides of these commercial packages, a number of special-purpose tools are available.

3.1.2 Gain

In practical terms, when we talk about gain, we mean the output power relative to the input power. The industry standard now uses 50 Ω; however, at the point where we have to match high-impedance devices such as monolithic crystal filters where the impedance jumps, things are not so obvious. The transformation equation that allows us to transform from one impedance to another is

$$m = \sqrt{\frac{R_2}{R_1}} \tag{3.1}$$

This correction factor, m, is required to do the transformation from R_1 (50 Ω) to R_2 (e.g., 1.2 kΩ). It needs to be understood that while the power gain remains the same, the voltage gain requires this correction factor. Since the power gain is expressed as 10 log A, A being the loop gain, 20 log A is the voltage gain, both expressed in dB. The reason for 10 versus 20 is that the power gain is always the voltage squared, which accounts for the 2. As reminder of these relationships

Distinction: Power ratios – voltage ratios

$$P = V^2/Z$$

$$n\mathrm{dB} = 10\ \log_{10}(P_2/P_1)\ \texttt{===========>}\ n = 10\ \log_{10} \frac{\frac{V_2^2}{Z_2}}{\frac{V_1^2}{Z_1}}\mathrm{dB}$$

$$Z_2 = Z_1\ \text{->}\ n\ \mathrm{dB} = 10\ \log_{10}(V_2^2/V_1^2) = 10\ \log_{10}(V_2/V_1)^2$$

$$\texttt{===>}\ n\ \mathrm{dB} = 20\ \log_{10}(V_2/V_1)\mathrm{dB}$$

The gain is typically defined under the condition of real resistive source and load terminations, which means that all the available power is being supplied to the ampli-fier, and likewise through proper matching at the output, all the available power is fed to the termination. This gain figure is typically measured at a single spot frequency, but since most amplifiers are not totally flat, one needs to specify the gain tolerances or gain flatness. A good number is x dB ± 0.5 dB gain variation. In the case of a narrowband amplifier, the gain typically is shaped like a tuned circuit or like a critically coupled bandpass filter frequency response while in the case of a real wideband application, the desired gain has some kind of ripple that is due to the various compensation components in the circuit. An extreme case of a wideband amplifier is a 6–18-GHz single-stage amplifier that uses some very clever matching. This is shown in the following example. Figure 3.8 shows the circuit diagram, and Figure 3.9 shows its frequency-dependent performance.

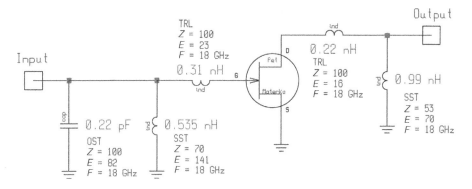

Figure 3.8 A 6–18-GHz FET amplifier.

Although this is a cute approach of generating a single-stage amplifier over such a wide bandwidth, in real life, it is not acceptable. The reason for this is that neither S_{11} nor S_{22} meet any useful specifications over the entire range. Typically, S_{11}/S_{22} values are only a few dB, while in reality more than 10 dB is required. We have chosen this example to show that gain and matching, depending upon the circuit topology, are connected. We get similar results for the noise figure whereby the best achievable noise figure (F_{min}) and the actually 50-Ω noise figure are only a few dB apart; however, its frequency dependence in a wideband stage is certainly not acceptable.

Another parameter associated with gain is the stability factor K of the circuit. A value of $K > 1$ is mandatory to guarantee stability together with magnitude of S_{11} and S_{22} being less than 1.

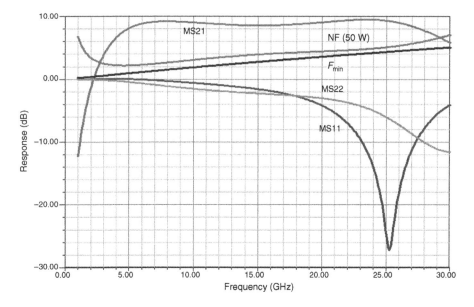

Figure 3.9 Frequency-dependent gain, matching, and noise performance of the Figure 3.8 amplifier.

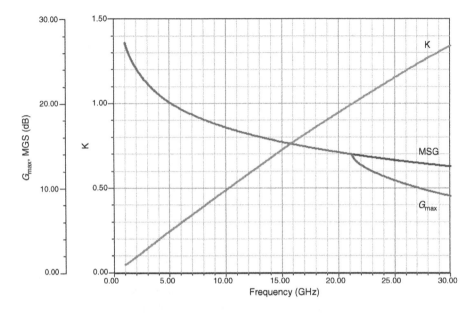

Figure 3.10 G_{max}, maximum stable gain (MSG), and K performance of the Figure 3.8 amplifier.

Figure 3.10 plots both the K stability factor, which for absolute stability must be >1, and maximum stable gain (MSG) as well as G_{max}. It turns out that MSG is a highly artificial definition of stability implies that input and output are terminated with pure resistances and then conjugately matched. There are four ways of achieving this resistive loading, either in the emitter (as a shunt resistor between gate/drain or base/collector, and finally in parallel to the input and output to ground). Needless to say, this is not a real operating mode because it deteriorates both noise and gain. In our case, the entity G_{max} is a function of S_{12} of the circuit, not just the transistor, and depends highly on S_{12}. If we bring the system close to oscillation, then G_{max} will increase at the expense of bandwidth. Unfortunately, some of these microwave definitions of gain (see Table 3.1) have become widely used in industry without always being useful. To evaluate these equations for S_{12} is not legitimate because other quantities like K become infinite, and the entire system of equations falls apart. The power gain with input and output conjugately matched, or the transducer power gain for arbitrary source and load matching, still give more insight in the system.

Definition for G_{max}: If $k > 1$, it is called G_{ma}, and if $k < 1$, it is called G_{ms}.

In the above example, we specifically stated that the input and output VSWR was totally unacceptable, and this was due to a one-stage design. Figure 3.11 shows an amplifier chip built around two transistors with standard feedback techniques applied; Figures 3.12 and 3.13 show its frequency-dependent gain, matching, stability, and noise performance. This amplifier was one of the early examples of GaAsFET wideband amplifiers, but it is interesting because while it has wild (and wide) gain swings, it exhibits a fairly constant input and output impedance or reflection coefficient. A further evaluation of the circuit shows that not only is the circuit unconditionally stable but also the maximum stable gain has an extremely high value based on essentially minimal S_{12}, the K factor is always significantly

Table 3.1 Nine Power Gains

Transducer power gain in 50-Ω system	$G_T = |S_{21}|^2$
Transducer power gain for arbitrary Γ_G and Γ_L	$G_T = \dfrac{(1 - |\Gamma_G|^2)|S_{21}|^2(1 - |\Gamma_L|^2)}{|(1 - S_{11}\Gamma_G)(1 - S_{22}\Gamma_L) - S_{12}S_{21}\Gamma_G\Gamma_L|^2}$
Unilateral transducer power gain	$G_{\mathrm{TU}} = \dfrac{|S_{21}|^2(1 - |\Gamma_G|^2)(1 - |\Gamma_L|^2)}{|1 - S_{11}\Gamma_G|^2|1 - S_{22}\Gamma_L|^2}$
Power gain with input conjugate matched	$G = \dfrac{|S_{21}|^2(1 - |\Gamma_L|^2)}{|1 - S_{22}\Gamma_L|^2(1 - |S_{11}'|^2)} = \dfrac{|S_{21}|^2}{1 - |S_{11}|^2}$ (for $\Gamma_L = 0$)
Available power gain with output conjugate matched	$G_A = \dfrac{|S_{21}|^2(1 - |\Gamma_G|^2)}{|1 - S_{11}\Gamma_G|^2(1 - |S_{22}'|^2)} = \dfrac{|S_{21}|^2}{1 - |S_{22}|^2}$ (for $\Gamma_G = 0$)
Maximum available power gain	$G_{ma} = \left|\dfrac{S_{21}}{S_{12}}\right| (k - \sqrt{k^2 - 1})$
Maximum unilateral transducer power gain	$G_{\mathrm{TU\,max}} = \dfrac{|S_{21}|^2}{(1 - |S_{11}|^2)(1 - |S_{22}|^2)}$
Maximum stable power gain	$G_{ms} = \dfrac{|S_{21}|}{|S_{12}|}$
Unilateral power gain	$U = \dfrac{1/2|S_{21}/S_{12} - 1|^2}{k|S_{21}/S_{12}| - \mathrm{Re}(S_{21}/S_{12})}$

higher than 1, and G_{max} follows, to some degree, the actual gain response of the amplifier. (Ultimately, designers gave up on simple RCL feedback amplifiers for such a wide bandwidth, replacing them with distributed amplifiers previously called traveling wave amplifiers.) The problem of gain and stability starts at audiofrequency amplifiers, where the

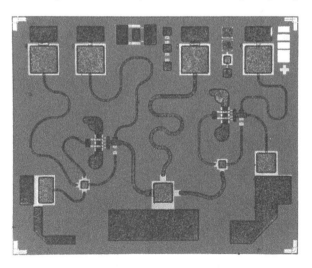

Figure 3.11 A two-stage, wideband feedback amplifier.

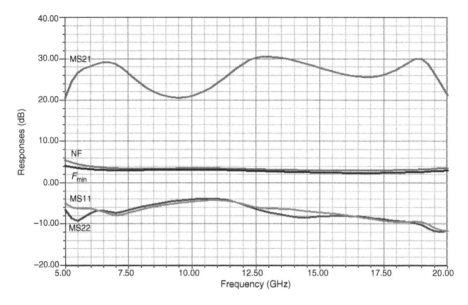

Figure 3.12 Frequency-dependent gain, matching, and noise performance for the Figure 3.11 amplifier.

motorboating phenomenon has been known for years, and the problem moves up in frequency with more modern technology, and is largely dependent not only on the device and the circuit but also on the layout. The motorboating effect has to do with ground loops and the fact that if the printed circuit traces from the power supply connection to the output power stage present too high a resistance, then at high currents the voltage will drop to a

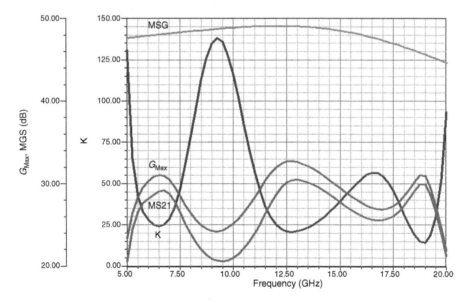

Figure 3.13 Frequency-dependent G_{max}, K, maximum stable gain, and gain performance for the Figure 3.11 amplifier.

point below useful operation, and as the volume is lowered, become stable again. In this case, the amplitude of the signal through the amplifier controls the stability of the output stage. Needless to say, this most unpleasant effect applies to high-power stages for all output transistors, where we use Class B or C operation, meaning that the dc operating current as a function of output does not remain constant.

In terms of gain, there is also an important gain characteristic called *differential gain* and *differential phase*. These expressions were first used in television circuits, where linearity in both areas is an absolute must. The amplitude of the signal determines the black and white contrast, and the phase determines the color. As an example, transmitting an image of one of our presidents playing golf in California to the New York area is done by many microwave hops. Any multipath reception, such as that caused by reflection from airplanes, may have little effect on the signal amplitude, but the resulting phase shift can transform a green lawn to a red lawn if a total shift of 180° occurs. Given a particular operating point, if the 1-dB compression point of multiple amplifiers in the chain is approached or exceeded, their differential gain, which is defined by the IEEE as "the difference between (a) the ratio of the output of a small high-frequency sine-wave signal at two different levels of a low-frequency signal on which it is superimposed, and (b) unity," may increase. This mode of saturation charges all the capacitances of the amplifier transistors, causing huge phase shifts that result in the known problem of changing colors. The same applies, of course, in test equipment amplifiers, in which such changes cannot be tolerated. A somewhat similar, but not quite the same, problem is spectral regrowth (evaluated in terms of adjacent channel power ratio [ACPR]), which is also due to nonlinearities. The question of linearity has already been covered in detail in Chapter 1. Needless to say, any overdrive condition also changes an amplifier's input and output SWR.

3.1.3 Noise Figure (NF)

Introduction   In this section, we look at the best way to describe noise in active devices (black boxes). Even when a two-port is linear, the output waveform will differ from the input, because of the failure to transmit all spectral components with equal gain (or attenuation) and delay. By careful design of the two-port, or by limitation of the bandwidth of the input waveform, such distortions can largely be avoided. However, noise generated within the two-port can still change the waveform of the output signal. In a linear passive two-port, noise arises only from the losses in the two-port; thermodynamic considerations indicate that such losses result in the random changes that we call noise. when the two-port contains active devices, such as transistors, there are other noise mechanisms that are present. A very important consideration in a system is the amount of noise that it adds to the transmitted signal. This is often judged by the ratio of the output signal power to the output noise power (S/N). The ratio of signal plus noise power to noise power $[(S + N)/N]$ is generally easier to measure, and approaches S/N when the signal is large.

In the evaluation of a two-port, it is important to know the amount of noise added to a signal passing through it. An important parameter for expressing this characteristic is the noise factor. The signal energy coming from a generator or antenna is amplified or attenuated in passing from the input to the output of a two-port, as is the noise that accompanies the input signal energy. A system generally includes a cascade of two-port networks that constitute one overall two-port that amplifies the signal to a high-enough power level for its intended use. The *noise factor* of a system is defined as the ratio of signal-to-noise ratios available at input and output:

$$F = \frac{(S/N)_{\text{input}}}{(S/N)_{\text{output}}} \geq 1 \tag{3.2}$$

The noise figure (or factor) of a receiver is an easily measured quantity that describes the signal-to-noise ratio reduction of that receiver.

When this ratio of powers is converted to decibels, it is generally referred to as the *noise figure* rather than noise factor. Various conventions are used to distinguish the symbols used for noise factor and noise figure. Here, we use F to represent the noise factor and NF to represent the noise figure, although the terms are usually used interchangeably.

For an amplifier with the power gain G, the noise factor can be rearranged as

$$F = \frac{S_i/N_i}{GS_i/G(N_i + N_a)} \tag{3.3}$$

when N_a is the additional noise power added by the amplifier referred to the input. This can be computed to be

$$F = 1 + N_a/N_i \tag{3.4}$$

The noise factor is often replaced by the noise figure (NF), which is defined in decibels as

$$NF = 10 \log_{10} F \tag{3.5}$$

In applications like satellite receivers, the noise factor becomes such a small number that it is inconvenient to work with. Many people have adopted the use of an equivalent noise temperature for a circuit to remedy this situation. Since the thermal noise power N available from a resistor at temperature T_e is

$$N = kT_e B \tag{3.6}$$

where k is Boltzmann's constant (1.38×10^{-23} J/K), T_e is the effective temperature in kelvins, and B is the bandwidth in Hertz. The equation above may be used to associate an effective noise temperature with circuits containing more than just thermal noise sources. This allows (3.4) to be written as

$$F = 1 + \frac{kT_e B}{kT_0 B} = 1 + \frac{T_e}{T_0} \tag{3.7}$$

where T_e is the effective noise temperature of the circuit and T_0 is the temperature of the generator resistor in Kelvins. The noise temperature T_e now characterizes our circuit noise contribution and can be directly related to the noise factor.

Assuming a reference noise temperature of 290 K ($-273 + 290 = 17°C$), let us determine the noise temperature of the system with a noise factor of 2.6 (4.15 dB):

$$T_e = (2.6 - 1)(290) = 464 \text{ K}$$

This temperature T_e should not be confused with the environmental operating temperature T_0. It is quite common to operate low-noise amplifiers with T_e below 100 K, an ambient temperature of 290 K.

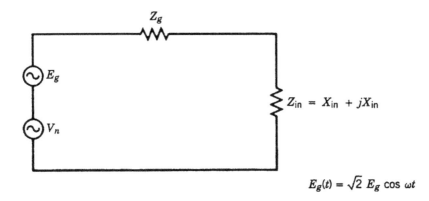

Figure 3.14 Combination of signal and noise voltages supplied to a complex termination.

Signal-to-Noise Ratio   Let us consider the signal-to-noise ratio of power delivered from a generator to a load as shown in Figure 3.14. The signal power delivered to the input is given by

$$S_{\text{in}} = P_{\text{in}} = \frac{E_g^2 \text{Re}(Z_{\text{in}})}{|Z_g + Z_{\text{in}}|^2} \tag{3.8}$$

where E_g is the rms voltage of the input signal supplied to the system, and the noise power supplied to the input is expressed by

$$N_{\text{in}} = \frac{\overline{v_n^2} \text{Re}(Z_{\text{in}})}{|Z_g + Z_{\text{in}}|^2} \tag{3.9}$$

where the noise power at the input is provided by the noise energy of the real part of Z_g. The input impedance Z of the system in the form $Z = R_{\text{in}} + jX_{\text{in}}$ is assumed to be complex.

The Johnson noise of a resistor [here $\text{Re}(Z_g)$], given by the mean-square voltage

$$\overline{v_n^2} = 4kTRB \tag{3.10}$$

where k (Boltzmann's constant) $= 1.38 \times 10^{-23}$ J/K, T is the absolute temperature of the resistor, and B is the bandwidth, is sufficiently small that the resistive component of impedance does not change. The available signal power from the generator has a lower limit, even if the signal is attenuated by the highest possible attenuation. The generator resistor acts as a Johnson noise generator, its power being

$$P_A = \frac{4kTBR}{4R} = kTB \tag{3.11}$$

where k is Boltzmann's constant, T is the absolute temperature, and B is the bandwidth. This power is the maximum available output power.

For an ambient temperature of 290 K, $kT = 4 \times 10^{-21}$ W/Hz. This expression is also given as $kT = -204$ dBW/Hz $= -174$ dBm/Hz $= -114$ dBm/MHz. We can combine (3.8) to (3.10) to obtain

$$\left(\frac{S}{N}\right)_{\text{in}} = \frac{E_g^2}{4kT\text{Re}(Z_g)B} \tag{3.12}$$

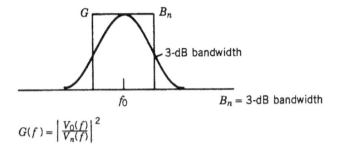

$$G(f) = \left| \frac{V_0(f)}{V_n(f)} \right|^2$$

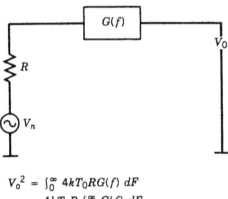

$$V_0^2 = \int_0^\infty 4kT_0 R G(f) \, dF$$
$$= 4kT_0 R \int_0^\infty G(f) \, dF$$
$$B_n = \frac{1}{G} \int_0^\infty G(f) \, dF$$
$$B_n = \text{noise bandwidth}$$

Figure 3.15 Graphical and mathematical explanation of the noise bandwidth from a comparison of the Gaussian-shaped bandwidth to the rectangular filter response.

This is the value of S/N contributed by the generator, which does not include the noise generated by the load, in this case $\mathrm{Re}(Z_{\mathrm{in}})$, which would need to be included in the measurement of the total S/N across the input impedance.

A critical parameter is the noise bandwidth, B_n, which is defined as the equivalent bandwidth, as shown in Figure 3.15. For reasons of group delay correction, most practical filters have round rather than sharp corners. The noise figure measurements shown later can be used to determine the "integrated" bandwidth, which is B_n.

An active system such as a combination of amplifiers and mixers will add noise to the input signals, and the noise factor that describes this is defined as the S/N ratio at the input to the S/N ratio at the output, which is always greater than unity [2]. In practice, a certain minimum signal-to-noise ratio is required for operation. For example, in a communication system, such a minimum is required for intelligible transmission, either voice or data. For high-performance TV reception, to provide a picture noise free to the eye, a typical requirement is for a 60-dB S/N. In the case of a TV system, a large dynamic range is required, as well as a very large bandwidth to reproduce all colors truthfully and all shades from high-intensity white to black. Good systems will have a bandwidth of 8 MHz or more.

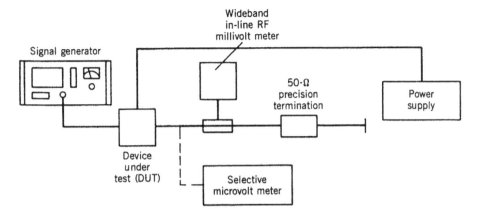

Figure 3.16   Test setup to measure signal-to-noise ratio.

Noise Figure Measurements   Some of the noise equations are based on mathematical models and physics. To understand some of these expressions, it is useful to look at a practical case of a system with amplifiers that has to be evaluated.

Let us look at Figure 3.16, which consists of a signal generator, the system or device under test (DUT), and a selective receiver with a built-in root-mean-square (rms) voltmeter to determine the signal and the noise voltage. It is necessary that the system have enough gain so that the noise voltage supplied by the generator will be indicated [3]

If we assume that our selective receiver is a video noise meter calibrated in rms voltage levels, we can perform two measurements. With an input termination connected to the TV system (typically, 75 Ω for cable TV, 50 Ω for satellite TV), the noise receiver/meter will read a value for proper termination that can easily be calculated. Since one-half of the mean-square noise voltage appears across the input,

$$v_{\text{in}} = \frac{v_n}{2} = \frac{\sqrt{4kTRB}}{2} \tag{3.13}$$

With $B = 10$ MHz, $T = 290$ K (T is always expressed in absolute temperature; $T^0 = -273°$C), and $k = 1.38 \times 10^{-23}$ J/K, then for $R = 75\Omega$

$$v_{\text{in}} = \frac{v_{\text{in}}}{2} = 1.73\mu V \tag{3.14}$$

where the rms noise voltage has been referred to the input port. We can verify this with our first measurement.

Now we increase the input voltage of the signal generator to a value that indicates a 60-dB *S/N* ratio at the output port. This should be about

$$E_g = \frac{v_n}{2}\sqrt{F} \times 1000 = 1.73\sqrt{F}\text{mV}$$

where F is the noise factor of the receiver. For a receiver noise factor of 10, we would obtain $E_g = 5.48$ mV (rms value). If the noise energy equivalent to a noise factor of F is assumed, we need $\sqrt{F}$ times more voltage. For a 60-dB ratio, this should be about $E_g = 1000 \times (v_n/2) \times \sqrt{F}$.

As they are done here, over a power range of 60 dB, the measurements can be performed over such a wide range only if special equipment is available. In cases where the internal detector of a piece of communications equipment is used, the signal-to-noise ratio measurements are performed over much smaller power ranges.

Let us assume that for the above-mentioned case ($F = 10$), we find a S/N ratio of 10 dB at the output for an input signal of 5.47 μV. By rewriting (3.7) as

$$E_g = \frac{v_n}{2}\sqrt{F} = \sqrt{kTRBF} \tag{3.15}$$

with F being the noise factor, we can solve for F with

$$F = \frac{P_s}{P_n} = \frac{E_g^2/R}{kTB} \tag{3.16}$$

While the input power from the thermal energy of the input termination resistor was $kTB = 4 \times 10^{-4}$ W, the input power required for the 10-dB S/N ratio was

$$P_s = \frac{(5.47 \times 10^{-6})^2}{75} = 3.98 \times 10^{-13}\text{W} \tag{3.17}$$

The noise factor is defined as the ratio P_s/P_n:

$$F = \frac{3.98 \times 10^{-13}}{4 \times 10^{-14}} = 10 \tag{3.18}$$

which is the proof.

This method is used more frequently at the 3-dB point, or double the input power if the dynamic range of the detector is small or only a linear indicator is available. Because of hum and other pickup, this is not an easy measurement. Using a signal generator is very expensive because in a laboratory or production environment, a wide frequency range requires several generators.

Another method is the use of a wideband noise generator. Modern gas discharge diodes or avalanche diodes are available that essentially provide white noise energy over a large frequency range. These microwave diodes typically have an output of 30 dB above kT when switched on and kT when switched off. To provide good matching at microwave frequencies, a 15-dB attenuator is cascaded. This means that the noise power of the source in the ON condition is about 15 dB above kT.

In the early 1960s, low-cost noise figure test equipment was built around vacuum diodes whose operating range was limited to 1200 MHz due to the resonance effects of the structure. Today the automatic noise gain analyzer offered current by Agilent and Eaton/AIL uses calibrated solid-state noise sources up to 26.5 GHz. It appears that the upper frequency limit has to do with matching, and the lower-frequency limit with $1/f$ noise.

Noisy Two-Port Description   Based on the convention by Rothe and Dahlke [4], any linear two-port can be in the form shown in Figure 3.17. This general case of a noisy two-port can be redrawn showing noise sources at the input and at the output. Figure 3.17b shows this in admittance form and Figure 3.17c in impedance form. The internal noise sources are assumed to produce very small currents and voltages, and we assume that linear two-port equations are valid. The internal noise contributions have been expressed by using external

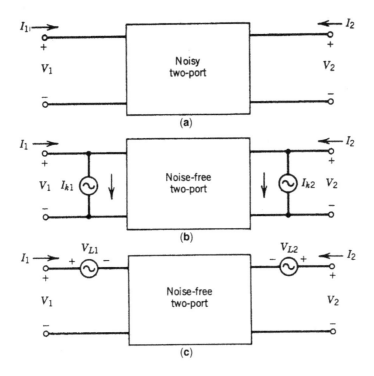

Figure 3.17   Noise linear two-ports: (a) general form; (b) admittance form; (c) impedance form.

noise sources:

$$I_1 = y_{11}V_1 + y_{12}V_2 + I_{k1}$$
$$I_2 = y_{21}V_1 + y_{22}V_2 + I_{k2} \tag{3.19}$$

$$V_1 = z_{11}I_1 + z_{12}I_2 + V_{L1}$$
$$V_2 = z_{21}I_1 + z_{22}I_2 + V_{L2} \tag{3.20}$$

where the external noise sources are I_{K1}, I_{K2}, V_{L1}, and V_{L2}.

Since we want to describe our noisy circuit in terms of the noise figure, the $ABCD$-matrix description will be more convenient since it refers both noise sources to the input of the two-port [5]. This representation is given below (note the change in direction of I_2):

$$V_1 = AV_2 + BI_2 + V_A$$
$$I_1 = CV_2 + DI_2 + I_A \tag{3.21}$$

where V_A and I_A are the external noise sources.

It is important to remember that all these matrix representations are interrelated. For example, the noise sources for the $ABCD$-matrix description can be obtained from the z-matrix representation shown in (3.20). This transformation is

$$V_A = -\frac{I_{k2}}{y_{21}} = V_{L1} - \frac{V_{L2}z_{11}}{z_{21}} \tag{3.22}$$

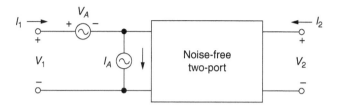

Figure 3.18 Chain-matrix form of linear noisy two-ports.

$$I_A = I_{k1} - \frac{I_{k2} y_{11}}{y_{21}} = -\frac{V_{L2}}{z_{21}} \tag{3.23}$$

The *ABCD* representation is particularly useful based on the fact that it allows us to define a noise temperature for the two-port referenced to input. The two-port itself (shown in Figure 3.18) is assumed to be noise free.

In the past, z and y parameters have been used, but in microwave applications, it has become common to use S-parameter definitions. This is shown in Figure 3.19. The previous equations can be rewritten in their new form using S parameters:

$$\begin{bmatrix} b_1 \\ b_2 \end{bmatrix} = \begin{bmatrix} S_{11} & S_{12} \\ S_{21} & S_{22} \end{bmatrix} \begin{bmatrix} a_1 \\ a_2 \end{bmatrix} + \begin{bmatrix} b_{n1} \\ b_{n2} \end{bmatrix} \tag{3.24}$$

There are different physical origins for the various sources of noise. Typically, thermal noise is generated by resistances and loss in the circuit or transistor, whereas shot noise is generated by current flowing through semiconductor junctions and vacuum tubes. Since these many sources of noise are represented by only two noise sources at the device input, the two equivalent noise sources are often a complicated combination of the circuit internal noise sources. Often, some fraction of V_A and I_A is related to the same noise source. This means that V_A and I_A are not independent in general. Before we can use V_A and I_A to

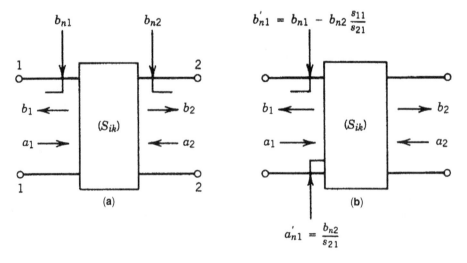

Figure 3.19 *S*-parameter form of linear noisy two-ports.

calculate the noise figure of the two-port, we must calculate the correlation between the V_A and I_A shown in Figure 3.18.

The noise source V_A represents all the device noise referred to the input when the generator impedance is zero, that is, when the input is short circuited. The noise source I_A represents all the device noise referred to the input when the generator admittance is zero, that is, the input in open circuited.

The correlation of these two noise sources considerably complicates analysis. By defining a correlation admittance, we can simplify the mathematics and get some physical intuition for the relationship between noise figure and generator admittance. Since some fraction of I_A will be correlated with V_A, we split I_A into correlated and uncorrelated parts as follows:

$$I_A = I_n + I_u \tag{3.25}$$

I_u is the part of I_A uncorrelated with V_A. Since I_n is correlated with V_A, we can say that I_n is proportional to V_A and the constant of proportionality is the correlation admittance.

$$I_n = Y_{\text{cor}} V_A \tag{3.26}$$

This leads us to

$$I_A = Y_{\text{cor}} V_A + I_u \tag{3.27}$$

The following derivation of noise figure will use the correlation admittance. Y_{cor} is not a physical component located somewhere in the circuit. Y_{cor} is a complex number derived by correlating the random variables I_A and V_A. To calculate Y_{cor}, we multiply each side of (3.27) by V_A^* and average the result. This gives

$$\overline{V_A^* I_A} = Y_{\text{cor}} \overline{V_A^2} \tag{3.28}$$

where the I_u term averaged to zero since it was uncorrelated with V_A. The correlation admittance is thus given by

$$Y_{\text{cor}} = \frac{\overline{V_A^* I_A}}{\overline{V_A^2}} \tag{3.29}$$

Often, people use the term "correlation coefficient." This normalized quantity is defined as

$$c = \frac{\overline{V_A^* I_A}}{\sqrt{\overline{V_A^2 I_A^2}}} = Y_{\text{cor}} \sqrt{\frac{\overline{V_A^2}}{\overline{I_A^2}}} \tag{3.30}$$

Note that the dual of this admittance description is the impedance description. Thus, the impedance representation has the same equations as above with Y replaced by Z, I replaced by V, and V replaced by I.

V_A and I_A represent internal noise sources in the form of a voltage source acting in series with the input voltage and a source of current flowing in parallel with the input current. This representation conveniently leads to the four noise parameters needed to the describe the noise performance of the two-port. Again using the Nyquist formula, the open-circuit voltage of a resistor at the temperature T is

$$\overline{V_A^2} = 4kTRB \tag{3.31}$$

This voltage is a mean-square fluctuation (or spectral density). It is the method used to calculate the noise identity. We could also define a noise equivalent resistance for a noise voltage as

$$R_n = \frac{\overline{V_A^2}}{4kTB} \tag{3.32}$$

The resistor R_n is not a physical resistor but can be used to simulate different portions of the noise equivalent circuit.

In a similar manner, a mean-square current fluctuation can be represented in terms of an equivalent noise conductance G_n, which is defined by

$$G_n = \frac{\overline{I_A^2}}{4kTB} \tag{3.33}$$

and

$$G_u = \frac{\overline{I_u^2}}{4kTB} \tag{3.34}$$

for the case of the uncorrelated noise component. The input generator to the two-port has a similar contribution

$$G_G = \frac{I_G^2}{4kTB} \tag{3.35}$$

with Y_G being the generator admittance and G_G being the real part. With the definition of F above, we can write

$$F = 1 + \left| \frac{I_A + Y_G V_A}{I_G} \right|^2 \tag{3.36}$$

The use of the voltage V_A and the current I_A has allowed us to combine all the effects of the internal noise sources.

We can use the previously defined (3.29) correlation admittance, $Y_{\text{cor}} = G_{\text{cor}} + jB_{\text{cor}}$, to simplify (3.36). First, we determine the total noise current

$$\overline{I_A^2} = 4kT(|Y_{\text{cor}}|^2 R_n + G_u)B \tag{3.37}$$

where R_n and G_u are defined in (3.32) and (3.34). The noise factor can now be determined by

$$F = 1 + \frac{G_u}{G_g} + \frac{R_n}{G_g}[(G_G + G_{\text{cor}})^2 + (B_G + B_{\text{cor}})^2] \tag{3.38}$$

$$F = 1 + \frac{R_u}{R_g} + \frac{G_n}{R_g}[(R_G + R_{\text{cor}})^2 + (X_G + X_{\text{cor}})^2] \tag{3.39}$$

The noise factor is a function of various elements, and the optimum impedance for the best noise figure can be determined by minimizing F with respect to generator reactance and resistance. This gives

$$R_{0n} = \sqrt{\frac{R_u}{G_n} + R_{\text{cor}}^2} \tag{3.40}$$

$$X_{0n} = -X_{\text{cor}} \tag{3.41}$$

and

$$F_{\min} = 1 + 2G_n R_{\text{cor}} + 2\sqrt{R_u G_n + (G_n R_{\text{cor}})^2} \tag{3.42}$$

(To distinguish between optimum noise and optimum power, we have introduced the convention $0n$ instead of the more familiar abbreviation *opt*.) At this point, we see that the optimum condition for minimum noise figure is not a conjugate power match at the input port. We can explain this by recognizing that the noise source V_A and I_A represent all the two-port noise, not just the thermal noise of the input port. We should observe that the optimum generator susceptance, $-X_{\text{corr}}$, will minimize the noise contribution of the two noise generators.

In rearranging for conversion to S parameters, we write

$$F = F_{\min} + \frac{g_n}{R_G}|Z_G - Z_{0n}|^2 \tag{3.43}$$

$$F = F_{\min} + \frac{R_n}{G_G}|Y_G - Y_{0n}|^2 \tag{3.44}$$

From the definition of the reflection coefficient,

$$\Gamma_G = \frac{Y_0 - Y_G}{Y_0 + Y_G} \tag{3.45}$$

and with

$$g_G = \frac{G_G}{Y_0} \tag{3.46}$$

$$r_n = \frac{R_n}{Z_0} \tag{3.47}$$

the normalized equivalent noise resistance

$$F = F_{\min} + \frac{4r_n|\Gamma_G - \Gamma_{0n}|^2}{g_G(1 - |\Gamma|^2)|1 + \Gamma_{0n}|^2} \tag{3.48}$$

$$r_n = (F_{50} - F_{\min})\frac{|1 + \Gamma_{0n}|^2}{4|\Gamma_{0n}|^2} \tag{3.49}$$

$$\Gamma_{0n} = \frac{Z_{0n} - Z_0}{Z_{0n} + Z_0} \tag{3.50}$$

The noise performance of any linear two-port can now be determined if the values of the four noise parameters, $F_{\min}$, $r_n = r_n/50$, and Γ_{0n} are known.

Figure 3.20 shows the noise factor of a high-frequency transistor as a function of B_g for $G_G = \text{constant}$ and as a function of G_g for $B_g = B_{\text{opt}}$.

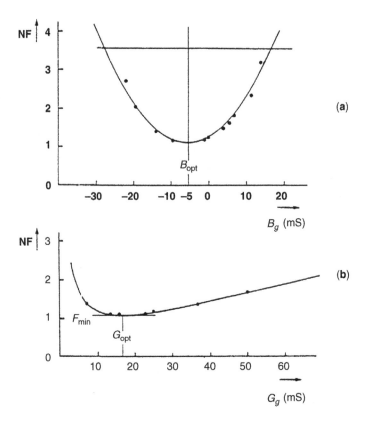

Figure 3.20 Noise factor in high-frequency BJTs for $f = 600\,\text{MHz}$: (a) as a function of B_g for $G_g =$ constant; (b) as a function of G_g for $B_g = B_{opt}$.

Noise Figure of Cascaded Networks   In a system with many circuits connected in cascade (Figure 3.21), we must consider the contributions of the various circuits. In considering the equivalent noise resistor R_n in series with the input circuit

$$F = \frac{R_G + R_n}{R_G} \tag{3.51}$$

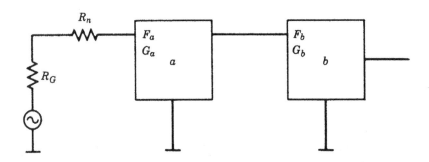

Figure 3.21 Cascaded noisy two-ports with the noise figures F_a and F_b and the gain figures G_a and G_b.

$$F = 1 + \frac{R_n}{R_G} \tag{3.52}$$

The excess noise added by the circuit is R_n/R_G.

Considering two cascaded circuits a and b, by definition, the available noise at the output of b is

$$N_{ab} = F_{ab}G_{ab}kTB \tag{3.53}$$

with B the equivalent noise bandwidth in which the noise is measured. The total available gain G is the product of the individual available gains, so

$$N_{ab} = F_{ab}G_aG_bkTB \tag{3.54}$$

The available noise from network a at the output of network b is

$$N_{a/b} = N_aG_b = F_aG_aG_bkTB \tag{3.55}$$

The available noise added by network b (its excess noise) is

$$N_{b/b} = (F_b - 1)G_bkTB \tag{3.56}$$

The total available noise N_{ab} is the sum of the available noise contributed by the two networks:

$$N_{ab} = N_{a/b} + N_{b/b} = F_aG_aG_bkTB + (F_b - 1)G_bkTB$$
$$= \left(F_a + \frac{F_b - 1}{G_a}\right)G_aG_bkTB \tag{3.57}$$

$$F_{ab} = F_a + \frac{F_b - 1}{G_a} \tag{3.58}$$

For any number of circuits, this can be extended to be

$$F = F_1 + \frac{F_2 - 1}{G_1} + \frac{F_3 - 1}{G_1G_2} + \frac{F_4 - 1}{G_1G_2G_3} + \cdots \tag{3.59}$$

When considering a long chain of cascaded amplifiers, there will be a minimum noise figure achievable for this chain. This is a figure of merit and was proposed by Haus and Adler [6]. It is calculated by rearranging (3.59).

$$(F_{tot})_{min} = (F_{min}) + \frac{F_{min} - 1}{G_A} + \frac{F_{min} - 1}{G_A^2} + \cdots + 1 \tag{3.60}$$

where F_{min} is the minimum noise figure for each stage and G_A is the available power gain of the identical stages. Using

$$\frac{1}{1 - X} = 1 + X + X^2 + \cdots \tag{3.61}$$

we find a quantity $(F_{\text{tot}} - 1)$, which is defined as the noise measure M. The minimum noise measure

$$(F_{\text{tot}})_{\min} - 1 = \frac{F_{\min} - 1}{1 - 1/G_A} = M_{\min} \tag{3.62}$$

refers to the noise of an infinite chain of optimally tuned, low-noise stages, so it represents a lower limit on the noise of an amplifier.

The minimum noise measure $M_{\min}$ is an invariant parameter and is not affected by feedback. It is somewhat similar to a gain-bandwidth product, in its use as a system invariant. The minimum noise measure is achieved when the amplifier is tuned for the available power gain and $\Gamma_G = \Gamma_{0n}$, given by (3.50).

Influence of External Parasitic Elements   Mounting an active two-port such as a transistor usually adds stray capacitance and lead inductance to the device, as shown in Figure 3.22. These external components consisting of transmission lines and parasitic reactances modify the noise parameters and the gain. Some researchers have published the results of these parasitic effects and have made manual computations or used some limited computer programs.

In a paper by Fukui [7], an attempt was made to determine the necessary equations, but the formulas are too involved even for pocket calculators. A more generic study by Iversen [8] is also very involved because of the various matrix manipulations, and is more suitable for a computer. Besser's paper in the *IEEE MTT-S* in 1975 [9] and Vendelin's paper [10] in the same issue have shown for the first time some practical results using computers and even optimization methods, using an early version of COMPACT. The intention of these investigations was to find feedback that modifies the device noise and scattering parameters such that a noise match could also provide a low-input VSWR. It can be seen from these discussions that some feedback, besides resulting in some gain reduction, may improve the noise matching at the input for a limited frequency range.

A more recent paper by Suter [11] based on a report by Hartmann and Strutt [12] has given a simple transformation starting from the S parameters and the noise parameters from common-source (or common-emitter) measurements. The noise parameters for the

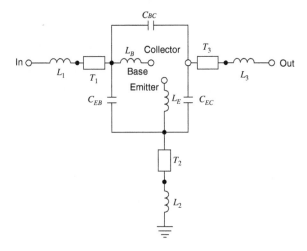

Figure 3.22   Equivalent circuit of the transistor package.

"packaged" device are calculated. This means that the parameters for the "new" device, including the common-gate (or common-base) case, are calculated. The equations are device dependent. They are valid for any active two-port.

A transformation matrix, n, may be used to combine the noise sources of the various circuit configurations. The transformation matrix parameters are given in Table 3.2 for series feedback, shunt feedback, and the common-gate (base) case, which will be important

Table 3.2 Transformation of Matrix Parameters

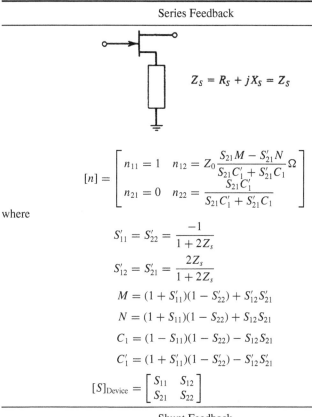

Series Feedback

$$Z_S = R_S + jX_S = Z_S$$

$$[n] = \begin{bmatrix} n_{11} = 1 & n_{12} = Z_0 \dfrac{S_{21}M - S'_{21}N}{S_{21}C'_1 + S'_{21}C_1}\,\Omega \\ n_{21} = 0 & n_{22} = \dfrac{S_{21}C'_1}{S_{21}C'_1 + S'_{21}C_1} \end{bmatrix}$$

where

$$S'_{11} = S'_{22} = \frac{-1}{1 + 2Z_s}$$

$$S'_{12} = S'_{21} = \frac{2Z_s}{1 + 2Z_s}$$

$$M = (1 + S'_{11})(1 - S'_{22}) + S'_{12}S'_{21}$$

$$N = (1 + S_{11})(1 - S_{22}) + S_{12}S_{21}$$

$$C_1 = (1 - S_{11})(1 - S_{22}) - S_{12}S_{21}$$

$$C'_1 = (1 + S'_{11})(1 - S'_{22}) - S'_{12}S'_{21}$$

$$[S]_{\text{Device}} = \begin{bmatrix} S_{11} & S_{12} \\ S_{21} & S_{22} \end{bmatrix}$$

Shunt Feedback

$$Z_P = R_P + jX_P = Z_P$$

$$[n] = \begin{bmatrix} n_{11} = \dfrac{S_{21}C'_2}{S_{21}C'_2 + S'_{21}C_2} & n_{12} = 0\,\Omega \\ n_{21} = \dfrac{1}{Z_0}\dfrac{S_{21}P - S'_{21}Q}{S_{21}C'_2 + S'_{21}C_2}S & n_{22} = 1 \end{bmatrix}$$

where

$$S'_{11} = S'_{22} = \frac{Z_p}{2 + Z_p}$$

(*continued*)

Table 3.2 (*Continued*)

$$S'_{12} = S'_{21} = \frac{2}{2 + Z_p}$$

$$P = (1 - S'_{11})(1 + S'_{22}) + S'_{12}S'_{21}$$

$$Q = (1 - S_{11})(1 + S_{22}) + S_{12}S_{21}$$

$$C_2 = (1 + S_{11})(1 + S_{22}) - S_{12}S_{21}$$

$$C'_2 = (1 + S'_{11})(1 + S'_{22}) - S'_{12}S'_{21}$$

$$[S]_{\text{Device}} = \begin{bmatrix} S_{11} & S_{12} \\ S_{21} & S_{22} \end{bmatrix}$$

Common Gate

$$[n] = \begin{bmatrix} n_{11} = \dfrac{2S_{21}}{-2S_{21} + C_4} & n_{12} = 0\,\Omega \\ n_{21} = \dfrac{1}{Z_0}\dfrac{C_3 C_4 - 4S_{12}S_{21}}{V(-2S_{21} + C_4)}S & n_{22} = -1 \end{bmatrix}$$

where

$$V = (1 + S_{11})(1 + S_{22}) - S_{12}S_{21}$$

$$C_3 = (1 - S_{11})(1 + S_{22}) + S_{12}S_{21}$$

$$C_4 = (1 + S_{11})(1 - S_{22}) + S_{12}S_{21}$$

$$[S]_{\text{Device}} = \begin{bmatrix} S_{11} & S_{12} \\ S_{21} & S_{22} \end{bmatrix} = \text{Common-source } S \text{ parameters}$$

for oscillator analysis. The transformation matrix gives the new four noise parameters as follows:

$$R'_n = R_n |n_{11} + n_{12}Y_{\text{cor}}|^2 + G_n |n_{12}|^2 \tag{3.63}$$

$$G'_n = \frac{G_n R_n}{R'_n} |n_{11}n_{22} - n_{12}n_{21}|^2 \tag{3.64}$$

$$Y'_{\text{cor}} = \frac{R_n}{R'_n}(n_{21} + n_{22}Y_{\text{cor}})(n^*_{11} + n^*_{12}Y^*_{\text{cor}}) + \frac{G_n}{R'_n}n_{22}n^*_{12} \tag{3.65}$$

A final transformation to the more common noise-parameter format given by (3.44) is still needed [10]:

$$F_{\text{min}} = 1 + 2R'_n(G'_{\text{cor}} + G'_{0n}) \tag{3.66}$$

$$R_n = R'_n \tag{3.67}$$

$$G_{0n} = \sqrt{\frac{G'_n}{R'_n} + G'^2_{cor}} \qquad (3.68)$$

$$B_{0n} = -B'_{cor} \qquad (3.69)$$

Figure 3.23 shows the noise figure as a function of external feedback for a low-noise microwave bipolar transistor, the AT-41435.

Bias-Dependent Noise Parameters Since we have elaborated on the effect of parasitics, it is now useful to point out that noise is also a strong function of bias point and frequency. Figure 3.24 shows the noise figure expressed in dB as a function of different dc currents and frequencies for a BFP420 BJT by Infineon. A change in the collector voltage will also affect the noise parameters, not quite as strongly as shown here, but still significantly. The reason for the turning point of the noise figure has to do with the fact that for lower currents, the f_t cutoff frequency has not reached its peak yet, and therefore the noise increases. The increase of the noise at higher currents is due to the Schottky noise calculated from the equation

$$P_n = 2Iq \qquad (3.70)$$

where P_n is the noise power, I is the saturation, collector, or drain current, and $q = $ charge of an electron (1.60×10^{-19} coulomb).

Table 3.3 shows the common-emitter noise parameters for the BFP420.

Noise Circles From the "Influence of External Parasitic Elements" section, we see that the noise factor is a strong function of the generator admittance (or admittance) presented to the input terminals of the noisy two-port. Noise tuning is the method to change the values of the input admittance to obtain the best noise performance. There is a range of values of input reflection coefficient over which the noise figure is constant. In plotting these points of constant noise figure, we obtain the so-called noise circles, which can be drawn on the Smith chart Γ_G plane [13]. Starting with the noise equation [see (3.44)] for a 50-Ω generator impedance, we find that

$$F_{50} = F_{min} + 4r_n \frac{|\Gamma_{0n}|^2}{|1 + \Gamma_{0n}|^2} \qquad (3.71)$$

We want to find the position of the reflection coefficient on the Smith chart, as in the case of the gain circles, for which F is constant. First, we rearrange (3.67) to read

$$r_n = (F_{50} - F_{min}) \frac{|1 + \Gamma_{0n}|^2}{4|\Gamma_{0n}|^2} \qquad (3.72)$$

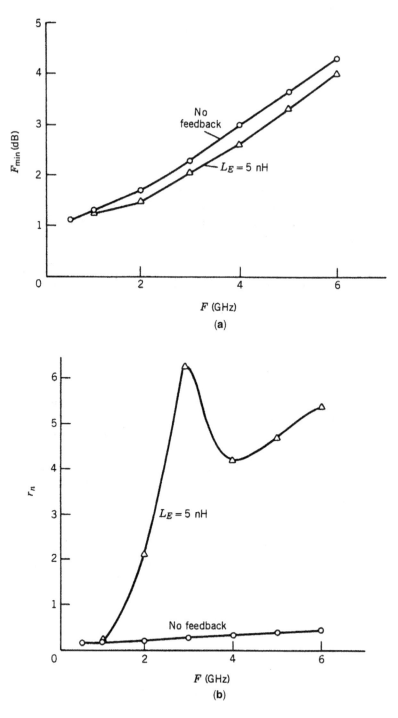

Figure 3.23 Noise parameters versus feedback for the AT-41435 silicon bipolar transistor: (a) F_{min} versus frequency and feedback; (b) r_n versus frequency and feedback; (c) Γ_{on} for AT-41435 versus frequency and feedback.

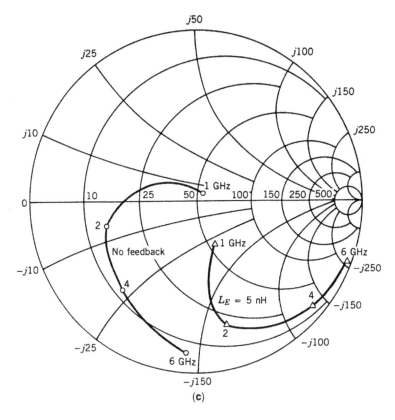

Figure 3.23 (*Continued*)

By introducing

$$N_i = \frac{F_i - F_{\min}}{4r_n}|1 + \Gamma_{0n}|^2 \tag{3.73}$$

we can find an expression for a circle of constant noise figure as introduced by Rothe and Dahlke [4, 13]. The center for the noise circle is

$$C_i = \frac{|\Gamma_{0n}|}{1 + N_i} \tag{3.74}$$

and the radius

$$r_i = \frac{\sqrt{N_i^2 + N_i(1 - |\Gamma_{0n}|^2)}}{1 + N_i} \tag{3.75}$$

with the definition of N used previously. However, if we consider only the minimum noise figure for a given device, we will not obtain the minimum noise figure for the multistage amplifier system. This was explained when the noise measure was introduced [see (3.25)]. Therefore, a better way to design the amplifier would be to use circles at constant noise measure instead of circles of constant noise figure. This was recently done by

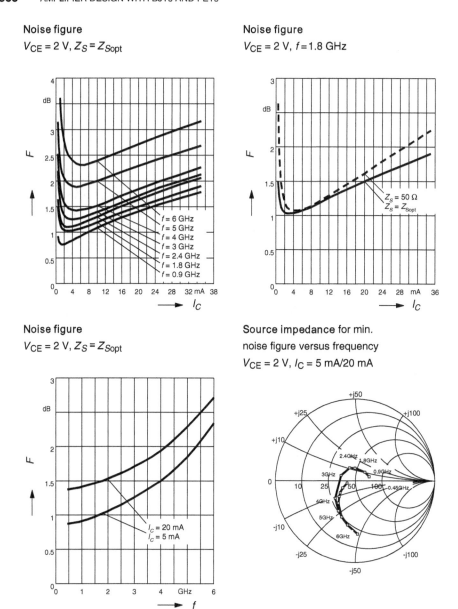

Figure 3.24 Noise figure minimum noise figure, noise figure, and source impedance for minimum noise figure for the Infineon BFP420 transistor. Courtesy Infineon Technologies.

Poole and Paul [14]. They derived the expressions for the noise measure circles as a function of S parameters, noise parameters, and Γ_G, using

$$G_A = \frac{|S_{21}|^2(1 - |\Gamma_G|^2)}{(1 - |S_{22}|^2) + |\Gamma_G|^2(|S_{11}|^2 - |\Delta|^2) - 2\mathrm{Re}(\Gamma_G C_1)} \tag{3.76}$$

Table 3.3 BFP420 Common-Emitter Noise Parameters

f	$F_{min}^{\,a}$	$G_a^{\,a}$	Γ_{opt}		R_N	r_n	$F_{50\Omega}^{\,b}$	$|S_{21}|^{2\,b}$
			MAG	ANG				
GHz	dB	dB			Γ	–	dB	dB

f	$F_{min}^{\,a}$	$G_a^{\,a}$	MAG	ANG	R_N	r_n	$F_{50\Omega}^{\,b}$	$|S_{21}|^{2\,b}$
$V_{CE}=2\,\text{V},\ I_C=5\,\text{mA}$								
0.9	0.90	20.5	0.28	41.0	8.7	0.17	1.02	20.3
1.8	1.05	15.2	0.20	82.0	6.7	0.13	1.11	15.8
2.4	1.25	13.0	0.20	124.0	5.5	0.11	1.32	13.5
3.0	1.38	12.1	0.22	−175.0	5.0	0.10	1.48	11.6
4.0	1.55	10.3	0.33	−157.0	5.5	0.11	1.83	9.1
5.0	1.75	8.6	0.45	−142.0	5.0	0.10	2.20	7.0
6.0	2.20	6.4	0.53	−123.0	15.0	0.30	3.30	5.3

[a]Input matched for minimum noise figure, output for maximum gain
[b]$Z_S = Z_L = 50\Omega$

where

$$C_1 = S_{11} - S_{22}^* \Delta$$
$$\Delta = S_{11}S_{22} - S_{12}S_{11}$$

In terms of the generator reflection coefficient, the noise measure can be expressed as

$$M = [(F_{min} - 1)(1 - |\Gamma_G|^2)|S_{21}|^2|1 + \Gamma_{0n}|^2 + 4r_n|S_{21}|^2|\Gamma_G - \Gamma_{0n}|^2]$$
$$\times \{|1 + \Gamma_{0n}|^2[|S_{21}|^2(1 - |\Gamma_G|^2) - (1 - |S_{22}|^2)$$
$$-|\Gamma_G|^2(|S_{11}|^2 - |\Delta|^2) + 2\text{Re}(\Gamma_G C_1)]\}^{-1} \tag{3.77}$$

Equation (3.77) can be shown to represent circles in the source reflection coefficient plane described by the following:

$$|\Gamma_G|^2 + |\Gamma_m|^2 - \Gamma_G \Gamma_m^* - \Gamma_G^* \Gamma_m = r_m^2 \tag{3.78}$$

The centers and radii of the constant-noise-measure circles are given by

$$C_m = \frac{M|1 + \Gamma_{0n}|^2 C_1^* + 4r_n|S_{21}|^2 \Gamma_{0n}}{M|1 + \Gamma_{0n}|^2 P + |S_{21}|^2(4r_n - W)} \tag{3.79}$$

$$r_m = \frac{\sqrt{M^2 M_a + MM_b + M_c}}{M|1 + \Gamma_{0n}|^2 P + |S_{21}|^2(4r_n - W)} \tag{3.80}$$

where

$$P = |S_{21}|^2 + |S_{11}|^2 - |\Delta|^2$$
$$Q = |S_{21}|^2 + |S_{22}|^2 - 1$$
$$W = |1 + \Gamma_{0n}|^2(F_{min} - 1)$$
$$M_a = |1 + \Gamma_{0n}|^4(PQ + |C_1|^2)$$
$$M_b = |1 + \Gamma_{0n}|^2|S_{21}|^2$$
$$\quad [-(4r_n|\Gamma_{0n}|^2 + W)P - (W - 4r_n)Q]$$
$$M_c = |S_{21}|^4|1 + \Gamma_{0n}|^2(F_{min} - 1)$$
$$\quad [W - 4r_n(1 - |\Gamma_{0n}|^2)]$$

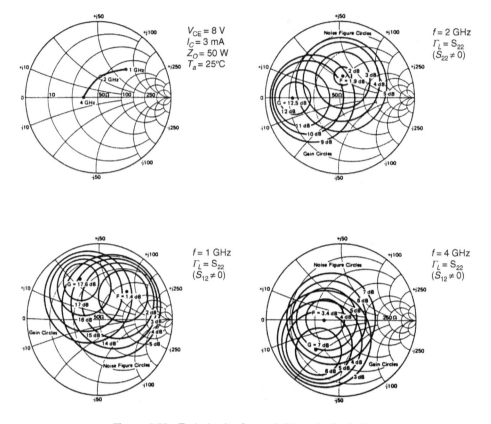

Figure 3.25 Typical noise figure circles and gain circles.

The value of the minimum noise measure can be found by considering the noise measure circle of zero radius, that is, set r_m equal to zero in (3.80). This results in

$$M_{\min} = \frac{-M_b \pm \sqrt{M_b^2 - 4M_a M_c}}{2M_a} \tag{3.81}$$

Equation (3.81) yields the same value of $M_{\min}$ as would have been obtained by using the immittance parameter equation given by Fukui [13] and can, therefore, be considered as the reflection coefficient plane analog of Fukui's expression. The elimination of the need for the parameter R_{eg}, however, results in a considerable simplification as compared with the earlier approach [13].

The value of the minimum noise measure is taken as the smallest nonnegative value of $M_{\min}$ given by (3.81). The source reflection coefficient that results in the minimum noise measure can now be obtained by employing (3.79):

$$\Gamma_{0m} = \frac{M_{\min}|1 + \Gamma_{0n}|^2 C_1^* + 4r_n |S_{21}|^2 \Gamma_{0n}}{M_{\min}|1 + \Gamma_{0n}|^2 P + |S_{21}|^2 (4r_n - W)} \tag{3.82}$$

The output reflection coefficients of the device, when Γ_{0m} is presented to the input port, is given by

$$S'_{22} = \frac{S_{22} - \Delta\Gamma_{0m}}{1 - S_{11}\Gamma_{0m}} \tag{3.83}$$

Noise Correlation in Linear Two-Ports Using Correlation Matrices   In the introduction to two-port noise theory, it was indicated that noise correlation matrices form a general technique for calculating noise in n-port networks. Haus and Adler have described the theory behind this technique [6]. In 1976, Hillbrand and Russer published equations and transformations that aid in supplying this method to two-port CAD [15].

This method is useful because it forms a base from which we can rigorously calculate the noise of linear two-ports combined in arbitrary ways. For many representations, the method of combining the noise parameters is as simple as that for combining the circuit element matrices. In addition, noise correlation matrices can be used to calculate the noise in linear frequency conversion circuits. The following is an introduction to this subject.

Linear, noisy two-ports can be modeled as a noise-free two-port with two additional noise sources. These noise sources must be chosen so that they add directly to the resulting vector of the representation, as shown in (3.84) and (3.85), and Figure 3.17.

$$\begin{bmatrix} I_1 \\ I_2 \end{bmatrix} = \begin{bmatrix} y_{11} & y_{12} \\ y_{21} & y_{22} \end{bmatrix} \begin{bmatrix} V_1 \\ V_2 \end{bmatrix} + \begin{bmatrix} i_1 \\ i_2 \end{bmatrix} \tag{3.84}$$

$$\begin{bmatrix} V_1 \\ V_2 \end{bmatrix} = \begin{bmatrix} z_{11} & z_{12} \\ z_{21} & z_{22} \end{bmatrix} \begin{bmatrix} I_1 \\ I_2 \end{bmatrix} + \begin{bmatrix} v_1 \\ v_2 \end{bmatrix} \tag{3.85}$$

where the i and v vectors indicate noise sources for the y and z representations, respectively. This two-port example can be extended to n-ports in a straightforward, obvious way.

Since the noise vector for any representation is a random variable, it is much more convenient to work with the noise correlation matrix. The correlation matrix gives us deterministic numbers to calculate with. The correlation matrix is formed by taking the mean value of the outer product of the noise vector. This is equivalent to multiplying the noise vector by its adjoint (complex conjugate transpose) and averaging the result:

$$\langle \overline{i}\overline{i}^+ \rangle = \begin{bmatrix} i_1 \\ i_2 \end{bmatrix} \begin{bmatrix} i_1^* & i_2^* \end{bmatrix} = \begin{bmatrix} \langle i_1 i_1^* \rangle & \langle i_1 i_2^* \rangle \\ \langle i_1^* i_2 \rangle & \langle i_2 i_2^* \rangle \end{bmatrix} = [C_y] \tag{3.86}$$

where the angular brackets denote the average value.

Note that the diagonal terms are the "power" spectrum of each noise source and the off-diagonal terms are complex conjugates of each other and represent the cross "power" spectrums of the noise sources. "Power" is used because these magnitude-squared quantities are proportional to power.

To use these correlation matrices in circuit analysis, we must know how to combine them and how to convert them between various representations. An example using y matrices will illustrate the method for combining two-ports and their correlation matrices. Given two matrices y and y', when we parallel them we have the same port voltages, and the terminal

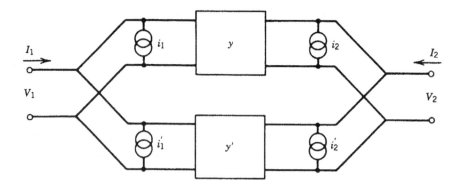

Figure 3.26 Parallel combination of two-ports using y parameters.

currents add (Figure 3.26):

$$I_1 = y_{11}V_1 + y_{12}V_2 + y'_{11}V_1 + y'_{12}V_2 + i_1 + i'_1$$
$$I_2 = y_{21}V_1 + y_{22}V_2 + y'_{21}V_1 + y'_{22}V_2 + i_2 + i'_2 \qquad (3.87)$$

or

$$\begin{bmatrix} I_1 \\ I_2 \end{bmatrix} = \begin{bmatrix} y_{11} + y'_{11} & y_{12} + y'_{12} \\ y_{21} + y'_{21} & y_{22} + y'_{22} \end{bmatrix} \begin{bmatrix} V_1 \\ V_2 \end{bmatrix} + \begin{bmatrix} i_1 + i'_1 \\ i_2 + i'_2 \end{bmatrix} \qquad (3.88)$$

Here, we can see that the noise current vectors add just as the y parameters add. Converting the new noise vector to a correlation matrix yields

$$\langle \bar{i}_{new} \bar{i}_{new}^+ \rangle = \left\langle \begin{bmatrix} i_1 + i'_1 \\ i_2 + i'_2 \end{bmatrix} [i_1^* + i_1'^* \quad i_2 i_2'^*] \right\rangle \qquad (3.89)$$

$$= \begin{bmatrix} \langle i_1 i_1^* \rangle + \langle i'_1 i_1'^* \rangle & \langle i_1 i_2^* \rangle + \langle i'_1 i_2'^* \rangle \\ \langle i_2 i_1^* \rangle + \langle i'_2 i_1'^* \rangle & \langle i_2 i_2^* \rangle + \langle i'_2 i_2'^* \rangle \end{bmatrix} \qquad (3.90)$$

The noise sources from different two-ports must be uncorrelated, so there are no cross products of different two-ports. By inspection, (3.90) is just the addition of the correlation matrices for the individual two-ports, so

$$[C_{ynew}] = [C_y] + [C'_y] \qquad (3.91)$$

The same holds true for g, h, and z parameters, but $ABCD$ parameters have the more complicated form shown below. If

$$[A_{new}] = [A][A'] \qquad (3.92)$$

then

$$[C_{Anew}] = [C_A] + [A][C_{A'}][A]^+ \qquad (3.93)$$

The transformation of one representation to another is best illustrated by an example. Let us transform the correlation matrix for a Y representation to a Z representation. Starting with

$$\begin{bmatrix} I_1 \\ I_2 \end{bmatrix} = [Y] \begin{bmatrix} V_1 \\ V_2 \end{bmatrix} + \begin{bmatrix} i_1 \\ i_2 \end{bmatrix} \tag{3.94}$$

we can move the noise vector to the left side and invert y:

$$\begin{bmatrix} V_1 \\ V_2 \end{bmatrix} = [Y^{-1}] \begin{bmatrix} I_1 - i_1 \\ I_2 - i_2 \end{bmatrix} = [Y^{-1}] \begin{bmatrix} I_1 \\ I_2 \end{bmatrix} + [Y^{-1}] \begin{bmatrix} -i_1 \\ -i_2 \end{bmatrix} \tag{3.95}$$

Since $(Y)^{-1} = (Z)$, we have

$$\begin{bmatrix} V_1 \\ V_2 \end{bmatrix} = [Z] \begin{bmatrix} I_1 \\ I_2 \end{bmatrix} + [Z] \begin{bmatrix} -i_1 \\ -i_2 \end{bmatrix} \tag{3.96}$$

so

$$= [Z] \begin{bmatrix} -i_1 \\ -i_2 \end{bmatrix} = [T_{yz}] \begin{bmatrix} -i_1 \\ -i_2 \end{bmatrix} \tag{3.97}$$

where the signs of i_1 and I_2 are superfluous since they will cancel when the correlation matrix is formed. Here, the transformation of the Y noise current vector to the Z noise voltage vector is done simply by multiplying by (Z). Other transformations are shown in Table 3.4.

To form the noise correlation matrix, we gain from the mean of the outer product:

$$\langle vv^+ \rangle = \begin{bmatrix} \langle v_1 v_1^* \rangle & \langle v_1 v_2^* \rangle \\ \langle v_1^* v_2 \rangle & \langle v_2 v_2^* \rangle \end{bmatrix} = [Z] \left\langle \begin{bmatrix} i_1 \\ i_2 \end{bmatrix} [i_1^* \quad i_2^*] \right\rangle [Z]^+ \tag{3.98}$$

or

$$[C_z] = [Z][C_y][Z]^+ \tag{3.99}$$

where

$$v^+ = [i_1^* \quad i_2^*][Z]^+ \tag{3.100}$$

This is called a congruence transformation. The key to all of these derivations is the construction of a correlation matrix from the noise vector, as shown in (3.90).

Table 3.4 Noise Matrix $T_{\alpha\beta}$ Transformation

		Original Form (α Form)					
		Y		Z		A	
Resulting form (β form)	Y	1	0	y_{11}	y_{12}	$-y_{11}$	1
		0	1	y_{21}	y_{22}	$-y_{21}$	0
	Z	z_{11}	z_{12}	1	0	1	$-z_{11}$
		z_{21}	z_{22}	0	1	0	$-z_{21}$
	A	0	A_{12}	1	$-A_{11}$	1	0
		1	A_{22}	0	$-A_{21}$	0	1

These correlation matrices may easily be derived from the circuit matrices of passive circuits with only thermal noise sources. For example,

$$[C_z] = 2kT\Delta f \text{Re}([Z]) \text{ and} \tag{3.101}$$

$$[C_y] = 2kT\Delta f \text{Re}([Y]) \tag{3.102}$$

The $2kT$ factor comes from the double-sided spectrum of thermal noise. The correlation matrix from the *ABCD* matrix may be related to the noise figure, as shown by Hillbrand and Russer [15]. We have

$$F = 1 + \frac{\bar{Y}[C_a]\bar{Y}^+}{2kT\text{Re}(Y_G)} \tag{3.103}$$

where

$$\bar{Y} = \begin{bmatrix} Y_G \\ 1 \end{bmatrix}$$

Expressing the noise factor in terms of the correlation matrix, here is a complete formula:

$$F = 1 + \frac{C_{22}^A(f) + 2\,\text{Re}\{Y_g(f)C_{12}^A(f)\} + |Y_g(f)|^2 C_{11}^A(f)}{2kT_0\text{Re}\{Y_g(f)\}} \tag{3.104}$$

Once we transform this in the *Y* parameter form, we obtain the following equation:

$$F(f) = F_{\min}(f) + \frac{R_n(f)|Y_{\text{opt}}(f) - Y_g(f)|^2}{\text{Re}\{Y_g(f)\}} \tag{3.105}$$

It should be noted that all these values are frequency dependent as expressed in this equation.

The *ABCD* correlation matrix can be written in terms of the noise-figure parameters as (double-sided spectrum)

$$[C_a] = 2kT \begin{bmatrix} R_n & \dfrac{F_0 - 1}{2} - R_n Y_{0n}^* \\[2mm] \dfrac{F_0 - 1}{2} - R_n Y_{0n} & R_n|Y_{0n}|^2 \end{bmatrix} \tag{3.106}$$

The noise correlation matrix method forms an easy and rigorous technique for handling noise in networks. This technique allows us to calculate the total noise for complicated networks by combining the noise matrices of subcircuits. It should be remembered that although noise correlation matrices apply to *n*-port networks, noise-figure calculations apply only to pairs of ports. The parameters of the C_a matrix can be used to give the noise parameters:

$$Y_{0n} = \sqrt{\frac{C_{ii*}}{C_{uu*}} - \left[\text{Im}\left(\frac{C_{ui*}}{C_{uu*}}\right)\right]^2} + j\,\text{Im}\left(\frac{C_{ui*}}{C_{uu*}}\right) \tag{3.107}$$

$$F_0 = 1 + \frac{C_{ui*} + C_{uu*}Y_{0n}^*}{kT} \tag{3.108}$$

$$R_n = C_{uu*} \tag{3.109}$$

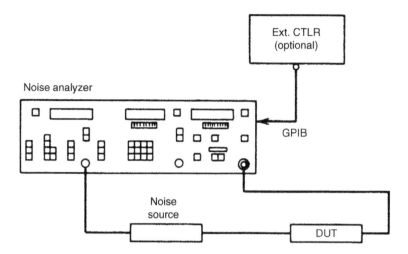

Figure 3.27 Noise figure measurement.

Noise Figure Test Equipment Figure 3.27 shows the block diagram of a noise test setup. It includes the noise source and the other components. The metering unit has a special detector that is linear and over a certain dynamic range measures linear power. The tunable receiver covers a wide frequency range (e.g., 10–1800 MHz) and controls the noise source. The receiver is a double-conversion superheterodyne configuration with sufficient image rejection to avoid double-sideband noise measurements that would give the wrong results.

These receivers are microprocessor controlled and the measurement is a two-step procedure. The first is a calibration step that measures the noise figure of the receiver system and a reference power level. Then the device under test (DUT) is inserted and the system noise figure and total output power are measured. The noise factor is calculated by

$$F_1 = F_{\text{system}} - \frac{F_2 - 1}{G_1} \tag{3.110}$$

and the gain is given by the change in output power from the reference level [16]. The noise of the system is calculated by measuring the total noise power with the noise source on and off. With the ENR (excess noise ratio) known [16],

$$F_{\text{system}} = \frac{\text{ENR}}{Y - 1} \tag{3.111}$$

The noise bandwidth is usually set by the bandwidth of the receiver, which is assumed to be constant over the linear range. The ENR of the noise source is given by

$$\text{ENR} = \frac{T_{\text{hot}}}{T_{\text{cold}}} - 1 \tag{3.112}$$

where T_{cold} is usually room temperature (290 K). This ENR number is about 15 dB for noise sources with a 15-dB pad and 5 dB for noise sources with a 25-dB pad. Since both gain and noise were stored in the initial calibration, a noise/gain sweep can be performed.

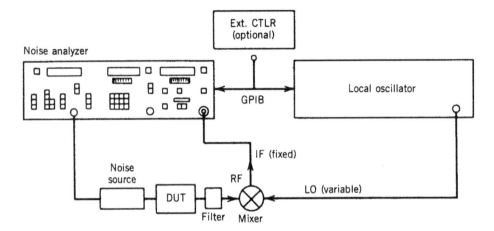

Figure 3.28 Single-sideband (SSB) noise figure measurements using an external mixer.

For frequencies above 1800 MHz, we can extend the range with the help of the external signal generators, as shown in Figure 3.28. As shown, a filter ahead of the external mixer reduces the noise energy in the image band. If the DUT has a very broad frequency range and has flat gain and noise over that range, a double-sideband (DSB) measurement is possible, with the image-rejection filter removed. However, a single-sideband (SSB) measurement is always more accurate [17].

How to Determine the Noise Parameters The noise figure of a linear two-port network as a function of source admittance may be represented by

$$F = F_{\min} + \frac{R_n}{G_G}[(G_{0n} - G_G)^2 + (B_{0n} - B_G)^2] \qquad (3.113)$$

where $G_G + jB_G$ = generator admittance presented to the input of the two-port, $G_{0n} + jB_{0n}$ = generator admittance at which optimum noise figure occurs, and R_n = empirical constant relating to the sensitivity of the noise figure to generator admittance, with dimensions of resistance.

It may be noted that for an arbitrary noise-figure measurement with a known generator admittance, equation (3.113) has four unknowns: $F_{\min}$, R_n, G_{0n}, and B_{0n}. By choosing four known values of generator admittance, a set of four linear equations are formed and the solution of the four unknowns can be found [18, 19]. Equation (3.113) may be transformed to

$$F = F_{\min} + \frac{R_n|Y_{0n}|^2}{G_G} - 2R_nG_{0n} + \frac{R_n|Y_G|^2}{G_G} - 2R_nB_{0n}\frac{B_G}{G_G} \qquad (3.114)$$

$$F = F_{\min} + \frac{R_n}{G_G}|Y_G - Y_{0n}|^2 \qquad (3.115)$$

Let

$$X_1 = F_{\min} - 2R_n G_{0n}$$
$$X_2 = R_n |Y_{0n}|^2$$
$$X_3 = R_n$$
$$X_4 = R_n B_{0n}$$

Then the generalized equation may be written as

$$F_i = X_1 + \frac{1}{G_{si}} X_2 + \frac{|Y_{si}|^2}{G_{si}} X_3 - 2\frac{G_{si}}{B_{si}} X_4 \qquad (3.116)$$

or, in matrix form

$$[F] = [A][X] \qquad (3.117)$$

and the solution becomes

$$[X] = [A]^{-1}[F] \qquad (3.118)$$

These parameters completely characterize the noise behavior of the linear two-port network. Direct measurement of these noise parameters by this method would be possible only if the receiver on the output of the two-port were noiseless and insensitive to its input admittance. In actual practice, the receiver itself behaves as a noisy two-port network and can be characterized in the same manner. What is actually being measured is the system noise figure of the two-port and the receiver. Thus, it becomes apparent that to do a complete two-port noise characterization, the gain of the two-port must be measured [20]. In addition, any losses in the input-matching networks must be carefully accounted for, because they add directly to the measured noise figure reading [21].

3.1.4 Linearity

Consistent with the elaborate discussion of linearity questions above, here is a quick refresher on what must be considered in amplifiers.

Dynamic Range, Compression, and IMD   An amplifier's dynamic range is related to the minimum discernible signal and the 1-dB compression point according to the equation

$$\mathrm{DR}_n = \frac{(n-1)[\mathrm{IP}_{n(\mathrm{in})} - \mathrm{MDS}]}{n} \qquad (3.119)$$

where DR is the dynamic range in decibels, n is the order, $\mathrm{IP}_{(\mathrm{in})}$ is the input intercept power in dBm, and MDS is the minimum detectable signal power in dBm. Once the 1-dB compression point is known, it is a fair assumption that the third-order intercept point, expressed in dBm, is 10 dB above this.

The intermodulation distortion gives the quality of an amplifier to withstand multiple signals without generating large intermodulation products or when using amplitude-modulated signals, cross-modulation. Figure 3.29 provides a quick reminder of IMD issues.

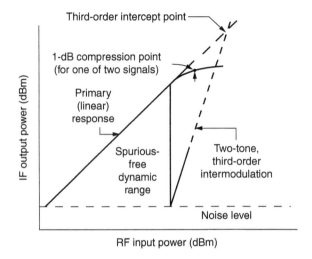

Figure 3.29 Amplifier linearity evaluation, including compression and two-tone IMD dynamic range. P_{-1dB} for a single-tone cannot be read directly from this graph because the values shown are the result of two equal-power tones.

In some amplifiers, we would like to have control over the gain. Although the better way to control gain while obtaining high linearity is a PIN-diode attenuator, special AGC circuits have become part of the amplifiers. The following shows a circuit of the SL610 made by Plessey that has an AGC input provision. The PIN-diode attenuator has essential the same noise figure as its attenuation. The AGC in multistage amplifiers is distributed over several stages, and there is a lesser correlation between noise and AGC than in pure passive attenuators.

Special Case of Linearity Requirements for Digital Modulation For cellular telephone systems, we use digital modulation formats, such as QPSK and $\pi/4$ DQPSK, that combine phase and amplitude modulation. Amplifiers handling such signals must be carefully characterized and designed if adequate amplitude and phase linearity are to be maintained.

Examples of Power Amplifiers: Looking into the Effects of Distortion This example presents modulation analysis using QPSK in a 2-GHz power amplifier (Figure 3.30). The simple amplifier was designed using the electronic Smith Tool of Serenade to determine the matching network for a narrow-band design. The example demonstrates setup of modulation analysis and available results.

Siemens among others provides integrated amplifiers based on the concept above, the most popular ones being the CGY96 and CGY94. For the purpose of measurements, these amplifiers were actually put on a breadboard. Figure 3.31 shows the big brother, the CGY96. It is a GaAs MMIC intended as a power amplifier for GSM class 4 phones with 3.2 W (35 dBm) at 3.5 V at an overall power added efficiency of 50%. Table 3.5 shows its electrical characteristics.

For the purpose of our experiments, we considered the CGY94 because it requires less external peripheral circuits. Table 3.6 shows its electrical characteristics. For those of us

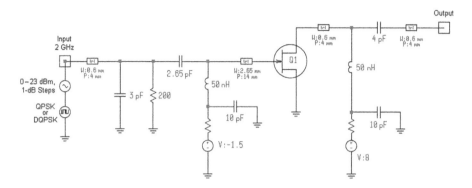

Figure 3.30 The power amplifier schematic.

interested in seeing part of the internal circuit, Figure 3.32 shows the chip. This two-stage amplifier uses an interesting internal feedback scheme. Its schematic is shown in Figure 3.33.

For the purpose of looking at the waveforms, we will continue with the single-stage amplifier (Figure 3.30) and then compare the measured versus predicted performance of this CGY94-based amplifier.

A modulation source is connected to the RF source at the input port of the amplifier. The RF source specifies the carrier power (or voltage or current) and the modulation source specifies the modulation format and properties of the modulated signal.

A brief review of the modulation source used in this example indicates the following properties.

Figure 3.31 CGY96 GaAs MMIC power amplifier.

Property	Value	Description
NB	128	Number of bits.
BR	1.2288E6	Bit rate.
N	8	Number of samples per bit.
BW_1	590.E3	One-sided bandwidth of the main channel in the PIB or P1IB calculation.
FS_1	740.E3	Start baseband frequency of the first adjacent channel.
BW_2	590.E3	One-sided bandwidth of the main channel in the P2IB calculation.
FS_2	1990.E3	Start baseband frequency of the second adjacent channel.
BW_3	590.E3	One-sided bandwidth of the main channel in the P3IB calculation.
FS_3	3240.E3	Start baseband frequency of the third adjacent channel.
M	4	Order of the signal space. $M = 4$ is QPSK.
DLY	0.0	Fractional bit delay.
IASC	1.0	I channel amplitude scale. Multiplier for the I waveform to model amplitude imbalance.
QASC	1.0	Q channel amplitude scale. Multiplier for the Q waveform to model amplitude imbalance.
LPF	3	Baseband low pass filter. 3 = Root raised cosine filter.
LPFC	665.E3	−3 dB cutoff frequency of LPF.
LPFN	3	Number of resonators if LPF = 1.
LPFR	0.35	Roll-off factor if LPF = 3 or LPF = 4.
FILE	N/A	File name of user-defined modulated signal file.

Table 3.5 CGY96 Electrical Characteristics

Characteristics	Symbol	Minimum	Typical	Maximum	Unit
Frequency range	f	880	–	915	MHz
Supply current $P_{in} = 0$ dBm	I_D	–	1.8	–	A
Supply current neg. voltage gener. $V_{aux} = 3.5\,V$	I_{aux}	–	10	–	mA
Gain (small signal)	G	–	40	–	dB
Power gain $P_{in} = 0$ dBm	G_p	–	35	–	dB
Output power $P_{in} = 0$ dBm, $V_{control} = 2.0$ V...2.5 V)	P_{out}	–	35	–	dBm
Overall power added efficiency $P_{in} = 0$ dBm	η	–	50	–	%
Dynamic range output power $V_{control} = 0.2...2.2$ V		–	80	–	dB
Harmonics $P_{in} = 0$ dBm	$H(2f_0)$	–	–40	–	dBc
	$H(3f_0)$	–	–43	–	dBc
	$H(4f_0)$	–	–44	–	dBc
Noise power in RX (935–960 MHz) $P_{in} = 0$ dBm, $P_{out} = 35$ dBm, 100 kHz RBW	N_{RX}	–	–81	–	dBm
Stability all spurious outputs < –60 dBc, VSWR load, all phase angles			10 : 1	–	–
Input VSWR			1.7 : 1	–	–

$T_A = 25°C$, $V_{neg} = -5$ V, $V_{control} = 2.2$ V; duty cycle 12.5%, $t_{on} = 577\mu s$

Table 3.6 CGY94 Electrical Characteristics

Characteristics	Symbol	Minimum	Typical	Maximum	Unit
Supply current $V_D = 3.0$ V; $P_{in} = 10$ dBm	I_{DD}	–	1.18	–	A
Negative supply current (normal operation)	I_G	–	2	–	mA
Shut-off current V_{TR} n.c.	I_D	–	400	–	μA
Negative supply current (shut off mode, V_{TR} pin n.c.)	I_G	–	10	–	μA
Gain $P_{in} = -5$ dBm	G	27.0	29.0	–	dB
Power gain $V_D = 3.6$ V; $P_{in} = 10$ dBm	G	22.8	23.6	–	dB
Output power $V_D = 3.0$ V; $P_{in} = 10$ dBm	P_0	31.5	32.3	–	dBm
Output power $V_D = 3.6$ V; $P_{in} = 10$ dBm	P_0	32.8	33.6	–	dBm
Output power $V_D = 5$ V; $P_{in} = 10$ dBm	P_0	34.5	35.5	–	dBm
Overall power added efficiency $V_D = 3.0$ V; $P_{in} = 10$ dBm	η	43	48	–	%
Overall power added efficiency $V_D = 3.6$ V; $P_{in} = 10$ dBm	η	42	47	–	%
Overall power added efficiency $V_D = 5$ V; $P_{in} = 10$ dBm	η	41	46	–	%
Harmonics ($P_{in} = 10$ dBm, CW) $2 f_0$	–	–	–49	–	dBc
$V_D = 3.6$ V; ($P_{out} = 33.1$ dBm) $3 f_0$	–	–	–45	–	dBc
Input VSWR $V_D = 3.6$ V;	–	–	1.5 : 1	2.0 : 1	–

$T_A = 25°$C, $f = 0.9$ GHz, $Z_S = Z_L = 50$, $V_D = 3.6$ V, $V_G = -4$ V, V_{TR} pin connected to ground, unless otherwise specified; pulsed with a duty cycle of 10%, ton $= 0.33$ ms

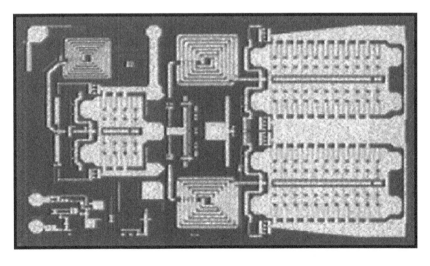

Figure 3.32   The CGY94 amplifier chip.

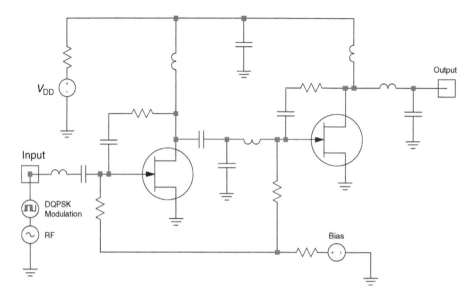

Figure 3.33 Schematic of the CGY94 GaAs MMIC power amplifier.

The modulation source is a QPSK stream with a bit rate of 1.2288 Mbits/s. A total of 128 bits will be analyzed, and each bit is sampled eight times to construct an accurate analog waveform. The BW and FS parameters are defined here to specify the main (in-band) channel and adjacent channel for ACPR calculations. Imbalance of the I & Q amplitudes and phases can be described by the IASC, QASC, and DLY parameters, if desired. The modulation signal is often filtered, and several types of filters are available. Here, a root-raised cosine filter is used with a cutoff frequency of 665 kHz and a roll-off factor of 0.35.

Analysis Figure 3.34 shows the compression characteristics of the amplifier as the available RF source power (a sinusoid) is swept. Note that P_{-1dB} is 11.5 dBm referred to the input and 28.5 dBm referred to the output. We will look at modulation characteristics in the linear region of operation at a source power of 0 dBm, and the nonlinear region at a source power of 20 dBm.

Next, we will view the voltage across the transistor to examine its behavior under compression. Figure 3.35 shows the drain-source voltage of the FET as the source power is swept. It is clear from this graph that clipping due to pinchoff and forward-gate conduction is limiting the performance of the device. A phase shift of the voltage waveform as power is increased is also apparent. The combination of the power compression and phase shift, or AM to PM conversion, will be used in the modulation analysis to determine the overall distortion of the modulated signal.

The output available from modulation analysis includes

- I & Q channel waveforms,
- eye and constellation diagrams,
- spectral plots, and
- ACPR, in-band and adjacent power.

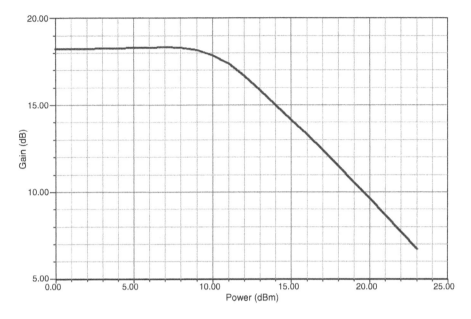

Figure 3.34 Single-tone RF power sweep analysis of the FET amplifier.

Figure 3.36 shows the eye diagrams for the 128-bit QPSK signal at two RF source powers, 0 dBm and 20 dBm. Operating linearly at 0 dBm input, the amplifier does not distort the eye and it remains wide open. Note that, in this example, the filter bandwidth does not produce any intersymbol interference, as witnessed by the open eye. At an input power of 20 dBm, the amplifier compresses the signal and distortion in the eye is evident.

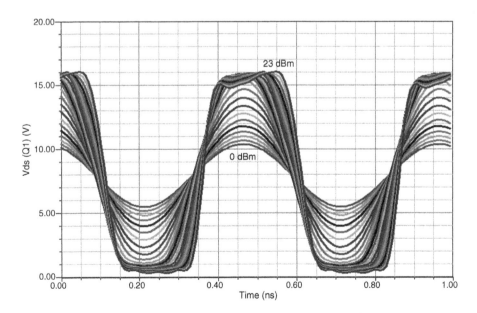

Figure 3.35 V_{DS} of the FET versus RF source power.

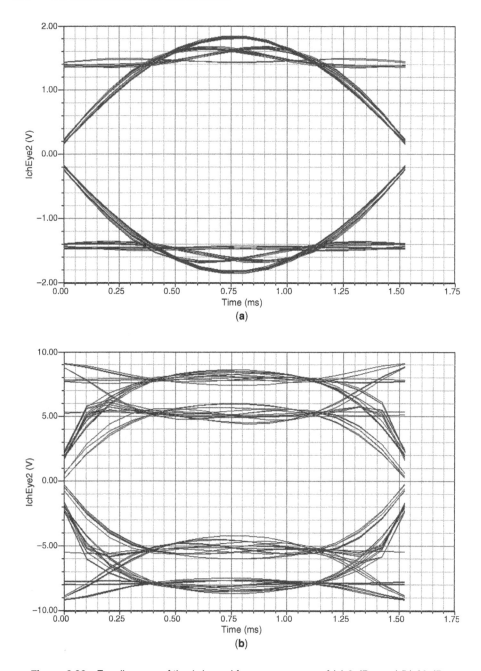

Figure 3.36 Eye diagrams of the *I* channel for source powers of (a) 0 dBm and (b) 20 dBm.

We can also view the modulation spectrum at these two power levels to investigate the intermodulation distortion and spectral regrowth that takes place. Figure 3.37 shows the spectral plots. The lower trace corresponds to 0 dBm source power and shows almost no regrowth, while the upper trace at 20 dBm source power shows considerable regrowth.

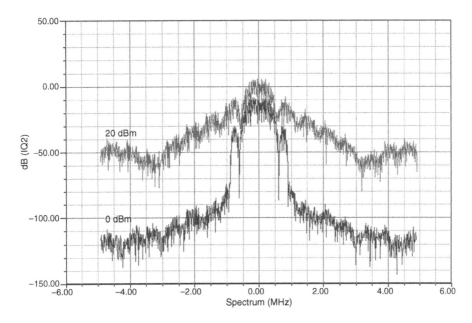

Figure 3.37 Spectrum of the modulation signal at the amplifier output. Source power is 0 dBm and 20 dBm.

The uncorrupted modulation source can also be shown. This corresponds to the source as a pure voltage without any circuit influence.

Now we evaluate the CGY94 board in a similar fashion. Figure 3.38 shows the waveform used for CAD purposes. Needless to say, such a signal cannot be produced by any signal generator. The one generated by an actual signal generator is shown in Figure 3.39. The wideband noise is around −80 dBm and the input level is about −25 dBm, consistent with the simulation signal. Looking at the output, Figure 3.40 shows the simulated signal with the expected increase in bandwidth. This is due to the nonlinearity of the amplifier and is consistent with Figure 3.37 for the single-stage amplifier. The actual measured output for the CGY94 shown in Figure 3.41 is quite close to its simulated output, indicating that the mathematical approach used for simulating is correct. This is only one example; in reality, many more waveforms can be analyzed and predicted using this method. Figure 3.41 shows essentially three distinct steps, which is sufficiently close to the response shown in the simulation. Needless to say, a dynamic range of 70 dB, while displayed in Figures 3.40 and 3.41, is not really necessary for good-quality transmission, and the discrepancy between the two pictures is in part due to the limited modeling quality of the active device as provided by the company that did the parameter extraction for this transistor.

Another useful view to aid in the understanding of the compression and distortion within the amplifier is to look at the magnitude of the complex waveform. Since the signal is composed of the in-phase and quadrature-phase components, it is represented as

$$s(t) = i(t) + j \cdot q(t) \tag{3.120}$$

The time-domain magnitude is written as

$$|s(t)| = \sqrt{i(t)^2 + q(t)^2} \tag{3.121}$$

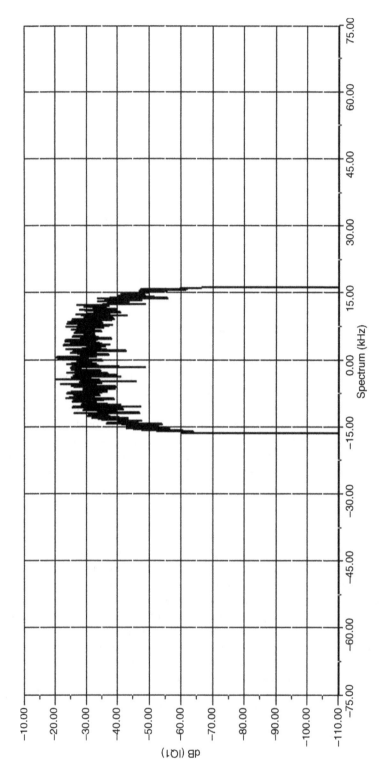

Figure 3.38 Simulated signal used for CAD analysis of the CGY94 amplifier.

406

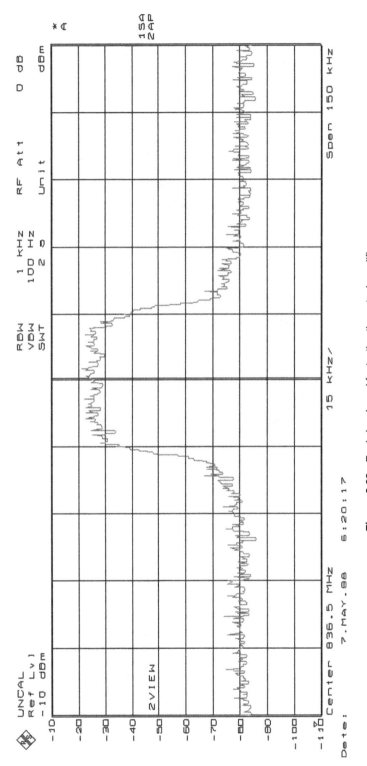

Figure 3.39 Real signal used for testing the actual amplifier.

407

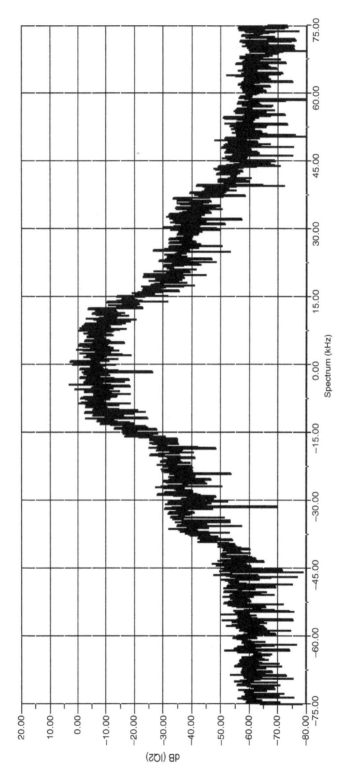

Figure 3.40 Spectral regrowth predicted by the simulation.

408

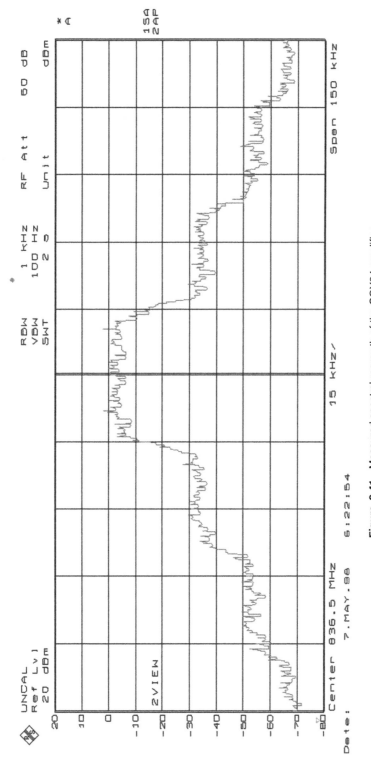

Figure 3.41 Measured spectral regrowth of the CGY94 amplifier.

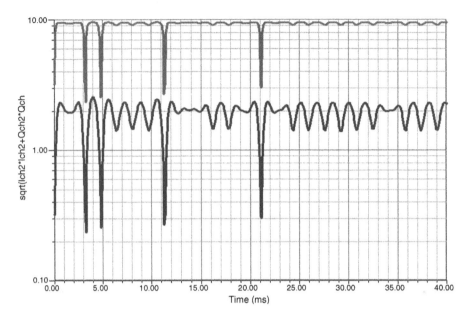

Figure 3.42 Time-domain magnitude of the complex modulation signal at 0 dBm and 20 dBm source powers.

Figure 3.42 shows the magnitude at 0 dBm and 20 dBm source powers. The X axis has been rescaled to 40 µs for improved viewing. Note the significant compression of the 20 dBm waveform at the higher signal levels. It is clear from this view that the signal is severely distorted.

We can also compute the adjacent channel power ratio, ACPR, against the RF power sweep. ACPR is the ratio of the adjacent channel power to the in-band channel power. The bandwidth (BW) and adjacent channel start frequency (FS) are used for this calculation. For accurate ACPR calculation, a large number of bits are needed. This example uses 128 bits so the computation time is short, but you should use 512 or more bits for a more accurate computation.

Figure 3.43 shows the ACPR as a function of RF source power. ACPR is nearly constant (and nonzero due to the gradual skirt of the baseband filter and inherent spillover to the adjacent channel) up to the P_{-1dB} compression point. As the intermodulation products spill into the adjacent channel and spectral regrowth occurs, the ACPR degrades.

π/4 DQPSK Circuit Analysis Next, we will examine the same amplifier using a π/4 DQPSK modulation source. The task is identical to the previous one except for the source and the number of bits. The number of bits for this project has been increased from 128 to 256. Figure 3.44 shows the eye diagrams from this modulation format, at 0 and 20 dBm source powers. The distortion and related intersymbol interference are clearly evident.

The constellation plot at 0 dBm is shown in Figure 3.45.

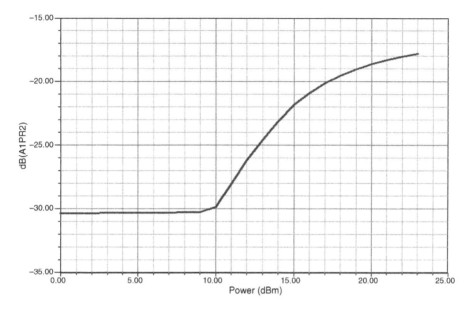

Figure 3.43 Adjacent channel power ratio as function of RF source power.

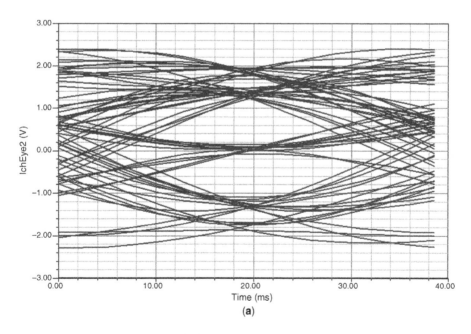

(a)

Figure 3.44 Eye diagrams for $\pi/4$ DQPSK at source powers of (a) 0 dBm and (b) 20 dBm. The distortion and related intersymbol interference are clearly visible.

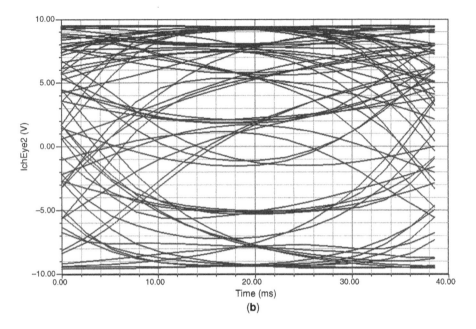

(b)

Figure 3.44   (*Continued*)

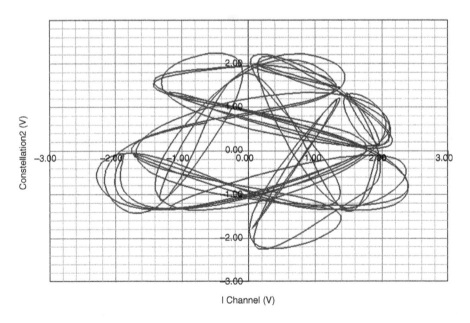

Figure 3.45   Constellation diagram for $\pi/4$ DQPSK at Port 2 and 0 dBm source power.

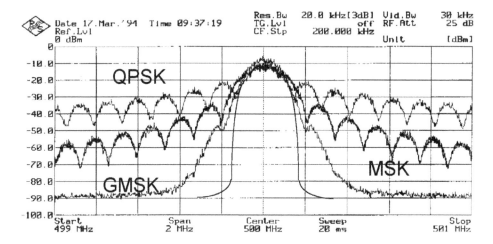

Figure 3.46   Spectrogram showing the characteristics of MSK, GMSK, and QPSK.

It may be interesting to take a look at the different waveforms currently in use. Figure 3.46 shows a spectrum analyzer picture of the most popular ones.

For reasons of time-division management, there is a gating used in the GSM standard to reduce the actual bandwidth of the spectrum. Figure 3.47 compares the GSM-signal with and without gating.

3.1.5 AGC

Many wireless systems consist of an up- or down-conversion using an intermediate frequency, typically 10% or less of the receiving frequency (FM broadcast IF has been selected at 10.7 MHz for the same reason).

We therefore will show two amplifiers with AGC and available data. The first one is a wideband amplifier-type SL610 that has been manufactured by Plessey; its current fate is unknown. However, since we put the circuit into the simulator it provides significantly insight in the operation of such AGC stages. Figure 3.48 shows the actual schematic of the SL610 entered in Ansoft's Serenade Design Environment for simulation.

The AGC action is derived from Q8 through Q5, which, together with Q4, forms a differential amplifier. Each consecutive stage (Q1, Q4, Q6) have increasing current, totaling about 20 mA, whereby the last transistor takes about 13 mA. Based on the feedback surrounding the transistors, the actual transistor characteristic is less important. We have shown this by putting much higher f_T devices in the circuit than the original design could have been made from at the time (30 years ago). The circuit is still interesting today because of its good large-signal-handling capability.

Figure 3.49 shows the frequency response of this amplifier, which is extremely close to the measured data. The good simulation also can be seen from comparing the published AGC curve with the one we found in the simulation. The bias of the first transistor remains essentially constant, resulting in a noise figure more independent of the AGC than the PIN-diode attenuator, and also stabilizes the input impedance as a function of AGC. Figure 3.50 shows the SL610's predicted AGC response.

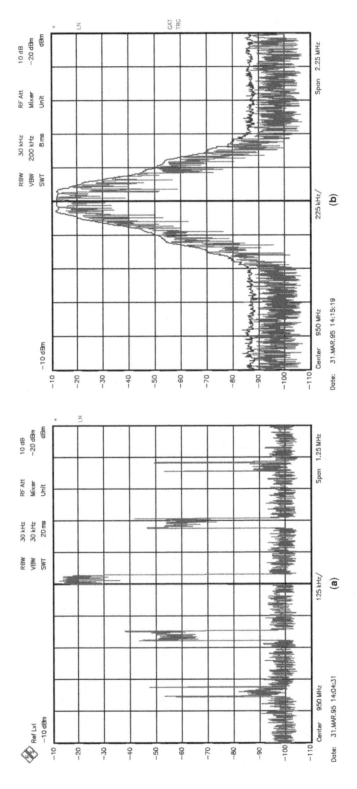

Figure 3.47 (a) Pulsed GSM signal without gating; (b) GSM signal with gating.

414

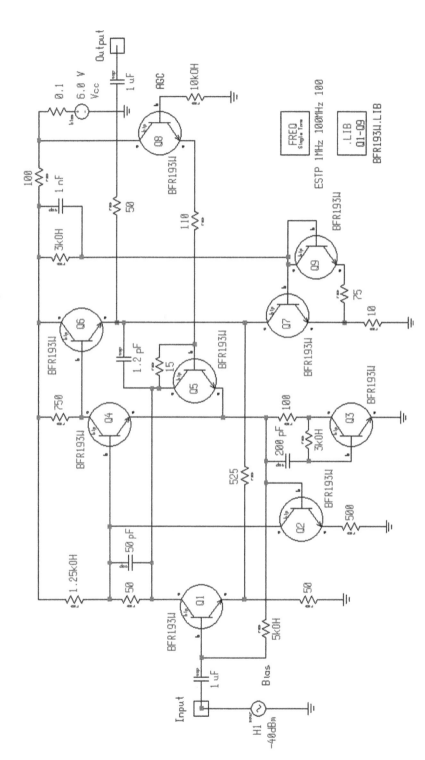

Figure 3.48 Schematic of the Plessey SL610 wideband amplifier with AGC. The 75-Ω resistor in the emitter of Q9 had to be added to obtain the appropriate dc current.

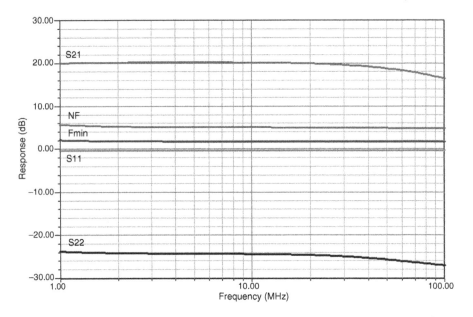

Figure 3.49 Simulated frequency response of the Plessey SL610 wideband amplifier IC with AGC = 0 V.

The same requirement is applicable for the wireless frequency range. Tables 3.7 and 3.8, and Figures 3.51–3.54 present specifications for the CGY121A GaAs MMIC by Infineon Technologies AG, formerly, the Semiconductor Group part of Siemens AG.

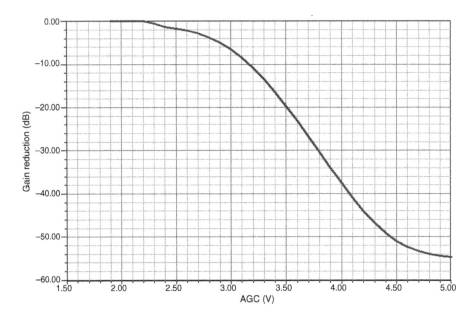

Figure 3.50 Simulated gain reduction versus AGC of the Plessey SL610 wideband amplifier IC.

Table 3.7 CGY121A Features

- Variable gain amplifier (MMIC amplifier) for mobile communication
- Typical gain control range over 50 dB
- Positive control voltage
- 50-Ω input and output matched
- Low power consumption
- Operating voltage range: 2.7–6 V
- Frequency range 800 MHz to 2.5 GHz

3.1.6 Bias and Power Voltage and Current (Power Consumption)

Biasing of transistors operating in Class A or B follows some standard rules. The only deviation from this tends to be when the operating voltage is very low, below 3 V, because then the luxury of voltage drops for good dc stability as a function of temperature disappears. (We will discuss low-voltage design in detail in Section 3.2.4.) In our simple example (Figure 3.55), we are operating from a 12-V power supply, and the first rule of thumb is to make sure that the voltage between the emitter and ground is at least 0.7 V. This is necessary to compensate the base–emitter junction threshold voltage, which decreases by about 2 mV/°C as the temperature increases. A larger voltage drop in the emitter–ground connection greatly reduces the influence. In this amplifier, which consists of an emitter, unbypassed resistor of 10 Ω in series with a 501-Ω resistor results in an overall voltage drop of 2.6 V. We could

Table 3.8 CGY121A Electrical Characteristics at 900 MHz[a] and 1.8 GHz[b]

Characteristics	Symbol	Minimum	typical	Maximum	Unit
Power gain $V_d = 3V$; $I = 45\,mA$; $V_{con} = 3V$	G	17	19	–	dB
Input return loss $V_d = 3V$; $I = 45\,mA$; $V_{con} = 3V$	RL_{in}	–	11	–	dB
Output return loss $V_d = 3V$; $I = 45\,mA$; $V_{con} = 3V$	RL_{out}	–	10	–	dB
Gain Control Range $V_{con} = 3\,V \ldots 0V$; $V_d = 3V$; $I = 45\,mA$	dG	48	53	–	dB
1dB gain compression $V_d = 3V$; $I = 45\,mA$; $V_{con} = 3V$	P_{1dB}	–	14	–	dBm
Power Gain $V_d = 3V$; $I = 45\,mA$; $V_{con} = 3V$	G	15.5	17.5	–	dB
Input return loss $V_d = 3V$; $I = 45\,mA$; $V_{con} = 3V$	RL_{in}	–	10	–	dB
Output return loss $V_d = 3V$; $I = 45mA$; $V_{con} = 3V$	RL_{out}	–	8.5	–	dB
Gain Control Range $V_{con} = 3\,V \ldots 0V$; $V_d = 3V$; $I = 45\,mA$	dG	48	53	–	dB
1dB gain compression $V_d = 3V$; $I = 45\,mA$; $V_{con} = 3V$	P_{1dB}	–	14	–	dBm

[a] $T_A = 25°C$, $f = 900$ MHz, $V_g = -4V$, $RS = RL = 50\,\Omega$ unless otherwise specified
[b] $T_A = 25°C$, $f = 1800$ MHz, $V_g = -4V$, $RS = RL = 50\,\Omega$ unless otherwise specified

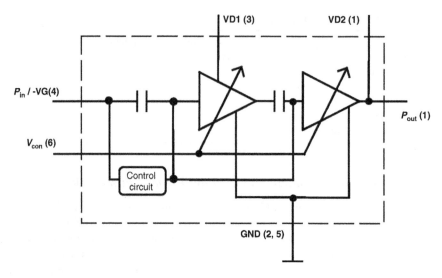

Figure 3.51 Functional block diagram of the gain-controllable CGY121A GaAs MMIC. Gain control is achieved by applying 0–3 V dc to the V_{con} pin.

reduce the supply voltage by about 1.9 V at the collector side and still maintain a safety margin of 0.7 V at the emitter–ground connection. The voltage 2.6 V, so to speak, is overkill.

On the collector side, for an RC type of amplifier, it is not a bad idea to set the collector voltage at about half the supply, which allows the RF swing to be half the supply voltage.

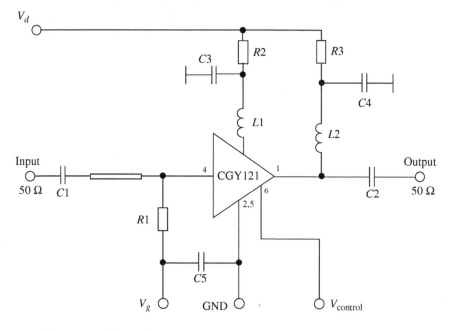

Figure 3.52 CGY121A 900-MHz application circuit. The values of the elements are: $C1 = C2 = 22$ pF, $C3 = C4 = 100$ nF, $C5 = 47$ nF, $L1 = 15$ nH, $L2 = 27$ nH, $R1 = 270\,\Omega$, $R2 = 12\,\Omega$, $R3 = 6.8\,\Omega$.

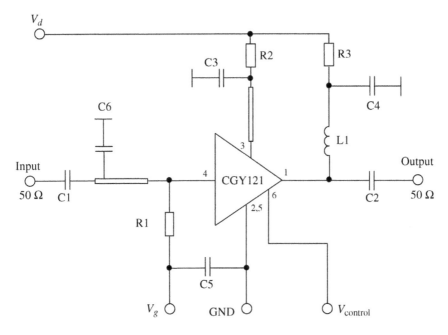

Figure 3.53 CGY121A 1.9-GHz application circuit. The values of the elements are: $C1 = C2 = 12\,pF$, $C3 = C4 = 100\,nF$, $C5 = 47\,nF$, $C6 = 1.2\,pF$, $L1 = 15\,nH$, $R1 = 270\,\Omega$, $R2 = 12\,\Omega$, $R3 = 2.7\,\Omega$.

In this case, we have violated the rule because of our 2.6-V drop. Assuming we had chosen a 0.7-V drop, then the collector–emitter voltage would be 6.3 V available for the swing. The choice of 5 mA for the transistor has come from a combination of good linearity and low noise figure. While the 10-Ω unbypassed resistor will deteriorate the noise figure somewhat, it reduces the emitter–current-dependent distortion. The base–emitter diffusion resistance equals 26 mV/5 mA, or roughly 5 Ω. The 10-Ω resistor is twofold larger, and therefore adds to the linearity and to the noise.

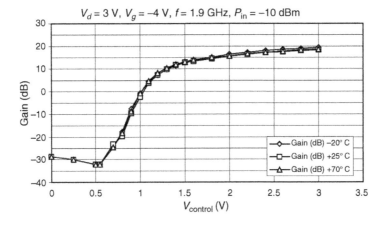

Figure 3.54 CGY121A gain versus $V_{control}$ at 1.9 GHz.

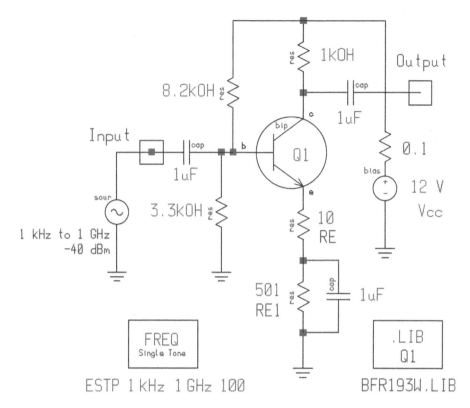

Figure 3.55 BJT amplifier example.

To determine the base bias resistor, it is a good assumption that the current going through the two resistors (in our case, 8.2 kΩ and 3.3 kΩ) is about 10% (or a slightly higher fraction) of the collector current. Since the dc current gain of the transistor is typically between 50 and 100, it is sufficient if the current through the resistive network is between 5 and 10 times higher than the dc bias current taken off the network. This type of bias scheme is only valid for Class A to AB$_1$. Later, we will see that for monolithic circuits, additional transistors will be used to generate the required voltage drops. Lately, combinations of NPN and PNP transistors have become popular to avoid any resistor in the emitter for dc biasing purposes. Figure 3.56 shows such a circuit. This of course invites thermal runaway. To minimize the circuitry, there are bias ICs, such as the Infineon BCR400W, that are now available. Figures 3.57 and 3.58 show its application in biasing BJTs and FETs, respectively; it can also be used to control PIN-diode bias current for TR antenna switching as shown in Figure 3.59.

To validate this, we have used the Serenade 8.0 CAD tool, which has a built-in bias- and temperature-dependent noise model for both BJTs and FETs. The SPICE-type dc analysis determines the actual currents, which were mentioned above. By maintaining the same bias and just bypassing both emitter resistors at the same time, the set of curves shown in Figure 3.60 will be obtained. As an exercise, we recommend the user to determine the optimum generator impedance for F_{min} with a 10-Ω resistor. Again, both the emitter current and the collector voltage affect the noise figure, the intermodulation, and f_T. If a transistor is used

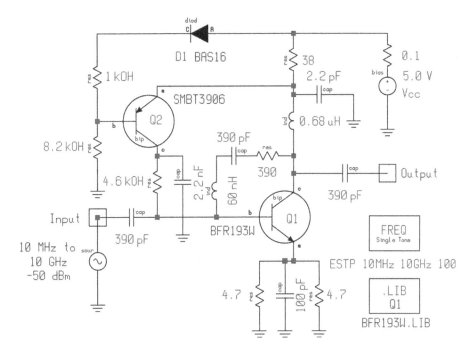

Figure 3.56 Wideband amplifier with active biasing. The 4.7-Ω resistors in the BFR193W emitter compensate the circuit's gain at low frequencies.

that is intended to be operated at significantly higher currents, then f_T will suffer while the other two parameters are less affected.

Since the dc input power translates into thermal dissipation, here is a plot of the noise figure as a function of temperature with and without emitter feedback. The heavy dc stabilization prevents a significant effect on the other parameters. The actual dc shift is from 5.0 to 5.2 mA for a temperature delta of 70 K.

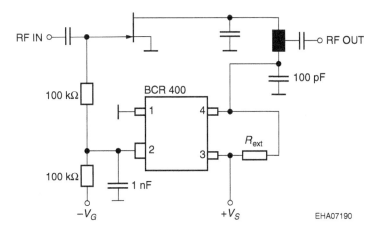

Figure 3.57 The Infineon BCR400 bias controller applied to a GaAsFET. Courtesy Infineon Technologies.

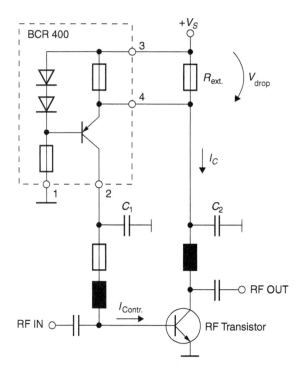

Figure 3.58   The Infineon BCR400 bias controller applied to a BJT. Courtesy Infineon Technologies.

As far as FETs are concerned, the enhancement types, typically PMOS, can use a similar type of bias, while the NMOS types complicate the designer's life as they require a negative gate voltage, specifically for low-noise operation, these devices are operated somewhat close to the pinchoff voltage. Putting a source resistor in place causes mostly headaches simply because the gain is so high that the transistors are always ready to find a frequency at which they love to oscillate. The transistor will look for every possible parasitic to turn the amplifier into an oscillator.

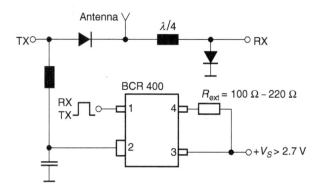

Figure 3.59   The Infineon BCR400 bias controller applied to TR switching. Courtesy Infineon Technologies.

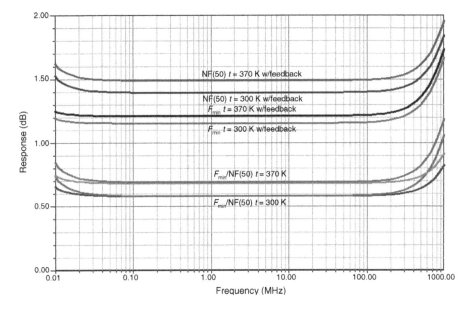

Figure 3.60 F_{min} and NF for the Figure 3.55 amplifier, with and without emitter feedback, at 300 and 370 K.

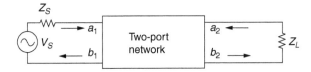

Figure 3.61 S parameters characterize a network (a two-port is shown) in terms of incident waves (a_1, a_2) and reflected waves (b_1, b_2). Figure 3.62 shows the flow graph for this network.

3.2 AMPLIFIER GAIN, STABILITY, AND MATCHING

Most semiconductor manufacturers describe their devices these days by providing either their S parameters (small-signal) or SPICE-type parameters (large-signal). One of our headaches in writing this book has been that many of the published models do not translate from large signal to small signal without giving unexpected or false results. Despite the availability of tools to do it right, many companies still can only provide data that are practically inadequate, and unfortunately the CAD manufacturer then gets the blame for results they are not really responsible for.

Tables 3.9 and 3.10 show typical data for a bipolar transistor (the BFP420 by Infineon). The next section presents a summary of the derivation and use of S parameters as published by Agilent and others.

3.2.1 Scattering Parameter Relationships

$$b_1 = s_{11}a_1 + s_{12}a_2 \tag{3.122}$$

Table 3.9 BFP420 Common-Emitter S Parameters

f (GHz)	S_{11} MAG	S_{11} ANG	S_{21} MAG	S_{21} ANG	S_{12} MAG	S_{12} ANG	S_{22} MAG	S_{22} ANG
$V_{CE} = 2$ V, $I_C = 20$ mA								
0.01	0.452	−2.3	37.62	178.3	0.0011	94.4	0.956	−0.6
0.1	0.447	−25.1	36.30	164.7	0.0068	82.5	0.941	−12.4
0.5	0.386	−101.1	23.41	121.0	0.0262	61.7	0.632	−47.2
1.0	0.378	−146.2	13.99	96.0	0.0395	57.8	0.395	−63.9
2.0	0.405	173.5	7.18	70.8	0.0664	54.0	0.222	−87.3
3.0	0.446	149.4	4.77	52.6	0.0949	47.1	0.133	−111.3
4.0	0.501	130.0	3.52	36.8	0.1206	38.5	0.133	−158.5
6.0	0.599	104.8	2.27	8.2	0.1646	18.9	0.196	142.0
8.0	0.700	78.5	1.51	−20.8	0.1800	−2.4	0.289	99.3
9.0	0.758	67.6	1.25	−34.4	0.1820	−13.0	0.379	84.1
10.0	0.800	62.0	1.04	−43.5	0.1800	−19.3	0.465	76.6
$V_{CE} = 2$ V, $I_C = 5$ mA								
0.01	0.790	−1.0	15.14	179.2	0.0012	83.4	0.988	−0.7
0.1	0.786	−11.6	14.98	171.8	0.0092	84.1	0.982	−6.5
0.5	0.702	−55.7	12.86	140.1	0.0398	62.8	0.857	−29.8
1.0	0.589	−99.1	9.63	112.6	0.0603	46.5	0.647	−48.6
2.0	0.507	−156.0	5.60	79.4	0.0798	34.6	0.401	−70.3
3.0	0.511	168.5	3.84	57.1	0.0957	29.8	0.267	−84.2
4.0	0.549	142.0	2.87	38.5	0.1121	25.1	0.207	−110.5
5.0	0.604	123.9	2.26	22.1	0.1285	19.4	0.150	−137.3
6.0	0.633	110.0	1.86	6.7	0.1442	13.1	0.173	−169.8

For more and detailed S- and Noise-parameters please see Internet:
http://www.siemens.de/Semiconductor/products/35/35.htm
Source: Courtesy Infineon Technologies.

Table 3.10 BFP420 Noise Parameters

f GH$_Z$	F_{min}^{a} dB	G_a^{a} dB	Γ_{opt} MAG	Γ_{opt} ANG	R_N Ω	r_n –	$F_{50\Omega}^{b}$ dB	$|S_{21}|^{2\,b}$ dB
$V_{CE} = 2$ V, $I_C = 5$ mA								
0.9	0.90	20.5	0.28	41.0	8.7	0.17	1.02	20.3
1.8	1.05	15.2	0.20	82.0	6.7	0.13	1.11	15.8
2.4	1.25	13.0	0.20	124.0	5.5	0.11	1.32	13.5
3.0	1.38	12.1	0.22	−175.0	5.0	0.10	1.48	11.6
4.0	1.55	10.3	0.33	−157.0	5.5	0.11	1.83	9.1
5.0	1.75	8.6	0.45	−142.0	5.0	0.10	2.20	7.0
6.0	2.20	6.4	0.53	−123.0	15.0	0.30	3.30	5.3

[a] Input matched for minimum noise figure, output for maximum gain
[b] $Z_S = Z_L = 50\,\Omega$
Source: Courtesy Infineon Technologies.

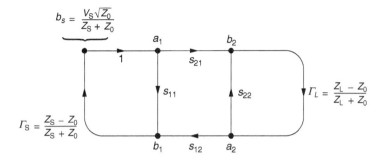

Figure 3.62 Flow graph for the network shown in Figure 3.61.

$$b_2 = s_{21}a_1 + s_{22}a_2 \tag{3.123}$$

Note that

$$s_{11} = \frac{b_1}{a_1} = \frac{\frac{V_1}{I_1} - Z_0}{\frac{V_1}{I_1} + Z_0} = \frac{Z_1 - Z_0}{Z_1 + Z_0} \tag{3.124}$$

and

$$Z_1 = Z_0 \frac{(1 + s_{11})}{(1 - s_{11})} \tag{3.125}$$

where $Z_1 = V_1/I_1$ is the input impedance at Port 1.

This relationship between reflection coefficient and impedance is the basis of the Smith Chart transmission-line calculator. Consequently, the reflection coefficients S_{11} and S_{22} can be plotted on Smith charts, converted directly to impedance, and easily manipulated to determine matching networks for optimizing a circuit design.

The above equations show one of the important advantages of S parameters. They are simply gains and reflection coefficients, both familiar quantities to engineers. By comparison, some of the Y parameters described earlier in the article are not so familiar. For example, the Y parameter corresponding to insertion gain S_{21} is the "forward transadmittance," Y_{21}. Clearly, insertion gain gives by far the greater insight into the operation of the network.

S parameters are simply related to power gain and mismatch loss, quantities that are often of more interest than the corresponding voltage functions:

$$|S_{11}|^2 = \frac{\text{Power reflected from the network input}}{\text{Power incident on the network input}} \tag{3.126}$$

$$|S_{21}|^2 = \frac{\text{Power delivered to a } Z_0 \text{ load}}{\text{Power available from } Z_0 \text{ source}} \tag{3.127}$$

$$|S_{12}|^2 = \text{Reverse transducer power gain with } Z_0 \text{ load and source} \tag{3.128}$$

$$|S_{22}|^2 = \frac{\text{Power reflected from the network output}}{\text{Power incident on the network output}} \tag{3.129}$$

From this we obtain the following.

Input reflection coefficient with arbitrary Z_L

$$S'_{11} = S_{11} + \frac{S_{12}S_{21}\Gamma_L}{1 - S_{22}\Gamma_L} \tag{3.130}$$

Output reflection coefficient with arbitrary Z_S

$$S'_{22} = S_{22} + \frac{S_{12}S_{21}\Gamma_S}{1 - S_{11}\Gamma_S} \tag{3.131}$$

Voltage gain with arbitrary Z_L and Z_S

$$A_V = \frac{V_2}{V_1} = \frac{S_{21}(1 + \Gamma_L)}{(1 - S_{22}\Gamma_L)(1 + S'_{11})} \tag{3.132}$$

$$\text{Power gain} = \frac{\text{Power delivered to load}}{\text{Power input to network}} :$$

$$G = \frac{|S_{21}|^2(1 - |\Gamma_L|^2)}{(1 - |S_{11}|^2) + |\Gamma_L|^2(|S_{22}|^2 - |D|^2) - 2\text{Re}(\Gamma_L N)} \tag{3.133}$$

$$\text{Available power gain} = \frac{\text{Power available from network}}{\text{Power available from source}}$$

$$G_A = \frac{|S_{21}|^2(1 - |\Gamma_S|^2)}{(1 - |S_{22}|^2) + |\Gamma_S|^2(|S_{11}|^2 - |D|^2) - 2\text{Re}(\Gamma_S M)} \tag{3.134}$$

$$\text{Transducer power gain} = \frac{\text{Power delivered to load}}{\text{Power available from source}} :$$

$$G_T = \frac{|S_{21}|^2(1 - |\Gamma_S|^2)(1 - |\Gamma_L|^2)}{|(1 - S_{11}\Gamma_S)(1 - S_{22}\Gamma_L) - S_{12}S_{21}\Gamma_L\Gamma_S|^2} \tag{3.135}$$

Unilateral transducer power gain ($S_{12} = 0$)

$$G_{Tu} = \frac{|S_{21}|^2(1 - |\Gamma_S|^2)(1 - |\Gamma_L|^2)}{|1 - S_{11}\Gamma_S|^2|1 - S_{22}\Gamma_L|^2} \tag{3.136}$$

$$= G_0 G_1 G_2$$

$$G_0 = |S_{21}|^2 \tag{3.137}$$

$$G_1 = \frac{1 - |\Gamma_S|^2}{|1 - S_{11}\Gamma_S|^2} \tag{3.138}$$

$$G_2 = \frac{1 - |\Gamma_L|^2}{|1 - S_{22}\Gamma_L|^2} \tag{3.139}$$

Maximum unilateral transducer power gain when $|S_{11}| < 1$ and $|S_{22}| < 1$

$$G_u = \frac{|S_{21}|^2}{|(1 - |S_{11}|^2)(1 - |S_{22}|)^2|}$$

$$= G_0 G_{1\,max} G_{2\,max} \tag{3.140}$$

$$G_{i\,max} = \frac{1}{1 - |S_{11}|^2} \quad i = 1, 2 \tag{3.141}$$

Constant-gain circles (unilateral case: $S_{12} = 0$):

• center of constant-gain circle is on line between center of Smith Chart and point representing S_{ii}^*
• distance of center of circle from center of Smith Chart:

$$r_i = \frac{g_i |S_{ii}|}{1 - |S_{ii}|^2(1 - g_i)} \tag{3.142}$$

• radius of circle:

$$\rho_i = \frac{\sqrt{1 - g_i}(1 - |S_{ii}|^2}{1 - |S_{ii}|^2(1 - g_i)} \tag{3.143}$$

where $i = 1, 2$, and

$$g_i = \frac{G_i}{G_{i\,max}} = G_i(1 - |S_{ii}|^2) \tag{3.144}$$

Unilateral figure of merit:

$$u = \frac{|S_{11} S_{22} S_{12} S_{21}|}{|(1 - |S_{11}|^2)(1 - |S_{22}|^2)|} \tag{3.145}$$

Error limits of unilateral gain calculation:

$$\frac{1}{(1 + u^2)} < \frac{G_T}{G_{Tu}} < \frac{1}{(1 - u^2)} \tag{3.146}$$

Conditions for absolute stability:
No passive source or load will cause a network to oscillate if a, b, and c are all satisfied:

a.
$$|S_{11}| < 1, |S_{22}| < 1 \tag{3.147}$$

b.
$$\left| \frac{|S_{12} S_{21}| - |M^*|}{|S_{11}|^2 - |D|^2} \right| > 1 \tag{3.148}$$

c.
$$\left| \frac{|S_{12} S_{21}| - |N^*|}{|S_{22}|^2 - |D|^2} \right| > 1 \tag{3.149}$$

where

$$D = S_{11}S_{22} - S_{12}S_{21} \qquad (3.150)$$

$$M = S_{11} - DS_{22}^* \qquad (3.151)$$

$$N = S_{22} - DS_{11}^* \qquad (3.152)$$

Condition that a two-port network can be simultaneously matched with a positive real source and load

$$K > 1 \qquad (3.153)$$

$$K = \frac{1 + |D|^2 - |S_{11}|^2 - |S_{22}|^2}{2|S_{12}S_{21}|} \qquad (3.154)$$

Source and load for simultaneous match:

$$\Gamma_{mS} = M * \left[\frac{B_1 \pm \sqrt{B_1^2 - 4|M|^2}}{2|M|^2} \right] \qquad (3.155)$$

$$\Gamma_{mL} = N * \left[\frac{B_2 \pm \sqrt{B_2^2 - 4|N|^2}}{2|N|^2} \right] \qquad (3.156)$$

where

$$B_1 = 1 + |S_{11}|^2 - |S_{22}|^2 - |D|^2 \qquad (3.157)$$

$$B_2 = 1 + |S_{22}|^2 - |S_{11}|^2 - |D|^2 \qquad (3.158)$$

Maximum available power gain. If $K > 1$

$$G_{A\,max} = \left| \frac{S_{21}}{S_{12}} (K - \sqrt{K^2 - 1}) \right| \qquad (3.159)$$

In conclusion, a two-port network (active or passive) is unconditionally stable when

$$K > 1$$

$$|S_{12}S_{21}| < 1 - |S_{11}|^2$$

and

$$|S_{12}S_{21}| < 1 - |S_{22}|^2$$

3.2.2   Low-Noise Amplifiers

A low-noise amplifier combines a low noise figure, reasonable gain, and stability without oscillation over its entire useful frequency range. These amplifiers are typically operated in Class A, which is characterized by a bias point more or less at the center of maximum current and voltage capability of the device used, and by RF current and voltages that are sufficiently small relative to the bias point that does not shift. This means that an amplifier operating at, say, 3 V and 1 mA for low-noise, high-gain operation will not be Class A if the RF current is larger than 0.1% of the current or the voltage swing is more than 0.1% of the current swing. These amplifiers can be designed using standard linear S parameters, which are (hopefully) available from the device manufacturer for the bias point of interest. If suitable S parameters are not available, we must resort to a CAD tool that allows us to determine a device's bias point as a function of its bias network and to generate small-signal S parameters for the device. With suitable S parameters in hand, and assuming the amplifier will be frequency selective, we then connect tuned circuits to the input and output to provide matching.

In designing amplifiers, the "power gain" of a circuit is frequently mentioned. This is actually deceptive because power gain is somewhat associated with dc power dissipation, and in considering it we are on the verge of discussing large-signal performance. Therefore, it would be better to define the power gain as a function of operating mode, such as "Class A power gain." We have also outlined that there is a difference between noise matching and gain matching. If we can be allowed to step back into the vacuum-tube era for a moment, the reader should be reminded that the difference between power gain and noise matching had to do with the grid-plate feedback capacitance which, via the Miller effect, showed up in parallel with the input. This actually detunes the input slightly off-center relative to the best gain. The reason for this is that as the feedback increases, Y_{12} or S_{12} starts playing a role, and the frequently assumed unilateral case no longer exists. The difference between noise matching and gain matching increases as a function of frequency. It can be brought closer together if appropriate reactive feedback is used. The two choices are: (1) increase the emitter or source inductance or (2) slightly increase the base-to-collector/gate-to-drain capacitance. Needless to say, the penalty for this is potential instability, and a sweep must be done from a few megahertz to several gigahertz to make sure the stage does not "take off."

The most efficiency this circuit provides is 50%; even this is a theoretical limit if one tries to keep distortion under control.

Design Guidelines   The design of a one-stage amplifier consists of finding

1. an input lossless matching network M_1, and
2. an output lossless matching network M_2

so that the maximum or desired transistor gain is achieved over the operating bandwidth of the amplifier. Usually, the common-emitter or common-source configuration is chosen for highest gain per stage. If the stability factor K is greater than unity, these two networks can be found to give the maximum available gain G_{max}. If the stability factor is less than or equal to unity, the amplifier could be terminated in a matching structure that causes oscillation, that is, G_{max} is infinite. This should be avoided by locating the regions of instability in the Γ_G and Γ_L planes. The input and output terminations (Γ_G and Γ_L) must be designed to avoid the instability regions. Usually, these unstable regions are near the conjugate match for S_{11} and S_{22}. Thus, a stable amplifier will require some input and/or output mismatch if K is less than or equal to unity.

There are at least two alternative approaches for potentially unstable amplifiers:

1. add resistive matching elements to make $K \geq 1$ and $G_{max} \approx G_{ms}$ and
2. add feedback to make $K \geq 1$ and $G_{max} \approx G_{ms}$.

For narrowband amplifiers, it is usually recommended to accept a transistor with $K < 1$, design the amplifier for a gain approaching G_{ms}, and to ensure that the Γ_G and Γ_L terminations provide stability at all frequencies, both inside and outside the amplifier passband.

The design of an amplifier would usually have the following specifications:

gain and gain flatness,

bandwidth and center frequency ($f_2 - f_1$, f_0),

noise figure,

linear output power,

input reflection coefficient (VSWR$_{in}$),

output reflection coefficient (VSWR$_{out}$),

bias voltage and current.

For small-signal amplifiers, the small-signal S parameters are sufficient to complete the design. After selecting an appropriate transistor based on these specifications, a one-stage amplifier design should be considered if sufficient gain can be achieved; otherwise, a two-stage amplifier should be designed.

The circuit topology should be chosen to allow dc bias for the transistor. Usually, RF short-circuited stubs are placed near the transistor to allow dc biasing. If the topology does not allow dc bias, a broadband bias choke or high-resistance bias circuit that does not affect the amplifier performance must be used.

The following steps can be tabulated for the design of a one-stage amplifier see Figure 3.63.

1. Select a transistor based on a datasheet description of the S parameters, noise figure, and linear output power.
2. Calculate K and G_{max} or G_{ms} versus frequency.
3. For $K > 1$, select the topologies that match the input and output (and allow dc biasing) at the upper band edge f_2. Ideally, this will give G_{max} and $S'_{11} = S'_{22} = 0$. Usually, $S_{12} = 0$ is assumed for the initial design. Next, the topology may be varied to flatten gain versus frequency at the expense of S'_{11} and S'_{22}
4. For $K < 1$, plot the regions of instability on the Γ_G and Γ_L planes and select topologies that partially match the input and output at the upper band edge and avoid the unstable regions. The gain will approach G_{ms} as an upper limit. Next, the topology may be varied to flatten gain versus frequency.
5. After finding initial M_1 and M_2, plot the amplifier S parameters versus frequency; make adjustments in topology until the specifications for gain, input reflection coefficient, and output reflection are satisfied. Also plot Γ_G and Γ_L versus frequency to verify amplifier stability.

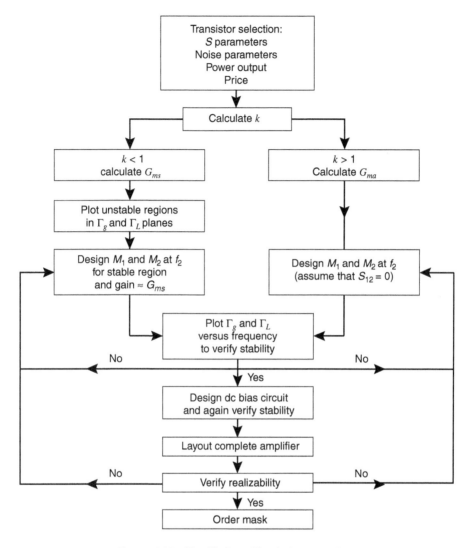

Figure 3.63   Simplified amplifier design procedure.

6. Design the dc bias circuit. Layout the elements of the complete amplifier and check realizability.

As a beginning point in an amplifier design, the circuit topologies may be designed with the assumption that S_{12} is zero. Later, an exact design procedure will be described that includes the topologies with terminations in the stable region.

CMOS Low-Noise Amplifiers   An optimal low-noise amplifier needs to be matched for lowest noise figure while providing low input return loss. In transistors, however, optimum power match differs from optimum noise match, and MOS transistors are no exception in this respect. Just adding matching networks to a MOS transistor therefore means to find a compromise between noise figure and return loss. This strategy will not yield the optimum performance the device can provide.

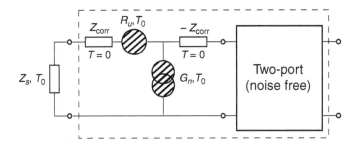

Figure 3.64 Representation of the noise of a two-port through two noise sources and a correlation impedance.

With the aim to provide excellent noise match and power match, one first needs to use a circuit topology that transforms the two to get closer together in the Smith Chart, or in the ideal case, to become identical.

This goal can be achieved since the physical origin of noise matching differs from that of power matching. In the latter case, reactive elements are added in order to transform the circuit's input impedance to be equal to the complex conjugate of the source impedance. The incoming power wave is no longer reflected, since resonant circuits compensate any mismatch in phase and amplitude.

Noise match, on the other hand, means to exploit the fact that the short-circuit noise current and the open-circuit noise voltage at the input of the device are partly correlated. Minimum noise figure is achieved by choosing a source impedance that forces the correlated part of the noise to cancel out.

In order to have a closer look on how noise-free feedback and matching impacts the noise performance, a generic two-port is considered as shown in Figure 3.64. The noise of the two-port is represented in terms of two uncorrelated sources, $\langle |v_n|^2 \rangle = 4kT_0BR_u$ and $\langle |i_n|^2 \rangle = 4kT_0BG_n$, and a correlation impedance Z_{corr}, as proposed by Rothe and Dahlke [4]. The noise factor of the two-port as a function of the source admittance $Z_s = R_s + jX_s$ is given by

$$F = 1 + \frac{R_u + G_n|Z_s + Z_{\text{corr}}|^2}{R_s} \tag{3.160}$$

This formulation is equivalent to the commonly used formula relying on the much more intuitive parameters minimum noise factor $F_{\min}$, optimum noise match Z_{opt}, and equivalent noise admittance G_n.

$$F = F_{\min} + \frac{G_n}{R_s}|Z_s - Z_{\text{opt}}|^2 \tag{3.161}$$

The parameter G_n in fact is identical in both descriptions, the other parameters can be calculated by

$$Z_{\text{opt}} = R_{\text{opt}} + jX_{\text{opt}}$$

$$X_{\text{opt}} = -X_{\text{corr}}$$

$$R_{\text{opt}} = \sqrt{\frac{R_u}{G_n} + R_{\text{corr}}^2}$$

$$F_{\min} = 1 + 2G_n(R_{\text{opt}} + R_{\text{corr}})$$

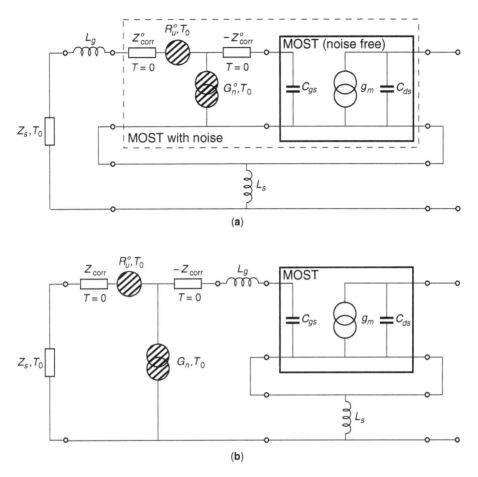

Figure 3.65 MOS transistor with source and gate inductances. (a) The noise of the transistor is described by the Rothe–Dahlke equivalent circuit. (b) The noise sources are transformed in order to describe the noise properties of the whole circuit.

The common approach to alter the noise matching is to introduce inductive series feedback at the source. Now, adding a lossless inductor at the source will not add any noise to the circuit. But a feedback will impact noise and signal differently, and in general alter F_{min}. However, due to the simplified equivalent circuit regarded here, F_{min} is not changed. For matching purposes, a series inductance is also connected to the gate, which should also be considered at this point. This configuration is shown in Figure 3.65. The generic two-port is now replaced by a basic MOS transistor equivalent circuit. For matching purposes, a series inductance is also connected to the gate, which should also be considered at this point. This configuration is shown in Figure 3.65. The generic two-port is now replaced by a basic MOS transistor equivalent circuit. The MOS transistor's noise is described by the Rothe–Dahlke description, not by its internal noise model. One might expect to see the internal FET noise current sources describing channel noise and induced gate noise. But this description is preferred since we are interested to see how the external inductances alter the noise four parameters, while the physics governing the noise is not in the focus.

The inductive feedback obviously has an impact on G_n and Z_{corr}. These parameters read [22]

$$G_n = G_n^o \cdot \left| \frac{Z_{21}}{Z_{21} + j\omega L_s} \right|^2$$

$$Z_{\text{corr}} = j\omega L_g + j\omega L_s \left(1 - \frac{Z_{11}}{Z_{21}} \right) + Z_{\text{corr}}^o \left(1 + \frac{j\omega L_s}{Z_{21}} \right)$$

where the superscript o denotes the original parameter.

Assuming the MOS transistor can be approximated by the simple equivalent circuit shown in Figure 3.65, the formulae can be simplified as follows:

$$G_n = G_n^o \cdot \left| \frac{g_m}{g_m + j\omega^3 L_s C_{gs} C_{ds}} \right|^2 \approx G_n^o$$

$$Z_{\text{corr}} = j\omega L_g + j\omega L_s \left(1 - \frac{j\omega C_{ds}}{g_m} \right) + Z_{\text{corr}}^o \left(1 + \frac{j\omega^3 L_s C_{gs} C_{ds}}{g_m} \right) \approx j\omega(L_g + L_s) + Z_{\text{corr}}^o$$

The optimum noise match in terms of optimum source impedance Z_{opt} can therefore be approximated by

$$Z_{\text{opt}} = Z_{\text{opt}}^o - j\omega(L_g + L_s) \tag{3.162}$$

The real part of Z_{opt} therefore remains unchanged, but its imaginary part is reduced. That is an important finding since it means that $\text{Re}(Z_{\text{opt}})$ is unchanged by the matching and feedback inductances. Optimizing this value to the needs of the circuit design therefore means to choose a suitable device and bias point.

Regarding the imaginary part of Z_{opt}^o, it is now required to know at least its sign, and its frequency dependence in order to see whether it can be compensated by the inductances or not. This calculation requires to consider the physics-based noise model of the MOS transistor, and quite a bit of calculations which, in all details, is found in the literature [23]. In the end, equation (3.162) can be rewritten approximating $\text{Im}(Z_{\text{opt}}^o)$ by the following:

$$Z_{\text{opt}} = \text{Re}(Z_{\text{opt}}^o) - m \frac{1}{j\omega C_{gs}} - j\omega(L_g + L_s) \tag{3.163}$$

with the parameter m in the range of 0.6 for long channel devices, while it is expected to approach 1 as gate width is scaled down. The imaginary part of the optimum noise match impedance is capacitive, which can very well be compensated by properly choosing the values of the inductances.

But noise match alone is not sufficient for a good LNA, it also reqires good input matching. The input impedance for the device with gate and source inductances, as shown in Figure 3.65, reads

$$Z_{\text{in}} = j\omega(L_s + L_s) + \frac{1}{j\omega C_{gs}} + \frac{g_m L_s}{C_{gs}} = j\omega(L_s + L_s) + \frac{1}{j\omega C_{gs}} + \omega_T L_s \tag{3.164}$$

with the transit frequency $\omega_T = 2\pi f_T = g_m / C_{gs}$.

The source inductance L_s introduces a real part in the input impedance, which is the main benefit of the inductive feedback. The formulae describing Z_{in} and Z_{opt} become similar in structure. With the designer's freedom to choose suitable values for inductance and device size, it becomes possible to bring these two impedances as close together as possible, allowing for simultaneous noise and power match. While $Re(Z_{in})$ is affected by ω_T, that depends on the transistor and how it is biased, the sum of the inductances $L_s + L_g$ can be used to compensate the capacitive component of Z_{opt} and Z_{in}.

In conclusion, the design of a CMOS LNA follows the following steps.

1. Choose a device of appropriate size that satisfies $Re(Z_{opt}) = Re(Z_s)$.
2. Select source inductance L_s and bias point to satisfy $\omega_T L_s = Re(Z_s)$.
3. Determine the value of the gate inductance L_g so that $Im(Z_{opt}) = Im(Z_s)$ holds.

In certain cases, especially if a very low-power design is required, it is not possible to choose the device size freely. In this case, usual matching techniques have to be applied instead of just compensating C_{gs} and forcing $Re(Z_{opt}) = Re(Z_s)$. In this case the inductive degeneration still has its benefit, since it allows in principle simultaneous power and noise matching, when $\omega_T L_s = Re(Z_{opt})$ is enforced. However, this is not always possible in practice. Z_{opt} increases with decreasing device size, requiring large inductance values. But if L_s excites a certain limit, noise figure is compromised. On one hand, large inducances come with higher parasitic resistances and therefore add thermal noise. On the other hand, as the inductive feedback increases, the transistor's internal feedback will also gain importance and degrade the minimum noise factor.

If it is practically not possible to select a source inductance, often an extrinsic gate-source capacitance C_{ex} is added. Regarding Z_{in}, it simply adds to C_{gs}, with the negative impact on the external transit frequency that now becomes $\omega_T' = g_m/(C_{gs} + C_{ex})$. But considering noise matching, this approach has the power to lower $Re(Z_{opt})$ [23]. In this case, $Re(Z_{opt})$ reads

$$Re(Z_{opt}) = \frac{A \cdot B}{\omega C_{gs} \left[A^2 \cdot B^2 + \left(\frac{C_{gs} + C_{ex}}{C_{gs}} + A \cdot |c|^2 \right)^2 \right]} \tag{3.165}$$

with $A = \alpha \sqrt{\delta/(5\gamma)}$, $B = \sqrt{1 - |c|^2}$. The parameters α, γ, δ refer to the noise sources of the FET [24]. The channel noise current $\langle |i_{nd}|^2 \rangle$ and the induced gate noise current $\langle |i_{gd}|^2 \rangle$ are given by

$$\langle |i_{ng}|^2 \rangle = 4kTB\delta \frac{\omega^2 C_{gs}^2}{5g_{d0}} \tag{3.166}$$

$$\langle |i_{nd}|^2 \rangle = 4kTB\gamma g_{d0} \tag{3.167}$$

where g_{d0} is the drain-source conductance at zero drain-source voltage V_{ds}. The typical values for the noise parameters are the following. For long-channel devices, γ starts at unity for zero V_{DS}, and approaches 2/3 in saturation. However, it increases at high V_{GS} and V_{DS}. For short-channel devices, it can exceed 2. The parameter δ is constant in long channel devices, its value is around 3/4. It is higher for short channel devices. The value α is defined as the ratio of g_m to g_{d0}, which is unity for long-channel devices and lower in the case of a short-channel device. The correlation coefficient c can be given for long channel devices, and its value is purely imaginary, $c = j0.395$.

Equation (3.165) shows that $Re(Z_{opt})$ is reduced by adding C_{ex}. This, therefore, is an adequate measure for the case in question. Lowering $Re(Z_{opt})$ reduces the requirement of a high value of $\omega_T L_s$. It thereby allows to reduce feedback, and it enables usage of very small, low-power, transistors that inherently have high impedance levels.

Thus, for a given device size, where $\mathrm{Re}(Z_{opt})$ is too high for the original approach, the design procedure is as follows.

1. Select a source inductance L_s and capacitance C_{ex} to obtain matching of $\mathrm{Re}(Z_s)$ and $\mathrm{Re}(Z_{opt})$ without compromising effective transit frequency (and, thus, gain) and noise.
2. Determine the value of the gate inductance L_g so that $\mathrm{Im}(Z_{opt}) = \mathrm{Im}(Z_s)$ holds.

The whole approach up to now, however, has a drawback. Introducing source and gate inductances is also a viable starting point in osciallator design, see Chapter 5. Another point is the low reverse isolation of a the single-transistor LNA that should be reduced also because of the Miller effect. Therefore, two CMOS transistors are commonly employed in cascode configuration, which will be described further in detail in Section 3.4.

Another issue in CMOS design is power dissipation. Being energy efficient is imperative for battery-powered mobile systems, which first impacts the choice of the transistor size as just discussed. But it also means that the supply voltage is constrained to a lower value [25].

The standard cascode is disadvantageous in this respect, as it requires the supply voltage to be twice the voltage V_{ds} required for a single transistor. The drain voltage of each of the transistors should be twice the overdrive voltage $V_{od} = (V_{gs} - V_t)$ in order to provide high gain. The supply voltage can therefore only be lowered if it is not applied to the full cascode stack, but instead each transistor is biased individually. This requires to decouple the RF path from the dc path, resulting in a topology known as folded cascode [26–28].

The folded cascode stage is shown in Figure 3.66. The supply voltage is fed in between the two transistors that are in fact dc-wise connected in parallel. CMOS offers the advantage that it provides NMOS and PMOS devices that require supply voltages of opposite sign. Thereby, the folded cascode transistors can be dc coupled. In the original approach based on bipolar technology, it was necessary to decouple the two transistors by means of capacitors and to bias them individually [26]. But in CMOS, the dc is supplied through an RF choke that provides an open to the RF signal. The RF therefore directly passes the two transistors in cascode configuration, without the need for any dc blocking capacitance in between.

The folded cascode is operated at lower supply voltages compared to the original cascode, but it still dissipates the same power. An approach to lower voltage and current at the same time is the complementary current-reuse architecture as shown in Figure 3.67 [29]. As in case of the folded cascode, dc and RF path is decoupled in this architecture. The two transistors are in series concerning dc, thus the second transistor "reuses" the current of the first. But for the RF signal, the structure acts as a two-stage amplifier. The inductance are employed like inductors to ground for interstage matching, together with the coupling capacitance. Althogh it looks like a cascode at the first glance, the complementary current-reuse stage works differently. Thus, it is not providing the benefits inherent to the cascode structure, like reduction of the miller capacitance, or improved stability. Its main advantage is that it provides the highest gain at lowest supply voltage. Due to the complementary MOS architecture, the required supply voltage is reduced by one overdrive voltage V_{od} as

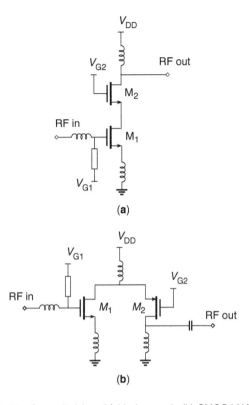

Figure 3.66 Cascode (a) and folded cascode (b) CMOS LNA circuit.

compared to the standard cascode architecture, without operating the transistors in weak inversion. Thereby, highest gain for for minimum dissipated power is achieved. The drawback of this architecture, however, is the need for quite a number of inductors that are not easily realized with high Q on wafer, and which also consume a lot of space.

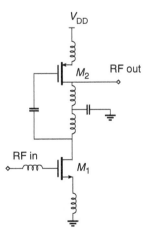

Figure 3.67 Complementary current-reuse CMOS LNA circuit.

Passive Elements Available in CMOS Technologies Monolithic-integrated CMOS processes use a larger number of interconnect layers that can be used to design passive elements such as capacitors, inductors, and transformers. Traditional MMIC processes only provide one metal layer, and transmission lines are either of the microstrip or coplanar waveguide (CPW) type. In either case, the metallization is high-conductivity electroplated gold, and the substrate is semiinsulating. Chips are often thinned down to $100\,\mu$m in order to suppress higher modes of the microstrip lines. Gold metallization, semiinsulating substrate, and the appropriate dimensions of the lines allow for optimum propagation of the electromagnetic waves along the transmission lines. CMOS is different and a rather hostile environment for RF. The standard silicon wafers are of low-resistivity type to safe on cost, and aluminum and copper are used instead of gold for the transmission lines. Therefore, it is to be expected that the losses due to the finite conductivity of the transmission line metal will be higher than on similar lines on GaAs, GaN, or InP processes. In addition, it is advisable to keep the electric field out of the substrate, as the substrate losses are significant and, if neglected, may render any RF design impossible.

On the other hand, there are some advantages of CMOS technologies besides the aspect of low cost in high-volume production. The high number of metal layers enable the custom design of various capacitor and inductor structures, a feature that is not available in the traditional GaAs world.

The metallization layer stack of a typical RFCMOS process is shown in Figure 3.68 [30]. This process provides six metallization layers, the lower four are made of copper, the upper ones are aluminum layers. Basically, the higher layers provide thicker metal. The layers are embedded in a low-loss dielectric, on top of the substrate. The present process uses two types of dielectric, first silicon oxide, and on top polyimide. The electric fields should be confined within these dielectric layers, as they are good isolators. Often, even a lower metallization layer is employed to shield the passive structures from the lossy substrate. Good RF transmission lines can be realized within the oxide if the RF ground is also realized within this metallization layer stack, forming a microstrip or stripline.

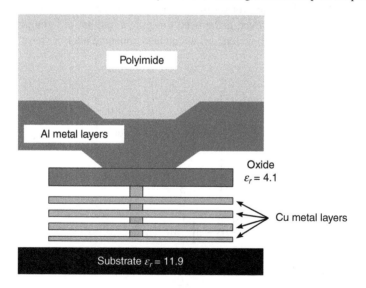

Figure 3.68 Metal-layer stack of a typical RFCMOS process (from Refs [30, 31], reprint with permission). © 2006 A. Vasylyev and W.A. Debski.

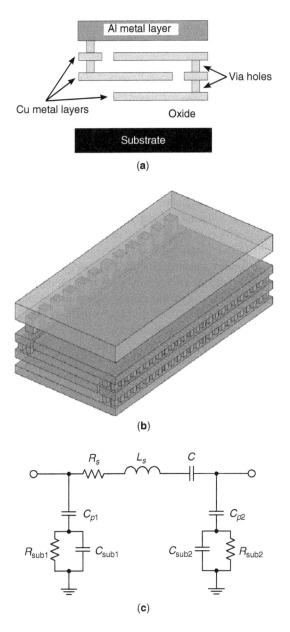

Figure 3.69 Parallel-plate capacitor realized in a typical RFCMOS process (from Refs [30, 31], reprint with permission). © 2006 A. Vasylyev and W.A. Debski.

Capacitance elements are designed straightforward in the form of parallel-plate capacitors. Figure 3.69a shows an example of such an approach using multiple layers to enhance the capacitance per square. In the example, four layers are used. The interconnection between the plates belonging to the same electrode are realized through via holes. Figure 3.69b shows the three-dimensional layout of the capacitor. The via holes are closely spaced and form a type of via fence in order to provide a good connection between the layers.

These structures, however, do not just behave like pure capacitances. The following effects are expected to be observed.

- The finite conductivity of the metal layers will lead to parasitic resistance.
- A parasitic capacitance between the metals and the substrate below will establish a lossy connection to ground.
- The structure has a certain dimension, giving rise to parasitic inductances at higher frequencies.

These effects are accounted for in the first-order equivalent circuit shown in Figure 3.69c. The capacitance C is the desired element, ballasted with the parasitic resistance R_s, and inductance L_s. The impact of coupling to substrate and substrate losses is described through $C_{p1,2}$, $R_{sub1,2}$ and $L_{sub1,2}$. The equivalent circuit reveals that this structure is a resonance circuit, not only the desired capacitance. Thus, it shows capacitive behavior only up to a certain frequency. Beyond this resonance frequency, it becomes inductive. Since the basic geometrical parameters for a certain technology are fixed, it is observed that the resonance frequency of the structure decreases with increasing size of the structure. Higher capacitance values will come with lower usable maximum frequencies.

Regarding lumped equivalent models like the one Figure 3.69c, it must be kept in mind that not only the device but also the model has an upper usable frequency. Even though the model consists of a number of elements, it only applies first-order approximations to model each of the physical effects. If the wavelength and structure size come into the same order of magnitude, distributed effects need to be considered. As a rule of thumb, such models can be expected to be reasonably accurate up to the resonance frequency, and definitely fail beyond. This rule of thumb is applicable to all first-order models of passive components, and even transistors.

The parallel-plate capacitances discussed so far provide a certain capacitance per square, depending mainly on the dielectric properties of the oxide isolation layers. These layers, however, are not desiged to provide high capacitance. Staggering multiple layers as shown in the figure yields higher capacitance, but also larger structures with the associated higher parasitic effects.

A number of processes therefore provide an extra layer of thin high-permittivity, high-breakdown dielectric to allow for metal-insulator-metal (MIM) capacitances, and an extra metal for the top electrode. This type of capacitances is commonly available in III–V processes, an can be integrated into RFCMOS processes at the cost of a few additional mask sets.

Figure 3.70 shows a schematic of a MIM capacitance within the layer stack, and also the associated equivalent circuit. The equivalent circuit looks a bit simpler than the one of the parallel plate capacitance before. Also the dimension will be much smaller due to the thin high-permittivity dielectric used. However, if the capacitance dimensions are increased, it will also show some parasitic inductive effect.

Theoretical limits of capacitance values and resonance frequencies were explored by Aparicio and Hajimiri [32]. In their work, a vast number of configurations to design capacitances is discussed. A main result is shown in Table 3.11. A third configuration, in addition to the types of capacitance discussed so far, regarded in this paper, is the vertical capacitance. The basic idea behind the vertical capacitance is simple. The achievable lateral dimensions are much smaller than the vertical dimensions, thus allowing for higher capacitance per size. While vertical dimensions are fixed by the dielectric layer thickness, the

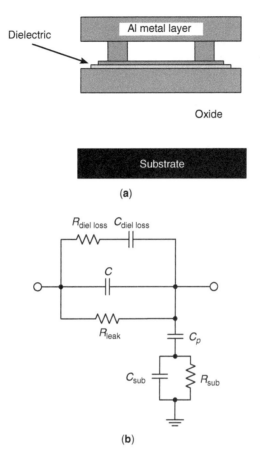

Figure 3.70 Metal–insulator–metal capacitor realized in a RFCMOS process (from Ref. [30], reprint with permission). Courtesy W.A. Debski.

lateral dimensions depend on how narrow via holes and metallization can be placed. Also, tolerances are lower in the lateral dimension that is structured by lithography compared to the vertical layers formed by deposition. However, in order to get a vertical capacitor plate, fences of vias are used that connect many metal layer strips. It depends on the individual technology whether its design rules allow for such a structure.

Table 3.11 Comparison of Perfomance Parameters for Different Types of CMOS Capacitances (from Ref. [32])

Structure	Capacitance Density (fF/μm^2)	Average Capacitance (pF)	Area (μm^2)	Standard Deviation σ (fF)	resonance frequency (GHz)	Q at 1 GHz	Breakdown Voltage (V)
Vertical plate	1.51	1.01	669.9	5.06	>40	83.2	128
Horizontal plate	0.20	1.09	5378.2	26.11	21	63.8	500
MIM	1.1	1.05	960.9			11	

Source: © 2006 Andriy Vasylyev.

The values shown in the table were obtained for a CMOS process featuring seven aluminum metal layers, four of which were used to fabricate the capacitances. The minimum lateral dimensions were a minimum 240 nm for distance between and for width of metal structures, and a metal and oxide thickness of 530 nm and 730 nm, respectively. Statistical data was obtained by measurements on different spots of two 8-in wafers. As expected, the vertical capacitor shows less tolerances, and higher capacitance, consuming less space and thus featuring higher resonance frequency. The values for the MIM capacitance are the typical values shown for comparison.

Inductances can be realized in monlithic processes through spiral lines. A fully symmetric structure is shown in Figure 3.71. Also this structure will show a number of parasitic effects.

- The finite conductivity of the metal layers will lead to parasitic resistance.
- A parasitic capacitance between the metals and the substrate below will establish a lossy connection to ground.
- The capacitance between the windings will lead to a parasitic parallel capacitance.

The equivalent-circuit model shown in Figure 3.71b provides a first-order model for the spiral inductance. It is quite similar to the model of the parallel-plate capacitor, at least regarding the impact of the lossy substrate that is modeled through $C_{ox1,2}$, $R_{sub1,2}$, and $C_{sub1,2}$. The parasitic capacitance turns out to be in parallel to the lossy inductance, which yields a parallel resonance circuit. There is, therefore, a maximum achievable inductance, depending on frequency and technology, since adding winding will increase the parasitic capacitance and loss resistance together with the inductance.

Variations and refinements of the equivalent circuit are possible, for example, the symmetrical structure shown in Figure 3.71c. It intends to enhance the validity of the equivalent circuit model somehow toward higher frequencies, by somehow distributing the coupling to the lossy substrate. However, even such a model fails close to or beyond the resonance frequency of the structure.

Transformers are important building blocks in many CMOS-integrated circuits. These can be realized by designing concentric spiral inductors. An example is shown in Figure 3.72. This transformer consists of one turn on the primary side and two turns on the secondary side.

An equivalent-circuit model of a transformer is slightly more complicated than that of a single inductor. For a structure providing a center connection, the basic model topology is shown in Figure 3.72b. It can be understood as four inductor models with additional coupling between the inductances. Although this model already consists of 30 elements, it still catches only first-order effects. Since this model is quite complicated, it can be reduced to the one shown in Figure 3.73. This model is, of course, very basic, as it only accounts for the self- and mutual inductance and line losses.

The accuracy of equivalent-circuit models is constrained to a certain frequency range, as already mentioned. An other issue is parameter extraction for such a model. It is quite tedious to determine a vast number of parameters basically from two-port measurements. In order to obtain physically significant model parameters, it is required to analytically develop a suitable algorithm. It commonly also requires to investigate dedicated test structures. On the positive side, it is in principle possible to develop parameterized models for a class of elements. It can be accouted for geometrical variations if it is known how the model parameters depend on geometry, and an appropriate formula is used to determine specific parameters.

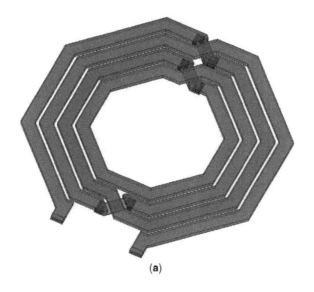

(a)

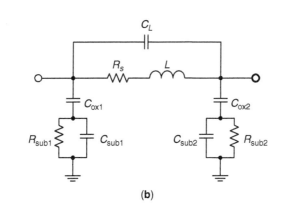

(b)

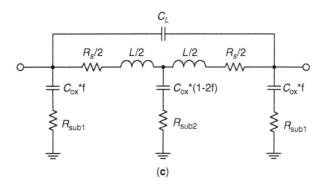

(c)

Figure 3.71 Spiral inductor realized in a RFCMOS process (from Ref. [30], reprint with permission). Courtesy W.A. Debski.

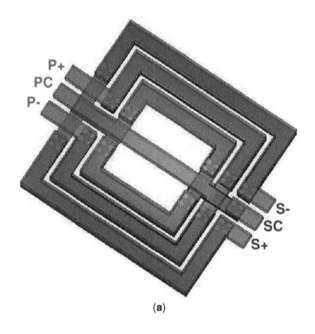

(a)

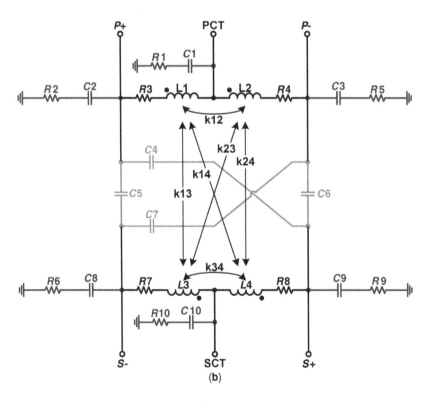

(b)

Figure 3.72 Transformer realized in a RFCMOS process (from Ref. [31], reprint with permission). © 2006 Andriy Vasylyev.

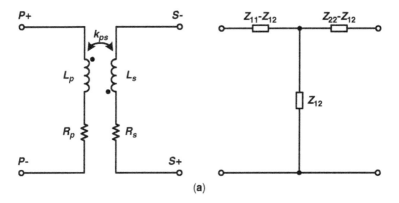

Figure 3.73 Simplified transformer equivalent-circuit model (from Ref. [31], reprint with permission). © 2006 Andriy Vasylyev.

A faster and, at least, at first sight, more accurate approach is to use measured S parameters of the passive models, or to rely on electromagnetic simulation. Both approaches promise to provide highly accurate descriptions of the device performance, since no assumptions restrict the validity of the model. The designer also saves the time required to determine the equivalent-circuit model; simulation becomes both faster and more accurate.

However, there must be a reason why equivalent-circuit models are still in use. The first is that not all circuit simulation tools calculate in frequency domain. S parameters are frequency-domain data. Using an S parameter model in a time-domain solver like SPICE requires to apply an inverse Fourier transform in order to determine the time-domain response of the model. First of all, this requires a certain simulator kernel that allows for this so-called convolution algorithm. Second, will the time-domain response be causal? This is definitely required for simulation, but ensuring it for just any set of S parameters requires elaborate mathematical treatment of the data. Thus, simulating the same circuit in time-domain and in frequency domain will show slightly different results. A model based on a lumped-element equivalent circuit, on the other hand, works with any circuit simulation tool.

It is another issue that S parameters are only known for the measured or simulated frequency range. The fundamental and the first harmonics is usually covered. But what about baseband and dc? Also it might be desirable in nonlinear harmonic-balance simulations to use many harmonics. If one leaves it to the simulation tool to extrapolate the data easily lead to nonphysical parameters. An equivalent-circuit model will not be accurate beyond a certain frequency as discussed above, but it will never become nonphysical and predict, for example, active behavior for an inductor. DC and baseband performance are also well covered.

Finally, there is basically no way to derive a scaled S parameter-based model. Only discrete measured or simulated structures are covered. Parameterized equivalent-circuit-based models have the capabilities to interpolate within certain geometrical limits. This allows to define the models behind the well-known pick-and-place symbols in circuit simulators that describe an element like a spiral inductors on the basis of a few electrical and geometrical parameters.

Design Examples Throughout this book, we always address the issue that we have to separate the battery-operated hand-held applications and the more stationary designs

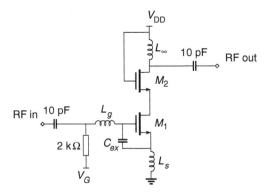

Figure 3.74 Schematic of the CMOS LNA. Both transistors are 130-nm NMOS, gate width 250 μm, and 20 gate fingers. Model parameters are given in Table 3.12. The values of the other circuit elements are $C_{ex} = 1.4$ pF, $L_s = 0.45$ nH, and $L_g = 2.7$ nH. The quality factor of these inductances is set to $Q = 6$ at the target frequency of 2.41 GHz. The biasing network and output matching are omitted for simplicity, $V_{DD} = 2$ V and $I_{DD} = 23$ mA.

like those for base stations. In the case of the hand-held or low-voltage applications, the dynamic range requirements are significantly less than the front-end of a base station requires.

130-nm CMOS Low-Noise Amplifier   As an example, we let us consider a LNA for 2.41 GHz applications that should be fabricated in a 130-nm CMOS technology. We choose devices with a gate width of 250 μm, and a bias of $V_{DD} = 1$ V, $I_{DD} = 20$ mA. A generic model for such a device can be found, for example, on the predictive technology model website [34]. The schematic of the circuit is given in Figure 3.74.

How simultaneous noise and power matching is achieved is shown in Figure 3.75. Starting point is the unmatched cascode, the respective values of S_{11}^* and Γ_{opt} at 2.4 GHz are indicated by the larger symbol. Obviously, the real part of the noise match impedance is higher than 50 Ω, which requires to either use a larger device or to start by adding an extrinsic gate-source capacitance C_{ex}. It is also obvious that the optimum noise and power match are quite far apart.

The impact of adding C_{ex} is shown by the bullet symbols. The capacitance values are increased by 100 fF from one symbol to the next, the highest value shown is $C_{ex} = 1.4$ pF, which is used in this example. Since noise and power match are still too far apart of each other, a source degeneration inductance L_s is added. In the example, L_s is increased in steps of 50 pH up to a maximum of 0.45 nH. The diamond symbols in the Smith charts show that indeed, S_{11} changes much faster than Γ_{opt}. Finally, a gate inductance L_s is added to obtain input matching; it is increased in steps of 300 pH up to a total value of $L_g = 2.7$ nH. In this simulation, it is accounted for inductor losses by setting the inductors' quality factor to $Q = 6$.

This practical example shows some discrepancies to the theory presented earlier. For example, it does not seem to be possible to achieve perfect simulatneous power and noise matching. But despite the fact that the theory was based on many simplifications, it proves to be applicable for real-life circuit design.

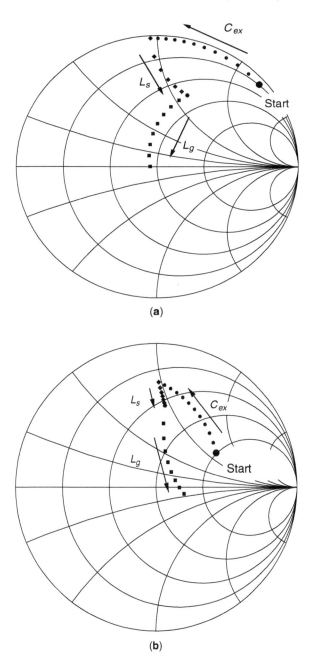

Figure 3.75 Impact of external gate-source capacitance C_p, source degradation inductance L_s, and gate inductance L_g on (a) complex conjugate of S_{11} and (b) optimum noise match Γ_{opt}.

Figure 3.76 presents the S parameters of the circuit after input matching. Noise figure and stability factor μ are shown in Figure 3.77. In these figures, the performance of the cascode LNA (solid lines) is compared with a single-transisor LNA (dashed lines). The input matching for the single-transistor LNA was achieved with similar component values,

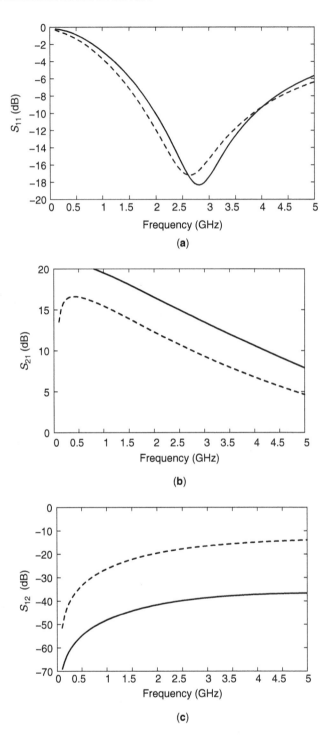

Figure 3.76 Simulated S parameters of the circuit shown in Figure 3.74 (solid lines), compared to the results obtained when the cascode is replaced by a single transistor (dashed lines). (a) S_{11}. (b) S_{21}. (c) S_{12}.

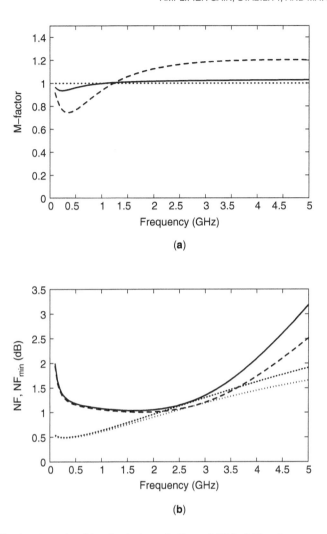

Figure 3.77 Simulated results of the circuit shown in Figure 3.74 (solid lines), compared to the results obtained when the cascode is replaced by a single transistor (dashed lines). (a) Stability factor μ. (b) Noise figures. The dotted lines denote the respective minimum noise figures.

that is, $C_{ex} = 1.4\,\text{pF}$, $L_s = 0.64\,\text{nH}$, $L_g = 2.2\,\text{nH}$. The benefit of the cascode configuration is obviously its better isolation, as S_{12} is about 20 dB lower than in the single-transistor case. The price is, however, a slightly higher achievable noise figure.

Other interesting design examples can be found online [33].

130-nm CMOS Transformer-Coupled Low-Noise Amplifier   An alternative approach to the inductively degenerated low-noise amplifier is the transformer-coupled LNA, a topology initially introduced as zwischenbasis-configuration by Cantz in 1953 [35]. The schematic is shown in Figure 3.78. The topology differs from the previously discussed LNA by the fact that the source and gate inductance now are mutually coupled. This feedback supports simultaneous noise and power matching.

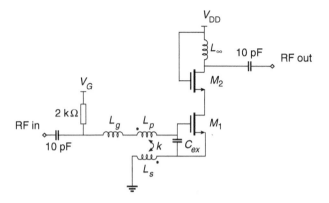

Figure 3.78 Schematic of the transformer-coupled CMOS LNA. Both transistors are 130-nm NMOS, gate width 250 μm, and 20 gate fingers. Model parameters are given in Table 3.12. The values of the other circuit elements are $C_{ex} = 1.5$ pF, the transformer parameters are $L_p = 0.8$ nH, $L_s = 0.2$ nH, and $k = 0.6$, and inductor losses are modeled by adding resistances of 1.1 Ω/0.1 pH to each of the inductances. $L_g = 1.7$ nH with a quality factor of $Q = 6$ at the target frequency of 2.41 GHz. The biasing network and output matching are omitted for simplicity, $V_{DD} = 2$ V and $I_{DD} = 23$ mA.

In the present case, the transformer was modeled by assuming a fixed ratio of 4:1 for the primary and secondary inductances L_s and L_p, respectively. Also the coupling factor was fixed to $k = 0.6$. In order to account for the transformer losses, it is assumed that each of the inductances has a parasitic resistance of 1.1 Ω/100 pH. The transformer together with the extrinsic gate-source capacitance C_{ex} was empoyed to bring noise and power match to the same impedance, with a real part of 50 Ω. The gate inductance L_g was finally used to transform the resulting input impedance to the 50 Ω source. This inductance is assumed to have a quality factor limited to $Q = 6$ at the target frequency of 2.41 GHz. The parameter values are $L_s = 200$ pH and $L_p = 0.8$ nH, $C_{ex} = 1.5$ pF, and $L_g = 1.7$ nH. The transistor model parameters are given in Table 3.12. The original transformer-coupled LNA relies on a single transistor instead of a cascode. Therefore, simulation results obtained with the same topology except for that the cascode is replaced by a single MOS transistor are included in the results shown in Figures 3.79 and 3.80. In case of the single-transistor amplifier, matching slightly differs. The element values are $L_s = 350$ pH, $L_p = 1.4$ nH, $C_{ex} = 1.2$ pF, and $L_g = 0.4$ nH.

The benefit of the transformer-coupled configuration over the traditional source degeneration topology is obviously the lower total inductance values that are required in order to achieve a good matching. In the present simulation, the main noise sources are not the transistors but the losses of the inductors.

NE68133 Matched Amplifier. We will first consider the design of a low-voltage, low-current, low-noise amplifier based on the NEC BJT NE68133. This device was chosen because it is unconditionally stable, has a minimum noise figure of about 1.2 dB at 7 mA, 15.5 dB gain, and a f_T of 8 GHz. While it is possible to start the design based on the published S and noise parameters, we will choose to involve also the nonlinear model as this allows us more freedom in selecting the bias point including the collector voltage. The actual topology of the amplifier is shown in Figure 3.81.

Table 3.12 130-nm NMOS BSIM4 Model Parameters Used to Simulate the LNA, as Published at the Predictive Technology Model Website (http://ptm.asu.edu/)

```
* Beta Version released on 2/22/06

* PTM 130nm NMOS

.model  nmos  nmos  level = 54

+version = 4.0          binunit = 1          paramchk= 1          mobmod  = 0
+capmod  = 2            igcmod  = 1          igbmod  = 1          geomod  = 1
+diomod  = 1            rdsmod  = 0          rbodymod= 1          rgatemod= 1
+permod  = 1            acnqsmod= 0          trnqsmod= 0

+tnom    = 27           toxe    = 2.25e-9    toxp    = 1.6e-9     toxm    = 2.25e-9
+dtox    = 0.65e-9      epsrox  = 3.9        wint    = 5e-009     lint    = 10.5e-009
+ll      = 0            wl      = 0          lln     = 1          wln     = 1
+lw      = 0            ww      = 0          lwn     = 1          wwn     = 1
+lwl     = 0            wwl     = 0          xpart   = 0          toxref  = 2.25e-9
+xl      = -60e-9
+vth0    = 0.3782       k1      = 0.4        k2      = 0.01       k3      = 0
+k3b     = 0            w0      = 2.5e-006   dvt0    = 1          dvt1    = 2
+dvt2    = -0.032       dvt0w   = 0          dvt1w   = 0          dvt2w   = 0
+dsub    = 0.1          minv    = 0.05       voff1   = 0          dvtp0   = 1.2e-010
+dvtp1   = 0.1          lpe0    = 0          lpeb    = 0          xj      = 3.92e-008
+ngate   = 2e+020       ndep    = 1.54e+018  nsd     = 2e+020     phin    = 0
+cdsc    = 0.0002       cdscb   = 0          cdscd   = 0          cit     = 0
+voff    = -0.13        nfactor = 1.5        eta0    = 0.0092     etab    = 0
+vfb     = -0.55        u0      = 0.05928    ua      = 6e-010     ub      = 1.2e-018
+uc      = 0            vsat    = 100370     a0      = 1          ags     = 1e-020
+a1      = 0            a2      = 1          b0      = 0          b1      = 0
+keta    = 0.04         dwg     = 0          dwb     = 0          pclm    = 0.06
+pdiblc1 = 0.001        pdiblc2 = 0.001      pdiblcb = -0.005     drout   = 0.5
+pvag    = 1e-020       delta   = 0.01       pscbe1  = 8.14e+008  pscbe2  = 1e-007
+fprout  = 0.2          pdits   = 0.08       pditsd  = 0.23       pditsl  = 2.3e+006
+rsh     = 5            rdsw    = 200        rsw     = 100        rdw     = 100
+rdswmin = 0            rdwmin  = 0          rswmin  = 0          prwg    = 0
+prwb    = 6.8e-011     wr      = 1          alpha0  = 0.074      alpha1  = 0.005
+beta0   = 30           agidl   = 0.0002     bgidl   = 2.1e+009   cgidl   = 0.0002
+egidl   = 0.8

+aigbacc = 0.012        bigbacc = 0.0028     cigbacc = 0.002
+nigbacc = 1            aigbinv = 0.014      bigbinv = 0.004      cigbinv = 0.004
+eigbinv = 1.1          nigbinv = 3          aigc    = 0.012      bigc    = 0.0028
+cigc    = 0.002        aigsd   = 0.012      bigsd   = 0.0028     cigsd   = 0.002
+nigc    = 1            poxedge = 1          pigcd   = 1          ntox    = 1

+xrcrg1  = 12           xrcrg2  = 5
+cgso    = 2.4e-010     cgdo    = 2.4e-010   cgbo    = 2.56e-011  cgdl    = 2.653e-10
+cgsl    = 2.653e-10    ckappas = 0.03       ckappad = 0.03       acde    = 1
+moin    = 15           noff    = 0.9        voffcv  = 0.02

+kt1     = -0.11        kt1l    = 0          kt2     = 0.022      ute     = -1.5
+ua1     = 4.31e-009    ub1     = 7.61e-018  uc1     = -5.6e-011  prt     = 0
+at      = 33000

+fnoimod = 1            tnoimod = 0

+jss     = 0.0001       jsws    = 1e-011     jswgs   = 1e-010     njs     = 1
+ijthsfwd= 0.01         ijthsrev= 0.001      bvs     = 10         xjbvs   = 1
+jsd     = 0.0001       jswd    = 1e-011     jswgd   = 1e-010     njd     = 1
+ijthdfwd= 0.01         ijthdrev= 0.001      bvd     = 10         xjbvd   = 1
+pbs     = 1            cjs     = 0.0005     mjs     = 0.5        pbsws   = 1
+cjsws   = 5e-010       mjsws   = 0.33       pbswgs  = 1          cjswgs  = 3e-010
+mjswgs  = 0.33         pbd     = 1          cjd     = 0.0005     mjd     = 0.5
+pbswd   = 1            cjswd   = 5e-010     mjswd   = 0.33       pbswgd  = 1
+cjswgd  = 5e-010       mjswgd  = 0.33       tpb     = 0.005      tcj     = 0.001
+tpbsw   = 0.005        tcjsw   = 0.001      tpbswg  = 0.005      tcjswg  = 0.001
+xtis    = 3            xtid    = 3

+dmcg    = 0e-006       dmci    = 0e-006     dmdg    = 0e-006     dmcgt   = 0e-007
+dwj     = 0.0e-008     xgw     = 0e-007     xgl     = 0e-008

+rshg    = 0.4          gbmin   = 1e-010     rbpb    = 5          rbpd    = 15
+rbps    = 15           rbdb    = 15         rbsb    = 15         ngcon   = 1
```

Source: Courtesy Nanoscale Integration and Modeling (NIMO) Group, ASU.

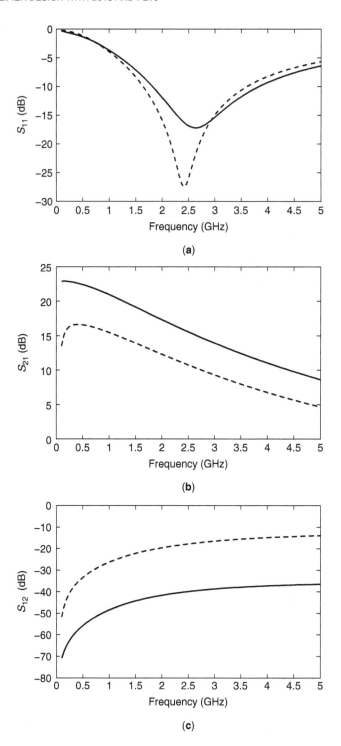

Figure 3.79 Simulated S parameters of the circuit shown in Figure 3.78 (solid lines), compared to the results obtained when the cascode is replaced by a single transistor (dashed lines). (a) S_{11}. (b) S_{21}. (c) S_{12}.

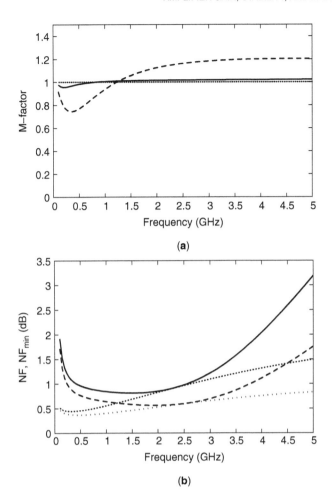

(a)

(b)

Figure 3.80 Simulated results of the circuit shown in Figure 3.78 (solid lines), compared to the results obtained when the cascode is replaced by a single transistor (dashed lines). (a) Stability factor μ. (b) Noise figures. The dotted lines denote the respective minimum noise figures.

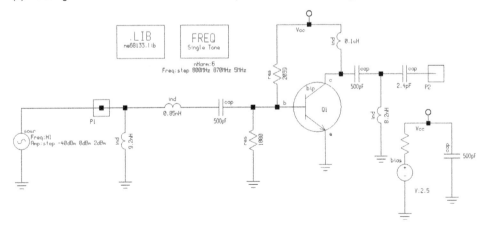

Figure 3.81 The NE68133 low-noise amplifier.

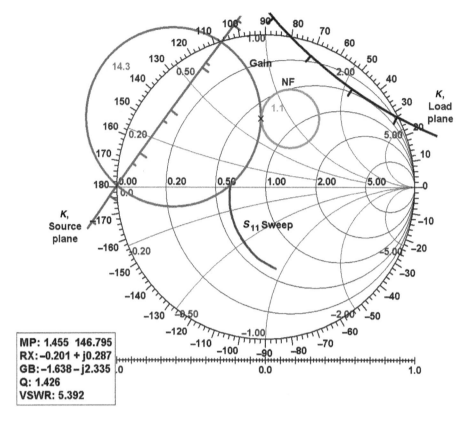

Figure 3.82   Display from the interactive Smith Tool, showing circles for gain, noise figure, and source and load-plane stability, as well as a sweep of S_{11} versus frequency. In synthesizing an input matching circuit, we choose a single working frequency and work from the chart's origin to the point of tangency between the gain and noise figure circles. See Figure 3.83.

Determining appropriate values for the input and output matching networks is essential to finalize the design. The next several figures illustrate matching-network syntheses undertaken with the interactive Smith Tool in Ansoft's Serenade Design Environment. We evaluate the point of tangency as well as stability circles in the source and load planes, indicating that our impedance match will result in a stable amplifier.

In Figure 3.82, the center circle (with 1.1 marker) refers to the noise circle for which the noise figure of 1.1 dB can be obtained. The left circle (with the 14.3 marker) corresponds to source-plane terminations that correspond to 14.3 dB of available gain. In order to get the required noise figure *and* gain, the input termination must be at the point of tangency of both circles. The arc touching the X axis close to 0.50 is the amplifier's S_{11} response from 800 to 875 MHz. The shallow arc that nearly bisects the 14.3 dB gain circle is actually a portion of the source-plane stability circuit; the arc at the top right is a portion of the load-plane stability circle. Because we will be terminating the amplifier input with a value inside the source-plane stability circle, and terminating the amplifier output with a value outside the load-plane stability circle, the amplifier will be unconditionally stable at the frequency at which the match is obtained.

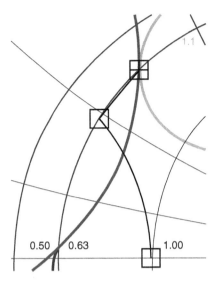

Figure 3.83 Input matching network extraction, assuming lossless matching components and no parasitic effects.

The next task is to obtain the values for the matching circuit that matches the input for the 1.1 dB noise figure. Since we do a noise matching will not get the best possible S_{11} matching (Figure 3.83). This translates into an inductor of 9.2 nH to ground and a series inductor of 0.85 nH based on the large-signal model of the NE68133 at the chosen bias point.

After we have determined our input matching network, it is time to find the output matching. First find termination point on the output plane that corresponds to 14.3 dB gain and 1.1-dB NF at the input plane. The output gain circle goes through the 1.1-dB NF circle at two points, and for reasons of easy matching, we selected the lower point. The next step is to find the conjugate reflection coefficient $\Gamma_L = S_{22}^*$. This is the lower square marker in the Smith diagram, see Figure 3.84. The remaining task is to find the matching network that brings this value to 50 Ω. By doing so, we first obtain a shunt inductance $L_{shunt} = 8.2$ nH and a series capacitor $C_{series} = 2.4$ pF. Again, these are the values applicable to the large-signal equivalent circuit.

The careful reader will have noticed that we have been avoiding the issue of small-signal exact matching using the manufacturer-supplied parameters. If this is done, the narrowband response shown in Figure 3.85 will be obtained. Therefore, we must optimize the circuit for a better frequency response to obtain the one shown in Figure 3.86.

After this optimization, we were well able to meet the specification being center band gain of 14 dB and a corresponding NF of 1.1 dB. To show the flatness of the gain, Figure 3.87 shows a gain variation in the range of 0.25 dB.

The next step is to look into the large-signal performance. The first order of business is to determine the third-order intercept point, which is based on a two-tone analysis. Given the fact that we run the device at low voltage and at a just-reasonable current, the actual intercept point of +13.5 dBm is quite a good number. Figure 3.88 shows the fundamental and third-order outputs. The crossover point between the two, as shown in Chapter 1, is defined as the third-order intercept point. If one would have chosen the third harmonic

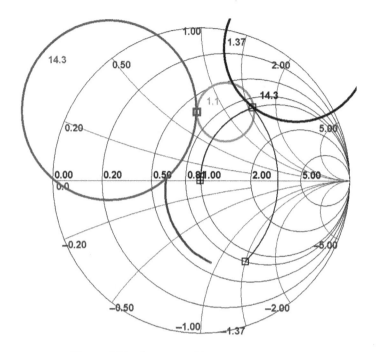

Figure 3.84   Output matching network extraction.

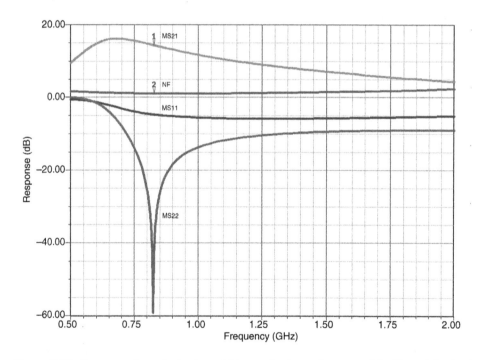

Figure 3.85   Frequency-dependent gain, matching, and noise performance of the Figure 3.81 circuit.

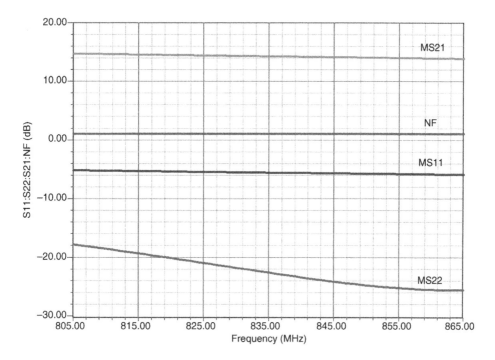

Figure 3.86 Optimized performance of input/output reflection, gain, and noise figure.

output rather than the fundamental, we would have shown the fifth-order intercept point. The second-order IMD product ($f_1 \pm f_2$) typically can be minimized by input selectivity.

How amplifier linearity affects its ability to preserve the characteristics of digital signals is a major concern in wireless design. Figures 3.89 and 3.90 show constellation diagrams with the amplifier handling a PSK signal at drive levels of -40 dBm and +10 dBm, respectively. As expected, we see significant distortion at the higher drive levels due to circuit nonlinearities.

BFP420 Matched Amplifier.   The NE68133 has an attractive f_T of 8 GHz, but if more gain is needed, one may play with the concept of using a much "hotter" device, such as the BFP420. We recall that the NEC device was unconditionally stable, and our first worry is how the Infineon BFP420 will behave.

A study of Figure 3.91 shows that indeed the NEC device had to be stable all over and therefore was a good basis for a risk-free design. Things change dramatically which the BFP420, for which the K factor is essentially less than 1 up to a corner frequency of 2.75

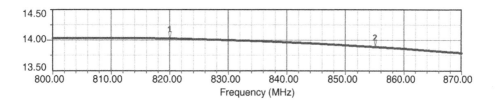

Figure 3.87 High-resolution display of gain as a function of frequency.

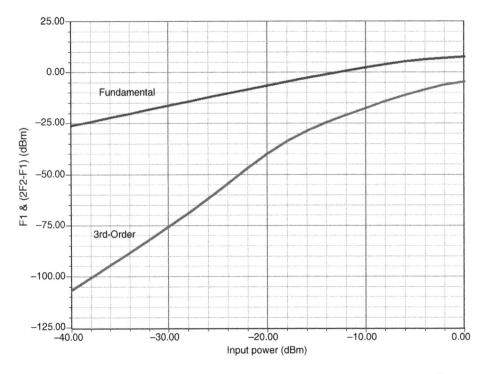

Figure 3.88 Simulated fundamental and third-order-IMD outputs of the NE68133 amplifier.

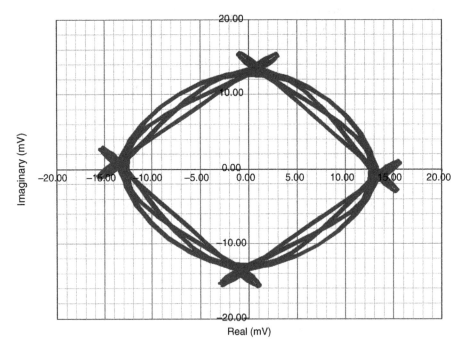

Figure 3.89 Output constellation at -40 dBm input.

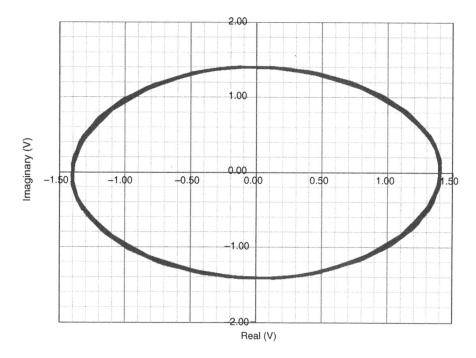

Figure 3.90   Output constellation at 10 dBm input.

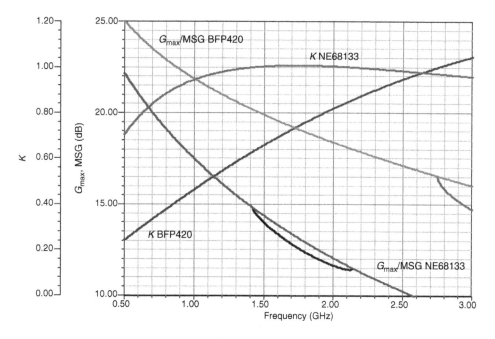

Figure 3.91   Comparison of K, G_{max}, and maximum stable gain (MSG) for the NE68133 and BPF420 BJTs.

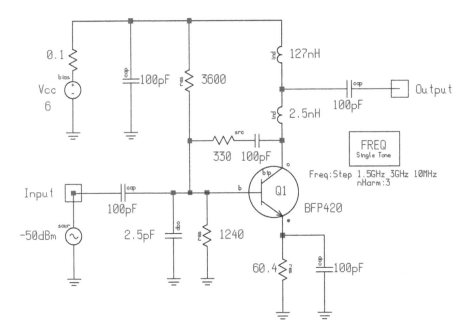

Figure 3.92   BFP420 amplifier with resistive voltage feedback from collector to base.

GHz, where the transistor becomes unconditionally stable. Around 1.75 GHz, the K factor for the NEC transistor barely touches 1. We explained earlier the fact that G_{max} deviating from the maximum stable gain is a more realistic number since the conditions for maximum stable gain involve feedback and conjugate matching.

To improve the stability of the BFP420, we will now set out to use the same topology as the previous amplifier but with a resistive feedback between base and collector (Figure 3.92). A good value for this is always somewhere around 200–500 Ω. It must be remembered that this voltage feedback also stabilizes the input and output impedance and reduces the Miller effect because of the feedback being resistive and more dominant.

While the first example covered a fairly narrow frequency band, we are now attempting to cover a much wider frequency range, specifically, from 1.5 to 3 GHz reasonable gain and noise. This, by definition, requires a much higher gain-bandwidth product, and the simple matching network at the input and output provides a nonoptimal input and output matching (Figure 3.93). The feedback introduces some additional noise. This application shows the limits of single-stage UHF/SHF amplifiers with feedback.

Narrowband BFP420 Amplifier   Our next attempt will be to design a narrowband input stage that combines very high selectivity with good noise figure. This can only be achieved by using two tuned single-stage resonators with capacitive/magnetic coupling. The following amplifier is tailored to have a center frequency of 900 MHz. The transistor will operate at 12 V/10 mA to have a sufficient dynamic range. The input selectivity is achieved by the use of two top-coupled tuned circuits, followed by a BFP420 (Figure 3.94). The first order of business is to determine the input tuned circuit. The tuning capacitance for the two tuned circuits is arbitrarily set to a manageable 2 pF (small values have too much tolerances and larger values will require smaller inductances, resulting in lower Q). The reactance of the

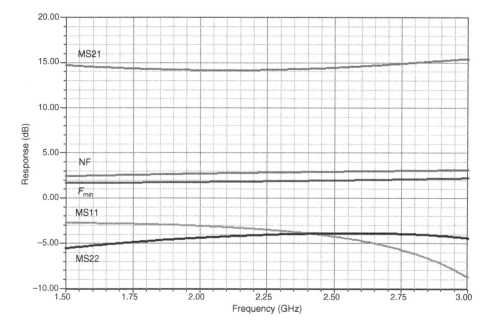

Figure 3.93 Frequency-dependent gain, matching, and noise performance of the BFP420 feedback amplifier.

first tuned circuit is

$$Z_C = \frac{1}{\omega C} = \frac{1}{2\pi \times 9 \cdot 10^8 \times 2 \cdot 10^{-12}} = 88.42 \, \Omega$$

The necessary inductance for resonance will be

$$\omega L = 88.42$$

or

$$L = 88.42/\omega = 88.42/\left(2\pi \times 9 \cdot 10^8\right) = 15.64 \, \text{nH}$$

We will make both tuned circuits identical, but with different loaded Q. By multiplying 88.42 times the loaded Q (20 for the input resonator and 50 for output resonator), the resulting parallel-resonant resistance is $88.42 \times 20 = 1768\Omega$ at 900 MHz (for the input resonator) and $88.42 \times 50 = 4421\Omega$ (for the output resonator). These values are achieved by loading the input with the 50-Ω source and by loading the output with the input of the transistor. Therefore, the input transformation ratio $m_1 = 1768.4/50 = 35.37$. At the output, the equivalent number $m_2 = 4421/50 = 88.42$. For the purpose of simulation, and because it is extremely difficult to find the physical location of the appropriate tap on an air-wound coil for the necessary impedance, we do this "on paper" by incorporating an ideal transformer. We are also going to show later the actual transmission-line-based filter and its modeling. This filter will have a higher insertion loss based on the PC-board material used.

Using our nonlinear simulator, in small-signal ac mode, we obtain the frequency response and noise figure shown in Figure 3.95. In actually building this circuit, there is

462

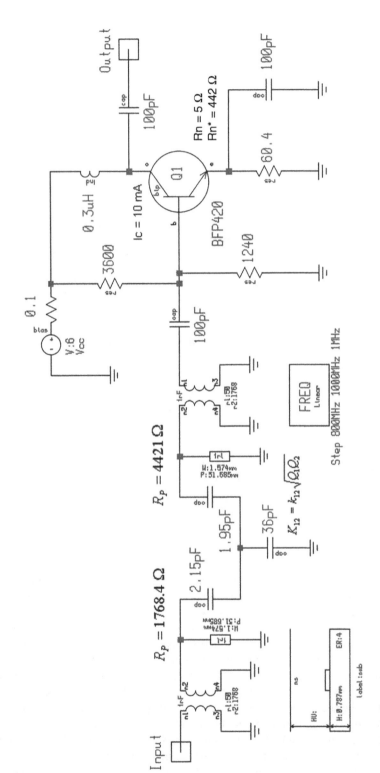

Figure 3.94 BFP420 amplifier with narrowband input filtering. Ideal transformers are used to match the filter's input and output to the input port and the BFP420 base, respectively.

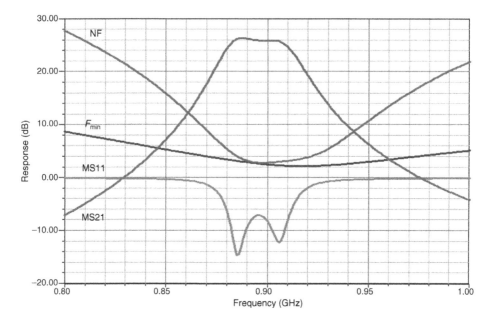

Figure 3.95 Frequency-dependent gain, matching, and noise performance of the narrowband BFP420 amplifier.

another problem besides these transformers—the top-coupling capacitor. We have used a grounded-T configuration to obtain a practically realizable value of 36 pF, whose lead inductance will not influence the circuit. (The actual coupling capacitance value for top coupling without the T configuration would have 0.1 pF. We would like to see this value repeatably realized.)

If the actual noise would be calculated relative to the output tuned circuit, the equivalent noise resistor R_n of 5 Ω would now be multiplied by m_2, being 442 Ω. Fortunately enough, the Serenade simulator has a highly accurate BJT noise model that is both bias- and temperature-dependent so the simulator will give us the answer directly.

We have not put any matching circuit at the output, and leave this task to the reader. In practice, the slight tilt in the MS_{21} curve in Figure 3.95 would be tuned out. Also note that the minimum noise figure (F_{min}) and the actual 50-Ω noise figure (NF) agree somewhat off center. This has to do with the Miller effect, which results in the well-known difference between noise matching and input-gain matching. By varying the coupling capacitor to move from the overcoupled response that gives a low noise figure to a critically coupled response ($K = 1$), we get significantly better selectivity, narrow bandwidth (at the obvious expense of the noise figure) as shown in Figure 3.96.

Since we somewhat compromised by not providing the reader with an actual input and output coupling that can be built, and further, because the inductance in question (15.26 nH) is hard to realize with a high Q, we now show the printed-circuit approach for the input filter. By using a Teflon-based material for the PC board, the losses would be less; however, the radiation would be more because the ε would go down to 2.1 from the value we used (4). It is also convenient to introduce this filter to show that at higher frequencies, where lumped elements are really no longer available, one has to switch to distributed elements. A list of

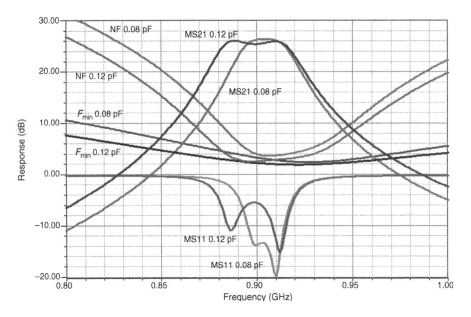

Figure 3.96   Comparison of filter responses with critical coupling (0.08 pF response) and overcoupling (0.12 pF response).

the commonly used distributed elements will be shown later; the transmission line and T junction are probably the most frequently such elements, followed by spiral and rectangular inductors, and bends.

The actual implementation of the input filter on a printed circuit board requires distributed elements (Figure 3.97). In practice, one has two parallel coupled lines with a tap somewhere toward the ground connection and, for modeling accuracy, we have to introduce a T junction with the appropriate impedance of 50 Ω for the input and a different impedance (if necessary) for the output. The coupling mechanism now is magnetic coupling, since no one can physically change the distance between the transmission lines by tuning some mechanical elements, one has to get it right the first time. This can only be achieved with a high-precision simulator. The reason why these simulators are so expensive is the investment of the number of man-years to produce valid models over a huge frequency and impedance range. Since we are now curious about the frequency response, we have plotted in Figure 3.98, the frequency response that shows a higher (expected) insertion loss. The PC-board material is partially responsible for this.

At the end of this chapter, we will have a chance to talk about passive elements as needed for microwave frequencies, and one should consider frequencies above 1500 MHz as microwave frequencies.

GaAsFET Feedback Amplifier   The implementations we have shown so far have been based on bipolar transistors. Figure 3.99 shows a single-stage feedback amplifier using a GaAsFET. The feedback is based around an LRC network and some attempts at matching at the input and output are made. Figure 3.100 shows its frequency-dependent gain, matching, and noise performance.

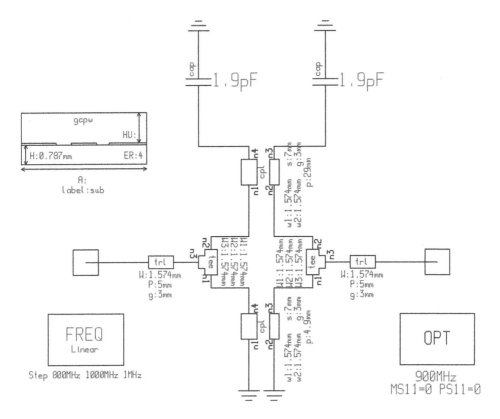

Figure 3.97   Input filter using distributed-element resonators and matching.

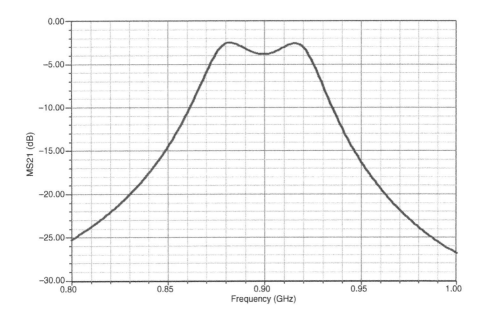

Figure 3.98   Gain versus frequency response for the distributed-element filter.

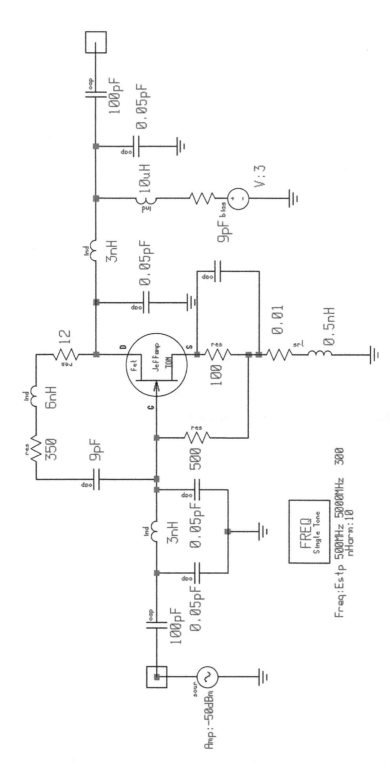

Figure 3.99 GaAsFET feedback amplifier.

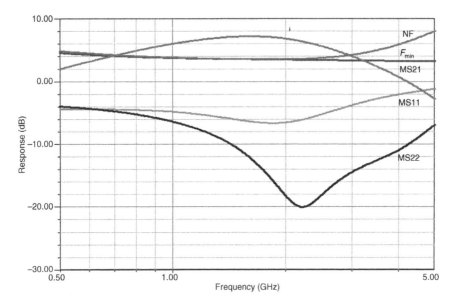

Figure 3.100   Frequency-dependent gain, matching, and noise responses for the GaAsFET amplifier.

3.2.3 High-Gain Amplifiers

In most cases, high-gain amplifiers receive an already amplified signal, which means that we are less concerned about the signal-to-noise ratio but remain concerned about distortion. To reduce distortion, we have two options: (1) increase the dc current in the active device, which always improves the current-related distortion, and (2) maybe add some voltage feedback, which reduces the voltage distortion. In simple terms again, only current feedback can reduce the current-related distortion of a square-law or exponential transfer characteristic and only voltage feedback can reduce the effect of overdrive as a function of the load, but only to some degree. Its main purpose remains to reduce the input impedance to useful values, specifically for FETs, since current feedback increases the input and output impedances. Some ultrasensitive circuits still require Class A operation, but now at higher current levels; these are typically CATV applications and there are several transistors available that can be operated at 5–15 V but at 60 mA or higher. Because of the higher power consumption, this excludes hand-held cordless or cellular telephones. The same circuit as used before (Figure 3.99) for Class A operation now can be operated at a higher current, which automatically increases the gain and the noise figure and reduces distortion. This is based on the fact that the transconductance (g_m) increases at higher currents. The second transistor in the TI circuit achieves this by being in reality a larger device, which could be modeled by several transistors, like the first stage, in parallel. This automatically increases the dc current by the scaling factor. The same is even more true for the output stage. Likewise, bipolar transistors can also be scaled. A visual inspection of some transistors in their packages will reveal that they are really not just one transistor, but several operating in parallel, each with its own ballast resistors (for bipolar transistors) or other means (for FETs).

Using tuned circuits rather than wideband matching, it is possible to go to Class AB or B operation. AB is really a hybrid between A and B, and Class B is probably better known. Figure 3.217 shows the Class B operating point. Most Class B stages are operated in push–pull because together with the tuned circuit, the one half of the cycle is supplied by the other transistor. The definition of Class B operation is a conducting angle of 90° (Figure 3.217). At this point, it should be noted that some older texts define the conducting angle as being twice the value we have defined in this chapter.

Without going into further details, the highest efficiency in Class B operation is 78.5%. Based on the fact that modern wireless applications generally use a hybrid of amplitude and angle modulation, pure Class C operation finds little application nowadays because of the severe nonlinear distortion it introduces. A Class B stage has less "power gain" as a Class A stage, but provides much more output power. We have already mentioned the problem associated with characterizing a Class A stage in terms of power gain, that is, the only time Kirchoff's equations are valid, meaning that the sum of all currents and voltages have to be zero. Because Kirchoff's equations do not consider nonlinearities and harmonic contributions, the very moment we move into considering medium and higher-power amplifiers, the nonlinearities will dominate and the performance of the stage is best analyzed with a high-quality circuit simulator with the appropriate dynamic range and good modeling capability. Without question, this also applies to Class C operation. As pointed out previously, the high-gain amplifier is optimized for gain rather than noise figure, and typically requires more collector or drain current. This can be easily seen by rebiasing the FET amplifier as shown above to a significantly higher value of 33 mA. This is accomplished by changing the source resistor down to 3 Ω. As a result of this, we obtain the gain, matching, and noise figure responses shown in Figure 3.101.

It may be a surprise to the reader that the actual noise figure now is less than that shown in Figure 3.100. This device is now operating at 33 mA; the reason for this has to do with the fact that if the transconductance is too low, the noise figure also increases, and the NF curve for the FET has a similar-looking current-dependent minimum—a curve that is frequently

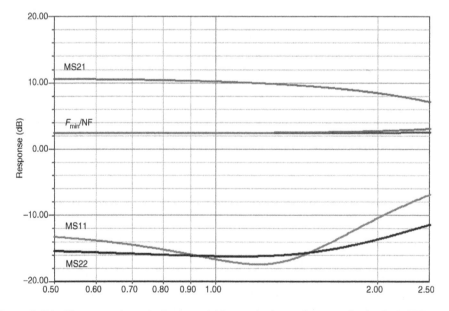

Figure 3.101 Frequency-dependent gain, matching, and noise performance for the GaAsFET amplifier of Figure 3.99 rebiased for a drain current of 33 mA.

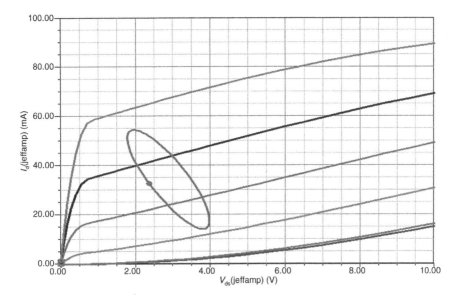

Figure 3.102   Dc *I–V* curves for the GaAsFET operating at $I_D = 33\,\text{mA}$, including the ac load line. The load line is an ellipse rather than a straight line for the reason described in the text.

not supplied by the manufacturer. This amplifier, now operating at 33 mA, should have quite good large-signal capabilities; we will examine this shortly.

Another interesting piece of information is the display of the dc *I–V* curves with the "load line." Since we have reactances in the circuit, we are not going to get a line, but rather an ellipse indicating the storage of energy. (The actual load line would be a line through the two foci of the ellipse.) This curve allows us to immediately see that at this operating point, the transistor is neither going into current saturation nor voltage limiting. The input power can be increased further until such a condition exists. Figure 3.102 shows the dc *I–V* curve for the amplifier shown in Figure 3.99.

Besides looking for the adjacent-channel power ratio (ACPR), the other traditionally important points are third-order intercept point and the equivalent of a multitone simulation as used for CATV application but relevant to the multisignal environment of wireless applications. Figure 3.103 shows the determination of the amplifier's third-order intercept point. It is most important for the reader to understand that very low values for input and output level have to be used to determine the crossover point. If this advice is not followed, totally erroneous numbers can occur.

Since an antenna supplies a large number of signals (this is true of hand-held radios as well as base-station sites) a three-tone standard has been adopted as a good approximation of multisignal operation. This standard is in accordance with the German DIN 45004E,3.3 three-tone measurement. The requirement is that the intermodulation distortion products at the output are better than 60 dB suppressed relative to the largest of the three tones. One tone is used as the reference, and the other two tones are 6 dB less. This measurement has been derived from television, whereby the video transmitter, based on the vestigial sideband power, is 6 dB above the two sidebands produced by the FM sound transmitter using an modulation index of 5. The result of a three-tone test can be seen in Figure 3.104.

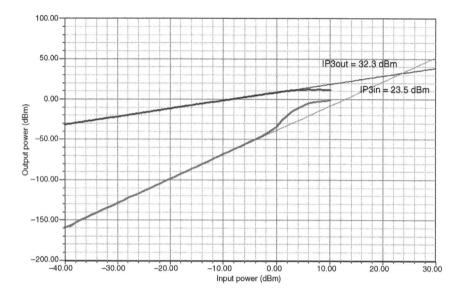

Figure 3.103   Determination of the third-order intercept point based on a two-tone measurement tones spaced at 5% of the operating frequency (1.8 GHz).

This amplifier clearly meets the DIN specification for linearity at input levels up to $-9/-15$ dBm at the input.

Bipolar transistors are also quite popular for this purpose. Figure 3.105 shows a single-stage amplifier with an intrinsic transistor and all the other "hidden" intrinsics, but normally not visible, added. The three diodes belong to those intrinsic elements, which are needed for accurate microwave modeling that the standard nonlinear models do not provide. The reader will see that we went through great detail to model all the necessary parts. This feedback amplifier, whose frequency response is shown in Figure 3.106, has been quite difficult to

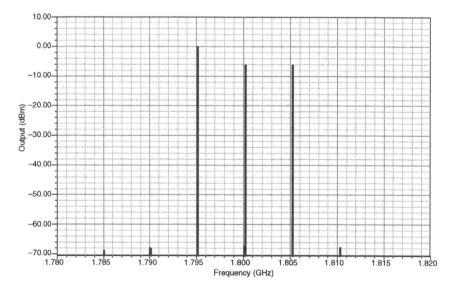

Figure 3.104   Simulated three-tone analysis of the GaAsFET amplifier with $I_D = 33$ mA.

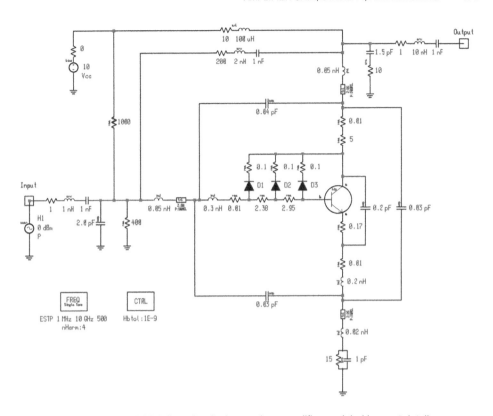

Figure 3.105 A high-linearity single-transistor amplifier modeled in great detail.

design, but there are monolithic circuits available where the higher gain-bandwidth product is being used to advantage.

Figure 3.106 shows the predicted frequency-dependent gain, matching, and noise figure at a bias point of 108 mA at 10 V. Using the same DIN three-tone test procedure, Figure 3.107 shows the expected good result.

Since this test requires two equal signals and one larger signal, we need to expect one IMD product larger than the rest. We have set our window for a dynamic range of 70 dB, and yes, in the left corner, we can see the product between the first two signals, while the other products are below the −50 dB level.

As a little task, Figure 3.108 shows a simple amplifier, based on a Infineon transistor, which is a feedback amplifier with fewer modeling details surrounding the transistor. Its associated frequency-dependent performance is shown in Figure 3.109. Interested readers are invited to improve its performance by using their skills and intuition.

Permitting higher currents and actually combining two transistors leads to our first MMIC. The following simulation is based on the MGA-64135.[1] Needless to say, there

---

[1] The Hewlett Packard MGA-64135 is now replaced by Avago's amplifiers MGA-81563 or MGA-82563, up to 6 GHz. Avago provides designers with an Agilent ADS design kit for these circuits that allow for detailed simulation. However, since the models are protected and do not allow us to even see the circuit topology, we keep the older, but detailed and not yet outdated circuit example.

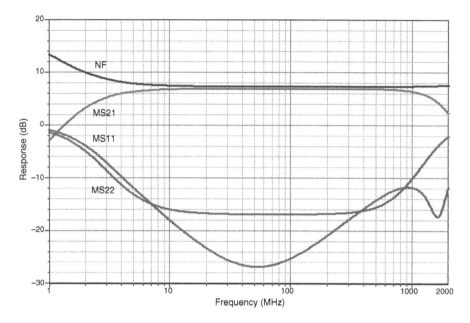

Figure 3.106 Simulated frequency-dependent gain, matching, and noise figure performance of the circuit shown in Figure 3.105.

are some compensation networks inside the MMIC and distributed elements that affect the frequency response. However, with this arrangement, it is possible to obtain an appreciable gain at our wireless frequencies. The noise figure, based on the feedback, however, is not too impressive. This is a problem that occurs with most feedback amplifiers, which try to

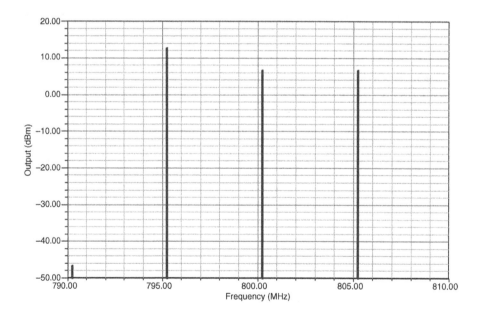

Figure 3.107 Simulated three-tone output spectrum of the Figure 3.105 amplifier.

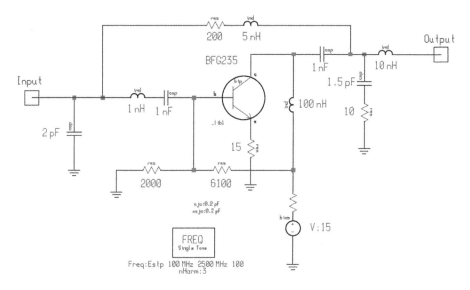

Figure 3.108 A high-linearity amplifier using the Infineon BFG235.

minimize the distortion and yet would like to keep the noise figure down. Figure 3.110 shows the schematic of the GaAs MMIC, Figure 3.111 shows its frequency-dependent gain, matching, and noise performance, Figure 3.112 shows its two-tone response, which is used to obtain the intercept point ($IP_{3,in} = 8.5$ dBm, $IP_{3,out} = 22$ dBm), and finally, Figure 3.113 shows the three-tone test results at an input RF level of $-21/-27$ dBm. While the

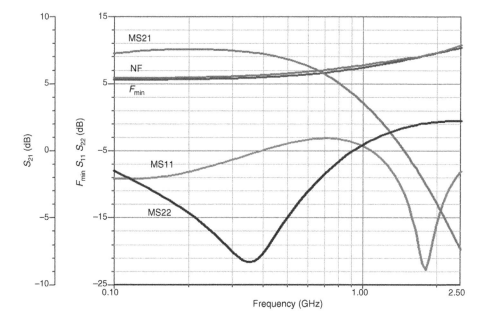

Figure 3.109 Frequency-dependent gain, matching, and noise performance of the Figure 3.108 amplifier.

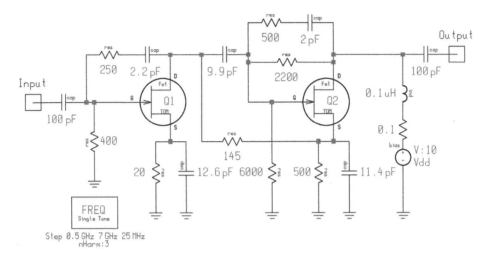

Figure 3.110 Simulation schematic for the MGA-64135 MMIC amplifier. Q_1 runs at 17.4 mA; Q_2 at 32.2 mA.

gain-bandwidth product is considerably larger than the single-stage amplifier we examined before, the intercept point and three-tone performance does not reach the same values. As a general rule, the dynamic range is directly proportional to the power dissipated in the devices, specifically, the current. As explained earlier several times, the first-order nonlinearity comes from the voltage-to-current transfer characteristic, such as the transconductance itself and can only be improved by a combination of resistive feedback in the emitter/source and

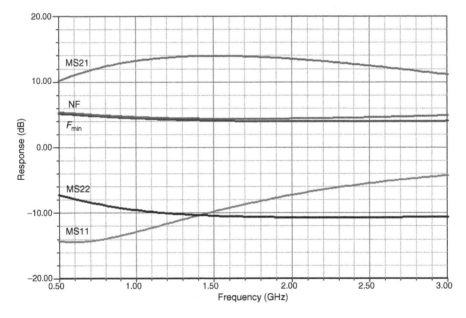

Figure 3.111 Frequency-dependent gain, matching, and noise-figure performance of the MGA-64135 MMIC amplifier.

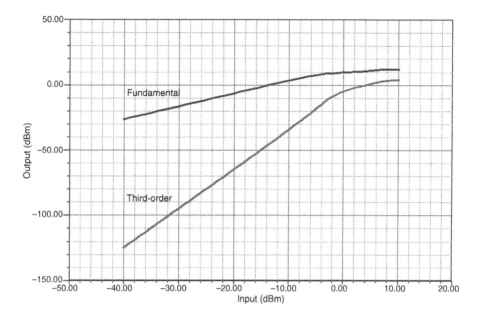

Figure 3.112 Simulated two-tone testing of the MGA-64135 MMIC amplifier. $IP_{3,in} = 8.5\,dBm$, $IP_{3,out} = 22\,dBm$.

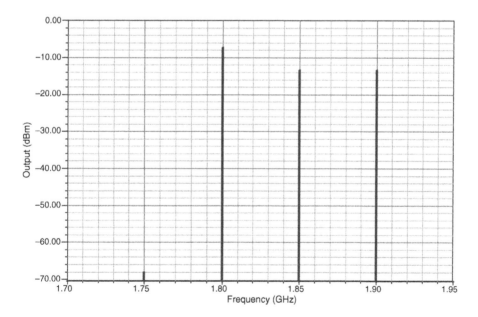

Figure 3.113 Simulated three-tone test of the MGA-64135 MMIC. The test tones were 1.80 GHz (–21 dBm), 1.85 GHz (–27 dBm), and 1.90 GHz (–27 dBm). The 1.75-GHz IMD product is down 60.5 dB relative to the 1.80-GHz signal.

by high currents. At these dc levels, the base–emitter (gate-source) junction gets linearized because the differential values are getting smaller compared to the parasitic loss resistances, including external dc/RF feedback. Once again, the voltage feedback reduces the voltage clipping but is initially more responsible for the input and output matching.

3.2.4 Low-Voltage Open-Collector Design [36–39][2]

Why R_C Acts Like a Source Resistor   An open-collector output allows a designer the flexibility to choose the value of the R_C resistor. Choosing this resistor value not only sets the dc bias point of the device but also defines the source impedance value. Figure 3.114 shows the ac model of the transistor. Converting the output structure by applying a Norton and Thevenin transformation, one can conclude that R_C becomes the source resistance. Thus, by choosing R_C to be equal to the load, maximum power transfer will then occur.

Figure 3.115 shows an active transistor with a collector and base resistor. From basic transistor theory, Equation A in Figure 3.115 is generated and has the same form as the general equation for a straight line ($y = mx + b$).

The slope of the dc load line is generated by the value of the collector resistor ($m = -1/R_C$) and is shown in Figure 3.116. For a given small-signal base current, the collector current is shown by the dotted curve.

The intersection of the dotted curve and the dc load line is called the *quiescent point* (*Q*-point) and dc bias determinant. The location of the *Q*-point is important because it determines whether the transistor is operated in the cutoff, active, or saturation region(s). In most cases, the *Q*-point should be in the active region because this is where the transistor acts like an amplifier.

Figure 3.117 shows the ac collector–emitter voltage (V_{CE}) output swing with respect to an ac collector current (I_C). Collector current is determined by the ac voltage presented to the input transistor's base (v_i) because it affects the base current (i_b), which then affects i_c. This is how the v_i is amplified and seen at the output. Recall that this is with no external load (R_{Load}) present at the collector. Since no load is present, the ac load line has an identical slope as the dc load line as seen in Equations A and B of Figure 3.115 ($m = -1/R_C$).

Open Collector with R_{Load}   A filter with some known input impedance is a typical load for the output of the transistor. For simplicity, we will assume a resistive load (R_{Load}) and neglect any reactance. Since this resistive load is used (see Figure 3.118), the ac output swing is measured at V_{Out} or V_{CE}.

A dc blocking capacitor is used between the R_{Load} and the V_{CE} output to assure that the *Q*-point is not influenced by R_{Load}. It is also necessary to avoid passing dc to the load in applications where the load is a SAW filter. However, R_{Load} will affect the ac load line, which is seen in Equation B in Figure 3.118. Note that the V_{CE} voltage swing is reduced; thus, V_{Out} signal is reduced (see Figure 3.119).

Since the value of R_C and R_{Load} affect the ac load line slope, the value chosen is important. The higher the impedance of R_{Load} and R_C, the greater the ac output swing will be at the output, which means more conversion gain in a mixer. This is due to the slope getting flatter, thus allowing for more output swing.

---

[2]Based on portions of the Philips Semiconductors/Signetics RF Communications Products Application Note AN1777, "Low-voltage front end circuits: SA601, SA620," August 20, 1997. Used with permission.

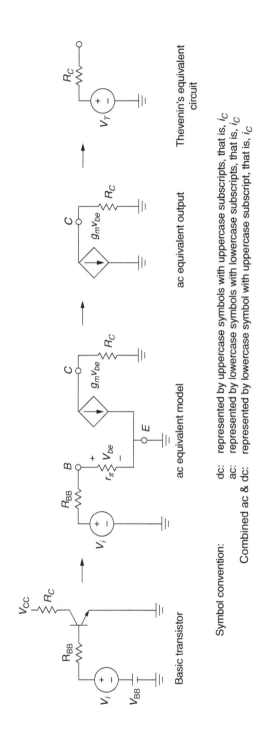

Basic transistor

ac equivalent model

ac equivalent output

Thevenin's equivalent circuit

Symbol convention:

dc: represented by uppercase symbols with uppercase subscripts, that is, i_C

ac: represented by lowercase symbols with lowercase subscripts, that is, i_c

Combined ac & dc: represented by lowercase symbol with uppercase subscript, that is, i_C

Figure 3.114 R_C as source resistor.

477

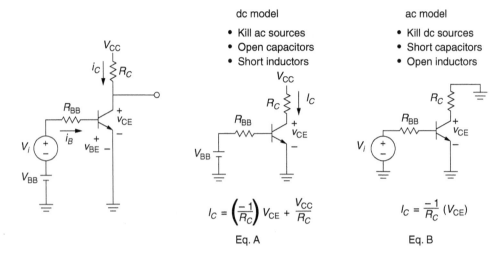

Figure 3.115 Basic transistor analysis.

Open-Collector with Inductor (L_C) Adding an inductor in parallel with R_C can increase the ac output signal V_{CE}. Figure 3.120 shows the dc and ac analysis of this circuit configuration. In Equation A of Figure 3.120, there is no R_C influence because the inductor acts like a short in the dc condition. This means the slope of the dc load is infinite and causes the Q-point to be centered around V_{CC}, thus moving it to the right of the curve. The ac load

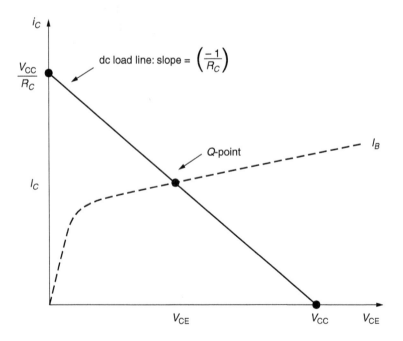

Figure 3.116 Load line and Q-point graph.

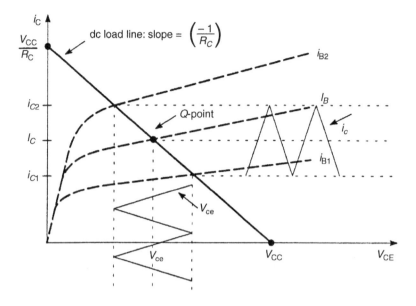

Figure 3.117   Graphical analysis for the circuitry in Figure 3.115.

line slope is set only by R_C because no load is present. Note that it has the same ac load line slope as the first condition in Equation B of Figure 3.115.

Referring to Figure 3.121, one might note that the base current (ac and dc) curves spread open as V_{CE} increases. This is caused by a noninfinite Early voltage (see Figure 3.122), which causes the collector current to be dependent on V_{CE}. Taking advantage of this nonideal condition, the peak-to-peak ac output swing V_{CE} can thereby be increased by moving the dc Q-point to the right due to the wider spreading between the curves corresponding to different base currents. Figure 3.123 combines Figures 3.117 and 3.121 to show the different ac output signals with different Q-points.

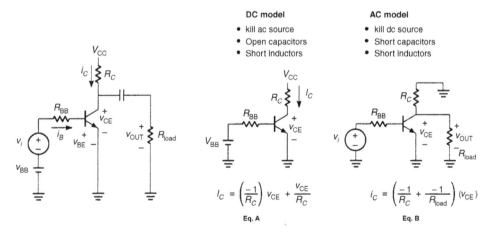

Figure 3.118   Basic transistor analysis with R_{load}.

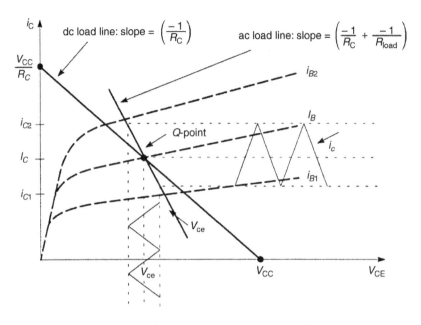

Figure 3.119 Graphical analysis for the circuitry in Figure 3.118.

Looking at the ac output level, one might ask how the V_{CE} peak voltage can exceed the supply voltage V_{CC}. Recall that the inductor is an energy-storing device ($v = L \, di/dt$). Therefore, total instantaneous voltage is V_{CC} plus the voltage contribution of the inductor.

Open-Collector with Inductor (L_C) and R_{Load}   Figure 3.124 shows the dc and ac analysis with the inductor and load resistor. Again, from the dc analysis, the inductor causes R_C to be nonexistence so the dc load line is vertical. In the ac analysis, the ac load-line slope is

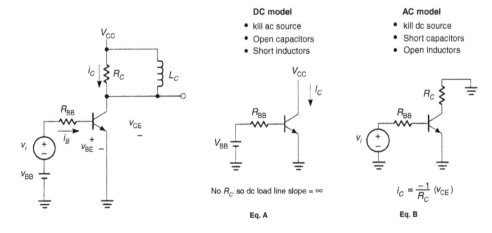

Figure 3.120 Basic transistor analysis with inductor added to collector.

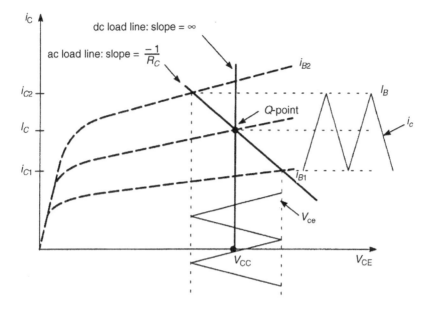

Figure 3.121 Graphical analysis for the circuitry in Figure 3.120.

influenced by both R_C and R_{Load} (see Equation B in Figure 3.124). The ac load-line slope is the same as the example of the open collector without the inductor. Figure 3.125 shows the response for the open collector with L_C and R_{Load}. Figure 3.126 combines Figures 3.119 and 3.125 showing the increase in ac output swing.

In conclusion, for the load line, R_C plays a role in setting up the bias-determined Q-point as well as the ac source impedance. However, when an inductor is placed in parallel with R_C, a different Q-point is set and the ac source impedance is altered. Moving the Q-point takes advantage of the transistor's nonideal i_c dependence on V_{CE} to get more signal output without having to change the base current. Since R_C is in parallel with R_{Load} in the ac condition, it influences the ac load line slope.

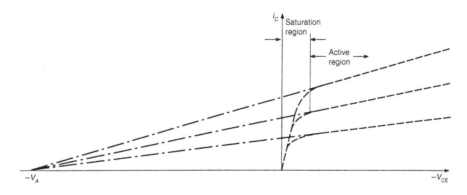

Figure 3.122 Graphical representation of Early voltage effect.

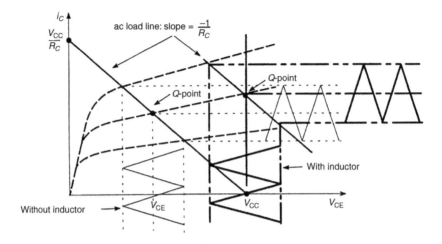

Figure 3.123 Comparison of open-collector circuit with inductor versus without inductor.

Flexible Matching Circuit [39] A useful variation of the open collector matching concepts previously outlined provided the capability of delivering equal power to two unequal resistive loads. This allows the power delivered to the load to be measured indirectly at another test point in the circuit where the impedance can be arbitrarily defined. If this impedance is defined to be 50 Ω, a spectrum analyzer can be easily placed directly into the circuit. This is an excellent troubleshooting technique and a valuable option to have available in high production environments.

Figure 3.127 shows the schematic for this flexible matching circuit. In this circuit, C_B functions only as a dc blocking capacitor and presents a negligible impedance at the frequency of interest. Recall from the previous open-collector matching discussions that, when R_C is placed in parallel with an inductor, it has no effect on the Q-point, but does influence

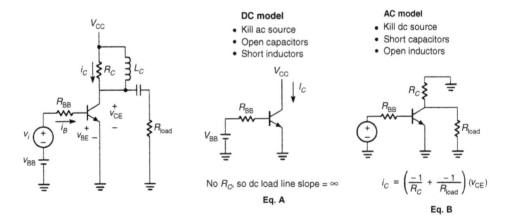

Figure 3.124 Basic transistor analysis with inductor and R_{Load}.

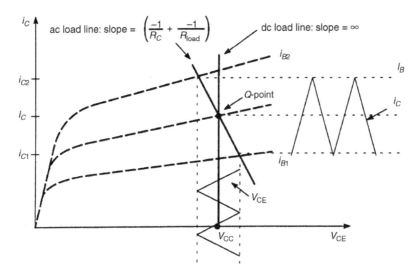

Figure 3.125 Graphical analysis for the circuitry in Figure 3.124.

the slope of the ac load line as

$$\text{Slope} = \frac{-1}{R_C} + \frac{-1}{R_{\text{Load}}} \tag{3.168}$$

The capacitor C_S functions not only as a dc blocking capacitor but also is chosen such that the impedance presented by the combination of L, R_C, and C_S is equal to R_{Load} for optimum power transfer. The analysis is done in the following manner. First, note that inductor L is connected to V_{CC}, which is an affective ac ground. So, L can be redrawn to ground. Next, R_C and C_S are converted to their parallel equivalent values as shown in Figure 3.128.

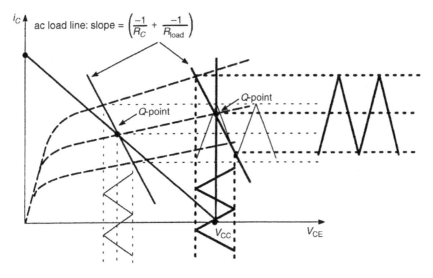

Figure 3.126 Comparison of open-collector R_{Load} circuit with inductor versus without inductor.

V_{CC}
dc blocking caps
L
C_S
A
R_C
C_B
B
R_{load}

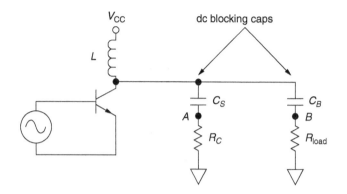

Figure 3.127 Flexible matching circuit.

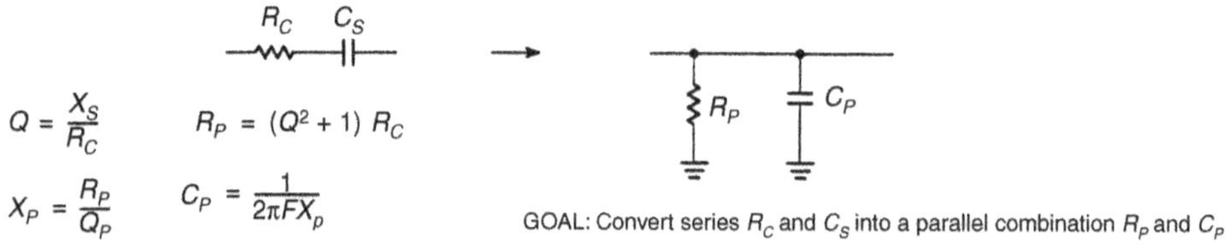

R_C C_S
$Q = \frac{X_S}{R_C}$
$R_P = (Q^2 + 1) R_C$
$X_P = \frac{R_P}{Q_P}$
$C_P = \frac{1}{2\pi F X_p}$
R_P
C_P
GOAL: Convert series R_c and C_s into a parallel combination R_p and C_p

Figure 3.128 Converting from series to parallel configuration.

GOAL: Find L_C and C_P such that R_P looks like R_{load} value at the resonal frequency
R_P
C_P
L

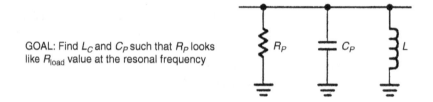

Figure 3.129 Converting from series to parallel configuration.

The resulting parallel LCR circuit is shown in Figure 3.129. At resonance, the parallel L, C_P combination will be an effective open circuit leaving only R_P. R_P is then simply chosen to be equal to R_{Load}.

Starting with the most simple transistor configuration and using R_C types of dc-coupled amplifier circuits, here is an overview of possible configurations for this application.

3.3 SINGLE-STAGE FEEDBACK AMPLIFIERS

This amplifier (Figure 3.130) is a standard minimal configuration amplifier with reduction of current distortion because of the emitter feedback. The voltage gain is

$$\text{VG} = \frac{R_L}{\frac{26\text{ mV}}{I_C} + R_e} \tag{3.169}$$

where R_L is the total load seen at the collector.

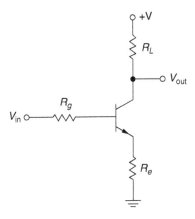

Figure 3.130 Single-stage amplifier with current feedback (provided by R_e).

The feedback amplifier in Figure 3.131 uses voltage feedback to "correct" the input and output impedance to be close to 50 Ω, but does not improve the current distortion. The voltage gain of this circuit is

$$VG = \frac{R_F}{R_g} \tag{3.170}$$

$$R_{in} = R_{out}$$
$$= R_F R_D \quad \text{with} \quad R_D = \frac{26\,\text{mV}}{I_C} \tag{3.171}$$

Example for $I_C = 10$ mA $\rightarrow R_D = 2.6\,\Omega$. Assuming $R_F = 200\,\Omega$, the input resistance $R_{in} = 22.63\,\Omega$. The gain is calculated by getting the inverse of $R_D = g_m = 0.39$ A/V. Once this is multiplied with the output impedance of 22.63 Ω, the resulting gain becomes 10 log (8.6) or 9.3 dB. The gain can be increased by either using a larger collector current or a larger feedback resistor. It should be reminded that this circuit does not improve the distortion caused by the base–emitter junction.

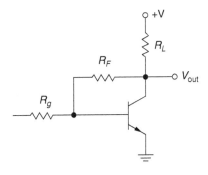

Figure 3.131 Single-stage amplifier with voltage feedback (provided by R_F).

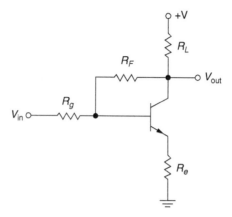

Figure 3.132 Single-stage amplifier with voltage and current feedback.

To reduce the current distortion as well, we use the circuit shown in Figure 3.132, which contains both types of feedback. In reality, the collector–base resistor is frequently split into an unbypassed section responsible for the feedback and a larger resistor responsible for the appropriate biasing. The example circuit, in Figure 3.133 is shown.

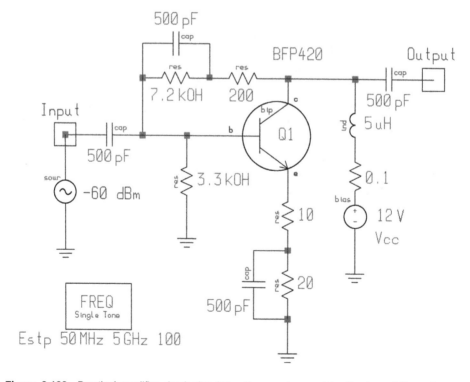

Figure 3.133 Practical amplifier circuit showing voltage and current feedback, and illustrating how reactances (in this case, 500-pF bypass capacitors) and multiple resistors can be used to provide different ac and dc resistances for the paths from collector to base and emitter to ground.

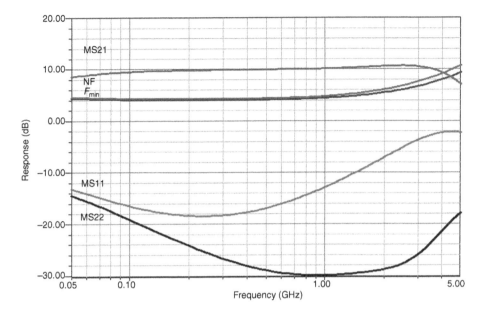

Figure 3.134 Frequency-dependent gain, matching, and noise performance of the circuit shown in Figure 3.133.

By using two of the BFP420 transistors in parallel, and balancing the transistors by using the appropriate emitter resistors (which, in total, will result in 10 Ω) plus the by-passed value, we obtain the frequency-dependent gain, matching, and noise responses shown in Figure 3.134. To determine the appropriate feedback resistor R_F, we recommend the formula

$$R_F = \frac{(Z_{\text{in}} Z_{\text{out}})}{R_E} \tag{3.172}$$

This equation requires some additional comments. The combined shunt and series feedback only work as long as a sufficient transconductance is provided. In other words, there must be enough open-loop gain that will be reduced to obtain more bandwidth. A necessary condition for this approach, which really comes from the frequency range of less than 30 MHz, is

$$Z_{\text{in}} = \frac{26\,\text{mV}}{I_C}\beta \gg R_F \tag{3.173}$$

This requires a minimum transconductance of

$$g_m = \frac{1 + S_{21}}{Z_{\text{in}}} \tag{3.174}$$

where S_{21} must be a negative number that reflects the 180° phase shift. Of course, life becomes interesting when S_{21} goes through 1, but it (hopefully) never becomes negative and larger than 1 at the same time.

Example: The BFP420 has been used here frequently. A 10-dB gain would require an S_{21} of 3.16 ($\sqrt{10}$); therefore, we need to inspect the datasheet that shows that the magnitude of S_{21} is 3.16 around 4.5 GHz, yet the phase shift has changed from 180° to around 20°. In order to design an amplifier, we now set

$$R_F = Z_{\text{in}} (1 + |S_{21}|) \tag{3.175}$$

which is valid until S_{21} becomes too small. For a given 10 dB or 3.16 for S_{21}, the resulting RF is 4.16 × 50, or 208 Ω. The transconductance under these conditions becomes

$$g_m = \frac{1 + |S_{21}|}{Z_0} \tag{3.176}$$

where $Z_0 = 50\,\Omega$ (the desired input impedance). g_m now becomes 83.2 mS. We now test the input and find

$$\frac{\beta}{g_m} = Z_{\text{in}} = \frac{50}{0.0832} = 600\,\Omega \tag{3.177}$$

Therefore, the emitter feedback resistor is not necessary to meet the previously mentioned condition whereby the input impedance needs to be higher than R_F. Assuming the transconductance now is 0.00832, or 10% of the previous value, the input impedance would be 60 Ω, no longer meeting the prescribed requirement. Therefore, we need an unbypassed emitter resistor that fulfills the equation

$$R_E = \frac{Z_0^2}{R_F} - \frac{1}{g_{m_{dc}}} \tag{3.178}$$

where

$$g_{m_{dc}} = \frac{1 - S_{21_{\max}}}{Z_0} \tag{3.179}$$

Again the reader should be cautioned that in the common-emitter circuit, there is 180° phase shift, and therefore $S_{21,\max}$ is a negative quantity, resulting in reality in

$$g_{m_{dc}} = \frac{1 + S_{21,\max}}{Z_0}$$

In our case,

$$g_{m_{dc}} = \frac{-S_{21}}{2Z_0} = \frac{38}{2 \times 50} = \frac{38}{100} = 0.38$$

so

$$R_E = \frac{50^2}{208} - \frac{1}{0.38} = 12 - 2.6 = 9.4\,\Omega \tag{3.180}$$

In practice, the designer would use a 10-Ω resistor.

Since the gain of the transistor continues to fall as a function of frequency, one can compensate to a degree the cutoff frequency by putting an inductance in series that follows the equation

$$L_F = \frac{R_F}{2\pi f_{3\,\text{dB}}} \tag{3.181}$$

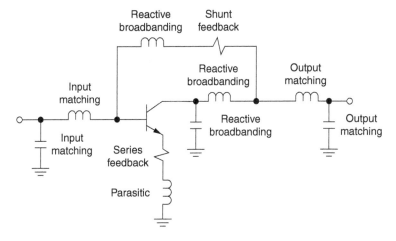

Figure 3.135 Elements of a feedback amplifier.

Assuming $f_{3dB} = 2$ GHz, the resulting series inductance $L_F = 16.6$ nH. Basically, the same applies to circuits where the second transistor is used as an emitter follower to feed the signal back from the collector to the base of the first transistor.

In summary (see Figure 3.135):

- Resistive negative feedback at lower frequencies reduces open-loop gain and simultaneously matches intput and output.
- Reactive elements set gain and match at high frequencies. Feedback is reduced at high frequencies.
- Emitter bypass capacitors are not used, as they would destroy the impedance match.

This is equally valid for FETs.

The next level of complexity can be reached by using two transistors with dc feedback.

3.3.1 Lossless or Noiseless Feedback

All the amplifiers shown above are based on a combination of resistive and reactive feedback. The very moment resistive feedback is used, the noise figure increases, as does the intercept point. As long as the intercept point increases overproportionally compared to the noise figure, this is an acceptable method. However, there are many cases in which an extremely low noise figure and a higher than normally achievable intercept point are required. To avoid this degradation of performance, Dr. Norton of Anzac developed a negative feedback structure based on the capabilities of the ferrite transformer. This patented technique (U.S. Patent Nos. 3,624,536 and 3,891,934, since expired), known as *lossless feedback*, provides lower noise figure and higher linear output than can be achieved using resistive feedback with the same transistor. Transformer feedback is used in which the transformer in the feedback network acts as a directional coupler. The coupling ratio between the input and output of the coupler determines the magnitude of the feedback, and hence the gain. Across the frequency band over which the transformer operates as intended, the feedback is negative due to the in-phase coupling and the inversion of the active device.

This technique was used in amplifiers ranging from 5 to 500 MHz, 20 to 1000 MHz, and 300 to 1800 MHz. The difficulty associated with this type of amplifier is the fact that it is based on transformers that use twisted pairs of wires as transmission lines. In some implementations used at higher frequencies, actual directional couplers were used at the expense of bandwidth; in these circuits, the coupling was determined by the spacing of the coupler lines rather than transformer turns ratio. These microstrip couplers have much narrower bandwidths than transformers and are fairly large.

Another headache with the lossless feedback design is the fact that many of these transistor configurations are used either in grounded base or grounded emitter using transistors that nowadays have f_Ts of 25 GHz and higher. Such devices are very hot, and to make sure one obtains freedom from oscillation over a wide frequency range is a continuing problem, even for the circuits we have shown so far. A similar clever technique has been implemented in the termination-insensitive mixer, which we will examine in the mixer chapter.

3.3.2 Broadband Matching

The circuitry above has enabled us to extend the frequency range of the amplifier and avoid the normal sharp cutoff. According to Fano [40], there is a limit to the bandwidth of matching. The concept is that impedance plus bandwidth determines the ability to match, since there is a Q of the circuit and there is a Q of the device. The following equations are valid:

$$|\rho_{\text{Bode}}| = e^{-\left(\frac{\pi Q_T}{Q_{\text{BP}}}\right)} \tag{3.182}$$

$$Q_T = \frac{\sqrt{f_1 f_2}}{f_2 - f_1} \tag{3.183}$$

$$Q_{\text{BP}} = \frac{\text{Im}(Z)}{\text{Re}(Z)} = \frac{X_s}{R_s} \tag{3.184}$$

Figure 3.136 shows an example. This combination results in a VSWR of 3.43, which for practical purposes is already too high. The workaround is either special circuitry (such as exponentially staggered impedance jumps in the matching network like 50 Ω > 20 Ω > 7 Ω > 3 Ω). This means we first transform the 50-Ω source of load impedance down to 3 Ω and then look for a matching network that matches the transistor output/input that is within a few ohms of 3 Ω. A detailed discussion of such matching can be found in the *Numerical Design* portion of Section 3.12.1 under the title "Broadband Matching Using Bandpass-Filter Networks—High-Q Case."

3.4 TWO-STAGE AMPLIFIERS

The following two popular configurations have been mostly used for low-frequency applications. The first one (Figure 3.137, implemented in the circuit shown in Figure 3.138) shows a nice high-frequency performance assuming that the emitter and collector currents are essentially the same. As to distortion products, this circuit behaves equivalently to the shunt impedance example from Figures 3.132 and 3.133, with the additional gain

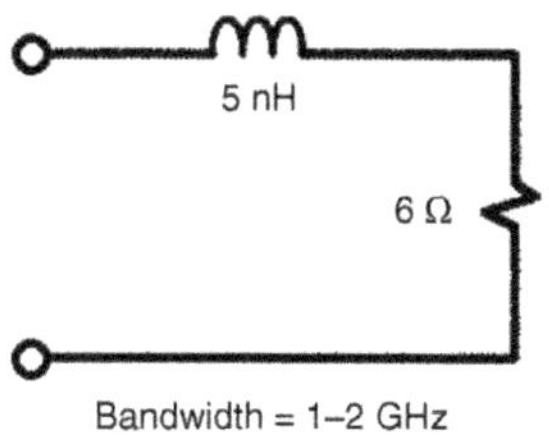

$$Q_{BP} = \frac{2 \pi \sqrt{(2 \text{ GHz}) (1 \text{ GHz}) (5 \text{ nH})}}{6 \, \Omega} = 7.40$$

5 nH

6 Ω

Bandwidth = 1–2 GHz

$$Q_T = \frac{\sqrt{(2 \text{ GHz}) (1 \text{ GHz})}}{(2 \text{ GHz}) - (1 \text{ GHz})} = 1.414$$

$$\left| \rho_{Bode} \right| = e^{-\left(\frac{\pi (1.414)}{7.40} \right)} = .549$$

$$VSWR_{Bode} = \frac{1 + .549}{1 - .549} = 3.43{:}1$$

Figure 3.136 Broadband matching example showing the Bode limit.

of the second transistor. Its frequency matching and noise performance are also shown in Figure 3.139.

The next two-stage circuit (Figure 3.140, implemented in Figure 3.141) uses the collector current of the second stage as the feedback to the first. Proper biasing of this circuit is quite difficult and interactive. Figure 3.142 shows the frequency-dependent gain, matching, and noise figure. We believe that this circuit can be optimized further, but leave such investigations to the reader.

A *cascode* amplifier is a special form of two-stage circuit. Although a cascaded pair acts like a single transistor, its merits are in its reduced feedback and result in higher gain-bandwidth product. Figure 3.143 shows a simple cascode application that assumes the

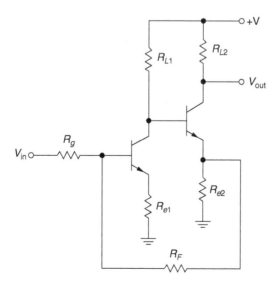

Figure 3.137 Two-stage amplifier with voltage feedback.

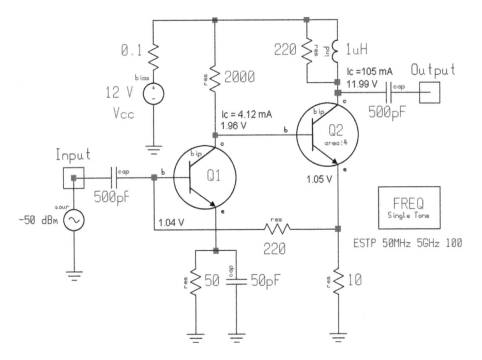

Figure 3.138   Implementation of the Figure 3.137 circuit.

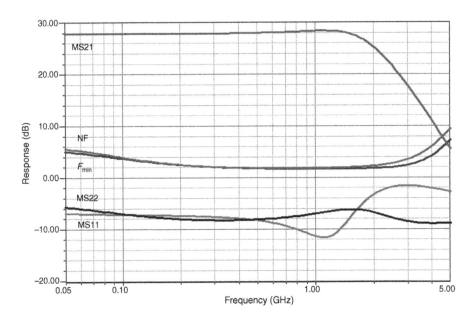

Figure 3.139   Frequency-dependent gain, matching, and noise performance of the circuit of Figure 3.138.

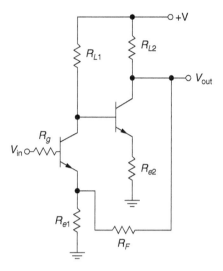

Figure 3.140 Two-stage amplifier with voltage feedback from Stage 2 to Stage 1, and current feedback in both stages.

presence of an output transformer that absorbs the output capacitance. Figure 3.144 shows this circuit's frequency-dependent gain, matching, and noise performance.

It has been obvious to us that most published wireless application circuits are evaluated on a no-input-selectivity basis; that is, their reported performance does not take into account

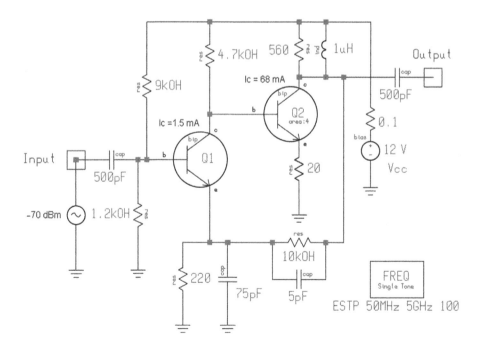

Figure 3.141 Simulated two-stage amplifier with feedback as shown in Figure 3.140.

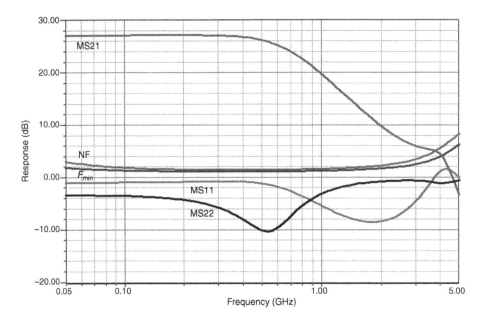

Figure 3.142   Gain, matching, and noise performance of the amplifier shown in Figure 3.141.

the filtering that is usually present in practical implementations. Evaluated in this way, a low-noise design may achieve a noise figure of less than 2 dB—1.2 dB or so is typical— but this is misleading because filters and connectors introduce insertion loss, resulting in a higher noise figure.

The cascode is a nice preamplifier combination. It combines (depending on the transistor) a low-noise figure with essentially no feedback. Since the output of a grounded-base transistor provides a high-impedance source, this configuration is ideal to work into the SAW filters frequently used in wireless applications at the operating frequency. Special transistor pairs are available for use in cascode; as can be seen from this example, not every transistor combination is ideal. This one gives a higher noise figure than wanted and at frequencies above 2 GHz tends to show possible instabilities. Both Infineon and Freescale are offering transistors appropriate for these cascodes. Mostly one would resort to transistors of less-exotic cutoff frequencies to guarantee stability. Typically, grounded-base/gate transistors tend to become oscillators, specifically because this configuration, like the emitter follower, allows a higher operating frequency while maintaining the same gain-bandwidth product, the actual gain for the last two circuit configurations is less. A grounded-base/gate stage has a high voltage gain and the emitter/source follower has power gain at low impedance levels. We invite our readers to experiment with the capacitor from the base of Q_1 to ground in Figure 3.143 to get an interesting education about these transistors and their willingness to oscillate.

The following is an example of a commercially available VHF to 2 GHz low-noise amplifier. As can be seen from Table 3.13, its use is ideal for the 900-MHz range providing good linearity and low noise.

This design is based on the cascode mentioned above, and because of high integration, uses many transistors for bias and level shifting.

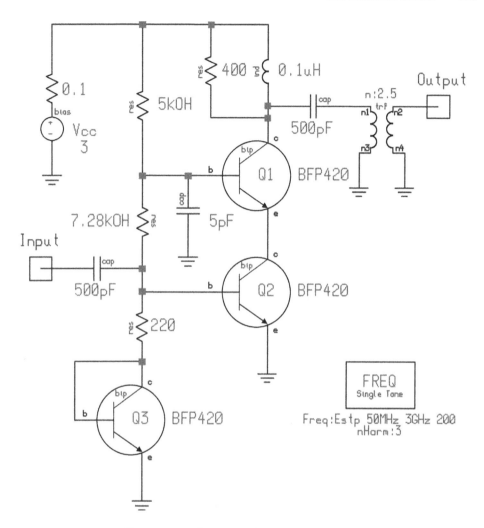

Figure 3.143 Schematic of the cascode amplifier.

Table 3.13 Freescale MBC13916 Features

- Usable frequency range 100–2500 MHz
- Small-signal gain 19 dB at 900 MHz and $V_{CC} = 2.7$ V
- $NF_{min} = 0.9$ dB at 900 MHz; 1.9 dB at 1.9 GHz
- P_{1dB} 2.5 dBm at 900 MHz and $V_{CC} = 2.7$ V
- OIP_3 16.5 dBm at 900 MHz
- Bias current 4.7 mA at $V_{CC} = 2.7$ V
- Supply voltage 2.7–5.0 V
- Device weight typically 0.00642 g, industry standard SOT-343R package

Source: Copyright of Freescale, Inc. 2012, used with permission.

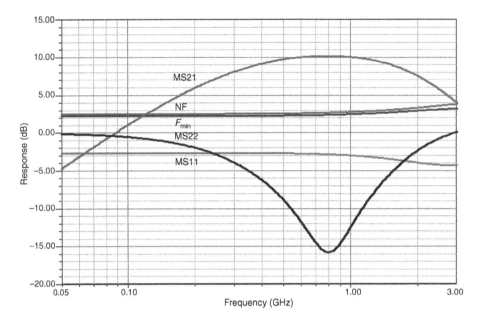

Figure 3.144 Frequency-dependent gain, matching, and noise performance of the cascode amplifier.

The 900-MHz application circuit in Figure 3.145 shows the required external components for a successful design. The matching circuit results in a slight degradation of the noise figure by 0.3 dB. A band-selection filter would be placed at the output of the amplifier, since it would deter the noise figure too much. The antenna is rarely in the vicinity of such strong transmitters so that such a strong input filter is required.

There are two more interesting applications for two-stage amplifiers. The one implemented in the Hewlett-Packard MSA-0735 MMIC has become a quite common one, and its usefulness up to higher frequencies depends mostly on the cutoff frequency of the transistors, their parasitics, and phase shift of the gain. The typical configuration is shown in Figure 3.146, and its frequency-dependent gain, match, and noise figure is shown in Figure 3.147. As we see an increase of S_{22} at higher frequencies, this indicates unwanted phase shift in the circuit. Some manufacturers hide the compensation circuits to overcome such peaking; most of the schematics shown by the manufacturers are really intended only to show operating principles, with proprietary details omitted.

Thanks to the properties of GaAs transistors, we can extend the frequency range significantly. The amplifier we are about to look at goes back several years ago and is described in great detail in Ref. [41]. Essentially, what we have as can be seen in Figure 3.148 is an FET version of Figure 3.141. The main goal of this circuit was to have an all-monolithic circuit, avoiding all ac coupling and accomplishing all matching and interaction with dc coupling. The transistor Q_1 is the main gain stage, which has the transistor Q_2 as an active load. The output impedance of Q_2 is $1/g_m$—typically, in the area of 50–200 Ω, depending on the biasing, which affects the transconductance. Transistor Q_3 has the same function as an emitter/source follower; again, its dc bias determines its transconductance and, therefore, its output impedance. (Given this, a device that has a transconductance of 20 mS, or close to it, would establish a very good output match to 50 Ω.) The drains of Q_1 and Q_5 are at

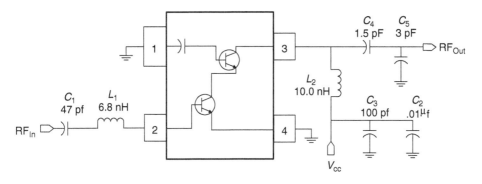

Electrical Characteristics for $V_{CC} = 2.7$ V $T_A = 25°C$, $f = 900$ MHz

Small signal gain	19 dB
Noise figure	1.25,dB
Reverse isolation	−42 dB
P_{1dB}	2.5 dBm
OIP_3	11 dBm

Bill of Materials

Component	Value	Case	Manufacturer	Comments
C_1	47 pF	0402	Murata	DC block input math
C_2	0.01 µs	0402	Murata	Low frequency bypass to improve IP3
C_3	100 pF	0402	Murata	RF bypass
C_4	1.5 pF	0402	Murata	DC block, output math
C_5	3.0 pF	0402	Murata	Output match, S_{22} improvement
L_1	6.8 nH	0402	Murata	input math
L_2	10 nH	0402	Toko	DC feedthrough, output math
L_3	5.6 nH	0402	Toko	Output math
Q_1	MBC13916	SOT343R	Freescale	SiGe cascode amp

Figure 3.145 A Freescale MBC13916 application circuit for 900 MHz, with typical performance and bill of materials. Copyright of Freescale, Inc. 2012, Used by Permission.

approximately 3.5 V. The diodes at the source of Q_3 are level shifters that must shift the dc voltage at the gate of Q_5 and at the drain of Q_4 to approximately −1.5 V relative to ground—the necessary bias condition for Q_1 and Q_5. The standard shunt feedback circuit would show resistive feedback between the drain of Q_1 (gate of Q_2) and the gate of Q_1; in the previous examples, we calculated the magic value of this part to be about 200 Ω. Adding in parallel to Q_1 the transistor Q_5 allows the designer to use a feedback scheme that isolates the feedback loop from the input. DC-wise, the two transistors are tied together via a 10-kΩ resistor, which can be included in the actual packaged device. Insufficient information was published to allow us to duplicate the design, considering all the intricacies involved (we somewhat object to the fact that even in the *IEEE Trans. Microwave Theory Tech.*, so little information is given about the actual implementation of this design—using space constraints as an excuse—that its educational value is usually minimal); it shows a new principle without revealing the design steps necessary to obtain a practical circuit.

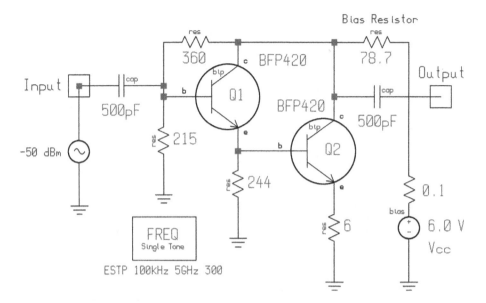

Figure 3.146   Schematic of the MSA-0735 MMIC amplifier. Copyright of Freescale, Inc. 2012, used with permission.

Several variations of this type of circuit are available; the literature is full of discussions about them. The magnitude of the feedback is set by the ratio of the width of Q_5 to that of Q_1. Typical values for the Q_5/Q_1 ratio range from 0.15–0.30. In simulation, sizing the devices can best be accomplished by using the scaling factor for the FET model.

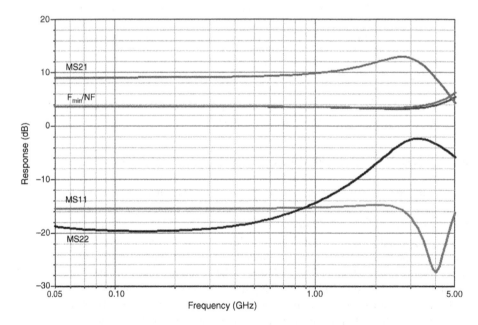

Figure 3.147   Frequency-dependent gain, matching, and noise performance of the MSA-0735 MMIC amplifier.

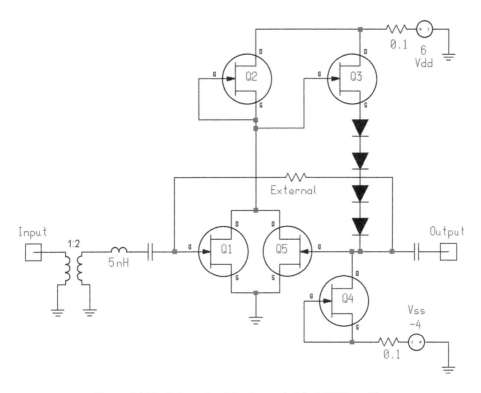

Figure 3.148 Schematic of the dc-coupled GaAsFET amplifier.

One of our questions was, "What would the perfect input termination for Q_1 be?" By using an ideal transformer and playing with its turns ratio (needless to say, we did this with CAD), we determined experimentally that relative to the gate, 200 Ω and a series inductor of 5–8 nH—the sort of details that designers often avoid telling their audiences—gave the ideal frequency response. Overall, we believe that the inventor of this circuit can be quite proud of the overall performance. The FETs we chose were taken from the Texas Instruments nonlinear foundry library; however, any similar device from another company would have worked equally well. The feedback loop used in this IC is an example of an approach frequently referred to as *active feedback*. The performance of this amplifier can be evaluated from Figure 3.149, which shows the frequency-dependent gain, matching, and noise figures (even including F_{min}).

3.5 AMPLIFIERS WITH THREE OR MORE STAGES

Most three-stage amplifiers are obtained by adding another buffer or power-gain stage to the above-shown principle designs. As an example, we decided to evaluate the recent Renesas Electronics μPC2749TB (uPC2749TB) IC, a silicon MMIC intended for application as a low-noise 1900-MHz amplifier operating at 3 V. Table 3.14 brief summarizes its electrical specifications.

In modeling the Renesas μ PC2749TB's three transistors, we chose to use our favorite, the BFP420, which, as we already pointed out, is not without headaches at the higher frequencies. Figure 3.150 shows the circuit diagram, which closely resembles the one pub-

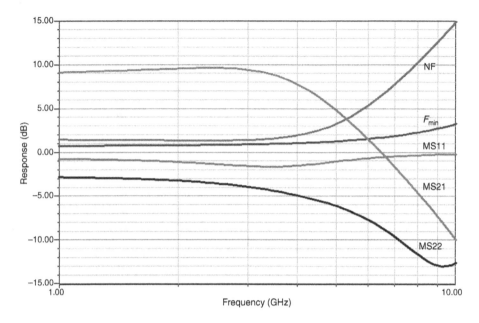

Figure 3.149 Frequency-dependent gain, matching, and noise performance of the dc-coupled GaAs-FET amplifier.

lished by Renesas; again, the only hard facts we had were the IC's basic circuit topology, gain, and dc current. Given those numbers, we staggered the current in the three devices by setting them at 0.64 mA for the first stage, 1.85 mA for the second stage, and 3 mA for the third stage. It should be noted that the two output transistors are not unlike the approach used for the Hewlett-Packard MSA-0735 (Figure 3.146). Taking (hopefully) all the right assumptions, we end up with a design that closely resembles the manufacturer's specifications for both power consumption, gain, and noise. This can be verified by looking at the results plotted in Figure 3.151.

Table 3.14 Renesas μPC2749TB Electrical Characteristics

		$T_A = +25°C$, $V_{CC} = 3.0\,\text{V}$, $Z_S = Z_L = 50\,\Omega$		
Symbol	Parameter	Test Conditions	Unit	Typical
I_{CC}	Circuit current	No signal	mA	6.0
G_P	Small-signal gain	$f = 1.9\,\text{GHz}$	dB	16.0
P_{sat}	Saturated output power	$f = 1.9\,\text{GHz}$, $P_{\text{in}} = -6\,\text{dBm}$	dBm	−6.0
NF	Noise figure	$f = 1.9\,\text{GHz}$	dB	4.0
		$f = 0.9\,\text{GHz}$		3.2
f_{corner}	Cutoff frequency	3 dB down below flat gain	GHz	2.9
ISOL	Isolation	$f = 1.9\,\text{GHz}$	dB	30
RL_{in}	Input return loss	$f = 1.9\,\text{GHz}$	dB	10
RL_{out}	Output return loss	$f = 1.9\,\text{GHz}$	dB	12.5
OIP_3	Output third-order intercept	$f_1 = 1.900\,\text{GHz}$, $f_2 = 1.902\,\text{GHz}$	dBc	−33

Source: Copyright by Renesas Electronics.

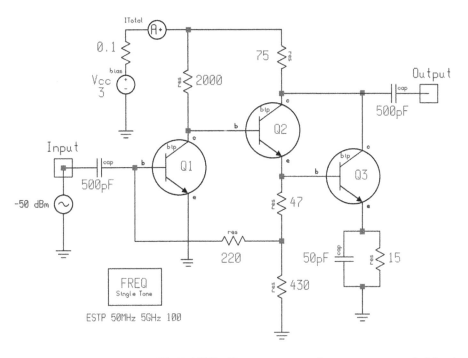

Figure 3.150 Circuit of the μPC2749 MMIC. The component values were not supplied by the manufacturer.

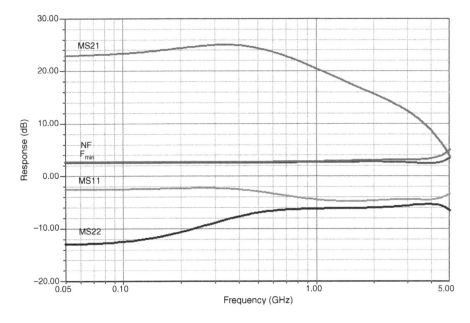

Figure 3.151 Simulated gain, matching, and noise performance of the Renesas μPC2749 MMIC.

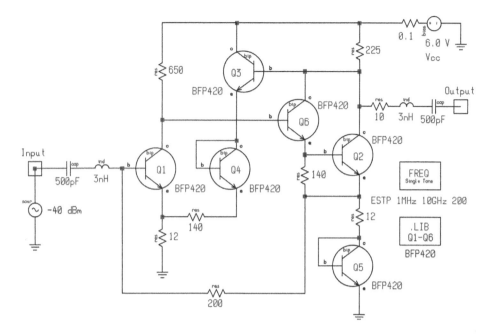

Figure 3.152 Schematic of the Philips NE/SA5204A amplifier IC entered for simulation.

The final example is a more complex derivative of the Philips NE/SA5204A amplifier, which was intended for standard 6-V operation and has a fairly "normal" circuit arrangement. This can be seen from Figure 3.152. A much more elaborate cousin of this is the low-voltage Renesas μPC2710 amplifier. According to its specifications, it operates at 5 V but has significantly more gain than the NE/SA5204. Table 3.15 shows its electrical specifications and Figure 3.153 shows its measured gain as published by the manufacturer.

The actual interior of the circuit is shown in Figure 3.154. It consists of the standard preamplifier and then a Darlington configuration with the collectors tied together. The middle section of the circuit acts as a power supply responsible for the bias relative to the

Table 3.15 Renesas μPC2710TB Electrical Characteristics

		$T_A = +25°C$, $V_{CC} = V_{out}$ 5.0 V, $Z_S = Z_L = 50\,\Omega$		
Symbol	Parameter	Test Conditions	Unit	Typical
I_{CC}	Circuit current	No signal	mA	22
G_P	Small-signal gain	$f = 0.5\,GHz$	dB	33
P_{sat}	Saturated output power	$f = 0.5\,GHz$, $P_{in} = -8$ dBm	dBm	13.5
NF	Noise figure	$f = 0.5\,GHz$	dB	3.5
f_{corner}	Cutoff frequency	3 dB down below flat gain	GHz	1.0
ISOL	Isolation	$f = 0.5\,GHz$	dB	39
RL_{in}	Input return loss	$f = 0.5\,GHz$	dB	6
RL_{out}	Output return loss	$f = 0.5\,GHz$	dB	12
ΔG_P	Gain flatness	$f_1 = 0.1–0.6\,GHz$	dB	±0.8

Source: Copyright by Renesas Electronics.

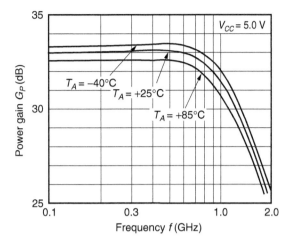

Figure 3.153 Manufacturer-supplied gain versus frequency for the Renesas μPC2710TB MMIC. Copyright Renesas Electronics.

emitter of the first transistor influencing the rest of the circuit as well. Since NEC (now Renesas) mentioned that they use their NESAT III process (f_T of 20 GHz at $V_{CE} = 3$) in producing the μPC2710T, and obtaining some information regarding it, we decided to use our standard BFP420 (f_T of 25 GHz; see the datasheet in Chapter 2) for the purpose of simulation.

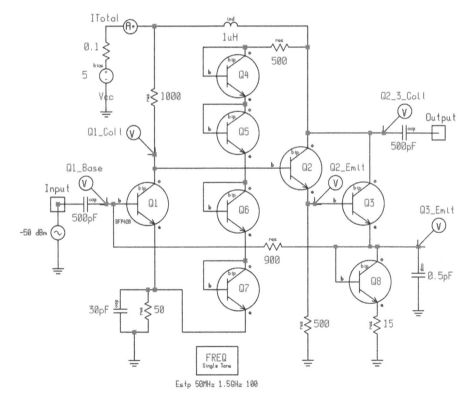

Figure 3.154 Schematic of the μPC2710T silicon MMIC entered for simulation.

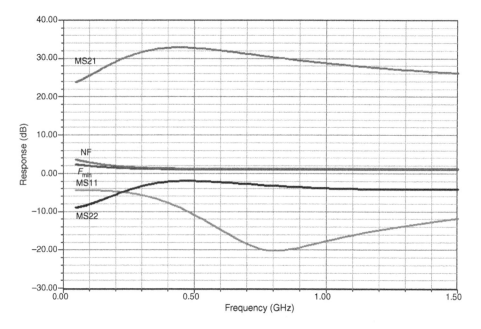

Figure 3.155 Simulated gain, match, and noise performance of the μPC2710T.

In our effort to simulate this circuit, we chose a combination of values that pretty much complies with the manufacturer's published specifications, but S_{22} is not close, as shown in Figure 3.155. We stopped at this point and invite our readers with CAD capabilities to provide the final touch in meeting all the published specifications for the circuit. We are particularly recommending such an effort because it very nicely shows the interaction of all the circuit elements in a fairly drastic way. Because of this, we admire the IC design even more, since the manufacturer has to sell a pretty consistent amplifier as a function of different production runs.

3.5.1 Stability of Multistage Amplifiers

Stability, needless to say, is a big issue, and the delay of the various stages adds to the phase shift, resulting in possible instabilities. The fact that S_{11} and/or S_{22} can become larger than 0dBm means there is the potential of oscillating, depending on the reactances available. If a cable of inappropriate length is added, hell will break loose. Figure 3.156 shows the dangerous peaking of our simulation of the Philips NE/SA5204A, in which the internal compensation was intentionally not assumed correctly. We will all agree that the high end performance of this amplifier can generate high anxiety. As the production process of the aging device becomes more modern—meaning that transistors with much higher f_T will be the basis for the production—it will become quite difficult to maintain a previously achieved stability. This, by the way, applies also to discrete transistors, which for reasons of economics and multiple manufacturers will be improved over time. Such evolution results in generally smaller base-spreading resistances and capacitances, and other parameter changes that, overall, cause headaches in manufacturing. We will leave the topic of three-stage amplifiers on this note.

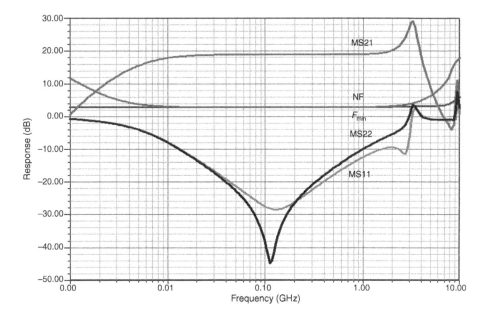

Figure 3.156   Simulated gain, match, and noise performance of the NE5204A IC.

3.6 A NOVEL APPROACH TO VOLTAGE-CONTROLLED TUNED FILTERS INCLUDING CAD VALIDATION [42]

Modern receivers control input stages as well as the oscillator band and frequency by electrical rather than mechanical means. Tuning is accomplished by voltage-sensitive capacitors (varactor diodes) and band switching by diodes with low forward conductance. Since the wireless band (essentially 400 MHz to 2.4 GHz) is so full of strong signals, the use of a tracking filter is desired as a solution to improve the performance and prevent second-order IMD products or other undesired overload effects. The dc control voltage needed for the filter can easily be derived from the VCO control voltage. There may be a small dc offset, depending on the IF used.

3.6.1 Diode Performance

The capacitance versus voltage curves of a varactor diode depend on the variation of the impurity density with the distance from the junction. When the distribution is constant, there is an "abrupt junction" and capacitance follows the law

$$C = \frac{K}{(V_d + V)^{1/2}} \tag{3.185}$$

where V_d is the contact potential of the diode and V is the applied voltage.

Such a junction is well approximated by an alloyed junction diode. Other impurity distribution profiles give rise to other variations, and the above equation is usually modified to

$$C = \frac{K}{(V_d + V)^n} \tag{3.186}$$

where n depends on the diffusion profile and $C_0 = K/V_d^n$.

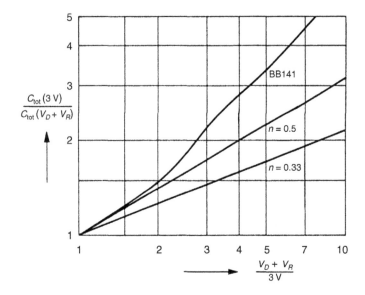

Figure 3.157 Voltage-dependent change of capacitance of different types of diodes. The BB141 is a hyperabrupt diode with $n = 0.75$.

A so-called graded junction, having a linear decrease in impurity density with the distance from the junction, has a value of n. This is approximated in a diffused junction.

In all cases, these are theoretical equations, and limitations on the control of the impurity pattern can result in a curve that does not have such a simple expression. In such a case, the coefficient n is thought of as varying with voltage. If the impurity density increases away from the junction, a value of n higher than 0.5 can be obtained. Such junctions are called hyperabrupt. A typical value for n for a hyperabrupt junction is about 0.75. Such capacitors are used primarily to achieve a large tuning range for a given voltage change. Figure 3.157 shows the capacitance-voltage variation for the abrupt and graded junctions as well as for a particular hyperabrupt junction diode. Varactor diodes are available from a number of manufacturers, such as Freescale, Infineon, and NXP. Maximum values range from a few to several hundred picofarads, and useful capacitance ratios range from about 5 to 15.

Figure 3.158 shows three typical circuits that are used with varactor tuning diodes. In all cases, the voltage is applied through a large resistor R_e or, better, an RF choke in series with a small resistor. The resistance is shunted across the lower diode and may be converted to a shunt load resistor across the inductance to estimate Q. The diode also has losses that

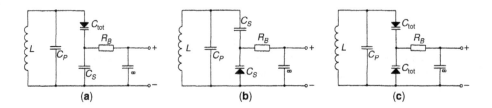

Figure 3.158 Various configurations to apply tuning diodes in a tuned circuit. Version (c) shows the lowest distortion.

may result in lowering the circuit Q at high capacitance. When the frequency is sufficiently high. This must be considered in the circuit design.

The frequent-dependent performance is not only determined by applying the dc tuning voltage to (3.186). If the RF voltage is sufficient to drive the diode into conduction on peaks, an average current will flow in the circuits of Figure 3.158, which will increase the bias voltage. The current is impulsive, giving rise to various harmonics of the circuit. Even in the absence of conduction, (3.186) deals only with the small-signal capacitance. When the RF voltage varies over a relatively large range, the capacitance changes. In this case, (3.186) must be changed to

$$\frac{dQ}{dV} = \frac{K}{(V + V_d)^n} \tag{3.187}$$

Here Q is the charge on the capacitor. When this relation is substituted in the circuit differential equation, it produces a nonlinear differential equation, dependent on the parameter n. Thus, the varactor may generate direct current and harmonics of the fundamental frequency. Unless the diodes are driven into conduction at some point in the cycle, the direct current must remain zero.

The current of Figure 3.158c can be shown to eliminate the even harmonics, and permits a substantially larger RF voltage without conduction than either circuit in Figure 3.158a or b. When $n = 0.5$, only second harmonic is generated by the capacitor, and this can be eliminated by the back-to-back connection of the diode pair. It has, integrating (3.187)

$$\begin{aligned} Q + Q_A &= \frac{K}{1 - n} (V + V_d)^{1-n} \\ &= \frac{C_v}{1 - n} (V + V_d) \end{aligned} \tag{3.188}$$

C_V is the value of (3.186) for applied voltage V, and Q_A is a constant of integration. By letting $V = V_1 + v$ and $Q = Q_1 + q$, where the lowercase letters represent the varying RF and the uppercase letters indicate the values of bias when RF is absent, thus follows

$$q + Q_1 + Q_A = \frac{K}{1 - n} [v + (V_1 + V_d)]^{1-n} \tag{3.189}$$

$$1 + \frac{v}{V'} = \left(1 + \frac{q}{Q'}\right)^{1/(1-n)} \tag{3.190}$$

where $V' = V_1 + V_d$ and $Q' = Q_1 + Q_A$. For the back-to-back connection of identical diodes, $K_{11} = K_{12} = K_1$, $V_1' = V_2' = V'$, $Q_1' = Q_2' = Q'$, $q = q_1 = -q_2$, and $v = v_1 - v_2$. Here, the new subscripts 1 and 2 refer to the top and bottom diodes, respectively, and v and q are the RF voltage across and the charge transferred through the pair in series. This notation obtains

$$\frac{v}{V'} \equiv \frac{v_1 - v_2}{V'} = \left(1 - \frac{q}{Q'}\right)^{1/(1-n)} - \left(1 - \frac{q}{Q'}\right)^{1/(1-n)} \tag{3.191}$$

For all n, this eliminates the even powers of q, hence even harmonics. This can be shown by expanding (3.191) in a series and performing term-by-term combination of the equal powers of q. In the particular case, $n = 1/2$, $v/V' = 4q/Q'$, and the circuit becomes linear.

The equations hold as long as the absolute value of v_1/V' is less than unity, so that there is no conduction. At the point of conduction, the total value of v/V' may be calculated by noticing that when $v_1/V' = 1$, $q/Q' = -1$, so $q_2/Q' = 1$, $v_2/V = 3$, and $v/V = -4$. The single-diode circuits conduct at $v/V' = -1$, so the peak RF voltage should not exceed this. The back-to-back configuration can provide a fourfold increase in RF voltage handling over the single diode. For all values of n, the back-to-back configuration allows an increase in the peak-to-peak voltage without conduction. For some hyperabrupt values of n, such that $1/(1-n)$ is an integer, many of the higher-order odd harmonics are eliminated, although only $n = 1/2$ provides elimination of the third harmonic. For example, $n = 2/3$ results in $1/(1 - n) = 3$. The fifth harmonic and higher odd harmonics are eliminated, and the peak-to-peak RF without conduction is increased eightfold; for $n = 3/4$, the harmonics 7 and above are eliminated, and the RF peak is increased 16 times. It must be noted in these cases that the RF peak at the fundamental may not increase so much, since the RF voltage includes the harmonic voltages.

Since the equations are only approximate, not all harmonics are eliminated, and the RF voltage at conduction, for the back-to-back circuit, may be different than predicted. For example, abrupt junction diodes tend to have n of about 0.46–0.48 rather than exactly 0.5. Hyperabrupt junctions tend to have substantial changes in n with voltage. The diode illustrated in Figure 3.157 shows a variation from about 0.6 at low bias to about 0.9 at higher voltages, with wiggles from 0.67 to 1.1 in the midrange. The value of V_d for varactor diodes tends to be in the vicinity of 0.7 V.

3.6.2 A VHF Example

The application of tuning diodes in double-tuned circuits has been found in TV tuners for many years. Figure 3.159 shows the circuit diagram.

The input impedance of 50 Ω gets transformed up to 10 kΩ. The tuned circuits consist of the 0.3-μH inductor and two sets of antiparallel diodes. By dividing the RF current in the tuned circuit and using several diodes instead of just one pair, intermodulation distortion is reduced.

The coupling between the two tuned circuits is tuned via the 6-nH inductor that is common to both circuits. This type of inductance is usually printed on the circuit board. The diode parameters used for this application were equivalent to the Infineon BB515 diode. The frequency response of this circuit is shown in Figure 3.160.

The coupling is less than critical. This results in an insertion loss of about 2 dB and a relatively steep passband sides. Once the circuit's large-signal performance (Figure 3.161) is seen, a third-order intercept point of about −2 dBm is not so unexpected. The reason for this poor performance is the high impedance (high L/C ratio), which provides a large RF voltage swing across the diodes. A better approach appears to be using even more diodes and at the same time changing the impedance ratio (L/C ratio).

3.6.3 An HF/VHF Voltage-Controlled Filter

The above example used a step-up procedure typically done by a tap at the input and output inductance. This method allows an impedance transformation; however, if one desires to change it into a series-tuned arrangement, it has to be done with a transformer. The large-signal conditions in the frequency range from 10 to 30 MHz on a medium- to large-sized antenna is equivalent to, if not worse than, the conditions for VHF operation. The only

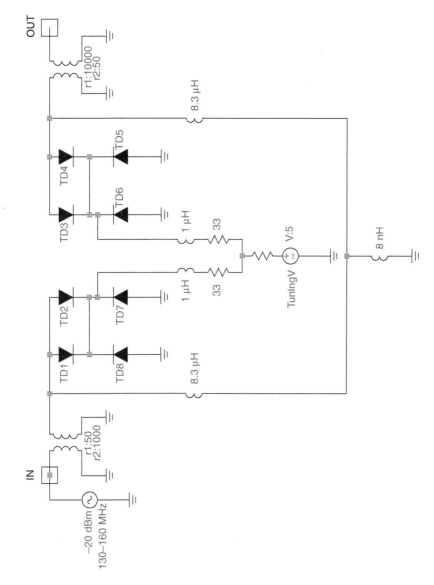

Figure 3.159 Double-tuned filter at 161 MHz using hyperabrupt tuning diodes. By using several parallel diodes, the IMD performance improves.

509

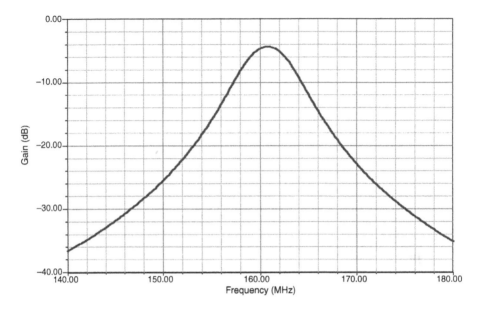

Figure 3.160   Frequency response of the tuned filter shown in Figure 3.159. The circuit is undercoupled (less than transitional coupling); $Q \times k < 1$.

exception would be line of sight into the transmitter, such as a tower in the middle of the city. Examples of such hostile conditions would be any large city such as Munich, New York, Miami, Chicago, and San Francisco, where the authors had significant experience with intermodulation distortion problems.

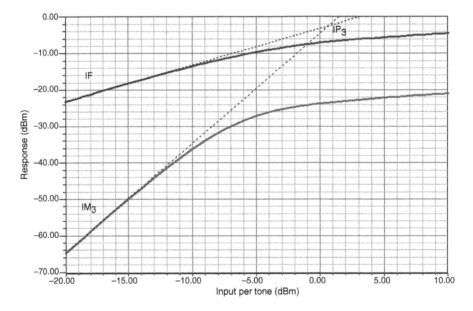

Figure 3.161   Prediction of intercept point of the double-tuned circuit shown in Figure 3.159. Note the compression of both the input signal and the IMD product.

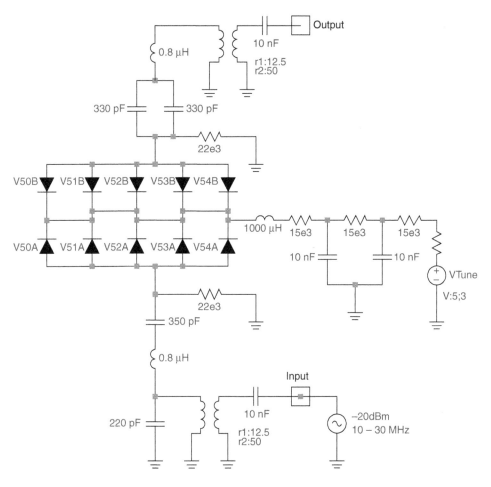

Figure 3.162   High dynamic range 10–20-MHz filter.

By translating the circuit into a low-impedance-drive arrangement, building it symmetrically, and reducing the tuning range somewhat by capacitors and many series diode pairs, the filter's large-signal performance was significantly improved. The tuning diodes had about 125 pF at 1 V dc per unit. Initial tests trying to use a high capacitance diode, such as the BB112, resulted in much higher IMD products. The B112 has a 1 V capacitance of 470 pF down to about 20 pF at 8 V. Standard diodes show about 30 pF at 1 V, while the diodes used in the experiment had a 125 pF at 1 V. These experimental diodes were made available from a well-known manufacturer for this evaluation. Figure 3.162 shows the circuit diagram.

A step-down transformer drives the tuned circuit with a source impedance of 12.5 Ω. The circuit is symmetrical. There are two 0.8-μH inductors and 2 × 5 diodes in the loop. The output of the circuit is transformed back up to 50 Ω. The tuning voltage is supplied via a heavily filtered arrangement. This circuit has actually been implemented in Rohde & Schwarz field strength meter equipment. They guarantee an intercept point of about +20 dBm. The selectivity of these circuits is quite reasonable and mostly intended to reduce second-order intermodulation distortion products by about 10 dB, at the same time having

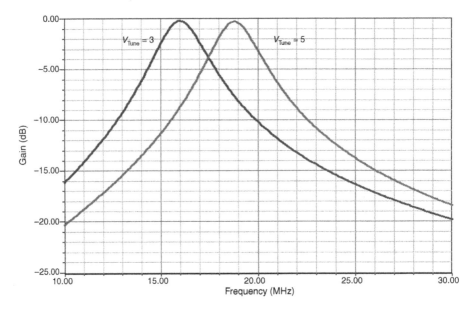

Figure 3.163   Frequency response of the filter shown in Figure 3.162. Filters of this type are intended to reduce second-order IMD by providing about 20 dB suppression at half the center frequency.

a high third-order intercept point. Figure 3.163 shows the selectivity curves at two different tuning voltages.

The most interesting number, however, is the third-order intercept point (already determined as about 20 dBm), which still had to be simulated. This sometimes sounds like a contradiction between once the measured values are available and are acceptable why one would want—or, rather, need—to do a simulation besides the necessity to develop a high-input intercept filter? It was desirable to validate the nonlinear models and prove that the above-mentioned equations will hold true. This type of simulation is now more difficult because the number of nonlinear elements went from four to ten, and some numerical problems, such as convergence difficulties, can be expected, and the effect of harmonic frequency cancellation (compensation) can also be seen. By using diode combinations that result in $1/(1 - n)$, n being an integer number, these IMD products can be drastically reduced. This implies that each of the five diode pairs has a selected value for n to meet this condition. Later, a final attempt will be made to improve the first VHF filter with the proper diode combinations.

Figure 3.164 shows the calculation of third-order intercept point for the arrangement shown in Figure 3.163. This number has now increased to +18 dBm. The differences between selection and measurement is approximately 2 dB. The reason for this more pessimistic value compared to the measured value probably has to do with slight variations of the exponent of the diodes' voltage-dependent capacitance. Therefore, the authors consider both the circuit performance as well as the simulation accuracy to be extremely good. This type of circuit, as mentioned earlier, has wide application in oscillator circuits. Mr. Danzeisen of Rohde & Schwarz was probably the first to have used this type of circuit by paralleling many diodes for improved performance.

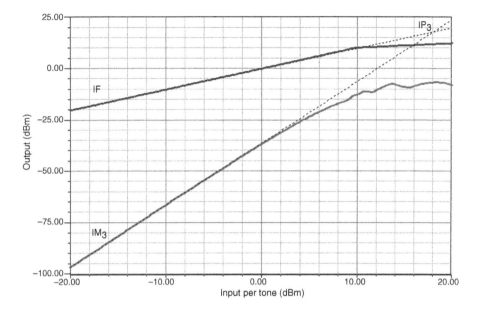

Figure 3.164 Prediction of third-order intercept point of the filter shown in Figure 3.162. The reason for the curves on the IMD plot lies in the interaction between the 10 nonlinear devices.

3.6.4 Improving the VHF Filter

By selecting the appropriate diode combinations and circuit modifications as shown in Figure 3.165, a significant IMD improvement of the Figure 3.159 circuit is obtained. A special shunt arrangement of several diodes with different exponents has been developed, which allows its value to be "adjusted." This circuit, for which a patent has been applied, showed

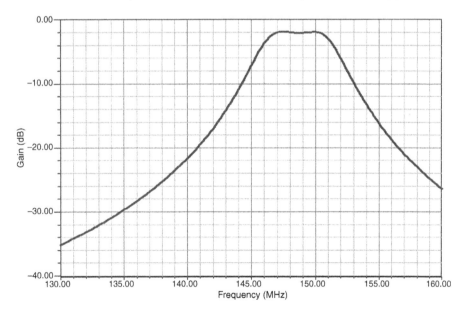

Figure 3.165 The frequency response of the improved circuit derived from Figure 3.159.

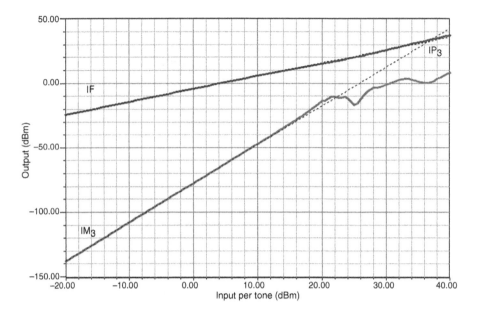

Figure 3.166 Predicted IMD performance of the improved version of Figure 3.159. This type of improvement is significant for all applications. A patent is in the process of being obtained for this; therefore, the circuit cannot be disclosed at present.

an improvement from −2 dBm to 32 dBm. The same high operating Q was maintained. Figure 3.166 shows the frequency response after the modification. Note that a coupling slightly greater than critical ($Q \times k = 1.1$) has been selected.

3.6.5 Conclusion

After explaining some of the nonlinearities in mathematical terms, we have given some examples of voltage-tuned circuits and discussed their large-signal performance. This section has also shown that modern CAD tools can accurately predict the performance of such circuits. The authors would like to thank Gregg Albrecht of Ansoft, Compact Division, for performing the actual simulations.

3.7 DIFFERENTIAL AMPLIFIERS

The differential amplifier (Figure 3.167) goes back many years and also was the first step from TTL to ECL. The differential amplifier is an emitter follower sinking its current into a grounded-base stage with a constant-current generator between the two emitters and the ground. The stage to ground needs to be properly biased so that the collector voltage does not increase unnecessarily above V_{BE}, but, on the other hand, has a differential output impedance at least two decades or higher than the input at the transistors above being driven into the emitter from this stage. Most of the modern integrated circuits in one way or another use this differential amplifier.

If the circuit is operated about its point of symmetry, the dc component in either current remains half of the emitter constant current for all symmetrical driving signals. In addition,

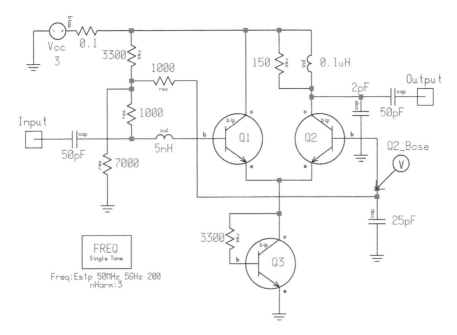

Figure 3.167 The differential amplifier schematic. The transistors are Infineon BFP420s. Figure 3.168, shows the circuit's frequency-dependent gain, matching, and noise performance.

for input signals that are aperiodic, even harmonics are not generated. This assumes that the stage is driven symmetrically at both inputs. As, in our case, the differential amplifier is also frequently used to have its second stage grounded, meaning that there is a base-to-ground capacitor, and the stage is asymmetrically fed. As the input voltage is increased, the differential stage becomes a limiter, which limits the current between plus and minus I_S, with I_S being the source current. The resulting output will become a square wave. This stage can also be used as a line receiver, translating a sine wave to an output logic suitable for driving ECL stages.

One can assume that the differential amplifier has a transconductance of

$$g_m = \frac{\alpha g_{in}}{2} \tag{3.192}$$

where α is the ratio of collector to emitter current. Within the limits of the output being a sine wave, this stage has very low distortion. The output current now is determined by

$$i = \frac{I_{in}}{2} \tanh \left(\frac{x}{2} \cos \omega t \right) \tag{3.193}$$

where ωt is the input frequency and I_{in} is the current forced by Q_3. The reason why the transconductance has to be divided by 2 comes from the fact that the current is split into two equal components for Q_1 and Q_2. Multiplying the new transconductance with the output load gives the voltage gain of this amplifier; its power gain can be obtained by multiplying this with $\sqrt{\text{output resistance} \div \text{load resistance}}$.

Having done the linear analysis and having learned that because of the hyperbolic tangent functions this amplifier can also be used as a line receiver, converting sinusoidal waves to square waves may be useful to take a closer look at these issues. If we take a CAD

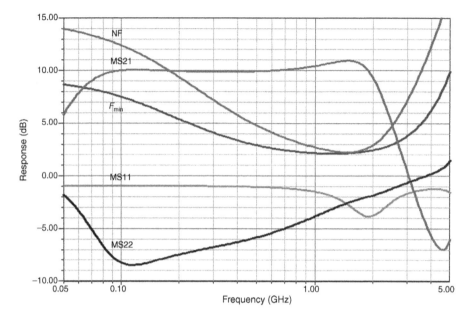

Figure 3.168 Frequency-dependent gain, matching, and noise performance of the differential amplifier.

oscilloscope, we see the resulting output waveform being a nonequal duty cycle output voltage that results in high harmonic content. The amount of harmonic content can be controlled by the drive power and by the actual bias. Figure 3.169 shows the resulting output waveform.

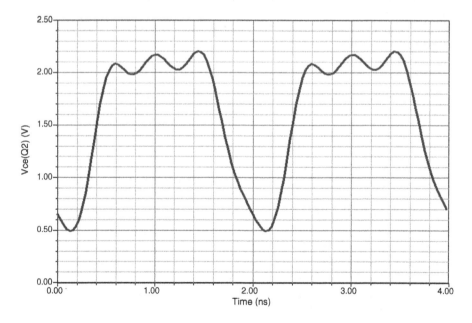

Figure 3.169 V_{CE} versus time for output transistor Q_2 with drive = 10 dBm.

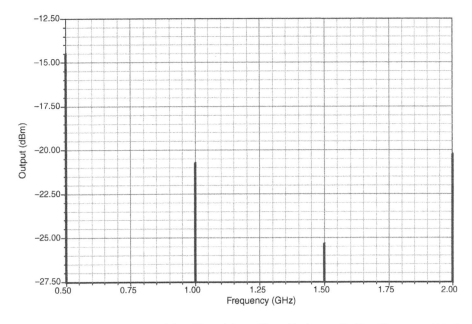

Figure 3.170 Output spectrum of the differential amplifier with drive = 10 dBm. The present biasing is not optimal for frequency multiplication.

If we now switch to a CAD spectrum analyzer, we can evaluate the harmonic contents (Figure 3.170). Figure 3.171 shows the output transistor's dc I–V curves and ac load line under these conditions. Based on the previously defined conducting angle of the differential amplifier, one can optimize for harmonic content. Therefore, it is logical at this point to

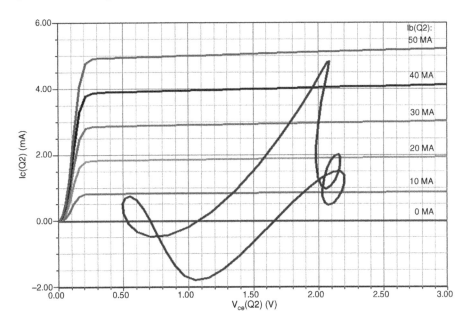

Figure 3.171 DC I–V curves and ac load line for output transistor Q_2 with drive = 10 dBm.

look at frequency multipliers, as they are sometimes needed. Finally, we are curious to see the so-called load line of the output transistor, which is, based on the interaction of all the stages, not looking the same way as the expected one, which we know from previous amplifier examples. Part of the reason for this is that for a certain voltage range, the collector voltage is in the negative region, and all kinds of saturation effects result in this surprising waveform. By reducing the input power, one would obtain a more "expected" load line. We invite readers with CAD capabilities to play with this example, as it gives a lot of insight into the operation of emitter-coupled logic (ECL) stages.

3.8 FREQUENCY DOUBLERS

Having seen that the harmonic content invites us to bias a stage from Class A into Class B or C if our goal is a harmonically rich current, we will now evaluate a single stage that will be biased close to cutoff and driven by a relatively low frequency (500 MHz). To obtain the best results, we will use a parallel notch filter in the output, followed by a high-pass matching circuit to "catch" the desired harmonic. The input frequency is heavily suppressed by the notch filter. Based on its finite Q, there is a limited range over which this frequency can be varied with constant subharmonic attenuation. Figure 3.172 will be our test circuit.

This circuit consists of a bias transistor, Q_2. This method has been discussed before and solves the temperature-dependent bias emitter thermal runaway problems through "temperature compensation." The purists among our readers may point out that the currents drawn by Q_1 and Q_2 are not quite identical (which would give 100% cancellation), but this approach shown is sufficient to do its job. Also, we have chosen a Class AB operation point, which is defined by Q_1's 1-mA dc standing current. The conducting angle has not been optimized for the second harmonic—a task that the interested reader can do by using the conducting angle relationship shown in the beginning of the amplifier chapter. To give readers with access to nonlinear CAD some homework, we have not set this at the optimum point.

The second transistor, here called Q_1 (because of its importance at RF) has a matching network at the input. This is recommended for optimum energy transfer. The matching condition can be validated by examining Figure 3.173. Looking at the higher frequency response, the input shows a possible trend of oscillation (negative loss); this could actually be eliminated by using a low-pass filter at the input instead of the high-pass filter used now. Findings of this type always make us nervous, and we hope that the nonlinear models are sufficiently accurate at higher frequencies to predict such behavior correctly. In the case of the FET doubler mentioned in Chapter 2, we experienced that the device modeling done by Compact (now Ansoft) gave the best results, the university answers were fairly close, and the device manufacturer data were worst. We have since become highly skeptical about device modeling unless we can validate it ourselves. Synergy Microwave has a complete set of test equipment in the laboratory, including network and spectrum analyzers with special software created by Rohde & Schwarz that can be used in addition to the Scout program to do in-house validation. This was achieved by the use of nonlinear optimization at a drive level of 0 dBm. One could say that this is similar to using a load–pull approach at the input, and then resorting to the large-signal S parameters obtained. Another note of caution: capacitors at these low values tend to have a very large percentage error; for instance, a

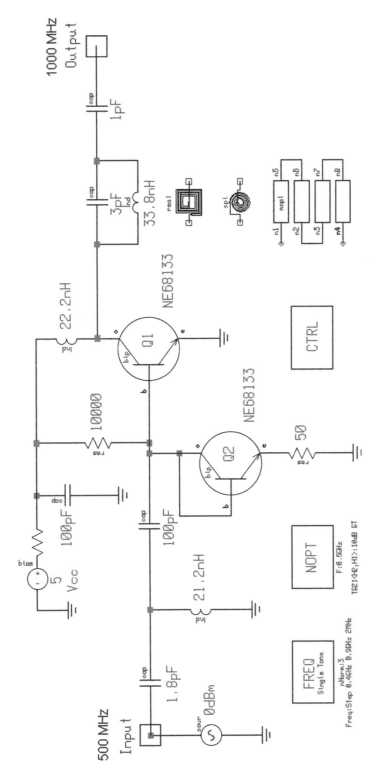

Figure 3.172 BJT frequency doubler schematic. A parallel-tuned trap in Q_1's collector attenuates fundamental feedthrough.

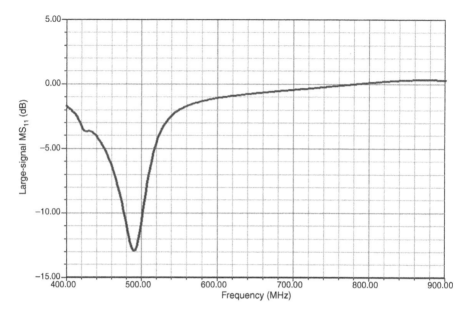

Figure 3.173 Large-signal S_{11} for the BJT doubler. Note the positive region above 800 MHz.

1-pF capacitor, if not selected carefully, can vary between 0.7 and 1.3 pF—a 50% error. Needless to say, this is not acceptable for useful production.

The output consists of a high-pass filter and a notch filter with the purpose of reducing the subharmonic significantly. We have again done this on purpose by leaving room for the reader to come up with a more inventive output circuit like a Cauer-type (same as elliptical) low-pass filter with discrete notches at higher harmonics of the input frequency, such as 1.5 GHz, 2 GHz, and so on. In addition, while in the previous biasing example looking at distributed elements, we have shown here three ways how to implement the inductance for the notch filter. This is equally applicable for other inductors in this value range. As frequency increases, this becomes more interesting, but one needs to remember that many materials especially silicon are extremely lossy. The same inductor approach has been used in Figure 5.61 of the oscillator chapter. On silicon, even the manufacturer becomes an issue. Figure 3.174 shows a top view of a "rectangular" inductor that actually approaches the spiral inductor; its implementation can be seen in Figure 3.175. Modeling this was a nightmare, and during our Compact days, we had to actually modify the electromagnetic simulator to accommodate this and get results where measures and simulation agreed. Figure 3.176 shows the measured and simulated Q as a function of frequency of a silicon-based inductor. In the case of other materials, such as PC board or GaAs, the models in the CAD tools are sufficient to predict the losses of the inductors. Following this initial exercise, Motorola (now Freescale) then developed an improved silicon spiral inductor; its frequency-dependent Q is shown in Figure 3.177. Note that this Q was measured and not simulated.

One of the reasons why we use a transistor multiplier is that we expect gain (because of the square-law characteristic, an FET would have been better tailored for a doubler); therefore, we have "measured" the gain of this circuit while driving it with 0 dBm. Another important factor is that because of the way the CAD tool displays it, the gain is reference

Figure 3.174 A 6.5-turn spiral inductor in silicon.

to the 500-MHz input. Mentioning this is important; otherwise, the reader may look for the 1 GHz gain, which does not exist. Figure 3.178 shows the doubler gain.

By now we are curious to see the doubler in action, and by reviewing Figure 3.179, we can see the output spectrum. The output spectrum validates the statement that the output

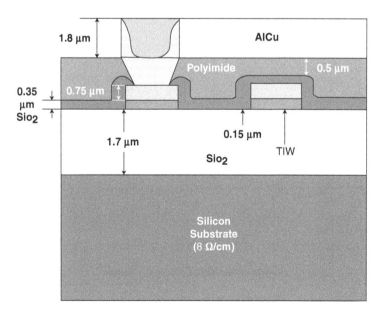

Figure 3.175 Cross section of the silicon inductor.

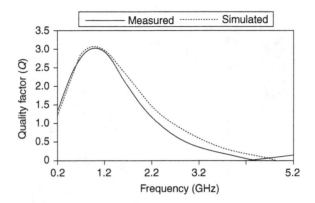

Figure 3.176 Measured and simulated Q versus frequency for the simulated inductor.

circuit, while doing a marvelous job of removing the 500-MHz input frequency, does not cure the problem of unwanted multiples of the input frequency. Again, a Cauer low-pass filter, while introducing some possible additional loss, should be considered for this circuit to have a bandpass response. On the other hand, many oscillators have very little harmonic suppression, and the same type of post-filtering needs to be considered. There are two-transistor solutions also available, such as push–push for frequency doubling (push–pull would eliminate the desired second harmonic quite well). Actually, two diodes could have also been used for doubling, but since it requires transformers and a post-amplifier, we have shown the preferred application. The only advantage of the diode application is that it is wideband, while this BJT circuit, as we have just learned, must be made narrowband for reasons of spectral purity.

Finally, we are curious to see the actual operating state of this amplifier, and this can be done by analyzing the load line (Figure 3.180) . Needless to say, it is complicated in

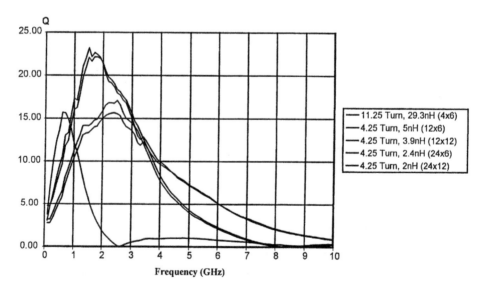

Figure 3.177 Q versus frequency for the Motorola improved inductor.

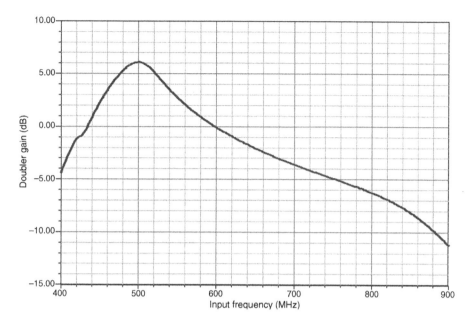

Figure 3.178 Frequency-dependent gain of the BJT doubler.

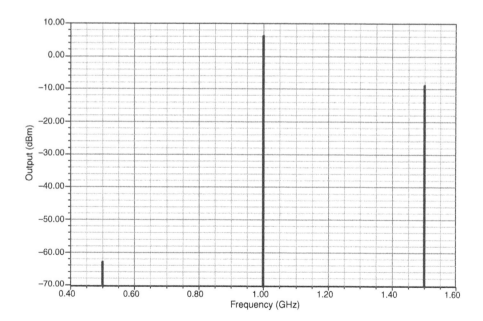

Figure 3.179 Output spectrum of the BJT doubler.

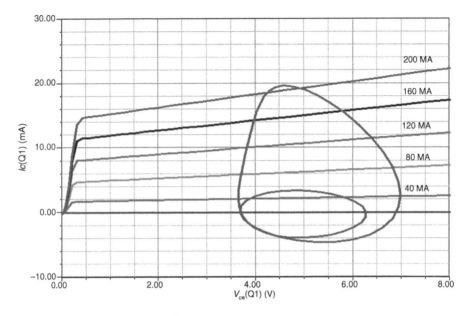

Figure 3.180 *I–V* curves and ac load line for RF transistor Q1 in the BJT frequency doubler.

a manner similar to that of the differential amplifier. Because of the stored energy in the various tuned circuits, it looks quite impressive, and would probably be difficult to predict without a nonlinear CAD tool.

3.9 MULTISTAGE AMPLIFIERS WITH AUTOMATIC GAIN CONTROL (AGC)

Having discussed the basic differential amplifier, we will now evaluate the performance of the MC1350/1490 video amplifier chip from Motorola.[3] Most of the circuitry in the chip is self-explanatory, which has been used for our simulation. Figure 3.181 shows the inner circuit of this chip, and Figure 3.182 shows a recommended application. By grounding one side of the input, we effectively get the same performance as the differential amplifier we just discussed, with one difference being that the input is now fed to the constant-current generator. To minimize the distortion at the two lower transistors, they are decoupled by a 66-Ω resistor. Figure 3.183 shows the frequency-dependent gain, matching, and noise figure. Since this particular device has been specifically designed for AGC purposes, we also show in Figure 3.184 its gain reduction versus applied AGC voltage.

3.10 BIASING

A lot has been written about biasing, and since the purpose of this book is not to focus too much on dc, the following examples including RF consequences have been selected

---

[3] Motorola is now Freescale, the video amplifiers are still available from Lansdale Semiconductors. Part numbers ML1350 and ML1490.

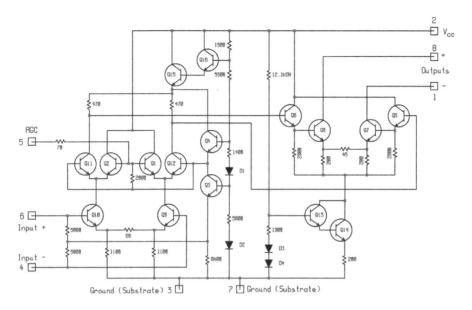

Figure 3.181 Schematic of the Motorola MC1350/1490 entered for simulation.

and considered sufficient. Early application reports by Motorola and others have provided a wealth of data. As a side effect of the introduction of the wireless era, we find that the engineers are so pressed for time that the number of complete application reports has gone toward zero. This is one of the reasons why we have tried to keep this book on a minimum mathematical basis (some is still needed), but provide the most practical assistance so the book does not become a turn-off. Figure 3.185 shows a textbook type of approach for RF.

In order to decouple the transistor from the biasing, we use RF chokes and dc decoupling (bypass) capacitors. This is the technique we have already used in the previous examples,

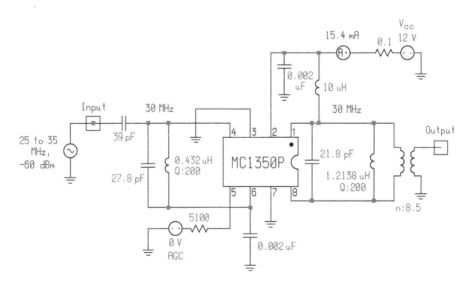

Figure 3.182 MC1350/1490 30-MHz application circuit.

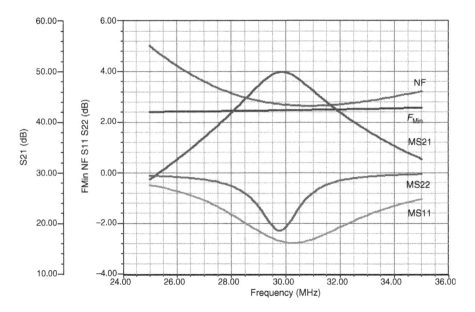

Figure 3.183   Frequency-dependent gain, matching, and noise performance of the MC1350/1490 in the circuit of Figure 3.182.

so one might ask the question, "What's new?" As frequency increases, these inductors are either not manufacturable or have such a low Q that their use becomes questionable. This is the point where one may introduce the so-called distributed elements.

Figure 3.186 shows the very same circuit but resorting to distributed rather than lumped elements. The elements we are introducing now (part of any good, up-to-date

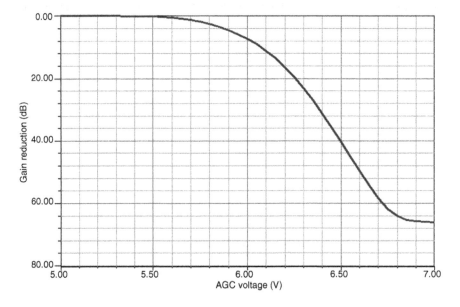

Figure 3.184   Gain reduction versus applied AGC voltage for the MC1350/1490 IC.

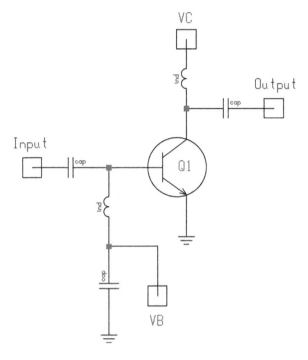

Figure 3.185 Simple BJT RF amplifier with lumped elements used for bypassing and dc blocking, and for feeds base bias and collector supply feeds.

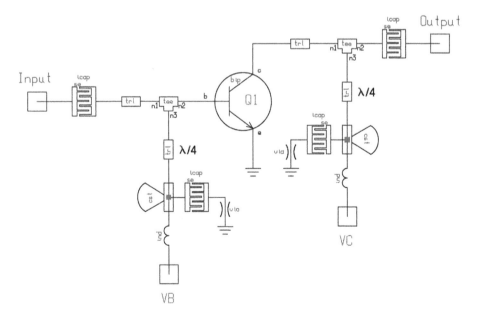

Figure 3.186 Simple BJT RF amplifier with distributed elements used for bypassing and dc blocking, and for base bias and collector supply feeds.

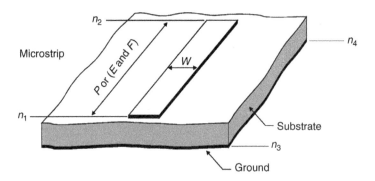

Figure 3.187 Transmission line in microstrip.

CAD tool; see element library for how to use their physical element descriptions) are as follows.

- *Transmission Line.* Any printed connection between two points on a circuit board is a transmission line (Figure 3.187). Its characteristic electrical impedance is a function of the square root of the dielectric constant (ε_r), the width, metallization, thickness, and height above substrate of the line, and the loss tangent of the substrate. Since lines frequently have to be laid out in the form of curved connections or have a bend in their direction, we have to add elements capable of describing the high-frequency consequences of such connections. Figures 3.188 and 3.189 show mitered and radial bend elements that perform this function.

- *T, Cross, and Y junction.* By the time a point like a collector or a base, or its FET equivalent, spreads out into connecting with other elements, we need additional modeling capability to describe *T* connections, crossings, and *Y* junctions. Figures 3.190–3.192 show the way in which these connections need to be modeled.

If the need exists, the standard inductances must be replaced with a transmission line whose length is λ/8 at the operating frequency. At higher frequencies, these transmission

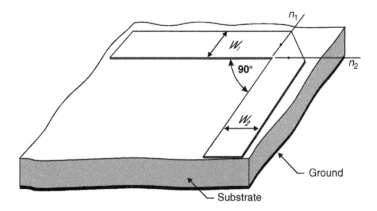

Figure 3.188 Mitered bend.

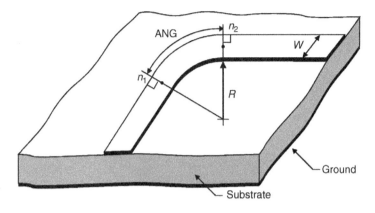

Figure 3.189   Radial bend.

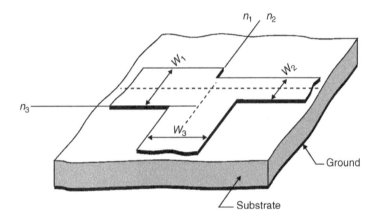

Figure 3.190   *T* junction.

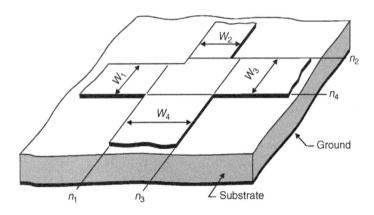

Figure 3.191   Cross.

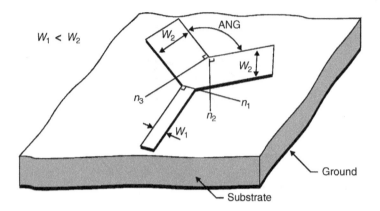

Figure 3.192 Y junction.

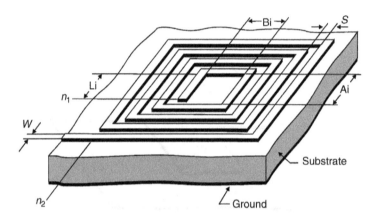

Figure 3.193 Rectangular inductor.

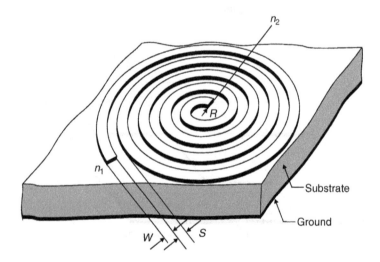

Figure 3.194 Spiral inductor.

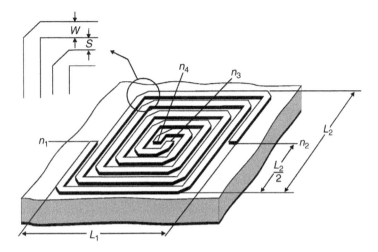

Figure 3.195   Transformer in microstrip.

lines, however, then go into $\lambda/4$ resonant mode and later become capacitive. This type of design makes it fairly narrowband. A way around this is the use of printed inductors, as shown in Figures 3.193 and 3.194.

These inductors have a self-resonant frequency similar to the transmission line mentioned above, but the safety margin is significantly higher.

Talking about printed inductors, a logical extension of this is the printed transmission-line-based transformer as shown in Figure 3.195. One can consider this as two interlaced rectangular inductors, and based on the substrate material, they are useful over a wide frequency range. Besides being used as a transformer, they can also be used to transit from unbalanced to balanced transmission provided that the difference in length from a connection point of view does not cause any problems (this is a layout issue).

A popular form of combining stages is the so-called Lange coupler (Figure 3.196) invented by the German Julius Lange. It is one of the major contributions in wideband applications.

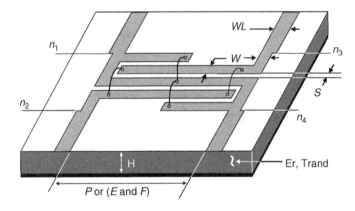

Figure 3.196   Four-strip Lange coupler.

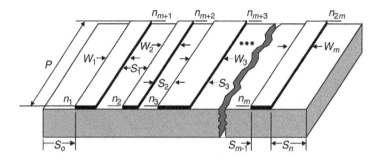

Figure 3.197   Multiple-coupled lines element in microstrip.

The useful application of the Lange coupler probably starts at 4 GHz. It consists of parallel transmission lines with the appropriate connections as shown. Lange couplers are typically built with four, six, and eight fingers. The Ansoft Serenade product has a Lange coupler synthesis program that can be used to gain more insight into this coupler's application. We assume that other modern software has similar capabilities.

Where meander-type of inductors are necessary, a neat way to implement and simulate them is to use the multiple coupled line element (Figure 3.197) of the Serenade product, which both fast and accurately calculates the behavior of the meander, including self-resonances and losses. We made use of this arrangement in our previous examples.

- *Interdigital Capacitors.* The issue of tolerances of small capacitors already has been brought up. The interdigital capacitor can be made on printed circuit board material as well as gallium arsenide, and if its dimensions are continuous with the transmission-line width it does not cause any abrupt changes in the impedance. This type of capacitor permits to obtain very small values. By the way, an alternative to this is the use of transmission lines being 3/8 λ. We have learned above that a transmission line below resonate frequency is inductive, goes into resonance, and then becomes capacitive. Again, bandwidth is also an issue. An interdigital capacitor consists of a number of parallel fingers as shown in Figure 3.198, and its capacitance can be varied by adjusting

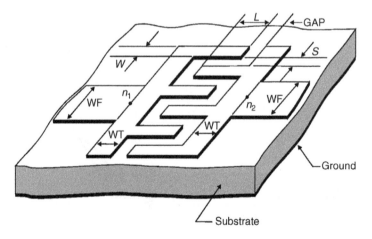

Figure 3.198   Interdigital capacitor.

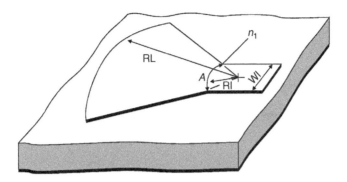

Figure 3.199   Radial stub.

the number of fingers and their spacing. The advantage of the interdigital capacitor compared to discrete components is its low variation in value.

- *Radial Stubs.* The radial stub (Figure 3.199) is not much different from a λ/4 resonator, but its bandwidth is much greater than a simple transmission line. This is another way to ground the "cold" side of a transmission line or part of a circuit that needs to be grounded for RF. Of course, the interdigital capacitor comes in a version that is a combination of a capacitor and a via hole.

- *Via Holes.* The "cold" end of the transmission line, being either considered an inductor or capacitor, needs to be connected to a sufficiently large copper backplane. One very efficient way to do this, especially if there is not enough copper left on the top of the board, is the use of via holes (Figure 3.200). One could theoretically generate a via hole with a rivet, but most manufacturing processes do not allow this; the normal solution is to use plated-through holes left open. In the PC boards, via holes are typically cylindrical; on substrates like GaAs, they may be conical.

- *Correction Elements.* Although the behavior of actual circuits proceeds regardless of our ability to measure and describe it, we do not enjoy this luxury in simulating circuit behavior in software. In a simulator, effects that are insufficiently described will be inaccurately simulated. For instance, segments of high-impedance transmission line (e.g., 120 Ω) are frequently used for dc feeds and RF chokes. In predicting the effect of a

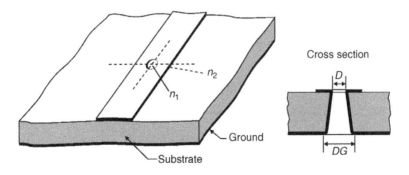

Figure 3.200   Via hole.

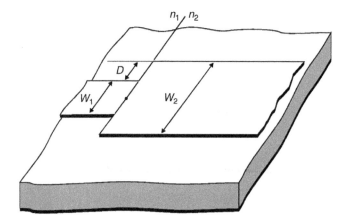

Figure 3.201 The STEP element tells the simulator to calculate the effects of joining transmission lines of differing characteristics.

transition from 120-Ω line to 50-Ω line, a simulator must be alerted to the discontinuity so that it can do the necessary mathematical corrections to account for the impedance jump. To do this, a specific circuit element, the STEP (Figure 3.201), must be inserted between the 120-Ω and 50-Ω line elements in the simulation circuit file. In addition to the substrate data, we characterize a STEP by providing the widths of its input and output lines. The element itself has no physical length.

A similar correction is necessary if a transmission line is used as a resonator or just "left open" at one end. Such a transmission line tends to radiate, and because of its high-impedance properties reacts differently as far as its electrical length is concerned. A zero-length one-port element, the OPEN (Figure 3.202), must be added to such a line for mathematically correct calculation.

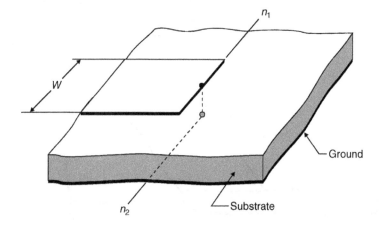

Figure 3.202 The OPEN element tells the simulator to calculate the effects of leaving the end of a transmission line unconnected.

We end this excursion into distributed elements here. It is most important to keep in mind that as frequency increases, we rapidly move into the area where we must consider all these distributed elements to achieve accurate simulations—even if doing so makes simulation a painful and time-consuming effort.

Finally, anyone who adventures in this area *must* obtain a foundry manual from the company that will build the integrated circuit or hybrid under design. Luckily, there are a rather high number of independent foundries in the market. This is partly a result of last decade's development where most of the major electronics companies spun off their semiconductor branch. Just naming the companies that provided the circuits used as examples in the first edition of this book: the semiconductor branches of Motorola, Siemens, NEC, Hewlett Packard, and Philips are now Freescale, Infineon, Renesas, Avago, and NXP, respectively. The global trend to outsource the semiconductor processes results in a number of companies that offer high-end foundry processes either in the III–V area, providing HEMTs and HBTs on GaAs or InP, or specialized to make silicon-based chips, in SiGe HBT, RF CMOS, or BiCMOS technology.

We mention the issue of foundries here again because each foundry has its own proprietary approach to modeling transmission line discontinuities. The availability of a foundry service somewhat eases the requirement that a designer be fully up to speed on the nuances of discontinuities, because a foundry's designer service will help customers account for all relevant parasitics or discontinuities in their designs. In addition, there are tables of S parameters for standard cells of either capacitors, resistors, or inductors. The designer may then be forced to adjust the circuit so that it will work with a particular inductance value or value of another component within the resolution of the table that describes these elements. Information on the active part such as diodes and transistors were already given in Chapter 2.

3.10.1 RF Biasing

Applying the knowledge we have just acquired about distributed elements, Figure 3.203 shows a somewhat exaggerated case of using the various elements. While it is consistent with the simple RF case we started with, modeling of this type is certainly necessary at higher frequencies.

Many but not all the elements we have just described are used, but we have not solved the question of the actual dc biasing.

3.10.2 dc Biasing

The following is a rehearsal of the hopefully known recommended way how to bias a transistor. We will start with the collector current (in our example, Figure 3.204, 10 mA). To separate RF and dc, we have a combination of a load resistor (R_L) and an inductance (L_C). Information about this operation was already given in Section 3.2.4. The emitter resistor should have a voltage drop about 0.7 V, and therefore, the emitter resistor is $0.7/I_C$. The next assumption is that the transistor has a dc current gain of 100, and therefore, we decide to have the bias resistor chain (R_1 and R_2) draw 10% of I_C, or 1 mA. Assuming a collector voltage of 5 V, the value of $R_1 + R_2$ has to be 5 kΩ to result in the required 1 mA. Because the base current now is 10% of the divider current, the base will draw 0.1 mA. The voltage across R_2 has to be the voltage drop across R_E (0.7 V) + V_{BE}, which is also 0.7 V, or 1.4 V.

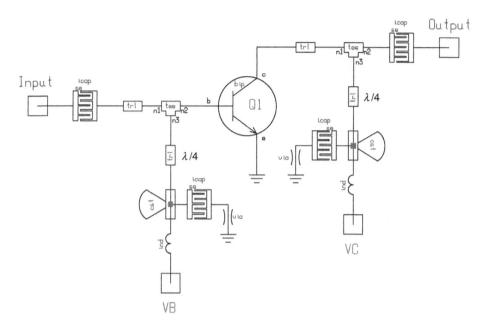

Figure 3.203 Simple BJT RF amplifier with distributed elements used for bypassing and dc blocking, and for base bias and collector supply feeds.

Given the 1 mA, $R_2 = 1.4/10^{-3} = 1.4\,k\Omega$. R_1, therefore, is $5\,k\Omega - 1.4\,k\Omega = 3.6\,k\Omega$. This concludes the calculation.

In the case of a dc beta of only 30, we need to raise I_{BIAS} from 1 to 3 mA since the base current now will be 0.3 mA. This changes the resistor values. $R_1 + R_2 = 5\,V/3\,mA = \approx 1.6\,k\Omega$. Following the same approach, we need $1.4/(3 \cdot 10^{-3}) = \approx 470\,\Omega$. This makes

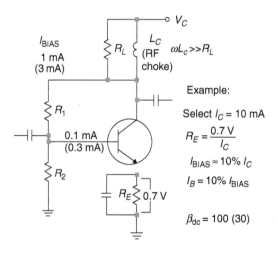

Figure 3.204 DC biasing example.

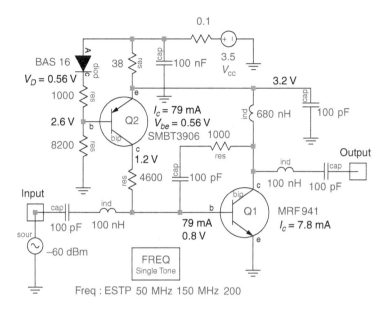

Figure 3.205   An RF amplifier with active biasing. (After the low-noise amplifier circuit in [43]).

for $R_1 = 1600\,\Omega - 470\,\Omega = \approx 1100\,\Omega$. These roundings were necessary because resistors can only be bought in certain resistance steps.

However, for low battery voltages, the 0.7 V between the emitter and the ground cannot be wasted. Therefore, a dc stabilization circuit, to which we have already referred, will be looked at next. Figure 3.205 shows a combination of an RF transistor, Q1, and a dc-stabilization transistor, Q2. The most important thing about this circuit is the fact that the difference between the supply voltage and the collector of Q1 is 0.3 V, which is approximately half of the voltage we had to "waste" across R_E in the above-mentioned circuit. Plus, the voltage drop is now in the collector, and given the fact that we feed Q1's collector through the inductor of a quasi-tuned circuit rather than through only a resistor, we get a much higher voltage swing. The base–collector diode temperature dependency of Q2 is compensated for with the silicon diode feeding the base. It is also fascinating to see that because of the low currents, both the voltage drop of the diode and V_{BE} of Q2 are about the same, and yet Q1, operating at a reasonable current, shows the expected 0.7–0.8 V V_{BE}. We have supplied all the necessary dc voltages to give the reader an incentive to calculate the mechanism of this stabilization. The key equation to this is to determine the current in the voltage divider feeding Q2, and going from there. On the other hand, having a nice CAD tool that, like SPICE, provides insight into the dc voltages, and allows the addition of dc voltage probes, makes life much easier. To prove that this circuit actually works, Figure 3.206 shows its frequency-dependent gain, matching, and noise-figure performance. It is based on a Motorola application that had excessive voltage drop for Q1, and which we have improved to be suitable for low-voltage operation [43].

Integrated active bias solutions are also available. Figure 3.207 shows the Infineon BCR400 active bias controlled applied to a BJT; it can also be applied to FETs and TR switching diodes, as we saw in Figures 3.57 and 3.59, respectively.

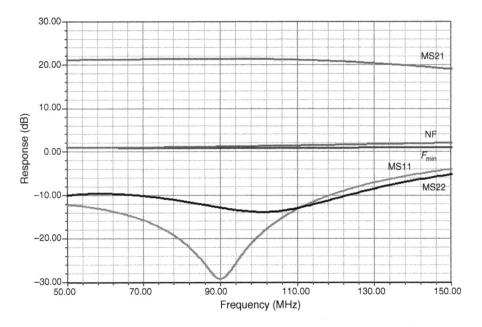

Figure 3.206 Frequency-dependent gain, matching, and noise responses for the amplifier in Figure 3.205.

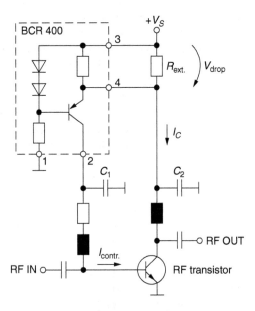

Figure 3.207 The Infineon BCR400 active bias controller can be applied to FETs as well as BJTs. Courtesy Infineon Technologies.

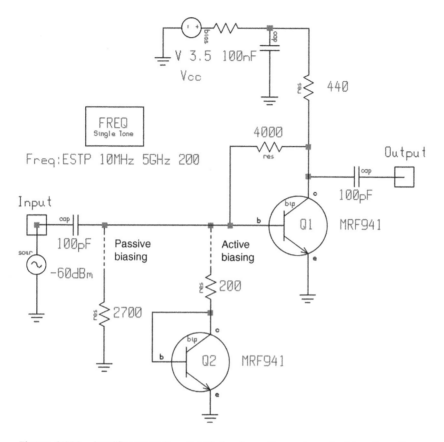

Figure 3.208 Amplifier stage ($I_C = 3\,\text{mA}$) showing active and passive bias alternatives.

3.10.3 dc Biasing of IC-Type Amplifiers

In integrated circuits, we do not have the unlimited flexibility as shown above. This results frequently in the use of constant-current sources or transistors being used as diodes for biasing purposes. Since our first suspicion is that this is going to cause a lot of noise in the IC, Figure 3.208 shows a test example.

As suspected, by choosing the same bias point of 3 mA, the noise figure using active bias is quite a bit higher than the passive one, a shown in Figure 3.209.

3.11 PUSH–PULL/PARALLEL AMPLIFIERS

Since it is practically impossible to obtain $S_{11} = S_{22} = 0$, using a combiner at the input and output reduces the problem. Figure 3.210 shows the 3-dB hybrid approach. It has a practical bandwidth limit of about 4:1; its advantage is that it is compact and ideal for cascading. If one stage opens, P_{OUT} and gain drop only 6 dB; however, it does not protect against load mismatch. A recommended distributed form is the Lange coupler (Figure 3.211).

To analyze this approach, Figure 3.212 shows the resulting impedances and necessary phase shifts. If a Wilkinson coupler [44] is used instead of the 3-dB hybrid, the bandwidth

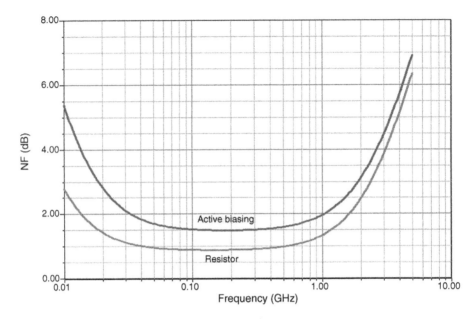

Figure 3.209 Frequency-dependent noise figure of the amplifier circuit shown in Figure 3.208. The active biasing curve is considerably noisier.

is reduced. The Wilkinson coupler is larger in size than the 3-dB hybrid but more easily realized, leading to lower cost. Figure 3.213 shows the actual impedances necessary for matching, with A1 and A2 being the actual amplifiers combined.

The solution for pushpull amplifiers resorts to a balun with the following results obtained.

Bandwidths of greater than one decade are achievable; the balun structure gives 4:1 impedance advantage; no isolation; $S_{11} \neq 0$, $S_{22} \neq 0$. Provides cancellation of even-order products.

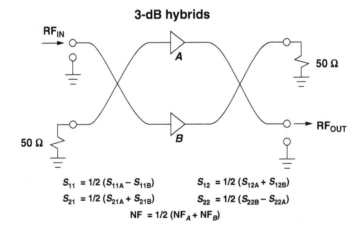

Figure 3.210 Combining power with 3-dB hybrid couplers.

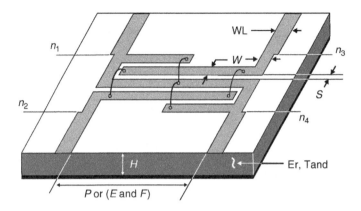

Figure 3.211   Four-strip version of the Lange coupler, a broadband 90° hybrid.

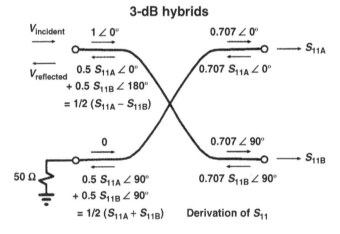

Figure 3.212   3-dB hybrids. This analysis method may be applied to any coupler.

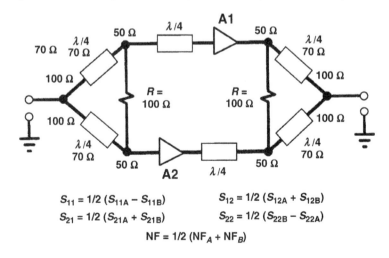

Figure 3.213   Wilkinson divider/combiners.

1) Balun

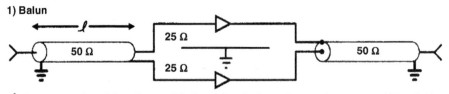

ℓ : Long enough to "de-reference" balanced end of transformer from ground (Typ $\lambda/8$)

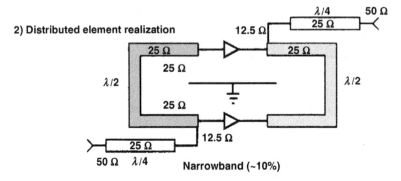

Figure 3.214 Push-pull amplifiers.

Since most amplifiers have 50-Ω rather than 25-Ω impedances, here is a circuit (Figure 3.214) that matches 25–50 Ω on both sides. Since the standard amplifier operates at 50 Ω rather than 25 Ω, an additional network may be necessary to transform the impedance up to 50 Ω again. A similar problem occurs at the amplifier output.

3.12 POWER AMPLIFIERS

Power amplifiers are devices that are tasked to transform as much dc power into RF/microwave power as possible. Depending upon the technology, devices are available that can produce several 100 W up at 2 GHz. The first concern regarding power amplification is the efficiency. As an example, the 100 W power amplifier running at an efficiency of 33% would draw 300 W from the grid and produce 200 W of heat. It is even worse than just that 2/3 of the energy is lost: lifetime of semiconductor devices decreases exponentially with temperature. Appropriate cooling is therefore imperative for successful power amplifier design.

The definitions for the efficiency can be given as the ratio of the desired RF output power P_{out} to the power that is delivered to the amplifier. The first is the drain (or collector) efficiency η that is just the ratio of P_{out} to the dc power consumed by the amplifier

$$\eta = \frac{P_{out}}{P_{dc}} \tag{3.194}$$

This definition for the efficiency neglects the input power P_{in} required to drive the amplifier. In the wireless and microwave range, power amplifiers might not have too much gain, and neglecting P_{in} yields too optimistic values. The power-added efficiency (PAE) accounts also for P_{in}:

$$PAE = \frac{P_{out} - P_{in}}{P_{dc}} \qquad (3.195)$$

This definition will yield negative values for PAE if $P_{out} < P_{in}$. But this is rarely an issue for an amplifier, and this definition is widely accepted. Sometimes, however, the following definition is preferred that is limited to 0% (no output power) to 100%:

$$PAE = \frac{P_{out}}{P_{dc} + P_{in}} \qquad (3.196)$$

The efficiency issue can be highlighted regarding Figure 3.215. A generic power amplifier consists basically of a transistor that converts the dc power provided by the power supply

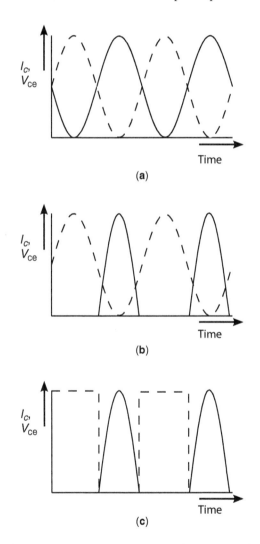

Figure 3.215 Sketch of collector current (solid lines) and collector voltage (dashed lines) of a power-amplifier transistor. (a) Class A operation. (b) Class C operation. (c) Class F operation.

into the output RF power. A filter is required to make sure that only power at the desired frequency is finally delivered to the load.

Figure 3.215a shows current and voltage at the collector of a bipolar transistor in class A operation. The RF power is delivered toward the load, as current and voltage are in antiphase. The dc power, however, is delivered toward the transistor where it is dissipated and converted to heat. In other words, the dissipated power is generated when collector current I_c and collector–emitter voltage V_{ce} exist at the same time:

$$P_{diss}(t) = i_c(t) \cdot v_{ce}(t) \tag{3.197}$$

An effective means to reduce the dissipated power is therefore to bias the transistor in a way that it only provides current pulses instead of a full sine wave, as shown in Figure 3.215b. How the current waveforms are obtained through proper biasing is shown in Figure 3.217. In order to distinguish the different modes of operation, the current conduction angle Θ is defined in Figures 3.217 and 3.218.

Power amplifier Classes A–C are defined in terms of the device conduction angle and/or the type of device operation:

Class A	"Linear" operation; 180° conduction angle
Class AB	"Linear" operation; conduction angle between 90° and 180°
Class B	"Linear" operation; 90° conduction angle
Class C	Fixed drive; less than 90° conduction angle

Since the transconductance of FETs at conducting angles of less than 180° becomes very small, the actual power gain sharply decreases compared to Class A operation while the power-added efficiency is optimized to values of 55% (Class B) or so. The dissipation of these devices is, of course, quite high, while in Class B operation this is even more of an issue. Power devices in Class C are dying to have big heatsinks to survive. One of the nice things about FETs is a much reduced thermal runaway effect compared to their bipolar brothers.

Figure 3.216 shows Class C operation. The output of the amplifier is a current pulse train that frequently leads to misunderstanding as to the calculation of its load. Given the fact that the amplifier is a current generator, it is silly to assume one has to do a conjugate matching looking backward into the amplifier. This is the condition for Class A operation. Unfortunately, this is still not understood by some engineers, but we hope we can clear this up today. The theoretical efficiency is 89.7% at a conducting angle of less than 180°; in practice, efficiencies more than 85% are not obtainable. (All these efficiencies are those of the device itself; losses in matching networks are not taken into account in these efficiency values.) Considering optimal gain and optimal output power, the conducting angle actually settles somewhere between 100° and 140°. Again, the power gain in Class C operation is less than in Class A and Class B operation.

While the equation $P_{OUT} = (V_{bat} - V_{sat})^2/(2R_L)$ is valid, it really applies only to power amplifiers in the sense that for Class A, high-gain operation R_L can be made significantly higher as the absolute power output is not the important item. While the supply voltage V_{bat} is always known, the saturation voltage V_{sat} has to be obtained by the load line dc analysis. Note that the RF saturation voltage is always higher than the dc. When looking at Figure 3.217, it becomes obvious that the conducting angle is 180°, which implies that the entire sine wave at the input is reproduced as an essentially undistorted sine wave at the output.

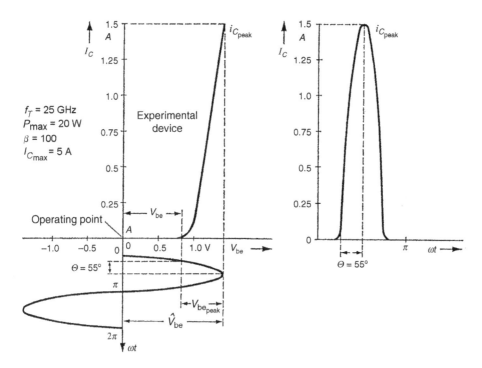

Figure 3.216 Output current as a function of drive voltage, taking into consideration the knee voltage of the base–emitter *p–n* junction. The output is a series of current pulses with a duty cycle determined by the conduction angle.

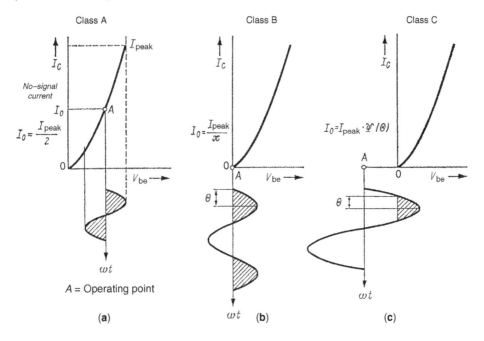

Figure 3.217 Definition of Class A, B, and C operation as mentioned above, including conduction angle (Θ). Figure 3.218 shows the determination of Θ.

We can now calculate the components of the distorted current in separating them into the dc component, the fundamental frequency, and its harmonics:

$$i_{cn} = i_{peak} \cdot \left[\frac{2\Theta}{2\pi[1 - \cos(\Theta)]} \cdot \{\text{Si}\,[(n-1)\Theta] + \text{Si}\,[(n+1)\Theta]\} - \right.$$
$$\left. - \frac{2\Theta\cos(\Theta)}{\pi[1 - \cos(\pi/2)]} \cdot \text{Si}\,(n\Theta) \right] \qquad (3.198)$$

With n being the index of the harmonic, starting with $n = 0$ at dc, and fundamental $n = 1$, the instantaneous current maximum i_{peak}, the function $\text{Si}(x) = \sin(x)/x$, and the conduction angle Θ. The equation is easily obtained from deriving the Fourier series for a clipped sine wave with dc offset, as shown in Figure 3.218. These harmonic components are a first-order approximation of the current waveform, but it is close enough to reality to explain the general concept. Figure 3.219 shows the relative amplitudes of the current harmonics as a function of conduction angle Θ. As expected, the harmonic content increases as the conduction angle decreases. Also the output power approaches zero with Θ, but peaks somewhere in Class AB. Figure 3.220 finally shows the emitter efficiency η calculated from equation (3.198), together with dc and fundamental power. It is obvious that increasing the efficiency by reducing the conduction angle comes at the expense of reduced output power. It finally leads to the trivial solution of 100% efficiency at no power.

It should be noted, however, that these amplifier classes are defined only under ideal conditions, for pure sinewaves of constant amplitude, and that the theoretical efficiencies are just drain/collector efficiencies not taking into account any kind of required drive power or parasitic or matching losses. Therefore, realistic PAE values are significantly lower, and the question of which class an amplifier is becomes questionable without a constant drive signal.

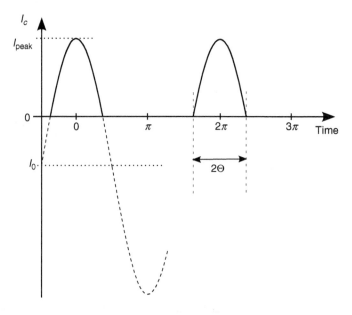

Figure 3.218   Definition of conduction angle Θ.

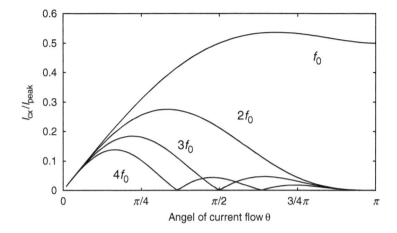

Figure 3.219 Function values for the fundamental, to fourth harmonic (f_0, $2f_0$, $3f_0$, and $4f_0$, respectively) versus device conduction angle.

Enhancing efficiency only by reducing conduction angle therefore seems not to be the best solution. There must be a way to obtain higher efficiency without degrading output power. The point is that up to now, it was not mentioned that the Classes A–C assume that the load is formed by a filter that shortens all higher harmonics of the current except for the fundamental. The purely sinusoidal shape of the voltage waveform is a result of this termination.

It has already been proposed quite early that appropriate harmonic termination can boost efficiency without the need to sacrifice fundamental power [45, 46]. A closer look at (3.198) reveals that for Class B, with $\Theta = \pi/2$, only the fundamental and even harmonics of the current exist at all. If only odd harmonics of the voltage would exist, losses could be minimized. Fortunately, the square-wave shape of the voltage waveform shown in Figure 3.215c has exactly this property. Its spectrum is proportional to $\mathrm{Si}(n\pi/2)$. Looking at the time domain waveforms, it is pretty obvious that power dissipation is zero since current and voltage do not overlap at any time. In the frequency domain, this means that the power

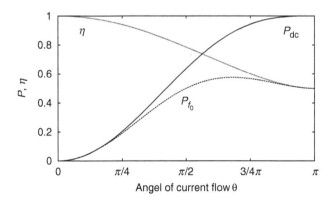

Figure 3.220 Values for the dc and fundamental frequency power, and resulting drain efficiency η versus device conduction angle.

drawn from the voltage supply equals the RF power at the fundamental delivered to the load. It is not easy to force current and voltage exactly into these shapes just by tuning the load. But what can be achieved is to approximate it by forcing zero current at even harmonics and zero voltage at odd harmonics, by terminating higher harmonics by open and short circuits, respectively. This is the approach taken in Class F amplifiers [47, 48]. Of course, in reality, higher harmonics than the third are usually neglected. Also there might be different optimum load impedances than exactly short or open. If the sequence of short and open circuits is reversed, for example, one speaks of inverse Class F or Class F^{-1}. For arbitrarily optimized reflective terminations, commonly the term "harmonically tuned amplifier" is used.

Class F	90° conduction angle, harmonics tuned in order to force the voltage waveform into square pulses at the times where the current is zero
Class inverse F	Basically Class F, but with voltage and current waveforms interchanged
Harmonically tuned	Individually optimized current conduction angle and reflective terminations for higher harmonics

It is necessary to consider that the output capacitance of the transistor is already part of the harmonic tuning circuit. Designs of optimized tuning circuits for ideal Class F operation taking into account up to the third harmonic are shown in Figures 3.221 and 3.222 [49]. The

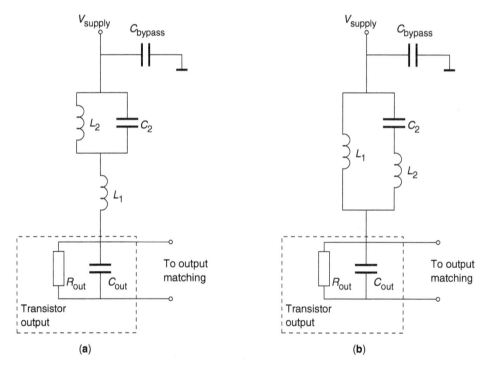

Figure 3.221 Class-F harmonic tuning circuits taking the transistor output capacitance into account, as proposed in Ref. [49]. The dimensions of the elements are: (a) $L_1 = 1/(6\omega^2 C_{out})$, $L_2 = 5/3 \cdot L_1$, $C_2 = 12/5 \cdot C_{out}$ and (b) $L_1 = 4/(9\omega^2 C_{out})$, $L_2 = 9/15 \cdot L_1$, $C_2 = 15/16 \cdot C_{out}$.

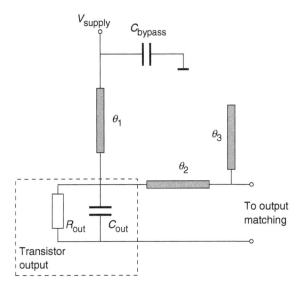

Figure 3.222 Class-F harmonic tuning circuits taking the transistor output capacitance into account, as proposed in Ref. [49]. The length of the transmission lines are: $\theta_1 = \pi/2$, $\theta_2 = 1/3\tan^{-1}\left[1/(3Z_0\omega_0 C_{out})\right]$ $\theta_3 = \pi/6$.

Class-F concept allows to boost efficiency compared to amplifiers without proper harmonic termination. But the ideal waveforms that provide an efficiency of $\eta = 100\%$ are only the theoretical limit. It would require to control all harmonics, while in practice, it is not feasible to go beyond the third harmonic. The theoretical efficiency limit for the case where only the lower harmonics are controlled is shown in Table 3.16. Practical PAE values will be even lower as no parasitic losses or other nonideal transistor behavior is regarded so far.

The power amplifiers so far all relied on the transistor as a controlled current source that generates a sine wave or parts of it. Another concept relies on the transistor as switch, just as in a high-efficiency dc–dc converter. A switch is either on or off, short or open circuit, respectively. Thus, it consumes ideally no power. The probably most famous concept relying on this principle is the Class-E amplifier.

This concept was proposed and patented by the Sokals in 1975 [50, 51]. The Class-E amplifier was derived to match the following conditions:

- account for device output capacitance
- minimize switching times

Table 3.16 Theoretical Maximum Efficiency of Class-F PAs [47]

n Harmonics of Voltage	$n = 1$	$n = 3$	$n = 5$	$n = \infty$
m harmonics of current				
$m = 1$	50%	58%	60%	64%
$m = 2$	71%	82%	85%	90%
$m = 4$	75%	87%	91%	96%
$m = \infty$	79%	91%	95%	100%

- voltage delay at switch turnoff
- zero voltage at switch turnon
- zero voltage slope at switch turnon
- flat top of voltage and current waveforms

These conditions are defined to make the concept useful in practical circuits. First, the condition *zero voltage at switch turnon* ensures that no capacitor is discharged when the transistor becomes a short circuit. Otherwise, all the stored energy would be lost. This of course requires to *account for device output capacitance*, too. During the event of switching, most of the power is lost. Thus, the output circuit is optimized to support *minimized switching times*. Finally, *voltage delay at switch turnoff* and *zero voltage slope at switch turnon* ensure low voltage while the transistor is switching, thereby reducing the dissipated power. *Flat top of voltage and current waveforms* has no impact on efficiency, but is required to protect the transistor from breakdown.

Figure 3.223 shows the idealized equivalent circuit of a first-order Class-E amplifier. The transistor is acting as a switch. The transistor's output capacitance is not neglected but absorbed into the capacitance C_1, which is a significant advantage of the Class-E concept.

In order to understand the principle of operation, we take a look at the currents and voltages shown in Figure 3.224. The currents in the collector (or drain) node are defined through the respective branches. The RF choke inductor L_1 forces I_{dc} to be a pure dc current. The series resonance of C_2 and L_2 guarantee that I_{RL} is a pure sinus signal at the fundamental frequency. The resulting current is a bias-shifted sinewave that is used to charge *and discharge* C_1 while the switch is open. Thus, I_{C1} has a positive and a negative portion, and as a result, the voltage V_T builds up *and returns to zero* again. After the voltage became zero, the switch is closed, and for the remaining time of the period, the current I_T flows through the transistor while the voltage is ideally zero. Therefore, by appropriate selection of the element values, either current or voltage at the transistor will be zero, and no charged capacitance is shortened when switching. Thus, the Class-E amplifier reaches the theoretical limit of $\eta = 100\%$. It is, however, usually not possible to actually see the switching of the current, even in circuit simulation. This is due to the fact that a major portion of C_1 might be the transistor's output capacitance. Hence, at the drain (or collector) node, the sum of I_T and part of I_{C1} is flowing and cannot be separated.

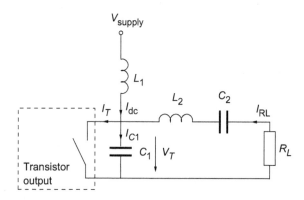

Figure 3.223 Circuit schematic of a first-order Class-E amplifier. The currents and voltages are shown in Figure 3.224.

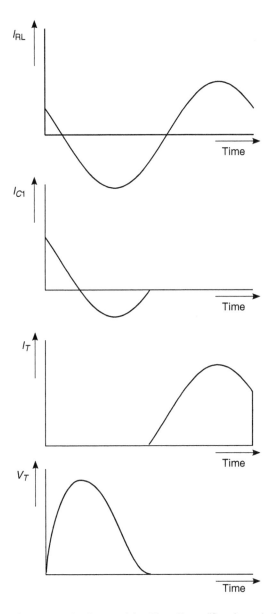

Figure 3.224 Currents and voltages of the Class-E amplifier shown in Figure 3.223.

It is of course necessary to choose the right circuit element values in order to achieve Class-E performance. Numerous works have been published that derive these values, for example, Refs. [50–54]. Also nonideal behavior and current conduction angles other than the standard 50% duty cycle were investigated. Values for the Class-E amplifier with a duty cycle of 50% are given in Table 3.17 [51]. With these relations between the different element values, an amplifier can be designed. It starts by choosing the dc voltage V_{CC}:

$$V_{CC} = \frac{BV_{CEV}}{3.65} SF \qquad (3.199)$$

Table 3.17 Dependence of Class-E Circuit Elements on Loaded Q and Output Power [51]

Q_L	$\frac{P \cdot R}{(V_{CC}-V_0)^2}$	$\omega C_1 R$	$\omega C_2 R$
∞	0.576801	0.18360	0
20	0.56402	0.19111	0.05313
10	0.54974	0.19790	0.11375
5	0.51659	0.20907	0.26924
3	0.46453	0.21834	0.63467
2.5	0.43550	0.22036	1.01219
2	0.38888	0.21994	3.05212
1.7879	0.35969	0.21770	∞

where BV_{CEV} is the transistor's breakdown voltage, and $SF \leq 0.8$ is a security factor. Since the maximum voltage, the transistor must withstand builds up when C_1 is charged, it is not limited to $2 \cdot V_{CC}$ as for Class-A amplifiers. In Class-E, the voltage peaks at values in excess of $3 \cdot V_{CC}$.

From a given V_{CC}, C_1, C_2, and R are determined from Table 3.17. This is an optimization procedure, as output power P should be maximized with realistic element values. Also, the transistor output capacitance constrains the minimum value for C_1. Table 3.17 also accounts for the saturation offset voltage V_0 that reduces the effective voltage swing. Finally, the value of L_2 is determined from the loaded Q of the series resonance circuit, $Q_L = \omega L_2/R$.

It is the promise of the Class-E concept that an efficiency close to 100% can be achieved without any kind of transistor modeling. Provided that the transistor behaves like a decent switch, no load–pull measurement and individual power matching optimization is required. The transistor just becomes a switch, and the rest is determined through the other circuit elements that are known a priori.

Regarding the different classes of amplifiers, it has to be noted that the definitions idealize the situation in a way that only a single frequency at a fixed amplitude is regarded. The issue of feedback or how to drive the amplifier is also not considered. But assuming a constant amplitude is much more severe. Modern communication standards use RF signals that do not fall into this category. It is rather common to have a bandwidth in excess of 10 MHz and an peak-to-average ratio of 10 dB. This means, that the average output power is only 10% of the peak power the amplifier must be able to provide. The high peak-to-average ratio means that, for example, optimizing an amplifier to optimum Class-C operation might not be feasible. For tuned concepts like the Class-E and Class-F approaches, also the relative bandwidth of modern signals is challenging. Class-E specifically is rather suited for narrowband PM/FM signals that inherently are of constant amplitude. Modern wireless power amplifiers, therefore, rely on linearization and efficiency boosting concepts like envelope tracking or the Doherty technique rather than on the pure single-transistor amplifier stage.

3.12.1 Example 1: 7-W Class C BJT Amplifier for 1.6 GHz

The following demonstrates how to design a 1.6-GHz, 7-W amplifier operating in Class C. We need to point out again that all modulation forms that have the information in both amplitude and phase require more linearity than Class C affords. Needless to say, this affects the efficiency; therefore, at a later point, we will also show a Class AB power amplifier. This is a tutorial and we are going to take a UHF example, and we will walk the reader through all the necessary problems arising from the design. The selection of the transistor is done by first looking at the f_T, which in our case should be at least 4–5 GHz. The first order of business is to define the point of operation for the output amplifier. To do this, we need to determine maximum voltage and current values supplied by the manufacturer. We have chosen a device that operates at 28 V and can sustain a peak collector current of 1.5 A. Figure 3.216 lets us determine the conducting angle Θ from the I_C/V_{BE} plane.

This figure shows on the left side the I_C/V_{BE} plane and the necessary voltage to drive the amplifier.

Initially, we assume that the base–emitter junction is operated without any bias ($V_{BE} = 0$). The dc I–V curve shows that the collector current starts at about 650 mV. The amplifier itself therefore is already in Class-C operation. For full output power, the drive voltage V_{BEpeak} has to be 1.14 V. Therefore, the conducting angle can be determined by

$$\cos \Theta = \frac{V_{be}}{\hat{v}_{be}} = \frac{0.65}{1.14} = 0.57$$

$$\cos 0.57 = 55°$$

V_{be} is the dc voltage at which the transistor starts conducting. From Figure 3.219 and equation (3.198), the fundamental current i_{c1} can be determined as

$$i_{c1} = 1.5\,A \times 0.36 = 0.54\,A$$

The average collector dc current yields

$$i_{c2} = 1.5 \times 0.21 = 0.315\,A$$

The collector saturation voltage $V_{ce(sat)}$, unfortunately is around 2–3 V; for the purpose of the calculation, we assume that the saturation voltage is 3 V, and therefore we can assume that the maximum collector ac voltage $\hat{v}_{ce} \approx (V_B - 3\,V) = 25\,V$.

The appropriate collector load has to be

$$R_c = \frac{\hat{v}_{ce}}{i_{c1}} = \frac{25\,V}{0.54\,A} = 46.3\,\Omega$$

The maximum output power is

$$P_{out} = \frac{\left(V_{ce} - V_{ce(sat)}\right)^2}{2R_c} = \frac{625}{92.6} = 6.75\,W$$

The only way to increase the power would be to have a smaller collector load at a higher current. In this family of Motorola transistors, there was no 10-W transistor; that is why we have chosen a 20-W experimental device, which for this purpose is overkill. The theoretical efficiency of the amplifier is defined as

$$\eta = \frac{P_{out}}{P_{in}} = \frac{6.75\,W}{0.315 \times 28} = 85\%$$

If the saturation voltage would be less, our efficiency would automatically increase since for the given load, the actual output power would increase (6.75 W → 7.5 W). Assuming that this stage would operate at 1900 MHz with a gain of roughly 22 dB, the required drive would be significantly less than the output power, or roughly 40 mW. For the purpose of the following calculations, we will assume that the drive power is 0.7 W. The total dissipation of the device now adds up to be

$$P_T = P_{\text{IN(dc)}} - P_{\text{OUT}} + P_{\text{DRIVE}}$$

$$P_T = (28 \times 0.315) - 6.75 + 0.7 = 2.77$$

Power-added efficiency is derived from this. Knowing the power dissipation, one can calculate a thermal resistance R_{therm} of the heatsink and the transistor:

$$R_{\text{therm}} = \frac{T_j - T_{\text{amb}}}{P_T}$$

where T_j is the device junction temperature and T_{amb} is the ambient temperature. For a maximum ambient temperature of 60°C and a maximum junction temperature of 150°C, the thermal resistance

$$R_{\text{therm}} = \frac{120 - 60}{2.77} = 21.7\,°\text{C/W}$$

R_{therm} is the sum of thermal resistance of the transistor and its package, and the thermal resistance between the transistor package and the heat sink, and the thermal resistance between the heat sink and the ambient air. Therefore

$$R_{\text{therm}} = R_{\text{th(case)}} + R_{\text{th(case/heatsink)}} + R_{\text{th(heatsink)}}$$

The required thermal resistance of the heatsink, which helps us to determine its size or surface area, is calculated from

$$R_{\text{th(heatsink)}} = 21.7 - 10 - 1.4 = 15.6°\text{C/W}$$

Assuming an aluminum heat sink (a 2-mm-thick plate), the required heatsink surface area A (cm^2) is

$$A = \frac{1}{\alpha R_{\text{th(heatsink)}}} : \qquad \alpha \approx 1.5 \frac{\text{mW}}{°\text{Ccm}^2}$$

Therefore, the required area is

$$A = \frac{1}{1.5 \times 0.0156} = 42.7\,\text{cm}^2$$

The next task is to match both the input and output to the (most likely) 50-Ω source and termination. The following section will guide us through how to do this. The input impedance Z_{11} (after some lengthy search) was found to be $1 + j10$ Ω. This means that the input consists of the base spreading resistance (approximately 1 Ω) and an input capacitance, which because of phase shift now turns out to be inductive. Since the popular method for input matching these days is a pi configuration, we are going to use a simple pi filter both for input and output matching. There are several equations in the literature that give the values for the two capacitors and inductors; most of them result in an incorrect resonance frequency. Since we have access to software tools, we decided to take three approximate values and

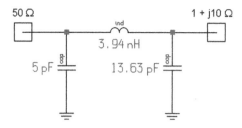

Figure 3.225 Input matching network for the amplifier used in this tutorial.

use the optimizer to give us the appropriate matching. The only remaining variable in such an approach is the bandwidth. If we specify $S_{21} = 1$ and $S_{11} = 0$ (100% energy transfer and perfect return loss), we will realize that this is possible only at one frequency. We therefore need to identify a frequency band or bandwidth over which an acceptable match is needed. As the bandwidth increases, the match for such a simple circuit will get poorer; there are not enough elements of freedom available. In this case, we took the center frequency plus/minus 100 MHz. Figures 3.225 and 3.226 show the actual circuit with the element values and the resulting input match. Likewise, we did the same thing at the output, but the actual resulting dynamic output capacitance is not known *a priori*. For low frequencies, the integral equation gives a value of $2C_{CE}$; however, practical tests in frequencies above 500 MHz show that the actual increase is only 1.3 times the dc output capacitance. Therefore, we decided to design the filter with no output capacitance assumed, and the input capacitance of about 5 pF needs to be reduced for proper reactive matching. Mathematically, this could result in a negative number, specifically if the resulting output capacitance exceeds the 5 pF value. To compensate for this, one needs to use an RF choke that tunes out the resulting dynamic output capacitance and then one can revert to the original 5 pF as calculated.

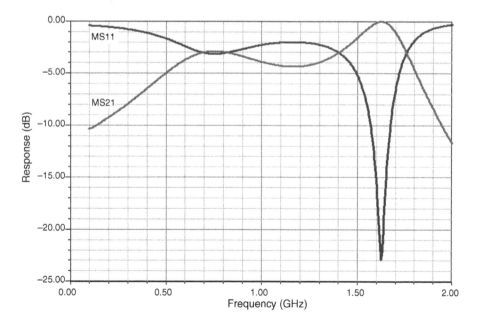

Figure 3.226 Frequency response of the input matching network.

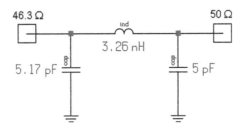

Figure 3.227   Output matching network for the amplifier.

The selection of the actual component is another issue, because these capacitors need to handle several amperes of RF current. It is not always easy to find capacitors with the appropriate value that can handle the current. The inductors can be printed. The paragraph following this section therefore will deal with the issue of how to translate the lumped values into distributed values, which will make the current handling much easier (we hope).

For this, we did the same optimizing approach by using a 50-Ω source and a 46.3-Ω load. Figures 3.27 and 3.28 show the frequency-dependent responses and the circuit with the appropriate values.

The next step is to connect the input and output matching network to an "actual" transistor. This is a point where difficulties become very high, since few companies provide SPICE-type parameters for transistors operating at currents more than 200 mA. Figures 3.229 and 3.230 show the overall schematic and frequency response of the actual amplifier with all filters connected. If more bandwidth is needed, the matching circuit must be increased in complexity, as will be shown in the following section.

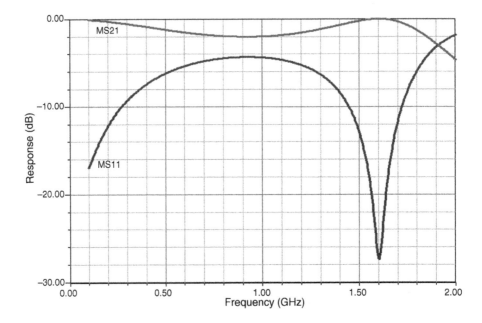

Figure 3.228   Frequency response of the output matching network.

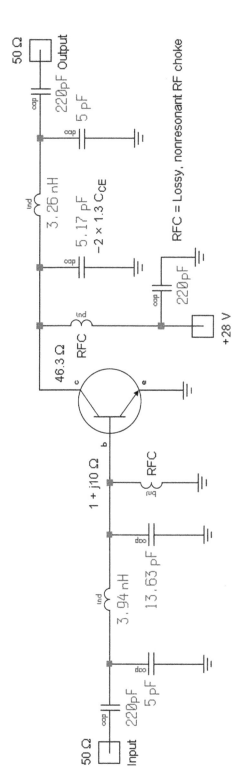

Figure 3.229 Actual schematic of our experimental 1.6-GHz amplifier. The RF chokes have enough RF losses not to be resonant and yet do not load the circuit. This is frequently accomplished either with ferrite beads or resistors in parallel with the chokes.

557

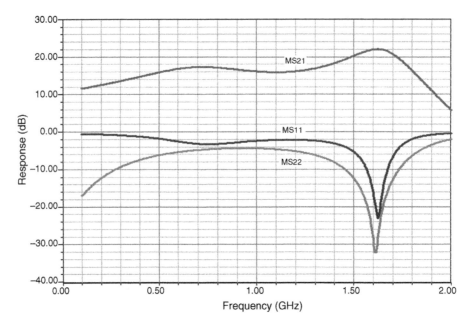

Figure 3.230 Overall frequency response of our experimental amplifier. Input and output matching, and gain, are shown as a function of frequency.

Now is a good time to look at the final results of our simulation, which is based on the various assumptions as outlined above. Figure 3.231 shows the dc I–V curve with a straight load line and no reactances. This was accomplished by using a wideband transformer instead of an output matching circuit. Figure 3.232, however, shows the "load line" that results when

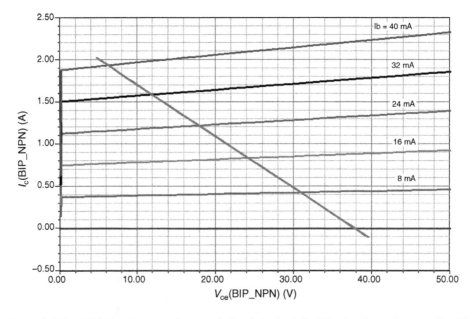

Figure 3.231 DC I–V curves assuming a perfect resistive load. No RF saturation voltage or other RF saturation effects are evident.

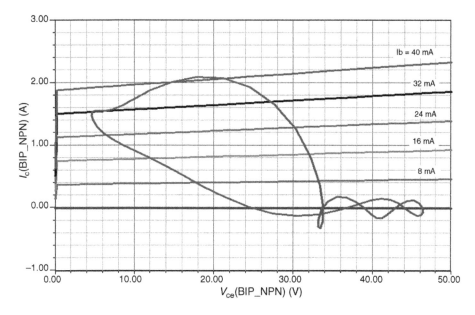

Figure 3.232   Load line with the output transformer replaced with a real matching circuit.

the transformer is replaced with a real matching network. Based on the harmonic contents and its improper termination, one can see how the output load line opens and shows ringing in the low-current area. If we expect to see the saturation voltage, we will be disappointed because the saturation voltage is frequency dependent, increases with frequency, and this

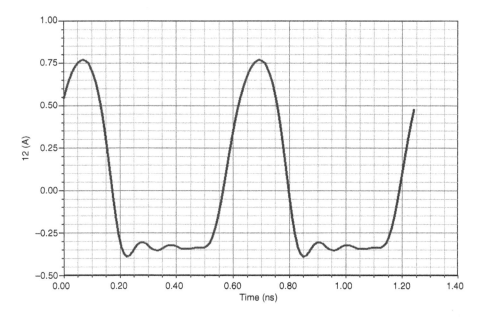

Figure 3.233   Output current as a function of time. The duty cycle is determined by the device conduction angle.

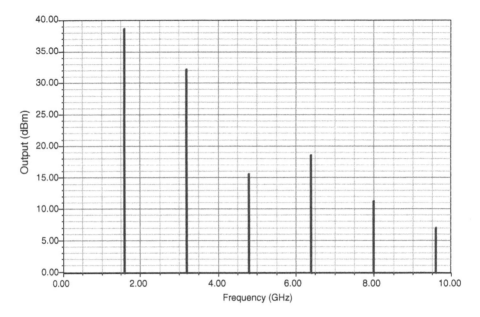

Figure 3.234   Power output at the collector prior to any filtering, assuming a real output termination.

figure would give an erroneous impression. Hence, we show dc *I–V* curves. To display the conduction angle, which we showed previously, we are going to look at the output current as a function of time at a drive level of +16.5 dBm (Figure 3.233). During the negative half, the transistor is not conducting as predicted, so this validates our assumptions. Next, we are interested in determining the power gain. Therefore, we look at the output spectrum for a drive level of +16.5 dBm (Figure 3.234). It shows an output level of 38.6 dBm. If we subtract the drive level (16.5 dBm) from this, we see that the power gain is 22.1 dB at this drive level. Since we are used to thinking in watts, Figure 3.235 shows the absolute output power and the harmonics prior to filtering. Figure 3.236 shows the output spectrum after replacing the transformer with a simple pi output network.

During the conduction period, there will be a voltage across the transistor 180° out of phase, as shown in Figure 3.237. Finally, to show the lack of linearity of a Class C amplifier, Figure 3.238 shows the amplifier's output power versus drive, and Figure 3.239 shows the amplifier's gain versus drive. When the transistor turns on, it is clearly evident.

A particular difficulty arises in the case of FETs, where the charge energy makes for a difference between dc and pulsed measurements. Figure 3.240 shows this. Some of the more modern transistors, such as SiGe types, may need some pulsing also, and as can be seen in Figure 3.240, even depending on the pulsing one may get different dc *I–V* curves. This area is still under investigation, and even the "experts" do not agree on all aspects of this. In real life, it turns out that a lot of experimentation is the answer. Anybody who thinks that a pure CAD approach will give the right design answers immediately will encounter many surprises! This is not a fault of the CAD tools—assuming that they are properly engineered—but rather the availability of appropriately "tweaked" models for high-power applications, and the willingness of device suppliers to standardize on certain models and to support them.

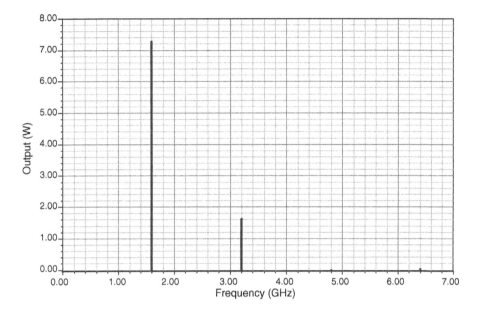

Figure 3.235   Available output power in watts, prior to any filtering, assuming a real output termination.

Because of the difficulties involved in cooling transistors under test and simulating RF pulse and CW conditions, depending on the application, manufacturers have so far stayed away from investing money in this area. The designer must therefore resort to application notes published by device manufacturers. Because even slight layer changes can have a large impact on the frequency response of a given device, application-note-based designs

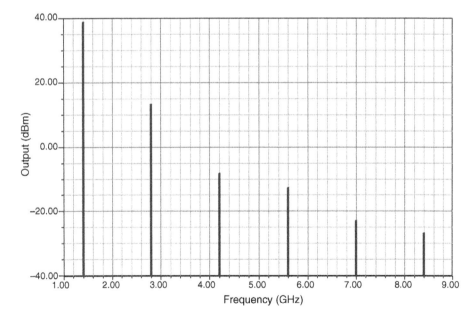

Figure 3.236   The output spectrum after replacing the transformer with a simple π output network.

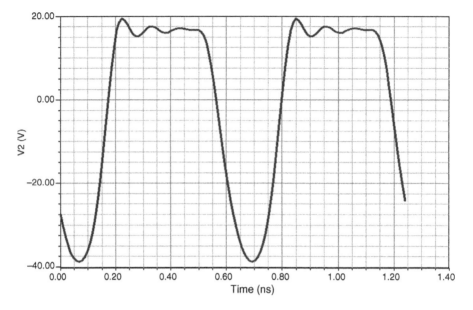

Figure 3.237 Output voltage at the collector. The voltage is asymmetrical compared to 0; this is important in the selection of a device capable of handling these voltages.

can run into difficulty if a device's fabrication has significantly evolved since the publication of its application notes.

Given the multitude of models outlined in Chapter 2, the CAD user is sometimes left hanging as to which is the best model to use. The latest indications are that the

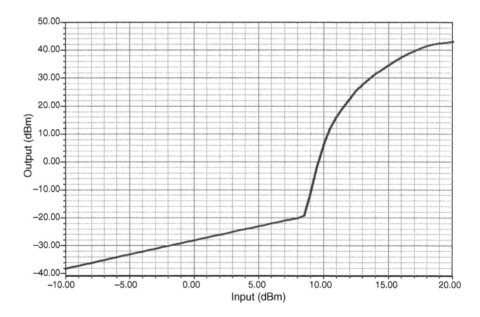

Figure 3.238 Output power versus input power for the Class C amplifier. Until the transistor turns on (at a drive level of about 8.5 dBm), only fed-through power appears at the output.

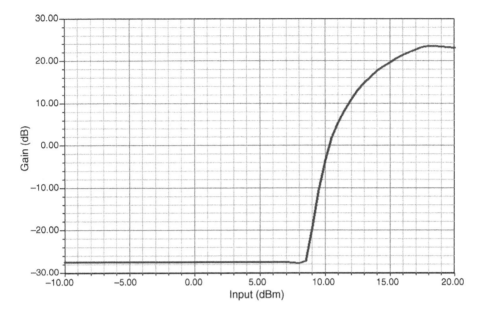

Figure 3.239 Amplifier gain as a function of drive. It is evident that one needs to cross the 0.65 V point for the transistor to become conductive and show gain. The initial amplification as a function of drive shows a very steep curve that then levels off and droops as the amplifier moves into saturation.

Most Exquisite Transistor Model (MEXTRAM) model, developed by the University of Delft, Holland, with Philips of Holland, shows the most promising result, specifically in the area of IMD products. While other late models have up to seven internal nodes, MEXTRAM fortunately uses only five. (The MEXTRAM model actually consists of three transistors, which makes modeling fun.) Power-added efficiency simulation is not available, since the MEXTRAM model was developed for small-signal applications, meaning circuits that operate at less than 100 mW of RF output power.

As if the absence of trustworthy high-power device models is not challenge enough, in developing this 1.6-GHz Class C amplifier example, we ran into another complication: using a CAD optimizer to simultaneously match circuit's input and output circuits failed because of the device's input impedance variation with drive. Every matching adjustment that improved the drive to the device resulted in an input-impedance change that reduced the drive to the device! In the end, the optimizer could not keep up. In actually building such a circuit, one would make the matching-network values adjustable on the initial breadboard and then, with the necessary input drive level applied, iteratively adjust the input and output matching for the desired output level. It appears that the CAD program could not handle such a degree of nonlinearity.

We then went to the approach of using large-signal S parameters at a drive level slightly less than 16 dBm, but this detuned the input filter, resulting in a poorer input match than the one initially predicted. Likewise, the output matching, which in a power stage must result in maximum output power rather than a conjugate match, was less impressive. The large-signal S parameters at the output for a power amplifier are somewhat dubious because Class C operation generates so many harmonics that the output termination will reflect a lot of energy and, depending on the output filter topology, different results can be expected.

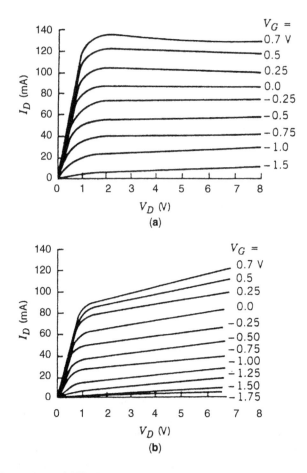

Figure 3.240 Comparison of FET drain current characteristics obtained from (a) pulsed *I–V* and (b) rectified sine-wave evaluation. The rectified sine-wave curves can be considered equivalent to curves obtained at dc.

We firmly believe that there will be further mathematical improvement in the CAD tools, and we very much invite our readers to use such a nonlinear example to test their tools, provided that a high-power model becomes available in this power range.

It has been rumored that the load-pull technique, rather than the application of large-signal *S* parameters, is a vehicle to solve such matching problems. However, the load-pull technique is really applied only to the fundamental frequency and does not deal with the issue of harmonics of the output current.

Actual measurements of amplifiers of this type show that our simulated efficiency and drive level come quite close to reality, but at a frequency somewhat offset from that chosen for the design. As an example, the input and output filters tuned at 1.4 GHz instead of 1.6 GHz using the large-signal *S* parameters. Again, an iterative process, similar to manually trimming a prototype, will give a similar result. This is consistent with our earlier statement that most final designs are really hand-tweaked, and that CAD can only bring us close, but

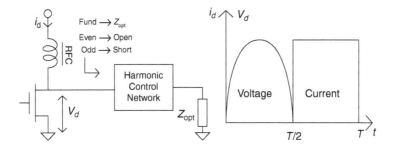

Figure 3.241   Current and voltage waveforms for an ideal inverse Class F PA.

not all the way, to a good solution. For passive problems, the CAD does a significantly better job; the nonlinear cases are the ones that still cause headaches.

3.12.2   Example: A Highly Efficient 3.5 GHz Inverse Class-F GaN HEMT Power Amplifier[4]

In the following, the design of an inverse Class-F amplifier is presented. It is a 3.5 GHz broadband high-efficiency design providing 12 W of output power. These specifications call for GaN HEMT technology since it provides higher cutoff frequencies compared to LDMOS while maintaining high power level.

3.12.2.1   Inverse Class-F PAs

All high-efficiency PAs are based on careful control of the harmonic content of the voltage and current waveforms at the transistor intrinsic terminals. The construction of an ideal inverse class-F PA consists in imposing a square and halfsinusoidal waveform for drain-current and drain-to-source voltage, respectively. Figure 3.241 presents the drain-current and drain-to-source voltage waveforms.

These idealized waveforms are obtained by control of the transistor load impedance at different harmonics. The optimum resistive load impedance must be matched to the system impedance (Z_c) at the fundamental frequency. Moreover, the output impedance seen by the device at even harmonics has to be open circuited because the half sine voltage wave only consists of even harmonics. On the other hand, all odd harmonic impedances must be short circuited in order to approximate a square drain current. According to [55], the second, $2 f_0$, and third, $3 f_0$, harmonics are most important for high-efficiency operation. In Ref. [56], it is shown that controlling these two harmonics is usually enough in practical microwave PAs.

Furthermore, controlling more harmonics increases circuit complexity without necessarily improving the performance [57]. On the contrary, manipulation of excessive harmonics

[4]Based on P. Saad, C. Fager, H. M. Nemati, H. Cao, H. Zirath, and K. Andersson: "A highly efficient 3.5 GHz inverse class-F GaN HEMT power amplifier" *International Journal Microwave Wireless Technology*, DOI:10.1017/S1759078710000395. © Cambridge University Press and European Microwave Association 2010, reprinted with permission.

may reduce the bandwidth of the PA. Thus, close to optimal, load impedances are given by

$$Z_L(f_0) = Z_{\text{opt}} \tag{3.200}$$

$$Z_L(2\ f_0) = 1 \tag{3.201}$$

$$Z_L(3\ f_0) = 0 \tag{3.202}$$

where f_0 is the fundamental frequency and Z_{opt} is the optimal load impedance at the fundamental frequency.

3.12.2.2 Design Methodology

In order to reduce the parasitics of the package and facilitate harmonic impedance optimization at the transistor output reference plane, a bare-die Cree CGH60015DE GaN HEMT is used.

In switched-mode PAs and in harmonically tuned amplifiers, the transistor operates in the on- and off-regions. A simplified transistor model optimized for this type of operation is developed and used in the PA design. The model is based on simplified expressions for the nonlinear currents and capacitances where focus is put on accurately predicting the high efficiency, on- and off-regions of the transistor characteristics. The model allows the intrinsic waveforms to be studied in the PA design and therefore allows a careful investigation of the transistor operation. Since the transistor is almost driven like a switch, diode models are added to the common HEMT model in order to accurately predict forward gate voltage and negative drain voltage conditions of a GaN transistor. The topology of this model is shown in Figure 3.242.

The first step to design the PA was to find the optimum input and output load conditions to maximize the output power and efficiency of the transistor. The procedure for optimization of the fundamental and harmonic impedances is summarized as follows:

1. Perform a fundamental load-pull/source-pull simulation to find the optimum fundamental load and source impedances for efficiency and output power.

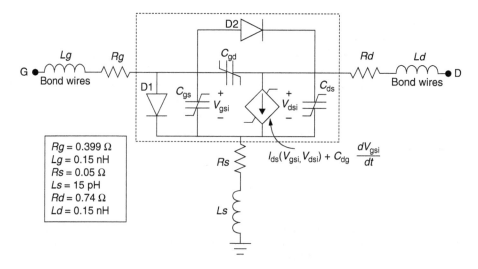

Figure 3.242 Large signal GaN HEMT equivalent circuit model.

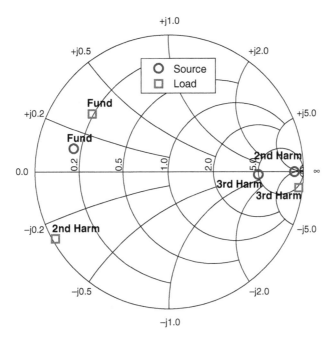

Figure 3.243 Simulated optimum harmonic loads at the transistor reference plane for maximum PAE.

2. Using the impedances found in the previous step, a harmonic load-pull simulation was performed to find the optimum second and third harmonic load and source impedances for high-efficiency operation. The obtained optimum source and load impedances at fundamental, second, and third harmonics that maximize the power-added efficiency (PAE) are shown in Figure 3.243.

Figure 3.244 shows the simulated intrinsic drain voltage and current waveforms of the transistor, corresponding to 80% PAE. The drain voltage waveform is a half-sinusoid whereas the drain current waveform is close to a square wave, which corresponds to the inverse class-F waveforms shown in Figure 3.241.

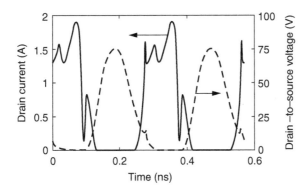

Figure 3.244 Simulated intrinsic current and voltage waveforms of the transistor resulting in 80% PAE at 3.5 GHz.

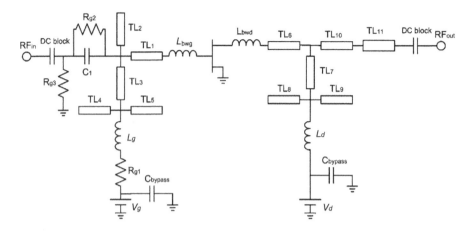

Figure 3.245 Circuit topology of the PA.

The circuit diagram of the designed inverse class-F PA is depicted in Figure 3.245. The space between the bonding pads on the chip and the PCB lines is reduced as much as possible in order to avoid narrowband and therefore sensitive harmonic matching. L_{bwg} and L_{bwd} are used in the circuit design to model the input and output bondwire inductances, respectively. Their values are estimated to 0.15 nH.

The input matching network consists of transmission lines TL_i; $i = 1 \ldots 5$, which are optimized to provide, at the input of the device, the optimum impedances obtained from the source/load pull simulations, see Figure 3.243.

The input matching network has been slightly modified in order to stabilize the PA. The Rollet stability factor (k) of the amplifier can be improved by increasing the real part of Z_{11}:

$$k = \frac{2 Re(Z_{11}) Re(Z_{22}) - Re(Z_{12} Z_{21})}{|Z_{12} Z_{21}|} \tag{3.203}$$

A 39-Ω series resistance R_{g2} is therefore added at the input of the amplifier to improve the stability in the high-frequency band. Further improvement in the amplifier stability in the low-frequency band can be achieved by reducing the low-frequency gain. The parallel resistance R_{g3}, set to 400 Ω, increases low-frequency stability by reducing the input impedance.

The output matching network consists of transmission lines TL_i; $i = 6 \ldots 11$, which provides the optimum fundamental, second, and third harmonics load impedances at the output of the device.

The values of the inductors L_g and L_d are equal to 28 and 8 nH, respectively. They are used to prevent the leakage of RF into the dc supply lines. The circuit was optimized for wideband operation to minimize the impact of mounting and manufacturing tolerances. It is important to note that the original design did not change significantly after the optimization.

Finally, Monte-Carlo simulations have been used to study the impact of components variability and uncertainty on the PA performance. Uncertainties introduced by the manufacturing process and the lumped components have been considered. The Monte-Carlo simulations have shown that the design is robust and not very sensitive to these variations.

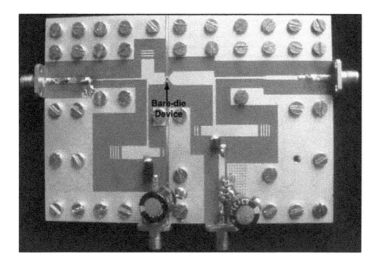

Figure 3.246 Fabricated inverse class-F PA. Size: $11 \times 8 \, cm^2$.

3.12.2.3 *Implementation and Measurement Results*

The inverse class-F PA was implemented on a Rogers 5870 substrate with $\epsilon_r = 2.33$ and thickness of 0.8 mm. Figure 3.246 shows a picture of the fabricated inverse class-F amplifier using the bare-die GaN HEMT device.

The implemented PA has been characterized by large signal and modulated measurements to study its performance.

Large Signal Measurements   All measurements were made using a continuous wave input signal generated by an Agilent E4438C microwave synthesized source boosted by a microwave driver amplifier and the relevant power levels were measured by a power meter (Agilent E4419B). A low-pass filter with cut-off frequency of 6 GHz has been added at the output of the PA to filter out the power at the harmonics. Hence, only the power at the fundamental frequency have been measured and used in evaluating the PA performance.

First the bias sensitivity was investigated by a gate bias, V_{GS} sweep. Figure 3.247 shows simulated and measured output power, P_{out}, and PAE versus V_{GS}. The measurements show that the performance of the PA is not very sensitive to the gate voltage. Results of the drain voltage, V_{DS}, sweep using 29 dBm input power are shown in Figure 3.248. It can be noticed that the output power can be further increased by increasing V_{DS}. The measured results agree well with simulations when the input drive level is high enough. From Figures 3.247 and 3.248, we conclude that the optimum bias for the PA is $V_{DS} = 28$ V and $V_{GS} = -2.5$ V.

A power sweep measurement has been performed at 3.5 GHz. The simulated and measured output powers versus input power are shown in Figure 3.249. As expected, good agreement between simulation and measurement results is obtained at high power levels, where the transistor is operated in a high-efficiency mode that the model was optimized for. The peak power level at the output of this amplifier is about 41 dBm or 12 W. Figure 3.250 shows the simulated and measured gain and PAE versus input power. A peak PAE of 78% is measured for an input drive level of 29 dBm. There is a good agreement between the simulated and measured results close to the peak efficiency operation.

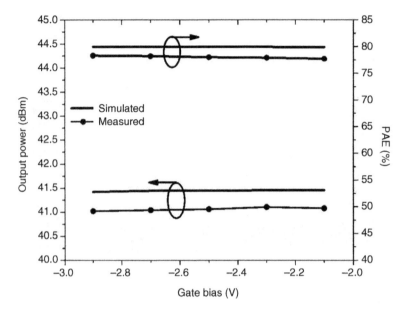

Figure 3.247   Simulated and measured output power and PAE versus gate bias.

The poor agreement between simulations and measurements at low input powers in Figures 3.249 and 3.250 is due to the fact that as the input power is reduced and the device is less overdriven, the harmonic contents of its output waveforms are no longer strong enough to be able to form high-efficiency switching conditions. Therefore, the device operation extends outside of the region where the model is optimized and the simulation results start

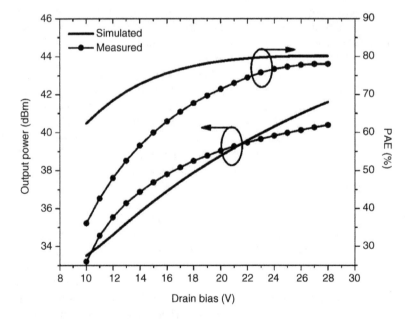

Figure 3.248   Simulated and measured output power and PAE versus drain bias.

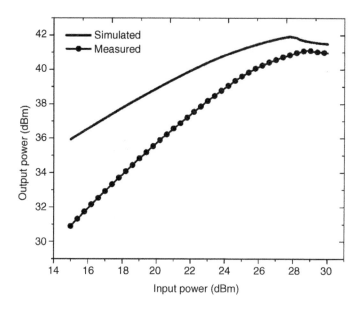

Figure 3.249 Simulated and measured output power versus input power.

to deviate from measurements. This is a fact that the PA designer should be aware of. The model is mainly optimized for switched mode or harmonically tuned overdriven operations where conventional models lack accuracy and fail to provide convergence in simulations.

The PA has been characterized versus frequency from 3 to 4 GHz, with a 29 dBm fixed input power drive level. The PAE and gain of the PA are plotted in Figure 3.251 and compared with simulations. A good agreement between simulation and measurements is observed at the PA operation frequency. A maximum gain and PAE of 12 dB and 78%, respectively,

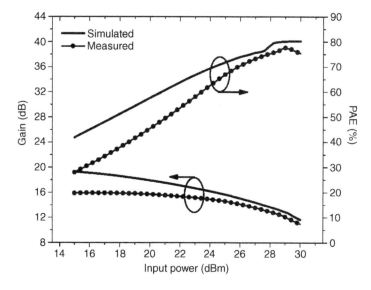

Figure 3.250 Simulated and measured PAE and gain versus input power level.

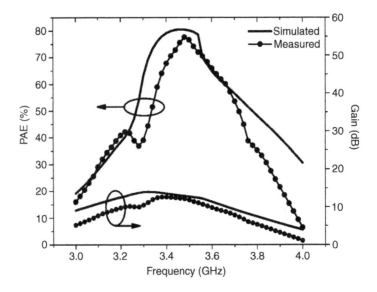

Figure 3.251   Simulated and measured PAE and gain versus frequency for 29 dBm input power.

are located at 3.5 GHz corresponding to a drain efficiency of 82% at this frequency. The amplifier exhibits higher than 50% PAE between 3.32 and 3.72 GHz, which corresponds to greater than 10% fractional bandwidth.

Input return loss have finally been measured under large signal conditions using 29 dBm input power, in the frequency band 3–4 GHz. Simulated and measured input return loss are shown in Figure 3.252, a return loss of 14 dB is obtained at 3.5 GHz and agrees well with simulations.

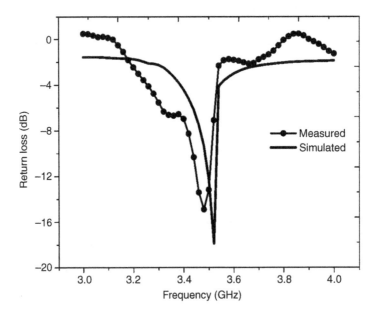

Figure 3.252   Simulated and measured large-signal input return loss.

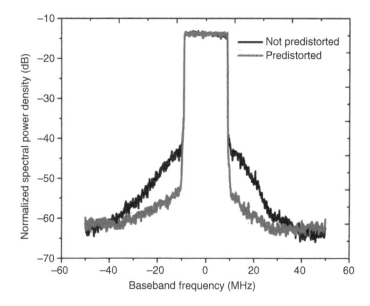

Figure 3.253 PA output signal spectrum of a 20 MHz LTE signal at center frequency of 3.5 GHz before and after DPD.

Modulated Measurements   Modulated measurements are used to evaluate the performance of the PA and show that the PA is linearizable to meet modern wireless communication system standards. In the experiment, a 20 MHz long-term evolution (LTE) signal with 11.2 dB peak-to-average ratio (PAR) and a 5 MHz WCDMA signal with 6.6 dB PAR are used. A relatively low average efficiency is expected when such signals with high PAR are used since the PA has to operate at large back-off most of the time.

The digital-predistortion (DPD) used, for both LTE and WCDMA signals, is the memory polynomial model with nonlinear 11 and memory depth 5 [59]. The measured output spectrum at 3.5 GHz of the LTE signal, before and after DPD for an average input power of 18 dBm, is shown in Figure 3.253.

The adjacent channel power ratio (ACLR) of the PA without DPD reaches −32 dBc with an average PAE of 35% whereas the ACLR of the PA with the DPD applied reaches −42 dBc at an average PAE of 30%.

Figure 3.254 shows the measured output spectrum at 3.5 GHz of the WCDMA signal before and after DPD. In this measurement, the average input power used is 19.4 dBm, so the PA was operating at 9.6 dB back-off. When the DPD is applied, the average PAE is decreased from 45% to 40% while 13 dB improvement in ACLR, from 234 to −47 dBc, is obtained.

Average PAE and ACLR without DPD for WCDMA and LTE signals are displayed versus average output power in Figure 3.255.

3.12.2.4 Conclusions
A high-efficiency inverse Class-F PA using a GaN HEMT has been presented. The design methodology is based on load-pull/source-pull and harmonic load-pull simulations. A bare-die device, instead of a packaged transistor, is used to minimize the influence of parasitics and therefore take full advantage of recent device technology improvements. The peak

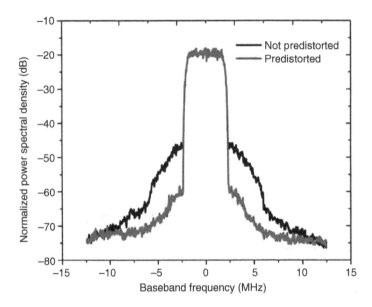

Figure 3.254   PA output signal spectrum of a 5 MHz WCDMA signal at center frequency of 3.5 GHz before and after DPD.

PAE of 78% with a power gain of 12 dB was achieved at an output power of 41 dBm at 3.5 GHz. A very broadband performance, with a power gain over 10 dB and PAE over 60%, was maintained over 300 MHz bandwidth. When DPD is used, modulated measurements demonstrate an average PAE of 30% and ACLR of –42 dBc for LTE signal. For WCDMA signal ACLR of –47 dBc and average PAE of 40% were obtained. These results represent state of the art for GaN PAs in this frequency range and demonstrate the success of the selected bare-die mounting, modeling, and circuit design methodologies used.

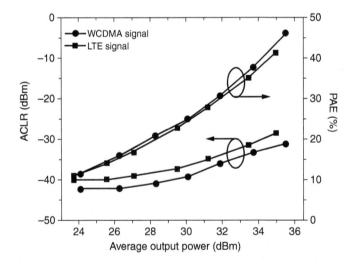

Figure 3.255   PA average performance with a power-swept 20 MHz LTE and 5 MHz WCDMA signals at center frequency of 3.5 GHz.

3.12.3 Linear Amplifier Systems[5]

In the past, solid-state linear RF power amplifiers have mainly been used for applications such as HF SSB and combined amplification of video and audio signals in TV transmitters. The linearity and efficiency requirements of these systems could often be met using relatively simple techniques, usually with manual alignment. Modern wireless systems increasingly make use of complex modulation schemes with demanding restrictions on spectral regrowth. The need to faithfully reproduce input signals (often in baseband I–Q format) and to keep the amplifier output within a strict emissions mask, while retaining high efficiency, has led to a resurgence of interest in linearization and efficiency enhancement techniques. Another application for the technologies has been in flexible multicarrier transmitters, where single channel amplifiers and cavity combiners are replaced by a single, wideband, high-power, ultra-linear amplifier. This section provides a brief review of the most popular linearization techniques and compares the trade-offs in terms of linearity, bandwidth, efficiency, and complexity. It must be stressed that the starting point for all the techniques described is always a carefully designed and repeatable power amplifier using well-characterized, reliable transistors. The techniques outlined below should not be used to compensate for poor frequency response, significant batch variations, poor ruggedness, and instability.

3.12.3.1 Class A/AB Operation and Power Back-Off

The most obvious, and certainly the simplest way to improve the linearity of an amplifier is to operate it in Class A and to reduce the operating power level (back-off) until the required linearity is obtained. In an ideal Class A amplifier, third-order IM products fall off with a 3rd power law, fifth order with a 5th power law, and so on; a few dB back-off can result in a significant reduction in IM products, especially the higher order products. This technique is often used in lower power circuits, and also for ultralinear applications such as UHF TV transmitters, where exceptional levels of back-off may be used (35 W of average RF output power from > 600 W of dc input). The price to be paid for the simplicity of this technique is poor efficiency and the need for overrated RF transistors (and additional combining networks if high power levels are required). Figure 3.256 shows the performance of an ultralinear 900 MHz amplifier constructed from a quadrature-combined pair of push–pull bipolar transistors, operating with a collector voltage of 26.5 V and a total supply current of 6 A.

The third-order IM distortion is very low, and falls off by around 8 dB for every 3 dB reduction in output power, which is slightly less than the theoretical figure of 9 dB per 3 dB. The higher order IM products at average powers below 2 W were almost immeasurable on a spectrum analyzer. The penalty for this linearity is exceptionally low efficiency of only 2.5% at 4 W output power. Wireless systems typically use 1–5 dB of back-off, with efficiencies in the range 10–40%. A more efficient alternative to Class A is Class AB operation, with bias typically set to between 10% and 50% of the Class A level. Linearity is obviously compromised by this approach and the effect of back-off becomes heavily dependent on the specific characteristics of the RF device technology. Some technologies such as LDMOS

---

[5]Based on "An overview of linear amplifier systems," Application note of UltraRF, copyright by Cree, reprint with permission.

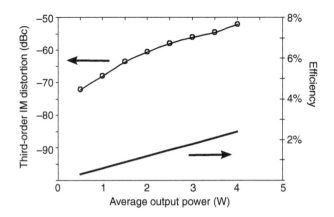

Figure 3.256 Class A amplifier with back-off.

demonstrate excellent Class AB linearity, but the rate of reduction of third-order IM with back-off is typically less than that in Class A and may reach a plateau at around −50 dBc.

3.12.3.2 RF Feedback

RF feedback is often used in low power amplifiers, but its application becomes limited as the operating frequency and output power increase. The finite time delay around the feedback loop reduces the bandwidth over which stable linearization can be achieved and makes feedback around multiple stages impractical. At high power levels, dissipation in the feedback network can become significant, forcing the use of high power resistors, which leads to cost increases and mechanical complexities. The other limitation is, of course, the fact that a single RF amplifier stage at UHF may only have a gain of around 10 dB and that it is only practical to reduce the gain (and hence the IMD) by a few dB.

3.12.3.3 Modulation Feedback

Modulation feedback techniques use some form of detection or demodulation to recover representations of the baseband modulating signal and the power amplifier output signal. The difference between the two signals provides error signals, which are then used to apply correction to the amplifier drive or control signals. Simple systems typically apply only amplitude feedback, while more advanced systems apply both amplitude and phase correction (in either polar or Cartesian format). Since the feedback is applied to the modulation rather than the RF signal, it is possible to achieve stable operation with a relatively large amount of feedback.

Envelope Feedback (Terman and Buss, 1940s) [60]   This is a very simple technique, which corrects for AM–AM distortion. It is also often used in transmitter AGC loops to compensate for power amplifier gain variations and to control pulse shaping in TDMA transmitters. A detector is used to demodulate the AM component of the PA output. This signal is fed back to a differential amplifier and compared with a detected sample of the input signal. The difference between the two signals represents the AM distortion of the amplifier and this signal is amplified, filtered, and used to modulate the driver stage of the power amplifier, see Figure 3.257.

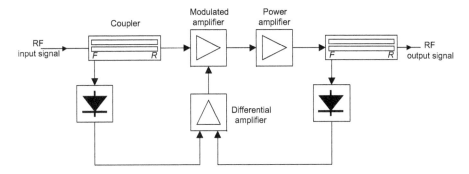

Figure 3.257 Envelope feedback.

The detectors must have wide dynamic range and accurate tracking, otherwise the loop gain and error signal accuracy will be signal dependent. This will lead to reduced correction and may even lead to an increase in higher order distortion products. The technique does not compensate for phase distortion, and, if there is significant delay in the signal processing, a phase difference may be created between the AM and PM components of the signals, degrading or imbalancing the correction process. In simple terms, this implies that the bandwidth of the correction circuitry must be 10 or more times wider than the bandwidth of the modulating signal envelope.

Envelope Elimination and Restoration (Kahn, 1950s) [61] This technique was initially developed for SSB and TV transmitters, although it is applicable to other modulation schemes. It should be noted that EE&R is not a linearization technique; it is an efficiency enhancement technique that can be used for linear amplification. Figure 3.258 shows a schematic of the system. The incoming RF signal is split into amplitude and phase components using a detector and a limiter, respectively. The phase component is amplified in a Class C amplifier. In the original (tube) design, the amplitude component was then reapplied in a final modulation stage. For solid-state designs it may be better, to apply the amplitude component directly at the Class C amplifier using a PWM supply modulator.

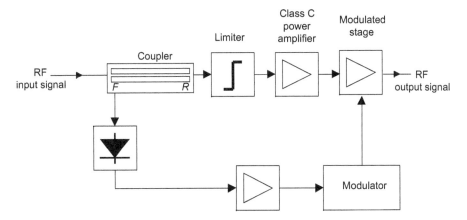

Figure 3.258 Envelope elimination and restoration.

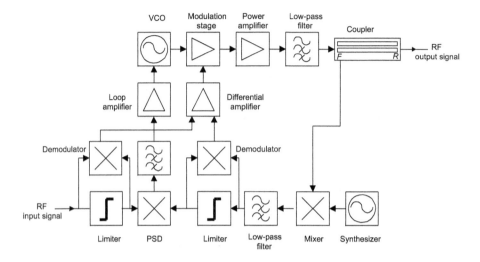

Figure 3.259   Polar loop (Type I).

Since the RF power amplifier is operated in Class C, very high efficiencies can be achieved, typically better than 50% for third-order IM products of –30 dBc. EE&R is not a feedback technique and imperfect components such as nonideal limiting or AM–PM conversion in the modulator will affect the level of distortion products at the output and may generate additional higher order products.

Polar Loop (Petrovic, 1970s) [62]   In some ways, polar loop is similar to EE&R, as the RF signal is resolved into amplitude and phase components. However, the implementation is considerably more complex and it includes the application of amplitude and phase modulation feedback around the power amplifier, see Figure 3.259. The power amplifier is operated in Class AB or, if combined with the modulation stage, in Class C. In one implementation (Type I), a VCO is operated at the transmit frequency and is used to apply both the phase modulation and the phase correction signals. An alternative implementation (Type II) separates the phase modulation and phase correction, reducing the required feedback bandwidth, and a third implementation uses a phase modulator rather than a VCO to apply the phase correction.

One of the limitations of the original design was the linearity and balance of the polar resolvers, although a custom design using modern IC techniques would help to reduce these effects. Another limitation is the wide bandwidth of the phase detector output, particularly during zero-crossings, leading to spurious outputs, particularly with 2-tone tests. Polar loop systems with efficiencies >40% have been reported (probably 50% if modern devices were used), with third-order IM products at −55 dBc. Polar loop techniques have been applied to very high power commercial medium wave transmitters. If the level of spurious products is reduced and sufficient feedback can be applied over the required bandwidth, they may be attractive for high-efficiency multicarrier amplification at VHF and UHF.

Cartesian Loop (Petrovic & Smith, 1980s) [63]   The Cartesian loop, in which the RF output signal is resolved into quadrature (*I* and *Q*) components, has become increasingly popular in the 1990s and some designs are now in volume productions. Since the input

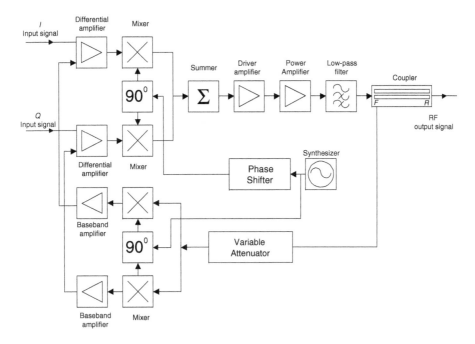

Figure 3.260 Cartesian loop.

signals to a Cartesian loop are at baseband, the Cartesian loop differs from Polar loop in that it is a linearized transmitter rather than a linearized amplifier, see Figure 3.260. The use of I/Q modulation makes it ideal for single-channel transmitters with complex digital or analogue modulation schemes, such as TETRA (400 MHz) or narrowband PMR/SMR (typically 220 or 800 MHz). As with polar loop, the use of modulation feedback permits the use of relatively large amounts of feedback, typically 40 dB for 5 kHz channel bandwidth systems and 30 dB for 25 kHz bandwidth systems. Cartesian loop may also be appropriate for wider bandwidth modulation such as GSM EDGE, where a reduced amount of feedback (around 10 dB) could be applied over a bandwidth of several hundred kibhertz.

The baseband I/Q input signals are applied through a differential amplifier and loop filter to a quadrature modulator and through the power amplifier stages to the output. A sample of the RF output is attenuated and fed to a quadrature demodulator, where it is compared with the baseband input signals to generate the drive signals to the modulator. The synthesizer is shared between the forward and reverse paths and is programmed to the center of the operating channel. The Cartesian loop is therefore acting as two orthogonal control loops in I and Q. There are several key factors that must be addressed in a Cartesian loop design: loop stability, dc offset, and "image generation."

- As the channel frequency is tuned across the operating band, the phase shift between the forward and reverse paths will change and instability sidebands or ears may appear. A programmable phase shifter is normally used to adjust the relative phase of the local oscillators to ensure that the feedback remains negative, with control either by dead-reckoning (look-up table) or by calibration procedures.

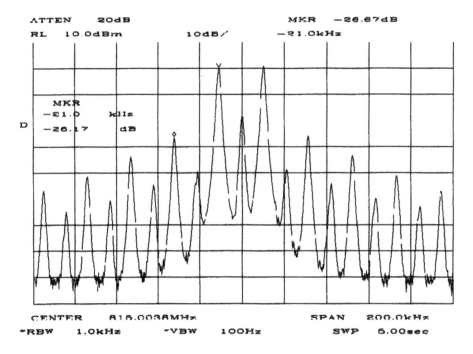

Figure 3.261 800 MHz Cartesian loop (open loop).

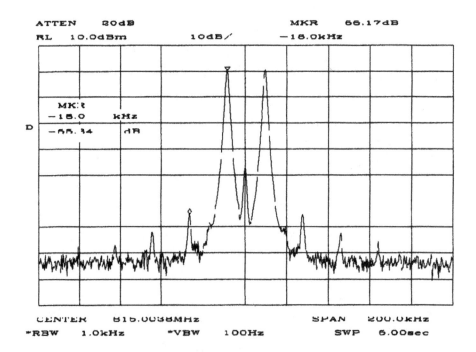

Figure 3.262 800 MHz Cartesian loop (closed loop).

- If high linearity is required, then the downconverter mixers must be operated at very low input levels and this can lead to dc offset (and noise) limitations. DC offset results in unwanted leakage of the local oscillator, increasing error vector magnitude (EVM). Typical solutions to the dc offset problem are the use of sample-and-hold circuits around the differential amplifiers or the deliberate introduction of suitable dc offsets at the I and Q inputs.
- In the early designs, the baseband I and Q signals were generated by phase shift networks, with poor phase quadrature and amplitude balance between the I and Q paths. This, combined with poor quadrature balance in the demodulator limited the orthogonality between the I and Q signals, leads to the generation of unwanted images of the wanted signal, and creates unwanted in-channel products and degrades the EVM. With the advent of low-cost DSP, generation of I and Q signals with very accurate amplitude and phase matching has become much simpler and the DSP can even be used to predistort the I and Q signals to compensate for inadequacies in the RF circuitry.

Innovative RF and software techniques enable >45 dB image and >50 dB carrier suppression to be reliably achieved in commercial products. It should be noted that even with this high level of image suppression, the Cartesian loop is not normally appropriate for multicarrier amplification. As with many of the other techniques described in this paper, designing a Cartesian loop system for low-cost, high-volume manufacture presents a non-trivial challenge. Although the loop itself is relatively simple, the local oscillator and down-converter must be extremely well shielded to prevent unwanted modulation by the PA output signal. The support circuitry such as phase shifters, power control attenuators, and monitor circuits for loop calibration can occupy as much PCB area as the actual loop itself, although Cartesian loop ASICs are now available. The most complex ASICs integrate the baseband input filtering and amplification, differential amplifiers, I–Q modulator, RF driver amplifier, variable attenuator, I–Q demodulator, baseband amplifiers, 90° phase shifters for modulators and demodulators, and a 360° variable phase shifter. The choice of power amplifier is critical for a successful Cartesian loop design. Class C amplifiers can be used but the open loop two-tone IMD3 is only around −15 dBc and large amounts of linearization feedback will be required. In addition, the variation of amplifier gain and phase shift with signal level will affect the loop gain and hence stability. Class AB PA designs are the normal choice and bipolar and MOSFET devices have been used successfully. Since the control loop also acts to maintain constant output power, operation at rated power into a range of load impedances and with different supply voltages should be taken into consideration.

3.12.3.4 Feedforward [65]

Feedforward linearization predates feedback [66] and is an inherently wideband and, superficially, a simple, elegant technique. It is a popular solution for the amplification of single IS-95 carriers or multiple AMPS, TDMA, or GSM carriers in cellular and PCS base stations. In a feedforward system, a sample of the distortion generated by the main amplifier is fed forward and is combined with the amplifier output in such a way that the amplifier distortion is (largely) cancelled, see Figure 3.263. Two loops are required: in the first loop, the undistorted reference signal (A) is subtracted from the distorted amplifier output (B), leaving a signal, which consists purely of distortion products (C). This signal is amplified

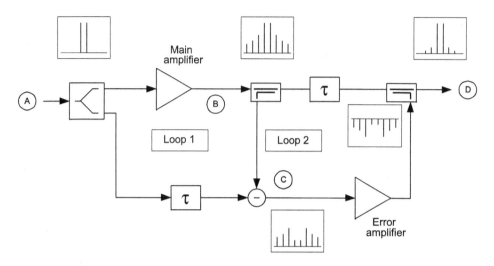

Figure 3.263 Feedforward amplifier.

by the error amplifier in the second loop and, using a directional coupler, is inserted in antiphase back into the main output path. The sampled distortion products will therefore cancel with the amplifier distortion, leaving the amplified wanted signal (D).

Signals traveling through an amplifier have an associated group delay, due to the transit time though the semiconductors and the delay through matching and interconnection networks. It is therefore necessary to introduce compensating time delay elements (often semi-rigid coaxial cables) into both loops, as any delay mismatch will lead to less than total cancelation. The loss in these lines can be significant, especially in loop 2. It is possible to use large diameter low-loss cable, but space limitations often result in a loss of around 1 dB at 2 GHz. The quality of cancelation (null-depth) is also limited by the frequency-dependent amplitude and phase characteristics of the amplifiers, and, to a lesser extent, of the passive components. This frequency dependence implies that the system will need to be optimized for a specific operating frequency, and that the required null-depth will have a limited RF bandwidth. It is usual to include amplitude and phase trimming networks both in loop 1 and loop 2; these can be manually adjusted or more commonly driven by automatic control systems. A variety of systems is in use, often including some form of pilot signal and complex signal processing, and consequently, many patents have been issued. Imperfect cancelation of the wanted signals in Loop 1 will lead to increased output power requirements in the Class A error amplifier, reducing system efficiency, and possibly adding additional distortion products generated in the error amplifier back into the main path. A typical Loop 1 cancelation target is 20–25 dB, and, it is arguable that little benefit is gained from canceling the main tones to a level lower than the most significant (usually 3rd order) distortion products. The null depth in loop 2 has a dB for dB impact on the output distortion cancelation, and again a typical target is 20–25 dB. Enhanced performance can be obtained by the use of multiple feedforward loops, although cost and complexity rapidly conspire to reduce the benefits.

The final directional coupler used for error signal reinsertion must have a relatively tight coupling factor; otherwise, the error amplifier power rating will need to be increased. A tight coupling factor leads to increased main path loss, which again degrades efficiency. Analysis of this dilemma for a typical system usually results in the use of an output coupler of around 8–10 dB coupling (0.5 dB or higher main line loss), and an error amplifier rated at approximately 25% of the main amplifier power. The combination of output coupler and delay-line loss typically results in around 1.5 dB power loss, and combined with the error amplifier dc power requirements leads to a typical dc to RF efficiency for a feedforward cellular amplifier of around 5%–10%. In addition, the isolation between the different sections is vital, particularly between the main amplifier output and the error path.

3.12.3.5 Predistortion

Predistortion techniques attempt to modify the incoming signal(s) to complement and therefore cancel the nonlinear effects in RF power amplifiers. Predistortion has been used with great success (mainly to correct third-order distortion) in TV transmitters (IF predistortion) and TWT amplifiers (RF predistortion). The technique is inherently stable, but the open-loop nature of simple predistorters means that external effects such as temperature changes will not be compensated unless some form of adaption is used.

Gain and Phase Compensation   The simplest predistorters use expansive networks to compensate for the gain compression experienced by a power amplifier as the operating point approaches the 1 dB compression point, see Figure 3.264 These networks, which are placed in series with the amplifier input, typically use diode-resistor networks and depending on the complexity can achieve 5–15 dB reduction in third-order IM distortion. An alternative approach is to place voltage-controlled attenuators (or amplifiers) and phase shifters in the input signal path, and to use the envelope of the RF signal to dynamically adjust the settings. The aim with these networks is to flatten-out the signal level-dependent compression of the amplitude (AM–AM) and phase (AM–PM) transfer characteristics. This system can work over a relatively wide bandwidth if the signal processing bandwidth is approximately 10 times wider than the modulation bandwidth. Typical improvements in third-order products are up to 10 dB, but as with all open loop correction systems, it is sensitive to changes such as temperature and amplifier gain, and some form of adaption is often added. The performance is also dependent on the tracking of the detection and control

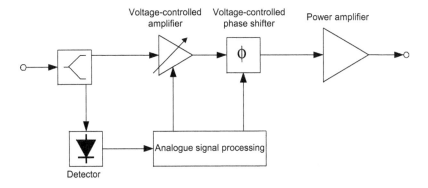

Figure 3.264   Predistortion (amplitude and phase compensation).

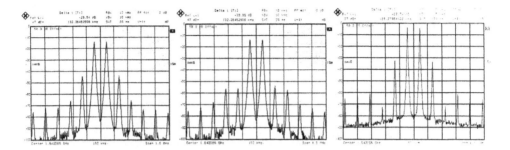

Figure 3.265 Nonlinear generator spectra: uncorrected amplifier, the amplifier with nulling optimized for third-order products, and nulling optimized for fifth-order products.

characteristics and it is challenging to achieve significant reduction of higher order products without fine tuning the control laws.

Non-Linear Generators   An alternative approach is to use nonlinear amplifiers to generate complementary IM distortion, which is used to cancel the power amplifier distortion. There are several ways of implementing this technique, with the simplest being a diode or transistor (often GaAs) network in series with the main signal path and this is mainly used to correct for third-order distortion. A more complex alternative is to implement the nonlinear generator using a low power transistor with similar distortion characteristics to the main amplifier and to apply feedforward predistortion. Practical implementations can be quite complex, as it is necessary to null-out the RF input signal from the NLG output (leaving just the distortion) and to match time delays, amplitudes, and phases in the main and secondary branches. It is possible, with care, to achieve 15 dB reduction in third-order IM products, although the optimum setting for third-order products may have little impact on higher order products (and may even increase them). Conversely, it is possible to achieve a reduction in higher order products at the expense of third and fifth-order products. The system is, of course, affected by temperature and gain changes unless some form of adaption is included. Figures 3.265 shows (from left to right) an uncorrected amplifier, the amplifier with nulling optimized for third-order products and nulling optimized for fifth-order products.

3.12.3.6 Baseband Predistortion [67]

Another approach for a complete transmitter is to use a DSP to apply the predistortion to the baseband (analog or digital) modulation signals prior to upconversion. The mapping-based predistorter uses a large look-up table to map the input *I/Q* signals to new predistorted values. This can provide excellent performance but may require many Mbits of memory and can be slow to converge and to adapt to changes. An alternative is the gain-based predistorter, which uses the envelope level to modify the complex output signal, with interpolation between stored values. The reduced memory requirements are traded-off for the increased computational requirements. A third alternative is analogue baseband predistortion in which the predistortion is applied to the baseband signals with analogue circuits, which may in turn, be controlled by a DSP controller. It is likely that baseband (adaptive) predistortion will become more widespread as the performance of DSPs and analog-digital converters increases and the cost and power consumption decreases.

Technique	RF Bandwidth	Linearity Improvement	Efficiency	Complexity/ Risk
Power back-off	Wide	Good	Low	Low
RF feedback	Narrow to moderate	Poor	Low	Moderate
Envelope feedback	Moderate	Low	Moderate	Low
EE&R	Narrow	Moderate	High	Moderate
Polar loop	Narrow to moderate	High	High	High
Cartesian loop	Narrow to moderate	High	High	Moderate to high
Feedforward nonadaptive	Wide	Moderate	Low	Moderate
Feedforward adaptive control	Wide	High	Low	High
Baseband predistortion	Moderate to wide	Moderate	High	Moderate to high
Nonlinear generator predistortion	Moderate to wide	Low	High	Moderate to high
Gain and phase predistortion	Moderate to wide	Low	High	Moderate

3.12.4 Impedance Matching Networks Applied to RF Power Transistors[6]

Introduction Some graphical and numerical methods of impedance matching will be reviewed here. The examples given will refer to high-frequency power amplifiers.

Although matching networks normally take the form of filters and therefore are also useful to provide frequency discrimination, this aspect will be considered only as a corollary of the matching circuit.

Matching is necessary for the best possible energy transfer from stage to stage. In RF-power transistors, the input impedance is of low value, decreasing as the power increases, or as the chip size becomes larger. This impedance must be matched either to a generator—of generally 50 Ω internal impedance—or to a preceding stage. Impedance matching has to be made between two complex impedances, which makes the design still more difficult, especially if matching must be accomplished over a wide frequency band.

Device Parameters

Input Impedance The general shape of the input impedance of RF-power transistors is as shown in Figure 3.266. It is a large-signal parameter, expressed here by the parallel combination of a resistance R_P and a reactance X_p [68].

The equivalent circuit shown in Figure 3.267 accounts for the behavior illustrated in Figure 3.266. With the presently used stripline or flange packaging most of the power devices for VHF low band will have their R_P and X_p values below the series resonant power f_s. The input impedance will be essentially capacitive.

Most of the VHF high-band transistors will have the series-resonant frequency within their operating range; that is, they will be purely resistive at one single frequency f_s, which the parallel-resonant frequency f_p will be outside.

---

[6]Based on Motorola application note AN-721 ("Impedance matching networks applied to RF power transistors"). Copyright of Motorola; used by permission.

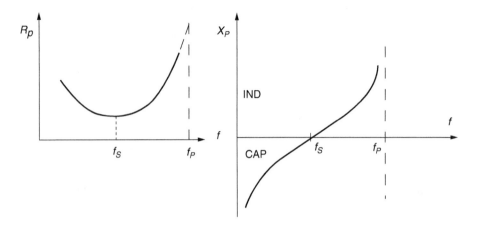

Figure 3.266   Input impedance of RF power transistors as a function of frequency.

Parameters for 1- or 2-GHz transistors will be beyond f_s and approach f_p. They show a high value of R_P and X_p with inductive character.

A parameter that is very often used to judge on the broadband capabilities of a device is the input Q or Q_{IN}, defined simply as the ratio R_P/X_p. Practically, Q_{IN} ranges around 1 or less for VHF devices and around 5 or more for microwave transistors.

Q_{IN} is an important parameter to consider for broadband matching. Matching networks normally are low-pass or pseudo-lowpass filters. If Q_{IN} is high, it can be necessary to use band-pass filter types matching networks and to allow insertion losses. But broadband matching is still possible. This will be discussed later.

Output Impedance   The output impedance of RF power transistors, as given by all manufacturers' datasheets, generally consists of only a capacitance C_{OUT}. The internal resistance of the transistor is supposed to be much higher than the load and is normally neglected. In the case of a relatively low internal resistance, the efficiency of the device would decrease by the factor

$$1 + R_L/R_T \tag{3.204}$$

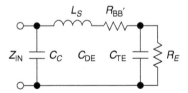

Figure 3.267   Equivalent circuit for the input impedance. R_E = emitter diffusion resistance; C_{DE}, C_{TE} = diffusion and transition capacitances, respectively, of the emitter junction; $R_{BB'}$ = base spreading resistance; C_C = package capacitance; L_S = base lead inductance.

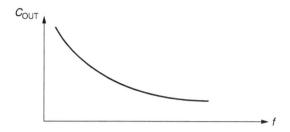

Figure 3.268 Output capacitance C_{OUT} as a function of frequency.

where R_L is the load resistance, seen at the collector–emitter terminals, and R_T the internal transistor resistance equal to

$$\frac{1}{\omega_T(C_{TC} + C_{DC})} \tag{3.205}$$

defined as a single parameter, where ω_T = transit angular frequency, and C_{TC} and C_{DC} = transition and diffusion capacitances, respectively, at the collector junction.

The output capacitance, C_{OUT}, which is a large-signal parameter, is related to the small-signal parameter C_{CB}, the collector–base transition capacitance.

Since a junction capacitance varies with the applied voltage, C_{OUT} differs from C_{CB} in that it has to be averaged over the total voltage swing. For an abrupt junction and assuming certain simplifications, $C_{OUT} = 2C_{CB}$.

Figure 3.268 shows the variation of C_{OUT} with frequency. C_{OUT} decreases partly due to the presence of the collector lead inductance, but mainly because of the fact that the base–emitter diode does not shut off anymore when the operating frequency approaches the transit frequency, f_T.

Output Load   In the absence of a more precise indication, the output load R_L is taken equal to

$$R_L = \frac{[V_{CC} - V_{CE(sat)}]^2}{2P_{out}} \tag{3.206}$$

with $V_{CE(sat)}$ equal to 2 or 3 V, increasing with frequency.

The above equation just express a well-known relation, but also shows that the load, in first approximation, is not related to the device, except for $V_{CE(sat)}$. The load value is primarily dictated by the required output power and the peak voltage; it is not matched to the output impedance of the device.

At high frequencies, this approximation becomes less exact, and for microwave devices the load that must be presented to the device is indicated on the datasheet.

Strictly speaking, impedance matching is accomplished only at the input. Interstage and load matching are more impedance transformations of the device input impedance and of the load into a value R_L (sometimes with additional reactive component) that depends essentially on the power demanded and the supply voltage.

Matching Networks   In the following, matching networks will be described by the order of complexity. These are ladder-type reactance networks.

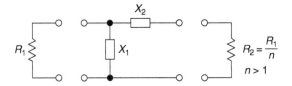

Figure 3.269   Two-reactance matching network.

The different reactance values will be calculated and determined graphically. Increasing the number of reactances broadens the bandwidth. However, matching networks consisting of more than four reactances are rare. Above four reactances, the improvement is small.

Numerical Design

Two-Resistance Networks   Resistance terminations will first be considered. Figure 3.269 shows the reactive L section and the terminations to be matched.

Matching or exact transformation from R_2 into R_1 occurs at a single frequency f_0. At f_0, X_1 and X_2 are equal to

$$X_1 = \pm R_1 \sqrt{\frac{R_2}{R_1 - R_2}} = R_1 \frac{1}{\sqrt{n-1}} \tag{3.207}$$

$$X_2 = \mp \sqrt{R_2(R_1 - R_2)} = R_1 \frac{\sqrt{n-1}}{1} \tag{3.208}$$

At f_0, $X_1 X_2 = R_1 R_2$. X_1 and X_2 must be of opposite sign. The shunt reactance is in parallel with the larger resistance.

The frequency response of the L section is shown in Figure 3.270, where the normalized current is plotted as a function of the normalized frequency.

If X_1 is capacitive and consequently X_2 is inductive, then

$$X_1 = -\frac{f_0}{f} R_1 \sqrt{\frac{R_2}{R_1 - R_2}} = -\frac{f_0}{f} R_1 \frac{1}{\sqrt{n-1}} \tag{3.209}$$

and

$$X_2 = -\frac{f}{f_0} \sqrt{R_2(R_1 - R_2)} = \frac{f}{f_0} R_1 \frac{\sqrt{n-1}}{1} \tag{3.210}$$

The normalized current absolute value is equal to

$$\left| \frac{I_2}{I_0} \right| = \frac{2\sqrt{n}}{\sqrt{(n-1)^2 \cdot \left(\frac{f}{f_0}\right)^4 - 2\left(\frac{f}{f_0}\right)^2 + (n+1)^2}} \tag{3.211}$$

where $I_0 = \frac{\sqrt{nE}}{2 \cdot R_1}$, and is plotted in Figure 3.270 [69].

If X_1 is inductive and consequently X_2 is capacitive, the only change required is a replacement of f by f_0 and vice-versa. The L section has low-pass form in the first case and high-pass form in the second case.

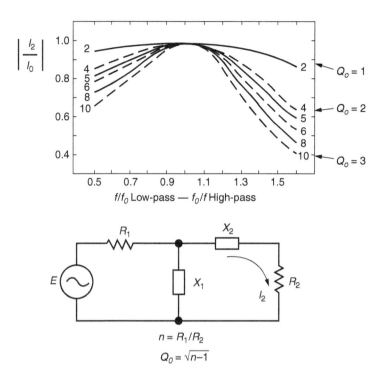

Figure 3.270 Normalized frequency response for the L section in low-pass or high-pass form.

The Q of the circuit at f_0 is equal to

$$Q_0 = \frac{X_2}{R_2} = \frac{R_1}{X_1} = \sqrt{n-1} \tag{3.212}$$

For a given transformation ratio n, there is only one possible value of Q. On the other hand, there are two symmetrical solutions for the network that can be either a low-pass filter or a high-pass filter.

The frequency f_0 does not need to be the center frequency, $(f_1 + f_2)/2$, of the desired band limited by f_1 and f_2. In fact, as can be seen from the low-pass configuration of Figure 3.270, it may be interesting to shift f_0 toward the high-band edge frequency f_2 to obtain a larger bandwidth w, where

$$w = \frac{2(f_1 + f_2)}{f_2 - f_1} \tag{3.213}$$

This will, however, be at the expense of poor harmonic rejection.

EXAMPLE: For a transformation ratio $n = 4$, it can be determined from the above relations:

Bandwidth w	0.1	0.3
Max insertion losses	0.025	0.2
X_1/R_1	1.730	1.712

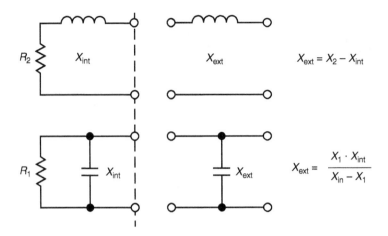

Figure 3.271   Termination reactance compensation.

If the terminations R_1 and R_2 have a reactive component X, the latter may be taken as part of the external resistance as shown in Figure 3.271. This compensation is applicable as long as

$$Q_{\text{INT}} = \frac{X_{\text{INT}}}{R_2} \text{ or } \frac{R_1}{X_{\text{INT}}} < n - 1 \tag{3.214}$$

Tables giving reactance values can be found in Refs [70] and [71].

Use of Transmission Lines and Inductors   In the preceding section, the inductance was expected to be realized by a lumped element. A transmission line can be used instead (Figure 3.272).

As can be seen from the computed selectivity curves (Figure 3.273) for the two configurations, transmission lines result in a larger bandwidth. The gain is important for a transmission line having a length $L = \lambda/4 (\Theta = 90°)$ and a characteristic impedance $Z_0 = \sqrt{R_1 R_2}$. It is not significant for lines short with respect to $\lambda/4$. One will notice that there is an infinity of solutions, one for each value of C, when using transmission lines.

Three-Reactance Matching Networks   The networks that will be investigated are shown in Figure 3.274. They are made of three reactances alternately connected in series and shunt.

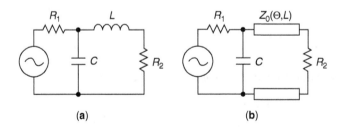

(a)   (b)

Figure 3.272   Use of a transmission line in the L section.

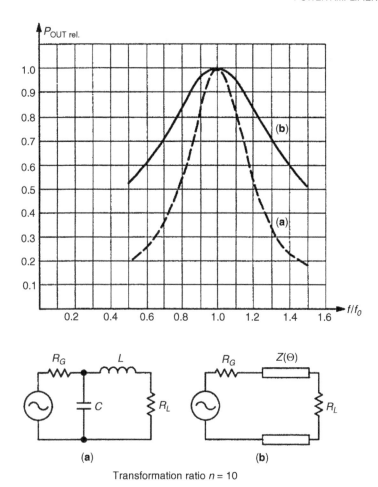

Transformation ratio $n = 10$

Figure 3.273 Bandwidth of the L section for $n = 10$ (a) with lumped constants and (b) with transmission line ($\lambda/4$).

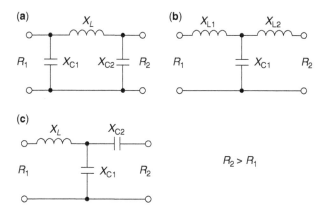

Figure 3.274 Three-reactance matching networks.

A three-reactance configuration allows the designer to make the quality factor (Q) of the circuit and the transformation ratio $n = R_2/R_1$ independent of each other and consequently to choose the selectivity between certain limits.

For narrowband designs, one can use the following formulas (see Ref. [72], where tables are given.

Network (a):

$$X_{C1} = R_1/Q \qquad Q \text{ must be first selected} \tag{3.215}$$

$$X_{C2} = R_2 \sqrt{\frac{R_1 R_2}{(Q^2 + 1) - \frac{R_1}{R_2}}} \tag{3.216}$$

$$X_L = \frac{QR_1 + (R_1 R_2/X_{C2})}{Q^2 + 1} \tag{3.217}$$

Network (b):

$$X_{L1} = R_1 Q \qquad Q \text{ must be first selected} \tag{3.218}$$

$$X_{L2} = R_2 \cdot B \tag{3.219}$$

$$A = R_1(1 + Q^2) \tag{3.220}$$

$$X_{C1} = \frac{A}{Q + B} \tag{3.221}$$

$$B = \sqrt{\frac{A}{R_2} - 1} \tag{3.222}$$

Network (c):

$$X_{L1} = Q \cdot R_1 \qquad Q \text{ must be first selected} \tag{3.223}$$

$$X_{C2} = A \cdot R_2 \tag{3.224}$$

$$A = \sqrt{\frac{R_1(1 + Q^2)}{R_2} - 1} \tag{3.225}$$

$$X_{C1} = \frac{B}{Q - A} \tag{3.226}$$

$$B = R_1 \cdot (1 + Q^2) \tag{3.227}$$

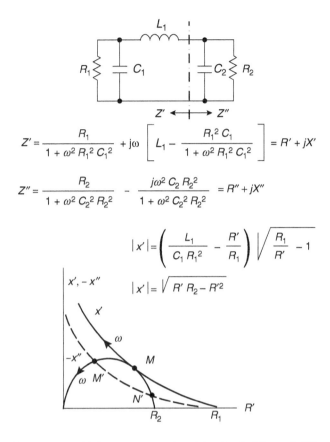

$$Z' = \frac{R_1}{1 + \omega^2 R_1^2 C_1^2} + j\omega \left[L_1 - \frac{R_1^2 C_1}{1 + \omega^2 R_1^2 C_1^2} \right] = R' + jX'$$

$$Z'' = \frac{R_2}{1 + \omega^2 C_2^2 R_2^2} - \frac{j\omega^2 C_2 R_2^2}{1 + \omega^2 C_2^2 R_2^2} = R'' + jX''$$

$$|X'| = \left(\frac{L_1}{C_1 R_1^2} - \frac{R'}{R_1} \right) \sqrt{\frac{R_1}{R'} - 1}$$

$$|X'| = \sqrt{R' R_2 - R'^2}$$

Figure 3.275 *Z*-plane representation of the circuit of Figure 3.274a.

The network that yields the most-practical component values should be selected for a given application.

The three-reactance networks can be thought of as being formed of an *L* section (two reactances) and of a compensation reactance. The *L* section essentially performs the impedance transformation, while the additional reactance compensates for the reactive part of the transformed impedance over a certain frequency band.

Figure 3.275 shows a representation in the *Z* plane of the circuit of Figure 3.274a split into two parts: $R_1 - C_1 - L_1$ and $C_2 - R_2$.

Exact transformation from R_1 into R_2 occurs at the points of intersection *M* and *N*. Impedances are then conjugate, or $Z' = R' + jX'$ and $Z'' = R'' + jX''$ with $R' = R''$ and $X' = -X''$.

The only possible solution is obtained when X' and $-X''$ are tangential to each other. For the dashed curve, representing another value of L_1 or C_1, a wider frequency band could be expected at the expense of some ripple inside the band. However, this can only be reached with four reactances as will be shown below.

With a three-reactance configuration, there are not enough degrees of freedom to permit $X' = -X''$ and simultaneously obtain the same variation of frequency on both curves from point M' to point N'. Exact transformation can, therefore, be obtained at only one frequency.

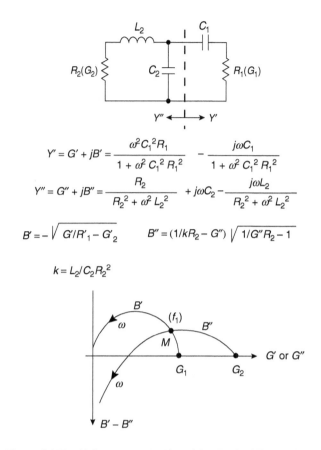

$$Y' = G' + jB' = \frac{\omega^2 C_1^2 R_1}{1 + \omega^2 C_1^2 R_1^2} - \frac{j\omega C_1}{1 + \omega^2 C_1^2 R_1^2}$$

$$Y'' = G'' + jB'' = \frac{R_2}{R_2^2 + \omega^2 L_2^2} + j\omega C_2 - \frac{j\omega L_2}{R_2^2 + \omega^2 L_2^2}$$

$$B' = -\sqrt{G'/R'_1 - G'_2} \qquad B'' = (1/kR_2 - G'')\sqrt{1/G''R_2 - 1}$$

$$k = L_2/C_2 R_2^2$$

Figure 3.276 *Y*-plane representation of the circuit of Figure 3.274c.

The values of the three reactances can be calculated by making

$$X' = -X'', \quad R' = R'', \quad \text{and} \quad \frac{dX'}{dR'} = -\frac{dX''}{dR''} \tag{3.228}$$

The general solution of these equations leads to complicated calculations. Therefore, computed tables should be used.

One will note from Figure 3.275 that the compensation reactance contributes somewhat to impedance transformation; that is, R' varies when going from complicated M to R_2.

The circuit of Figure 3.274b is dual with respect to the first one and gives exactly the same results in a *Y*-plane representation. The circuit of Figure 3.274c is somewhat different since only one intersection M exists as shown in Figure 3.276. Narrower frequency bands must be expected from this configuration. The widest band is obtained for $C_1 = \infty$.

Again, if one of the terminations has a reactive component, the latter can be taken as a part of the matching network, provided that it is not too large (see Figure 3.273).

Four-Reactance Networks   Four-reactance network are used essentially for broadband matching. The networks that will be considered in the following consist of two two-reactance sections in cascade. Some networks have pseudo-lowpass-filter character, others

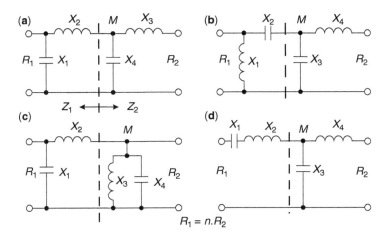

Figure 3.277 Four-reactance networks.

have bandpass-filter character. In principle, the former show narrower bandwidth since they extend the impedance transformation to very low frequencies unnecessarily, while the latter ensure good matching over a wide frequency band around the center frequency only (see Figure 3.279).

The two-reactance sections used in the networks in Figure 3.277 have either transformation properties or compensation properties. Impedance transformation is obtained with one series reactance and one shunt reactance. Compensation is made with both reactances in series or in shunt. If two cascaded transformation networks are used, transformation is accomplished partly by each one.

With four-reactance networks there are two frequencies, f_1 and f_2, at which the transformation from R_1 into R_2 is exact. These frequencies may also coincide. For the network in Figure 3.277b, for instance, at point M, R_1 or R_2 is transformed into $\sqrt{R_1 R_2}$ when both frequencies fall together. At all points (M), Z_1 and Z_2 are conjugate if the transformation is exact.

In the case of Figure 3.277b, the reactances are easily calculated for equal frequencies:

$$X_1 = \sqrt{\frac{R_1}{\sqrt{n} - 1}} \tag{3.229}$$

$$X_2 = R_1 \sqrt{\frac{\sqrt{n} - 1}{n}} \tag{3.230}$$

$$x_1 \cdot X_4 = R_1 \cdot R_2 = X_2 \cdot X_3 \tag{3.231}$$

$$X_3 = \frac{R_1}{\sqrt{n(\sqrt{n} - 1)}} \tag{3.232}$$

$$X_4 = \frac{R_1}{n} \sqrt{\sqrt{n} - 1} \tag{3.233}$$

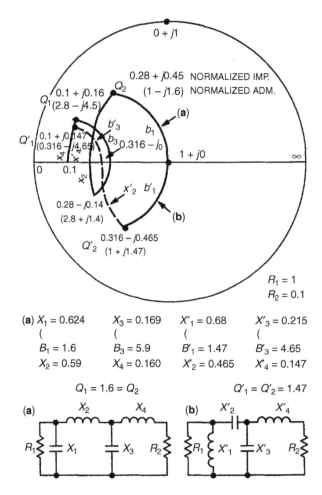

Figure 3.278 Transformation paths for networks (a) and (b).

For the network in Figure 3.277a, normally, at point (M) Z_1 and Z_2 are complex. This pseudo-lowpass filter has been computed elsewhere [70]. Many tables can be found in the literature for networks of four and more reactances having Chebyshev character or maximally flat response [70, 71, 73].

Figure 3.278 shows the transformation path from R_1 to R_2 for networks (a) and (b) in Figure 3.277 on a Smith Chart.

Case (a) of Figure 3.278 has been calculated using tables in Ref. [71]. Case (b) has been obtained from the relationship given above for $X_1 \cdots X_4$. Both apply for a transformation ratio equal to 10 and for $R_1 = 1$. There is no simple relationship for $X'_1 \cdots X'_4$ of network (b) if f_1 is made different from f_2 for larger bandwidth. Figure 3.279 shows the respective bandwidths of network (a) and (b) for the circuits shown in Figure 3.278.

If the terminations include a reactive component, the computed values for X_1 or X_4 may be adjusted to compensate for this.

For configuration (a), it can be seen from Figure 3.278 that, in the considered case, the Qs are equal to 1.6. For configuration (b), Q'_1, which is equal to Q'_2, is fixed for each

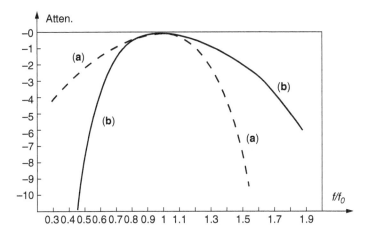

Figure 3.279 Selectivity curves for networks (a) and (b) of Figure 3.278.

transformation ratio

$$
\begin{array}{c|c|c|c|c|c}
n & 2 & 4 & 8 & 10 & 16 \\
\hline
Q'_1 = Q'_2 & 0.65 & 1 & 1.35 & 1.46 & 1.73
\end{array}
\qquad Q' = \sqrt{\sqrt{n} - 1} \qquad (3.234)
$$

The maximum value of reactance that the terminations may have for use in this configuration can be determined from the above values of Q'.

If R_1 is the load resistance of a transistor, the internal transistor resistance may not be equal to R_1. In this case, the selectivity curve will be different from the curves given in Figure 3.279. Figure 3.280 shows the selectivity for networks (a) and (b) when the source resistance R_1 is infinite. From Figure 3.280, it can be seen that network (a) is more sensitive to changes in R_1 than network (b).

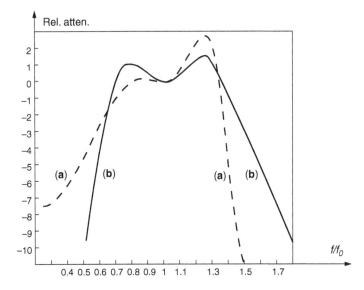

Figure 3.280 Selectivity curves for networks (a) and (b) of Figure 3.278 with infinite R_1.

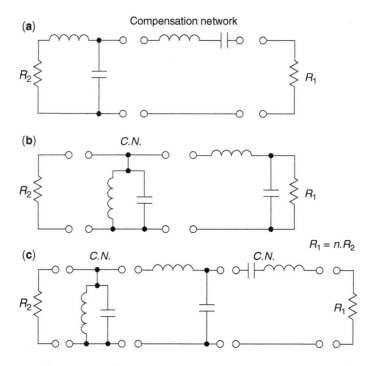

Figure 3.281 Compensation networks used with an *L* section.

As mentioned earlier, the four-reactance network can also be thought of as two cascaded two-reactance sections: one used for transformation and the other for compensation. Figure 3.281 shows commonly used compensation networks, together with the associated *L* section.

The circuit of Figure 3.281a can be compared to the three-reactance network shown in Figure 3.274c. The difference is that capacitor C_2 if that circuit has been replaced by an *LC* circuit. The resulting improvement can be seen by comparing Figure 3.282c with Figure 3.276.

By adding one reactance, exact impedance transformation is achieved at two frequencies. It is now possible to choose component values such that the point of intersection M' occurs at the same frequency f_1 on both curves and simultaneously that N' occurs at the same frequency f_2 on both curves. Among the infinite number of possible intersections, only one allows the achievement of this.

When M' and N' coincide in M, the new $dX'/df = dX''/df$ condition can be added to the condition $X' = -X''$ (for three-reactance networks) and similarly $R' = R''$ and $dR'/df = dR''/df$.

Again, a general solution of the above equations leads to still more complicated calculations than in the case of three-reactance networks. Therefore, tables are preferable [70, 71, 73].

The circuit of Figure 3.281b is the dual the circuit of Figure 3.279a and does not need to be treated separately. It gives exactly the same results in the *Z* plane. Figure 3.281c shows a higher-order compensation requiring six reactive elements.

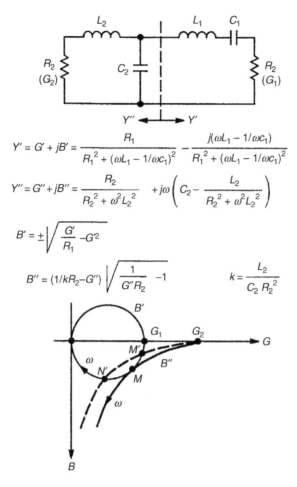

$$Y' = G' + jB' = \frac{R_1}{R_1{}^2 + (\omega L_1 - 1/\omega c_1)^2} - \frac{j(\omega L_1 - 1/\omega c_1)}{R_1{}^2 + (\omega L_1 - 1/\omega c_1)^2}$$

$$Y'' = G'' + jB'' = \frac{R_2}{R_2{}^2 + \omega^2 L_2{}^2} + j\omega\left(C_2 - \frac{L_2}{R_2{}^2 + \omega^2 L_2{}^2}\right)$$

$$B' = \pm\sqrt{\frac{G'}{R_1} - G'^2}$$

$$B'' = (1/kR_2 - G'')\sqrt{\frac{1}{G''R_2} - 1} \qquad k = \frac{L_2}{C_2 R_2{}^2}$$

Figure 3.282 *Y*-plane representation of the circuit of Figure 3.281a.

The above-discussed matching networks employing compensation circuits result in narrower bandwidths than the former solutions (see "Four-Reactance Networks," above) using two transformation sections. A matching arrangement with higher-order compensation, such as in Figure 3.281c, is not recommended. Better use can be made of the large number of reactive elements by using them all for transformation.

When the above configurations are realized using short portions of transmission lines, the equations or the usual tables no longer apply. The calculations must be carried out on a computer because of their complexity. However, a graphical method can be used (see the next section) that will consist essentially in tracing a transformation path on the *Z–Y* chart using the computed lumped element values and replacing it by the closest path obtained with distributed constants. The bandwidth change is not significant as long as short portions of lines are used.

Matching Networks Using Quarter-Wave Transformers   At sufficiently high frequencies, where λ/4-long lines of practical size can be realized, broadband transformations can easily be accomplished by the use of one or more λ/4 sections. Figure 3.283 summarizes

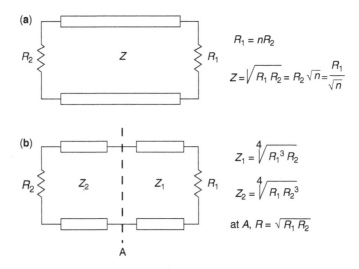

Figure 3.283 Transformation networks using $\lambda/4$-long transmission lines.

the main relations for (a) one-section and (b) two-section transformation. A compensation network can be realized using a $\lambda/2$-long transmission line. Figures 3.284 and 3.285 show the selectivity curves for different transformation ratios and section numbers.

Exponential Lines Exponential lines have largely frequency-independent transformation properties. The characteristic impedance of such lines varies exponentially with their length l:

$$Z = Z_0 \cdot e^{kl} \tag{3.235}$$

where k is a constant, but these properties are preserved only if k is small.

Broadband Matching Using Bandpass-Filter Networks—High-Q Case The above circuits are applicable to devices having low input or output Q, if broadband matching

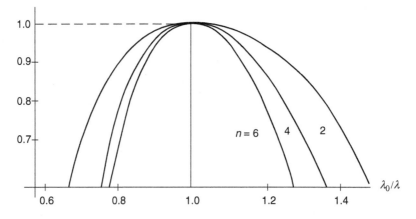

Figure 3.284 Selectivity curves for two $\lambda/4$-section networks at different transformation ratios.

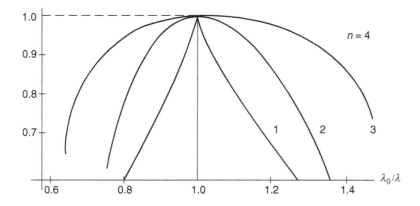

Figure 3.285 Selectivity curves for one, two, and three $\lambda/4$ sections.

is required. Generally, if the impedances to be matched can be represented for instance by a resistor R in series with an inductor L (sometimes a capacitor C) within the band of interest and if L is sufficiently low, the latter can be incorporated into the first inductor in the matching network. This is also valid if the representation consists of a shunt combination of a resistor and a reactance.

Practically, this is feasible for Qs around 1 or 2. For higher Qs or for input impedances consisting of a series or parallel resonant circuit (see Figure 3.267), as it appears to be for large bandwidths, a different treatment must be followed.

Let us first recall that, as shown by Bode and Fano [74, 75], limitations exist on the impedance matching of a complex load. In the example of Figure 3.286, the load to be matched consists of a capacitor C and a resistor R in shunt.

The reflection coefficient between the transformed load and generator is equal to

$$\Gamma = \frac{Z_{IN} - R_g}{Z_{IN} + R_g} \tag{3.236}$$

When $\Gamma = 0$, there is perfect matching; when $\Gamma = 1$, there is total reflection. The ratio of reflected to incident power is

$$\frac{P_r}{P_i} = |\Gamma|^2 \tag{3.237}$$

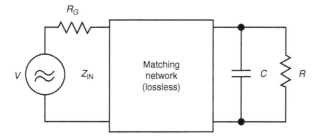

Figure 3.286 General matching conditions.

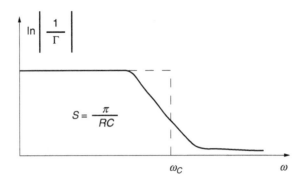

Figure 3.287   Representation of the Bode equation.

The fundamental limitation on the matching takes the form

$$\int_{\omega=0}^{\infty} \ln\left(\frac{1}{|\Gamma|}\right) d\omega \leq \frac{\pi}{RC} \quad \text{Bode equation} \tag{3.238}$$

and is represented in Figure 3.287.

The meaning of the Bode equation is that the area S under the curve cannot be greater than π/RC, and therefore, if matching is required over a certain bandwidth, this can only be done at the expense of less power transfer within the band. Thus, power transfer and bandwidth appear as interchangeable quantities.

It is evident that the best utilization of the area S is obtained when $|\Gamma|$ is kept constant over the desired band ω_c and made equal to 1 over the rest of the spectrum. Then

$$|\Gamma| = e^{-\frac{\pi}{\omega_C RC}} \tag{3.239}$$

within the band and no power transfer happens outside. A network fulfilling this requirement cannot be obtained in practice as an infinite number of reactive elements would be necessary.

If the attenuation is plotted versus frequency for practical cases, one may expect to have curves like the ones shown in Figure 3.288 for a low-pass filter having Chebyshev character.

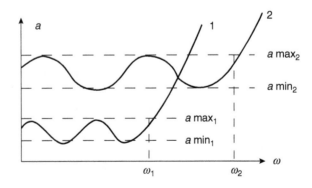

Figure 3.288   Attenuation versus angular frequency for different bandwidths with the same load.

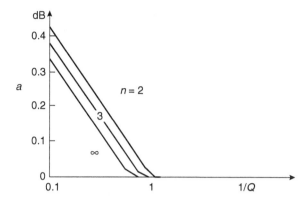

Figure 3.289 Insertion losses as a function of $1/Q$.

For a given complex load, an extension of the bandwidth from ω_1 to ω_2 is possible only with a simultaneous increase of the attenuation a. This is especially noticeable for Qs exceeding 1 or 2. (see Figure 3.289).

Thus, devices having relatively high input Qs are usable for broadband operation, provided the consequent higher attenuation or reflection introduced is acceptable.

The general shape of the average insertion losses or attenuation a (neglecting the ripple), the attenuation decreases if the number n of the network element increases. But above $n = 4$, the improvement is small. For a given attenuation a and bandwidth, the larger the n the smaller the ripple; for a given attenuation, the larger the n the larger the bandwidth.

Computations show that for $Q < 1$ and $n \leq 3$, the attenuation is below 0.1 dB (approximately). The impedance transformation ratio is not free here. The network is a true low-pass filter. For a given load, the optimum generator impedance will result from the computation.

Before impedance transformation is introduced, a conversion of the low-pass prototype into a bandpass-filter network must be made. Figure 3.290 summarizes the main relations for this conversion. r is the conversion factor. For the bandpass filter, Q_{INmax}, or the maximum possible input Q of the device to be matched, has been increased by the factor r (from Figure 3.290, $Q'_{INmax} = rQ_{INmax}$).

Impedance inverters will be used for impedance transformation. These networks are suitable for insertion into a bandpass filter without affecting the transmission characteristics.

Figure 3.291 shows four impedance inverters. It will be noticed that one of the reactances is negative and must be combined in the bandpass network with a reactance of at least equal positive value. Insertion of the inverter can be made at any convenient place [69, 76].

When using the bandpass filter for matching the input impedance of a transistor, reactances $L_1'C_1'$ should be made to resonate at ω_0 by addition of a convenient series reactance.

As stated above, the series combination of R_0, L_1' and C_1' normally constitutes the equivalent input network of a transistor when considered over a large bandwidth. This is a good approximation up to about 500 MHz.

In practice, the normal procedure for using a bandpass-filter matching network will be the following.

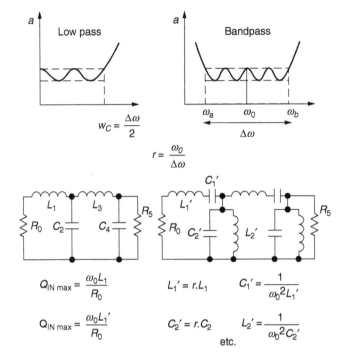

Figure 3.290 Conversion from low-pass filter into bandpass filter.

1. For a given bandwidth center frequency and input impedance of a device to be matched—for example, 50 Ω—first determine Q'_{IN} from the data sheet as

$$\frac{\omega_0 L'_1}{R_0} \tag{3.240}$$

after having eventually added a series reactor for centering.

2. Convert the equivalent circuit $R_0 L'_1 C'_1$ into a low-pass prototype $R_0 L_1$ and calculate Q_{IN} using the formulas of Figure 3.290.

3. Determine the other reactance values from tables [70] for the desired bandwidth.

4. Convert the element values found by Step 3 into series- or parallel-resonant circuit parameters.

5. Insert the impedance inverter in any convenient place.

In the above discussions, the gain rolloff has not been taking into account. This is of normal use for moderate bandwidths (e.g., 30%). However, several methods can be employed to obtain a constant gain within the band despite the intrinsic gain decrease of a transistor with frequency. Tables have been computed elsewhere [77] for matching networks approximating 6 dB/octave attenuation versus frequency.

Another method consists of using the above-mentioned network and then adding a compensation circuit as shown in Figure 3.292.

Resonance ω_b is placed at the high edge of the frequency band. Choosing Q correctly, rolloff can be made 6 dB/octave. The response of the circuit of Figure 3.292 is

Inverter   Equivalent to:

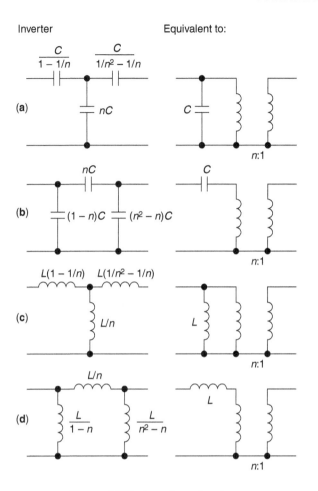

Figure 3.291 Impedance inverters.

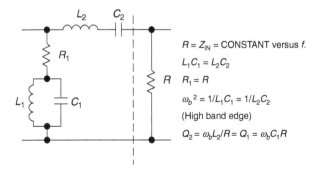

$R = Z_{IN}$ = CONSTANT versus f.

$L_1 C_1 = L_2 C_2$

$R_1 = R$

$\omega_b{}^2 = 1/L_1 C_1 = 1/L_2 C_2$

(High band edge)

$Q_2 = \omega_b L_2 / R = Q_1 = \omega_b C_1 R$

Figure 3.292 Roll-off compensation network.

expressed by

$$\frac{1}{1 + Q^2 \left(\dfrac{\omega}{\omega_b} - \dfrac{\omega_b}{\omega} \right)^2} \tag{3.241}$$

where $\omega < \omega_b$. This must be equal to ω/ω_b for 6 dB/octave attenuation.

At the other band edge a, exact compensation can be obtained if

$$Q = \frac{\left(\dfrac{\omega_b}{\omega_a} \right)^2 - 1}{\dfrac{\omega_a}{\omega_b} - \left(\dfrac{\omega_b}{\omega_a} \right)^2} \tag{3.242}$$

There are several methods available for synthesis of matching networks. They should not be confused with filters because they can be either high-pass, low-pass, or bandpass, and their main purpose is to match one complex impedance to another. In many cases, one of the terminations happens to be 50 Ω and real, but this is just a "lucky" case of the general approach. An excellent paper on this [78] refers to the "real frequency" technique. Another equally important publication is the one by Cuthbert [79], since followed by his book [80]. Two more good references are [81] and [82].

3.12.5   Example 2: Low-Noise Amplifier Using Distributed Elements

In our previous examples, the amplifiers were built around lumped elements. Taking the fact that a transmission line length of less than λ/4 is an inductor and 3/4 λ is a capacitor. The whole amplifier can be built around microstrip technology.

If anybody wonders why we are going to show a low-noise amplifier in the middle of the power section, the reason is that we will select a low-power Infineon transistor which, being operated Class A, has an intercept point of +30 dBm and can easily achieve a noise figure of less than 2 dB. Such an amplifier is attractive when feeding an input stage from an antenna that sees many signals and curtails the bandwidth with an input filter of good selectivity that needs a good output termination, which in many cases works against the achievement of a good noise figure because when designing for best noise figure the input impedance of the transistor stage is far from 50 Ω.

First, we will start with the textbook approach and use the Smith diagram to obtain the best noise figure the device is capable of, and see what happens. Figure 3.293 shows the actual approach to input and output matching.

We start our design by querying the Smith Tool about the minimum noise figure and the gain that comes with it. We mark the point of the lowest noise figure (labeled Γ_{opt}). The small noise circle yields the location for Γ_{opt} and tells us the minimum noise figure is going to be 1.91 dB. (By the way, this assumes no further losses, even in soldering the transistor to the PC board.) Then we construct the input gain circle (G1) that touches Γ_{opt} and turns out to represent 13.47 dB of available gain.

The next step is to find a matching network between Γ_{opt} and the 50-Ω source impedance. Immediately, there is a conflict; if we look at S_{11}, which is the input reflection coefficient shown in the Smith diagram, it is on the other side of the complex plane and sufficiently far away, indicating that there is going to be quite a difference between noise matching and

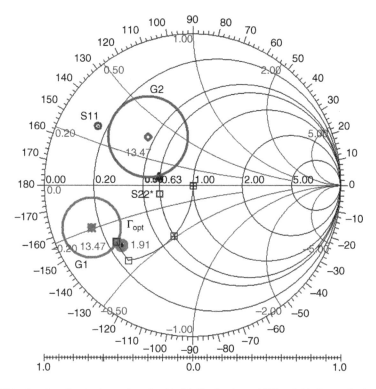

Figure 3.293 Input and output matches for 13.47 dB of gain and NF = 1.91 dB as displayed by the interactive Smith Tool.

power-gain matching. We have two options for matching the input: either we start from the origin and move toward the position of Γ_{opt}, or we start with Γ_{opt} and go to 50 Ω. Depending on the direction in which we do this, this can result in different topologies, such as a high-pass or low-pass filter. The high-pass filter is obtained with starting with Γ_{opt}; the low-pass topology is obtained when starting at the 50-Ω point. The Smith Tool then provides us with the corresponding values of capacitors and inductors after the transformation has been accomplished.

The final step is to do the same type of matching at the output. We start with S_{22}, and either select S_{22}^* (the complex conjugate for highest gain) or $S_{22\text{load}}$. As a rule, the second one is always a lower impedance (in terms of the real part) than the conjugate match. This graphic procedure is somewhat of a trial and error. As to the topology, a high-pass match should be preferred at the input and a low-pass match at the output. The reason for this is that the high-pass filter will tend to protect the stage from the enormous number of signals present below 1 GHz; using a low-pass filter at the output suppresses unwanted harmonics that the amplifier may generate. Figure 3.294 shows one of the approaches we like simply best because it gives a good input termination and we were trying for a simultaneous match for best noise figure and power gain.

The way we were able to achieve this was to take advantage of the existing feedback in the transistor and select a less-than-optimal output termination but gaining at the input. The way this works is, taking advantage of the Miller effect, we rotate the phase angle of the feedback with the load to move the optimum noise and gain matching points closer. The danger of this

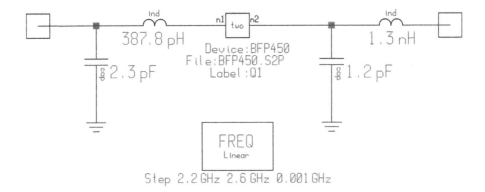

Figure 3.294 A 2.4-GHz amplifier using a BFP450 transistor and lumped-element input and output matching.

is that this condition is now highly sensitive to both input and output termination. The output *must* see 50 Ω to maintain this good input match, but since the second transistor stage can now be optimized for best input match, a two-stage combination can be configured to meet this requirement. Also let us not forget that the device is running 50 mA and as far as bipolar transistors is concerned, this is probably the best amplifier combination we have seen for getting the highest dynamic range and lowest noise figure, providing a high dynamic range. The resulting gain, matching, and noise figure characteristics are shown in Figure 3.295.

Now we turn to using distributed elements and at the same time use a shunt feedback at the output; to be specific, we will load it with 150 Ω. The reason for this is to be able

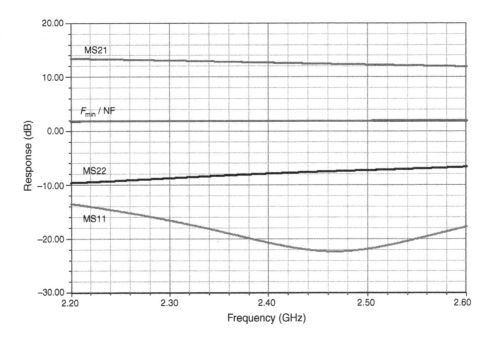

Figure 3.295 Frequency-dependent gain, matching, and noise characteristics of the matched high-dynamic-range BFP450 2.4-GHz amplifier.

Input   Output

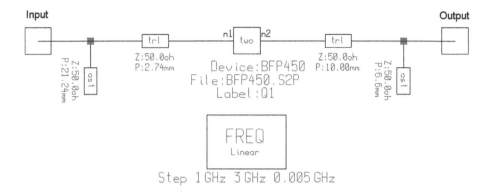

Step 1 GHz 3 GHz 0.005 GHz

Figure 3.296 Distributed-element input and output matching for the 2.4-GHz BFP450 amplifier.

to get still a good compromise at the input that will depend less on the output and at the same time improve the S_{22} matching at the output. The procedure to doing the matching now offers only certain degrees of freedom as a transmission line while doing the matching with lumped elements allows the choice of series or parallel inductors or capacitors. In the case of transmission line, we have open stubs, shorted stubs, and transmission lines. This may result in somewhat more complex matching schemes. On the positive side, they can be easily trimmed, are reproducible, and certainly cost less than exotic capacitors, which need to have small tolerances and be able to tolerate high RF currents.

The resulting input and output matching is shown in Figure 3.296, and its frequency response in Figure 3.297. We were able to maintain a good input and output match with an insignificant deterioration of the noise figure.

Since this is a high-dynamic-range amplifier, we also show its predicted two-tone response in Figure 3.298. The manufacturer specifies an $IP_{3,in}$ of 29 dB for a bias point of $V_{CE} = 3$ and $I_C = 50$ mA with $Z_S = Z_{Sopt}$ and $Z_L = Z_{Lopt}$ at 1.8 GHz. Significantly, however, the manufacturer-supplied nonlinear BFP450 library we used exhibits considerably different characteristics at $V_{CE} = 4$ and $I_C = 50$ than those reflected in the corresponding S parameters supplied by the same manufacturer. In our circuit, the S parameters resulted in about 13.5 dB of gain (based on a compromise between matching for optimum noise and maximum gain) versus about 9.5 dB of gain for the nonlinear library (match optimized for maximum gain). The difference between the 2.4 GHz F_{min} values returned for the two modeling approaches is similarly striking: 1.90 dB for the bare S parameter set and 7.76 dB for the bare nonlinear library biased to $V_{CE} = 4$ and $I_C = 50$ mA. Although the foundry supplied model for this transistor surely has been improved since the time of writing the first edition of this book, we keep this result as a warning not to trust any model without doing any validation.

Having fully evaluated the amplifier at 2.4 GHz as a working example, we note that this frequency is used by radio amateurs for some applications. We will therefore supply examples for 432 and 1296 MHz, popular spot frequencies in the amateur 70- and 23-cm bands, respectively. Figures 3.299 and 3.300 show the 432-MHz version's schematic and frequency-dependent responses. Figures 3.301 and 3.302 show the schematic and frequency-dependent responses for the 1296-MHz version.

Using a bipolar transistor that operates at much less current, the noise figure can be improved at the expense of the dynamic range/intercept point. It is possible to reduce the

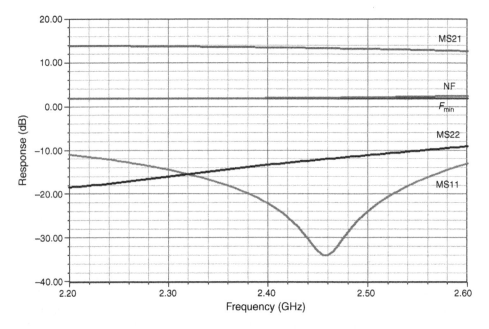

Figure 3.297 Predicted frequency-dependent gain, matching, and noise performance for the BFP450 amplifier with distributed-element matching.

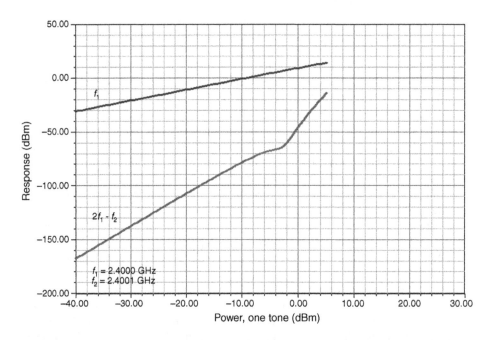

Figure 3.298 Predicted two-tone response of the BFP450 amplifier with the circuit adjusted for maximum gain (9.5 dB): $IP_{3,in} \approx 28.5$ dBm and $IP_{3,out} \approx 38$ dBm. That this version's gain is significantly lower than that of the S-parameter-based equivalent illustrates the need for better accuracy in parameter extraction.

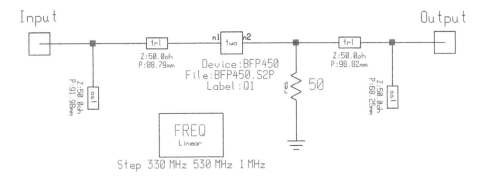

Figure 3.299 Schematic of BPF450 amplifier for 432 MHz. Because the BFP450's $f_T = 25$ GHz, at this lower frequency we must shunt the device output with a 50-Ω resistor to maintain stability.

noise figure by about 1 dB or so, but the reduction in intercept point will be much more dramatic. Another choice is the use of GaAsFETs for this application, but their matching in comparable circuits has been described as a nightmare simply because the devices have too much gain and very high input impedances (essentially 0.2 pF in series with 3 Ω), combined with reasonable output impedances in the vicinity of 240 Ω. Here is a wide field for engineers to experiment and look at price/performance and performance/real estate issues. GaAsFETs also tend to require more than 30–40 mA to be alive, and even the low operating voltage of 4–5 V makes the device take a lot of dc power.

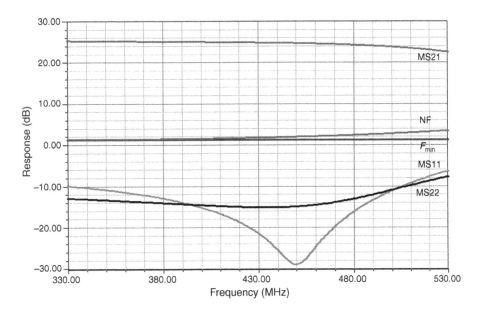

Figure 3.300 Frequency-dependent gain, matching, and noise for the BPF450 amplifier at 432 MHz. As with the circuit in Figure 3.299, the input and output network values have been chosen to achieve a low NF and good input matching at the expense of the output match.

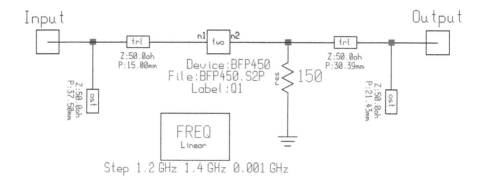

Figure 3.301 Schematic of the BPF450 amplifier for 1296 MHz. A loading resistor is still necessary to ensure stability, but because the transistor exhibits less gain at this frequency than at 432 MHz, we can increase the loading resistor to 150 Ω.

3.12.6 Example 3: 1-W Amplifier Using the CLY15

The following is an application showing a fairly wideband amplifier with +30 dBm (1 W) output. We need to point out that in accordance with the various digital modulation modes used with this amplifier, this is not continuous operation, but a much less than 1:1 duty cycle operation. The device would probably not survive operating at full output under CW conditions. Since there is no useful large-signal model available at the time of this writing, we will design this Class A amplifier using large dc bias S parameters. In the actual design,

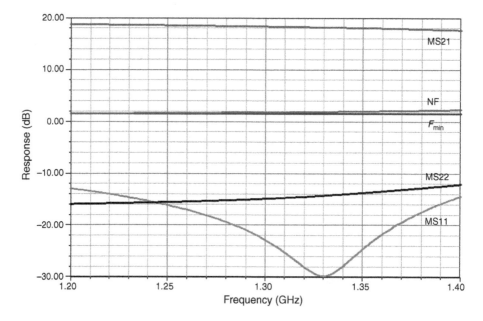

Figure 3.302 Frequency-dependent gain, matching, and noise responses of the BFP450 amplifier at 1296 MHz. As with the two previous versions, the input and output network values have been chosen to achieve a low NF and good input matching at the expense of the output match.

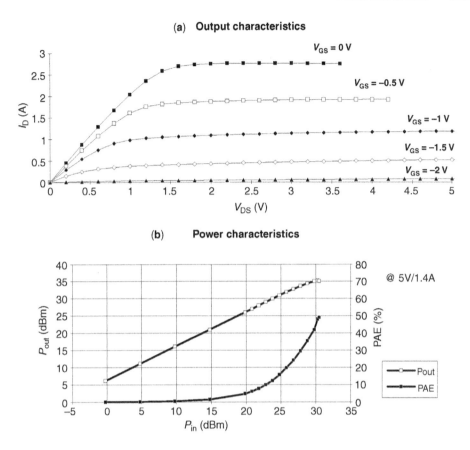

Figure 3.303 Output and power characteristics for the CLY15 medium-power GaAsFET.

there will be some cut-and-try, but we find that in this type of application this is still the best way of handling.

The CLY15 (gate width 16 mm) is the largest of a series of Siemens GaAsFETs. Its device family also includes the CLY2 (gate width 2 mm), CLY5 (4 mm), and CLY10 (8 mm). One of the most critical parameters will be the saturation voltage; from the datasheet of the CLY15, we can see that at 1.5 A the saturation voltage is approximately 1 V. This drop of 1 V includes losses that result from other parasitic elements; the dc I–V curve (Figure 3.303a) indicates a saturation voltage of 0.75. Since we aim for 1 W output power, the resistive portion of the load will be $(5 - 1)^2/(1 \times 2) = 8\Omega$. This assumes a supply voltage of 5 V.

The next step is to design the input and output matching network. Previewing the S parameters, it will be noticed that they are essentially inductive, which means that the input match, using lumped elements, will result in a combination of capacitors both at the input and the output. Figures 3.304 and 3.305 show the way the input and output match were done on the Smith diagram resulting in lumped matching elements. By inspecting the lumped input matching, here is what we find. We start with S_{11}, which is in the lower part of the Smith diagram, and determine its conjugate match, S_{11}^*. We already pointed out that this device is inductive, which means in order to match it from the conjugate point we need

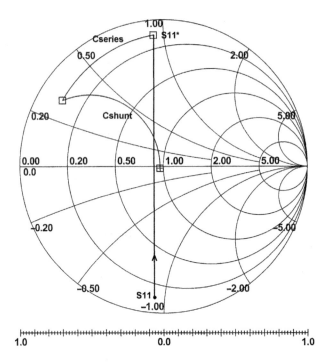

Figure 3.304 Input match for S_{11}^* using lumped elements.

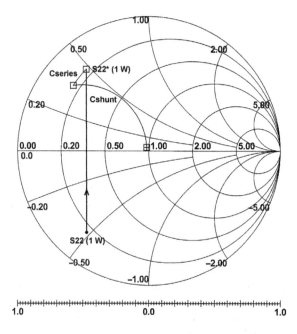

Figure 3.305 Output match for S_{22}^* (1 W) using ideal lumped elements.

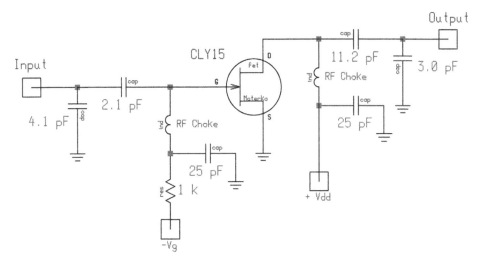

Figure 3.306   The complete amplifier using the CLY15 medium-power GaAsFET, with lumped-element input and output matching.

two capacitors. The first one is a series capacitor, as identified in the Smith diagram, and then finally a shunt capacitor to transform the impedance to 50 Ω. A close inspection of the Smith diagram shows that the resulting point is not quite on S_{11}; it is an iterative process to exactly find the crossover point, yet the input matching of better than 20 dB is more than acceptable.

As far as the output is concerned, we use the same technique by moving from S_{22} to S_{22}^*. S_{22} has been determined from the required output power; as we showed above, the real portion is 8 Ω. Using the lumped elements, the matching technique is identical; we first need to use a series capacitor and then a shunt capacitor for the transformation. The arc for the series capacitor this time is much longer, resulting in a series capacitance of 11.2 pF instead of the 2.1-pF capacitance used at the input. The actual schematic is shown Figure 3.306. The drawback of the lumped elements is that it is not practical to buy components that can handle the heavy RF currents that flow in the input and output networks. Taking the values obtained from the lumped input and output matching approach, Figure 3.307 shows the resulting frequency response.

The next step is to provide the same matching condition using distributed elements, which of course have the advantage that they can be printed on the PC board. Instead of using ideal elements, we used a loss tangent of $2 \cdot 10^{-3}$, or a Q of 500. The manufacturer of the R4000 material (Rogers) provided this specification. Now we start with the same approach by moving from S_{11} to S_{11}^*. From there, we use an open stub with a transmission-line impedance of 50 Ω, followed by a transmission line to the point in the left lower corner, and finally, a shorted transmission-line stub allows the 50-Ω matching. For this amplifier, we are not looking for noise matching but for input and output power matching. Figure 3.308 shows this match on the Smith chart.

At the output, the procedure again is similar: We move from the S_{11} for 1 W to a conjugate value, then use an open stub that acts like a capacitor until we cross the real axis; finally, a $\lambda/4$ transmission line is used to match from approximately 75–50 Ω. Figure 3.309 shows

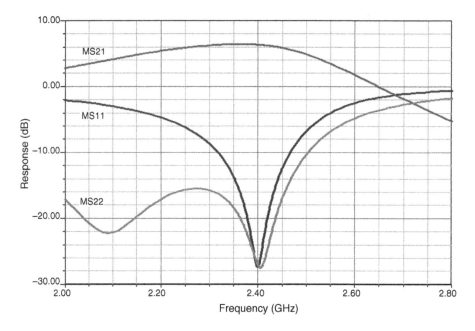

Figure 3.307 Frequency-dependent gain and matching responses of the CLY15 amplifier using ideal lumped elements.

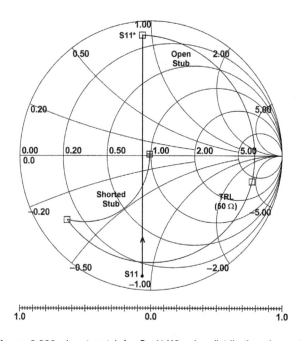

Figure 3.308 Input match for S_{11}^* (1 W) using distribution elements.

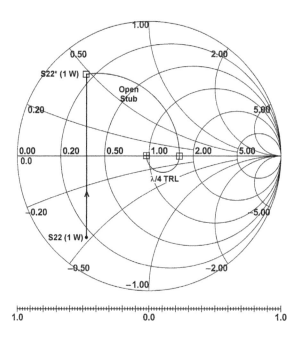

Figure 3.309 Output match for S_{22}^* (1 W) using distributed elements.

this match on the Smith chart. This concludes the distributed matching design for the CLY15 amplifier.

Finally, the complete circuit using distributed elements is shown in Figure 3.310. From a dc point of view, we took advantage of the incoming shorted stub by using a transmission line of the appropriate length and a bypass capacitor to ground. This allows us to feed the gate with the required negative voltage without interfering with the frequency response of the amplifier. When evaluating the total frequency response (Figure 3.311), we noticed that this circuit seems to have a higher Q than the lumped equivalent. Again, to obtain a wider bandwidth, more transformation stages will be needed. We leave this task to the interested reader.

The 1-dB bandwidth of the amplifier using distributed, lossy elements is 120 MHz. If a wider bandwidth is required, matching networks with more elements are required. Because of the gain-bandwidth product, the peak gain for such a wider-bandwidth arrangement will be less than this narrowband example.

Having used a lumped to distributed transformation, we next present a power amplifier that uses coaxial lines, baluns, and other transformation steps capable of handling the increased power requirement.

3.12.7 Example 4: 90-W Push–Pull BJT Amplifier at 430 MHz

The last complete application we want to take a look at is a 90-W-output push–pull amplifier using two TRW transistors (that are no longer made—an all-too-frequent occurrence after a design is complete, resulting in a scramble for replacements). Figure 3.312 shows the circuit diagram of this UHF transistor amplifier, and Figure 3.313 shows its mechanical design. The amplifier requires 28 V dc and will draw approximately 6 A, requiring 15 W of drive

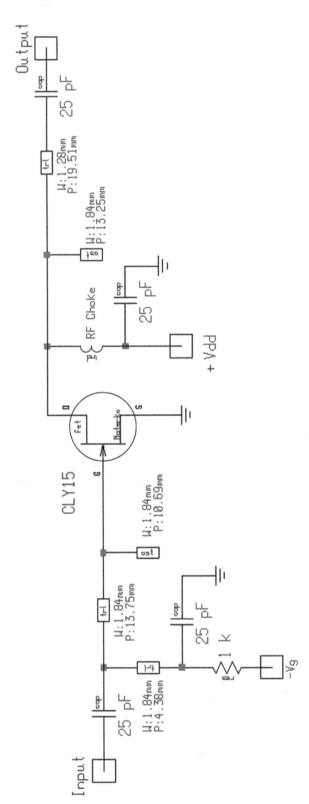

Figure 3.310 The complete amplifier using the CLY15 medium-power GaAsFET, with distributed-element input and output matching.

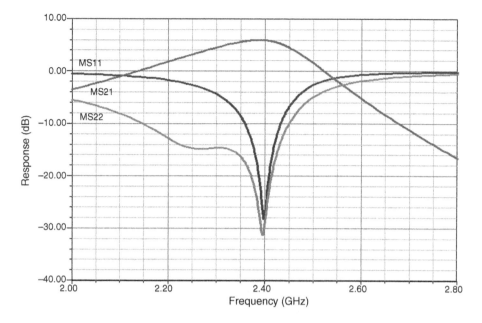

Figure 3.311 Frequency-dependent gain and matching using distributed, lossy elements.

at 438 MHz. This amplifier is a Class AB-type amplifier, which means its application is not limited to FM and other angle-modulated signals, but accommodates even SSB (single sideband) operation.

The drive power coming from the source connected to the input N connector (15 W) is fed through a short piece of coaxial cable to the PC board sitting on top of the heat sink. To change from single ended to push–pull, a 50-Ω coaxial cable is coiled up with sufficient electrical length and then feeds two symmetrical paths through amplifier. The RC section of 10 Ω in parallel with a 220-pF capacitor stabilizes the amplifier at lower frequencies, while at higher frequencies the 10-Ω resistor has no contribution. There are two independent 4:1 transformers, one for each section; they are built using 25-Ω coaxial cable with a set of two binocular ferrite cores to force the transformation. At this point, the dc voltage is also fed to the transistors. The stabilization circuit consists of a temperature monitor using the p–n junction of a transistor, side mounted on the heat sink, and a combination of operational amplifiers and adjustments. This guarantees proper, temperature-independent biasing. The 4:1 input baluns result in 12.5 Ω impedance or 3.125 Ω per transistor. Since the input impedance at this drive level is somewhere in the vicinity of $[(1\ldots2) + j(3\ldots5)]\ \Omega$. The final matching is accomplished by an RC section whereby the inductance is actually printed and the shunt capacitor across is soldered from base to base as close to the transistor bodies as possible. A ceramic-dielectric variable capacitor in parallel with this allows fine tuning.

The output consists of an inductor from collector to collector that tunes out the capacitive reactance and makes the transistor output impedance real. We then follow the same pattern by taking a 1:4 transformer, but this time built using 10-Ω rigid line instead of 25-Ω rigid line. Together with the output capacitance, prior to the output balun, this is a medium-bandwidth matching arrangement that transforms up to 50 Ω. The collector impedance per collector had to be $(28 - 2)^{2/2/90}$, resulting in roughly 3.8 Ω—not that different from the input match at the balun point. This way, the matching networks at the input and output

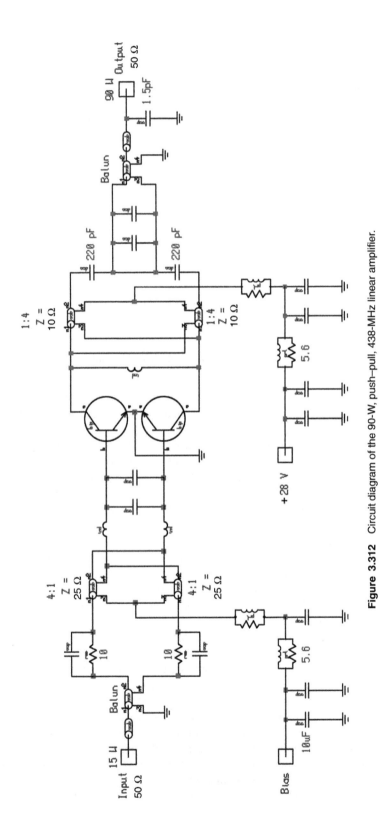

Figure 3.312 Circuit diagram of the 90-W, push–pull, 438-MHz linear amplifier.

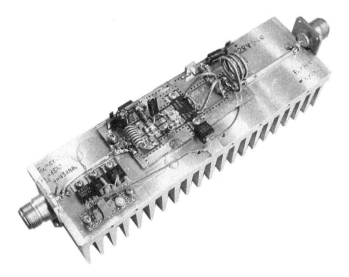

Figure 3.313 Mechanical assembly of the 90-W amplifier shown in Figure 3.312.

can be identical. At the output of the matching network, another coiled 50-Ω coaxial cable serves as balun; because it must operate at the 90 W level, larger-diameter line was used than in the input balun. Finally, another section of rigid line is used to connect the output balun on the PC board to the output N connector. The output capacitor again helps in the transformation.

Both the base and collector voltage supplies are heavily filtered using either ferrite-bead-based RF chokes and heavy wire inductors wound on low-impedance resistors. A fairly large number of these amplifiers were actually built and are still in use.

3.12.8 Quasiparallel Transistors for Improved Linearity

Some applications require really low intermodulation distortion. One of the authors (Rohde), during his tenure at RCA Government Systems Division, and his team developed a parallel-device topology that is somewhat similar to the conventional parallel approach but shows vastly increased linearity. In creating higher-power devices, manufacturers parallel several transistors in one package, connecting them in a clever way so the overall drive, dissipation, and output power are equally divided between all the devices. Special nickel–chromium ballasting resistors in the many fingers of the emitter connections of the multiple internal devices equalize any differences among the transistors while increasing the saturation voltage only by a small amount. In our high-linearity design, we used two transistors in parallel, one being biased somewhere between Class A and B, and the second one somewhere between B and C (Figure 3.314). The exact bias voltages and operating points had to be found through experiment. In practice, Q1, the "initially on" transistor, is driven to a point close to that at which its IMD products would increase. Q2 is biased such that at this same drive level it adds to the output power.

We frequently referred to this configuration as "turbocharged." The IMD products at full output power were down 40–45 dB, compared to 26–30 dB for a single transistor. While such an arrangement shows superior IMD performance, because of the somewhat abrupt kick-in of the second transistor, the transconductance response is semismooth, a characteristic that

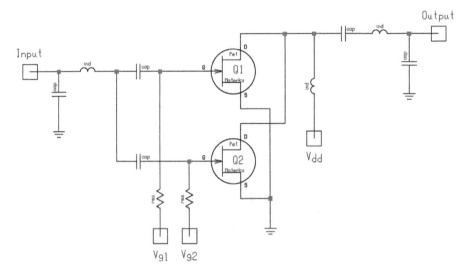

Figure 3.314 The quasiparallel amplifier topology looks like a standard parallel amplifier. Biasing the devices to different operating points nets an improvement in IMD performance.

might make alert readers think that IMD will occur at this point. On the other hand, the transconductance for Q1 in effect saturates or stays constant, so when the transconductance responses of both transistors are overlaid, the result is a smooth response overall.

This technique has been validated up to 100 W and up to at least 50 MHz. Its drawbacks include the need for two transistors with their associated cost, additional real estate, and higher input and output capacitances. Nonetheless, considering that MOS transistors capable of operation to 1 GHz were unavailable when this technique was developed, we believe that it should be revisited.

The Class D and higher operating modes require a large number of additional components and will make the efficiency higher (selective in frequency), but not necessarily reduce the IMD products. A further point of consideration is that it is incorrect to assume that reducing the input level to a given Class A will necessarily reduce its IMD products. As with tubes, the closer one gets to the knee, even in Class A operation, the worse the IMD. (This phenomenon was particularly troublesome in systems using 1-kW amplifiers up to 150 MHz; underdrive with the intent of getting better IMD performance produced just the opposite result.) If a wide power range is considered, we recommend using a bias scheme that includes sufficient intelligence to find the best operating point.

Another hot amplifier issue is that of IMD as a function of load impedance. If an amplifier is used at output power considerably below its design value, its load impedance will be too low and, depending on the circuit configuration, the result may be additional IMD.

3.12.9 Distribution Amplifiers

It is not uncommon that a signal from a particular source like an antenna has to be distributed at several places. A condominium is a typical place where from a main signal source the energy has to be split to provide a larger number of users—enough voltage or power to produce a noise-free picture. An amplifier capable of doing this is a *distribution* amplifier, not to be confused with a *distributed* (or traveling-wave) amplifier, such as that shown

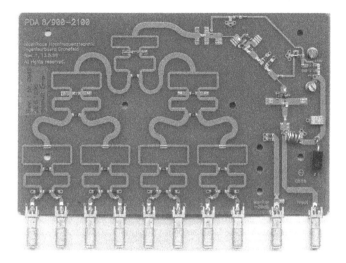

Figure 3.315 Distribution amplifier showing the use of multiple Wilkinson power dividers.

earlier in this chapter in Figure 3.6. An interesting means of achieving this is to use multiple Wilkinson couplers following an amplifier. Figure 3.315 shows an arrangement that makes heavy use of multiple power dividers to provide multiple outputs. Theoretically, the reverse is also possible; if we want to collect energy from a number of sources and combine them, as on a cable, we can use Wilkinson couplers at the *input* of a gain block. Note, however, that depending the level and number of the combined signals, this may put enormous strain on the linearity of the amplifier. What we want to avoid is lot of distortion that, in the case of television, will show up as color confetti in the noise and signal.

3.12.10 Stability Analysis of a Power Amplifier

One of the hidden pitfalls in narrowband circuit design is that a circuit may be unstable outside the frequency range of interest. Circuits containing structures built from distributed elements, which exhibit multiple resonances, can particularly susceptible to such instabilities. In this example, we analyze the stability of a simple FET power amplifier designed using small-signal concepts and the interactive Smith Tool utility in Ansoft's Serenade Design Environment. The evaluation shows how K factor analysis alone is not sufficient to check for stability of a design.

Small-Signal ac Analysis We begin by taking the Figure 3.316 amplifier, which was designed for about 16 dB of gain at 2.2 GHz. Due to the very low, capacitive input impedance of the FET and the proximity of the source-plane gain circles and stability circles to the edge of the Smith chart, the matching network attempts to transform 50 Ω to an impedance of approximately $(2.5 + j15)\Omega$. The impedance lies close to the stability circles at 2 GHz, but remains in a stable region. The bias point selected (-3.5 V) is chosen for Class AB operation. Higher linear gain can be achieved by reducing the gate voltage to -2.0 V; at this bias point, the amplifier gain will increase to about 19 dB.

First, we will do a small-signal ac analysis and examine the S parameters, and K and B_1 factors [see (3.154) and (3.157), respectively] from 1 to 6 GHz—well beyond the bandwidth

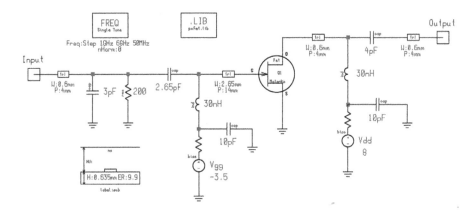

Figure 3.316 2.2-GHz amplifier design for stability analysis.

of the design. Figure 3.317 shows the circuit's frequency-dependent gain and matching characteristics.

If we examine B_1 and the stability factor K, we see that the amplifier is not unconditionally stable in the 2.2-GHz passband as shown in Figure 3.318. K dips slightly below 1, although B_1 remains above zero. In practice, a slight change in design would be required to bring K above 1 and remove this potential instability. However, for this example, this is of no consequence, as we shall see below.

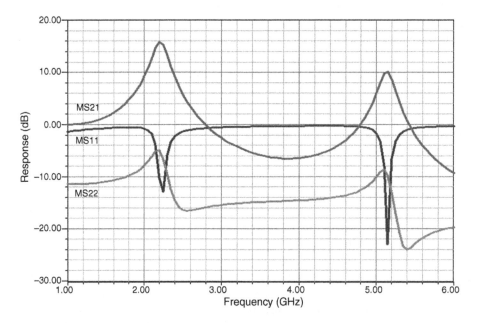

Figure 3.317 Frequency-dependent gain and matching characteristics of the amplifier. Although the circuit is designed to 2.2-GHz operation, the periodic responses of its distributed-element-based input and output matching networks result in a second gain peak near 5.1 GHz. As we shall see, however, instabilities may result at frequencies that cannot be intuited from this graph.

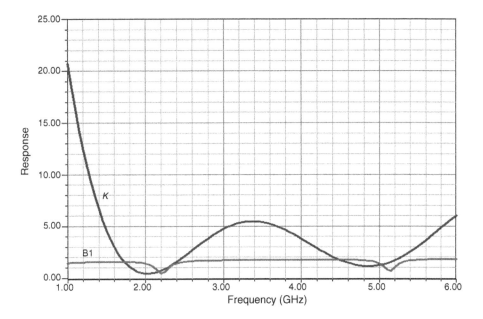

Figure 3.318 K and B_1 for the 2.2-GHz amplifier.

Nyquist Stability Analysis   Now we will utilize a Nyquist stability analysis to check the stability of the amplifier. To fully realize the advantages of a Nyquist stability analysis, we must sweep frequency over a very wide range. We want to be able to pick up instabilities at *any* frequency, not just near the design frequency. In this example, we sweep from 1 kHz to 20 GHz. A start frequency of 1 kHz is chosen so we can determine the starting point of the Nyquist plot and we ensure we pick up any low-frequency resonances. The stop frequency is chosen past the f_{max} of the transistor (where we know it cannot oscillate).

The analysis results are shown in polar plot form in Figure 3.319 and in magnitude-angle form in Figure 3.320. The polar graph shows the real and imaginary parts of the system determinant as frequency is swept. The magnitude-angle graph shows the magnitude and cumulative angle (no cut at ±180°) of the system determinant as frequency is swept. The key to interpreting these graphs lie in the following statement drawn from the Nyquist criterion.

> **The circuit will be unstable if the Nyquist plot encircles the origin in a clockwise direction.**

The term "unstable" refers to existence of natural frequencies lying in the right half of the complex frequency plane (RHP). These natural frequencies on the RHP will cause oscillations. To encircle the origin in a clockwise direction means that the Nyquist plot will travel in a clockwise path, cross the negative real axis, and continue to completely surround the origin. If the path appears to encircle the origin but then "unravels" itself in a counter-clockwise direction, the origin is not actually encircled. Figure 3.319 shows that the origin is not encircled and therefore this circuit is stable.

The magnitude-angle plot can assist you to determine if a clockwise encirclement has actually been made by recognizing the following.

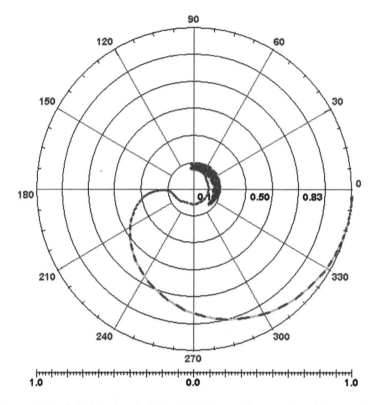

Figure 3.319 Polar Nyquist plot for the 2.2-GHz amplifier covering 1 kHz to 20 GHz.

- To travel in a clockwise direction on a polar chart means to decrease phase angle. On a magnitude-angle plot, the phase will become more negative as frequency is increased (although it may not be monotonic).
- If the path crosses the negative real axis on the polar chart, it will cross $-180°$ on the magnitude-angle plot.
- Encirclement will occur if the phase angle continues to decrease to, say, $-360°$. If "unraveling" occurs, the phase angle will not stay below $-360°$. Figure 3.320 clearly shows that the phase angle decreases to about $-180°$, but does not decrease any more. Rather, the angle returns to a positive value and oscillates between about $-40°$ and $+100°$. Therefore, an encirclement did not occur, and the circuit is stable.

An Unstable Case Let us now change the gate bias to -2 V—a value that increases the amplifier gain enough to cause instability. Figures 3.321 and 3.322 shows the results. From Figure 3.321, we can see that the origin is encircled in a clockwise direction. Figure 3.322 confirms this since the phase angle goes through $-180°$ and continues without unraveling.

Approximate Frequency of Oscillation The frequency where the Nyquist plot crosses $-180°$ is often the approximate frequency of the unstable right-half plane zeros and provides an estimate of the potential oscillation. For this example, there is a sharp crossing at 200 MHz that is characteristic of a circuit resonance. Note that the frequency is not related to a

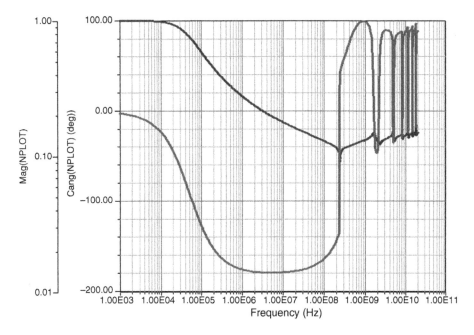

Figure 3.320 Magnitude and cumulative angle Nyquist plot for the 2.2-GHz amplifier covering 1 kHz to 20 GHz.

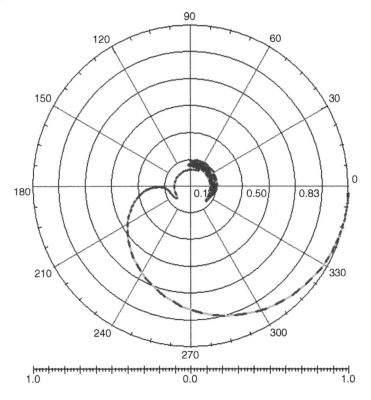

Figure 3.321 Polar Nyquist plot of the unstable amplifier.

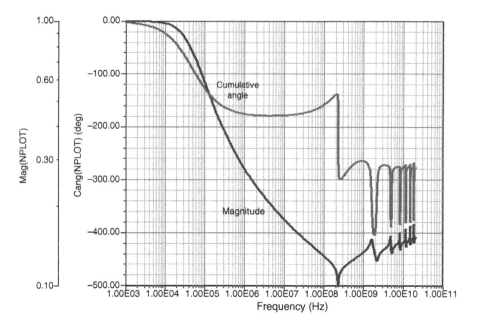

Figure 3.322 Magnitude-cumulative angle Nyquist plot of the unstable amplifier.

resonant frequency in a rigorous mathematical fashion, but it has been our experience that this frequency approximates the resonant frequency. To confirm the circuit's readiness for oscillation in the vicinity of this frequency, we evaluate it with Serenade's Oscillator Design Aid from 10 MHz to 1 GHz. The Design Aid finds an initial resonant frequency near 230.6 MHz—not too far from the $-180°$ crossing frequency of 200 MHz of the Nyquist plot. Figure 3.323 shows this point.

Oscillator analysis finds a final oscillation frequency of 228.48 MHz. Figure 3.324 shows the resulting output spectrum at Port 2. At this point, we decided to do a quick jump into Chapter 5 and take advantage of the CAD capability to predict the resulting phase noise, which is shown in Figure 3.325. As we compare this with other measured results, the approximately 228-MHz oscillator has a poor phase noise since the same phase noise is achievable at 1 GHz and higher. The reader needs to be reminded that this is an unwanted effect, but all of us who have built amplifiers will have experienced either low-frequency oscillations (motorboating) or very high-frequency oscillation, sometimes caused by the components in the immediate vicinity of the transistor itself and found somewhere above 3 GHz. This case is really a demonstration case; a closer look reveals that the transconductance required for this condition is unrealistically high, but regardless it is a good example to demonstrate the stability issue.

Where Does Oscillation Begin? An interesting question to solve now is, "What gate bias voltage is needed for oscillation to begin?" Ansoft's Serenade Design Environment includes oscillator synchronous stability analysis, which uses a modification of the harmonic-balance technique to determine critical points, such as the voltage needed to "turn on" an oscillator. Mathematically, this is called a *Hopf bifurcation* because the behavior of the circuit changes dramatically from a dc steady state to an oscillatory response. Figure 3.326 shows this effect

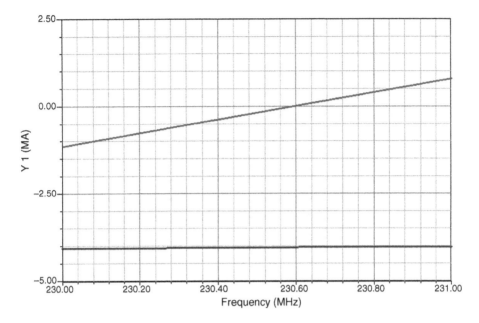

Figure 3.323 Graph from Serenade Oscillator Design Aid analysis of the amplifier. A resonant frequency is indicated where the imaginary part of the test current equals zero and the real part is negative.

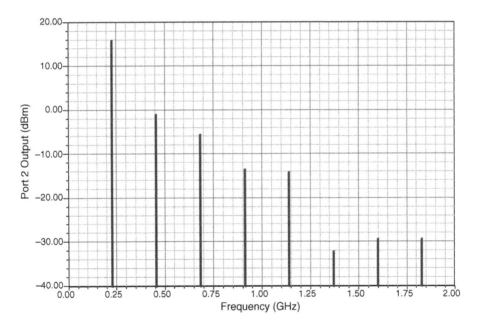

Figure 3.324 Predicted output spectrum of the unintended oscillator.

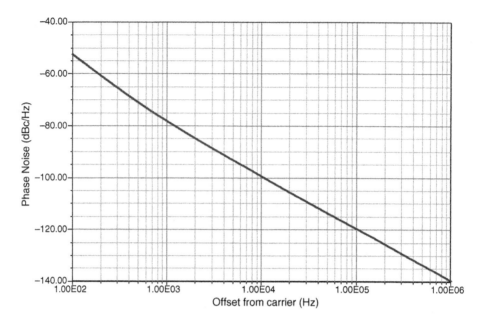

Figure 3.325   Predicted phase noise of the oscillating amplifier.

by plotting output power versus gate bias. The turning point at −3.2 V (22 mW) is clearly visible, as is the bifurcation at −3.1 V (0 mW output).

Figure 3.326 can be interpreted as follows.

1. If the bias voltage is increased from, say, −4.0 V, the circuit will begin to oscillate at the bifurcation voltage and the output power will jump to 26 mW.

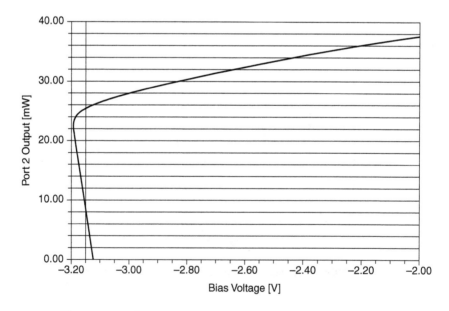

Figure 3.326   Output power versus gate bias for the unstable amplifier.

2. If the circuit is oscillating and the voltage is decreased from -2.0 V, the circuit will cease oscillating at the turning voltage of -3.2 V.

3. The branch between the turning point and the bifurcation point is unstable and cannot be realized in practice. This forms a hysteresis loop around points $(-3.1$ V, 0 mW), $(-3.1$ V, 26 mW), $(-3.2$ V, 22 mW), and $(-3.2$ V, 0 mW).

This amplifier stability analysis section is based on the stability analysis examples provided by Ansoft for the Serenade Design Environment.

An interesting example that combines both CAD and the feed-forward amplifier technique can be found in the Serenade Design Environment system simulator examples manual. This technique has been under evaluation for the last 2 years, but our feeling is that not enough measured data at frequencies above 500 MHz with reliable performance information has yet been made available in the literature. Along these lines, work done at Ansoft under a U.S. Air Force contract has provided very good insight into high-efficiency amplifiers at microwave frequencies [83].

REFERENCES

1. U. L. Rohde, A. M. Pavio, and R. A. Pucel, "Accurate noise simulation of microwave amplifiers using CAD," *Microwave Journal*, Vol. 31, No. 12, December 1988, pp. 130–141.

2. H. T. Friis, "Noise figures of radio receivers," *Proceedings of the IRE*, Vol. 32, July 1944, pp. 419–422.

3. "IRE standards on methods of measuring noise in linear twoports," *Proceedings of the IRE*, Vol. 48, January 1960, pp. 60–68.

4. H. Rothe and W. Dahlke, "Theory of noisy fourpoles," *Proceedings of the IRE*, Vol. 44, June 1956, pp. 811–818.

5. "IRE standards on methods of measuring noise in linear twoports, 1959," *Proceedings of the IRE*, Vol. 48, No. 1, January 1960, pp. 60–68. doi: 10.1109/JRPROC.1960.287380

6. H. A. Haus and R. B. Adler, *Circuit Theory of Linear Noisy Networks*. Cambridge, MA: MIT Press, 1959, and New York: Wiley, 1959.

7. H. Fukui, "The noise performance of microwave transistors," *IEEE Transactions on Electron Devices*, Vol. ED-13, March 1966, pp. 329–341.

8. S. Iversen, "The effect of feedback on noise figure," *Proceedings of the IEEE*, Vol. 63, March 1975, pp. 540–542.

9. L. Besser, "Stability considerations of low-noise transistor amplifiers with simultaneous noise and power match," *IEEE MTT-S International Microwave Symposium Digest*, 1975, pp. 327–329.

10. G. D. Vendelin, "Feedback effects on the noise performance of GaAs MESFETs," *IEEE MTT-S International Microwave Symposium Digest*, 1975, pp. 324–326.

11. W. A. Suter, "Feedback and parasitic effects on noise," *Microwave Journal*, February 1983, pp. 123–129.

12. K. Hartman and M. J. O. Strutt, "Changes in four noise parameters due to general changes in linear two-port circuits," *IEEE Transactions on Electron Devices*, Vol. ED-20, No. 10, October 1973, pp. 874–877.

13. H. Fukui, "Available power gain, noise figure, and noise measure of two-ports and their graphical representation," *IEEE Transactions on Circuit Theory*, Vol. CT-13, June 1966, pp. 137–142.

14. C. R. Poole and D. K. Paul, "Optimum noise measure terminations for microwave transistor amplifiers," *IEEE Transactions on Microwave Theory and Techniques*, Vol. MTT-33, No. 11, November 1985, pp. 1254–1257.

15. H. Hillbrand and P. H. Russer, "An efficient method for computer aided noise analysis of linear amplifier networks," *IEEE Transactions on Circuits and Systems*, Vol. CAS-23, No. 4, April 1976, pp. 235–238.

16. W. E. Pastori, "Topics on noise," Eaton Electronics Instrumentation Division, (EID) Seminar Notes, Los Angeles, 1983.

17. W. E. Pastori, "Image and second stage corrections resolve noise figure measurement confusion," *Microwave Systems News*, May 1983, pp. 67–86.

18. R. Q. Lane, "The determination of device noise parameters," *Proceedings of the IEEE*, Vol. 57, August 1969, pp. 1461–1462.

19. R. Q. Lane, "A microwave noise and gain parameter test set," *IEEE International Solid-State Circuits Conference*, 1978, pp. 172–173.

20. N. Kuhn, "Accurate and automatic noise figure measurements with standard equipment," *IEEE MTT-S International Microwave Symposium Digest*, 1980, pp. 425–427.

21. E. W. Strid, "Measurement of losses in noise-matching networks," *IEEE Transactions on Microwave Theory and Techniques*, Vol. MTT-29, March 1981, pp. 247–252.

22. W. Dahlke, "Transformation der Rauschkennwerte durch Netzwerke," in *Die Telefunken-Röhre*, H. Rothe, ed. Vol. 33, October 1956, pp. 30–45.

23. T.-K. Nguyen, C.-H. Kim, G.-J. Ihm, M.-S. Yang, and S.-G. Lee, "CMOS low-noise amplifier design optimization techniques," *IEEE Transactions on Microwave Theory and Techniques*, Vol. 52, No. 5, May 2004, pp. 1433–1442, and "Comments on 'CMOS low-noise amplier design optimization techniques'," *IEEE Transactions on Microwave Theory and Techniques*, Vol. 52, No. 7, July 2004, p. 3155.

24. A. van der Ziel, *Noise in Solid State Devices and Circuits*. New York: Wiley, 1986.

25. L.-H. Lu and H.-S. Chen, "Lower the voltage for CMOS RFIC," *IEEE Microwave Magazine*, February 2010, pp. 70–77.

26. T. Manku, G. Beck, and E. J. Shin, "A low-voltage design technique for RF integrated circuits," *IEEE Transactions Circuits Systems-II: Analog and Digital Signal Processing*, Vol. 45, No. 10, pp. 1408–1413, October 1998.

27. T. K. K. Tsang and M. N. El-Gamal, "Gain and frequency controllable sub-1 V 5.8 GHz CMOS LNA," in *IEEE International Circuits System Symposium*, Vol. 4, May 2002, pp. IV-795–IV-798.

28. H.-H. Hsieh, J.-H. Wang, and L.-H. Lu, "Gain-enhancement techniques for CMOS folded cascode LNAs at low-voltage operation," *IEEE Transactions Microwave Theory Technology*, Vol. 56, August 2008, pp. 1807–1816.

29. H.-H. Hsieh and L.-H. Lu, "Design of ultra-low-voltage RF front ends with complementary current-reused architectures," *IEEE Transactions Microwave Theory Technology*, Vol. 55, July 2007, pp. 1445–1458.

30. W. A. Debski, "Multi-band adaptive WLAN receivers in 0.13 μm CMOS," Ph.D. thesis, Brandenburg University of Technology, Cottbus, Germany, 2007. Available at http://opus.kobv.de/btu/volltexte/2007/368/.

31. A. Vasylyev, "Integrated RF power amplier design in silicon-based technologies," Ph.D. thesis, Brandenburg University of Technology, Cottbus, Germany, 2006. Available at http://opus.kobv.de/btu/volltexte/2006/20/.

32. R. Aparicio and A. Hajimiri, "Capacity limits and matching properties of integrated capacitors," *IEEE Journal Solid-State Circuits*, Vol. 37, No. 3, March 2002, pp. 384–393.

33. J. P. Silver, *RF, RFIC & Microwave theory, design*. Available at http://www.odyseus.nildram.co.uk/.

34. *The Predictive Technology Model (PTM)*. Available at http://ptm.asu.edu/

35. R. Cantz, "Hochfrequenzverstärkung mit trioden," *Telefunken-Röhre*, Vol. 30, 1953, pp. 52–69.

36. C. Bowick, *RF Circuit Design*. Indianapolis, IN: Howard F. Sams & Co, 1982.

37. A. S. Sedra and K. C. Smith, *Microelectronic Circuits*. Chicago, IL: Saunders College Publishing, 1991.

38. P. R. Gray and R. G. Meyer, *Analysis and Design of Analog Integrated Circuits*, 2nd ed. New York: Wiley, 1984.

39. S. H. Lee, A. K. Wong, and M. G. Wong, *A Current-Combiner Circuit for Better Mixer Conversion Gain*, Philips Semiconductors applications document.

40. F. M. Fano, "Theoretical limitations in the broadband matching arbitrary impedances," *Journal of Franklin Institute*, Vol. 249, 1950, pp. 57–85 and 139–155.

41. R. Goyal, ed., "Wideband amplifiers," *High-Frequency Analog Integrated-Circuit Design*. New York: Wiley, 1995, Chapter 6.

42. S. Georgi and U. L. Rohde, "A novel approach to voltage controlled tuned filters using CAD validation," *Microwave Engineering Europe*, October 1997, pp. 35–42.

43. *Analog/Interface ICs Device Data*, Vol. II, Q4/96, DL128, Rev 6, Motorola, Inc., MC13158 Wideband FM IF Subsystem specifications, pp. 8-308–8-329.

44. E. J. Wilkinson, "An *N*-way power divider," *IRE Transactions Microwave Theory Technology*, Vol. MTT-8, January 1960, p. 116.

45. V. J. Tyler, "A new high-efficiency high-power amplifier," *Marconi Revision*, Vol. 21, 1958, pp. 96–109.

46. D. M. Snider, "A theoretical analysis and experimental confirmation of the optimally loaded and overdriven RF power amplifier," *IEEE Transactions on Electron Devices*, Vol. ED-14, December 1967, pp. 851–857.

47. F. H. Raab, "Class-F power amplifiers with maximally flat waveforms," *IEEE Transactions on Microwave Theory Technology*, Vol. 45, November 1997, pp. 2007–2012.

48. S. Gao, "High-efficiency class-F RF/microwave amplifiers," *IEEE Microwave Magazine*, February 2006, pp. 40–48.

49. A. V. Grebennikov, "Circuit design technique for high efficiency Class F amplifiers," in *IEEE MTT-S International Microwave Symposium Digest*, Vol. 2, June 2000, pp. 771–774.

50. N. O. Sokal and A. D. Sokal, "Class E: A new class of high-efficiency tuned single-ended switching power amplifiers," *IEEE Journal of Solid-State Circuits*, Vol. SC-10, June 1975, pp. 168–176.

51. N. O. Sokal, "Class-E RF power amplifiers," *QEX*, 2001, pp. 9–20.

52. F. H. Raab, "Idealized operation of the class E tuned power amplifier," *IEEE Transactions on Circuits Systems*, Vol. CAS-24, December 1977, pp. 725–735.

53. M. K. Kazimierczuk and K. Puczko, "Exact analysis of class E tuned power amplifier at any Q and switch duty cycle," *IEEE Transactions on Circuits System*, Vol. CAS-34, February 1987, pp. 149–159.

54. D. J. Kessler and M. K. Kazimierczuk, "Power losses and efficiency of class-E power amplifier at any duty ratio," *IEEE Transactions on Circuits and Systems I: Regular Papers*, Vol. 51, No. 9, September 2004, pp. 1675–1689.

55. F. Raab, "Class-E, Class-C, and Class-F power amplifiers based upon a finite number of harmonics," *IEEE Transactions on Microwave Theory Technology*, Vol. 49, No. 8, August 2001, pp. 1462–1468.

56. R. Negra and W. Bachtold, "Lumped-element load-network design for class-E power amplifiers," *IEEE Transactions on Microwave Theory Technology*, Vol. 54, No. 6, June 2006, pp. 2684–2690.

57. R. Negra, F. Ghannouchi, and W. Bachtold, "Study and design optimization of multiharmonic transmissionline load networks for class-E and class-F K-Band MMIC power amplifiers," *IEEE Transactions Microwave Theory Technology*, Vol. 55, No. 6, June 2007, pp. 1390–1397.

58. H. Nemati, C. Fager, M. Thorsell, and H. Zirath, "High-efficiency LDMOS power-amplifier design at 1 GHz using an optimized transistor model," *IEEE Transactions Microwave Theory Technology*, Vol. 57, No. 7, July 2009, pp. 1647–1654.

59. J. Kim and K. Konstantinou, "Digital predistortion of wideband signals based on power amplifier model with memory," *Electronic Letters*, Vol. 37, No. 23, 2001, pp. 1417–1418.

60. F. E. Terman and R. R. Buss, "Some notes on linear and grid-modulated radio frequency amplifiers," *Proceedings of IRE*, Vol. 29, 1941, pp. 104–107.

61. L. R. Kahn, "Single sideband transmissions by envelope elimination and restoration," *Proceedings of IRE*, Vol. 40, 1952, pp. 803–806.

62. V. Petrovic and W. Gosling, "Polar loop transmitter," *Electronics Letters*, Vol. 15, 1979, pp. 1706–1712.

63. V. Petrovic and C. N. Smith, "The design of VHF polar loop transmitters," in *IEE Communications Conference*, Vol. 82, 1982, pp. 148–155.

64. S. M. Whittle, "A practical cartesian loop transmitter for narrowband linear modulation PMR systems," in *IEE Colloquium on Linear Amplifiers and Transmitters*, April 1994.

65. H. Seidel, "A microwave feedforward experiment," *Bell System Technical Journal*, November 1971, pp. 2879–2916.

66. H. S. Black, "Translating system," US patent 1,686,792, February 1925.

67. A. R. Mansell and A. Bateman, "Practical implementation issues for adaptive predistortion transmitter linearisation," in *IEE Colloquium on Linear RF Amplifiers and Transmitters*, April 1994.

68. *Systemizing RF Power Amplifier Design*, Motorola Application Note AN-282A.

69. W. E. Everitt and G. E. Anner, *Communication Engineering*. New York: McGraw-Hill.

70. G. L. Matthaei, L. Young, and E. M. T. Jones, *Microwave Filters, Impedance-Matching Networks and Coupling Structures*. New York: McGraw-Hill.

71. G. L. Matthaei, "Tables of Chebyshev impedance-transforming networks of low-pass filter form," *Proceedings of IEEE*, August 1964.

72. *Matching Network Designs with Computer Solutions*, Motorola Application Note AN-267.

73. E. G. Cristal, "Tables of maximally flat impedance-transforming networks of low-pass filter form," *IEEE Transactions on Microwave Theory Technology*, Vol. MTT 3, No. 5, September 1965.

74. H. W. Bode, *Network Analysis and Feedback Amplifier Design*. New York: D. Van Nostrand Company.

75. R. M. Fano, "Theoretical limitations on the broadband matching of arbitrary impedances," *Journal of the Franklin Institute*, January–February 1950.

76. J. H. Horwitz, "Design wideband UHF-power amplifiers," *Electronic Design*, Vol. 11, May 24, 1969.

77. O. Pitzalis and R. A. Gilson, "Tables of impedance matching networks which approximate prescribed attenuation versus frequency slopes," *IEEE Transactions on Microwave Theory Technology*, Vol. MTT-9, No. 4, April 1971.

78. Anthony N. Gerkis, "Broadband impedance matching using the 'real frequency' network synthesis technique," *Applied Microwave and Wireless*, July/August 1998, pp. 26–36.

79. T. Cuthbert, "Broadband impedance matching methods," *RF Design*, August 1994.

80. T. Cuthbert, *Broadband Direct-Coupled and Matching RF Networks*, TRCPEP publications, Grenwood, AR, 1999.

81. G. D. Vendelin, A. M. Pavio, and U. L. Rohde, *Microwave Circuit Design Using Linear and Nonlinear Techniques*. New York: Wiley, 1990, pp. 279–283.

82. A. Potter, "HP RF compiler automates schematic capture and extends capabilities of circuit synthesis," *Applied Microwave and Wireless*, Vol. 11, No. 6, June 1999, pp. 106–118.

83. A. Kain, *Final Report for an Integrated CAD Tool for High-Power, High-Efficiency RF and Microwave Amplifiers*, STTR Phase II, contract no. F33615-95-C-174, Department of the Air Force, Small Business Technology Transfer (STTR) Program.

84. K. K. Clarke and D. T. Hess, *Communication Circuits: Analysis and Design*. Reading, MA: Addospm-Wesley, 1971.

FURTHER READING

A. Pascht, J. Fischer, and M. Berroth, "A CMOS low noise amplifier at 2.4 GHz with active inductor load," *IEEE Topical Meeting on Silicon Monolithic Integrated Circuits in RF Systems*, Ann Arbor, MI, 2001.

P. L. D. Abrie, *The Design of Impedance-Matching Networks for Radio-Frequency and Microwave Amplifiers*. Norwood, MA: Artech House, 1985.

A. Ambrózy, *Electronic Noise*. New York: McGraw-Hill, 1982.

A. Cappy, "Noise modeling and measurement techniques," *IEEE Transactions on Microwave Theory and Techniques*, Vol. MTT-36, January 1988, pp. 1–10.

H. J. Carlin, "A new approach to gain-bandwidth problems," *IEEE Transactions on Circuits and Systems*, Vol. CAS-23, April 1977, pp. 170–175.

H. J. Carlin and P. Amstutz, "On optimum broad-band matching," *IEEE Transactions on Circuits and Systems*, Vol. CAS-23, April 1977, pp. 401–405.

H. J. Carlin and J. J. Komiak, "A new method of broad-band equalization applied to microwave amplifiers," *IEEE Transactions on Microwave Theory and Techniques*, Vol. MTT-27, February 1979, pp. 93–99.

H. J. Carlin and B. Yarman, "A simplified real frequency technique applied to broad-band multistage amplifiers," *IEEE-MTT*, Vol. 30, No. 12, December 1982.

H. J. Carlin and B. S. Yarman, "The double matching problem: analytic and real-frequency solutions," *IEEE Transactions on Circuits and Systems*, Vol. CAS-30, No. 1, January 1983, pp. 15–28.

R. S. Carson, *High Frequency Amplifiers*. New York: Wiley, 1975.

G. Caruso and M. Sannino, "Computer-aided determination of microwave two-port noise parameters," *IEEE Transactions on Microwave Theory and Techniques*, Vol. MTT-26, September 1978, pp. 639–642.

J. A. Eisenberg, "Designing amplifiers for optimum noise figure," *Microwaves*, Vol. 13, April 1974, pp. 36–41.

R. M. Fano, "Theoretical limitations on the broadband matching of arbitrary impedances," *Journal of the Franklin Institute*, Vol. 249, January 1960, pp. 57–83 and February 1960, pp. 139–155.

H. Fukui, *Low-Noise Microwave Transistors and Amplifiers*. New York: IEEE Press, 1981.

C. Gentili, *Microwave Amplifiers and Oscillators*. New York: McGraw-Hill, 1987.

S. Georgi and U. L. Rohde, "A novel approach to voltage-controlled tuned filters using CAD validation," *Microwave Engineering Europe*, October 1997, pp. 35–42.

G. Gonzales, *Microwave Transistor Amplifier Analysis and Design*. Englewood Cliffs, NJ: Prentice-Hall, 1984.

H. A. Haus and R. B. Adler, "Optimum noise performance of linear amplifiers," *Proceedings of IRE*, Vol. 46, August 1958, pp. 1517–1533.

S. A. Maas, "FET models for volterra-series analysis," *Microwave Journal*, May 1999, pp. 260–270.

D. J. Mellor and J. G. Linvill, "Synthesis of interstage networks and prescribed gain versus frequency slopes," *IEEE Transactions on Microwave Theory and Techniques*, Vol. MTT-23, No. 12, December 1975, pp. 1013–1020.

S. Mercer, "Exact simulation of LNAs reduces design cycle time," *Applied Microwave and Wireless*, January 1999, pp. 74–82.

R. P. Meys, "A wave approach to the noise properties of linear microwave devices," *IEEE Transactions on Microwave Theory and Techniques*, Vol. MTT-26, January 1978, pp. 34–37.

C. K. S. Miller, W. C. Daywitt, and M. G. Arthur, "Noise standards, measurements and receiver noise definitions," *Proceedings of the IEEE*, Vol. 55, June 1967.

B. J. Minnis, *Designing Microwave Circuits by Exact Synthesis*. Boston, MA: Artech House, 1996.

R. Minton, *Design of Large Signal VHF Transistor Power Amplifiers*, RCA Application Note SMA-36, July 1964.

R. Minton and H. C. Lee, *Frequency Multiplication Using Overlay Transistors*, RCA Application Note SMA-40, September 1965.

M. Mitama and H. Katoh, "An improved computational method for noise parameter measurement," *IEEE Transactions on Microwave Theory and Techniques*, Vol. MTT-27, June 1979, pp. 612–615.

T. Mukaihata, "Applications and analysis of noise generation in N-cascaded mismatched two-port networks," *IEEE Transactions on Microwave Theory and Techniques*, Vol. MTT-16, September 1968, pp. 699–708.

W. W. Mumford and E. H. Scheibe, *Noise Performance Factors in Communications Systems*. Dedham, MA: Horizon House, 1968.

D. E. Norton, "High dynamic range transistor amplifier using lossless feedback," *Microwave Journal*, May 1976, pp. 53–57.

D. E. Norton, "Strong signal survivability design for lossless feedback amplifiers," *Proc. 11th Annual European Microwave Conference*.

T. Y. Otoshi, "The effect of mismatched components on microwave noise-temperature calibrations," *IEEE Transactions on Microwave Theory and Techniques*, Vol. MTT-16, September 1968, pp. 675–686.

P. Penfeld, "Wave representations of amplifier noise," *IRE Transactions on Circuit Theory*, March 1962, pp. 84–86.

U. L. Rohde, J. C. Whitaker, and T. T. N. Bucher, "Receiver system planning," *Communications Receivers: Principles and Design*, 2nd ed. New York: McGraw-Hill, 1997, Chapter 3.

C. Rorabaugh, *Circuit Design and Analysis—Featuring C Routines*. New York: McGraw-Hill, 1992.

M. Sannino, "On the determination of device noise and gain parameters," *Proceedings of the IEEE*, Vol. 67, September 1979, pp. 1364–1366.

A. van der Ziel, *Noise in Solid State Devices and Circuits*. New York: Wiley, 1986.

D. Vondran, "Noise figure measurement: corrections related to match and gain," *Microwave Journal*, March 1999, pp. 22–38.

D. F. Wait, "Thermal noise from a passive linear multiport," *IEEE Transactions on Microwave Theory and Techniques*, Vol. MTT-16, September 1968, pp. 687–691.

B. S. Yarman and H. J. Carlin, "A simplified 'real frequency' technique applied to broad-band multistage microwave amplifiers," *IEEE Transactions on Microwave Theory and Techniques*, Vol. MTT-30, No. 12, December 1982, pp. 2216–2222.

4

MIXER DESIGN

4.1 INTRODUCTION

Radiocommunication requires that we shift a baseband information signal to a frequency or frequencies suitable for electromagnetic propagation to the desired destination. At the destination, we reverse this process, shifting the received radiofrequency signal back to baseband to allow the recovery of the information it contains. This frequency-shifting function is traditionally known as mixing; the stages that perform it, as mixers. Any device that exhibits amplitude-nonlinear behavior can serve as a mixer, for example, as we saw in Section 1.7.2, nonlinear distortion results in the production, from the signals present at the input of a device, of signals at new frequencies. Even a rusty screw or bolt on an antenna element can act as a mixer, producing unwanted IMD products that appear at the receiver input.

Although mixers are equally important in wireless transmission and reception, traditional mixer terminology favors the receiving case because mixing was first applied as such in receiving applications. Thus, the signal to be frequency-shifted is applied to the mixer's RF port, and the frequency-shifting power or voltage (from a local oscillator [LO]) is applied to the mixer's LO port, resulting in two outputs at the mixer's intermediate-frequency (IF) port. If the wanted IF is lower in frequency than the RF signal, the mixer is a downconverter; if the wanted IF is higher than the RF, the mixer is an upconverter. Converter may also be used as a term for a single stage that simultaneously acts as mixer and LO.

For a given RF signal, an ideal mixer with a perfect LO (i.e., an LO with no harmonics and no noise sidebands) would produce only two IF outputs: one at the frequency sum of the RF and LO, and another at the frequency difference between the RF and LO. Filtering can be used to select the desired IF product and reject the unwanted one, which is sometimes referred to as the IF image.

RF/Microwave Circuit Design for Wireless Applications, Second Edition. Ulrich L. Rohde and Matthias Rudolph.
© 2013 John Wiley & Sons, Inc. Published 2013 by John Wiley & Sons, Inc.

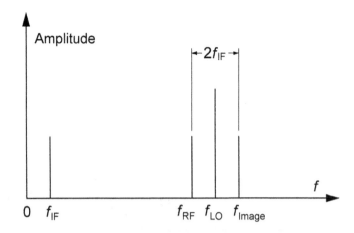

Figure 4.1   Relationship between a mixer's image and desired-signal responses. The image is 2 f_{IF} away from the desired signal.

The simultaneous generation of LO+RF and LO−RF outputs results not from a departure of mixer performance from the ideal, but from the mathematics of mixing itself. Another unavoidable mixing artifact, the RF image response, also results from the mathematics of mixing rather than mixer nonideality. Just as a given RF/LO combination produces two IF outputs (LO+RF and LO−RF, the IF and IF image), the mixer will produce output at the desired IF (LO+RF or LO−RF) in response to two possible RF inputs: one at LO+IF and another at LO−IF (Figure 4.1). The undesired response, the RF image (traditionally referred to merely as the image), is 2 f_{IF} removed from the desired response. Even if no man-made signals exist at the RF image frequency, reducing a mixer's RF image response can be important because noise at that frequency, including that produced by circuitry between the mixer and antenna, will still be mixed to the desired IF, degrading the signal-to-noise ratio. Filtering and phasing techniques can be used to reduce the RF or IF image responses— filtering if the image is sufficiently removed from the desired response for filtering to provide the necessary rejection, phasing if the desired and image responses are insufficiently spaced for filtering to work, as in the case of a double-conversion receiver in which signals at a high first IF (e.g., 50–70 MHz) must be converted to a very low first IF, such as 25 kHz.

The output of every real mixer includes a vast number of additional unwanted products, including noise, the fundamentals of the mixer's RF and LO signals and their harmonics, and the sums and differences of the RF and LO and their harmonics. Intermodulation distortion between multiple signals present at the RF port and IF output resulting from the mixing to IF of LO noise-sideband energy by strong adjacent signals (reciprocal mixing, Section 1.7.2), further complicate a mixer's output spectrum and may compromise system performance.

All mixers are multipliers in the sense that the various new outputs they produce can be described mathematically as the multiplicative products of their inputs. From an im-plementation standpoint, however, a given mixer circuit can be characterized as additive or multiplicative depending on how RF and LO signals are applied to it. Additive mixing occurs when the RF and LO signals are applied to the same input port, as in Figures 4.2 and 4.3. Multiplicative mixing occurs when the RF and LO signals are applied to separate ports, as in Figure 4.4. As a rule, multiplicative mixers afford better isolation between their LO

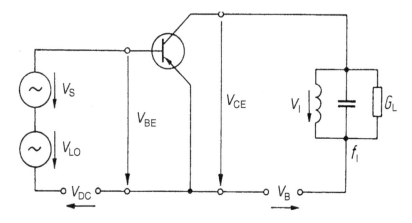

Figure 4.2 Additive mixing in a BJT [1].

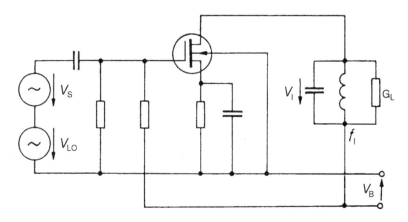

Figure 4.3 Additive mixing in a single-gate MOSFET [1].

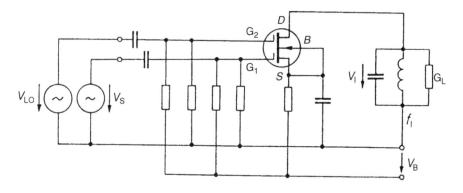

Figure 4.4 Multiplicative mixing in a dual-gate MOSFET. A dual-gate device is actually two single-gate devices in series [1].

and RF ports than additive mixers, and this enhanced interport isolation is their principal merit. Multiplicative mixing does not in itself suppress unwanted products; the spurious response of a basic multiplicative mixer cell is poor unless it is used in a push–pull or quad configuration.

Let us now consider the basic theory of mixers. Mixing is achieved by the application of two signals to a nonlinear device. Depending upon the particular device, the nonlinear characteristic may differ. However, it can generally be expressed in the form

$$I = K (V + v_1 + v_2)^n \tag{4.1}$$

The exponent n is not necessarily an integer, V may be a dc offset voltage, and the signal voltages v_1 and v_2 may be expressed as $v_1 = V_1 \sin(\omega_1 t)$ and $v_2 = V_2 \sin(\omega_2 t)$.

When $n = 2$, (4.1) may then be written as

$$I = K [V + V_1 \sin (\omega_1 t) + V_2 \sin (\omega_2 t)]^2 \tag{4.2}$$

This assumes the use of a device with a square-law characteristic. A different exponent will result in the generation of other mixing products, but this is not relevant for a basic understanding of the process. Expanding (4.2)

$$I = K[V^2 + V_1^2 \sin^2 (\omega_1 t) + V_2^2 \sin^2 (\omega_2 t) + 2V \, V_1 \sin (\omega_1 t)$$
$$+ 2V \, V_2 \sin (\omega_2 t) + 2V_2 \, V_1 \sin (\omega_2 t) \sin (\omega_1 t)] \tag{4.3}$$

The output comprises a direct current and a number of alternating current contributions. We are interested only in that portion of the current that generates the IF; so, if we neglect those terms that do not include both V_1 and V_2, we may write

$$I_{\text{IF}} = 2KV_1 V_2 \sin (\omega_1 t) \sin (\omega_2 t)$$
$$I_{\text{IF}} = KV_2 V_1 \{\cos [(\omega_2 - \omega_1) t] - \cos [(\omega_2 + \omega_1) t]\} \tag{4.4}$$

This means that at the output, we have the sum and difference signals available, and the one of interest may be selected by the IF filter.

4.2 PROPERTIES OF MIXERS

4.2.1 Conversion Gain/Loss

Even though a mixer works by means of amplitude-nonlinear behavior in its device(s), we generally want (and expect) it to act as a linear frequency shifter. The degree to which the frequency-shifted signal is attenuated or amplified is an important mixer property. Conversion gain can be positive or negative; by convention, negative conversion gains are often stated as conversion loss.

In the case of a diode (passive) mixer , the insertion loss is calculated from the various loss components:

$$\text{Loss (dB)} = \text{Conversion loss} + \text{Transformer loss}$$
$$+ \text{Losses due to harmonic generation} + \text{Diode loss} \tag{4.5}$$

In the case of a doubly balanced mixer, we must add the transformer losses (on both sides) and the diode losses as well as the mixer sideband conversion, which accounts by definition,

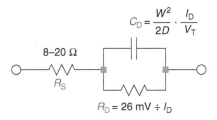

Figure 4.5 Equivalent circuit of a mixer diode.

for 3 dB. Ideally, the mixer produces only one upper and one lower sideband, which results in the 3 dB loss compared to the input signal. Also, the input and output transformers add about 0.75 dB on each side, and of course there are the diode losses because of the series resistances of the diodes.

Figure 4.5 shows the equivalent circuit of a diode. It consists of a series (loss) resistor R_S and a time-variable electronic resistor, typically called the diffusion resistance, R_D, which equals $26\,\mathrm{mV}/I_D$, and a capacitance C_D shunting R_D. C_D can be found from

$$C_D = \frac{W^2}{2D} \frac{I_D}{V_T} \tag{4.6}$$

where D is the diffusion constant, a material-dependent value, and W is the physical width. The average value for R_D is somewhere between the calculated value of $26\,\mathrm{mV}/I_D$ and some leakage current, simply because it is generated by a rectification mechanism, which turns the LO power into an RF current and then into a combination of dc and RF currents.

We can calculate the diode loss according to

$$\text{Diode loss (dB)} = \log_{10}\left(\frac{50 + (2 \times R_S)}{50}\right) \tag{4.7}$$

Assuming that $R_S = 8\,\Omega$, the diode loss for a diode-ring mixer

$$\text{Diode loss (dB)} = \log_{10}\left(\frac{50 + (2 \times 8)}{50}\right) = 0.5\,\text{dB} \tag{4.8}$$

From (4.5), the insertion loss for this mixer is therefore

Loss (dB) = 3 dB (conversion loss) + 1.5 dB (transformer loss)

  + 1 dB (losses from harmonic generation) + 0.5 dB (diode loss)

= 6 dB   (4.9)

This assumes mixing at the fundamental frequency. Sometimes the diode loss resistor R_S is as high as 25 Ω per arm (as in a MOSFET switch), or 50 Ω total. This now results in

Loss (dB) = 3 dB + 1.5 dB + 1 dB + 3 dB = 8.5 dB (insertion loss)   (4.10)

Since the value of R_S is partially determined by the threshold voltage of the diode and the diode diffusion resistance, R_D, a wide range of values can be noticed for different drive levels and mixer topologies. Figure 4.5 shows a shunt capacitance C_D, the so-called diode diffusion capacitance. When the diode is conducting, the influence of this nonlinearity is frequency dependent, which adds to the insertion loss. In this discussion, we have not considered its frequency dependency. At wireless frequencies, modern Schottky diodes, also frequently called hot-carrier diodes, are operated far from their cutoff frequency, resulting in less than 1 dB of additional losses. There are also mixers with special circuitry to terminate the IF image. This is done with a diplexer circuit or equivalent circuitry. This makes insertion loss values as low as 4 dB possible; however, the large-signal condition or intercept point suffers.

4.2.2 Noise Figure

Like any network, a mixer contributes noise to the signals its frequency shifts. The degree to which a mixer's noise degrades the signal-to-noise ratio (see Section 1.7.1) of the signals its frequency shifts is evaluated in terms of noise factor and noise figure as discussed in Section 1.7.1.

4.2.2.1 Passive Mixer

For a long time, the literature has stated that the noise figure of a passive mixer, which is pretty much independent of its circuit arrangement, is equal to the mixer's insertion loss. But this neglects the influence of the white-noise contribution of the mixer's diode(s). This is ironic, considering that RF noise generators were long based on thermionic diodes operated in saturation (the 5722 noise diode was a popular type, as in the Rohde & Schwarz SKTU). Such a diode's noise-power output can be readily determined from its saturation current. On the other hand, all Schottky diodes, while conducting, generate white noise that follows the same principle as above. This fact has been practically recognized only by a few companies that make modern noise-measurement equipment. Modern noise-measurement devices measure the noise figure of a system by "hot and cold" technique, an approach based on knowledge of the absolute noise energy emitted under hot conditions (conductance). This method has the advantage that it can be used up to several tens of gigahertz, while the old vacuum-tube-based noise generators ran out of steam at around 1 GHz due to the inability to match the tube to the 50 Ω termination. This was typically accomplished by connecting a 50 Ω resistor between anode and ground (without dc connection), followed by a low-pass filter, which would match the tube capacitance and other parasitics to the required termination of 50 Ω, purely resistive.

In reality, we can take the loss calculation from above and add the Schottky noise generated by the diodes as they are driven by the local oscillator.

If a Schottky diode is used as a noise generator in the conductive mode, it generates a continuous frequency spectrum, possibly up to several gigahertz. There is a mathematical relationship between the noise power spectrum emitted by the diode and the time-averaged current of this diode, which generates the noise. If the noise source impedance is set (typically 50 Ω), the available noise power can be calculated according to

$$I_R = \sqrt{2e \times I_s \times \Delta f} \tag{4.11}$$

where

$e = 1.6 \times 10^{-19}$ coulombs; I_s = saturation current of the diode; Δf = effective noise bandwidth.

For $S_{11} = 0$ or proper termination of this circuit ($R_G = R_{term}$),

$$P_R = \left(\frac{I_R}{2}\right)^2 \times R_i$$

$$= \frac{e}{2} \times I_s \times \Delta f \times R_i \qquad (4.12)$$

Calculated at a bandwidth of 1 Hz

$$\frac{P_R}{\Delta f} = \frac{e}{2} \times I_s \times R_i \qquad (4.13)$$

If

$$\frac{P_R}{\Delta f} = kT_0 \times F \qquad (4.14)$$

the noise factor becomes

$$F = \frac{e \times I_s \times R_i}{2kT_0} \qquad (4.15)$$

If the values for e and kT_0 are inserted,

$$F = 20 \times I_D \times \frac{26\,\text{mV}}{I_D} + \frac{R_s + R_G}{2R_G} \qquad (4.16)$$

4.2.2.2 Example

Assume the passive mixer mentioned above with its 6 dB insertion loss is considered and a dc current of 15 mA results as a function of the LO drive. Since I_D gets canceled, the noise factor (F) of the diode portion equals

$$F = 20 \times I_D \times \frac{26\,\text{mV}}{I_D} + \frac{R_s + R_G}{2R_G}$$

$$= 0.52 + 0.58$$

$$= 1.1 \qquad (4.17)$$

The noise figure, NF, is $10 \log F$, or 0.413 dB. Now this number and the insertion loss must be added. The resulting noise figure would be 6.413 dB. This is consistent with published measurement data.

4.2.2.3 Exact Mathematical Nonlinear Approach

The exact noise factor of a real mixer is computed by the formula

$$F = \frac{N_0(\omega_{IF}) + kT_0}{K_B T_0 G_{T_c}(\omega_{RF})} \qquad (4.18)$$

where

$N_0(\omega_{IF})$ = total noise power (per unit bandwidth) delivered to the IF load at intermediate frequency; K_B = Boltzmann's constant; T_0 = reference temperature (290 or 300 K is commonly used); $G_{T_c}(\omega_{RF})$ = transducer conversion gain from ω_{RF} to ω_{IF}.

Let us now further elaborate on (4.18). We may write

$$N_0(\omega_{IF}) = N_s(\omega_{IF}) + N_{INT}(\omega_{IF}) + N_L(\omega_{IF}) + kT_0 \qquad (4.19)$$

where N_S is noise generated by the RF source resistance and transferred to the IF load through frequency conversion, N_{INT} is noise generated internally to the mixer, and N_L is noise generated by the IF termination. If the source resistance is held at temperature T_0, N_S will basically originate from noise generated at the RF and image frequencies, which are transferred to the IF with approximately the same conversion gain, plus a relatively small contribution transferred from other sidebands with a smaller conversion gain. We may then write synthetically

$$N_s (\omega_{IF}) = 2aK_B T_0 G_{T_c} (\omega_{RF}) \tag{4.20}$$

where a is a coefficient slightly larger than 1. $N_{INT}(\omega_{RF})$ is generated by transformer losses, by the diode Schottky noise and by the diode resistive parasitics, and in principle may take on any value; in particular, it may be zero if both the transformers and the diodes are ideal (i.e., if the latter are pure nonlinear resistors). As for $N_L(\omega_{IF})$, by Nyquist's theorem the IF load resistor R_L may be described as a noiseless resistor in series with a noise voltage source whose mean-square voltage (per unit bandwidth) is

$$|V_L|^2 = 4K_B T_L R_L \tag{4.21}$$

where T_L is the IF termination temperature. If the IF load is driven by a source with an output impedance $Z_{out}(\omega_{IF})$, the noise power actually delivered to the load will obviously be

$$N_{out} = \frac{4K_B T_L R_L^2}{|Z_{out} (\omega_{IF}) + R_L|^2} \tag{4.22}$$

In addition to N_{out}, the thermal noise originating from the IF termination delivered to the IF load at ω_{IF} will also include contributions from other sidebands that are back converted by the mixer nonlinearities with a relatively small conversion gain. Thus, we may write

$$N_L (\omega_{IF}) = \frac{4bK_B T_L R_L^2}{|Z_{out} (\omega_{IF}) + R_L|^2} \tag{4.23}$$

where b is slightly larger than 1. If we now introduce the mixer conversion loss, namely

$$L_C = \frac{1}{G_{T_c} (\omega_{RF})} \tag{4.24}$$

and combine (4.18) with (4.19), (4.20), and (4.23), we finally get the noise factor expression

$$F = 2a + \frac{N_{INT} (\omega_{IF}) + kT}{K_B T_0} L_C + \frac{4bT_L R_L^2}{T_0 |Z_{out} (\omega_{IF}) + R_L|^2} L_C \tag{4.25}$$

In the normal region of operation of the mixer (sufficient LO drive), we may assume

$$Z_{out} (\omega_{IF}) \quad R_L \tag{4.26}$$

so that (4.25) becomes

$$F = 2a + \frac{N_{INT} (\omega_{IF}) + kT}{K_B T_0} L_C + \frac{bT_L}{T_0} L_C \tag{4.27}$$

Multiplying the $\log_{10}$ of F by 10 gives us the exact mixer noise figure in dB.

Table 4.1 Noise Figure and Conversion Gain Versus LO Power for a Diode DBM

LO Power (dBm)	NF (dB)	Conversion Gain (dB)
−10.0	45.3486	−45.1993
−8.0	32.7714	−32.5264
−6.0	19.8529	−19.2862
−4.0	12.1154	−11.3228
−2.0	8.85188	−8.05585
0.0	7.26969	−6.51561
2.0	6.42344	−5.69211
4.0	5.85357	−5.15404
6.0	5.50914	−4.84439
8.0	5.31796	−4.66871
10.0	5.19081	−4.54960
12.0	5.08660	−4.45887
14.0	4.99530	−4.38806
16.0	4.91716	−4.33322
18.0	4.85920	−4.29407
20.0	4.82031	−4.26763

The method just discussed is the basis for the far noise calculation for oscillators as mentioned in the oscillator chapter, in which it is referred to as conversion noise. This includes the AM-to-PM conversion noise.

Table 4.1 shows how the noise figure and conversion gain vary with LO power for a generic diode DBM (Figure 4.6). "Starving" a diode mixer by decreasing its LO drive rapidly degrades its performance in all respects.

4.2.2.4 Differential CMOS Mixer

Let us consider the noise generated in a differential switching CMOS mixer. Its topology is shown in Figure 4.7. We will follow the derivation of the individual origins of noise as detailed in Ref. [2]. The discussion will focus on 1/f noise for the reason that this kind of noise exists only in relative narrow bands at dc or, if upconverted, as noise sidebands of a carrier. It is interesting to see at which frequencies these noise sidebands will arise and why. Also, in direct conversion (zero IF), the baseband 1/f noise will be the governing type of noise. But of course, the mechanisms controlling 1/f noise affect white noise in the same manner. The difference is, however, that the white noise is constant over frequency, therefore, it does not change anything if it is translated in frequency.

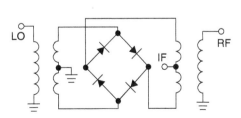

Figure 4.6 Generic diode DBM.

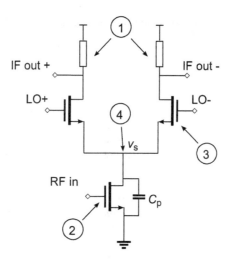

Figure 4.7 Differential CMOS mixer, indicating the origins of noise. (1) Load noise, (2) transconductance noise, (3) direct switching noise, (4) indirect switching noise.

This investigation aims at providing a basic understanding of the noise mechanisms present in an ideal differential switching mixer. Imperfections such as simultaneous conduction of both transistors during switching will cause additional noise. For simplicity, only a single-balanced mixer is discussed. There is, however, no principal difference to double-balanced mixers, so that the theory is easily extended to this case. Only switching-type mixers will be discussed, but also a Gilbert cell mixer will show qualitatively the same effects.

The following four noise mechanisms are present in this type of mixer: load noise, transconductance noise, and direct and indirect switching noise. These four effects will now be discussed. In Figure 4.7, it is indicated by the numbers (1–4) where these mechanisms are located. Throughout this section, the transistor noise is modeled by a single equivalent voltage noise source at the gate of the respective transistor. Figure 4.8 shows the resulting output $1/f$ noise spectra caused by each of the mechanisms.

Load Noise   The load of the differential mixer adds noise to the signal. This noise is simply superimposed to the signal at the output of the mixer. Minimizing this noise is done by selecting appropriate load transistors, or even resistors that provide no $1/f$ noise.

Transconductance Noise   The current source transistor is used to modulate the switched current by the RF signal. The switching is then responsible for the mixing of the RF signal. However, any noise generated in this transistor will also be mixed by the differential pair just as any other signal. The good news concerning $1/f$ noise is, however, that it is not at the same frequency as the RF signal, but it is in baseband. It gets therefore upconverted to the LO frequency and to its odd harmonics. Under ideal conditions, the $1/f$ noise caused by the transconductance transistor is shifted out of the baseband and should not degrade SNR at the mixer output.

Direct Switching Noise   After discussing the current source and the load, it is time to discuss the $1/f$ noise of the switching transistors. Their noise is not just mixed, it impacts the

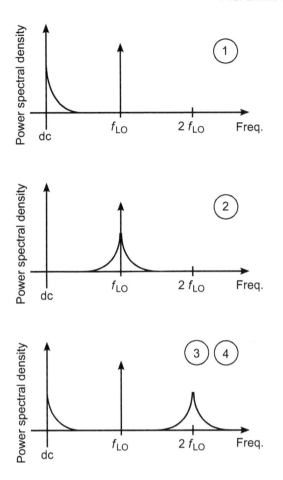

Figure 4.8 Contributions of the four basic noise sources to the output mixer noise. (1) Load noise, (2) transconductance noise, (3) direct switching noise, (4) indirect switching noise.

switching. Let us consider the ideal case. The switching transistors are assumed to perform the transition from on to off instantaneously when the gate voltage crosses a certain value. With no noise, the switching times would be exactly each half cycle of the LO signal.

Now consider additional $1/f$ noise at the gate, it is a slow signal that is superimposed to the LO switching signal. The LO signal will be just slightly increased or decreased, and the switching events are consequently shifted in time, a bit earlier or later. Thus, the noise causes jitter.

In order to quantify the noise contribution, we need to describe the difference between the switched current with jitter and without, as shown in Figure 4.9. If the two signals are simply subtracted, one obtains a series of pulses with the following properties

- The pulses are located at the switching times.
- The amplitude is either positive or negative, twice the switched current.
- The width (and the sign) of the pulse depends on the ratio between the magnitude of the gate noise voltage $V_n(t)$ at the switching time and the slope S of the LO voltage applied to the gate. Figure 4.10 highlights this relation.

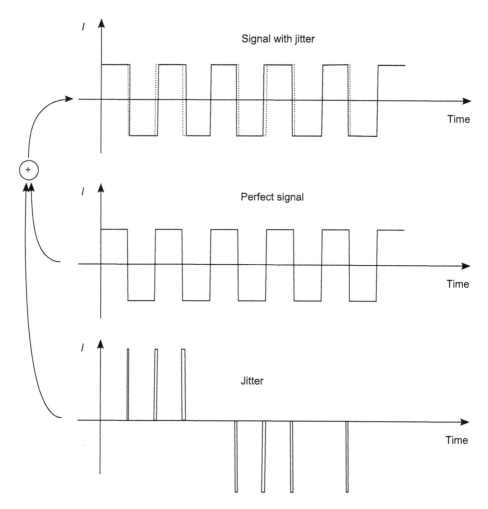

Figure 4.9   Direct switching noise: the LO signal with jitter can be understood as an undisturbed signal superimposed with a pulse train.

Thus, the jitter can be understood as a sampled replica of the original $1/f$ noise, that is, additionally scaled according to the signal's slope S. The sampling frequency equals twice the LO frequency since the jitter occurs at every switching event, thus twice per LO period.

The jitter noise current normalized for one period is therefore

$$I_{n,o} = \frac{2}{T_{LO}} \cdot 2I \cdot \frac{v_n}{S} \tag{4.28}$$

With the period of the LO signal $T_{LO} = 1/f_{LO}$, the switched current I, the noise voltage at the switching times v_n, and the slope of the LO gate voltage crossing zero, S.

For a sinusoidal signal, $T_{LO} \cdot S$ is given by $4\pi A$, where a factor of 2 accounts for the fact that the noise is compared to a differential signal. the baseband noise spectrum thereby becomes

$$i_{n,o}(f) = \frac{I}{\pi A} \cdot v_n(f) \tag{4.29}$$

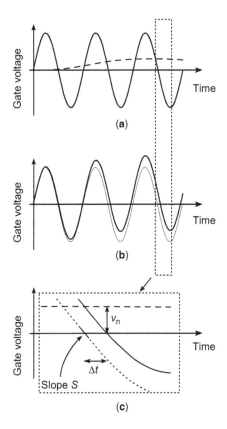

Figure 4.10 (a) Ideal LO gate voltage (solid line), $1/f$ noise voltage (dashed line); (b) actual (solid line) and ideal (dotted line) LO gate voltage; (c) the time-offset (jitter) at a zero crossing is calculated from the noise voltage and the slope of the LO signal.

Due to the sampling of this signal, this spectrum appears at baseband, and as noise sidebands around all even harmonics of the LO frequency.

In the ideal case, there is no interaction with the other noise sources. But if the mixer is unbalanced due to an offset voltage, noise of the transconductance transistor can leak through and add to the jitter. It affects the output noise spectrum in the same manner as the switching noise discussed before.

Indirect Switching Noise The direct switching noise can be reduced by fast switching. Applying an ideal square-wave voltage to the gates would suppress this kind of noise completely. However, $1/f$ noise of the switching transistors adds to overall noise in a second way that only depends on LO frequency and FET capacitances.

Let us assume ideal switching, and investigate the voltage at the virtual ground. Under this assumption, it will be constant. However, if $1/f$ noise exists at the switching transistor's gate, things change. Due to the balanced structure, it is allowed just to regard the unsymmetrical noise. At one transistor gate, the $1/f$ noise will be switched between two states: v_n if the MOST is conducting, and 0 if the MOST is turned off. However, due to the differential structure, the virtual ground potential will anticipate this imbalanced switching, and will also shift by v_n when the respective transistor is on.

The noise signal is stochastic in nature, but much slower than the LO signal. Let us treat it like a switched voltage of slowly varying amplitude. The relevant time constants in our case are the switching edges, as these will enable a current through the capacitance C_p associated with the transconductance transistor.

$$I_{n,o} = \frac{2}{T_{LO}} \int_0^{T_{LO}/2} C_p \cdot \left[\frac{d}{dt} v_s(t)\right] dt \qquad (4.30)$$

which leads to the noise contribution

$$i_{n,o} = \frac{2}{T_{LO}} C_p \cdot v_n \qquad (4.31)$$

Concerning the spectrum, the same observations are made as for the direct switching noise. The current bursts at the virtual ground change their polarity depending on whether the transistor is switched on or off. But regarding the load, this change in polarity is compensated by the differential pair, that routes one burst through one branch and the next through the other. Thus, at the output, again $1/f$ noise appears, sampled with twice the LO frequency.

Anyway, the indirect switching noise should usually be much lower than direct switching noise. Direct conversion noise can be reduced by fast switching, but switching noise can never be suppressed completely.

4.2.2.5 SSB Versus DSB Noise Figure

In radiocommunication, we typically use mixers as "single-sideband" devices; that is, we are interested in only one of the two possible RF inputs ($f_{LO} + f_{IF}$ or $f_{LO} - f_{IF}$) that produce output at the desired IF. The unused input is the RF image or simply image. A mixer's RF image response can be important even if no man-made signals are present at f_{image} because noise there, including that produced by the mixer's RF-port termination, will be converted to the IF, possibly compromising the system's signal-to-noise ratio. As a result of this and the IEEE definition of noise figure [3], which, for a mixer, considers only the noise associated with a system's principal frequency transformation, some controversy exists about the definition and measurement of mixer noise figure. IEEE's NF definition for mixers assumes no noise contribution at f_{image}, not even from the mixer's RF-port termination at that frequency; yet, by convention (and out of necessity, since an absolutely noiseless, $T = 0$ f_{image} termination is unavailable on the average test bench) the noise figures we measure always include some noise at f_{image}, even if, as a result of filtering, it arises only from the mixer's RF-port termination. The NF measurements and simulations presented in this book reflect this conventional SSB NF scenario. Stephen Maas provides in-depth material on these issues, including the importance of double-sideband NF for some systems, and the limitations of NF as a figure of merit, in Ref. [4].

4.2.3 Linearity

4.2.3.1 1 dB Compression Point

Like other networks, a mixer is amplitude nonlinear above a certain input level; above this point, the output level fails to track input level changes proportionally. This figure of merit, $P_{-1\,dB}$, identifies the single-tone input-signal level at which the output of the mixer has fallen 1 dB below the expected output level. The 1 dB compression point in a conventional double-balanced diode mixer is approximately 6 dB below the LO power. For lower distortion mixers, it is usually 3 dB below the LO power.

4.2.3.2 1 dB Desensitization Point

This specification is another figure of merit similar to the 1 dB compression point. However, the 1 dB desensitization point refers to the level of an interfering (undesired) input signal that causes a 1 dB decrease in nominal conversion gain for the desired signal. For a diode-ring DBM, the 1 dB desensitization point is usually 2–3 dB below the 1 dB compression point.

4.2.3.3 Dynamic Range

The dynamic range of any RF/wireless system can be defined as the difference between the 1 dB compression point and the minimum discernible signal (MDS). These two points are specified in units of power (dBm), giving dynamic range in dB. When the RF input level approaches the 1 dB compression point, harmonic and intermodulation products begin to interfere with the system performance. High dynamic range is obviously desirable, but cost, power consumption, system complexity, and reliability must also be considered.

4.2.3.4 Harmonic Intermodulation Products (HIP)

These are spurious products that are harmonically related to the f_{LO} and f_{RF} input signals.

$$\mathrm{HIP} = M f_{LO} + N f_{RF} \qquad (4.32)$$

Table 4.2 shows relative harmonic intermodulation product levels for a high-level diode DBM.

4.2.3.5 Intermodulation Distortion (IMD)

Nonlinearities in the mixer devices give rise to intermodulation distortion products whenever two or more signals are applied to the mixer's RF port. Testing this behavior with two (usually closely spaced) input signals of equal magnitude can return several figures of merit depending on how the results are interpreted. A mixer's third-order output intercept point ($IP_{3,out}$) is defined as the output power level where the spurious signals generated by $(2 f_{RF1} \pm f_{RF2}) \pm f_{LO}$ and $(f_{RF1} \pm 2 f_{RF2}) \pm f_{LO}$ are equal in amplitude to the desired output signal as shown in Figure 4.11.

The third-order input intercept point, $IP_{3,in}$—IP_3 referred to the input level—is of particularly useful value and is the most commonly used mixer IMD figure of merit. $IP_{3,in}$ can be calculated according to

$$\mathrm{IP}_{n,in} = \mathrm{IMR} \div (n - 1) + \text{input power (dBm)} \qquad (4.33)$$

Table 4.2 Typical Spurious Responses of High-Level Double-Balanced Mixer (Decibels Below $f_{LO} \pm f_{RF}$ Response)

RF Input Signal Harmonics		f_{LO}	$2 f_{LO}$	$3 f_{LO}$	$4 f_{LO}$	$5 f_{LO}$	$6 f_{LO}$	$7 f_{LO}$	$8 f_{LO}$
$8 f_{RF}$	100	100	100	100	100	100	100	100	100
$7 f_{RF}$	100	100	97	102	95	100	100	100	90	100
$6 f_{RF}$	100	100	92	97	95	100	100	95	100	100
$5 f_{RF}$	90	90	84	86	72	92	70	95	70	92
$4 f_{RF}$	90	90	84	97	86	97	90	100	90	92
$3 f_{RF}$	75	75	63	66	72	72	58	86	58	80
$2 f_{RF}$	70	70	72	72	70	82	62	75	75	100
f_{RF}	60	60	0	35	15	37	37	45	40	50
		60	60	70	72	72	62	70	70

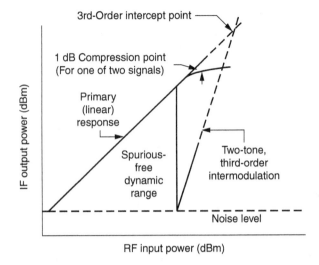

Figure 4.11 Mixer linearity evaluation, including compression and two-tone IMD dynamic range. $P_{-1\,\text{dB}}$ for a single-tone cannot be read directly from this graph because the values shown are the result of two-equal-tone drive.

where IMR is the intermodulation ratio (the difference in dB between the desired output and the spurious signal, and n is the IM order—in this case, 3). In a conventional diode double-balanced mixer, $IP_{3,\text{in}}$ is approximately 14 dB above the single tone 1 dB compression point ($P_{-1\,\text{dB}}$)—approximately 8 dB greater than the local oscillator power. As will be seen later, this does not apply to feedback active mixers. They have their own agenda, above a particular input level, their IMD products increase almost exponentially.

Although designers are usually more concerned with odd-order IM performance, second-order IM can be important in wideband systems (systems that operate over a 2:1 or greater bandwidth) as discussed in Section 1.7.2.

4.2.4 LO Drive Level

A mixer's specifications are usually guaranteed at a particular LO drive level, usually specified as a dBm value that may be qualified with a tolerance. Insufficient LO drive degrades mixer performance; excessive LO drive degrades performance and may damage mixer devices. Commercially available diode mixers are often classified by LO drive level. For example, a "Level 17" mixer requires 17 dBm of LO drive.

4.2.5 Interport Isolation

In a mixer, isolation is defined as the attenuation in dB between a signal input at any port and its level as measured at any other port. High isolation numbers are desirable. Figure 4.12 shows LO–IF and LO–RF isolation versus frequency for a triple-balanced diode DBM. Isolation is dependent mainly on transformer and physical symmetry, and device balance, but the level of signals applied to the mixer also plays a role, as shown in Figure 4.13.

4.2.6 Port VSWR

The load presented by a mixer's ports to the outside world can be of critical importance to a designer. For example, high LO-port VSWR may result in inefficient use of available

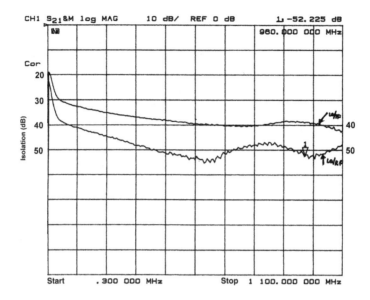

Figure 4.12 LO–IF and LO–RF isolation versus frequency for a high-level triple-balanced diode mixer. The periodic roughness of the traces is a measurement-system artifact, proving that even sophisticated FFT-based instruments are not perfect.

LO power, resulting in LO starvation (underdrive) that degrades the mixer's performance. Figure 4.14 shows LO-port VSWR versus frequency for a high-level diode DBM with two values of LO power. Like interport isolation, port VSWR can vary with the level of the signal applied.

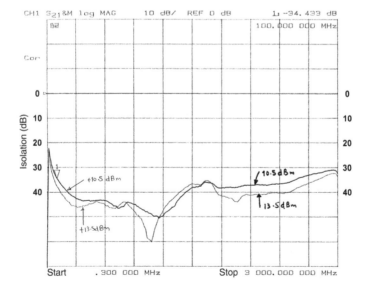

Figure 4.13 LO–IF isolation versus frequency and LO drive level for a high-level diode DBM.

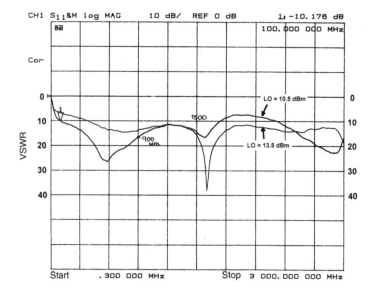

Figure 4.14   LO-port VSWR versus frequency for a high-level diode DBM.

4.2.7   dc Offset

Isolation between ports plays a major role in reducing dc offset in a mixer. Like isolation, dc offset is a measure of the unbalance of the mixer. In phase-detector and phase-modulator applications, dc offset is a critical parameter.

4.2.8   dc Polarity

Unless otherwise specified, mixers with dc output are designed to have negative polarity when RF and LO signals are of equal phase.

4.2.9   Power Consumption

Circuit power consumption is always important, but in battery-powered wireless designs it is critical. Mixer choice may be significant in determining a system's power consumption, sometimes in ways that seem paradoxical at first glance. For instance, a passive mixer might seem to be a power smart choice because it consumes no power—until we factor in the power consumption of the circuitry needed to provide the (often considerable) LO power a passive mixer requires. If a mixer requires a broadband resistive termination that will be provided by a postmixer amplifier operating at a high standing current, the power consumption of the amplifier stage must be considered as well. Evaluating the suitability of a given mixer type to a task therefore requires a grasp of its ecology as well as its specifications.

4.3   DIODE MIXERS

Passive mixers based on diode switches are common in base-station applications, where their high dynamic range and 50 Ω port impedances overcome objections to their inherent conversion loss, unsuitability to integration, and relatively high LO-power requirement. As we will discuss later, FET-based passive mixers are overcoming some of these limitations in mobile wireless applications.

4.3.1 Single-Diode Mixer

Figure 4.15 shows the schematic of a simplistic single-diode mixer. The LO and RF signals are applied in series to the diode, with no effort made to match the sources to the diode or isolate the sources from each other. The LO and RF signals are both present in the diode simultaneously, so the mixing performed is additive. The LO signal switches the diode on and off, gating the RF signal to the circuit's output at the LO frequency. In this simple circuit, the output port's 50 Ω resistance serves as the diode's ac load and dc return. Because we want the diode to operate as a linear switch with respect to the RF signal, the diode should turn on hard and turn off as completely as possible, and the LO, not the RF source, should switch the diode on and off. The upper frequency limit of the circuit will depend on how rapidly the diode can switch between the on and off states. The LO can be only so strong before the diode's dissipation limits are exceeded; the RF source can be only so strong relative to the LO before it begins to play a role in switching the diode on and off. The LO must therefore be considerably stronger than the RF source—20 dB or more for low-distortion applications.

Figure 4.16 shows how the single-diode mixer's conversion gain and noise figure vary with applied LO power. Figure 4.17 shows the mixer's output spectrum.

The following several pages present sample data on diodes suitable for mixer service in Tables 4.3 and 4.4 and the full manufacturer's data sheet on the Infineon BAT15-99.

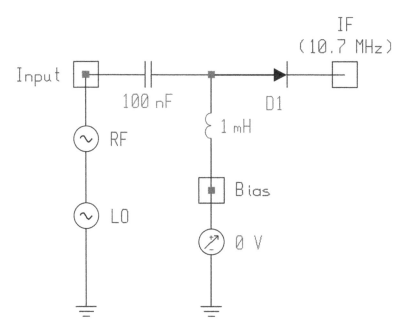

Figure 4.15 Schematic of the single-diode mixer. The circuit's 50 Ω output termination serves as the diode's dc return, as well as completing the circuit for RF, LO, and IF.

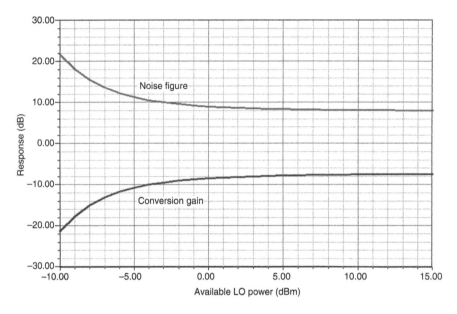

Figure 4.16 Conversion gain and noise figure versus LO power for the single-diode mixer. The values reported are worse than those predicted by theory because RF, LO, and IF matching have not been attempted. In this analysis, LO = 310.7 MHz (−10 to 25 dBm), RF = 300 MHz (−50 dBm), and IF = 10.7 MHz.

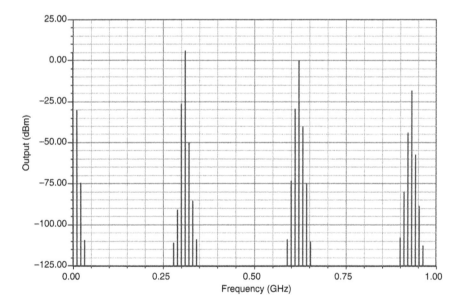

Figure 4.17 Output spectrum of the single-diode mixer. Three LO harmonics and three LO sidebands were used in this analysis. In this analysis, LO = 310.7 MHz (13 dBm), RF = 300 MHz (−20 dBm), and IF = 10.7 MHz.

Table 4.3 Example RF Schottky Diodes (Infineon)

Type	Maximum Ratings V_R (V)	I_F (mA)	Characteristics $(T_A = 25°C)$ C_T (pF)	V_F at I_F V	mA	Configuration	Package
BAT15-02LRH	4.0	110.0	0.26	0.23	1	Single	TSLP-2-7
BAT15-03W	4.0	110.0	0.26	0.23	1	Single	SOD323
BAT15-04R	4.0	110.0	0.26	0.23	1	Single	SOT23
BAT15-04W	4.0	110.0	0.26	0.23	1	Dual	SOT323
BAT15-05W	4.0	110.0	0.26	0.23	1	Dual	SOT323
BAT15-099	4.0	110.0	0.26	0.23	1	Single	SOT143-4-1
BAT17	4.0	130.0	0.55	0.34	1	Single	SOT23
BAT17-04	4.0	130.0	0.55	0.34	1	Dual	SOT23
BAT17-04W	4.0	130.0	0.55	0.34	1	Dual	SOT323
BAT17-05	4.0	130.0	0.55	0.34	1	Dual	SOT23
BAT17-05W	4.0	130.0	0.55	0.34	1	Dual	SOT323
BAT17-06W	4.0	130.0	0.55	0.34	1	Dual	SOT323
BAT17-07	4.0	130.0	0.75	0.34	1	Dual	SOT143-4-1
BAT24-02LS	4.0	110.0	0.21	0.23	1	Single	TSSLP-2-1

Source: Courtesy Infineon Technologies.

Table 4.4 Example Schottky Detector Diodes (Infineon)

Type	Maximum Ratings V_R (V)	I_F (mA)	Characteristics $(T_A = 25°C)$ C_T (pF)	V_F at I_F V	mA	R_O at $V_F = 0$ V kΩ	Configuration	Package
BAT62-02L	40.0	20.0	0.35	0.58	2	225.0	Single	TSLP-2-1
BAT62-02W	40.0	20.0	0.35	0.58	2	225.0	Single	SCD80
BAT62-03W	40.0	20.0	0.35	0.58	2	225.0	Single	SOD323
BAT62-07W	40.0	20.0	0.35	0.58	2	225.0	Dual	SOT343
BAT63-02	3.0	100.0	0.65	0.19	1	30.0	Single	SC79
BAT63-07W	3.0	100.0	0.65	0.19	1	30.0	Dual	PG-SOT343-4
BAT68	8.0	130.0	0.75	0.318	1	—	Single	SOT23
BAT68-04	8.0	130.0	0.75	0.318	1	—	Dual	SOT23
BAT68-04W	8.0	130.0	0.75	0.318	1	—	Dual	SOT323
BAT68-06	8.0	130.0	0.75	0.318	1	—	Dual	PG-SOT23-3
BAT68-06W	8.0	130.0	0.75	0.318	1	—	Dual	SOT323

Source: Courtesy Infineon Technologies.

BAT15...

Silicon Schottky Diodes

- Low barrier type for DBS mixer applications up to 12 GHz, phase detectors and modulators
- Low noise figure
- Pb-free (RoHS compliant) package

BAT15-02LRH   **BAT15-04W**   **BAT15-05W**   **BAT15-099**   **BAT15-099R**
BAT15-03W   **BAT15-099LRH**

ESD (Electrostatic discharge) sensitive device, observe handling precaution!

Type	Package	Configuration	L_S(nH)	Marking
BAT15-02LRH	TSLP-2-7	single, leadless	0.4	NP
BAT15-03W	SOD323	single	1.8	white P
BAT15-04W	SOT323	series	1.4	S8s
BAT15-05W	SOT323	common cathode	1.4	S5s
BAT15-07LRH	TSLP-4-7	parallel pair, leadless	0.4	NP
BAT15-098LRH	TSLP-4-7	anti-parallel pair, leadless	0.4	B
BAT15-099	SOT143	anti-parallel pair	2	S5s
BAT15-099R	SOT143	cross-over ring	2	S6s
BAT15-099LRH	TSLP-4-7	anti-parallel pair, leadless	0.4	S5

Source: Courtesy Infineon Technologies.

   **BAT15...**

Maximum Ratings at T_A = 25°C, unless otherwise specified

Parameter	Symbol	Value	Unit
Diode reverse voltage	V_R	4	V
Forward current	I_F	110	mA
Total power dissipation	P_{tot}		mW
BAT15-02LRH, -099LRH $T_S \le 76$ °C		100	
BAT15-03W, $T_S \le 70$ °C		100	
BAT15-04W, $T_S \le 68$ °C		100	
BAT15-05W, $T_S \le 65$ °C		100	
BAT15-099, $T_S \le 48$ °C		100	
BAT15-099R, $T_S \le 67$ °C		100	
Junction temperature	T_j	150	°C
Operating temperature range	T_{op}	-55 ... 150	
Storage temperature	T_{stg}	-55 ... 150	

Thermal Resistance

Parameter	Symbol	Value	Unit
Junction - soldering point[1]	R_{thJS}		K/W
BAT15-02LRH, -099LRH		≤ 780	
BAT15-03W		≤ 795	
BAT15-04W		≤ 820	
BAT15-05W		≤ 850	
BAT15-099		≤ 1020	
BAT15-099R		≤ 830	

[1]For calculation of R_{thJA} please refer to Application Note Thermal Resistance

Source: Courtesy Infineon Technologies.

BAT15...

Electrical Characteristics at T_A = 25°C, unless otherwise specified

Parameter	Symbol	Values			Unit
		min.	typ.	max.	
DC Characteristics					
Breakdown voltage $I_{(BR)}$ = 100 µA	$V_{(BR)}$	4	-	-	V
Forward voltage I_F = 1 mA I_F = 10 mA	V_F	 0.16 0.25	 0.23 0.32	 0.32 0.41	
Forward voltage matching[1] I_F = 10 mA	ΔV_F	-	-	20	mV
AC Characteristics					
Diode capacitance V_R = 0 V, f = 1 MHz, all other types V_R = 0 V, f = 1 MHz, BAT15-099R	C_T	 - -	 - -	 0.35 0.5	pF
Differential forward resistance I_F = 10 mA / 50 mA	R_F	-	5.5	-	Ω

[1]ΔV_F is the difference between lowest and highest V_F in a multiple diode component.

Source: Courtesy Infineon Technologies.

Diode capacitance $C_T = f(V_R)$
f = 1MHz

Reverse current $I_R = f(V_R)$
T_A = Parameter

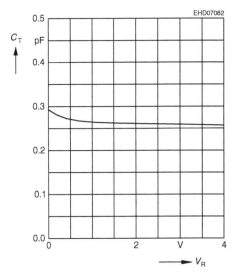

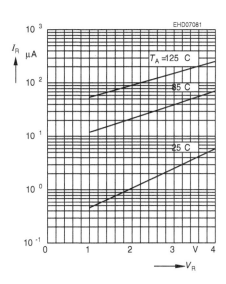

Forward current $I_F = f(V_F)$
T_A = Parameter

Forward current $I_F = f(T_S)$
BAT15-02LRH, BAT15-099LRH

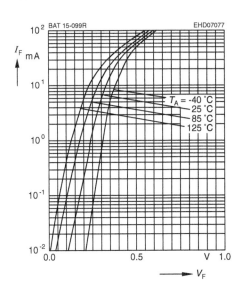

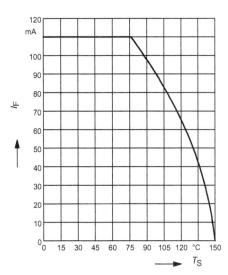

Source: Courtesy Infineon Technologies.

Forward current $I_F = f(T_S)$
BAT15-099R

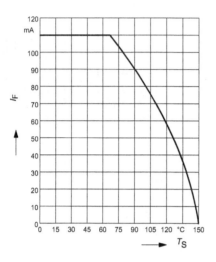

Permissible Puls Load $R_{thJS} = f(t_p)$
BAT15-02LRH, BAT15-099LRH

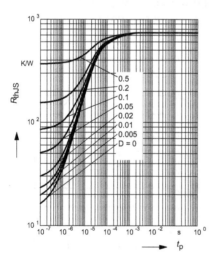

Permissible Pulse Load
$I_{Fmax}/I_{FDC} = f(t_p)$ BAT15-02LRH,
BAT15-099LRH

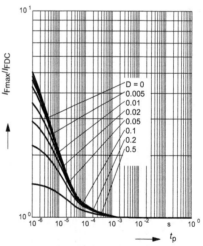

Source: Courtesy Infineon Technologies.

S_{11}-Parameters for BAT15-099

Typical impedance characteristics (with external bias I and Zo = 50Ω)

f	I = 0.02 mA		I = 0.05 mA		I = 0.1 mA		I = 0.2 mA		I = 0.5 mA	
GHz	MAG	ANG	MAG	ANG	MAG	ANG	MAG	ANG	MAG	ANG
1	0.94	-16.4	0.84	-16.6	0.77	-16.4	0.59	-17.2	0.19	-16.7
2	0.93	-33.8	0.88	-33.8	0.77	-34.5	0.58	-35.2	0.15	-36.1
3	0.92	-53.8	0.86	-54.5	0.75	-54.1	0.58	-56.1	0.13	-64.8
4	0.91	-74.3	0.84	-75.3	0.72	-76.4	0.51	-78.4	0.11	-104.8
5	0.91	-96.6	0.84	-97.6	0.72	-99.1	0.53	-102.3	0.15	-135.7
6	0.91	-115.4	0.84	-116.7	0.73	-118.7	0.53	-122.9	0.18	-160.9
7	0.91	-131	0.84	-132.3	0.73	-134.1	0.54	-138.1	0.2	-168.8
8	0.91	-143	0.84	-144.5	0.73	-146.8	0.55	-150.5	0.81	179.4
9	0.91	-155.6	0.83	-150.2	0.71	-159.7	0.53	-163.9	0.18	179.4
10	0.9	-167.3	0.83	-169.7	0.71	-178.8	0.51	-175.8	0.14	151.2
11	0.89	175.5	0.8	172.6	0.7	170	0.45	164.9	0.09	105.5
12	0.88	175.5	0.76	146.5	0.62	142.8	0.39	134.2	0.14	43.6

S_{11} = (f, I) BAT15-099

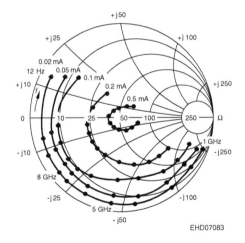

Source: Courtesy Infineon Technologies.

The simple single-diode mixer circuit shown in Figure 4.15 is intended only as an illustration of the basic behavior of diode mixer behavior. A practical single-diode mixer would include filtering at its RF, LO, and IF ports—RF filtering for image rejection, reduced LO radiation, and optimum matching of the RF source to the diode; LO filtering to keep RF out of the LO and optimally match the LO to the diode, and IF filtering to optimally match the diode to its IF (and IF image) load, preferably while providing some rejection of the mixer's unwanted outputs, the strongest (and most potentially troublesome) of which is the LO signal. Detail on the design of single-diode mixers can be found in Ref. [5].

4.3.2 Single-Balanced Mixer

Figure 4.18 shows the schematic of a two-diode (single-balanced) mixer. Unlike the single-diode mixer, it performs multiplicative mixing because its RF and LO signals are applied to different ports. In this more commonly seen two-diode mixer configuration, a balanced transformer drives the diodes out of phase for the LO and in phase for signals present at the RF port. Figure 4.19 shows how this mixer's conversion gain and noise figure vary with applied LO power. Figure 4.20 shows how the mixer's conversion gain and noise figure vary with frequency for a constant LO power.

The two-diode mixer is used mostly in the frequency range above 1 GHz in a manner akin to a phase discriminator, using step-recovery diodes in the LO feed for enhanced harmonic mixing. Such mixers are mainly used in medium-cost spectrum analyzers or microwave receivers up to several tens of gigahertz, with the necessary transformers and baluns printed on the circuit board.

With perfectly matched diodes and perfect transformer and constructional symmetry, no LO energy arrives at the IF and RF ports, and there is only slight attenuation between the RF and IF ports. Both building and computer modeling such a mixer is impossible: building, because perfectly matched diodes and perfect transformer and constructional symmetry cannot be achieved in practice; computer modeling, because floating-point mathematics runs out of gas in handling the infinite amplitude spread involved in calculating the perfect cancellation of the LO signal as it travels to the RF and IF ports. Figure 4.21 compares

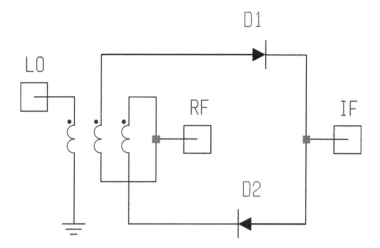

Figure 4.18 Schematic of the two-diode (also known as single-balanced) mixer.

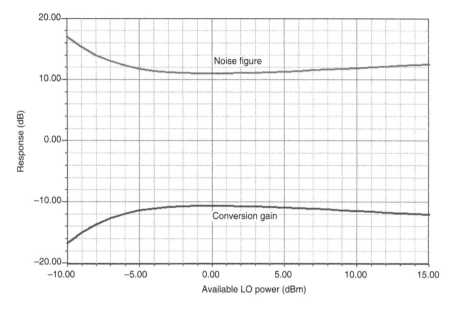

Figure 4.19 How the nonideal mixer's conversion gain and noise figure vary with available LO power. In this analysis, LO = 500 MHz (13 dBm), RF = 500.455 MHz (−20 dBm), and IF = 455 kHz.

the mixer's quasi-ideal port-to-port isolation (matched diodes, a perfect transformer, no stray inductances and capacitances, and a 10 mΩ resistors connected from port to port) and nonideal port-to-port isolation (slightly mismatched diodes and 0.5 pF between the upper terminal of the middle winding and ground).

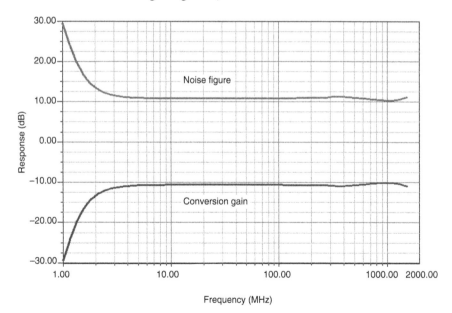

Figure 4.20 How the nonideal two-diode mixer's conversion gain and noise figure vary with frequency for a constant LO power. In this analysis, LO = 1–1500 MHz (2 dBm), RF = 1.455–1500.455 MHz (−40 dBm), and IF = 455 kHz.

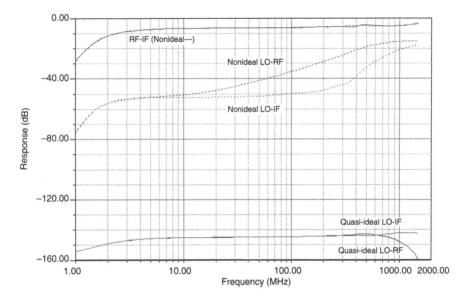

Figure 4.21 Port-to-port isolation of the quasi-ideal (identical diodes and no stray capacitance) and nonideal (slightly mismatched diodes and 0.5 pF of stray capacitance between the upper terminal of the middle transformer winding and ground) two-diode mixer. In this analysis, LO = 1–1500 MHz (2 dBm), RF = 1.455–1500.455 MHz (−40 dBm), and IF = 455 kHz.

Figure 4.22a and b shows the mixer's output spectrum for the quasi-ideal and nonideal cases, respectively.

4.3.2.1 Subharmonically Pumped Single-Balanced Mixer

Figure 4.23 shows a single-balanced mixer with a difference: antiparallel diode pairs take the place of single diodes, the RF and IF are buffered from each other only by filtering, and the LO is applied at 1/2 the frequency necessary to provide the desired frequency conversion. The RF–IF isolation is limited to that provided by the seriesed input and output filtering, but the LO–IF isolation is higher at f_{LO}, and much higher at $2 f_{LO}$, than that achievable with a double-balanced mixer (DBM) with the LO signal at $2 f_{LO}$ (Figure 4.24). Although the example shown is for an upconverting HF receiver, this technique finds application well into the microwave range as the basis for *I/Q* modulators, in which carrier leakage must be reduced to a level difficult to achieve with conventional DBMs [6–11].

4.3.3 Diode-Ring Mixer

Adding two more diodes and another transformer to the singly balanced mixer results in a double-balanced mixer (DBM) as shown in Figure 4.25. A DBM's frequency response is largely determined by the frequency response of its transformers, which act as transmission lines. The low-frequency limit is determined by the inductance of the transformer windings, the reactance of which, at the lowest frequency of interest, should be at least four times the impedance at which the transformer operates. The upper frequency limit is determined mainly by the degradation of the transformers' transmission-line behavior at higher frequencies, although the increasing importance of diode capacitance also plays a role.

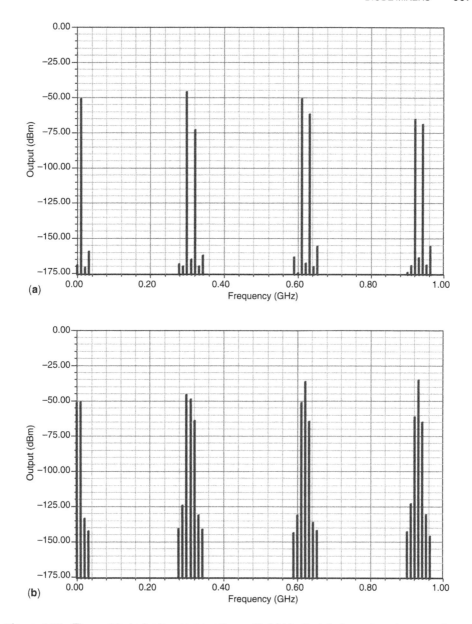

Figure 4.22 The nonideal mixer's output spectrum with (a) identical diodes and no stray capacitance and (b) slightly mismatched diodes and 0.5 pF of stray capacitance between the upper terminal of the middle transformer winding and ground. In these analysis, LO = 310.7 MHz (2 dBm), RF = 300.0 MHz (−40 dBm), and IF = 10.7 MHz. Four LO harmonics and three LO sidebands were used.

A DBM's interport isolation is determined by the symmetry of its transformers, diodes, and physical construction. In practice, the effects of diode mismatch can be minimized by using a dual diode, such as—in the case of the BAT15—the BAT15-099, the maximum VF spread between the diodes in which it is specified as 20 mV.

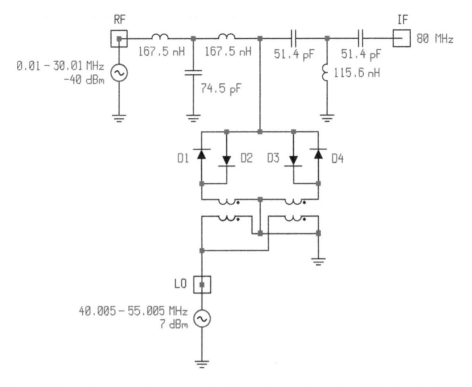

Figure 4.23 A subharmonically pumped single-balanced mixer using antiparallel diode pairs. The LO operates from 40.005 to 55.005 MHz to mix 0.01–30.01 MHz RF to an IF of 80 MHz.

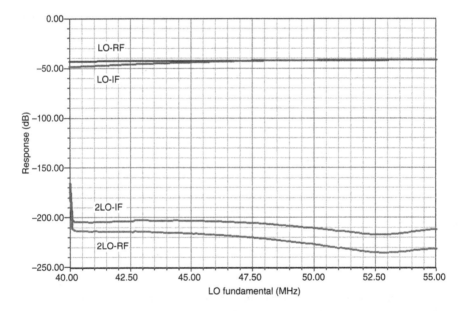

Figure 4.24 Simulated interport isolation of the subharmonic SBM. For realism, the diodes and transformers are slightly mismatched.

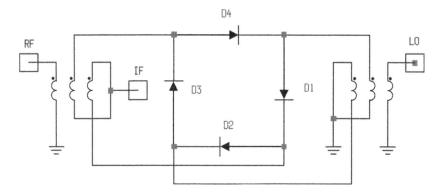

Figure 4.25   Schematic of the diode-ring doubly balanced mixer.

Figure 4.26 shows how the DBM's conversion gain and noise figure vary with applied LO power. Figure 4.27 shows how the DBM's conversion gain and noise figure vary with frequency for an LO power of 7 dBm, with quasi-ideal and nonideal balance. Figure 4.28 shows how the DBM's port-to-port isolation differs with quasi-ideal and nonideal balance for an LO power of 7 dBm. Figure 4.29 shows how the DBM's RF- and LO-port return loss varies with frequency; the sharp peak corresponds to a resonance caused by one of the stray capacitances added to simulate less-than-ideal balance in the modeled mixer.

Figure 4.30 shows the DBM's output spectrum. Figure 4.31 shows the DBM's output waveform over 50 cycles of the LO signal.

Two-tone testing the DBM allows us to characterize its IP_3 figures of merit. Figure 4.32 shows the nonideal DBM's IF and IM_3 responses for LO powers of -5, 1, 7, and 13 dBm. Figure 4.33 details how the DBM's IP_3 increases with LO drive, and Figure 4.34 shows the

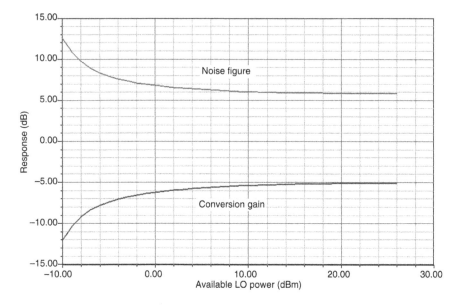

Figure 4.26   DBM conversion gain and noise figure versus LO power. In this analysis, LO = 310.7 MHz (-10 to 26 dBm), RF = 300 MHz (-40 dBm), and IF = 10.7 MHz.

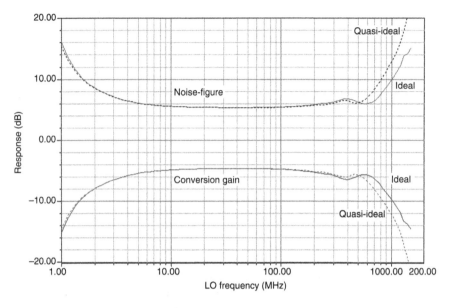

Figure 4.27 This plot of conversion gain and noise figure versus frequency for quasi-ideally and nonideally balanced versions of the same DBM reveals that balance plays a relatively minor role in the CG and NF performance achieved. In these analyses, the LO (−7 dBm) sweeps from 1 to 1500 MHz and the RF (−40 dBm) sweeps from 1.455 to 1500.455 MHz to produce an IF of 455 kHz.

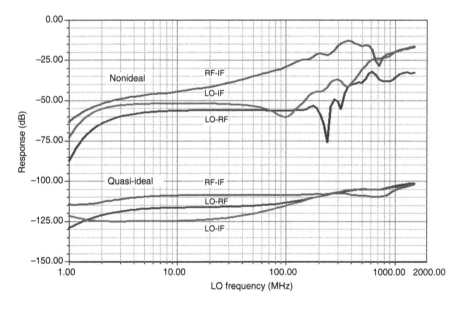

Figure 4.28 Interport isolation for DBMs with quasi-ideal and nonideal balance. In these analyses, the LO (−7 dBm) sweeps from 1 to 1500 MHz and the RF (−40 dBm) sweeps from 1.455 to 1500.455 MHz to produce an IF of 455 kHz.

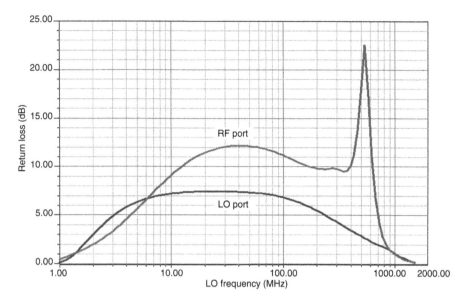

Figure 4.29 Return loss versus frequency for the DBM's RF and LO ports. The sharp peak results from a stray resonance.

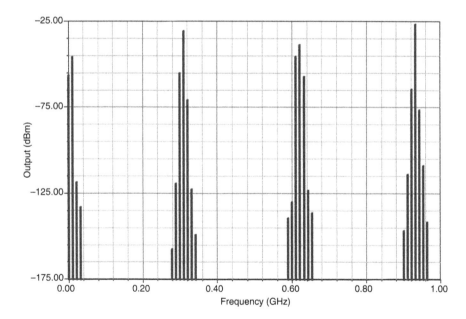

Figure 4.30 Output spectrum of a nonideally balanced DBM. In this analysis, LO = 310.7 MHz (−7 dBm) and RF = 300 MHz (−40 dBm) for an IF of 10.7 MHz. Four LO harmonics and three LO sidebands were used.

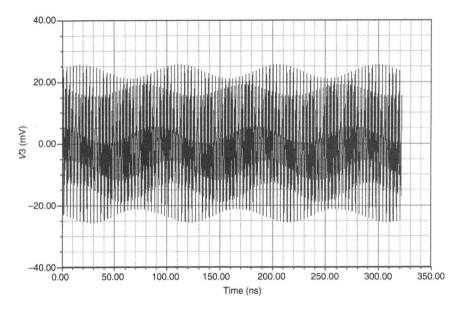

Figure 4.31 IF-port voltage waveform of the DBM over 100 cycles of the LO signal. The 310.7 MHz LO and 10.7 MHz IF components are clearly evident.

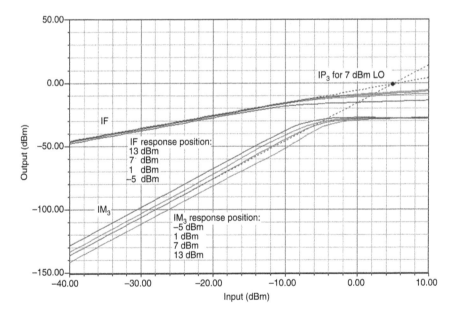

Figure 4.32 Diode DBM IF and IM_3 output versus RF power for four LO-drive levels. The responses for LO = 7 dBm have been extrapolated to show IP_3. Within limits, varying a mixer's LO drive affects its linear IF output relatively little while significantly affecting IMD (see Figure 4.33). For all four analyses, LO = 310.7 MHz, RF_1 = 300.0 MHz (−40 to 10 dBm), and RF_2 = 300.3 MHz (−40 to 10 dBm); four LO harmonics and three LO sidebands were used.

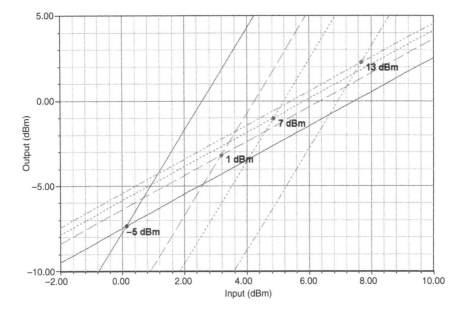

Figure 4.33 Extrapolating the responses for all four LO levels represented in Figure 4.32 shows how varying a diode DBM's LO drive shifts its third-order intercept point. Although these curves indicate that the simulated mixer's IP_3 generally increases with LO drive, the improvement in IP_3 is not as great as we might expect. The reason for this is that these four analyses, as well as the other diode-mixer analyses in this chapter, were done using diode models with a threshold voltage (V_J) of 0.23. If high-level diodes with a V_J of about 0.8 V had been used, IP_3, out for the 13 dBm LO case shown here would increase to +13 dBm. $IP_{3,in}$ for the 13 dBm LO case would turn out to be 13 dBm + insertion loss = 13 dBm + 7 dB = 20 dBm. The issue of diode damage aside, attempting to increase IP_3 merely by driving a low- or medium-barrier diode harder eventually results in diminishing returns. High-barrier diodes are essential in getting the best IP_3 performance with high LO drive.

desired IF outputs and close third-order spurs near 10.7 MHz. Figure 4.35 shows the DBM's output voltage over 100 cycles of the LO signal, and Figure 4.36 shows the anode–cathode voltage of one of the ring's diodes, also over 100 LO cycles, both under two-tone IMD test conditions.

A double-balanced mixer, unless it is termination insensitive, is extremely sensitive to nonresistive termination. This is because the transmission-line transformers do not operate properly when they are not properly terminated, and the reflected power generates high voltage across the diodes. This effect results in much higher distortion levels than in a properly terminated transformer.

4.3.3.1 Termination-Insensitive Mixer
Figure 4.37 shows a mixer circuit that tolerates a fairly high VSWR at its output without significant degradation of its third-order IM performance [12].

4.3.3.2 Phase Detector
Theoretically, any mixer with a dc-coupled IF port can be used as a phase detector. When two signals of equal frequency are applied simultaneously to the reference and incoming signal ports, the phase detector produces a dc output at the IF port proportional to the cosine of the phase difference (Figure 4.38).

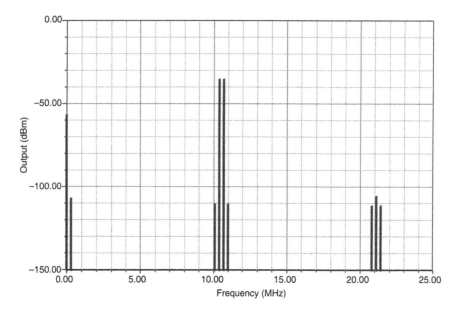

Figure 4.34   The DBM's output in the 11 MHz region during two-tone testing. The third-order products are clearly visible above and below the desired output signals. The test conditions for this analysis are those for Figure 4.32 with LO = 13 dBm.

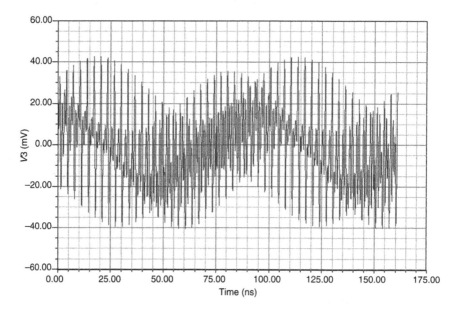

Figure 4.35   The DBM's IF-output voltage over 50 cycles of the LO signal. The test conditions for this analysis are those for Figure 4.32 with LO = 13 dBm.

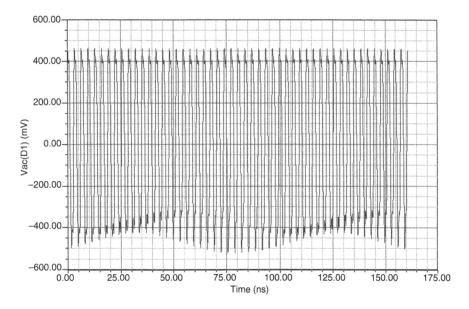

Figure 4.36 The anode–cathode voltage of one of the DBM's diodes, also over 50 LO cycles under two-tone IMD test conditions. The test conditions for this analysis are those for Figure 4.32 with LO = 13 dBm.

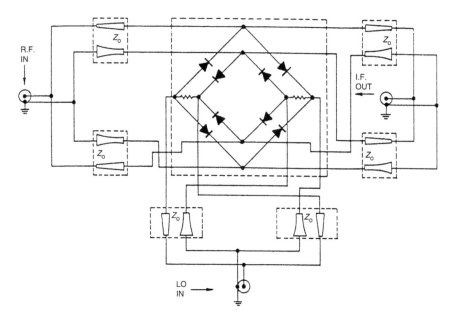

Figure 4.37 Example of a termination-insensitive mixer from Ref. [12].

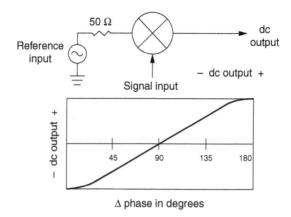

Figure 4.38 A mixer with a dc-coupled IF port can be used as a phase detector.

4.3.3.3 Binary Phase-Shift Keying (BPSK) Modulator

Binary phase modulation occurs when a positive and negative signal current shifts the RF carrier between 0 and 180°. Figure 4.39 shows a double-balanced mixer operating as a BPSK modulator.

4.3.3.4 Quadrature Phase-Shift Keying (QPSK) Modulator

A typical QPSK modulator consists of two biphase modulators, a 90° divider and a 0° power combiner as shown. Data inputs at the control ports will cause the carrier to shift between 0, 90, 180, and 270° as shown in Figure 4.40.

4.3.3.5 Quadrature IF Mixer

A quadrature IF mixer produces two IF outputs in phase quadrature. Its basic structure consists of two double-balanced mixers, a 90° splitter and 0° splitter. The basic block diagram is shown in Figure 4.41.

4.3.3.6 Image-Reject Mixer

The image-reject mixer consists of a basic quadrature IF mixer with an additional 90° hybrid at the IF ports as shown in Figure 4.42. The primary function is to differentiate between the real signal and the image signal. This type of device is especially useful in applications where the desired RF signal and image are so close in frequency that rejecting the image with filtering is not practical.

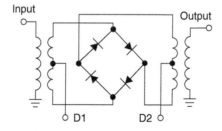

Figure 4.39 A diode-ring mixer as phase modulator.

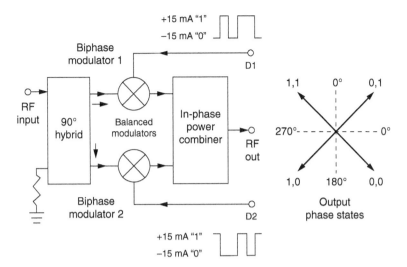

Figure 4.40 Two biphase modulators form the basis for a QPSK modulator.

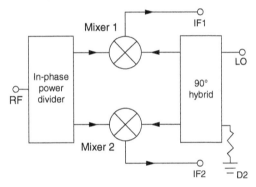

Figure 4.41 Quadrature mixer.

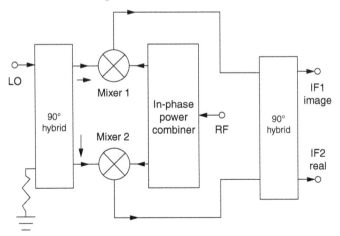

Figure 4.42 An image-reject mixer uses phasing to differentiate between its LO+IF and LO−RF IF outputs.

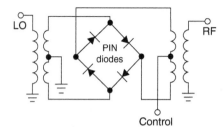

Figure 4.43   A diode DBM can be used as a dc-controlled attenuator if PIN diodes are used instead of Schottky devices in its ring.

4.3.3.7   Diode Attenuator/Switch

A ring of PIN diodes can be used as electronic attenuators by applying variable forward bias to the diodes (Figure 4.43). Maximum attenuation is achieved when the current at the control port is zero. The maximum attenuation is the isolation between the input and output port. Minimum attenuation (insertion loss) is achieved when the IF port current is 20 mA.

4.3.3.8   Single-Sideband (SSB) or In-Phase/Quadrature (I/Q) Modulator

SSB or *I/Q* modulators are useful in discriminating and removing the lower sideband (LSB) or upper sideband (USB) generated during frequency conversion, especially when the sidebands are very close in frequency and attenuation of one of the sidebands cannot be achieved with filtering. This is the case with audio and video modulation, where signals from dc to 10 MHz must be converted to a higher frequency that is appropriate for transmission. In such cases, both sidebands will be very close in frequency to the carrier frequency. With an *I/Q* modulator, one of the sidebands is easily canceled or attenuated along with its carrier.

   Attenuation of the carrier has been the most troublesome aspect in the design of passive *I/Q* modulators. Isolation between the local oscillator port and the RF port of the mixers, which is the main parameter in determining carrier rejection, is usually insufficient at frequencies above 200 MHz.

   I/Q modulator designs have basically comprised two double-balanced mixers (Figure 4.44). The mixers are fed at the LO ports by a carrier phase-shifted through a 90° hybrid.

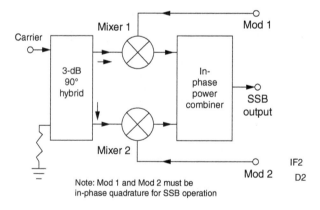

Figure 4.44   An SSB modulator matches two high-frequency mixers, a 90° hybrid, and an in-phase power combiner to produce an SSB output signal.

Thus, the carrier signal's relative phase is $0°$ to one mixer and $90°$ to the other mixer. Modulation signals are fed externally in phase quadrature to the two mixers' IF ports. The mixers' modulated output signals are combined through a two-way, in-phase power divider/combiner.

The circuit forms a phase-cancellation network to one of the sidebands and a phase-addition network to the other sideband. The carrier is somewhat attenuated and is directly dependent on the inherent LO–RF isolation of the mixers and the modulating signal level. In industry-standard *I/Q* modulators, USB suppression results when the first modulation port (MOD 1) is fed with a signal that is $90°$ in advance of the signal feeding the second modulation port (MOD 2). Opposite phasing can be arranged by changing the internal phase polarity of the mixers or by interchanging the $90°$ hybrid output ports to the LO ports of the mixers.

The phase and amplitude imbalances between the various components used in the manufacturing of the *I/Q* modulators must be tightly maintained for optimum SSB rejection. Matching of the two mixers for conversion loss and insertion phase is extremely critical, since differences in these parameters will add to amplitude- and phase-imbalance errors. The $90°$ hybrid in the LO port must be in nearly perfect phase quadrature.

Phase- and amplitude-imbalance errors adversely affect sideband suppression (Figure 4.45). In most cases, a typical passive *I/Q* modulator operates with a carrier input level of $+10$ dBm, which is required to drive the diodes in the mixers to operate in the linear range. The dynamic range of these mixers can be significantly improved by using diodes with a higher barrier height. The LO signal in this case must be increased in order to drive these diodes into conduction in their linear range.

Carrier rejection is also a problem when designing an SSB modulator, since only a few decibels of suppression can be achieved in standard high-frequency models. In the past, the major contributor to carrier suppression was the inherent LO–RF isolation through the mixers. Unfortunately, this isolation is usually poor at cellular frequencies (800–1000 MHz), where at least 25 dB of carrier rejection is necessary. In some cases, designers feed a small amount of dc into the IF ports to control the carrier rejection, which complicates the driver circuitry and calls for temperature compensation when operating at different temperatures.

As an example, an SSB modulator is assumed to operate with $+10$-dBm LO drive with each modulating signal at -10 dBm and in phase quadrature to each other when applied to the modulating ports (MOD 1 and MOD 2). The result will be a modulated signal at -16 dBm, assuming 6 dB conversion loss.

For 20 dB carrier rejection with respect to the desired modulated signal, the carrier must be at -36 dBm, which translates to LO–RF isolation of 46 dB.

By employing a subharmonic approach, the performance of SSB modulators can be extended beyond the limits of conventional designs as reported by Joshi in Ref. [8]. The approach is based on the use of subharmonic mixers in place of fundamental-frequency mixers and is applicable from about 140 to 3000 MHz. Subharmonic mixers use antiparallel diode pairs in their construction [13–15]. Matched antiparallel diode pairs used in single-ended or single-balanced mixer configurations cancel even-order intermodulation products (such as $2 f_{LO} \times 2 f_{RF}$, and $3 f_{LO} \times 3 f_{RF}$) at all ports.

Single-ended mixers lack the port-to-port isolation needed for SSB modulator applications. Odd-order products of the RF and LO frequencies (even $f_{LO} \times$ odd f_{RF}) and (odd $f_{LO} \times$ even f_{RF}) appear on all ports, requiring extensive filtering for satisfactory performance. For a single-balanced mixer, even harmonics of the LO combining with odd

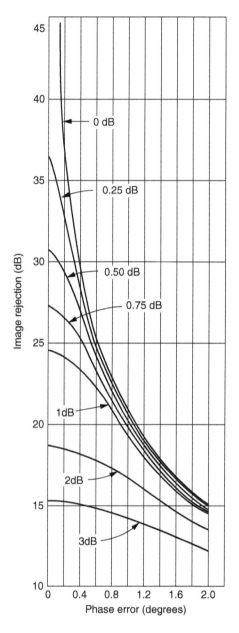

Figure 4.45 The level of SSB rejection improves as the phase and amplitude imbalance performance of an SSB modulator improves.

harmonics of the RF appear at the IF port, whereas odd harmonics of the LO combining with even harmonics of the RF appear at the RF and IF ports. This assumes that a balanced transformer is placed at the LO port, which is a logical choice due to the fact that the highest level signal appears at the LO port. Since the desired odd-order IF products appear at both the RF and IF ports, a need arises for a diplexing network to isolate the RF and IF signals.

The subharmonic modulator design provides a unique way to isolate the RF and IF signals. A single-balanced harmonic mixer offers good LO–RF and LO–IF isolation but poor RF–IF isolation. Fortunately, harmonically related signals are spaced well apart in the frequency spectrum, simplifying filtering of harmonically related signals.

Harmonic mixing also works well with low LO power levels, with somewhat lower 1 dB compression on the RF port than with fundamental-frequency mixing. The ability to operate with LO frequencies that are a fraction of the carrier frequency (1/2, 1/4, 1/6, etc.) significantly reduces the cost of an LO source, especially at higher frequencies. Also, using lower frequency LO sources helps avoid the signal-leakage problems inherent with higher frequency LO sources. Minimizing signal leakage, especially at higher frequencies, becomes expensive and bulky. Subharmonic mixing offers several advantages:

- The technique offers the ability to operate at LO frequencies that are 1/2, 1/4, or 1/6 of the carrier frequency. For example, for an IF of 100 MHz at an RF of 2 GHz, the LO can be $(2000\pm100) \div 2 = 950$ or 1050 MHz.
- The LO's even harmonics are strongly attenuated.
- The filtering requirements for fundamental frequency and odd harmonic signals of the LO are not critical.
- The cost of generating the LO is reduced due to the fact that the LO frequency need only be a fraction of the carrier frequency.

As an example of the performance improvements possible with the subharmonic mixers, units were evaluated at both cellular (935–960 MHz) and PCN/PCS (1.8–1.9 GHz) bands. For a conventional SSB modulator at 1.9 GHz fed with +10 dBm modulation signals, carrier rejection is barely 10 dB (Figure 4.46).

Sideband rejection can be improved by tuning, but the carrier rejection is controlled by the LO–RF isolation of the double-balanced mixers. Conventional double-balanced mixers with high isolation at cellular and PCN bands are very expensive and large when special techniques are used to improve LO–RF isolation. In contrast, the subharmonic nature of

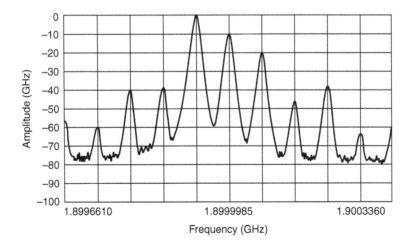

Figure 4.46 This plot of carrier and sideband rejection was measured for a conventional SSB modulator operating at 1.9 GHz.

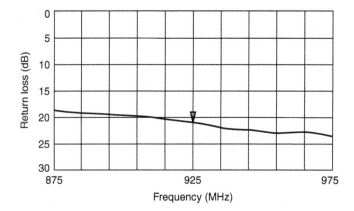

Figure 4.47   The SSB modulator's return loss as measured at the local oscillator port.

the new approach allows the use of lower frequency, less-expensive components in the modulators' construction.

The subharmonic modulators offer an improvement of more than 15 dB in carrier suppression compared to the conventional approach.

The measured VSWR (return loss) at the LO and RF ports is better than 1.50:1 (Figures 4.47 and 4.48). Measurements made on a cellular-band SSB modulator reveal carrier rejection on the order of 40 dB. Typical insertion loss is 7 dB while sideband rejection is 30 dB (Figure 4.49).

By the virtue of harmonic mixing, even-order mixing products are attenuated by about 30 dB with respect to the desired modulated output signal. The fundamental-frequency feedthrough into the output port is approximately 5 dB lower than the desired modulated signal, whereas the fourth harmonic mixing with the modulating signal is approximately 10 dB lower. Typical loss for fourth-harmonic mixing is 17–19 dB while maintaining 30 dB of carrier rejection.

Since harmonically related products are well-spaced in frequency, filtering undesired signals is relatively inexpensive using standard octave-bandwidth filters. Low-cost commercial

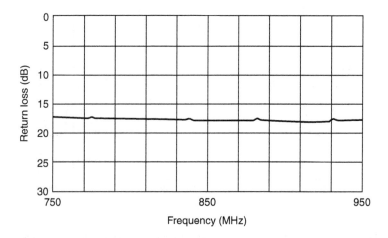

Figure 4.48   The novel harmonic SSB modulator's return loss as measured at the RF port.

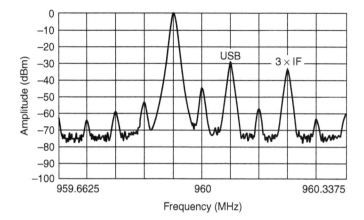

Figure 4.49 This plot of carrier and sideband rejection as measured for the novel harmonic SSB modulator operating at cellular frequencies.

bandpass filters typically offer better than 40–50 dB attenuation of unwanted harmonic signals. Constant-impedance bandpass filters offering good impedance match at desired stop-bands can also be used in cases where harmonically related products require impedance termination within a system.

The subharmonic modulator design is easily applied at custom frequencies. Conversion of an SSB modulator with output frequency corresponding to twice the LO frequency to one with output corresponding to four times the LO frequency requires only one component change, in the form of a signal-combining network at the modulator's output. Although the conversion loss of the fourth-harmonic LO component mixing with the modulating signal is in the vicinity of 18 dB, the cost of generating the LO is drastically reduced with the subharmonic modulator. In spite of higher signal loss, the carrier rejection is still at least 30 dB at the fourth harmonic, and harmonically related products can be eliminated with an inexpensive filter.

4.3.3.9   Triple-Balanced Mixer
Two diode rings can be combined to form a double-balanced mixer, or triple-balanced mixer (TBM), as shown in Figure 4.50. Triple-balanced mixers achieve higher dynamic range and interport isolation than double-balanced designs at the expense of LO power and increased complexity and size.

Figure 4.51 shows the Figure 4.50 circuit's interport isolation with the circuit configured in a less than ideally balanced form, with small variations in transformer-winding inductance and diode parameters introduced for more realistic modeling. Note that the mixer's interport isolation generally increases with frequency, rather than decreasing with frequency as with the DBM (Figure 4.28).

4.3.3.10   Rohde and Schwarz Subharmonically Pumped DBM
Figure 4.52 shows an example of a complete microwave diode mixer used as the input stage of a spectrum analyzer. Its LO is applied at 13.2 GHz, 1/3 the frequency necessary to mix the 40.1 GHz input signal to the 500 MHz IF, with distortion in the diodes providing the 3 × frequency multiplication. The circuit's conversion gain is −18.8 dB; its noise figure, 20.5 dB.

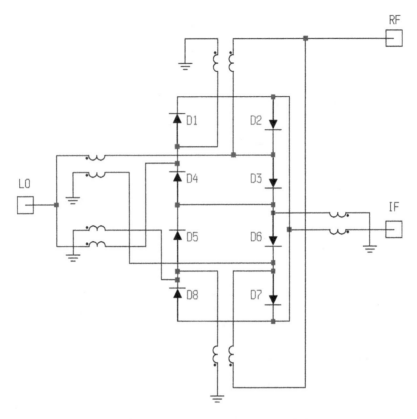

Figure 4.50 A triple-balanced diode mixer. A limitation of this configuration is that the internal dc common connections associated with its RF and LO transformers disallow usable IF response down to dc.

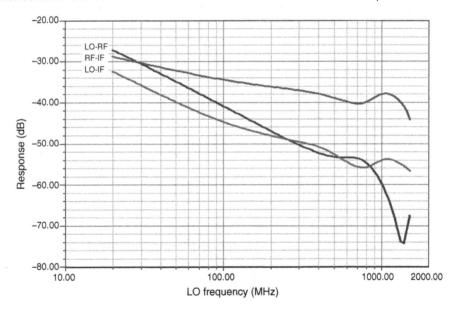

Figure 4.51 The triple-balanced mixer offers improved high-frequency isolation over a standard DBM.

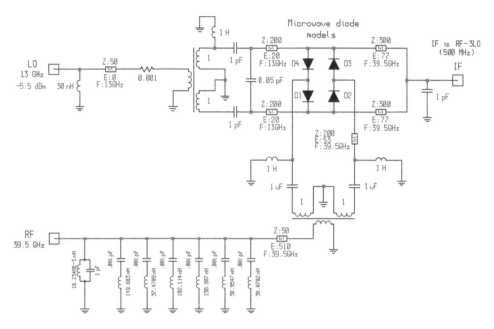

Figure 4.52 Schematic of the spectrum-analyzer mixer. The LO is applied at 1/3 the frequency necessary to mix the RF signal to the 500 MHz IF.

The subharmonic-drive technique exemplified by this mixer is important for another reason in addition to LO isolation. Depending on the application, directly generating a sufficiently phase-quiet LO signal at the higher wireless frequencies and on up into the microwaves may not be feasible. In such cases, injecting the LO at a subharmonic of the desired LO frequency may provide better phase noise performance than a fundamental LO even though the phase noise of a frequency-multiplied source increases by $\log_{10} n$ dB, where n is the multiplication factor.

4.4 TRANSISTOR MIXERS

Diode mixers are lossy, termination-sensitive, and require considerable LO power. Their attractiveness to designers of highly integrated wireless products is further reduced by their dependence on transformers for balance and port-to-port isolation. Transistor mixers are therefore used where high integration and reduced current drain are paramount—that is, in most non-base-station wireless applications.

Until 10 years ago, single BJTs were commonly used in a simple additive-mixing arrangement, an example of which appears in Figure 4.53. Such a circuit behaves like a combination of a preamplifier and single-diode mixer. Single-BJT mixers were used in early AM–FM radios until 1980, and unfortunately in some handheld 2 m and 70 cm ham equipment, to achieve the lowest possible power consumption. By definition, however, the presence of such a mixer also destroys any possibility of achieving a high intercept point. In addition to this, many combination frequencies occur despite elaborate input filtering. Although single-BJT mixers can exhibit considerable conversion gain (>10 dB), their dynamic range is restricted and their port-to-port isolation is poor. BJT mixers in today's

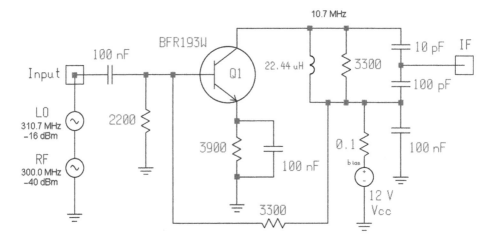

Figure 4.53   A single-BJT mixer.

competitive wireless designs use multiple transistors, and are based almost exclusively on the Gilbert multiplier cell.

4.4.1 BJT Gilbert Cell

The classic active mixer (Figure 4.54), conceived by Barrie Gilbert in 1967, is the basis for most active mixers used in wireless products today. Figure 4.55 shows a Gilbert cell mixer implementation for analysis. Figure 4.56 shows its conversion gain and noise figure versus LO power. Figure 4.57 shows its IM_3 and IP_3 responses. As Gilbert wrote a 1994 monograph [16]

> This circuit is attractive because (a) it can be monolithically integrated with other signal-processing circuitry; (b) it can provide conversion gain, whereas a diode-ring mixer always has conversion loss; (c) it requires very low power to drive the LO port; (d) it provides excellent isolation between the signal ports; and (e) it is far less fussy about load-matching.

The two major advantages afforded by the Gilbert cell are port isolation (it should really be called a multiplicative mixer) and its significant reduction of even-order frequency combinations. As with the single-transistor mixer, the Gilbert cell does not particularly shine with high intercept point and low current consumption. Another major drawback is that it requires a high-impedance output, which gives the user headaches because most of the better filters are in 50 Ω technology, and only very simple monolithic filters offer input and output impedances around 1 kΩ. Impedances on this order can result in significant crosstalk in high-density circuit boards.

The reason the Gilbert cell can be so linear for small input signal levels is twofold. It is a differential amplifier and it makes use of the tanh function, which is linear over a wide range around the zero-crossing point [17]. Attempts have been made to increase the intercept point. The Plessey SL6440 IC was a special version of the Gilbert cell that achieved high $IP_{3,in}$ (+30 dBm) at the expense of dc current.

Our suspicion, however, is that a combination of a modern double-balanced passive mixer using medium- to high-level diodes followed by a feedback FET amplifier, or even one of

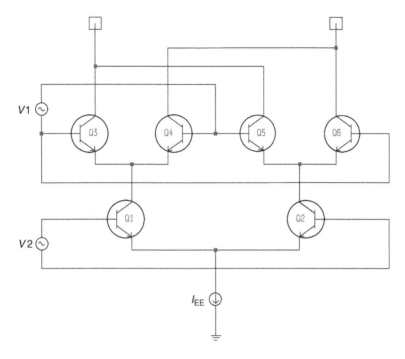

Figure 4.54   The basic Gilbert multiplier cell.

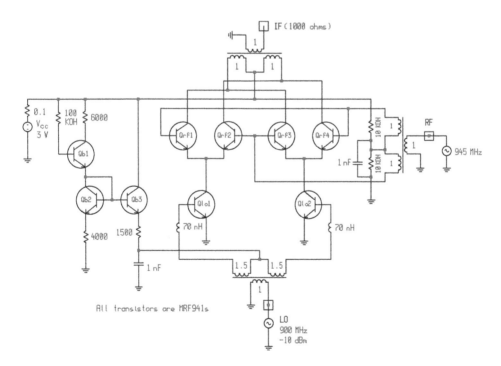

Figure 4.55   Gilbert cell mixer validation circuit.

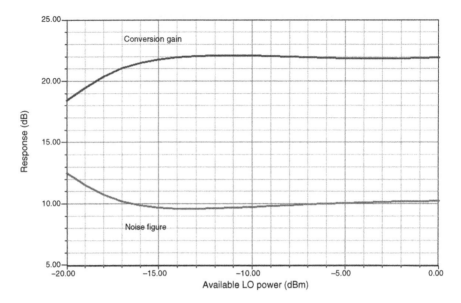

Figure 4.56 Conversion gain and noise figure versus LO power for the Gilbert cell mixer. In this analysis, LO = 900 MHz (−20 to 0 dBm) and RF = 945 MHz (−50 dBm) for an IF of 45 MHz.

the late CATV transistors with f_Ts of more than 25 GHz, would allow the achievement of a significantly higher intercept point. As pointed out in our amplifier chapter, CATV transistors are now available that combine a noise figure of less than 1 dB noise figure with a $IP_{3,in}$ of more than 30 dBm. Built in push–pull, structures using such transistors can

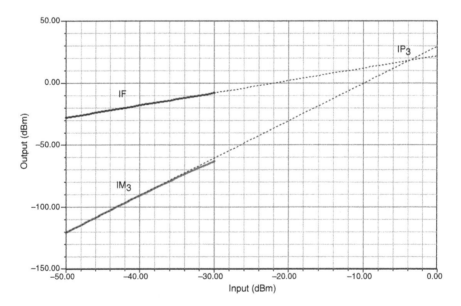

Figure 4.57 IM$_3$ and IP$_3$ responses for the BJT Gilbert cell mixer with RF signals at 945 and 946 MHz. In this analysis, LO = 900 MHz (−10 dBm), RF$_1$ = 945 MHz (−50 to −30 dBm), and RF$_2$ = 946 MHz (−50 to −30 dBm). Four LO harmonics and three LO sidebands were used.

achieve an $IP_{2,in}$ of 70 dBm, matching the numbers for the diode DBM. In the amplifier chapter, we learned that feedback always improves the dynamic range, mostly at the expense of gain, since the gain-bandwidth product gets reduced. Motorola has recently come out with a clever extension of its high-level preamplifier and modified it to become a mixer.

4.4.2 BJT Gilbert Cell with Feedback

Motorola's MC13143 low-power 2.4 GHz mixer IC uses a patented topology consisting of a class-AB-biased Gilbert cell augmented by feedback. Its linearity can be programmed via an external current source to achieve an $IP_{3,in}$ of 20 dBm at the expense of additional supply current.

The MC13143 contains 29 active transistors. Figure 4.58 shows the core of the MC13143 circuit reduced to its essentials and configured for computer analysis per Motorola's MC13143 test circuit. Figure 4.59 shows the circuit's conversion gain and noise figure versus LO drive, and Figure 4.60 shows the circuit's typical IF, IM_3, and IP_3 responses,

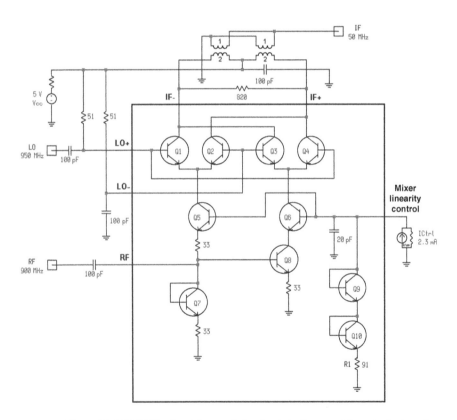

Figure 4.58 Motorola's MC13143 mixer IC uses dc feedback for improved linearity that can be programmed by applying a control current (0–2.3 mA) to its mixer linearity control pin. Per Motorola, an $IP_{3,in}$ of 20 dBm may be achieved with a control current of 2.3 mA, at the expense of approximately 7 mA of additional supply current. In this validation circuit, R1 (the resistor in Q10's emitter) has been adjusted for a total current drain (supply + linearity control) of 8 mA.

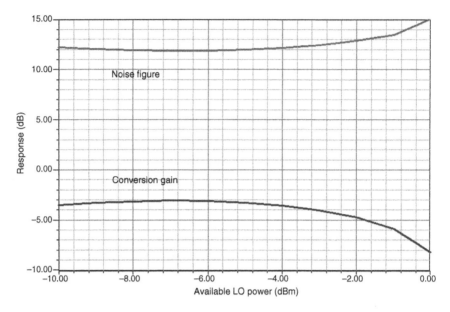

Figure 4.59 Calculated conversion gain and noise figure versus LO drive for the Motorola MC13143 IC operating at $V_{CC} = 2V$. Motorola's specifications for the typical values of these characteristics are -5.0 and $12\,dB$, respectively, at an LO drive level of $0\,dBm$. The RF signal is at 900 MHz and the LO is at 950 MHz, for an IF of 50 MHz. In this analysis, LO = 950 MHz (-10 to $5\,dBm$) and RF = 901 MHz ($-50\,dBm$) for an IF of 49 MHz.

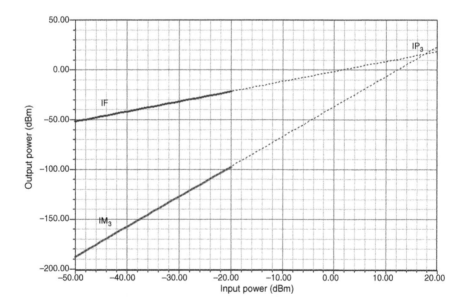

Figure 4.60 Calculated typical IF, IM_3, and IP_3 responses for the Motorola MC13143 mixer configured as shown in Figure 4.58. In this analysis, LO = 950 MHz ($-5\,dBm$), RF_1 = 900 MHz (-50 to $-20\,dBm$), RF_2 = 901 MHz (-50 to $-20\,dBm$), and V_{CC} = 5 V. Four LO harmonics and three LO sidebands were used.

Order this document by MC13143/D

 **MOTOROLA**

Ultra Low Power DC - 2.4 GHz Linear Mixer

The MC13143 is a high compression linear mixer with single–ended RF input, differential IF output and differential LO inputs which consumes as little as 1.8 mW. A new circuit topology is used to achieve a high third order intermodulation intercept point, high linearity and high 1.0 dB output compression point while maintaining a linear 50 Ω input impedance. It is designed for Up or Down conversion anywhere from dc to 2.4 GHz.

Ultra Low Power: 1.0 mA @ V_{CC} = 1.8 to 6.5 V

- Wide Input Bandwidth: DC–2.4 GHz
- Wide Output Bandwidth: DC–2.4 GHz
- Wide LO Bandwidth: DC–2.4 GHz
- High Mixer Linearity: $P_{i1.0\ dB}$ = 3.0 dBm

Linearity Adjustment of up to IP_{3in} = 20 dBm

- 50 Ω Mixer Input
- Single–Ended Mixer Input
- Double Balanced Mixer Operation
- Differential Open Collector Mixer Output

MC13143

ULTRA LOW POWER DC – 2.4 GHz LINEAR MIXER

SEMICONDUCTOR
TECHNICAL DATA

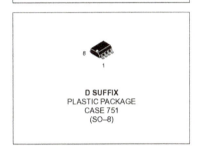

D SUFFIX
PLASTIC PACKAGE
CASE 751
(SO–8)

ORDERING INFORMATION

Device	Operating Temperature Range	Package
MC13143D	T_A = –40 to 85°C	SO–8

MAXIMUM RATINGS (T_A = 25°C, unless otherwise noted.)

Rating	Symbol	Value	Unit
Power Supply Voltage	V_{CC}(max)	7.0	Vdc
Junction Temperature	T_{Jmax}	150	°C
Storage Temperature Range	T_{stg}	–65 to 150	°C

NOTE: ESD data available upon request.

PIN CONNECTIONS

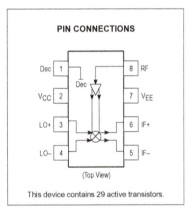

Dec	1	8	RF
V_{CC}	2	7	V_{EE}
LO+	3	6	IF+
LO–	4	5	IF–

(Top View)

This device contains 29 active transistors.

MC13143

RECOMMENDED OPERATING CONDITIONS

Rating	Symbol	Min	Typ	Max	Unit
Power Supply Voltage	V_{CC}	1.8	–	6.0	Vdc

DC ELECTRICAL CHARACTERISTICS ($T_A = 25°C$, $V_{CC} = 3.0$ V, $f_{RF} = 1.0$ GHz, Pin = –25 dBm.)

Characteristic	Symbol	Min	Typ	Max	Unit
Supply Current (Lin Control Current = 0)	I_{CC1}	–	1.0	–	mA
Supply Current (Lin Control Current = 1.6 mA)	I_{CC2}	–	4.1	–	mA

AC ELECTRICAL CHARACTERISTICS ($T_A = 25°C$, $V_{CC} = 3.0$ V, $f_{RF} = 1.0$ GHz, Pin = –25 dBm.)

Characteristic	Symbol	Min	Typ	Max	Unit
Mixer Voltage Conversion Gain ($R_P = R_L = 800$ Ω)	VG_C	–	9.0	–	dB
Mixer Power Conversion Gain ($R_P = R_L = 800$ Ω)	PG_C	–3.5	–2.6	–1.5	dB
Mixer Input Return Loss	Γin_{mx}	–	–20	–	dB
Mixer SSB Noise Figure	NF_{SSB}	–	14	15	dB
Mixer 1.0 dB Compression Point (Mx Lin Control Current = 1.6 mA)	$Pin_{-1.0\ dB}$	–1	0	–	dBm
Mixer Input Third Order Intercept Point ($d_f = 1.0$ MHz, $I_{control} = 1.6$ mA)	$IP3_{in}$	–	16	–	dBm
LO Drive Level	LO_{in}	–	–5.0	–	dBm
LO Leakage to Mixer IF Outputs	P_{LO-IF}	–	–33	–25	dB
Mixer Input Feedthrough Output	P_{RFm-IF}	–	–25	–	dB
LO Leakage to Mixer Input	P_{LO-RFm}	–	–40	–25	dB
Mixer Input Leakage to LO	P_{RFm-LO}	–	–35	–	dB

Figure 1. Test Circuit

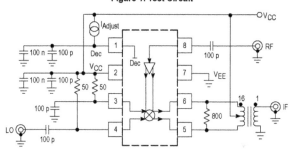

MC13143

TYPICAL PERFORMANCE CURVES

Figure 2. Power Conversion Gain and Supply Current versus Supply Voltage

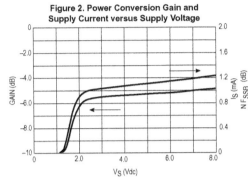

Figure 3. Noise Figure and Gain versus LO Power

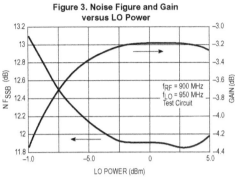

Figure 4. Mixer Input Return Loss versus RF Input Frequency

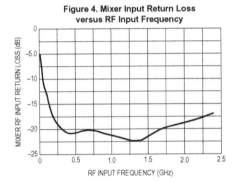

Figure 5. Power Conversion Gain and Supply Current versus RF Input Power

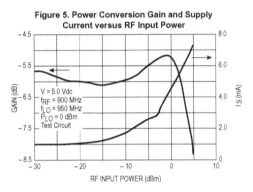

Figure 6. Noise Figure and Gain versus RF Frequency

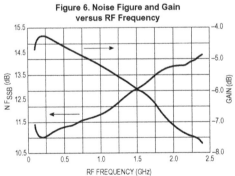

Figure 7. IIP3, Gain, Supply Current versus Mixer Linearity Control Current

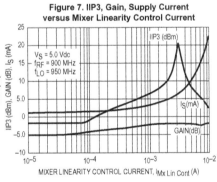

MC13143
CIRCUIT DESCRIPTION

General

The MC13143 is a double–balanced Mixer. This device is designated for use as the frontend section in analog and digital FM systems such as Wireless Local Area Network (LAN), Digital European Cordless Telephone (DECT), PHS, PCS, GPS, Cellular, UHF and 800 MHz Special Mobile Radio (SMR), UHF Family Radio Services and 902 to 928 MHz cordless telephones. It features a mixer linearity control to preset or auto program the mixer dynamic range, an enable function and a wideband IF so the IC may be used either as a down converter or an up converter.

Current Regulation

Temperature compensating voltage independent current regulators provide typical supply current at 1.0 mA with no mixer linearity control current.

Mixer

The mixer is a unique and patented double–balanced four quadrant multiplier biased class AB allowing for programmable linearity control via an external current source. An input third order intercept point of 20 dBm may be achieved. All 3 ports of the mixer are designed to work up to 2.4 GHz. The mixer has a 50 Ω single–ended RF input and open collector differential IF outputs (see Internal Circuit Schematic for details). The linear gain of the mixer is approximately –5.0 dB with a SSB noise figure of 12 dB.

Local Oscillator

The local oscillator has differential input configuration that requires typically –10 dBm input from an external source to achieve the optimal mixer gain.

Figure 8. MC13143 Internal Circuit*

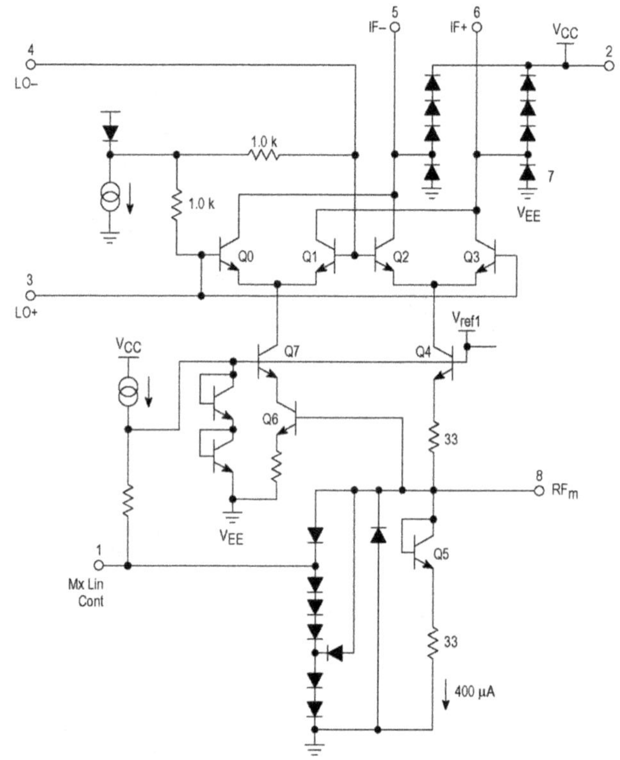

NOTE: * The MC13143 uses a unique and patented circuit topology.

MC13143
APPLICATIONS INFORMATION

Evaluation PC Board

The evaluation PCB is very versatile and is intended to be used across the entire useful frequency range of this device. The PC board is laid out to accommodate all SMT components on the circuit side (see Circuit Side Component Placement View).

Component Selection

The evaluation PC board is designed to accommodate specific components, while also being versatile enough to use components from various manufacturers. The circuit side placement view is illustrated for the components specified in the application circuit. The Component Placement View specifies particular components that were used to achieve the results shown in the typical curves and tables.

Mixer Input

The mixer input impedance is broadband 50 Ω for applications up to 2.4 GHz. It easily interfaces with a RF ceramic filter as shown in the application schematic.

Mixer Linearity Control

The mixer linearity control circuit accepts approximately 0 to 2.3 mA control current. An Input Third Order Intercept Point, IIP3 of 20 dBm may be achieved at 2.3 mA of control current (approximately 7.0 mA of additional supply current).

Local Oscillator Inputs

The differential LO inputs are internally biased at $V_{CC} - 1.0 \, V_{BE}$; this is suitable for high voltage and high gain operation.

For low voltage operation, the inputs are taken to V_{CC} through 51 Ω.

IF Output

The IF is a differential open collector configuration which is designed to use over a wide frequency range for up conversion as well as down conversion.

Input/Output Matching

It is desirable to use a RF ceramic or SAW filter before the mixer to provide image frequency rejection. The filter is selected based on cost, size and performance tradeoffs. Typical RF filters have 3.0 to 5.0 dB insertion loss. The PC board layout accommodates both ceramic and SAW RF filters which are offered by various suppliers such as Siemens, Toko and Murata.

Interface matching between the RF input, RF filter and the mixer will be required. The interface matching networks shown in the application circuit are designed for 50 Ω interfaces.

Differential to single–ended circuit configuration is shown in the test circuit. 6.0 dB of additional mixer gain can be achieved by conjugately matching the output of the MiniCircuits transformer to 50 Ω at the desired IF frequency. With narrowband IF output matching the mixer performance is 3.0 dB gain and 12 dB noise figure (see Narrowband 49 and 83 MHz IF Output Matching Options). Typical insertion loss of the Toko ceramic filter is 3.0 dB. Thus, the overall gain of the circuit is 0 dB with a 15 dB noise figure.

Figure 9. Narrowband IF Output Matching with 16:1 Z Transformer and LC Network

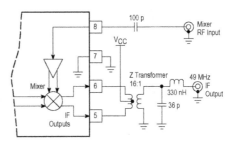

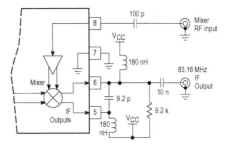

MC13143

Figure 10. Circuit Side Component Placement View

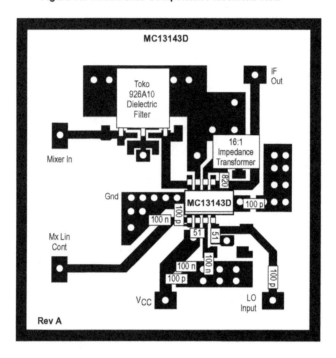

NOTES: 926.5 MHz preselect dielectric filter is Toko part # 4DFA–926A10; the 4DFA (2 and 3 pole SMD type) filters are available for applications in cellular and GSM, GPS, DECT, PHS, PCS and ISM bands at 902–928 MHz, 1.8–1.9 GHz at 2.4–2.5 GHz.

The PCB also accommodates a surface mount RF SAW filter in an eight or six pin ceramic package for the cellular base and handset frequencies. Recommended manufacturers are Siemens and Murata.

The PCB may also be used without a preselector filter; AC coupled to the mixer as shown in the test circuit schematic. All other external circuit components shown in the PCB layout above are the same as used in the test circuit schematic.

16:1 broadband impedance transformer is mini circuits part #TX16–R3T; it is in the leadless surface mount "TX" package. For a more selective narrowband match, a lowpass filter may be used after the transformer. The PCB is designed to accommodate lump inductors and capacitors in more selective narrowband matching of the mixer differential outputs to a single–ended output at a given IF frequency.

The local oscillator may also be driven in a differential configuration using a coaxial transformer. Recommended sources are the Toko Balun transformers type B4F, B5FL and B5F (SMD component).

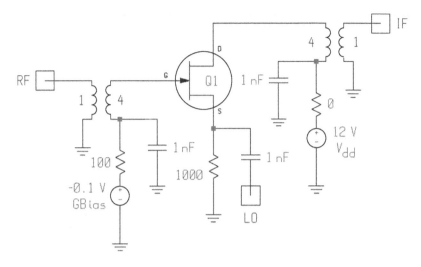

Figure 4.61 An additive JFET mixer cell using gate RF injection and source LO injection.

both as calculated by Ansoft Corporation's Serenade 8.0 circuit simulator. A reproduction of the MC13143 datasheet, used by permission, follows Figure 4.61.

In considering the approach used in the MC13143, one has to remember that feedback around many stages gives an IMD ripple, which means that the higher order IPs are no longer a straight-line calculation. Above a certain level, the 3 dB/dB law fails, and as a result of this, the IMD characteristic can be as bad as 10 or 20 dB/dB—almost like a break-down point.

4.4.3 FET Mixers

FETs, used in active and passive circuits, are still a popular solution for low-power inte-grated mixers up to 100 MHz (passive quad), 1 GHz (dual-gate MOSFET). Cost-sensitive applications may require the use of a single-gate FET. These types of single-gate microwave mixers are frequently used to validate the quality of software but the user keeps forgetting this is a quality of the model in question and not of the simulator. A good example of bad modeling is Figure 7.88 in Ref. [18]. Unfortunately we have neither seen the same circuit analyzed with modern tools nor do the authors of the paper cited [18] provide sufficient details of the circuit and the transistor to do this ourselves.

Figure 4.61 shows a basic, active FET mixer. Like the single-diode and single-BJT mixers, it is additive, although it is common to lessen RF–LO interaction by injecting one signal at the gate and the other at the source as shown. Single-gate GaAs and LDMOS FETs are also used in this arrangement with gate biasing as appropriate to the device type. Figure 4.62 shows a dual-gate FET in an additive mixing configuration.

The linearity of FETs is based on the fact that a FET follows a square law and therefore the first derivative, its transconductance, is supposed to be constant. This is valid within a wide amplitude range. FET mixers have a noise figure similar to those found in bipolar mixers: typically around 6–9 dB, depending on the configuration. A FET mixer's intercept point is subject to load-impedance variations, with a purely resistive termination providing the best

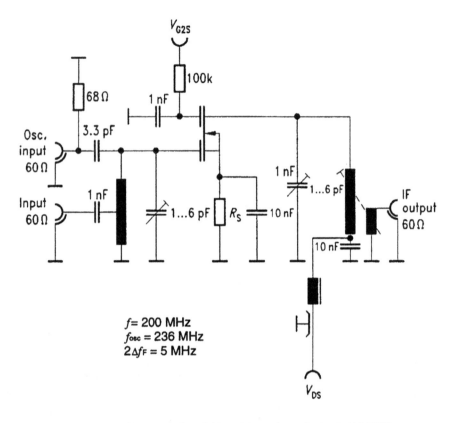

$f = 200$ MHz
$f_{osc} = 236$ MHz
$2\Delta f_F = 5$ MHz

Figure 4.62 Test circuit for additive mixing using a dual-gate MOSFET.

case. Terminating the mixer with a filter is problematic in that a filter looks purely resistive only within its 3 dB passband; in the transition band and beyond, the filter impedance rises rapidly and the mixer intercept point goes down. The best means devised so far to minimize this effect is to configure the FET output as a high-pass filter, using capacitive coupling from the output-tuned circuit to a 50 Ω bandpass filter, as shown in Figure 4.63. Outside its passband, this impedance inverter acts more like a short circuit and maintains IMD products at a reasonable level. The alternative to this is the popular diplexer, which requires more components and has more insertion loss. The high-pass configuration has been barely mentioned in the literature.

The active FET mixer achieves gain at the expense of intercept point; the difference can be as much as 20 dB. As an example, the passive MOSFET quad mixer, which is to be referenced, achieves an intercept point of greater than 40 dBm while similar active silicon FETs barely make 30 dBm. This 30 dBm performance is possible only at fairly low frequencies, and requires careful selection and adjustment of the dc bias point of all transistors of the circuit. This performance also requires that the FETs be operated grounded-gate, which limits their gain. The other drawback of these active arrangements is that higher impedances at the output reduce the third-order intercept point overproportionally; as an example, a factor of two in impedance change can cost as much as 6 dB intercept point. In these discussions, we assume that the IMD products are generated solely by the

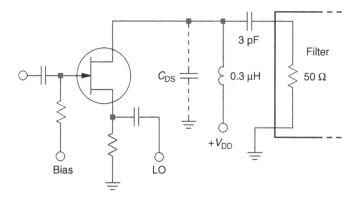

Figure 4.63 FET mixer output coupling that minimizes the effects of filter reactance outside the filter passband.

active device, and not by any passive devices, such as transformers. Actually, there are also some capacitors whose linearity varies with RF current—another unpleasant effect that is frequently overlooked.

On the other hand, one can use any FET as a passive device similar to a diode mixer, in which the source-drain channel gets switched on and off. This impedance modulation is somewhat similar to a diode mixer, but the gate electrode is isolated from both source and drain. It nonetheless falls in the category of additive mixers because there is sufficient interaction between gate and source, although the impedance at the gate changes significantly less than in an additive diode mixer. Implementation is a challenge in that building a high-performance passive FET mixer requires a pair or a quad of mixer cells that are sufficiently matched to suppress even-order IMD products. Intercept points, depending on the LO drive, vary between $+20$ and $+45$ dBm. The lower number is more applicable for microwave and RF frequencies, while the $+45$ dBm level is easily obtainable between 5 and 30 MHz. It is necessary to remember that the FET is a voltage-driven switch, and LO matching becomes an issue: How do you generate 30 V peak-to-peak across a few picofarads over a wide frequency range? Physical layout can also be critical: Symmetry is not only important for the active devices, but also for the input, output, and LO circuit. We have seen cases where fractions of an inch in different lengths leading to the gate electrodes has cost up to 10 dB of IP_3—just because of the resulting lack of symmetry.

Figure 4.64 shows the circuit of Figure 4.61 modified to work with an LDMOS FET, the harmonic-balance version of the SPICE level 3 model. Figure 4.65 shows the LDMOS circuit's conversion gain and noise figure versus frequency. The "noisy" traces reflect the model's numerical problems. This model, which works up to about 0.8 μm and has been validated against the Motorola data and SPICE results, has already been made mathematically continuous but still shows the inability to properly model the real RF transistor. This is why more research needs to be done on RF and microwave applications for MOS in the modeling area. As can be seen from the references, several attempts have been made to improve the model, but the major headache remains with parameter extraction, and as the models get more complex, the modeling quality does not seem to improve as a linear function. The reader should be reminded of the load–pull measurement as outlined earlier.

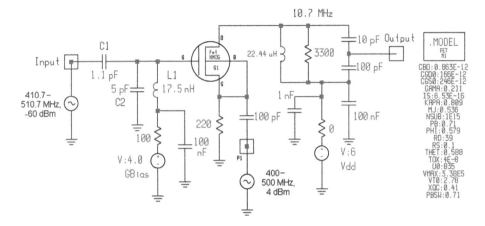

Figure 4.64   The N-JFET circuit of Figure 4.61 modified to work with an LDMOS FET.

Dual-Gate MOS/GaAs Mixers   Significantly better LO–RF isolation can be obtained by applying the RF to LO signals separate gates of a dual-gate FET (Figure 4.66). The result is multiplicative mixing with fewer constraints on LO and RF filtering.

4.4.4   MOSFET Gilbert Cell

The trend to go to smaller voltages has created some CMOS implementations of the Gilbert cell. While this may allow us to integrate reasonable mixers on the same chip, nobody should

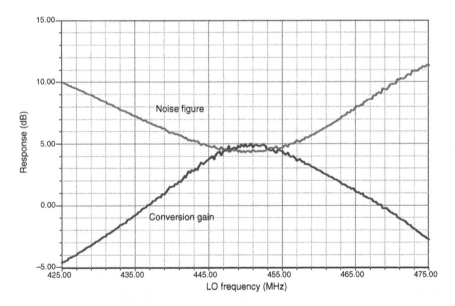

Figure 4.65   The LDMOS FET mixer's calculated conversion gain and noise figure. Although the "noisy" traces reflect numerical problems in the device model, the results can be put to good use in circuit design.

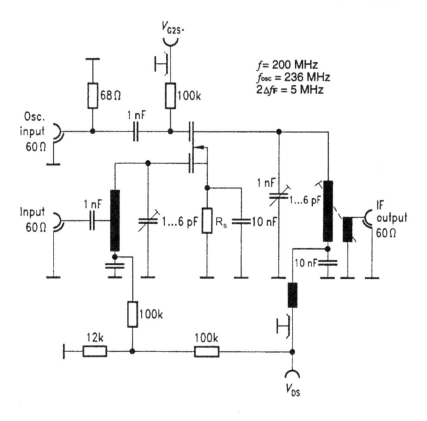

$f = 200$ MHz
$f_{osc} = 236$ MHz
$2\Delta f_F = 5$ MHz

Figure 4.66   Test circuit for multiplicative mixing using a dual-gate MOSFET.

expect any miracles from these mixers. They are frequently starved in operating voltage and current and rarely fare better than a single diode mixer with proper drive level. On the other hand, the symmetry reduces some of the unwanted spurious frequencies and because of the high impedance, the actual RF power level is much less than the diode would need. There are so many CMOS transistors available in different processes that we cannot provide information about the general behavior but this "low cost" solution, as pointed out, is really only attractive because one can stay in the same technology while building the amplifier and other stages. Because of the flicker corner frequency of MOS, the noise figure of the mixer will be high compared to other devices, but not as bad as its GaAsFET brothers. In any case, to combine reasonably performance and simplicity, one needs to have a preamplifier that reduces the noise figure as well as the third-order intercept point of the mixer by the amount of preamplification.

4.4.5 GaAsFET Single-Gate Switch—Resistive Mixer

It is a drawback of the diode mixers that the mixing process is performed by a one-port device. Local oscillator, RF, and IF signals are all present at the same port and need to be carefully separated through filters or balanced topologies. It is especially critical to suppress the LO signal on the RF line, since the signals are closely spaced in frequency and the LO is at a much higher power level. Replacing the diode by a FET offers a number of advantages. Good RF to LO isolation is achieved when the LO signal is applied at the gate, while the port for RF and IF signals is at the drain.

The FET in a FET switching or so-called resistive mixer is biased at a drain-source voltage of 0 V, therefore it operates in the resistive region. The LO at the gate modulates the resistance of the channel, and ideally, switches it fast between fully open and fully closed. The input signal (RF or IF) are fed to source or drain, where the output signal is generated (IF or RF). Besides the improved RF to LO isolation, the resistive mixer has the advantage that it is inherently a low-power circuit, since no dc supply is required at the drain. Still, the conversion loss is reasonable. The concept is relatively easy to realize for almost any frequency range, relatively low LO power is required, and the circuit provides good noise and linearity properties.

The good linearity comes from the fact that the resistive mixer rather acts like a switch, which is much more linear regarding RF input power than active mixers that use the inherently nonlinear transconductance. Intercept points of around 40 dBm are common in this type of mixer. It should be noted, that highest linearity can only be achieved if the source of the mixing transistors is directly connected to ground. For example, a record IP_3 of 53 dBm was achieved for such a "vertical" double-balanced mixer by Synergy Corp. on the basis of a standard TriQuint process. In contrast, if the differential mixer concept is realized by connecting the FETs in form of a ring—as usual for diode mixers—a typical maximum IP_3 of 30 dBm cannot be exceeded. If the sources are floating, it is practically impossible to control the gate-source voltage. High RF voltage swings, therefore, easily drive the gate into conducting state, an unwanted effect that degrades mixer performance and possibly device lifetime.

Before this type of mixer is discussed in more detail, a word of caution regarding simulating it. It is often observed that a simulation of a resistive mixer is not sufficiently accurate. The cause is not found in the numerics of the simulation tools, it can be the transistor model. Standard transistor models are most accurate for the common transistor operation, that is, the active region. The performance of a passive transistor is often not considered during model development. Thus, before designing a resistive mixer, one needs to verify that the models used are specified for switching operation at zero volts bias.

One of the pioneers of ICs for the wireless market is TriQuint, and we have been working with Wes Hayward of TriQuint in circuit analysis and validation. In doing so, we used some commercial and experimental circuits. In the amplifier chapter, we have already pointed out one amplifier that shows good agreement between measured and predicted data. By the time one has realized the entire circuit, including its associated biasing problems, even these circuits tend to grow. Figure 4.67 shows the circuit, which consists of an oscillator/amplifier, switching mixer, and a differential IF postamplifier.

Since the highest accuracy is needed to model these type of circuits, we found that only very recent simulators, such as Serenade 8 from Ansoft, can handle such a large number of FETs without barking at the user. Initially, we had all kinds of difficulties with these circuits because of the TriQuint TOM model, which had been developed for SPICE, and its IF/ELSE statements did not make it mathematically continuous. The type of GaAsFETs used by TriQuint were both regular FETs and enhancement FETs, which require a positive voltage at the gate. Further details of this should be looked up in the foundry manual by TriQuint. In the modeling chapter, we have shown some comparison between different models and their prediction accuracy. Such converters need to resort to inductors in GaAs technology, and compared to other parts they become quite big (Figure 4.68).

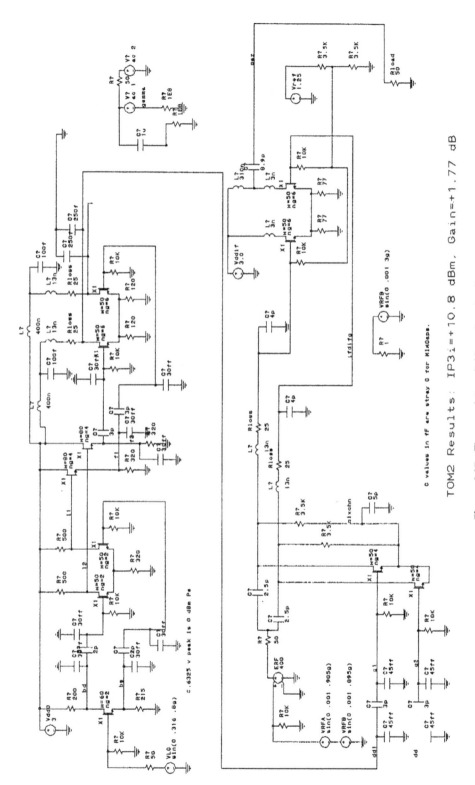

TOM2 Results: IP3i=+10.8 dBm, Gain=+1.77 dB

0 values in fF are stray C for MIMCaps.

Figure 4.67 The complete switching mixer circuit.

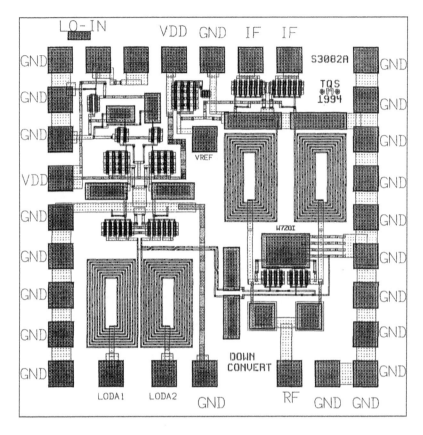

Figure 4.68 Physical layout of the GaAs oscillator/mixer/postamplifier IC, showing the considerable real estate required for integrated inductors.

In order to reduce some of the complexity in modeling and yet deal with the problems of mixers, here is a simpler mixer which is the combination of a set of switching transistors and a postamplifier for the IF. The reader may notice that the small inductances of 15 nH have an extremely high loss modeled by a series resistor of 18 Ω. The first to recommend this type of switching mixers was probably Stephen Maas. The circuit diagram itself (Figure 4.69) is self-explanatory, but again we need to point out the complexity it ends up with.

The GaAsFET mixer probably the best performance on the market today is the TriQuint model CMY210 (Figure 4.70). A reproduction of its datasheet, used by permission, follows the schematic.

The CMY210 circuit is a clever arrangement of a switching type of mixer with an automatic level control arrangement that provides the mixer with a reasonably constant drive level following the LO amplifier. Similar to cases we have shown, there is a clear relationship between the local oscillator power (invested dc power) and large-signal performance. This mixer externally has a series-tuned circuit at the input that allows the UHF frequency to pass, and has a parallel-tuned circuit that keeps the UHF frequency from getting into the IF stage. The resulting conversion loss of 5.5 dB is consistent with the example we presented

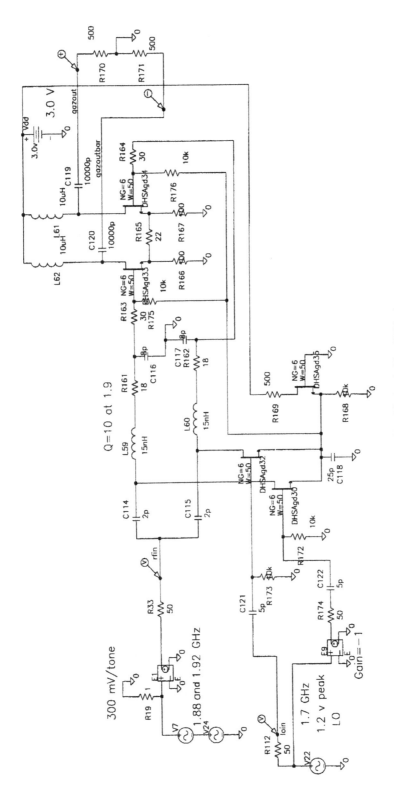

Figure 4.69 Simplified switching FET mixer.

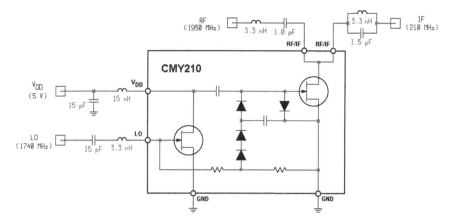

Figure 4.70 The TriQuint CMY210 IC includes a shunt switch mixer and AGC-equipped LO amplifier for more constant LO drive. Despite its simplicity, this mixer achieves a typical $IP_{3,in}$ of 25 dBm. Courtesy of TriQuint Semiconductor.

before of 6 dB, and while the noise figure also is stated as 5.5 dB, that value is deceiving because it was determined using a 120 MHz IF. If an IF of 10 MHz or less was used, the noise figure would be tremendously worse. This is due to the frequently mentioned flicker noise contribution.

4.4.5.1 Noise in Resistive Mixers [19]

A mixer's noise performance is not as easy to comprehend as it might be in a linear circuit. The nonlinear operation of the transistor or diode is a prerequisite for the mixer to operate at all. The signal is shifted in frequency, therefore we need to consider the relation between RF and IF noise. And finally, we do not simply have dc currents driving our noise sources and only small-signal RF. In this case, dc, if applied at all, might be smaller than the LO signal's amplitude.[1]

However, let us have a look at the different mechanisms one after the other. First, consider the nonlinear behavior and the many frequencies involved. This is not a real issue, if we assume for a moment that all noise sources are known. Noise currents and voltages are stochastical signals, but still signals that propagate through an electrical circuit like any other small-signal electrical signal would do. That is a standard task for any circuit simulator. Starting with an initial noise correlation matrix describing the noise sources, it needs to be calculated how these small-signal quantities are converted to the other frequencies. The result is a number of noise correlation matrices that each describe the noise associated with one mixing product. The total mixer noise is obtained by adding all these contributions.

The second issue is the noise sources. If we first consider white noise only, the issue becomes simpler in a resistive mixer. With the transistor acting as a controlled resistor, only thermal noise is observed. There is no excess noise due to pn junctions or hot electrons in the channel. Thus, it can be assumed that the noise figure of a resistive mixer follows the same rule as the noise figure of any passive mixer as discussed in Section 4.2.2.

---

[1] Based on Ref. [20]. Figures reprinted with permission. The mixer noise modeling approach was also published in Refs. [21–23].

CMY210
Datasheet

Ultra-Linear Mixer with Integrated LO Buffer

Description

The CMY210 is an all port, single-ended, general purpose up- and down-converter.
It combines small conversion losses and excellent intermodulation characteristics with a low demand of LO- and DC-power.
The internal level controlled LO-buffer enables good performance over a wide range of LO level inputs.
The mixer configuration allows RF or IF feed to either pin 1 or 6 and thus requires a frequency separation circuit on the pc board.

Applications

- Up- or Down-Converters

- Mobile Phones Receivers

- WLAN Receivers

- Mobile Phone or WLAN Basestations

Features

- Very High Input IP3 of 24 dBm typical

- Very Low LO Power demand of 0 dBm typical; Wide input range

- Wide LO Frequency Range: <500 MHz to >2.5 GHz

- Single-Ended Ports

- RF- and IF-Port Impedance 50 ohms

- Operating Voltage Range: <3 to 6 V

- Very Low Current Consumption: 6 mA typical

- All Gold Metalization

Package Outline, MW-6

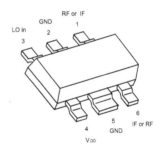

For additional information and latest specifications, see our website: www.triquint.com   2
Revision E, September 12, 2006

CMY 210 Datasheet

Maximum Ratings

Parameter	Port	Symbol	Value		Unit
			min	max	
Supply Voltage	4	V_{DD}	0	6	V
DC-Voltage at LO Input	3	V_3	-3	0,5	V
DC-Voltage at RF-IF Ports [2]	1, 6	$V_{1,6}$	- 0,5	+ 0,5	V
Power into RF-IF Ports	1, 6	$P_{in,RF}$		17	dBm
Power into LO Input	3	$P_{in,LO}$		10	dBm
Channel Temperature		T_{Ch}		150	°C
Storage Temperature		T_{stg}	-55	150	°C

Thermal Resistance

Channel to Soldering Point (GND)	R_{thChS}	≤100	K/W

1) For detailed dimensions see page 7.
2) For DC test purposes only, no DC voltages at pins 1, 6 in application

Source: © TriQuint Semiconductor, reprint with permission.

CMY 210 Datasheet

Electrical Characteristics

Test conditions: T_a = 25°C; V_{DD}= 3V, see test circuit; f_{RF} = 808MHz;
f_{LO} = 965MHz; P_{LO} = 0dBm; f_{IF} = 157MHz, unless otherwise specified:

Parameter, Test Conditions	Symbol	min	typ	max	Unit
Operating Current	I_{op}	-	6.0	8.0	mA
Conversion Loss	L_c	-	5.7	7.0	dB
SSB Noise Figure	F_{ssb}	-	6.0	-	dB
2 Tone 3rd Order IMD P_{RF1} = P_{RF2} = -3dBm f_{RF1} = 806MHz; f_{RF2} = 810MHz; f_{LO} =965MHz	d_{IM3}	-	54	-	dBc
3rd Order Input Intercept Point	$IP3_{in}$	20	24	-	dBm
P_{-1dB} Input Power	P_{-1dB}	-	14	-	dBm
LO Leakage at RF/IF-Port (1,6)	$P_{LO\ 1,6}$	-	-8	-	dBm

Applications Information
Test circuit / application example

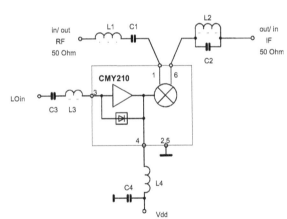

Notes for external elements:

L1, C1: Filter for upper frequency;
C2, L2: Filter for lower frequency;
each filter is a throughpath for the desired frequency (RF or IF) and isolates the other frequency (IF or RF) and its harmonics.
These two filters must be connected to pin 1 and pin 6 directly.
Parasitic capacitances at the ports 1 and 6 must be as small as possible.
L4 and C4 are optimized by indicating lowest I_{op} at used LO-frequency; same procedure for L3. The ports 1, 3 and 6 must be DC open.

Lumped element values for 800MHz test and application circuit:

f LO	F RF	F IF	L1	C1	L2	C2	L3	C3	L4	C4
MHz	MHz	MHz	nH	pF	nH	pF	nH	pF	nH	pF
965	808	157	8.2	3.9	8.2	3.3	6.8	47	15	33

CMY 210 Datasheet

Applications Information (cont)
PCB-Layout for 800MHz test and application circuit:

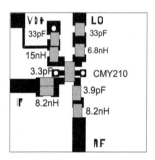

Actual size

Typical lumped element values for different RF-frequencies:

f RF	L1	C1	L2	C2
MHz	nH	pF	nH	pF
400	12	15	12	12
450	12	12	12	10
900	8.2	3.9	8.2	3.3
1500	3.3	2.7	3.3	2.2
1800	3.3	2.2	3.3	1.8
2000	3.3	1.8	3.3	1.2
2400	1.8	2.7	1.8	1.5

Typical lumped element values for different LO-frequencies:

f LO	L3	C3	L4	C4
MHz	nH	pF	nH	pF
500	15	82	47	82
750	6.8	33	22	33
800	6.8	33	18	33
950	6.8	27	15	27
1100	6.8	27	12	27
1400	6.8	22	6.8	22
1600	6.8	18	4.7	18
1800	6.8	15	3.3	15
2000	6.8	12	2.2	12
2100	6.8	12	1.8	12
2300	4.7	12	1.2	12

For additional information and latest specifications, see our website: www.triquint.com
Revision E, September 12, 2006

5

CMY 210 Datasheet

General description and notes

The CMY 210 is an all port single ended general purpose Up- and Down-Converter.
It combines small conversion losses and excellent intermodulation characteristics with a low
demand of LO- and DC-power.
The internal level controlled LO-Buffer enables a good performance over a wide LO level range.
The internal mixers principle with one port RF and IF requires a frequency separation at pin 1 and 6
respectively.

Note 1:
Best performance with lowest conversion loss is achieved when each circuit or device for the
frequency separation meets the following requirements:

Input Filter: Throughpass for the signal to be mixed; reflection of the mixed signal and the harmonics
of both.

Output Filter: Throughpass for the mixed signal and reflection of the signal to be mixed and the
harmonics of both.

The impedance for the reflecting frequency range of each filter toward the ports 1 and 6 should be as
high as possible.
In the simplest case a series- and a parallel- resonator circuit will meet these requirements but also
others as appropriate drop in filters or micro stripline elements can be used.
The two branches with filters should meet immediately at the package leads of the port 1 and 6.
Parasitic capacitances at these ports must be kept as small as possible.
The mixer also can be driven with a source- and load impedance different to 50Ω, but performance
will degrade at larger deviations.

Note 2:
The LO-Buffer needs an external inductor L4 at port 4; the value of inductance depends on the
LO frequency. It is tuned for minimum I_{op} consumption into port 4.
At lower LO frequencies it can be reduced by an additional capacitor C5.

Note 3:
The LO Input impedance at Port 3 can be matched with a series inductor. It also can be tuned for a
minimum current I_{op} into port 4. C3 is a DC blocking capacitor.
Since the input impedance of port 3 can be slightly negative, the source reflection coefficient should
be kept below 0.8 ($Z_0 = 50$ Ω).

The Conversion Noise Figure Fssb is corresponding with the value of Conversion Loss L_c. The LO
signal must be clean of noise and spurious at the frequencies $f_{LO} \pm f_{IF}$.

Source: © TriQuint Semiconductor, reprint with permission.

CMY 210 Datasheet

Electrical Characteristics (cont)

Operating Current $I_{op} = f(P_{LO})$
$V_{DD} = 3V$
$f_{LO} = $ Parameter

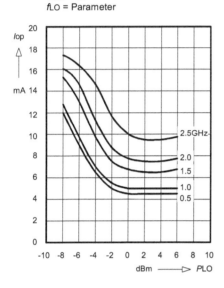

Conversion Loss $L_c = f(P_{LO})$
$V_{DD} = 3V; f_{IF} = 120MHz$
$f_{LO} = $ Parameter

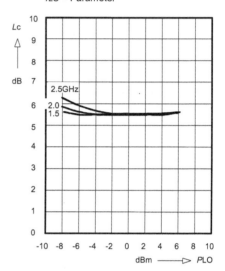

Conversion Loss $L_c = f(V_{DD})$
$P_{LO} = 0dBm$
$f_{LO} = 1500MHz; f_{IF} = 120MHz$

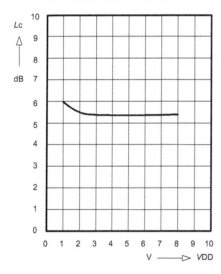

Third Order IP3 $IP3in = f(P_{LO})$
$P_{in} = 2 \times -3dBm; f_{IF} = 40/45MHz$
$V_{DD} = 3V; f_{LO} = $ Parameter;

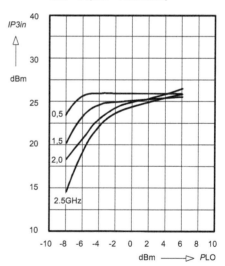

For additional information and latest specifications, see our website: www.triquint.com   7
Revision E, September 12, 2006

Things get a bit more involved when it comes to $1/f$ noise. The low-frequency noise becomes an issue when the $1/f$ noise corner is higher than the IF frequency, or if the RF is directly converted into baseband. In this case, $1/f$ noise is the dominant noise. This is obviously the case in baseband, but the noise also gets upconverted and forms sidebands at the harmonics of the LO signal. Since the spacing between RF signal sidebands and LO equals the IF frequency, it can be assumed that the upconverted $1/f$ noise is also dominant at all other RF frequencies. And again, the mixing process is straightforward to calculate or simulate.

The complicated issue in resistive mixer noise calculation is the question, how the currents control the transistor's noise sources. Classical theory considers only dc current. The result is that no $1/f$ noise is expected at all, since, the circuit topology fixes V_{DS} at 0 V and no dc current is flowing. Unfortunately, measurement reveals that this optimistic guess is wrong.

In order to predict the $1/f$ noise in a resistive mixer accurately, it is necessary that the transistor model implements cyclostationary noise sources. Cyclostationary noise sources are controlled by all harmonics of the large-signal current, not only by the dc component. Thereby, even without dc current flowing, noise sidebands appear at all LO frequencies. The traditional approach of implementing a noise source, on the other hand, assumes that only the DC current is driving the $1/f$ noise. If this assumption would hold, no $1/f$ noise would be observed in a resistive mixer since no DC current is flowing. Cyclostationary sources, on the other hand, assume that large-signal current of any frequency drives the noise source. As a result, the device generates $1/f$-shaped noise side band at all large-signal frequencies. This noise gets downconverted and shows up as baseband $1/f$ noise, even in absence of DC current. An introduction to the cyclostationary noise theory is found in [24].

The $1/f$ noise performance of resistive mixers was investigated at the circuit setup shown in Figure 4.71. It is basically a Fujitsu HEMT SHC40LG on a circuit board, in a configuration that allows for a detailed investigation of the transistor noise performance. The whole configuration is shielded in order to suppress coupling of low-frequency noise. An important

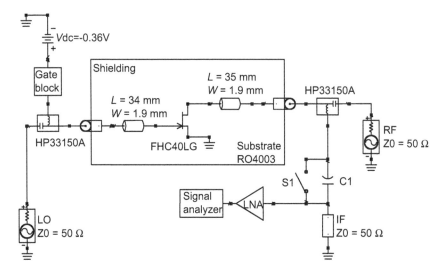

Figure 4.71 FET resistive mixer setup used to investigate the noise performance. From Ref. [20].

detail of this setup is the switch S1 that allows to short circuit the capacitance C1. Closing the switch enables dc current to flow through the transistor, which will impact the noise performance as we will see later on.

In order to minimize the resistive mixer noise, it is necessary to make sure that the noise performance is not additionally degraded, for example, through the following mechanisms:

1. High LO amplitude potentially drives the gate in forward conduction mode. The result is additional shot noise and excess $1/f$ noise of the schottky junction. It is, however, advisable to prevent this from happening as the gate is not meant to draw current and driving it in forward direction could possibly reduce lifetime. Anyway, monitoring that no gate dc current flows prevents these problems.

2. The LO itself will have phase noise. In the investigation, it is necessary to sort out which part of the noise is generated by the mixer, and what is mixed LO noise.

3. If the RF power is significant, it will add to the $1/f$ noise, due to the same mechanisms that are responsible for the noise caused by the LO signal.

Measured and simulated single-sideband $1/f$ noise at 1 KHz is shown in Figure 4.72 as a function of LO frequency. The LO power is set to 0 dBm. Simulations are shown as straight lines and symbols refer to measured points. The higher values were obtained when the switch S1 was open, so that no dc current can flow. Lower $1/f$ noise is observed with S1 closed, providing a dc path through the FET.

When the switch is open, classical noise models would not predict any $1/f$ noise at all, since its power level would be controlled by DC current only. No DC current, no $1/f$ noise. This example, therefore, is a proof of the cyclostationary noise source concept. The simulations were performed assuming that the $1/f$ noise sources are to be modeled this way. But still, assuming cyclostationaty noise sources alone cannot fully explain the measured $1/f$ noise. It is expected that any of the harmonics of the LO signal generates $1/f$-noise sidebands. But the dc component is nonexistent as long as the switch is open.

An important mechanism that is responsible for the baseband noise observed in this case is the cross-coupling of the LO signal from the gate to the channel. Figure 4.73 shows a sketch highlighting the coupling paths. LO leaks through C_{gd}, and in packaged FETs also through C_{gs}, if L_s is significantly large. This cross-coupled LO signal undergoes a mixing

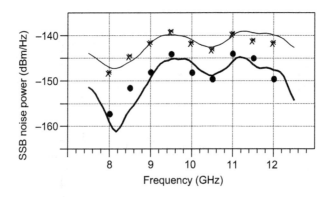

Figure 4.72 SSB noise power density at 1 KHz and LO power 0 dBm. Simulations: lines, measurement: symbols. Thin lines and crosses refer to the condition without dc current, thick lines and bullets refer to the circuit setup with dc path. From Ref. [20].

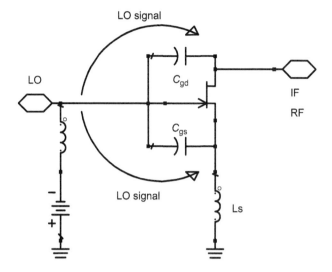

Figure 4.73 The LO signal couples through the capacitances C_{gs}, C_{gd} and the source inductance L_s into the RF and IF path. From Ref. [20].

process as any other signal in the channel does: it is downconverted to dc. And indeed, a dc current is observed when the switch is closed. Opening it can suppress dc, but it cannot change the fact that the LO fundamental noise sideband is downconverted with the cross-coupled signal. In short, the reason for the $1/f$ noise at baseband is the fact that the LO signal causes $1/f$-type noise sidebands, and that the cross-coupled LO signal downconverts these sidebands.

The impact of LO power level is shown in Figure 4.74 for a LO frequency of 9 GHz, obtained under the same conditions such as the previous graph. The $1/f$ noise first increases with LO power, due to decreased conversion loss and increasing cross-coupled power, and finally saturates.

Getting back to Figure 4.72, it seems that the $1/f$ noise can be reduced by providing a dc path to the mixing FET. At first sight, this seems to be a strange behavior, since a dc current causes additional noise. One would guess that the measured noise is increased. In fact, the

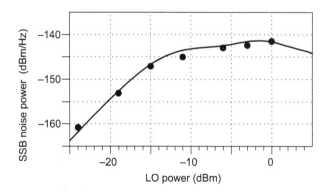

Figure 4.74 Dependence of SSB noise power density at 1 KHz on LO power, and an LO frequency of 9 GHz. Line: simulation, bullets: measurement. From Ref. [20].

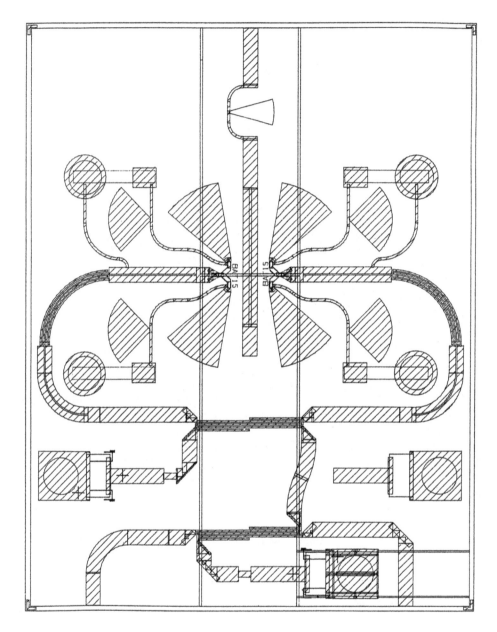

Figure 4.75 Layout of the 20 GHz diode mixer.

dc current causes $1/f$ noise that is superimposed to the downconverted noise. The noise at a certain offset frequency, however, is correlated between the sidebands at all harmonics. The dc current's noise obviously cancels the downconverted noise out. Depending on the type of GaAsFET, around 5 dB of improvement are to be expected, which can even be improved to around 10 dB improvement by forcing very quite dc currents through the device.

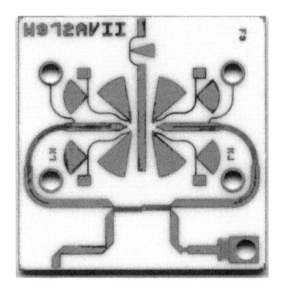

Figure 4.76 The 20 GHz diode–mixer circuit board.

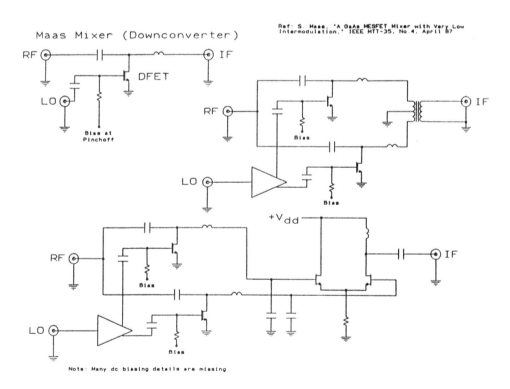

Figure 4.77 Example mixer circuit topologies.

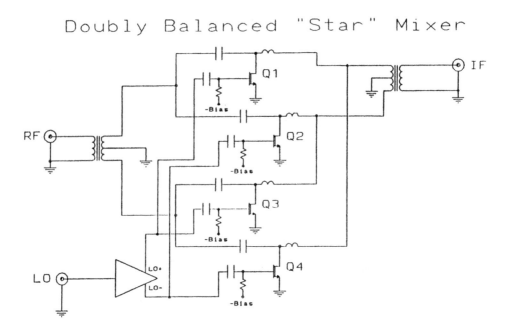

Figure 4.78   Example mixer circuit topologies.

For readers who are interested in what mixers look like when pushed well into the microwave region, we present the layout for a 20 GHz diode mixer. Figure 4.75 presents its layout, which includes radial stubs, circular stubs, and a Lange coupler and Figure 4.76 shows the layout implemented on a ceramic substrate.

As further food for thought and experimentation, Figures 4.77–4.87 present mixer and related circuit notes as presented by Wes Hayward in TriQuint Semiconductor's GaAs Design Class.

HA Mixer Family Measurements

The mixers are all identical except for FET width. Mixers used the HA
process and are half micron gate length. Bias was adjusted to best gain
at +6 LO. The LO baluns were half wave lines of semi-rigid coax.
F-LO=1.65, RF=1.9, and IF=0.25 GHz.
Input power is -13 or -10 dBm RF.
All measurements show down conversion results.

FET Width	P-av RF	P-LO dBm	P-out at IF	IMDR dB	Conv Gain dB	IP$_3$-out dBm	IP$_{3in}$ dBm
50	-13	6	-31	44.7	-11.7	-2.35	9.35
50	-13	10	-30.5	53.5	-11.2	2.55	13.75
50	-10	10	-27.8	49	-11.5	3	14.5
100	-10	6	-25.7	41.5	-9.4	1.35	10.75
100	-10	10	-24.8	48	-8.5	5.5	14.0
100	-10	14	-24.5	68.8	-8.2	16.2	24.4
200	-10	6	-24.3	40.5	-8	2.25	10.25
200	-10	10	-23.7	45	-7.4	5.1	12.5
200	-10	14	-23.5	60.8	-7.2	13.2	20.4
200	-10	17	-24	60.7	-7.7	12.65	20.3
400	-10	6	-24	42.8	-7.7	3.7	11.4
400	-10	10	-23.5	52	-7.2	8.8	16.0
400	-10	14	-23.3	67.7	-7	16.85	23.85
400	-10	17	-23.2	57.5	-6.9	11.85	18.75
50	-10	6	-30.8	56	-14.5	3.5	18.0
50	-10	6	-28.6	42.5	-12.3	-1.05	11.25
50	-10	10	-28	49.2	-11.7	2.9	14.6
50	-10	14	-28.2	60.8	-11.9	8.5	20.4
50	-10	17	-28.3	63.5	-12	9.75	21.75
50	-10	17	-28	62.8	-11.7	9.7	21.4
100	-10	6	-29.5	42.5	-13.2	-1.95	11.25
100	-10	10	-25	47.2	-8.7	4.9	13.6
100	-10	14	-24.7	63.7	-8.4	13.45	21.85
100	-10	16	-25.5	62	-9.2	11.8	21.0

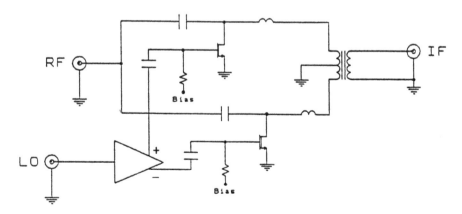

Figure 4.79 Example mixer circuit topologies.

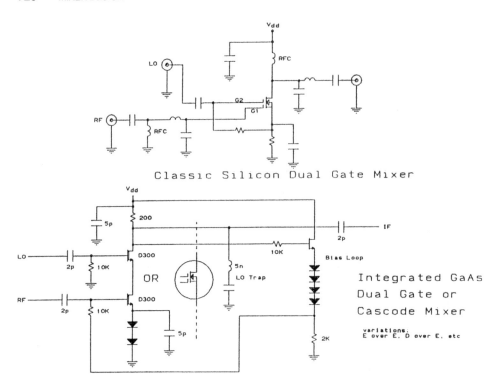

Figure 4.80 Example mixer circuit topologies.

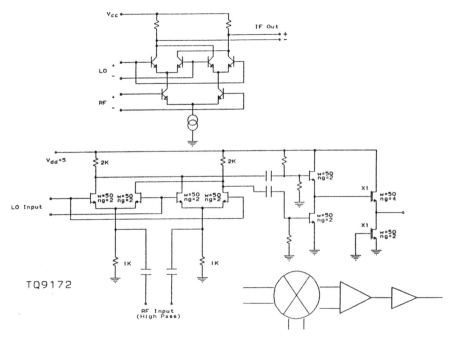

Figure 4.81 Example mixer circuit topologies.

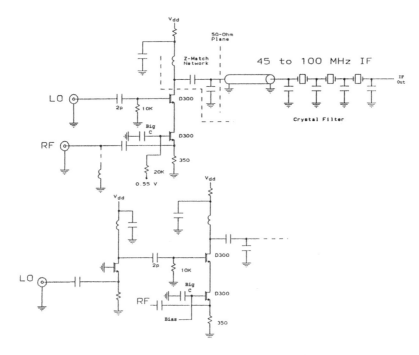

Figure 4.82 Example mixer circuit topologies.

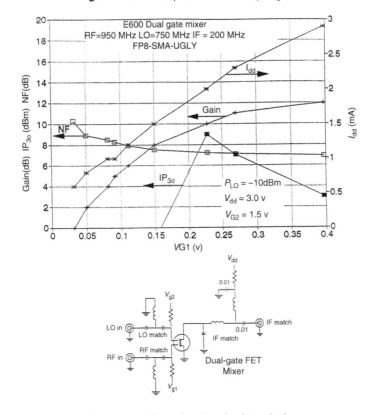

Figure 4.83 Example mixer circuit topologies.

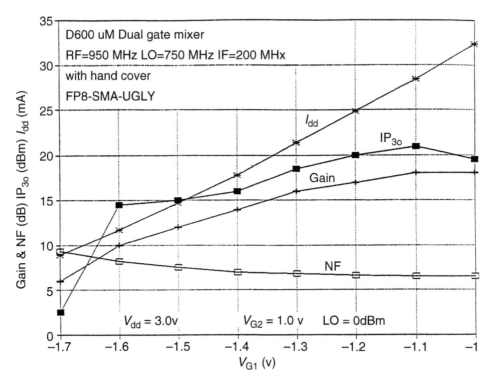

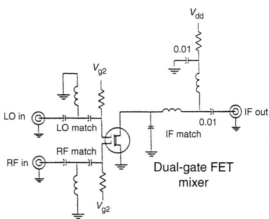

Figure 4.84   Example mixer circuit topologies.

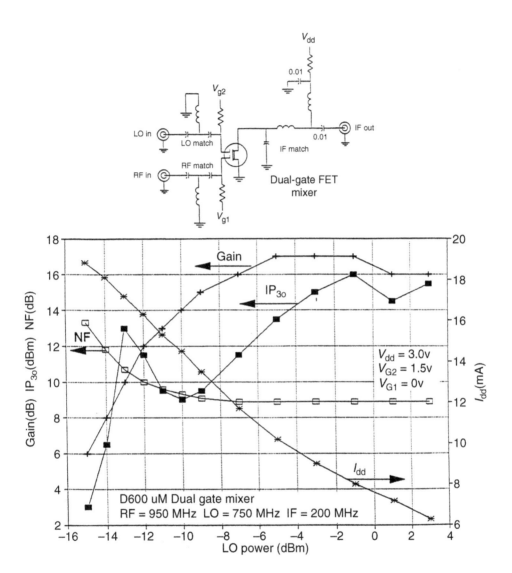

Figure 4.85 Example mixer circuit topologies.

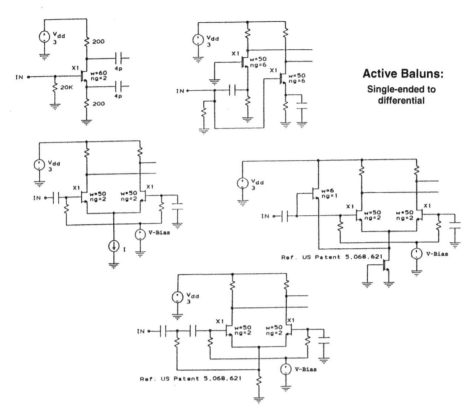

Figure 4.86   Example mixer circuit topologies.

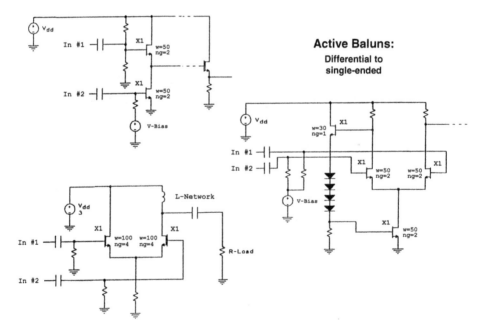

Figure 4.87   Example mixer circuit topologies.

REFERENCES

1. O. Zinke and H. L. Hartnagel, *Lehrbuch der Hochfrequenztechnik, Band 2: Elektronik und Signalverarbeitung.* New York: Springer Verlag, 1987, pp. 429–430.

2. H. Darabi and A. A. Abidi, "Noise in RF-CMOS mixers: a simple physical model," *IEEE Transactions on Solid State Circuits*, Vol. 35, No. 1, pp. 15–25, 2000.

3. IRE Subcommittee 7.9 on Noise, "Description of noise performance in receiving systems," *Proc. IRE*, Vol. 51, 1963, p. 436.

4. S. A. Maas, *Microwave Mixers*, 2nd ed. Boston: Artech House, 1993, pp. 153–171.

5. O. Zinke and H. Brunswig, *Lehrbuch der Hochfrequenztechnik*, 3rd ed. New York: Springer Verlag, Vol. 2, 1987, p. 435.

6. C. Qian, *Subharmonic Mixers Simplify Digital-Modulation Systems*, Microwaves & RF, 1998, pp. 62–70.

7. K. Itoh et al., "Even-harmonic quadrature modulator with low vector modulation error and low distortion for microwave digital radio," *Microwave Theory and Techniques Society Symposium Digest*, pp. 967–970, 1996.

8. S. R. Joshi, *Novel I/Q Modulators Mix Cellular Signals*, Microwaves & RF, Vol. 51, 1994, pp. 119–123. Also see S. R. Joshi, Modulator with Harmonic Mixers, Synergy Microwave Corporation, Paterson, NJ, US patent 5,416,449, May 16, 1995.

9. M. V. Schneider, R. Trambarulo, A. Gnauck, and M. J. Gans, "High-speed millimeter-wave modulator/demodulator," *Microwave Theory and Techniques Society Symposium Digest*, pp. 611–614, 1990.

10. J. Charamiec, "Subharmonically pumped Schottky diode single-sideband modulator," *IEEE Transactions on Microwave Theory and Techniques*, Vol. 26, No. 9, pp. 635–638, 1978.

11. I. Doyle, *A Simplified Subharmonic I/Q Modulator*, Applied Microwave & Wireless, 1998, pp. 34–40.

12. P. Will, "Broadband Doubly Balanced Mixer Having Improved Termination Insensitivity Characteristics," Adams-Russell Co., Inc., Waltham, MA, US patent 4,224,572, September 23, 1980.

13. J. T. Lee, "Balanced subharmonic mixers," *Microwave Journal*, August 1983.

14. D. Neuf, "Fundamental versus harmonic mixing," *Microwave Journal*, 1984.

15. B. Henderson, "Full-range orthogonal circuit mixers reach 2 to 26 GHz," *Microwave Systems News*, January, 1982.

16. B. Gilbert, *Demystifying the Mixer*, Self-published monograph, 1994, p. 27.

17. K. K. Clarke and D. T. Hess, *Communication Circuits: Analysis and Design*, Reading, MA: Addison-Wesley, 1971, p. 115.

18. G. D. Vendelin, A. M. Pavio, and U. L. Rohde, *Microwave Circuit Design Using Linear and Nonlinear Techniques*, 2nd ed. Wiley, 2005.

19. G. K. Tie and C. S. Aitchison, "Noise figure and associated conversion gain of a microwave MESFET gate mixer," *Conference Proc. of the 13th Eur. Microw. Conf.*, 1983.

20. M. Margraf, Niederfrequenz-Rauschen und Intermodulationen von resistiven FET-Mischern [PhD dissertation] Berlin Institute of Technology, 2004.
Available at http://opus.kobv.de/tuberlin/volltexte/2004/572/pdf/margraf_michael.pdf

21. M. Margraf and G. Boeck, "A New Scaleable Low Frequency Noise Model for Field-Effect Transistors Used in Resistive Mixers," *IEEE MTT-S Intl. Microwave Symp. Dig.*, 2003, Vol.1, 2003, pp. 559–562.

22. M. Margraf and G. Boeck, "Analysis and modeling of low-frequency noise in resistive FET mixers," *IEEE Transactions on Microwave Theory and Techniques*, Vol. 52, no. 7, 2004, pp. 1709–1718.

23. M. Margraf and G. Boeck, "1/f Noise optimum for field-effect transistors in single-ended resistive mixers," *Proc. 33rd Eur. Microw. Conf.*, 2003. Vol. 3, 2003, pp. 1015–1017.

24. M. Rudolph, F. Bonani, "Low-Frequency Noise in Nonlinear Systems", *IEEE Microwave Mag.*, Vol. 10, no. 1, Feb. 2009, pp. 84–92.

FURTHER READING

P. L. D. Abrie, *The Design of Impedance-Matching Networks for Radio-Frequency and Microwave Amplifiers*, Norwood: Artech House, 1985.

R. Goyal, *High-Frequency Analog Integrated Circuit Design*, New York: Wiley, 1995.

W. Hayward and S. Taylor, "Compensation Method and Apparatus for Enhancing Single Ended to Differential Conversion," TriQuint Semiconductor, Inc., Beaverton, OR, US patent 5,068,621, November 26, 1991.

R. G. Meyer, "Intermodulation in high-frequency bipolar transistor integrated-circuit mixers," *IEEE Journal of Solid-State Circuits*, Vol. SC-21, No. 4, pp. 534–537, 1986.

5

RF/WIRELESS OSCILLATORS

5.1 INTRODUCTION OF FREQUENCY CONTROL

Practically, all modern telecommunication equipment and test equipment uses frequency-control techniques based on frequency synthesis, the production of the many frequencies involved in a radiocommunication system's modulation, transmission, reception, and demodulation functions through the combination and mathematical manipulation of a very few input frequencies. Although the frequency synthesis techniques now ascendant in wireless systems—phase-locked loops and, less commonly, direct digital synthesizers—are fundamentally different, all are ultimately based on RF oscillators. This chapter covers oscillator theory, evaluation, and design.

Two types of oscillators are needed in a phase-locked-loop system (Figure 5.1). One, typically a crystal oscillator, generates the synthesizer's reference signal. As Figure 5.1 reflects, the reference oscillator may be built into the synthesizer IC in highly integrated systems. The other oscillator, a voltage-controlled oscillator (VCO), is varied in frequency by the system to produce the synthesizer's output signal. Although designing good oscillators remains somewhat like black magic or a special art, we will shown that mathematics and CAD tools, applied in conduction with practical experience, can keep the oscillator design process well under control.

5.2 BACKGROUND

The oscillator was probably first discovered by people who wanted to build an amplifier back in the vacuum-tube days. Its modern equivalent is shown in Figure 5.2. Here we see an active device, a bipolar transistor or field-effect transistor, with a tuned input and output circuit. Under ideal circumstances, the voltage at the input is 180° out of phase and the

RF/Microwave Circuit Design for Wireless Applications, Second Edition. Ulrich L. Rohde and Matthias Rudolph.
© 2013 John Wiley & Sons, Inc. Published 2013 by John Wiley & Sons, Inc.

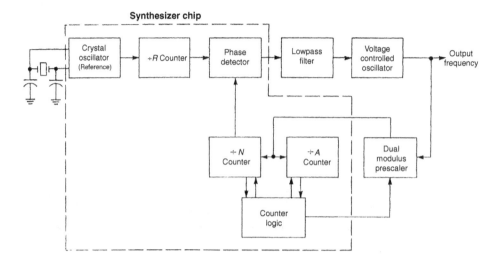

Figure 5.1 Block diagram of a modern, integrated frequency synthesizer. In this case, the designer has control over the VCO and the loop filter; the reference oscillator is part of the chip. In most cases (up to 2.5 GHz), the dual-modulus prescaler is inside the chip.

amplifier is "stable." Because of the unavoidable feedback capacitance, a certain amount of energy is transferred from the output to the input at about 90° out of phase. If one of the tuned circuits is detuned to the point where the phase shift is −180°, oscillation will occur. This means that the small energy from the output will be amplified, brought back to the input in phase and further amplified, causing the stage to "takes off" and oscillate. Unless there is some mechanism to limit the amplitude of the oscillation at the input or output, its amplitude can theoretically increase to a level sufficient to destroy the device through breakdown effects or thermal runaway.

The criterion for oscillation was first described by Barkhausen (see equation 5.6 in the next section). The conditions for oscillation exist when the feedback gain is high enough to cancel all losses and the difference between forward gain and reverse gain is less than zero. This also implies that oscillator circuits (Figure 5.2) provide a negative resistance, which is responsible for oscillation. The following mathematical treatment explains this.

5.3 OSCILLATOR DESIGN

A phase-locked loop generally has two oscillators: the oscillator at the output frequency and the reference oscillator. The reference oscillator at times can be another loop that is being mixed in, and the VCO is controlled by either the reference or the oscillator loop. The VCO is one of the most important parts in a phase-locked-loop system, the performance of which is determined only by the loop filter at offsets inside the loop bandwidth, and only by the quality of the VCO design at offsets outside the loop bandwidth.

To some designers, VCO design appears to be magic. Shortly, we will go through the mathematics of the oscillator and some of its design criteria, but the results have only limited meaning. This is due to component tolerances, stray effects, and most of all, nonlinear performance of the oscillator device, which can be modeled with only a certain degree of

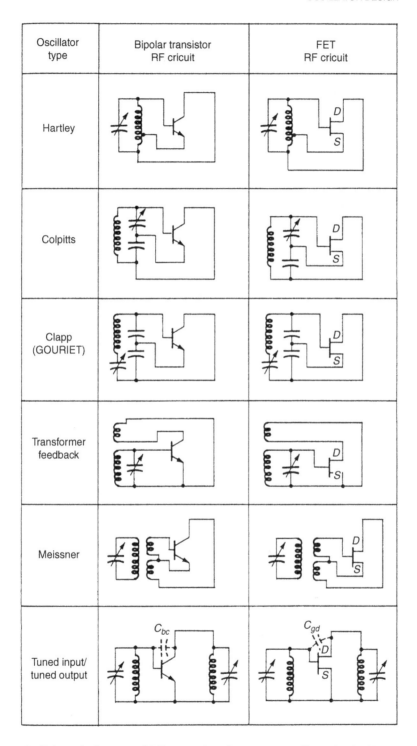

Oscillator type	Bipolar transistor RF cricuit	FET RF cricuit
Hartley		
Colpitts		
Clapp (GOURIET)		
Transformer feedback		
Meissner		
Tuned input/ tuned output		

Figure 5.2 Schematic diagrams of RF connections for common oscillator circuits (dc and biasing circuits not shown) [1].

accuracy. However, after building oscillators for awhile, a certain feeling will be acquired for how to do this, and certain performance behavior will be predicted on a rule-of-thumb basis rather than on precise mathematical effort. For reasons of understanding, we will deal with the necessary mathematical equations, but we consider it essential to explain that these are only approximations.

5.3.1 Basics of Oscillators

An electronic oscillator is a device that converts dc power to a periodic output signal (ac power). If the output waveform is approximately sinusoidal, the oscillator is referred to as sinusoidal. There are many other oscillator types normally referred to as relaxation oscillators. For application in frequency synthesizers, we will explore only sinusoidal oscillators for reasons of purity and noise-sideband performance.

All oscillators are inherently nonlinear. Although the nonlinearity results in some distortion of the signal, linear analysis techniques can normally be used for the analysis and design of oscillators. Figure 5.3 shows, in block diagram form, the necessary components of an oscillator. It contains an amplifier with frequency-dependent forward loop gain $G(j\omega)$ and a frequency-dependent feedback network $H(j\omega)$.

The output voltage is given by

$$V_0 = \frac{V_i G (j\omega)}{1 + G (j\omega) H (j\omega)} \tag{5.1}$$

For an oscillator, the output V_0 is nonzero even if the input signal $V_i = 0$. This can only be possible if the forward loop gain is infinite (which is not practical), or if the denominator

$$1 + G (j\omega) H (j\omega) = 0 \tag{5.2}$$

at some frequency ω_0. This leads to the well-known condition for oscillation (the Nyquist criterion), where at some frequency ω_0

$$G (j\omega_0) H (j\omega_0) = -1 \tag{5.3}$$

That is, the magnitude of the open-loop transfer function is equal to 1:

$$|G (j\omega_0) H (j\omega_0)| = 1 \tag{5.4}$$

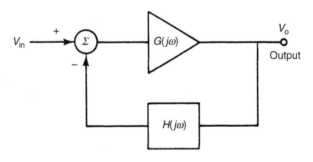

Figure 5.3 Block diagram of an oscillator showing forward and feedback loop components.

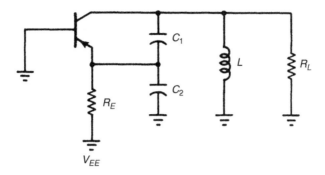

Figure 5.4 Oscillator with capacitive voltage divider.

and the phase shift is 180°:

$$\arg\left[G\left(j\omega_0\right) H\left(j\omega_0\right)\right] = 180° \qquad (5.5)$$

This can be more simply expressed as follows. If in a negative-feedback system, the open-loop gain, has a total phase shift of 180° at some frequency ω_0, the system will oscillate at that frequency provided that the open-loop gain is unity. If the gain is less than unity at the frequency where the phase shift is 180°, the system will be stable, whereas if the gain is greater than unity, the system will be unstable.

This statement is not correct for some complicated systems, but it is correct for those transfer functions normally encountered in oscillator design. The conditions for stability are also known as the Barkhausen criterion, which states that if the closed-loop transfer function is

$$\frac{V_0}{V_i} = \frac{\mu}{1 - \mu\beta} \qquad (5.6)$$

the system will oscillate provided that $\mu\beta = 1$. This is equivalent to the Nyquist criterion, the difference being that the transfer function is written for a loop with positive feedback. Both versions state that the total phase shift around the loop must be 360° at the frequency of oscillation and the magnitude of the open-loop gain must be unity at that frequency.

The following analysis of the relatively simple oscillator shown in Figure 5.4 illustrates the design method. The linearized (and simplified) equivalent circuit of Figure 5.4 is given in Figure 5.5. h_{rb} has been neglected and $1/h_{ob}$ has been assumed to be much greater that the load resistance R_L and is also ignored.

Note that the transistor is connected in the common-base configuration, which has no voltage phase inversion (the feedback is positive), so the conditions for oscillation are

$$\left|G\left(j\omega_0\right) H\left(j\omega_0\right) = 1\right| \qquad (5.7)$$

and

$$\arg\left[G\left(j\omega_0\right) H\left(j\omega_0\right)\right] = 0° \qquad (5.8)$$

The circuit analysis can be greatly simplified by assuming that

$$\frac{1}{\omega\left(C_2 + C_1\right)} \ll \frac{h_{ib}R_E}{h_{ib} + R_E} \qquad (5.9)$$

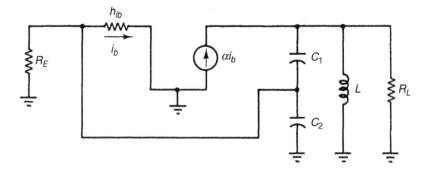

Figure 5.5   Linearized and simplified equivalent circuit of Figure 5.4.

and also that the Q of the load impedance is high. In this case, the circuit reduces to that of Figure 5.6, where

$$V = \frac{V_0 C_1}{C_1 + C_2} \tag{5.10}$$

and

$$R_{eq} = \frac{h_{ib} R_E}{h_{ib} + R_E} \left(\frac{C_1 + C_2}{C_1} \right)^2 \tag{5.11}$$

Figure 5.7 shows how the input impedance of an oscillator circuit—in this case, a BJT Colpitts oscillator minus its resonator—satisfies this condition for oscillation.

Then the forward gain

$$G(j\omega) = \frac{h_{fb}}{h_{ib}} Z_L = \frac{\alpha}{h_{ib}} Z_L \tag{5.12}$$

and

$$H(j\omega) = \frac{C_1}{C_1 + C_2} \tag{5.13}$$

where

$$Y_L = \frac{1}{Z_L} = \frac{1}{j\omega L} + \frac{1}{R_{eq}} + \frac{1}{R_L} + \frac{1}{j\omega C} \tag{5.14}$$

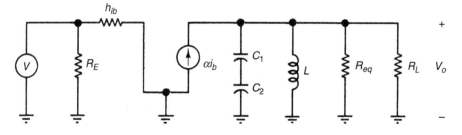

Figure 5.6   Further simplification of Figure 5.4, assuming high-impedance loads.

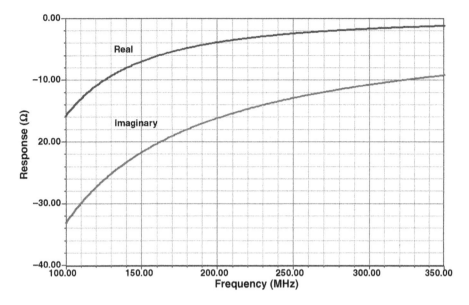

Figure 5.7 Calculated input impedance of a 200-MHz BJT Colpitts oscillator with its resonator removed. (Figure 5.46 shows an equivalent graph with the resonator present.)

A necessary condition for oscillation is that

$$\arg\left[G\left(j\omega\right)H\left(j\omega\right)\right] = 0° \tag{5.15}$$

Since H does not depend on frequency in this example, if $\arg[G(j\omega)H(j\omega)]$ is zero, the phase shift of the load impedance Z_L must also be zero. This occurs only at the resonant frequency of the circuit

$$\omega_0 = \frac{1}{\sqrt{L\left[C_1 C_2/\left(C_1 + C_2\right)\right]}} \tag{5.16}$$

At this frequency

$$Z_L = \frac{R_{eq} R_L}{R_{eq} + R_L} \tag{5.17}$$

and

$$GH = \frac{h_{fb}}{h_{ib}}\left(\frac{R_{eq} R_L}{R_{eq} + R_L}\right)\frac{C_1}{C_1 + C_2} \tag{5.18}$$

The other condition for oscillation is the magnitude constraint that

$$G\left(j\omega\right)H\left(j\omega\right) = \frac{\alpha}{h_{ib}}\left(\frac{R_{eq} R_L}{R_{eq} + R_L}\right)\frac{C_1}{C_1 + C_2} = 1 \tag{5.19}$$

Three-Reactance Oscillators   Although the block-diagram formulation of the stability criteria is the easiest to express mathematically, it is frequently not the easiest to apply since it is often difficult to identify the forward loop gain $G(j\omega)$ and the feedback ratio $H(j\omega)$ in

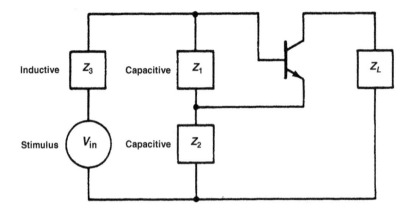

Figure 5.8 Generalized circuit for an oscillator using an amplifier model. Z_3 is inductive even with a capacitor in series with it.

electronic systems. A direct analysis of the circuit equations is frequently simpler than the block diagram interpretation (particularly for single-stage amplifiers). Figure 5.8 shows a generalized circuit for an electronic amplifier.

By inspecting Figure 5.8, one can see the three necessary reactances and the load. All other known circuits are derivatives of this by rotation. The small-signal equivalent circuit is given in Figure 5.9 (where h_{re} has been neglected).

Normally, h_{oe} can also be assumed sufficiently small and can be neglected. The loop equations are then

$$G(j\omega) H(j\omega) = \frac{\alpha}{h_{ib}} \left(\frac{R_{eq} R_L}{R_{eq} + R_L} \right) \frac{C_1}{C_1 + C_2} = 1 \tag{5.20}$$

$$0 = -I_1 Z_1 + I_B (h_{ie} + Z_1) \tag{5.21}$$

For the amplifier to oscillate, the currents I_b and I_1 must be nonzero even when $V_1 = 0$. This is only possible if the system determinant

$$\Delta = \begin{vmatrix} Z_3 + Z_1 + Z_2 & \beta Z_2 - Z_1 \\ -Z & h_{ie} + Z_1 \end{vmatrix} \tag{5.22}$$

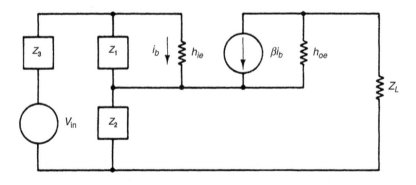

Figure 5.9 Small-signal equivalent circuit of Figure 5.8.

is equal to 0. That is,

$$(Z_3 + Z_1 + Z_2)(h_{ie} + Z_1) - Z_1^2 + \beta Z_1 Z_2 = 0 \tag{5.23}$$

which reduces to

$$(Z_1 + Z_2 + Z_3)h_{fe} + Z_1 Z_2 \beta + Z_1(Z_2 + Z_3) = 0 \tag{5.24}$$

Assume for the moment that Z_1, Z_2, and Z_3 are purely reactive impedances. (It is easily seen that equation 5.24 does not have a solution if all three impedances are real.) Since both the real and imaginary parts must be zero, equation (5.24) is equivalent to the following equations if

$$h_{ie}(Z_1 + Z_2 + Z_3) = 0 \tag{5.25}$$

and

$$Z_1[(1 + \beta)Z_2 + Z_3] = 0 \tag{5.26}$$

Since β is real and positive, Z_2 and Z_3 must be of opposite sign for equation (5.26) to hold. That is,

$$(1 + \beta)Z_2 = -Z_3 \tag{5.27}$$

Therefore, since h_{ie} is nonzero, equation (5.25) reduces to

$$Z_1 + Z_2 - (1 + \beta)Z_2 = 0 \tag{5.28}$$

or

$$Z_1 = \beta Z_2 [C_1 < C_2; L_1 > L_2] \tag{5.29}$$

Thus, since β is positive, Z_1 and Z_2 will be reactances of the same kind. If Z_1 and Z_2 are capacitors, Z_3 is an inductor and the circuit is as shown in Figure 5.10. It is referred to as a Colpitts oscillator, named after the person who first described it.

If Z_1 and Z_2 are inductors and Z_3 is a capacitor as illustrated in Figure 5.11, the circuit is called a Hartley oscillator.

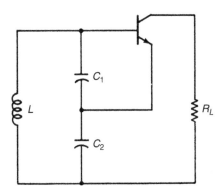

Figure 5.10 Colpitts oscillator.

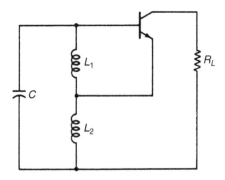

Figure 5.11 Hartley oscillator.

5.3.1.1 Example 1

Design a Colpitts circuit to oscillate at 200 MHz, using a transistor (operating at $I_C = 6\,\text{mA}$ dc) that has an input impedance

$$\text{Mag}(Z_{11}) = \frac{26\,\text{mV}}{I_C} \beta = 282\,\Omega\,(h_{ie} \hat{=} Z_{11})$$

and

$$\beta_{RF} = 65; \qquad \beta_{OSC} = 10 \qquad \text{(large-signal condition)}$$

The oscillating transistor operates under large-signal conditions, generating harmonics in addition to the fundamental. Most of the transistor's current gain will come from dc biasing. β_{OSC}, its large-signal beta at the fundamental frequency of oscillation, must therefore be considerably less than β_{RF}, the small-signal, single-tone value, as the transistor's available current gain, β_0, is distributed among the dc, fundamental and harmonic components present ($\beta_0 = \beta_{DC} + \beta_{OSC} + \beta_{OSC,F2} + \beta_{OSC,F3}\cdots$).

Solution. For the Colpitts circuit, Z_1 and Z_2 are capacitive reactances and Z_3 is an inductive reactance. Let $Z_2 = -j1.6\,\Omega$; then (equation 5.27)

$$Z_3 = -(1 + \beta)\,Z_2 = -(1 + 10) \times -1.6 \approx j17\,\Omega$$

and (equation 5.25)

$$Z_1 = -(Z_2 + Z_3) \approx -j17\,\Omega$$

At 200 MHz, these impedances correspond to component values of $C_1 = 50\,\text{pF}$, $C_2 = 500\,\text{pF}$, and $L = 30\,\text{nH}$. The oscillator will probably work with C_2 values larger than 500 pF, but at the loss of safety margin for tolerances in production, beta, temperature, and phase-noise performance. The completed circuit, except for biasing, is shown in Figure 5.12.

If Z_1, Z_2, Z_3, or h_{ie} is complex, the preceding analysis is more complicated, but the conditions for oscillation can still be obtained from equation (5.24). For example, if, in

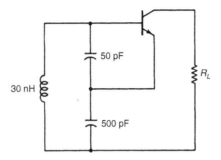

Figure 5.12 Design example of a Colpitts oscillator.

the Colpitts circuit, there is a resistor R in series with $L(Z_3 = R + j\omega L)$, equation (5.24) reduces to two equations

$$h_{ie}\left(\omega L - \frac{1}{\omega C_1} - \frac{1}{\omega C_2}\right) - \frac{R}{\omega C_1} = 0 \tag{5.30}$$

and

$$h_{ie}R - \frac{1 + \beta}{\omega^2 C_1 C_2} + \frac{1}{\omega C_1}\omega L = 0 \tag{5.31}$$

Define

$$C_1' = \frac{C_1}{1 + R/h_{ie}} \tag{5.32}$$

The resonant frequency at which oscillations will occur is found from equation (5.30) to be

$$\omega_0 = \frac{1}{\sqrt{L\left[C_1' C_2 / \left(C_1' + C_2\right)\right]}} \tag{5.33}$$

and for oscillations to occur $\text{Re}(h_{ie}) = \text{Re}(Z_{11})$ must be less than or equal to

$$R_E(h_{ie}) \le \frac{1 + \beta}{\omega_0^2 C_1 C_2} - \frac{L}{C_1} \tag{5.34}$$

If R becomes too large, equation (5.32) cannot be satisfied and oscillations will stop. In general, it is advantageous to have

$$X_{C_1} X_{C_2} = \frac{1}{\omega^2 C_1 C_2} \tag{5.35}$$

with $X \hat{=} 1/\omega C$ as large as possible, since R can be large. However, if C_1 and C_2 are too small (large X_{C_1} and X_{C_2}), the input and output capacitances of the transistor, which shunt C_1 and C_2, respectively, become important. A good, stable design will always have C_1 and C_2 much larger than the transistor capacitances they shunt.

5.3.1.2 Example 2

In Example 1, will the circuit still oscillate if the inductor now has a $Q_u = 100$? If the transistor input capacitance is 5 pF, what effect will this have on the system?

Solution.   In Example 1, $X_L = 17\,\Omega$, $C_1 = 50\,\text{pF}$, and $C_2 = 500\,\text{pF}$. Since $C_1 = 50\,\text{pF}$, adding 5 pF in parallel will change the equivalent C to 55 pF. As the inductor $Q_u = 100$, the equivalent resistance R in series with the lossless inductor is

$$R = \frac{17}{100} = 0.17\,\Omega$$

The new resonant frequency can be determined from equations (5.32) and (5.33)

$$C_1' = \frac{C_1}{1 + R/h_{ie}} = \frac{55}{1 + (0.17/280)} \approx \frac{55}{1.0006} = 54.96\,\text{pF}$$

and

$$f_0 = \frac{1}{2\pi L \left[C_1' C_2 / \left(C_1' + C_2 \right) \right]^{1/2}} \simeq +191.56\,\text{MHz}$$

The effect of the finite inductor Q causes a negligible change in the oscillating frequency compared to the effect of the transistor input capacitance, which reduces the resonant frequency 4.2%. Because the large-signal beta, β_{OSC}, is 10, there is enough loop gain to maintain oscillation.

5.3.1.3 Two-Port Oscillator

Although equations (5.25) and (5.26) can be used to determine the exact expressions for oscillation, they are often difficult to use and add little insight into the design process. An alternative interpretation, although not as accurate, will now be presented. It is based on the fact that an ideal tuned circuit (infinite Q), once excited, will oscillate infinitely because there is no resistance element present to dissipate the energy. In the actual case where the inductor Q is finite, the oscillations die out because energy is dissipated in the resistance. It is the function of the amplifier to maintain oscillations by supplying an amount of energy equal to that dissipated. This source of energy can be interpreted as a negative resistor in series with the tuned circuit. If the total resistance is positive, the oscillations will die out, while the oscillation amplitude will increase if the total resistance is negative. To maintain oscillations, the two resistors must be of equal magnitude. To see how a negative resistance is realized, the input impedance of the circuit in Figure 5.13 will be derived.

If h_{oe} is sufficiently small ($h_{oe} \ll 1/R_L$), the equivalent circuit is as shown in Figure 5.13. The steady-state loop equations are

$$V_{\text{in}} = I_{\text{in}} \left(X_{C_1} + X_{C_2} \right) - I_b \left(X_{C_1} - \beta X_{C_2} \right) \tag{5.36}$$

$$0 = -I_{\text{in}} \left(X_{C_1} \right) + I_b (X_{C_1} + h_{ie}) \tag{5.37}$$

After I_b is eliminated from these two equations, Z_{in} is obtained as

$$Z_{\text{in}} = \frac{V_{\text{in}}}{I_{\text{in}}} = \frac{(1+\beta)\,X_{C_1} X_{C_2} + h_{ie} \left(X_{C_1} + X_{C_2} \right)}{X_{C_1} + h_{ie}} \tag{5.38}$$

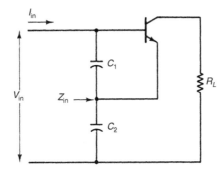

Figure 5.13   Calculation of input impedance of the negative-resistance oscillator.

If $X_{C_1} \ll h_{ie}$, the input impedance is approximately equal to

$$Z_{\text{in}} \approx \frac{1+\beta}{h_{ie}} X_{C_1} X_{C_2} + \left(X_{C_1} + X_{C_2}\right) \tag{5.39}$$

$$Z_{\text{in}} \approx \frac{-g_m}{\omega^2 C_1 C_2} + \frac{1}{j\omega \left[C_1 C_2 / (C_1 + C_2)\right]} \tag{5.40}$$

That is, the input impedance of the circuit shown in Figure 5.14 is a negative resistor,

$$R = \frac{-g_m}{\omega^2 C_1 C_2} \tag{5.41}$$

in series with a capacitor,

$$C_{\text{in}} = \frac{C_1 C_2}{C_1 + C_2} \tag{5.42}$$

which is the series combination of the two capacitors.

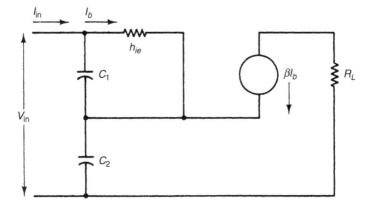

Figure 5.14   Equivalent small-signal circuit of Figure 5.13.

With an inductor L (with the series resistance R_S) connected across the input, it is clear that the condition for sustained oscillation is

$$R_S = \frac{g_m}{\omega^2 C_1 C_2} \tag{5.43}$$

and the frequency of oscillation

$$f_0 = \frac{1}{2\pi\sqrt{L\left[C_1 C_2/(C_1 + C_2)\right]}} \tag{5.44}$$

This interpretation of the oscillator readily provides several guidelines that can be used in the design. First, C_1 should be as large as possible so that

$$X_{C_1} \ll h_{ie} \tag{5.45}$$

and C_2 is to be large so that

$$X_{C_2} \ll \frac{1}{h_{oe}} \tag{5.46}$$

When these two capacitors are large, the transistor base-to-emitter and collector-to-emitter capacitances will have a negligible effect on the circuit's performance. However, equation (5.43) limits the maximum value of the capacitances since

$$r \le \frac{g_m}{\omega^2 C_1 C_2} \le \frac{G}{\omega^2 C_1 C_2} \tag{5.47}$$

where G is the maximum value of g_m. For a given product of C_1 and C_2, the series capacitance is at maximum when $C_1 = C_2 = C_m$. Thus, equation (5.47) can be written as

$$\frac{1}{\omega C_m} > \sqrt{\frac{r}{G}} \tag{5.48}$$

This equation is important because it shows that for oscillations to be maintained, the minimum permissible reactance $1/\omega C_m$ is a function of the resistance of the inductor and the transistor's mutual conductance, g_m.

An oscillator circuit known as the Clapp circuit or Clapp–Gouriet circuit is shown in Figure 5.15. This oscillator is equivalent to the one just discussed, but it has the practical advantage of being able to provide another degree of design freedom by making C_0 much smaller than C_1 and C_2.

It is possible to use C_1 and C_2 to satisfy the condition of equation (5.47) and then adjust C_0 for the desired frequency of oscillation ω_0, which is determined from

$$\omega_0 L - \frac{1}{\omega_0 C_0} - \frac{1}{\omega_0 C_1} - \frac{1}{\omega_0 C_2} = 0 \tag{5.49}$$

5.3.1.4 *Amplitude Stability*

Linearized analysis of the oscillator is convenient for determining the frequency but not the amplitude of the oscillation. The Nyquist stability criterion defines the frequency of oscillation as the frequency at which the loop phase shift is 360°, but it says nothing about the

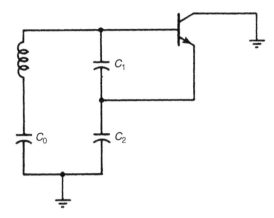

Figure 5.15   Circuit of a Clapp oscillator.

oscillation amplitude. If no provisions are taken to control the amplitude, it is susceptible to appreciable drift. Two frequently used methods for controlling the amplitude are operating the transistor in the nonlinear region or to use a second stage for amplitude limiting. For the single-stage oscillator, amplitude limiting is accomplished by designing an unstable oscillator, that is, the loop gain is made greater than 1 at the frequency at which the phase shift is 180°. As the amplitude increases, the β of the transistor decreases, causing the loop gain to decrease until the amplitude stabilizes. This is a self-limiting oscillator. There are nonlinear analysis techniques predicting the amplitude of oscillation, but their results are approximate except in idealized cases, forcing the designer to resort to an empirical approach.

An example of a two-stage emitter-coupled oscillator is shown in Figure 5.16. In this circuit, amplitude stabilization occurs as a result of current limiting in the second stage. This circuit has the additional advantage that it has output terminals that are isolated from the feedback path. The emitter signal of Q_2, having a rich harmonic content, is normally used as output. Harmonics of the fundamental frequency can be extracted at the emitter of Q_2 by using an appropriately tuned circuit. Note that the collector of Q_2 is isolated from the feedback path.

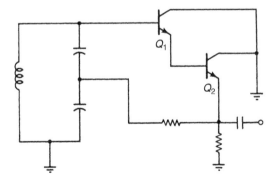

Figure 5.16   Two-stage, emitter-coupled oscillator.

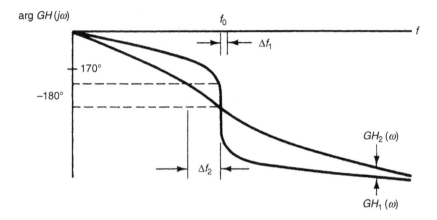

Figure 5.17   Phase plot of two open-loop systems with different resonator Qs.

5.3.1.5   Phase Stability

An oscillator has a frequency or phase stability that can be considered in two separate parts. First, there is long-term stability, in which the frequency changes over a period of minutes, hours, days, weeks, or even years. This frequency stability is normally limited by the circuit components' temperature coefficients and aging rates. The other part, short-term frequency stability, is measured in terms of seconds. One form of short-term instability is due to changes in phase of the system; here the term phase stability is used synonymously with frequency stability. It refers to how the frequency of oscillation reacts to small changes in phase shift of the open-loop system. It can be assumed that the system with the largest rate of change of phase versus frequency $d\phi/df$ will be the most stable in terms of frequency stability. Figure 5.17 shows the phase plots of two open-loop system used in oscillators. At the system crossover frequency, the phase shift is $-180°$. If some external influence causes a shift in phase—say, it adds $10°$ of phase lag—the frequency will change until the total phase shift is again $0°$. In this case, the frequency will decrease to the point where the open-loop phase shift is $170°$. Figure 5.17 shows that Δf_2, the change in frequency associated with the $10°$ change in phase of GH_2, is greater than the change in frequency Δf_1, associated with the open-loop system GH_1, whose phase is changing more rapidly near the open-loop crossover frequency.

This qualitative discussion illustrates that $d\phi/df$ at $f = f_0$ is a measure of an oscillator's phase stability. It provides a good means of quantitatively comparing the phase stability of two oscillators. Consider the simple parallel tuned circuit shown in Figure 5.18.

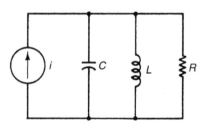

Figure 5.18   Parallel tuned circuit for phase-shift analysis.

For this circuit, the two-port is

$$\frac{V_0\,(j\omega)}{I\,(j\omega)} = \frac{R}{1 + jQ\left[(\omega/\omega_0) - (\omega_0/\omega)\right]} \tag{5.50}$$

where

$$\omega_0 = \frac{1}{\sqrt{LC}} \qquad \text{and} \qquad Q = \frac{R}{\omega_0 L} \tag{5.51}$$

The circuit phase shift is

$$\arg\frac{V_0}{I} = \theta = \tan^{-1}Q\left(\frac{\omega}{\omega_0} - \frac{\omega_0}{\omega}\right) \tag{5.52}$$

and

$$\frac{d\theta}{d\omega} = \frac{1/Q}{1/Q^2 + \left[\left(\omega^2 - \omega_0^2\right)/\omega_0\omega\right]^2}\frac{\omega^2 + \omega_0^2}{(\omega_0\omega)^2} \tag{5.53}$$

at the resonant frequency ω_0

$$\left.\frac{dQ}{d\omega}\right|_{\omega=\omega_0} = \frac{2Q}{\omega_0} \tag{5.54}$$

The frequency stability factor is S_F defined as the change in phase $d\phi/d\omega$ divided by the normalized change in frequency $\Delta\omega/\omega_0$. That is

$$S_F = 2Q \tag{5.55}$$

where S_F is a measure of the short-term stability of an oscillator. Equation (5.54) indicates that the higher the circuit Q, the higher the stability factor. This is one reason for using high-Q circuits in oscillator circuits. Another reason is the ability of the tuned circuit to filter out undesired harmonics and noise.

The two oscillator types shown so far can be considered two-port oscillators because one port is responsible for oscillation and the other port supplies the output power. This type of output supposedly has better isolation (although this is not true at higher frequencies) but has a higher noise floor. A better way of extracting energy is shown in Figure 5.19.

These types of oscillators are essentially what is referred to as two-port oscillators. The name comes from the fact that the oscillating device has two ports, with the output energy taken from the output port, typically the collector or drain, while the tuned circuit is at the input port. For years, it was widely held that this type of design provides better isolation between the tuned circuit and termination; unfortunately, because of the Miller effect, this is not true. At the higher frequencies, changes in output-port loading affect the input port, resulting in pulling, a detuning effect that increases with frequency. In addition, taking the output from the collector or the drain results in a poorer signal-to-noise ratio than that achievable by extracting energy from the tuned circuit itself. The only real advantages the collector output provides are more power and higher efficiency, but definitely at the expense of phase noise, which is often referred to as SSB phase noise because it is expressed in terms

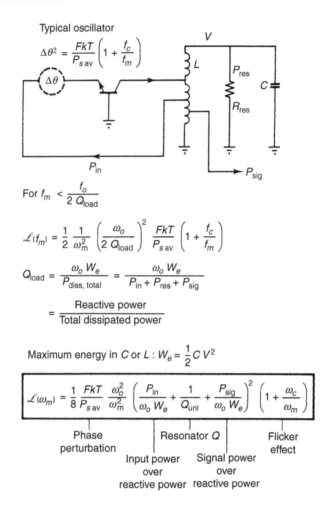

Typical oscillator

$$\Delta\theta^2 = \frac{FkT}{P_{s\,av}}\left(1 + \frac{f_c}{f_m}\right)$$

For $f_m < \dfrac{f_o}{2\,Q_{load}}$

$$\mathcal{L}_{(f_m)} = \frac{1}{2}\,\frac{1}{\omega_m^2}\left(\frac{\omega_o}{2\,Q_{load}}\right)^2\frac{FkT}{P_{s\,av}}\left(1 + \frac{f_c}{f_m}\right)$$

$$Q_{load} = \frac{\omega_o\,W_e}{P_{diss,\,total}} = \frac{\omega_o\,W_e}{P_{in} + P_{res} + P_{sig}}$$

$$= \frac{\text{Reactive power}}{\text{Total dissipated power}}$$

Maximum energy in C or L: $W_e = \dfrac{1}{2}C\,V^2$

$$\mathcal{L}_{(\omega_m)} = \frac{1}{8}\,\frac{FkT}{P_{s\,av}}\,\frac{\omega_o^2}{\omega_m^2}\left(\frac{P_{in}}{\omega_o\,W_e} + \frac{1}{Q_{unl}} + \frac{P_{sig}}{\omega_o\,W_e}\right)^2\left(1 + \frac{\omega_c}{\omega_m}\right)$$

Phase perturbation   Resonator Q   Flicker effect

Input power over reactive power   Signal power over reactive power

Figure 5.19 Diagram for a feedback oscillator illustrating the principles involved, and showing the key components considered in the phase-noise calculation and its contribution.

of the decibel ratio of the noise power in a single (the upper or lower) noise sideband, in a 1-Hz bandwidth centered at a specified frequency offset from the oscillator carrier, to the carrier power. Figure 5.20 compares the phase noise of an oscillator with the energy taken off the collector and the tuned circuit; the difference is highly visible.

5.4 OSCILLATOR CIRCUITS

5.4.1 Hartley

Figure 5.21 shows the Hartley oscillator configuration, in which feedback is obtained by tapping the resonator. In practice, the resonator can consist of a single tapped inductor or two separate, magnetically uncoupled inductors, with feedback obtained at their junction.

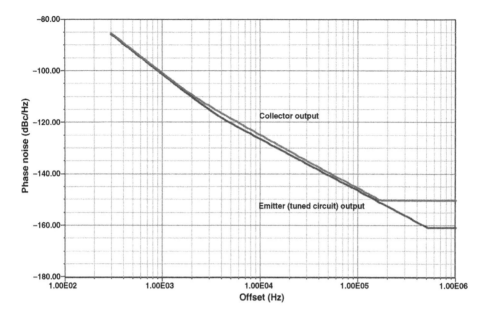

Figure 5.20   Phase noise with oscillator output taken from collector (upper trace) versus emitter (lower trace) of a BJT oscillator.

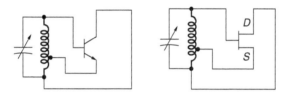

Figure 5.21   Hartley oscillator.

5.4.2 Colpitts

Figure 5.22 shows the Colpitts oscillator configuration, in which feedback is obtained via a capacitive voltage divider.

5.4.3 Clapp–Gouriet

Figure 5.23 shows the Clapp–Gouriet oscillator. Like the Colpitts, the Clapp–Gouriet obtains its feedback via a capacitive voltage divider; unlike the Colpitts, an additional capacitor

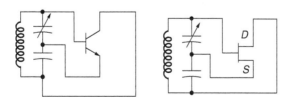

Figure 5.22   Colpitts oscillator.

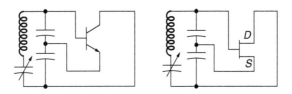

Figure 5.23   Clapp–Gouriet oscillator.

series-tunes the resonator. The Pierce oscillator, a configuration used only with crystals, is a rotation of the Clapp–Gouriet oscillator in which the emitter is at RF ground.

5.5 DESIGN OF RF OSCILLATORS

5.5.1 General Thoughts on Transistor Oscillators

An estimate of the noise performance of an oscillator is

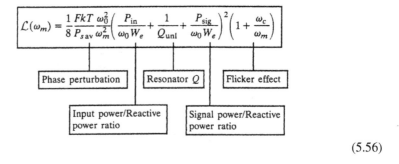

(5.56)

We just state this here without having dived into oscillator noise theory itself. (Also see Figure 5.19.) This equation is based on work done by Dieter Scherer of Hewlett-Packard about 1978. He was the first to introduce the flicker effect to the Leeson equation by adding the AM-to-PM conversion effect, which is caused by the nonlinear capacitance of the active devices. This equation must be further expanded as

$$\mathcal{L}(f_m) = 10\log\left\{\left[1 + \frac{f_0^2}{(2 f_m Q_{\text{load}})^2}\right]\left(1 + \frac{f_C}{f_m}\right)\frac{FkT}{2P_{\text{sav}}} + \frac{2kTRK_0^2}{f_m^2}\right\}$$ (5.57)

where $\mathcal{L}(f_m)$ = ratio of sideband power in 1-Hz bandwidth at f_m to total power in dB, f_m = frequency offset, f_0 = center frequency, f_c = flicker frequency, Q_{load} = loaded Q of the tuned circuit, F = noise factor, $kT = 4.1 \times 10^{-21}$ at 300 K (room temperature), P_{sav} = average power at oscillator output, R = equivalent noise resistance of tuning diode (typically 200 Ω to 10 kΩ), and K = oscillator voltage gain.

The following table shows the flicker corner frequency f_C as a function if I_C for a typical small-signal microwave BJT. $I_{C(\text{max})}$ of this transistor is about 10 mA.

I_C (mA)	f_C (kHz)
0.25	1
0.5	2.74
1	4.3
2	6.27
5	9.3

Source: Motorola

Note that f_C, which is defined by AF and KF in the SPICE model, increases with I_C. This gives us a clue about how f_C changes when a transistor oscillates. As a result of the bias-point shift that occurs during oscillation (see Section 5.3.1.1), an oscillating BJT's average I_C is higher than its small-signal I_C. KF is therefore higher for a given BJT operating as an oscillator than for the same transistor operating as a small-signal amplifier. This must be kept in mind when considering published f_C data, which is usually determined under small-signal conditions without being qualified as such. Determining a transistor's oscillating f_C is best done through measurement: operate the device as a high-Q UHF oscillator (we suggest using a ceramic-resonator-based tank in the vicinity of 1 GHz), and measure its close-in (10 Hz to 10 kHz) phase noise versus offset from the carrier. f_C will correspond to a slight decrease in the slope of the phase noise versus offset curve. Generally, f_C varies with device type as follows: silicon JFETs, 50 Hz and higher; microwave RF BJTs, 1–10 kHz (as above); MOSFETs, 10–100 kHz; GaAsFETs, 10–100 MHz. Figure 5.24 shows the phase noise of oscillators using different semiconductors and resonators.

Equation (5.57) is based on Rohde et al. [2]. The additional term introduces a distinction between a conventional oscillator and a VCO. Whether the voltage- or current-dependent capacitance is internal or external makes no difference; it simply affects the frequency.

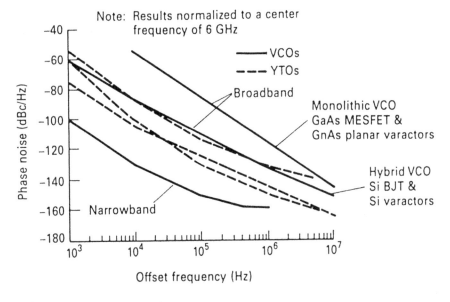

Figure 5.24   Phase noise of oscillators using different semiconductors and resonators.

For a more complete expression for a resonator oscillator's phase noise spectrum, we can write

$$
\begin{aligned}
s_\varphi (f_m) = {} & \frac{\alpha_R F_0^4 + \alpha_E \left(\frac{F_0}{2Q_L}\right)^2}{f_m^3} \\[2mm]
& + \frac{\left(\frac{2G\,FkT}{P_0}\right)\left(\frac{F_0}{2Q_L}\right)^2}{f_m^2} \\[2mm]
& + \frac{2\alpha_R Q_L F_0^3}{f_m^2} + \frac{\alpha_E}{f_m} + \frac{2G\,FkT}{P_0}
\end{aligned}
\tag{5.58}
$$

where G = compressed power gain of the loop amplifier, F = noise factor of the loop amplifier, k = Boltzmann's constant, T = temperature in Kelvins, P_0 = carrier power level (in watts) at the output of the loop amplifier, F_0 = carrier frequency in hertz, f_m = carrier offset frequency in hertz, $Q_L (= \pi F_0 \tau_g)$ = loaded Q of the resonator in the feedback loop, and α_R and α_E = flicker-noise constants for the resonator and loop amplifier, respectively.

In frequency synthesizers, we have no use for LC oscillators that do not include a tuning diode or diodes, but it may still be of interest to analyze the low-noise, fixed-tuned LC oscillator first and later make both elements, inductor and capacitor, variable.

Later, we will show the performance changes if we utilize the two possible ways of getting coarse and fine tuning in oscillators:

1. use of tuning diodes, and
2. use of switching diodes.

We will spend some time looking at the effects that switching and tuning diodes have in a circuit because they will ultimately influence the noise performance more strongly than the transistor itself.

The reason is that the noise generated in tuning diodes will be superimposed on the noise generated in the circuit, while switching diodes have losses that cause a reduction of circuit Q. The selection of the proper tuning and switching diodes is important, as is the proper way of connecting them. As both types are modifications of the basic LC oscillator, we start with the LC oscillator itself.

Early signal generators as were offered by several companies (namely, Rohde & Schwarz, Hewlett-Packard, Boonton Electronics, or Marconi), if they are not synthesized, use an air-dielectric variable capacitor or, as in the case of one particular Hewlett-Packard generator, the Model HP8640, a tuned cavity.

Tuning here is accomplished by changing the value of an air-dielectric variable capacitor or changing the mechanical length of a quarter-wave resonator.

Using the equations shown previously, it is fairly easy to calculate the performance of oscillators and understand how they work, but this does not necessarily optimize their design. For LC oscillator applications in which noise performance is crucial, the oscillator shown in Figure 5.25, used in the famous Rohde & Schwarz SMDU, is the state of the art. Its noise performance is equivalent to the noise found in the cavity-tuned oscillator used in Hewlett-Packard's popular HP8640 signal generator, and because of the unique way a tuning diode is coupled to the circuit, its modulation capabilities are substantially superior to any of the signal generators currently offered. To develop such a circuit from design equations is not

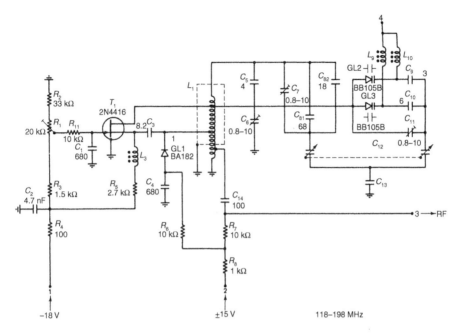

Figure 5.25 A 118- to 198-MHz oscillator from the Rohde & Schwarz SMDU signal generator.

possible. This circuit is the result of many years of experience and research and looks fairly simple. The grounded-gate FET circuit provides the best performance because it fulfills the important requirements of (5.56).

The tuned circuit is not connected directly to the drain, but the drain is put on a tap of the oscillator section. Therefore, the actual voltage across the tuning capacitor is higher than the supply voltage, and thus the energy stored in the capacitor is much higher than in a circuit connected between the gate electrode and ground, as it is in a normal Colpitts oscillator. In addition, the high output impedance of the FET does not load the circuit, which also provides a reduced noise contribution. Since this oscillator is optimized for best frequency-modulation performance in the FM frequency range, it becomes apparent that it fits the requirements of low-distortion stereo modulation.

For extremely critical blocking measurements in the 2-m region (140–160 MHz), the noise specification at 20 kHz off the carrier is of high importance, while the peak modulation typically does not exceed 5 kHz. Figure 5.25 shows the schematic of the oscillator section optimized for this frequency range.

Because of f_T limitations, it has been found experimentally that these LC oscillators, if based on silicon JFETs, should not be used above about 500 MHz; rather, a doubler stage should be employed. Analyzing the signal generators currently on the market, we find that their highest baseband typically ranges from 200 to 500 MHz, with frequency doubling used to 1000 MHz. As can be seen in Figure 5.26, the mechanical layout of such an oscillator is extremely compact.

5.5.2 Two-Port Microwave/RF Oscillator Design

A common method for designing oscillators is to resonate the input port with a passive high-Q circuit at the desired frequency of resonance. It will be shown that if this is achieved

Figure 5.26 Photograph of the helical-resonator system from the Rohde & Schwarz SMDU signal generator.

with a load connected on the output port, the transistor is oscillating at both ports and is thus delivering power to the load port. The oscillator may be considered a two-port structure, with M_3 being the lossless resonating port and M_4 provides lossless matching such that all the external RF power is delivered to the load. See Figure 5.27. The resonating network has been described. Nominally, only parasitic resistance is present at the resonating port, since a high-Q resonance is desirable for minimizing oscillator noise. It is possible to have loads at both the input and the output ports if such an application occurs, since the oscillator is oscillating at both ports simultaneously.

The simultaneously oscillation conditions is proved as follows. Assume that the oscillation condition is satisfied at port 1:

$$1/S'_{11} = \Gamma_G \tag{5.59}$$

Thus,

$$S'_{11} = S_{11} + \frac{S_{12}S_{21}\Gamma_L}{1 - S_{22}\Gamma_L} = \frac{S_{11} - D\Gamma_L}{1 - S_{22}\Gamma_L} \tag{5.60}$$

$$\frac{1}{S'_{11}} = \frac{1 - S_{22}\Gamma_L}{S_{11} - D\Gamma_L} = \Gamma_G \tag{5.61}$$

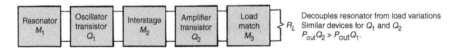

Figure 5.27 Buffered oscillator design.

By expanding equation (5.61), we find

$$\Gamma_G S_{11} - D\Gamma_L \Gamma_G = 1 - S_{22}\Gamma_L$$
$$\Gamma_L(S_{22} - D\Gamma_G) = 1 - S_{11}\Gamma_G$$
$$\Gamma_L = \frac{1 - S_{11}\Gamma_G}{S_{22} - D\Gamma_G} \tag{5.62}$$

Thus,

$$S'_{22} = S_{22} + \frac{S_{12}S_{21}\Gamma_G}{1 - S_{11}\Gamma_G} \tag{5.63}$$

$$\frac{1}{S'_{22}} = \frac{1 - S_{11}\Gamma_G}{S_{22} - D\Gamma_G} \tag{5.64}$$

Comparing equations (5.62) and (5.64), we find

$$1/S'_{22} = \Gamma_L \tag{5.65}$$

which means that the oscillation condition is also satisfied at port 2; this completes the proof. Thus, if either port is oscillating, the other port must be oscillating as well. A load may appear at either or both ports, but normally the load is in Γ_L, the output termination. This result can be generalized to an n-port oscillator by showing that the oscillator is simultaneously oscillating at each port:

$$\Gamma_1 S'_{11} = \Gamma_2 S'_{22} = \Gamma_3 S'_{33} = \cdots = \Gamma_n S'_{nn} \tag{5.66}$$

Before concluding this section on two-port oscillator design, the buffered oscillator shown in Figure 5.27 must be considered. This design approach is used to provide the following.

1. A reduction in loading–pulling, which is the change in oscillator frequency when the load reflection coefficient changes.
2. A load impedance that is more suitable to wideband applications ($k < 1$).
3. A higher output power from a working design, although the higher output power can also be achieved by using a larger oscillator transistor.
4. This type of oscillator is not optimized for low phase noise. However, buffered oscillator designs are quite common in wideband YIG applications, where changes in the load impedance must not change the generator frequency.

Two-port oscillator design may be summarized as follows.

1. Select a transistor with sufficient gain and output-power capability for the frequency of operation. This may be based on oscillator data sheets, amplifier performance, or S-parameter calculation.
2. Select a topology that gives $k < 1$ at the operating frequency. Add feedback if $k < 1$ has not been achieved.
3. Select an output load-matching circuit that gives $|S'_{11}| > 1$ over the desired frequency range. In the simplest case, this could be a 50-Ω load.

Table 5.1 *S* **parameters and stability factors** ($V_{CE} = 15\,\text{V}$, $I_C = 25\,\text{mA}$)

$L_B = 0$	$L_B = 0.5\,\text{nH}$
$S_{11} = 0.94\ \angle 174°$	$1.04\ \angle 173°$
$S_{21} = 1.90\ \angle -28°$	$2.00\ \angle -30°$
$S_{12} = 0.013\ \angle 98°$	$0.043\ \angle 153°$
$S_{22} = 1.01\ \angle -17°$	$1.05\ \angle -18°$
$k = -0.09$	-0.83

4. Resonate the input port with a lossless termination so that $\Gamma_G S_{11}' = 1$. The value of S_{22}' will be greater than unity with the input properly terminated.

In all cases, the transistor delivers power to a load and the input of the transistor. Practical considerations of realizability and dc biasing will determine the best design.

For both bipolar and FET oscillators, a common topology is the common-base or common-gate, since a common-lead inductance can be used to raise S_{22} to a large value, usually greater than unity even with a 50-Ω generator resistor. However, it is not necessary for the transistor S_{22} to be greater than unity, since the 50-Ω generator is not present in the oscillator design. The requirement for oscillation is $k < 1$; then resonating the input with a lossless termination will ensure that $|S_{11}'| > 1$.

A simple example will clarify the design procedure. A common-base bipolar transistor (HP2001) was selected to design a fixed-tuned oscillator at 2 GHz. The common-base *S*-parameters and stability factor are given in Table 5.1.

Using the load circuit in Figure 5.28, we see that the reflection coefficients are

$$\Gamma_L = 0.62\ \angle 30°$$

$$S_{11}' = 1.18\ \angle 173°$$

Thus, a resonating capacitance of $C = 20\,\text{pF}$ resonates the input port. In a YIG-tuned oscillator, this reactive element could be provided by the high-Q YIG element. For a dielectric-resonator oscillator (DRO), the puck would be placed to give $\Gamma_G \approx 1.0\ \angle -173°$.

Another two-port design procedure is to resonate the Γ_G port and calculate S_{22}' until $|S_{22}'| > 1$, then design the load port to satisfy. This design procedure is summarized in Figure 5.29.

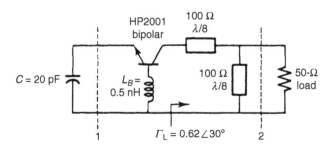

Figure 5.28   Oscillator example at 2 GHz.

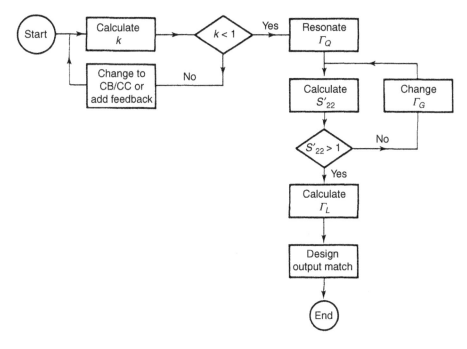

Figure 5.29 Oscillator design flow chart.

An example using this procedure at 4 GHz is given in Figure 5.30 using an AT41400 silicon bipolar chip in the common-base configuration with a convenient value of base and emitter inductance of 0.5 nH. The feedback parameter is the base inductance, which can be varied if needed.

Since a lossless capacitor at 4 GHz of 2.06 pF gives $\Gamma_G = 1 - 0\angle -137.7°$, this input termination is used to calculate S'_{22}, giving $S'_{22} = 0.637\angle 44.5°$. This circuit will oscillate into any passive load. Varying the emitter capacitor about 20° on the Smith chart to 1.28 pF gives $S'_{22} = 1.16\angle -5.5°$, which will oscillate into a load of $\Gamma_L = 0.861\angle 5.5°$. The completed lumped element design is given in Figure 5.31.

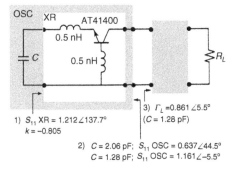

1) S_{11}, XR = 1.212∠137.7°
   $k = -0.805$

2) C = 2.06 pF; S_{11} OSC = 0.637∠44.5°
   C = 1.28 pF; S_{11} OSC = 1.161∠–5.5°

3) Γ_L =0.861 ∠5.5°
   (C = 1.28 pF)

Figure 5.30 A 4-GHz lumped-resonator oscillator using the AT41400.

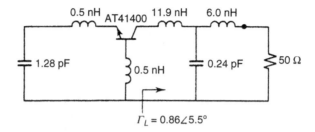

Figure 5.31   Completed lumped-resonator oscillator (LRO).

5.5.3 Ceramic-Resonator Oscillators

An important application for a new class of resonators called ceramic resonators (CRs) has emerged for wireless applications. The CRs are similar to rigid coaxial cable, where the center conductor is connected at the end to the outside of the cable. These resonators are generally operating in quarter-wavelength mode and their characteristic impedance is approximately $10\,\Omega$. Because their coaxial assemblies are made from a high-ε low-loss material with good silver plating throughout, the electromagnetic field is internally contained and therefore results in very little radiation. These resonators are therefore ideally suited for high-Q, high-density oscillators. The typical application for this resonator is VCOs ranging from not much more than 200 MHz up to about 3 or 4 GHz. At these high frequencies, the mechanical dimensions of the resonator become too tiny to offer any advantage. One of the principal requirements is that the physical length is considerably larger than the diameter. If the frequency increases, this requirement can no longer be met.

Manufacturers supply ceramic resonators on a prefabricated basis. Figures 5.32 and 5.33 show the standard round/square packaging available and the typical dimensions for a ceramic resonator. Figure 5.34 shows a ceramic-resonator oscillator built on a ceramic substrate.

Figure 5.32   Ceramic resonators.

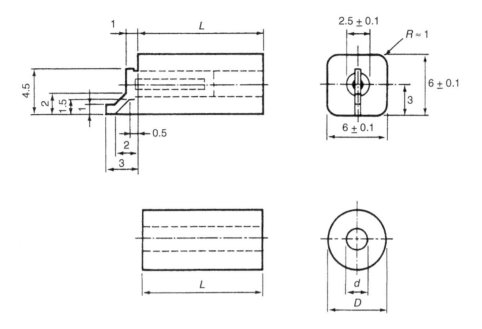

Figure 5.33 Standard round/square ceramic resonator packaging and dimensions.

The available material has a dielectric constant of 88 and is recommended for use in the 400- to 1500-MHz range. The next higher frequency range (800 MHz to 2.5 GHz) uses an ε of 38, while the top range (1–4.5 GHz) uses an ε of 21. Given the fact that ceramic resonators are prefabricated and have standard outside dimensions, the following quick calculation applies.

Figure 5.34 Modern ceramic resonator oscillator for hand-held cellular phones.

Relative dielectric constant of resonator material	$\epsilon_r = 21$	$\epsilon_r = 38$	$\epsilon_r = 88$
Resonator length (mm)	$l = \frac{16.6}{f}$	$l = \frac{12.6}{f}$	$l = \frac{8.2}{f}$
Temperature coefficient (ppm/°C)	10	6.5	8.5
Available temperature coefficients	-3 to $+12$	-3 to $+12$	-3 to $+12$
Typical resonator Q	800	500	400
Frequency range	1–4.5 GHz	800–2500 MHz	400–1500 MHz

5.5.3.1 Calculation of Equivalent Circuit

The equivalent parallel-resonant circuit afforded by a ceramic resonator has a resistance at resonance of

$$R_p = \frac{2\,(Z_0)^2}{R^*l} \tag{5.67}$$

where $Z_0 = $ characteristic impedance of the resonator, $l = $ mechanical length of the resonator, and $R^* = $ equivalent resistor due to metallization and other losses.

As an example, one can calculate

$$C^* \frac{2\pi\varepsilon_0\varepsilon_R}{\log_e(D/d)} = 55.61 \times 10^{-12} \frac{\varepsilon_R}{\log_e(D/d)} \tag{5.68}$$

and

$$L^* \frac{\mu_R\mu_0}{2\pi} = \log_e\left(\frac{D}{d}\right) = 2 \times 10^{-7} \log_e\left(\frac{D}{d}\right) \tag{5.69}$$

$$Z_0 = 60\,\Omega \frac{1}{\sqrt{\varepsilon_R}} \log_e\left(\frac{D}{d}\right)$$

$$= 60\frac{1}{\sqrt{88}} \log_e\left(\frac{6}{2.5}\right)$$

$$= 5.6\,\Omega \tag{5.70}$$

A practical example for $\varepsilon_r = 88$ and 450 MHz is

$$C_p = \frac{C^*l}{2} = 49.7 \text{ pF} \tag{5.71}$$

$$L_p = 8L^*l = 2.52 \text{ nH} \tag{5.72}$$

$$R_p = 2.5 \text{ k}\Omega \tag{5.73}$$

Figure 5.35 shows the schematic of a ceramic-resonator-based oscillator. Figures 5.36 and 5.37 show the simulated and measured phase-noise of the oscillator.

By using ceramic-resonator-based oscillators in conjunction with miniature synthesizer chips, it is possible to build extremely small phase-locked loop systems for cellular telephone operation. Figure 5.38 shows one of the smallest currently available PLL-based synthesizers manufactured by Synergy Microwave Corporation. Because of the high-Q resonator, these types of oscillators exhibit extremely low phase noise. Values of better than -150 dBc/Hz at 1 MHz off the carrier are achievable. The ceramic resonator reduces the oscillator's sensitivity to microphonic effects and proximity effects caused by other components.

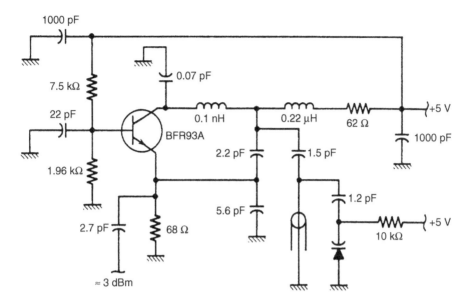

Figure 5.35 Schematic of a high-performance ceramic-resonator-based oscillator that can be used from 500 MHz to 2 GHz.

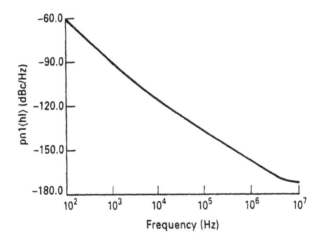

Figure 5.36 Simulated phase noise of an NPN bipolar 1 GHz ceramic-resonator-based oscillator.

5.5.4 Using a Microstrip Inductor as the Oscillator Resonator [3–5][1]

A high-Q microstrip resonator can be used to improve an oscillator's phase-noise performance. The following section, based on information in the Philips Semiconductors

---

[1]Based on portions of the Philips Semiconductors/Signetics RF Communications Products Application Note AN1777, "Low-Voltage Front End Circuits: SA601, SA620," August 20, 1997. Available at http://ics.nxp.com/support/documents/interface/pdf/an1777.pdf. Used with permission.

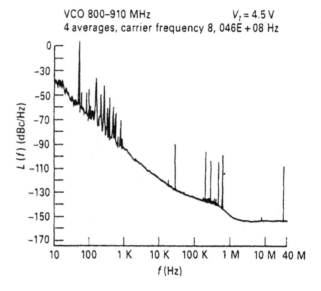

Figure 5.37   Measured phase noise of a ceramic-resonator-based oscillator.

application note AN1777, describes the application of such a resonator in a BJT differential oscillator.

It can be seen from (5.56) that the phase noise is proportional to the term

$$\frac{1}{4Q_{L^2}} \tag{5.74}$$

Figure 5.38   Miniature PLL-based synthesizer manufactured by Synergy Microwave Corporation.

This means that for any 10% increase in Q, we get something like 20% improvement in phase noise, which makes it most desirable to increase the Q. One way to reduce the loading of the resonator is to tap the device down on the resonator. The underlying oscillator circuit is based on the differential type of oscillator found in ICs.

5.5.4.1 Increasing Loaded Q

Figure 5.39 shows the oscillator section of the NXP SA620 low-voltage front-end IC with a conventional second-order parallel-tuned tank circuit configured as the external resonator.

From the basic equations for parallel resonance, the resonator's loaded Q is given by

$$Q_L = R_p \left[\frac{C_T}{L_T}\right]^{1/2} \tag{5.75}$$

where

$$R_p = R_t || R_C = \left[\left(\frac{1}{R_T}\right) + \left(\frac{1}{R_C}\right)\right]^{-1} \tag{5.76}$$

represents the net shunt tank resistance appearing across the network at resonance. R_T represents losses in the inductance L_T and capacitance C_T (almost always dominated by inductor losses), and R_C is the active circuit's load impedance at resonance. Improving the quality of the tank components L_T and C_T (i.e., improving their Q) will increase R_T, but since R_C is low in the case of the SA620, the resulting increase in R_P is relatively small. We can increase R_P by decoupling Z_C from the tank circuit (note that R_C is the real part of Z_C at resonance). Decoupling this impedance by using either tapped-L or tapped-C tank configurations is possible. Inspecting of the circuit shows that dc biasing is necessary

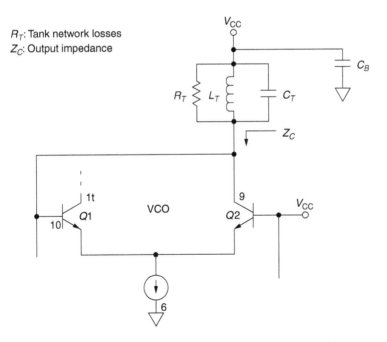

Figure 5.39   Oscillator portion of NXP SA620 front-end IC with external resonator.

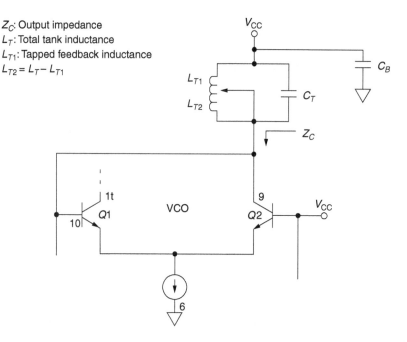

Figure 5.40 Oscillator portion of SA620 IC with tapped-microstrip resonator.

for pins 9 and 10 (osc1 and osc2, respectively), so a tapped-C approach would require shunt-feeding these pins to V_{CC}. Thus, the most practical way is to employ a tapped-L network.

Figure 5.40 shows the basic tapped-L circuit. The effective multiplied shunt impedance at resonance across C_T is given by

$$\hat{R}_p = R_p \left[\left(\frac{L_{T2}}{L_{T1}} \right) + 1 \right]^{1/2} \tag{5.77}$$

with a corresponding increase in loaded Q

$$\hat{Q}_L = Q_L \left[\left(\frac{L_{T2}}{L_{T1}} \right) + 1 \right]^{1/2} \tag{5.78}$$

Feedback is caused by an RF voltage appearing across L_T, coupled through bypass capacitor C_B and its ground return to the cold end of the current source at pin 6. The significance of this path increases with frequency. At 900 MHz, it becomes very critical in obtaining stable oscillations and must be kept as short as possible. Attention to layout detail is pivotal here!

5.5.4.2 High-Q Microstrip Inductor
Realizing a stable tapped-L network using lumped surface-mount inductors is impractical due to their low unloaded Q and finite physical size. Another approach uses a high-Q short microstrip inductor. The conventional approach to microstrip resonators treats them as a length of transmission line terminated with a short that reflects back an inductance impedance that simulates a lumped inductor. The approach presented here deviates from

this technique by specifically exploiting the very high unloaded Q attainable with short microstrip inductors where we design for a large C/L ratio by making the microstrip a specific, but short, length. Resonance is achieved by replacing the short with whatever capacitance is needed for proper resonance. One equation for the inductance of a strip of metal having a length ℓ (mils) and width w (mils) is

$$L = \left[5.08 \cdot 10^{-3}\right] \ell \left[\ln\left(\frac{\ell}{w}\right) + 1.193\left(\frac{w}{\ell}\right)\right] \qquad (5.79)$$

where L is in nH/mil. This technique results in a much shorter strip of metal for a given inductance. One intriguing property of microstrip inductors is that they are capable of very high unloaded Qs. An equation describing the quality factor for a "wide" microstrip is

$$Q_C = 0.63h \left[\sigma f_{\text{GHz}}\right]^{1/2} \qquad (5.80)$$

where h is the dielectric thickness in centimeters, σ is the conductivity in S/m, and f is frequency in GHz. This equation predicts an unloaded Q exceeding 700 when silver-coated copper is used. Also, wide microstrip lines are defined as those whose strip width w to height h ratio is approximately greater than 1, that is, $w/h > 1$. The parallel capacitor formed by the metal strip over the ground plane should also be included and may be calculated by the classic formula and included in the total needed to resonate with the inductance result found from equation (5.79). This "strip" capacitance is given by

$$C_S = K\varepsilon_0\varepsilon_R \left[\frac{w\ell}{h}\right] \qquad (5.81)$$

where C_S is in farads, $\varepsilon_0 = 8.86$ pF/cm is the permittivity of free space, and ε_r is the relative dielectric constant of the substrate. A "fringe factor," K, is included to account for necessary fringing (estimated to be 2%–15%). These equations are meant to provide insight into circuit behavior and should, therefore, be applied cautiously to specific applications.

5.5.4.3 UHF VCO Using the Tapped-Inductor Differential Oscillator at 900 MHz

The basic electrical circuit of this configuration is shown in Figure 5.41. Feedback occurs when sufficient RF voltage develops across the tap inductances L_{T1} and L_{T2} (due to tap 1 and tap 2, respectively); note that inductance is reckoned from the cold end of the tank at node A. Taps 1 and 2 should be as close as possible to pins 9 and 10, respectively. Also node A (cold end of tank) and node B (pin 6) must be as physically close together and exhibit as low an impedance as possible to discourage parasitic oscillations. This "inner loop," composed of L_{T2}, the inductance of connections between taps 1 and 2 and pins 9 and 10, respectively, and "stray" low-Q inductance between nodes A and B, creates a parasitic loop that will support oscillation that no longer depends on the full tapped-L tank circuit. Oscillation based on this low-Q loop typically occurs well above 1.2 GHz and has been observed as high as 1.6 GHz, and exhibits very poor phase-noise performance.

Another factor also affects this unwanted parasitic oscillation. As the full tank circuit's unloaded Q decreases, the circuit becomes conditionally stable and eventually favors the parasitic loop exclusively. This occurs because the loaded Q of the tapped microstrip affects the magnitude of the RF voltage appearing across the entire tank. Thus, feedback at taps 1 and 2 is reduced as the Q decreases. This increases the likelihood of the parasitic inner loop

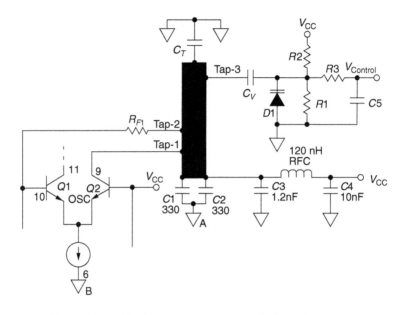

Figure 5.41   Differential oscillator with tapped-microstrip resonator.

controlling the oscillator, since its feedback voltage is largely independent of the microstrip tank Q.

Equation (5.80) shows that microstrip inductors are capable of very high unloaded Qs. This depends on a number of physical factors: frequency, dielectric thickness and quality, and the skin-depth conductance of the metal strip itself. For example, at 900 MHz, the unloaded Q of a silver-coated microstrip line is in the vicinity of 750. Coating the same line with lead (sloppy soldering can do it) reduces the Q to less than 230. If the unloaded Q is too low, the loaded Q also drops. The circuit becomes conditionally stable, and now may oscillate either at the higher parasitic-loop frequency or the desired lower frequency at which the entire microstrip tank resonates. Components should be connected by narrow traces closely connected to component that leads to avoid soldering on the microstrip proper.

R_{F1} introduces necessary losses to control the inner-loop parasitic oscillation. Its size should be made as small as possible consistent with stability, startup, and good phase-noise performance. Since R_{F1} also decreases feedback, the oscillator's output falls as R_{F1} increases. Typically, experimental values from 4 to 30 Ω have proven sufficient.

Stability as a function of V_{CC} is a good way to assess conditional stability. Provided the microstrip tank has a high loaded Q, the oscillator's stability should be independent of V_{CC} down to less than 2.5 V or so. When loaded Q decreases sufficiently to favor the inner-loop parasitic, conditional stability will occur. This can readily be observed by increasing V_{CC} from about 1.0 V and noting whether the VCO frequency jumps back and forth unpredictably. Mixer and LO port loading can also affect this condition and should be considered.

Dimensions for the microstrip resonator are based on several criteria. The strip must be "wide": its width-to-height ratio must be on the order of 1 or greater. Minimum strip length seems to be about 200 mils for 62-mil-thick board. L/C ratios somewhere about 500 have yielded quite good results. Note that we usually specify the "L/C" ratio because it is always greater than 1, even though in a parallel resonant circuit it appears in the Q equation as C/L. Thus, decreasing the "L/C ratio" by making the microstrip shorter helps the loaded inductor Q, and the circuit loaded Q_L, increase (see equation (5.75)). However, the effect of the lower Q of C_T, even when using a high-Q surface-mount type, decreased the expected larger increase in the net loaded circuit Q. Thus, the expected increase in Q_L may not be fully realizable. Assuming a 62-mil-thick board with FR5 dielectric, a strip 50 by 300 mils gave very good experimental results where C_T was about 4.3 pF, with oscillation occurring from about 950 to 1000 MHz with various test boards. A shorter inductor designed to increase the C/L ratio requiring 7.5 pF experimentally resulted in only about 1–2 dB phase-noise improvement at 950 MHz. The reason for this is that as the microstrip inductor degenerates to a plain inductor, all of its advantages are lost.

Tap 1 may be anywhere between the cold end of the tank (where C_1 and C_2 are located) and tap 2. Tap 2 yields good results at about 1/3 the strip length. Making it too close to the cold end in an effort to increase Q_L will result in loss of control over the inner-loop parasitic. To keep Q_L as high as possible, C_T should be a high-Q surface-mount type. Differences greater than 5 dB in SSB phase noise have been observed between a generic NP0 surface-mount capacitor and a high-Q NP0 surface-mount capacitor.

Tap 3 can be at the end of the microstrip when C_T is connected. However, the tuning diode inevitably will cause a decrease in overall loaded Q, typically resulting in a phase noise increase of as much as 5 dB. Moving it some distance down from the high end of the tank will decrease this effect and yield better results. A good starting point is about 1/4 down or 3/4 of the length from the cold end of the tank. C_1 and C_2 are paralleled lower-value capacitors to yield a better low-impedance ground return to pin 6 (node B) since relatively large RF currents flow through them. Note that as Q_L increases, the peak circulating RF current will also increase, as will the RF voltage at the hot end of the microstrip. At 900 MHz, good results have been obtained when both are about 300 pF. The RFC was chosen to be approximately series resonant around 900 MHz, and constitutes the series feed path for dc biasing. It may be possible to neglect this component entirely, provided that the PC-board connection to the cold end of the tank is very close to RF ground. Finally, note that shielding of the entire microstrip may be necessary to meet FCC Part 15 emission limitations (where applicable). Figure 5.42 shows one application of the tapped-microstrip oscillator with the NXP SA620 front-end IC.

5.5.5 Hartley Microstrip Resonator Oscillator

G. Smithson and S. Fitz of Plextek Communications Technology Consultants have reported on the construction of Hartley microstrip-resonator oscillators using thick-film techniques on ceramic (alumina) substrates for high-volume wireless products. The resonator in one such oscillator (Figure 5.43) exhibited an unloaded Q of approximately 80. Table 5.2 summarizes its phase noise versus tuning range with different tuning-diode configurations.

5.5.6 Crystal Oscillators

Most wireless applications do not require ultra-high frequency stability that mandates the use of a primary standard, such as one based on cesium, or a secondary standard, such as one based on rubidium. For very stable oscillators at fixed or only slightly variable frequencies up to a few hundred megahertz, oscillators using piezoelectric resonators are generally used. A piezoelectric material transduces mechanical stress (deformation) to electrical stress

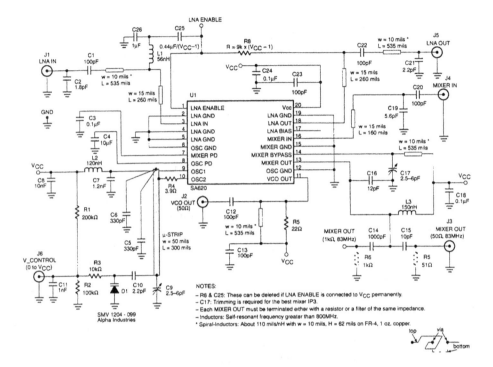

Figure 5.42   NXP SA620 application board schematic showing the tapped microstrip resonator.

(voltage), and vice versa. In noncritical applications up to 10 MHz, a piezolectric ceramic may be used. The Q and temperature stability of such materials is relatively poor, however, so quartz is the resonator material of choice for temperature-stable, low-noise piezoelectric oscillators.

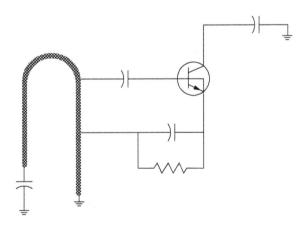

Figure 5.43   "Walking stick" microstrip Hartley resonator as used by Plextek in the 800–1500 MHz range. Below 800 MHz, the resonator's physical size becomes impractical; above 1500 MHz, the transistor's parasitic reactances make the Hartley configuration problematic.

Table 5.2   Achieved Oscillator Phase Noise Versus Tuning Range

Tuning Range (0.5–3 V)	Tuning Range (1–3 V)	Phase Noise at 20 kHz Offset
23%	16%	−100 dBc/Hz
16%	11%	−103 dBc/Hz
10%	8%	−106 dBc/Hz

Just as an electrical resonator consisting of distributed L and C exhibits multiple electrical resonances, a physical piece of piezolectric material, depending upon its shape, exhibits multiple mechanical resonances. Fundamental resonance occurs at the frequency at which the crystal is one-half wavelength long; overtone resonances occur near odd multiples (3, 5, 7, and so on) of the fundamental. Figure 5.44 shows the electrical equivalent of a crystal. Table 5.3 presents examples of approximate parameters for several crystal types. Appropriate shaping and placement of the crystal's connecting electrodes, combined with suitable oscillator circuit design and adjustment, allow the selection or ehancement of a particular resonance for excitation.

Figure 5.45 shows a typical crystal oscillator circuit configuration, and the associated table supplies recommended values for the elements. This is in line with the chip manufacturers' recommendation for their built-in oscillator circuits. Figure 5.46 shows how this oscillator's input impedance varies around resonance; Figure 5.47 shows its simulated phase noise. Placing a small resistance in series with the crystal (Figure 5.48) allows us to use the crystal as a high-Q filter, substantially improving the oscillator's spectral purity relative to that obtainable at the transistor emitter (Figure 5.49).

Some applications demand ultralow-phase-noise crystal oscillators. Figures 5.50 and 5.51 show an appropriate circuit and its phase noise with and without the current-supplying hot-carrier diode (HCD) used for noise reduction.

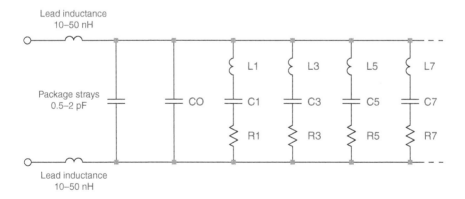

Figure 5.44   Electrical equivalent of a piezoelectric crystal resonator, showing fundamental (1) and overtone (3, 5, 7, ...) resonances. (Courtesy of Roger Clark, Vectron International.)

Table 5.3 Approximate Crystal Parameters

Crystal Q Approximations ($f_{\text{resonance}}$ in MHz)

Nonprecision fundamental crystals:

Worst: $Q = \frac{2 \times 10^6}{f}$ Better: $Q = \frac{5 \times 10^6}{f}$

Precision crystals processed for high Q:

SC: $Q = \frac{12 \times 10^6}{f}$ AT Cut: $Q = \frac{10 \times 10^6}{f}$

Crystal Resistance (r) for 10-MHz, Third-Overtone Example Using Precision Crystal Approximation (f in MHz and C_1 in pF)

$$r = \frac{(2\pi f C_1)}{Q}$$

$$r = 6.964$$

Typical crystal parameters

10-MHz fundamental: $C_1 = 0.02$ pF, $C_0 = 3.6$ pF, $Q = 350{,}000$,
$r = 2.5\ \Omega$, $L = 0.012665$ H

50-MHz third overtone: $C_1 = 0.0026$ pF, $C_0 = 4.212$ pF, $Q = 100{,}000$,
$r = 13\ \Omega$, $L = 0.003897$ H

100-MHz fifth overtone: $C_1 = 0.0005$ pF, $C_0 = 2.25$ pF, $Q = 80{,}000$,
$r = 40\ \Omega$, $L = 0.005066$ H

155-MHz high-frequency fundamental: $C_1 = 0.005$ pF, $C_0 = 1.5$ pF,
$Q = 20{,}000$, $r = 12\ \Omega$, $L = 0.002108$ H

Courtesy of Roger Clark, Vectron International.

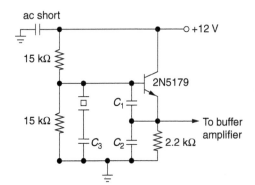

	3 MHz	6 MHz	10 MHz	20 MHz	30 MHz
C_1 (pF)	330	270	180	82	43
C_2 (pF)	430	360	220	120	68
C_3 (pF)	39	43	43	36	32
Crystal C_L (pF)	32	32	30	20	15

C_3 is the approximate value needed to set occillator on frequency using indicated crystal C_L.

Figure 5.45 A 3–30-MHz Colpitts crystal oscillator.

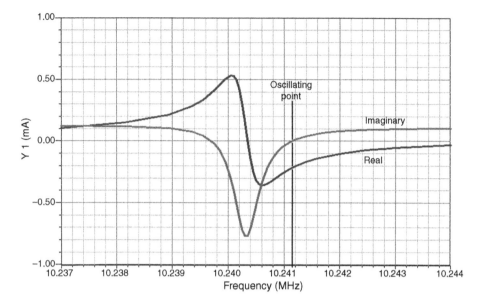

Figure 5.46 As with every oscillator, the crystal oscillator's input impedance is negative and real at resonance. This graph plots the real and imaginary components of the test voltage injected by the Oscillator Design Aid tool in Ansoft Serenade 8.0. Compare this graph with Figure 5.7, which plots an oscillator's input impedance with the resonator absent.

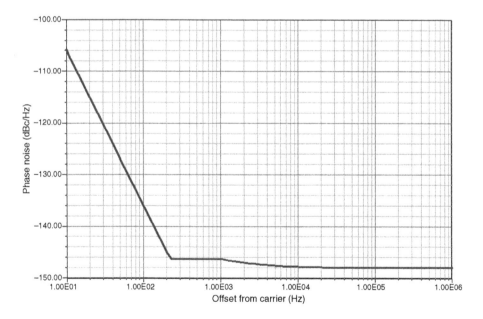

Figure 5.47 Simulated phase noise of the crystal oscillator.

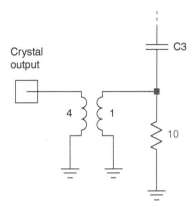

Figure 5.48 This alternative output tap uses the crystal's high selectivity for improved harmonic reduction.

5.5.7 Voltage-Controlled Oscillators

A schematic of a ceramic-resonator-based VCO (Figure 5.35) has already been shown in the section on ceramic-resonator oscillators. To tune the oscillator within the required range, tuning diodes are used. These diodes are often called varactors or voltage-sensitive diodes. By way of approximation, we can use the equation

$$C = \frac{K}{(V_R + V_D)^n} \tag{5.82}$$

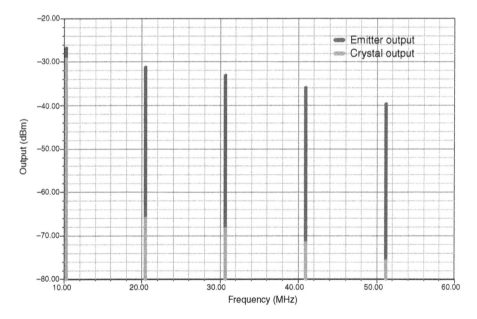

Figure 5.49 Comparison of the crystal oscillator's simulated output spectrum as obtained at the transistor emitter (dark gray, fundamental −26.7 dBm) and crystal (light gray, fundamental −29.1 dBm). At the emitter, the second harmonic is down only 4.3 dB relative to the carrier; at the crystal, the second harmonic is down 36 dB relative to the carrier.

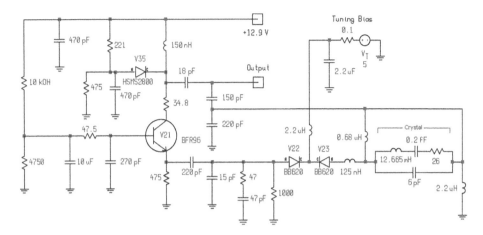

Figure 5.50 100 MHz VCXO suitable for ultralow-phase-noise applications. The HSMS2800 hot-carrier diode provides noise reduction.

wherein all constants and all parameters determined by manufacturing process are contained in K. The exponent is a measure of the slope of the capacitance/voltage characteristic and is 0.5 for alloyed diodes, 0.33 for single-diffused diodes, and (on average) 0.75 for tuner diodes with a hyperabrupt pn junction [6, 7]. Figure 5.52 shows the capacitance–voltage characteristics of an alloyed, a diffused, and a tuner diode.

Recently, an equation was developed that, although purely formal, describes the practical characteristic better than equation (5.82):

$$C = C_0 \left(\frac{A}{A + V_R} \right)^m \tag{5.83}$$

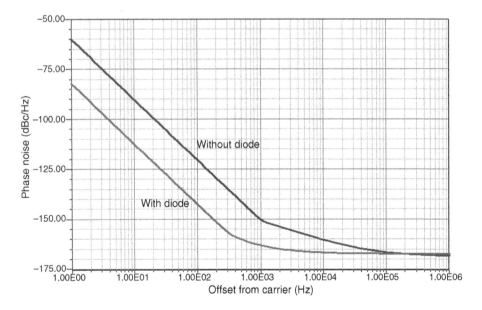

Figure 5.51 Phase noise of the VCXO with and without the noise-reduction diode.

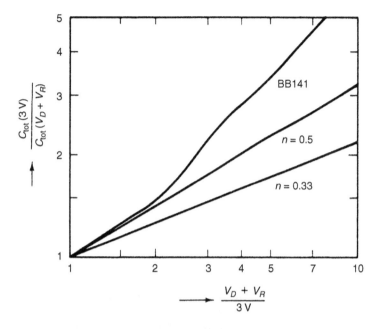

Figure 5.52 Capacitance/voltage characteristic for an alloyed capacitance diode ($n = 0.33$), a diffused capacitance diode ($n = 0.5$), and a wide-range tuning diode (BB141).

wherein C_0 is the capacitance at $V_R = 0$, and A is a constant whose dimension is a volt. The exponent m is much less dependent on voltage than the exponent n in equation (5.84).

The operating range of a capacitance diode or its useful capacitance ratio,

$$\frac{C_{\max}}{C_{\min}} = \frac{C_{\text{tot}}(V_{R\min})}{C_{\text{tot}}(V_{R\max})} \tag{5.84}$$

is limited by the fact that the diode must not be driven by the alternating voltage superimposed on the tuning voltage, either into the forward mode or the breakdown mode. Otherwise, rectification would take place, shifting the bias of the diode and considerably affecting its figure of merit.

There are several manufacturers of tuning diodes. Freescale is a typical supplier in this country; Infineon of Germany and NXP also provide good diodes. Table 5.4 and Figure 5.53 contain information for tuning diodes useful for our applications.

5.5.8 Diode-Tuned Resonant Circuits

5.5.8.1 Tuner Diode in Parallel-Resonant Circuit

Figures 5.54 and 5.55 illustrate two basic circuits for the tuning of parallel resonant circuits by means of capacitance diodes. In the circuit diagram of Figure 5.54, the tuning voltage is applied to the tuning diode via the bias resistor, R_B. Series connected from the tuning diode to the top of the tank circuit is capacitor C_S, which completes the circuit for alternating current but isolates the cathode of the tuner diode from the coil and thus from the negative terminal of the tuning voltage. Moreover, a fixed parallel capacitance C_P is provided. The decoupling resistor preceding the bias resistor is large enough to be disregarded in the following discussion. Since for high-frequency purposes the biasing resistor is connected in

Table 5.4 Example Tuning Diodes (Infineon)

Type	Maximum Ratings				Characteristics ($T_A = 25°C$)						Package	Chip Code
	V_R (V)	I_F (mA)	C_T (pF)	V_R (V)	C_T (pF)	V_R (V)	C Ratio	I_R (nA)	V_R (V)		Package	Chip Code
BB439	30	20	29	3	5.0	25	5.8	≤20	28		SOD-323	Q
BB535	30	20	18.7	1	2.1	28	8.9	≤10	30		SOD-323	A
BB545	30	20	20	1	2.0	28	10.0	≤10	30		SOD-323	C
BB639	30	20	38.3	1	2.65	28	14.5	≤10	30		SOD-323	K
BB639C	30	20	39	1	2.55	28	15.3	≤10	30		SOD-323	B
BB640	30	20	69	1	3.05	28	22.6	≤10	30		SOD-323	N
BB804 (dual)	20	50	44.75	2	26.15	8	1.71	≤20	16		SOT-23	E
BB814 (dual)	20	50	44.75	2	20.8	8	2.15	≤20	16		SOT-23	O
BB833	30	20	9.3	1	0.75	28	12.4	≤20	30		SOD-323	F
BB914 (dual)	20	50	43.75	2	18.7	8	2.34	≤20	16		SOT-23	P
BBY51-03W	7	20	5.30	1	3.10	4	1.75	≤10	6		SOD-323	H
BBY52-03W	7	20	1.75	1	1.25	4	1.40	≤10	6		SOD-323	I
BBY53-03W	6	20	5.30	1	2.40	3	2.20	≤10	4		SOD-323	L

Source: Courtesy Infineon Technologies.

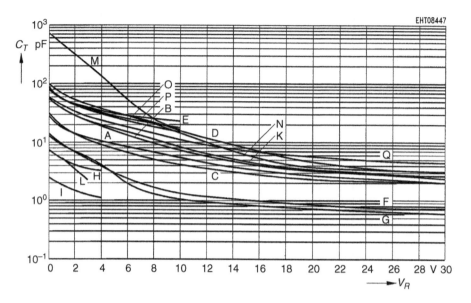

Figure 5.53 Diode capacitance C_T versus reverse voltage V_R for the diodes shown in Table 5.4. The curve labels correspond to the chip codes called out in the table.

parallel with the series capacitor, it is transformed into the circuit as an additional equivalent shunt resistance R_c. For the parallel loss resistance transformed into the circuit, we have the expression

$$R_c = R_B \left(1 + \frac{C_{\text{tot}}}{C_S}\right)^2 \tag{5.85}$$

If in this equation the diode capacitance is replaced by the resonator circuit frequency ω, we obtain

$$R_c = R_B \left[\frac{\omega^2 L C_S}{\omega^2 L \left(C_S + C_p\right) - 1}\right]^2 \tag{5.86}$$

In the circuit of Figure 5.55, the resonant circuit is tuned by two tuning diodes, which are connected in parallel via the coil for tuning purposes, but series connected in opposition

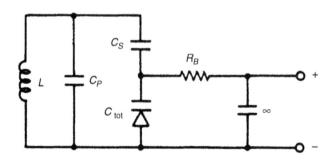

Figure 5.54 Parallel-resonant circuit with tuner diode and bias resistor in parallel to the diode.

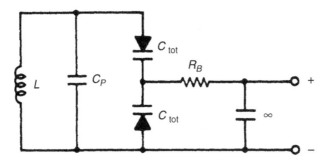

Figure 5.55 Parallel-resonant circuit with two tuning diodes.

for high-frequency signals. This arrangement has the advantage that the capacitance shift caused by the ac modulation acts in opposite directions in these diodes, and therefore cancels itself. The bias resistor R_B, which applies the tuning voltage to the tuning diodes, is transformed into the circuit at a constant ratio across the entire tuning range. Given two identical, loss-free tuning diodes, we obtain the expression

$$R_c = 4R_B \tag{5.87}$$

5.5.8.2 Capacitances Connected in Parallel or in Series with the Tuning Diode

Figure 5.54 shows that a capacitor is usually in series with the tuner diode, in order to close the circuit for alternating current and, at the same time, to isolate one terminal of the tuning diode from the rest of the circuit with respect to direct current, so as to enable the tuning voltage to be applied to the diode. If possible, the value of the series resistor C_S will be chosen such that the effective capacitance variation is not restricted. However, in some cases—for example, in the oscillator circuit of receivers in whose intermediate frequency is of the order of magnitude of the reception frequency—this is not possible, and the influence of the series capacitance will then have to be taken into account. By connecting the capacitor C_S, assumed to be lossless, in series with the diode capacitance C_{tot}, the tuning capacitance is reduced to the value

$$C^* = C_{tot} \frac{1}{1 + C_{tot}/C_S} \tag{5.88}$$

The Q of the effective tuning capacitance, taking into account the Q of the tuning diode, increases to

$$Q^* = Q \left(1 + \frac{C_{tot}}{C_S}\right) \tag{5.89}$$

The useful capacitance ratio is reduced to the value

$$\frac{C^*_{max}}{C^*_{min}} = \frac{C_{max}}{C_{min}} \frac{1 + C_{min}/C_S}{1 + C_{max}/C_S} \tag{5.90}$$

wherein C_{max} and C_{min} are the maximum and minimum capacitances of the tuner diode.

On the other hand, the advantage is gained that, due to the capacitive potential division, the amplitude of the alternating voltage applied to the tuner diode is reduced to

$$\hat{v}^* = \hat{v}\, \frac{1}{1 + C_{tot}/C} \tag{5.91}$$

so that the lower value of the tuning voltage can be smaller, and this results in a higher maximum capacitance C_{max} of the tuning diode and a higher useful capacitance ratio. The influence exerted by the series capacitor, then, can actually be kept lower than equation (5.89) would suggest.

The parallel capacitance C_P that appears in Figures 5.54 and 5.55 is always present, since wiring capacitances are inevitable and every coil has its self-capacitance. By treating the capacitance C_P, assumed to be lossless, as a shunt capacitance, the total tuning capacitance rises in value and, if C_S is assumed to be large enough to be disregarded, we obtain

$$C^* = C_{tot}\left(1 + \frac{C_p}{C_{tot}}\right) \tag{5.92}$$

The Q of the effective tuning capacitance, derived from the Q of the tuning diode, is

$$Q^* = Q\left(1 + \frac{C_p}{C_{tot}}\right) \tag{5.93}$$

or, in other words, it rises with the magnitude of the parallel capacitance. The useful capacitance is reduced as

$$\frac{C_{max}^*}{C_{min}^*} = \frac{C_{max}}{C_{min}} \frac{1 + C_p/C_{max}}{1 + C_p/C_{min}} \tag{5.94}$$

In view of the fact that even a comparatively small shunt capacitance reduces the capacitance ratio considerably, it is necessary to ensure low wiring and coil capacitances in the layout stage of the circuit design.

5.5.8.3 Tuning Range

The frequency range over which a parallel-resonant circuit (according to Figure 5.54) can be tuned by means of the tuning diode depends on the useful capacitance ratio of the diode and on the parallel and series capacitances present in the circuit. The ratio is

$$\frac{f_{max}}{f_{min}} = \sqrt{\frac{1 + \frac{C_{max}}{C_p(1 + C_{max}/C_S)}}{1 + \frac{C_{max}}{C_p(C_{max}/C_{min} + C_{max}/C_S)}}} \tag{5.95}$$

In many cases, the value of series capacitor can be chosen to be large enough for its effect to be negligible. In that case, equation (5.95) is simplified as follows:

$$\frac{f_{max}}{f_{min}} = \sqrt{\frac{1 + C_{max}/C_p}{1 + C_{min}/C_p}} \tag{5.96}$$

From this equation, the diagram shown in Figure 5.56 is computed. With the aid of this diagram, the tuning diode parameters required for tuning a resonant circuit over a stipulated frequency range (i.e., the maximum capacitance and the capacitance ratio) can

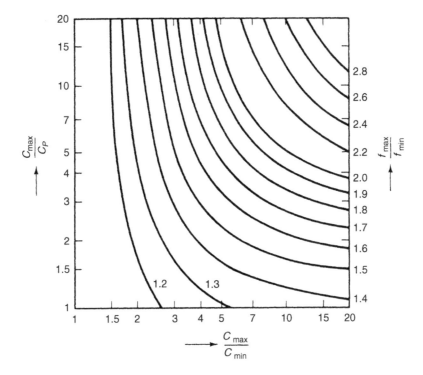

Figure 5.56 Diagram for determining capacitance ratio and maximum capacitance.

be determined. Whenever the series capacitance C_S cannot be disregarded, the effective capacitance ratio is reduced according to equation (5.90).

5.5.8.4 Tracking
When several tuned circuits are used on the same frequency, diodes must be selected for perfect tracking.

5.5.9 Practical Circuits

5.5.9.1 Oscillators with Coarse and Fine Tuning
After so much theory, it may be nice to take a look at some practical circuits, such as the one shown in Figure 5.57. This oscillator is being used in the Rohde & Schwarz ESN/ESVN40 field-strength meter, and in the HF1030 receiver produced by Cubic Communications, San Diego. This circuit combines all the various techniques shown previously. A single diode is being used for fine tuning a narrow range of less than 1 MHz; coarse tuning is achieved with the antiseries diodes.

Several unusual properties of this circuit are apparent as follows.

1. The fine tuning is achieved with a tuning diode that has a much larger capacitance than that of the coupling capacitor in the circuit. The advantage of this technique is that the fixed capacitor and the tuning diode form a voltage divider whereby the voltage across the tuning diode decreases as the capacitance increases. For larger values of the

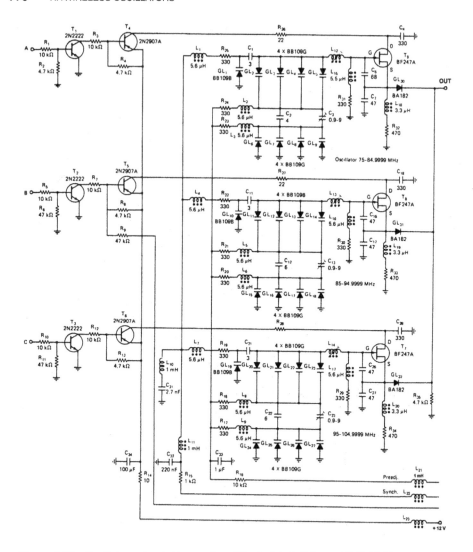

Figure 5.57   Oscillator and switching section of the Rohde & Schwarz ESH2/ESH3 test receiver.

capacitance of the tuning diode, the Q changes and the gain K_0 increases. Because of the voltage division, the noise contribution and loading effect of the diode are reduced.

2. In the coarse-tuning circuit, several tuning diodes are used in parallel. The advantage of this circuit is a change in LC ratio by using a higher C and storing more energy in the tuned circuit. There are no high-Q diodes available with such large capacitance values, and therefore preference is given to using several diodes in parallel rather than one tuning diode with a large capacitance, normally used only for AM (medium-frequency broadcast) tuner circuits.

We have mentioned previously that, despite this, the coarse-tuning circuit will introduce noise outside the loop bandwidth, where it cannot be corrected. It is therefore preferable

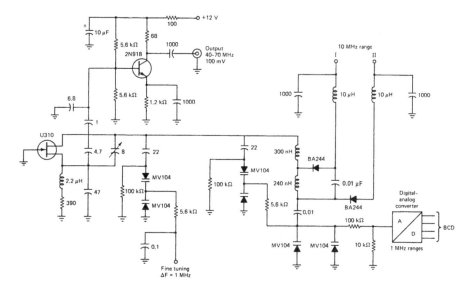

Figure 5.58 40–70-MHz VCO with two coarse-steering ranges and a fine-tuning range of 1 MHz.

to incorporate switching diodes for segmenting ranges at the expense of switching current drain.

Figure 5.58 shows a circuit using a combined technique of tuning diodes for fine- and medium-resolution tuning, and coarse tuning with switching diodes.

5.5.9.2 DC-Coupled Oscillator

An interesting circuit that was cut out of an IC was supplied to us by David Lovelace of Motorola. Figures 5.59 and 5.60 show the circuit and its phase noise, respectively. As a result of the constraints of integration, this BJT oscillator has no base bias circuitry in the conventional form. As a fairly noisy performer, it ultimately was not used in production.

5.5.9.3 Siemens Colpitts Oscillator

A more elaborate approach that is essentially a standard Colpitts oscillator with a lot of detailed modeling is the oscillator of Figure 5.61, supplied to us by Siemens. Considering its simplicity, its noise performance is good, as shown in Figure 5.62. Modeling this circuit involved the consideration of via holes and coupled lines. Figure 5.63 shows an actual implementation of this oscillator. The value of external emitter resistance (R7 in Figure 5.61) plays a critical role in determining a common-emitter Colpitts oscillator's upper frequency limit as described in Ref. [8].

5.5.9.4 DC-Stabilized Oscillator

For high-performance synthesizers, such as the Rohde & Schwarz series SMG/SMH, a dc-stabilized oscillator with fairly wide tuning range may be used. Figure 5.64 shows such an oscillator, the measured phase noise of which is shown in Figure 5.65. It should be noted that the output loop is magnetically coupled and therefore takes advantage of the inherent selectivity of the tuned circuit.

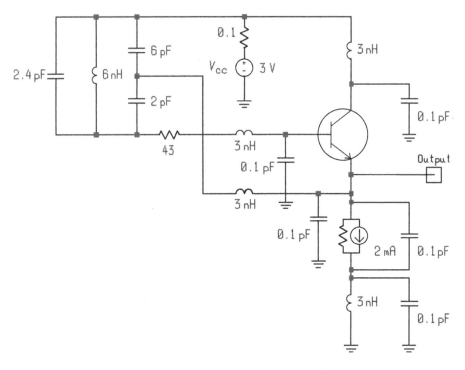

Figure 5.59 Schematic of the dc-coupled oscillator.

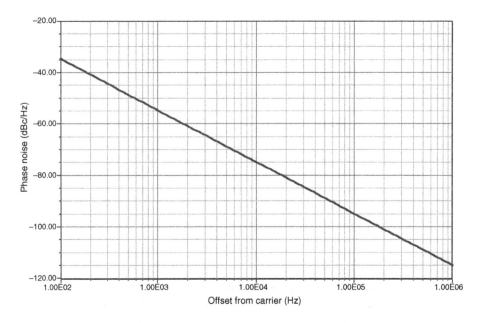

Figure 5.60 Phase noise of the dc-coupled oscillator operating at 897 MHz.

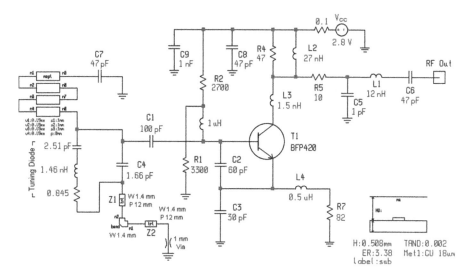

Figure 5.61 Schematic of the Siemens VCO.

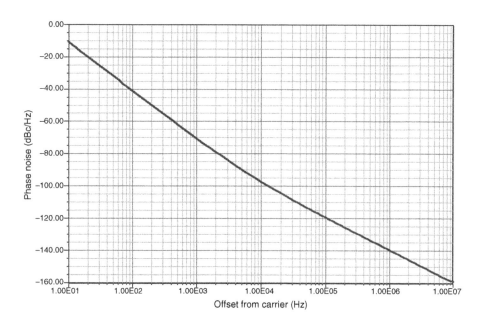

Figure 5.62 Simulated phase noise of the VCO. The frequency of oscillation is 1.052 GHz.

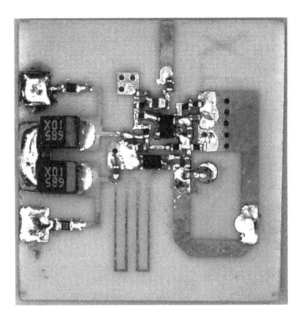

Figure 5.63 Implementation of the Siemens VCO.

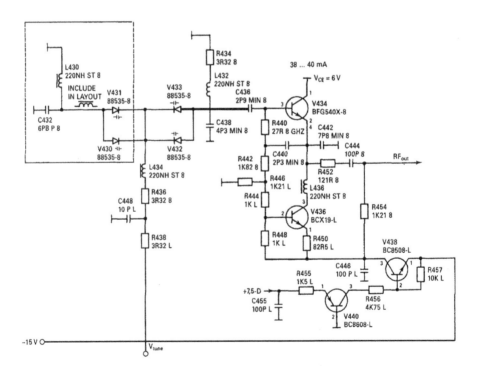

Figure 5.64 A recommended very low-phase-noise VCO that operates in the 1 GHz region. The V436 stage is a constant-current generator responsible for improved phase-noise performance.

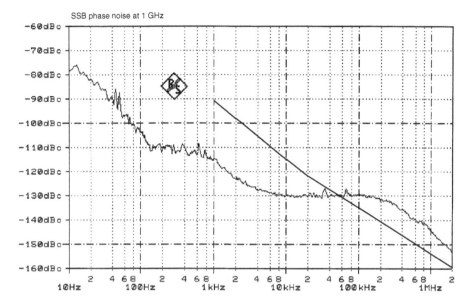

Figure 5.65 Phase noise of the low-noise VCO. The "ledges" from 100 Hz to 1 kHz and 10–100 kHz reflect the noise-cleanup effects of a dual-loop synthesizer; the superimposed line from 1 kHz to 2 MHz shows the oscillator's noise characteristic with both loops unlocked. The closed-loop response is noisier than the open-loop response at offsets above 60 kHz because a wide-loop bandwidth is used to achieve a faster switching speed.

5.6 NOISE IN OSCILLATORS

In transmitters, oscillator noise can result in adjacent-channel interference and modulation errors; in receivers, oscillator noise can result in demodulation errors, and degraded sensitivity and dynamic range. The specification, calculation, and reduction of oscillator noise is therefore of great importance in wireless system design.

5.6.1 Linear Approach to the Calculation of Oscillator Phase Noise

Since an oscillator can be viewed as an amplifier with feedback (Figure 5.66), it is helpful to examine the phase noise added to an amplifier that has a noise figure F. With F defined by Ref. [9]

$$F = \frac{(S/N)_{\text{in}}}{(S/N)_{\text{out}}} = \frac{N_{\text{out}}}{N_{\text{in}} G} = \frac{N_{\text{out}}}{G k T B} \tag{5.97}$$

$$N_{\text{out}} = F G k T B \tag{5.98}$$

$$N_{\text{in}} = k T B \tag{5.99}$$

where N_{in} is the total input noise power to a noise-free amplifier. The input phase noise in a 1-Hz bandwidth at any frequency $f_0 + f_m$ from the carrier produces a phase

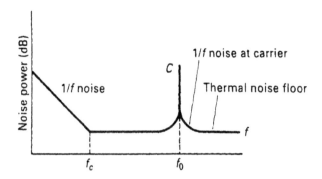

Figure 5.66 Noise power versus frequency of a transistor amplifier with an input signal applied.

deviation given by (Figure 5.67)

$$\Delta\theta_{\text{peak}} = \frac{V_{n\text{RMS1}}}{V_{\text{avsRMS}}} = \sqrt{\frac{FKT}{P_{\text{avs}}}} \qquad (5.100)$$

$$\Delta\theta_{1\text{RMS}} = \frac{1}{\sqrt{2}}\sqrt{\frac{FkT}{P_{\text{avs}}}} \qquad (5.101)$$

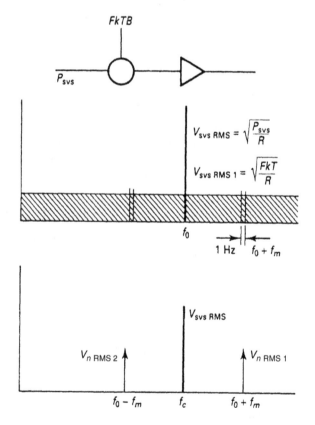

Figure 5.67 Phase noise added to carrier.

Since a correlated random phase-noise relation exists at $f_0 - f_m$, the total phase deviation becomes

$$\Delta\theta_{\text{RMStotal}} = \sqrt{FkT/P_{\text{avs}}} \qquad (5.102)$$

The spectral density of phase noise becomes

$$S_\theta(f_m) = \Delta\theta^2_{\text{RMS}} = FkTB/P_{\text{avs}} \qquad (5.103)$$

where $B = 1$ for a 1-Hz bandwidth. Using

$$kTB = -174\,\text{dBm/Hz} \qquad (B = 1) \qquad (5.104)$$

allows a calculation of the spectral density of phase noise that is far removed from the carrier (i.e., at large values of f_m). This noise is the theoretical noise floor of the amplifier. For example, an amplifier with $+10$ dBm power at the input and a noise figure of 6 dB gives

$$S_\theta(f_m > f_c) = -174\,\text{dBm} + 6\,\text{dB} - 10\,\text{dBm} = -178\,\text{dBm} \qquad (5.105)$$

Only if P_{out} is > 0 dBm, we can expect $\mathcal{L}$ (signal-to-noise ratio) to be greater than 174 dBc/Hz (1-Hz bandwidth). For a modulation frequency close to the carrier, $S_\theta(f_m)$ shows a flicker or $1/f$ component, which is empirically described by the corner frequency f_c. The phase noise can be modeled by a noise-free amplifier and a phase modulator at the input as shown in Figure 5.68.

The purity of the signal is degraded by the flicker noise at frequencies close to the carrier. The spectral phase noise can be described by

$$S_\theta(f_m) = \frac{FkTB}{P_{\text{avs}}}\left(1 + \frac{f_c}{f_m}\right) \qquad (B = 1) \qquad (5.106)$$

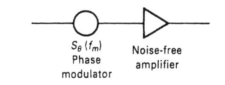

$S_\theta(f_m)$
Phase
modulator

Noise-free
amplifier

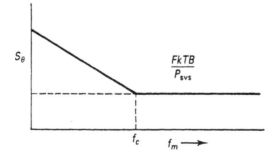

Figure 5.68 Phase noise modeled by a noise-free amplifier and phase modulator.

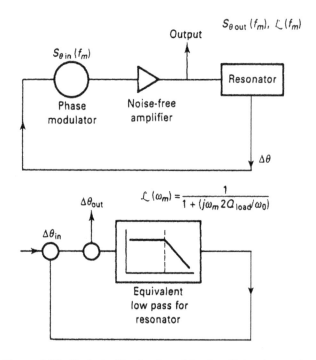

Figure 5.69   Equivalent feedback models of oscillator phase noise.

No AM-to-PM conversion is considered in this equation. The oscillator may be modeled as an amplifier with feedback as shown in Figure 5.69. The phase noise at the input of the amplifier is affected by the bandwidth of the resonator in the oscillator circuit in the following way. The tank circuit or bandpass resonator has a low-pass transfer function

$$\mathcal{L}(\omega_m) = \frac{1}{1 + j(2Q_L\omega_m/\omega_0)} \tag{5.107}$$

where

$$\omega_0/2Q_L = B/2 \tag{5.108}$$

is the half bandwidth of the resonator. These equations describe the amplitude response of the bandpass resonator; the phase noise is transferred unattenuated through the resonator up to the half bandwidth.

The closed-loop response of the phase feedback loop is given by

$$\Delta\theta_{\text{out}}(f_m) = \left(1 + \frac{\omega_0}{j2Q_L\omega_m}\right)\Delta\theta_{\text{in}}(f_m) \tag{5.109}$$

The power transfer becomes the phase spectral density

$$S_{\theta\text{out}}(f_m) = \left[1 + \frac{1}{f_m^2}\left(\frac{f_0}{2Q_L}\right)^2\right]S_{\theta\text{in}}(f_m) \tag{5.110}$$

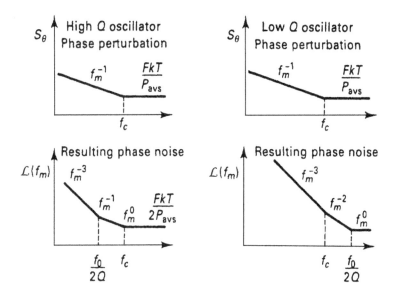

Figure 5.70 Equivalent feedback models of oscillator phase noise.

where $S_{\theta\text{in}}$ was given by equation (5.106). Finally, $\mathcal{L}(f_m)$ is

$$\mathcal{L}(f_m) = \frac{1}{2}\left[1 + \frac{1}{f_m^2}\left(\frac{f_0}{2Q_L}\right)^2\right] S_{\theta\text{in}}(f_m) \qquad (5.111)$$

This equation describes the phase noise at the output of the amplifier (flicker corner frequency and AM-to-PM conversion are not considered). The phase perturbation $S_{\theta\text{in}}$ at the input of the amplifier is enhanced by the positive phase feedback within the half bandwidth of the resonator, $f_0/2Q_L$.

Depending on the relation between f_c and $f_0/2Q_L$, there are two cases of interest, as shown in Figure 5.70. For the low-Q case, the spectral phase noise is unaffected by the Q of the resonator, but the $\mathcal{L}(f_m)$ spectral density will show a $1/f^3$ and $1/f^2$ dependence close to the carrier. For the high-Q case, a region of $1/f^3$ and $1/f$ should be observed near the carrier. Substituting equation (5.106) in equation (5.111) gives an overall noise of

$$\mathcal{L}(f_m) = \frac{1}{2}\left[1 + \frac{1}{f_m^2}\left(\frac{f}{2Q_L}\right)^2\right]\frac{FkT}{P_{\text{avs}}}\left(1 + \frac{f_c}{f_m}\right)$$

$$= \frac{FkTB}{2P_{\text{avs}}}\left[\frac{1}{f_m^3}\frac{f^2 f_c}{4Q_L^2} + \frac{1}{f_m^2}\left(\frac{f}{2Q_L}\right)^2 + \left(1 + \frac{f_c}{f_m}\right)\right] \text{ dBc/Hz} \quad (5.112)$$

Examining equation (5.112) gives the four major causes of oscillator noise: the upconverted $1/f$ noise or flicker FM noise, the thermal FM noise, the flicker phase noise, and the thermal noise floor, respectively.

Q_L (loaded Q) can be expressed as

$$Q_L = \frac{\omega_0 W_e}{P_{\text{diss,total}}} = \frac{\omega_0 W_e}{P_{\text{in}} + P_{\text{res}} + P_{\text{sig}}} = \frac{\text{Reactive power}}{\text{Total dissipated power}} \qquad (5.113)$$

where W_e is the reactive energy stored in L and C,

$$W_e = \frac{1}{2}CV^2 \tag{5.114}$$

$$P_{res} = \frac{\omega_0 W_e}{Q_{unl}} \tag{5.115}$$

The limitation of this equation is that the loaded Q in most cases has to be estimated and the same applies to the noise factor. In practice, the Leeson formula as given above is well suited to help interpreting phase noise of an oscillator, and it enables to determine qualitatively possible directions to improve the design. But noise factor F of the transistor in the circuit is different from the noise factor of the transistor alone, the same applies to the loaded Q and to the output power.

These quantities can be determined using a microwave harmonic-balance simulator, which is based on the noise modulation theory (published by Rizzoli) [10].

But a good analytical estimate is highly desirable since it allows to spot promising options for phase-noise reduction. Analytically determining the parameters, however, requires certain restrictions and assumptions:

1. The formulas below were derived for a Colpitts oscillator with a BJT.
2. The circuit only accounts for the first-order effects. Parasitic effects are omitted, if they can be regarded as second order. These simplifications, however, are required to obtain a human-readable result.
3. The derivation is based on a time-domain large-signal analysis of the oscillator. This calculation is quite tedious, so the derivation of the formulas will not be given in detail. The approach basically assumes that one sinusoidal signal of significant amplitude exists. Neglecting higher harmonics of the oscillation frequency simplifies matters in a way that the calculation can be handled analytically. This approach is based on what was published by Kurokawa for coupled oscillators [11] and was extended by Rohde [12] to the present case.
4. Similar results could be obtained for other oscillator topologies and transistor technologies.

The following equations, based on the equivalent circuit shown in Figure 5.71, are the exact values for P_{sav}, Q_L, and F which are needed for the Leeson equation.

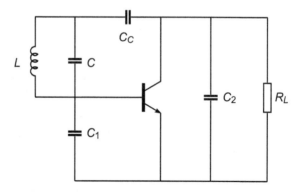

Figure 5.71   Equivalent circuit of the Colpitts oscillator.

The power can be given as

$$P_{\text{sav}} = \frac{V_{ce}^2}{4} \times \frac{Q_L^2 R_L}{\frac{Q_L^2}{\omega_0^2}\left(\frac{C_1+C_2}{C_1 C_2}\right)^2 + \omega_0^2 L^2} \tag{5.116}$$

To calculate the loaded Q_L, we have to consider the unloaded Q_0 and the loading effect of the transistor. There we have to consider the influence of the BJT's $Y_{21}^{\dagger}$. The inverse of this is responsible for the loading and reduction of the Q.

$$Q_L = \frac{Q_0 \times Q^*}{Q_0 + Q^*} \tag{5.117}$$

with

$$Q^* = \frac{\omega_0 \times |1/Y_{21}^{\dagger}|(C_1 + C_2)}{1 - \omega_0^2 C_1 L} \tag{5.118}$$

Finally, the noise factor F can be derived. This noise calculation, while itself uses a totally new approach, is based on the general noise calculations such as the one shown by Hawkins [13] and Hsu and Snapp [14]. An equivalent procedure can be found for FETs rather than bipolar transistors.

$$F = 1 + X_C \left[r_b \frac{1}{2r_e\beta}\left(r_b + \frac{1}{X_C}\right)^2 + \frac{r_e}{2} + \frac{1}{2r_e}(r_b + X_C)^2\left(\frac{f_0}{f_T}\right)^2 \right] \tag{5.119}$$

with

$$X_C = \frac{C_2 C_C}{(C_1 + C_2)C_1 r_e} \tag{5.120}$$

In this formula, β denotes the BJT's common-emitter current gain, f_T is its transit frequency, and r_e and r_b denote the emitter and base resistances, respectively.

This equation is extremely significant because it covers most of the causes of phase noise in oscillators. (AM-to-PM conversion must be added; see equation 5.56.)

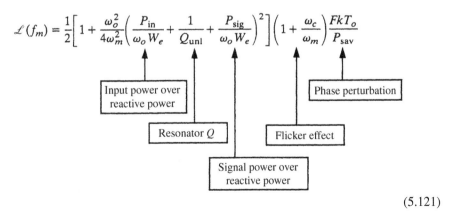

$$\mathcal{L}(f_m) = \frac{1}{2}\left[1 + \frac{\omega_o^2}{4\omega_m^2}\left(\frac{P_{\text{in}}}{\omega_o W_e} + \frac{1}{Q_{\text{unl}}} + \frac{P_{\text{sig}}}{\omega_o W_e}\right)^2\right]\left(1 + \frac{\omega_c}{\omega_m}\right)\frac{FkT_o}{P_{\text{sav}}}$$

(5.121)

To minimize the phase noise, the following design rules apply.

1. Maximize the unloaded Q.
2. Maximize the reactive energy by means of a high RF voltage across the resonator and obtain a low LC ratio. The limits are set by breakdown voltages of the active devices and the tuning diodes, and the forward-bias condition of the tuning diodes. If L becomes too high, the circuit degenerates into a squegging oscillator. Using lower L allows larger capacitance, which allows us to swamp the oscillating device's internal, nonlinear capacitance with external, linear capacitance, reducing the phase-noise degradation caused by the nonlinear capacitance. The ceramic-resonator oscillator best illustrates this approach, which may be hard to accomplish in discrete circuits because high-Q inductors with values above 1 nH cannot be built unless high-Q transmission lines or resonators are used.
3. Avoid saturation at all cost, and try to either have limiting or automatic gain control (AGC) without degradation of Q. Isolate the tuned circuit from the limiter or AGC circuit. Use tuning diodes in antiseries configurations to avoid forward bias.
4. Choose an active device with the lowest possible noise figure and flicker corner frequency. The noise figure of interest is the noise figure obtained with the actual impedance at which the device is operated. Using FETs rather than BJTs, it is preferable to deal with the equivalent noise voltage and noise currents rather than with the noise figure, since they are independent of source impedance. The noise figure improves as the ratio between source impedance and equivalent noise resistance increases. In addition, in a tuned circuit, the source impedance changes drastically as a function of the offset frequency, and this effect has to be considered. For low phase-noise operation, use a medium-power transistor. If you need your output power to be achieved at 6–9 mA, select a transistor with an I_{Cmax} of 60–90 mA. Also avoid an f_T greater than 3–5× the operating frequency. To make transistors that are stable across their full frequency ranges, manufacturers add circuitry that makes the flicker corner frequency higher as f_T increases.

   The following transistors have the lowest noise figure.

   • BFG67 by NXP
   • BFR106/92; BFP405/420 by Infineon
   • 2SC3356 by Renesas

   Among the lowest-noise JFETs are the U310, 2N4416/17, and 2N5397.

   In designing MMICs, one must resort to HEMTs. They have a fairly high flicker frequency, which is typically around 10 MHz but can go up to as high as 100 MHz.

5. Phase perturbation can be minimized by using high-impedance devices such as FETs, where the signal-to-noise ratio of the signal voltage relative to the equivalent noise voltage can be made very high. This also indicates that in the case of a limiter, the limited voltage should be as high as possible.

6. Choose an active device with low flicker noise. The effect of flicker noise can be reduced through RF feedback. An unbypassed emitter resistor of 10–30 Ω in a bipolar circuit can improve the flicker noise by as much as 40 dB.

   The proper bias point of the active device is important, and precautions should be taken to prevent modulation of the input and output dynamic capacitance of the active device, which will cause amplitude-to-phase conversion, and therefore introduce noise.

7. The energy should be coupled from the resonator rather than from another portion of the active device so that the resonator limits the bandwidth.

Equation (5.121) assumes that the phase perturbation and the flicker effect are the limiting factors, as practical use of such oscillators requires that an isolation amplifier must be used.

In the event that energy is taken directly from the resonator and the oscillator power can be increased, the signal-to-noise ratio can be increased above the theoretical limit of 174 dBc/Hz, due to the low-pass-filter effect of the tuned resonator. However, since this is mainly a theoretical assumption and does not represent a real-world system, this noise performance cannot be obtained. In an oscillator stage, even a total noise floor of -170 dBc/Hz is rarely achieved.

What other influences do we have that cause an oscillator's noise performance to degrade? So far, we have assumed that the Q of the tuned circuit is really determined only by the LC network and the loading effect of the transistor. In synthesizer applications, however, we find it necessary to add a tuning diode. The tuning diode has a substantially lower Q than that of a mica capacitor or even a ceramic capacitor. As a result of this, the noise sidebands change as a function of the additional loss. This is best expressed as form of adjusting the value for the loaded Q in equation (5.113).

There seems to be no precise mathematical way of predetermining the noise influence of a tuning diode, but the following approximation seems to give proper results.

$$\frac{1}{Q_{T\text{load}}} = \frac{1}{Q_{\text{load}}} + \frac{1}{Q_{\text{diode}}} \qquad (5.122)$$

The tuning diode is specified to have a cutoff frequency $f_{\max}$, which is determined from the loss resistor R_S and the value of the junction capacitance as a function of voltage (e.g., measured at 3 V). This means that the voltage determines the Q and, consequently, the noise bandwidth.

5.6.2 Phase-Noise Analysis Based on the Feedback Model

In order to optimize the phase noise of an oscillator, it is helpful to know the contribution of every single physical noise source. The fast way to achieve this goal is to rely on circuit simulation software. A number of simulations can be performed switching all noise sources off except for one, and the resulting simulated phase noise equals the contribution of the noise source in question to overall phase noise.

Analytical formulas, on the other hand, give much more insight into the mechanisms behind and can support analytical minimization of phase noise that is always to be preferred over numerical optimization.

First thing we need for an analytical approach is a transfer function. The input signal $X(j\omega)$ in this case would be the noise source's current or voltage. The output $Y(j\omega)$ is defined as the noise voltage at the load resistor. No matter where the noise source is located in the circuit, it is still an oscillator that can be described by the feedback system shown in Figure 5.72.

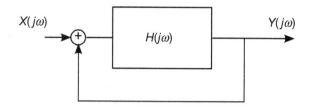

Figure 5.72 Feedback loop.

The transfer function of the closed-loop system is given by

$$TF(j\omega) = \frac{Y(j\omega)}{X(j\omega)} = \frac{H(j\omega)}{1 + H(j\omega)} \tag{5.123}$$

At the oscillation frequency ω_0, the oscillation condition is fulfilled, and it holds

$$H(j\omega)|_{\omega=\omega_0} = -1 \tag{5.124}$$

The transfer function thus is singular at the oscillation frequency, as expected. But phase noise is observed close to the carrier, at a frequency $\omega = \omega_0 + \Delta\omega$. Thus, approximating H by a Taylor series gives

$$H(j\omega)|_{\omega=\omega_0+\Delta\omega} = H(j\omega_0) + \frac{dH(j\omega)}{d\omega}\bigg|_{\omega=\omega_0} \cdot \Delta\omega \tag{5.125}$$

Inserting this approximation into the formula for *TF* yields

$$TF(j\omega_0 + \Delta\omega) = \frac{Y(j\omega_0 + \Delta\omega)}{X(j\omega_0 + \Delta\omega)} \approx \frac{-1}{\frac{dH(j\omega)}{d\omega}\big|_{\omega=\omega_0} \cdot \Delta\omega} \tag{5.126}$$

This formula describes how $1/f$ or white noise is transferred to become noise sidebands close to the carrier. The open-loop transfer function $H(j\omega)$, of course, needs to be defined and calculated differently for each location where a noise source is found in the circuit.

When dealing with noise powers, it is of course required to use the square of the transfer function, as it is defined in terms of voltages and currents:

$$|TF(j\omega_0 + \Delta\omega)|^2 = \left| \frac{-1}{\frac{dH(j\omega)}{d\omega}\big|_{\omega=\omega_0} \cdot \Delta\omega} \right|^2 \tag{5.127}$$

As a practical example, a Colpitts oscillator will be regarded, as discussed in greater detail in Ref. [12]. The equivalent circuit is shown in Figure 5.73. The bipolar transistor is represented in the figure by a noise-free two-port and its base resistance. The following noise sources are present in the oscillator:

- thermal noise associated with the losses of the resonator, $\langle|i_{nr}|^2\rangle = 4kTB/R_p$, with the Boltzmann constant k, the actual temperature T, the bandwidth B, and the loss resistance of the resonator, R_p;

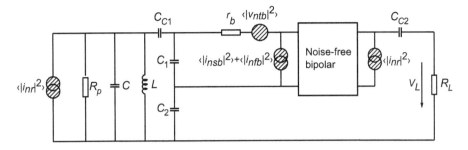

Figure 5.73 Schematic of the oscillator circuit with noise sources.

- thermal noise associated with the transistor's base resistance, $\langle|v_{ntb}|^2\rangle = 4kTBr_b$, with the base resistance R_b;
- transistor shot noise at the base, $\langle|i_{nsb}|^2\rangle = 2eBI_b$, with the electron charge e, and the base current I_b;
- transistor flicker ($1/f$) noise at the base, $\langle|i_{nfb}|^2\rangle = K_f BI_b^{AF}/f_m$, with the parameters K_f and AF and the baseband frequency f_m;
- transistor shot noise at the collector, $\langle|i_{nc}|^2\rangle = 2eBI_c$, with the collector current I_c.

As a further assumption, we consider the baseband noise to be completely upconverted to the oscillation frequency $f_0 = \omega_0/(2\pi)$. This assumption has no effect on the white noise sources, but shifts the flicker noise from $K_f I_b^{AF}/f_m$ in baseband to the noise sidebands around $K_f I_b^{AF}/\Delta f$.

The phase-noise contributions can finally be given by the following formulas [12]. The formulas consist of two parts. The squared magnitude term gives the transfer function, as defined above in equation (5.126). The leading term on the right-hand side gives the noise spectral power density of the physical source, as introduced above. Please note that the bandwidth B is omitted in this description, since the phase noise is given in dBc/Hz.

The phase-noise contribution of the resonator losses is given by

$$PN_{inr}(\omega_0 + \Delta\omega) = \frac{4kT}{R_p} \left| \frac{1}{2} \left(\frac{1}{2j\omega_0 C_{eff}} \right) \left(\frac{\omega_0}{\Delta\omega} \right) \right|^2 \tag{5.128}$$

The following equation gives the phase-noise contribution of the transistor's base resistance:

$$PN_{vntb}(\omega_0 + \Delta\omega) = 4kTr_b \left| \frac{1}{2} \left(\frac{C_1 + C_2}{C_2} \right) \left(\frac{1}{2jQ} \right) \left(\frac{\omega_0}{\Delta\omega} \right) \right|^2 \tag{5.129}$$

The transistor's base shot noise and flicker noise sources are located at the same branch of the equivalent circuit, thus the transfer functions are identical.

$$PN_{insb}(\omega_0 + \Delta\omega) = 2eI_b \left| \frac{1}{2} \left(\frac{C_2}{C_1 + C_2} \right) \left(\frac{1}{2j\omega_0 Q C_{eff}} \right) \left(\frac{\omega_0}{\Delta\omega} \right) \right|^2 \tag{5.130}$$

$$PN_{infb}(\omega_0 + \Delta\omega) = \frac{K_f I_b^{AF}}{f_\Delta} \left| \frac{1}{2} \left(\frac{C_2}{C_1 + C_2} \right) \left(\frac{1}{2j\omega_0 Q C_{eff}} \right) \left(\frac{\omega_0}{\Delta\omega} \right) \right|^2 \tag{5.131}$$

Finally, the phase-noise component caused by the collector shot noise reads

$$PN_{insc}(\omega_0 + \Delta\omega) = 2eI_c \left| \frac{1}{2} \left(\frac{C_1}{C_1 + C_2} \right) \left(\frac{1}{2j\omega_0 QC_{eff}} \right) \left(\frac{\omega_0}{\Delta\omega} \right) \right|^2 \quad (5.132)$$

The transfer functions rely on an effective capacitance C_{eff} that is defined as

$$C_{eff} = C + \frac{C_1 C_2}{C_1 + C_2}$$

For the given circuit, these equations allow for an estimation of the main contributors to the oscillator phase noise.

5.6.3 AM-to-PM Conversion

We went into more detail in dealing with the mechanism and influence of tuning diodes in the voltage-controlled oscillators section, where we evaluated the various methods of building voltage-tunable oscillators using tuning diodes. In this section, we limit ourselves to practical results.

The loading effect of the tuning diode is due to losses, and these losses can be described by a resistor in parallel with the tuned circuit.

It is possible to define an equivalent noise $R_{a,eq}$ that, inserted in Nyquist's equation,

$$V_n = \sqrt{4kT_0 R\Delta f} \quad (5.133)$$

where $kT_0 = 4.2\times10^{-21}$ at about 300 K, R is the equivalent noise resistor, and Δf is the bandwidth, determines an open-circuit noise voltage across the tuning diode. Practical values of $R_{a,eq}$ for carefully selected tuning diodes are in the vicinity of 200 Ω to 50 kΩ. If we now determine the noise voltage $V_n = \sqrt{4 \times 4.2 \times 10^{-21} \times 10,000}$, the resulting voltage value is 1.296×10^{-8} V $\sqrt{Hz}$.

This noise voltage generated from the tuning diode is now multiplied with the VCO gain K_0, resulting in the rms frequency deviation

$$\left(\Delta f_{rms} \right) = K_0 \times \left(1.296 \times 10^{-8} V \right) \qquad \text{in 1-Hz bandwidth} \qquad (5.134)$$

To translate this into an equivalent peak phase deviation,

$$\theta_d = \frac{K_0\sqrt{2}}{f_m} \left(1.296 \times 10^{-8} \right) \qquad \text{rad in 1-Hz bandwidth} \qquad (5.135)$$

or for a typical oscillator gain of 100 kHz/V,

$$\theta_d = \frac{0.00183}{f_m} \qquad \text{rad in 1-Hz bandwidth} \qquad (5.136)$$

For $f_m = 25$ kHz (typical spacing for adjacent-channel measurements for FM mobile radios), the $\theta_c = 7.32\times10^{-8}$. This can be converted now into the SSB signal-to-noise ratio as

$$\mathcal{L}(f_m) = 20 \log_{10} \frac{\theta_c}{2} = -149 \text{ dBc/Hz} \qquad (5.137)$$

For the typical oscillator gain of 10 MHz/V found in wireless applications, the resulting phase noise will be 20 dB worse [10 log (10 MHz ÷ 100 kHz)]. However, the best tuning diodes, like the BB104, have an R_n of 200 Ω instead of 10 kΩ, which again changes the picture. According to equation (5.133), with $kT_0 = 4.2 \times 10^{-21}$, the resulting noise voltage will be

$$V_n = \sqrt{4 \times 4.2 \times 10^{-21} \times 200}$$
$$= 1.833 \times 10^{-9} \; V\sqrt{Hz} \tag{5.138}$$

From equation (5.114), the equivalent peak phase deviation for a gain of 10 MHz/V in a 1-Hz bandwidth is then

$$\theta_d = \frac{1 \times 10^7 \sqrt{2}}{f_m} \left(1.833 \times 10^{-9} \right) \; \text{rad} \tag{5.139}$$

or

$$\theta_d = \frac{0.026}{f_m} \quad \text{rad in 1-Hz bandwidth} \tag{5.140}$$

With $f_m = 25 \, \text{kHz}$, $\theta_c = 1.04 \times 10^{-6}$. Expressing this as phase noise

$$\mathcal{L}(f_m) = 20 \log_{10} \frac{\theta_c}{2} = -126 \; \text{dBc/Hz} \tag{5.141}$$

The phase-noise value in equation (5.137) is that typically achieved in the Rohde & Schwarz SMDU or with the Hewlett-Packard 8640 signal generator, and considered state of the art for a free-running oscillator. It should be noted that both signal generators use a slightly different tuned circuit; the Rohde & Schwarz generator uses a helical resonator, whereas the Hewlett-Packard generator uses an electrically shortened quarter-wavelength cavity. Both generators are mechanically pretuned, and a tuning diode with a gain of about 100 kHz/V is used for AFC and frequency-modulation purposes. It is apparent that, because of the nonlinearity of the tuning diode, the gain is different for low dc voltages than for high dc voltages. The impact of this is that the noise varies with the tuning range.

If this oscillator had to be used for a synthesizer, its 1-MHz tuning range would be insufficient; therefore, a way had to be found to segment the band into the necessary ranges. In VCOs, this is typically done with switching diodes that allow the proper frequency bands to be selected. These switching diodes insert in parallel or series, depending on the circuit, with additional inductors or capacitors, depending on the design.

In low-energy-consuming circuits, the VCO frequently is divided into a coarse-tuning section using tuning diodes and a fine-tuning section with a tuning diode. In the coarse-tuning range, this results in very high gains, such as 1–10 MHz/V, for the diodes. The noise contribution of those diodes is therefore very high and can hardly be compensated by the loop. For low-noise applications, which automatically mean higher power consumption, it is unavoidable to use switching diodes.

Let us now examine some test results. Referring back to equation (5.106), Figure 5.74 shows the noise sideband performance as a function of Q, whereby the top curve with $Q_L = 100$ represents a somewhat poor oscillator and the lowest curve with $Q_L = 100,000$ probably represents a crystal oscillator where the unloaded Q of the crystal was in the vicinity of 3×10^6. Figure 5.75 shows the influence of flicker noise.

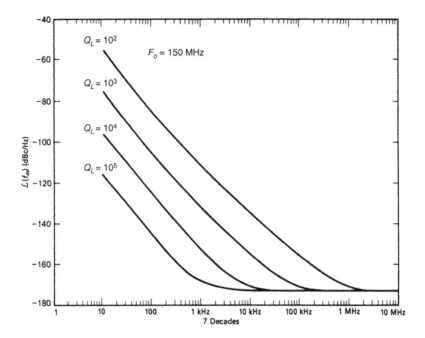

Figure 5.74   Noise sideband of an oscillator at 150 MHz as a function of the loaded Q of the resonator.

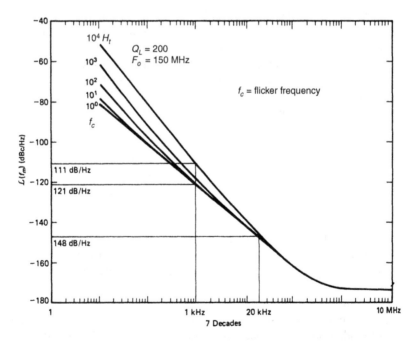

Figure 5.75   Noise-sideband performance as a function of the flicker frequency ω_C varying from 10 Hz to 10 kHz.

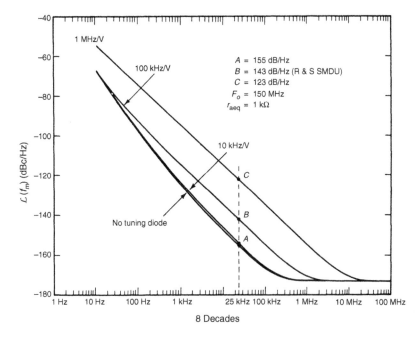

Figure 5.76 Noise-sideband performance of an oscillator at 150 MHz, showing the influence of various tuning diodes.

Corner frequencies of 10 Hz to 10 kHz have been selected, and it becomes apparent that around 1 kHz the influence is fairly dramatic, whereas the influence at 20 kHz off the carrier is not significant. Figures 5.76 and 5.77 show the influence of the tuning diodes on a high-Q oscillator.

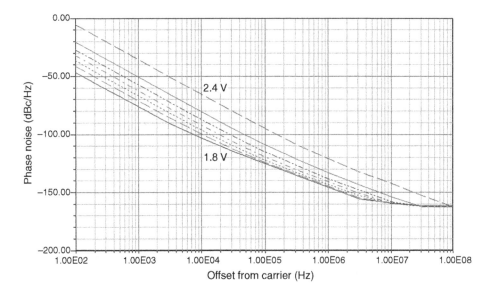

Figure 5.77 Noise-sideband performance of an oscillator at 147 MHz, showing the influence of a tuning diode operated at bias voltages from 1.8 to 2.4 V in 0.1-V increments.

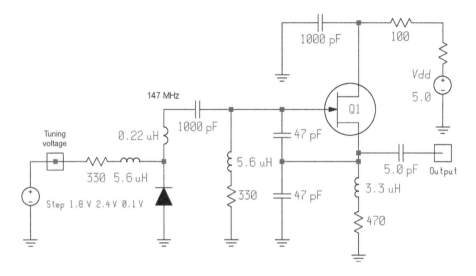

Figure 5.78 The VCO schematic. Because the tuning diode is in series with the resonator, it passes all of the tank's RF current and plays a disproportionately large role in determining the circuit's phase-noise performance.

Curve A in Figure 5.76 uses a lightly coupled tuning diode with a K_0 of 10 kHz/V; the lower curve is the noise performance without any diode. As a result, the two curves are almost identical, which can be seen from the somewhat smeared form of the graph. Curve B shows the influence of a tuning diode at 100 kHz/V and represents a value of −143 dBc/Hz from −155 dBc/Hz, already some deterioration. Curve C shows the noise if the tuning diode results in a 1-MHz/V VCO gain, and the noise sideband at 25 kHz has now deteriorated to −123 dBc/Hz. These curves speak for themselves.

Figure 5.77 shows how the phase noise of a 147-MHz VCO varies with the bias applied to its tuning diode. Because a tuning diode's Q increases with the reverse bias applied to it, we expect a VCO's phase-noise performance to improve, even if only slightly, as its tuning voltage is increased. Yet Figure 5.77 illustrates the reverse of this relationship. Inspecting this VCO's circuit (Figure 5.78), we see that the diode is in series with the circuit's 220-nH resonator. All the RF current flowing through the resonator must flow through the relatively low-Q diode, and even small decreases in the diode's capacitance will significantly modify the tank's LC ratio.

Graphing the diode's ac load line (Figure 5.79), we see that as the tuning voltage increases (i.e., as the tuning diode capacitance decreases), the RF voltage across the diode becomes disproportionately high. We also note that at every value of tuning voltage, the RF voltage across the diode goes positive (relative to the anode) during part of each cycle.

Modifying the oscillator tank to add capacitance in parallel with the tuning diode (Figure 5.80) reduces the tuning diode's effect on tank Q and LC ratio, and therefore its contribution to phase noise. With this arrangement, we now see the expected relationship between tuning bias and phase noise: the higher the tuning bias, the lower the phase noise (Figure 5.81). Figure 5.82 shows that the diode's ac load line is now virtually unchanged across the tuning-voltage range. The RF voltage across the diode still goes positive (relative to the anode) during part of each cycle.

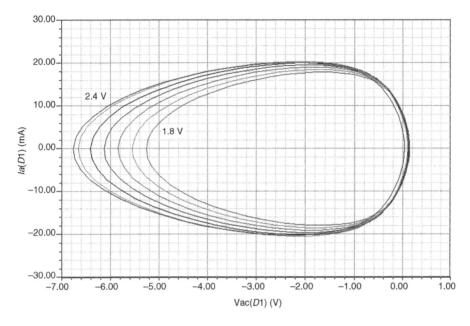

Figure 5.79 Graphing the tuning diode's ac load line reveals that increasing the diode's tuning bias (decrease its capacitance) results in a disproportionally large increase in the RF voltage across the diode. At all tuning voltages, the RF voltage across the diode is slightly positive (relative to the anode) during part of each cycle.

Figure 5.83 shows one more modification to the VCO tank. Now, the tuning diode is lightly coupled to the hot end of the tank, in parallel with the resonator. The result is phase-noise performance that is essentially unchanged across the 1.8–2.4 V tuning-voltage range. Figure 5.84 compares this circuit's phase-noise performance with the worst-case performance of the tank configurations in Figures 5.78 and 5.80. Figure 5.85 compares

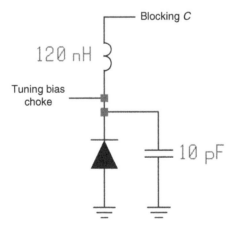

Figure 5.80 Adding capacitance across the tuning diode reduces the effect of its nonlinearities on phase-noise performance. We have also revalued the resonator to 120 nH to maintain oscillation near 147 MHz.

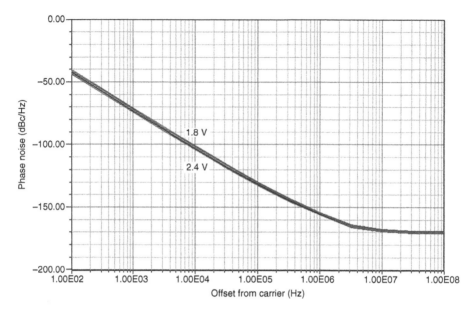

Figure 5.81   Now that the tuning diode passes only part of the tank's RF current, varying its capacitance does not result in the large change in phase-noise performance shown in Figure 5.77. Also note that the phase-noise versus tuning-bias relationship has reversed relative to that shown in Figure 5.77: the highest tuning voltage now corresponds to the lowest phase noise.

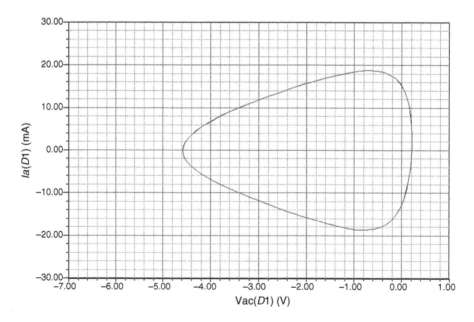

Figure 5.82   AC load line for the tuning diode in Figure 5.80. Now the voltage swing across the diode is virtually the same at all tuning voltages. The RF voltage across the diode still goes positive (relative to the anode) during part of the cycle.

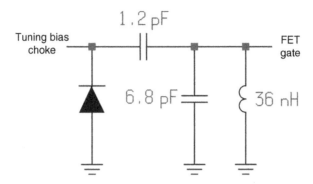

Figure 5.83 Tank circuit reconfigured per Figure 5.54 to lightly couple the tuning diode in parallel with the resonator, with the resonator inductance again decreased to maintain oscillation near 147 MHz. The 1000-pF capacitor and gate choke of Figure 5.78 (5.6 μH in series with 330 Ω) are no longer needed.

the diode ac load lines that correspond to the phase-noise curves in Figure 5.84. The best phase-noise performance corresponds to the diode ac load line that exhibits the narrowest voltage swing and never goes positive relative to the anode.

It is important to keep in mind that we have not gained something for nothing in achieving the best phase-noise response in Figure 5.84. As we reduced the diode's phase-noise contribution, we also reduced the oscillator's tuning range—from 7.1% (Figure 5.78 circuit) to 1.8% (Figure 5.80 circuit) to 0.046% (Figure 5.83). Simultaneously, optimizing phase-noise performance and tuning range requires a combination of techniques. First, we would maximize diode Q by using as high a tuning voltage as feasible. Second, we would use antiseries-connected diodes to minimize the RF voltage swing across them (Figure 5.55).

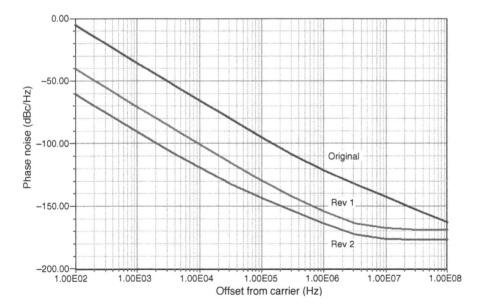

Figure 5.84 Comparison of worst-case phase-noise performance obtained with the tank configurations of Figures 5.78 (Original), 5.80 (Rev 1), and 5.83 (Rev 2).

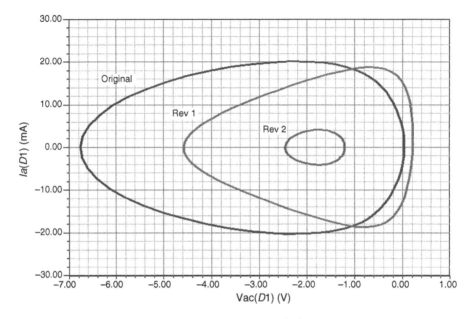

Figure 5.85 Comparison of the tuning diode's ac load line for the worst-case phase-noise curves of Figure 5.82.

Both of these techniques work against the achievement of wide tuning ranges by minimizing the capacitance swing the diodes can contribute. As a result of this, and because hyperabrupt diodes are not available for the frequencies at which wireless VCOs operate, we would therefore use multiple antiseries diode pairs in parallel.

It is of interest to compare various oscillators. Figure 5.86 shows the performance of a 10-MHz crystal oscillator, a 40-MHz *LC* oscillator, the HP8640 cavity-tuned oscillator at 500 MHz, the 310–640-MHz switched-reactance oscillator of the HP8662 oscillator, and a 2–6-GHz YIG oscillator at 6 GHz.

5.6.4 Numerically Optimized Oscillators

We will look at two examples where we compare predicted and measured data. Figure 5.87 shows the abbreviated circuit of a 10-MHz crystal oscillator. It uses a high-precision, high-Q crystal made by companies such as Bliley. Oscillators like this are intended for use as frequency and low-phase-noise standards. In this case, the circuit under consideration is part of the HP3048 phase-noise measurement system.

Figure 5.88 shows the measured phase noise of this HP frequency standard and Figure 5.89 shows the phase noise predicted using the mathematical approach outlined above.

In corporation with Motorola, we also analyzed an 800-MHz VCO. In this case, we also did the parameter extraction for the Motorola transistor. Figure 5.90 shows the circuit, a Colpitts oscillator that uses RF feedback in the form of a 15-Ω resistor and a capacitive voltage divider consisting of 1 pF between the BJT's base and the feedback resistor, and 1 pF between the feedback resistor and common. Also, the tuned circuit is loosely coupled to this part of the transistor circuit. Figure 5.91 shows a comparison between predicted and

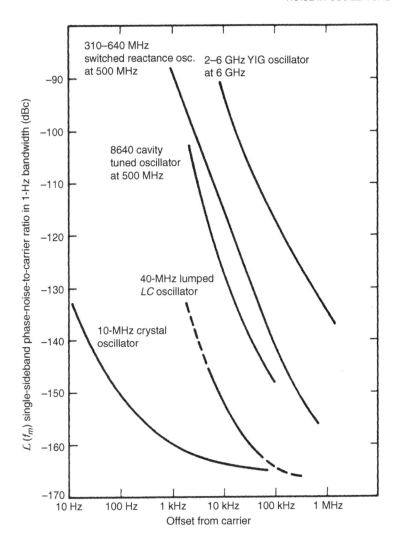

Figure 5.86 Comparison of noise-sideband performances of a crystal oscillator, *LC* oscillator, cavity-tuned oscillator, switched-reactance oscillator, and YIG oscillator.

measured phase noise for this oscillator. Figures 5.92–5.94 show the oscillator's predicted output waveform, spectrum, and dc I-V and ac load line responses, respectively.

5.6.4.1 Phase-Noise Optimization

By allowing the simulator's optimizer to vary the oscillator's feedback and dc operating point, phase noise can be improved. The values that were allowed to vary were the values of the capacitors in the feedback voltage divider, the value of a negative-feedback resistor (starting value, 15 Ω) that reduces AM-to-PM conversion, and the emitter resistor, which changes the transistor's dc bias and therefore affects its bias-dependent flicker noise. Figure 5.95 shows the improvement of the 800-MHz VCO as previously shown. While the close-in phase noise can be improved drastically by approximately 32 dB, the far-out noise

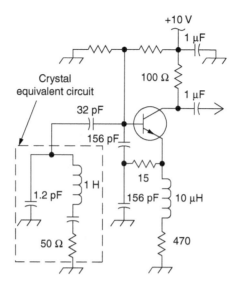

Figure 5.87 Abbreviated circuit of a 10-MHz crystal oscillator.

at 20 MHz and beyond is deteriorated. This is due to the resistive feedback, which reduces AM-to-PM conversion [27–29].

In testing an oscillator's resistance to frequency pushing (frequency shift with supply-voltage changes), variations in phase noise can be observed. Phase noise is also bias-

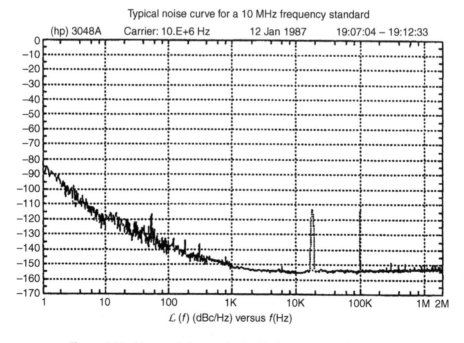

Figure 5.88 Measured phase noise for this frequency standard by HP.

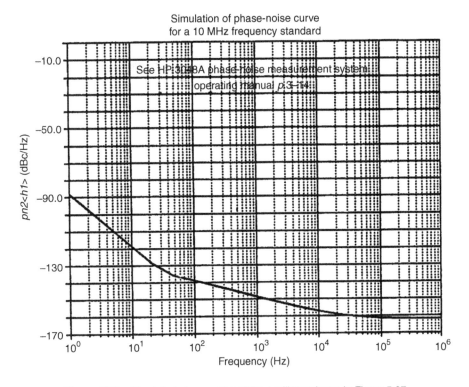

Figure 5.89   Simulated phase noise of the oscillator shown in Figure 5.87.

dependent, and this fact can be used for phase-noise optimization. Figure 5.96 shows the phase noise of an oscillator as its supply voltage is varied.

5.7 OSCILLATORS IN PRACTICE

5.7.1 Oscillator Specifications

Although we have tended to emphasize phase noise as a critical indicator of oscillator quality, phase noise is only one of a number of oscillator performance criteria that must be considered in wireless system design. In addition to phase noise, the following characteristics are used by commercial companies to describe oscillator performance.

Frequency pushing characterizes the degree to which an oscillator's frequency is affected by its supply voltage. A sudden current surge caused by activating a transceiver's RF power amplifier may produce a spike on the VCO's dc power supply and a consequent frequency jump. Like tuning sensitivity (below), pushing is specified in frequency/voltage form, and is tested by varying the VCO's dc supply voltage (typically ±1 V) with its tuning voltage held constant.

> *Harmonic Output Power.* The harmonic content is measured relative to the output power. Typical values are 20 dB or more suppression relative to the fundamental. This suppression can be improved by additional filtering.

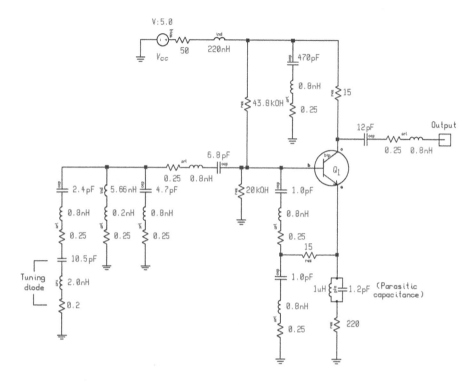

Figure 5.90 Colpitts oscillator that uses RF negative feedback between the emitter and capacitive voltage divider. To be realistic, we have also used real components rather than ideal ones. The suppliers for the capacitors and inductors provide some typical values for the parasitics. The major changes are 0.8 nH and 0.25 Ω in series with the capacitors. The same thing applies for the main inductance, which has a parasitic connection inductance of 0.2 nH in series with a 0.25-Ω resistance. These types of parasitics are valid for a fairly large range of components assembled in surface-mount applications. Most engineers model the circuit only by assuming lossy devices, not adding these important parasitics. One of the side effects we have noticed is that the output power is more realistic and, needless to say, the simulated phase noise agrees quite well with measured data. This circuit can also serve as an example for modeling amplifiers and mixers using surface-mount components.

Output Power. The output power of the oscillator, typically expressed in dBm, is measured into a 50-Ω load. The output power is always combined with a specification for flatness or variation. A typical specification would be 0 dBm ±1 dB.

Output Power as a Function of Temperature. All active circuits vary in performance as a function of temperature. An oscillator's output power over a temperature range should vary less than a specified value, such as 1 dB.

Post-Tuning Drift. After a voltage step is applied to the tuning diode input, the oscillator frequency may continue to change until it settles to a final value. This post-tuning drift is one of the parameters that limits the bandwidth of the VCO input.

Power Consumption. This characteristic conveys the dc power, usually specified in milliwatts and sometimes qualified by operating voltage, required by the oscillator.

Sensitivity to Load Changes. To keep costs down, many wireless applications use a VCO alone, without the buffering action of a high reverse-isolation amplifier stage. In such applications, frequency pulling, the change of frequency resulting from partially

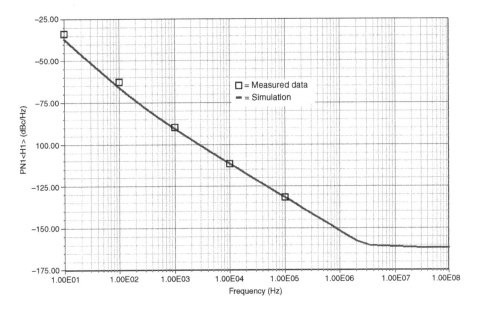

Figure 5.91 Comparison between predicted and measured phase noise for the oscillator shown in Figure 5.90.

reactive loads, is an important oscillator characteristic. Pulling is commonly specified in terms of the frequency shift that occurs when the oscillator is connected to a load that exhibits a nonunity VSWR (such as 1.75, usually referenced to 50-Ω), compared to the frequency that results with a unity-VSWR load (usually 50-Ω). Frequency

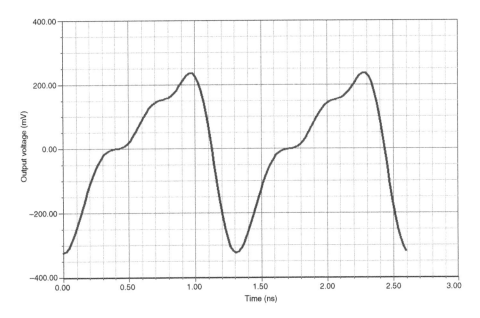

Figure 5.92 Predicted output waveform for the oscillator shown in Figure 5.90.

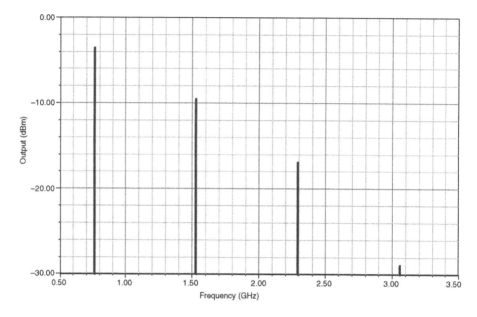

Figure 5.93   Predicted output spectrum for the oscillator shown in Figure 5.90.

pulling must be minimized, especially in cases where power stages are close to the VCO unit and short pulses may affect the output frequency. Such feedback may make phase locking impossible.

Spurious Outputs. A VCO's spurious output specification, expressed in decibels, enumerates the strength of unwanted and nonharmonically related components relative

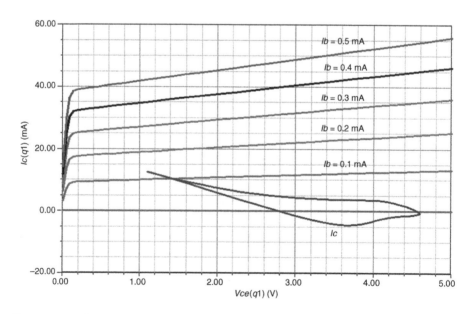

Figure 5.94   AC load line and dc I-V curves for the transistor in the oscillator shown in Figure 5.90.

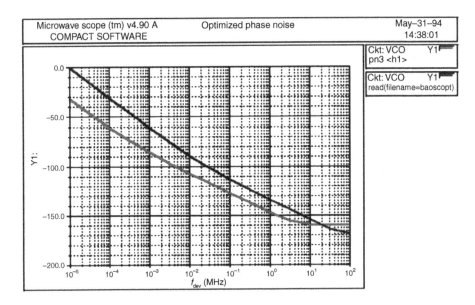

Figure 5.95   Phase noise of the 800-MHz VCO before and after optimization.

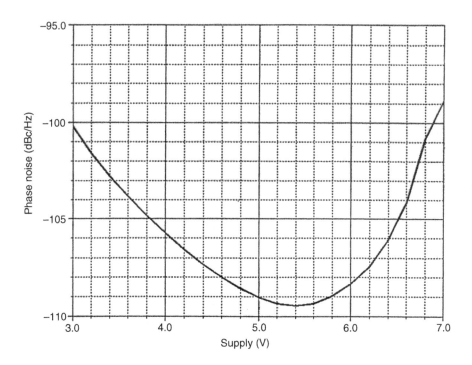

Figure 5.96   Calculated phase noise as a function of the supply voltage for a BJT oscillator. The graph shows a distinct minimum for a particular bias, which could be translated into a collector current change. Such phase-noise minimization could also be accomplished by changing just the voltage applied to the bias portion of the oscillator.

to the oscillator fundamental. Because a stable, properly designed oscillator is inherently clean, such spurs are typically introduced only by external sources in the form of radiated or conducted interference. See harmonic output power.

Temperature Drift. Although the synthesizer is responsible for locking and maintaining the oscillator's frequency, the VCO's frequency change as a function of temperature is a critical parameter and must be specified. Its value varies between 10 kHz/°C to several hundred kHz/°C depending on the center frequency and tuning range.

Tuning Characteristic. This specification shows the relationship, depicted as a graph, between a VCO's operating frequency and the tuning voltage applied. Ideally, the correspondence between operating frequency and tuning voltage is linear. See tuning linearity.

Tuning Linearity. For stable synthesizers, a constant deviation of frequency versus tuning voltage is desirable. It is also important to make sure that there are no breaks in the tuning range—for example, that the oscillator does not stop operating with a tuning voltage of 0 V. See tuning characteristic.

Tuning Sensitivity, Tuning Performance. This datum, typically expressed in megahertz per volt (MHz/V), enumerates how much a VCO's frequency changes per unit of tuning-voltage change.

Tuning Speed. This characteristic is defined as the time necessary for the VCO to reach 90% of its final frequency on the application of a tuning-voltage step. Tuning speed depends on the internal components between the input pin and tuning diode—including, among other things, the capacitance present at the input port. The input port's parasitic elements determine the VCO's maximum possible modulation bandwidth.

5.7.2 More Practical Circuits

Because of the need for integration, we will present three IC oscillators of increasing complexity. The circuit in Figure 5.97 is found in early Siemens ICs. Figure 5.98 shows its phase-noise performance.

Figure 5.99 shows the basic push–pull oscillator topology underlying the circuit in Figure 5.97. The omission of the current source and active bias components improves the simpler circuit's phase-noise performance over that of the circuit shown in Figure 5.97. Figure 5.100 compares the phase noise of this basic circuit to the integrated version in Figure 5.97.

Figure 5.101 is for those who speculate on the possibility of replacing the BJTs in Figure 5.99 with LDMOS FETs or JFETs. Figure 5.102 compares its phase noise to that of the BJT case. The rule of thumb is that FETs do not have enough gain for high-Q oscillation in oscillators above 400 MHz, and bipolar transistors are a better choice. Because of their higher flicker noise contribution, GaAsFETs should be considered only at frequencies above 4 or 5 GHz.

Figure 5.103 shows a circuit based on two differential amplifiers and frequently found in Gilbert cell IC applications. Figure 5.104 shows its phase-noise performance.

Figure 5.105 is a validation circuit that demonstrates a comparison of measured phase-noise performance versus its prediction by circuit simulation software. The circuit is based on the core of Motorola's MC12148 ECL oscillator IC, the topology of which is derived from the superseded MC1648. Motorola specifies the MC12148's typical phase-noise performance as −90 dBc/Hz at an offset of 25 kHz. The close agreement between the measured

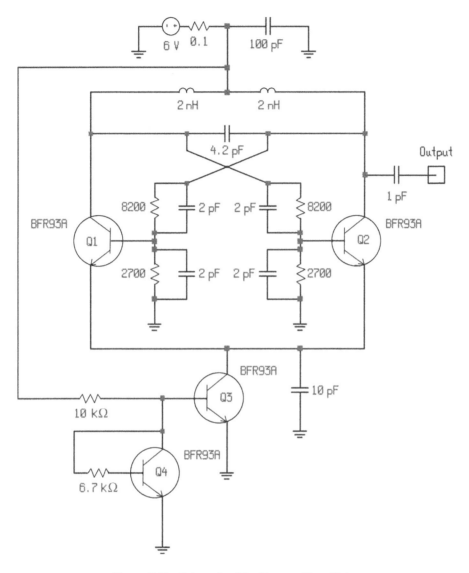

Figure 5.97   Schematic of the Siemens IC oscillator.

and simulated results highlights the value of state-of-the-art CAD tools in wireless oscillator design (Figure 5.106).

Figure 5.106 presents the results of two simulations: one with ideal L and C components and another with realistic values specified for these reactances. In using a circuit simulator to validate and compare oscillator topologies, it is important to start with ideal components because we are interested in comparing the inherent phase-noise performance of various circuits—particularly the role played by the nonlinear reactances in active devices. Because the Qs of practical components are so low in the microwave range (very generally, 200–300 for capacitors and 60–70 for inductors), the inherent merits of one topology, or one device, over another would be masked by loading effects if we conducted our investigations using

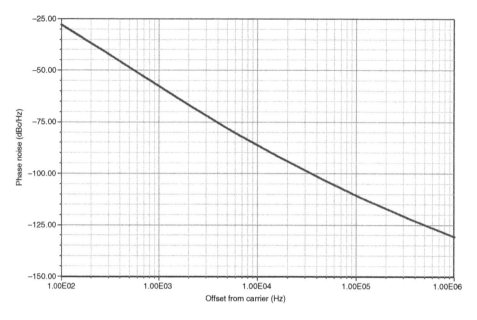

Figure 5.98   Phase noise of the Siemens IC oscillator. The frequency of oscillation is 1.051 GHz.

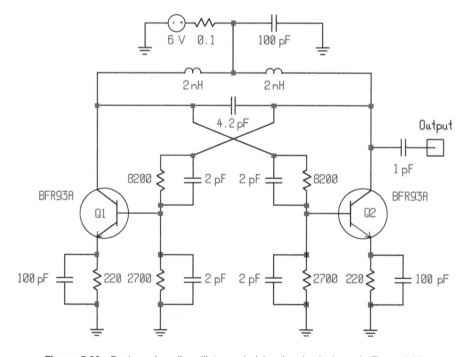

Figure 5.99   Basic push–pull oscillator underlying the circuit shown in Figure 5.97.

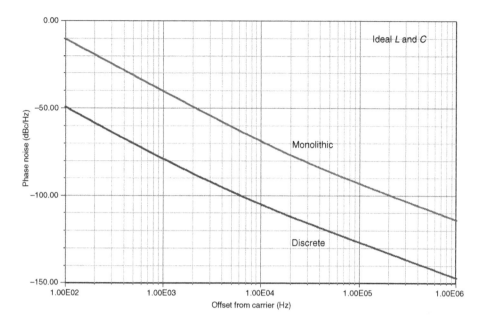

Figure 5.100 The basic version of the oscillator (discrete curve) is a better phase-noise performer than the integrated version (monolithic curve). The frequency of oscillation is 1.039 GHz.

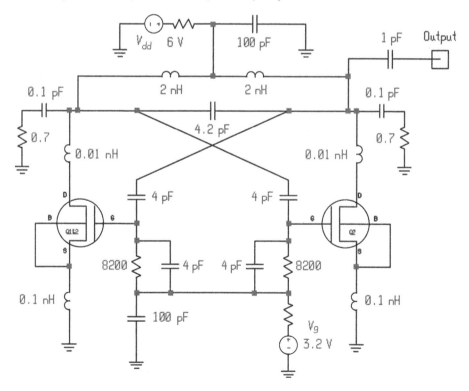

Figure 5.101 1.04-GHz push–pull oscillator using LDMOS FETs instead of BJTs. (LDMOS FET parameters courtesy of David K. Lovelace, Motorola, Inc.)

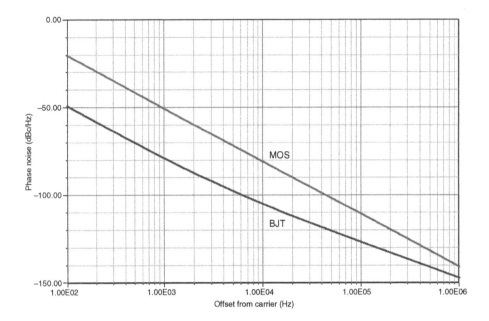

Figure 5.102 Phase-noise comparison of the BJT and MOSFET oscillators.

nonideal reactances. Once initial evaluation is completed, realistic Qs should be specified for increased accuracy in phase-noise, output-power, and output-spectrum simulation.

Finally, as evidence of how moving from lumped to distributed techniques can improve oscillator performance at frequencies where LC tanks become problematic, Figure 5.107 compares, for a simulated BJT Colpitts oscillator operating at 2.3 GHz, the difference in phase-noise performance obtainable with a resonator consisting of an ideal 2-nH inductor and a 1/4-λ transmission line (11 Ω, 90° long at 2.6 GHz, attenuation 0.1 dB/m) with the transistor biased by constant-current and constant-voltage sources.

5.7.2.1 Silicon/GaAs-Based Integrated VCOs and Possible Difficulties

Due to process variations, the center frequency of the VCO cannot be repeated without trimming. Trimming can be avoided by making the VCO gain large enough to compensate for tank circuit variations, but this degrades phase-noise performance and increases the VCO susceptibility to unwanted spurs. Example: A VCO with a gain of 10 MHz/V has a phase noise of −80 dBc/Hz at 10 kHz while a VCO with a gain of 75 MHz/V has a phase noise of only −60dBc/Hz at 10 kHz. All but the tank inductors are integrated.

Tank inductors are generally not integrated for reasons of size and Q. The area of a 1-nH inductor is approximately 42,000 mm^2, which is $\approx 3.5\%$ of a typical IC; the area of a 10-nH inductor is approximately 167,000 mm^2, which is $\approx 14\%$ of a typical IC. For GaAs substrate, the resulting Q would be about 5; for silicon, as we already saw, 20 is already a very high value.

The customer can use PCB traces for the required inductance instead of external surface mount inductors (cost: $0.03/each, most expensive passive component). It must be pointed out that even for a nonintegrated approach, the higher VCO gain is noisier by definition. This was explained in Section 5.6.3. These problems are not unique to silicon; they apply to GaAs also. Figure 5.108 shows the size relationship of on-chip capacitors and inductors

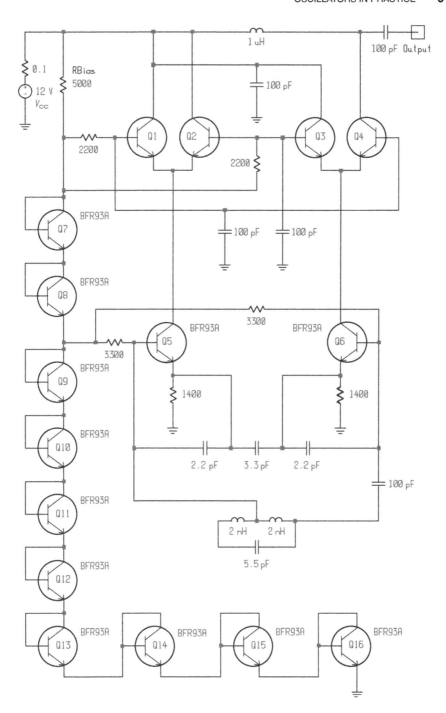

Figure 5.103 1.022-GHz IC oscillator based on two differential amplifiers. With an RF signal fed to the bases of Q1–Q4 and Q2–Q3 in push–pull, and IF output extracted from the collectors of Q1–Q3 and Q2–Q4 in push–pull, the circuit would act as a Gilbert cell converter. In the circuit shown, we have modified the differential amplifiers' collector circuitry to give single-ended output and avoid cancellation of the oscillator signal.

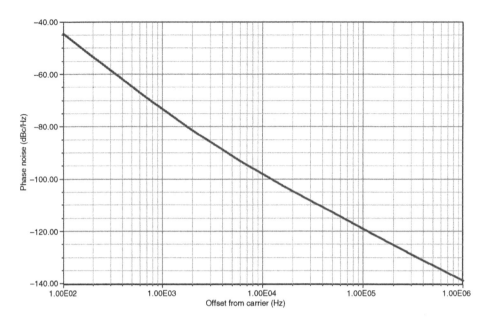

Figure 5.104   Phase noise of the two-differential-amplifier oscillator circuit.

compared to the actual transistor; therefore, on-chip inductors should really be avoided because of reasons of space and Q. Just imagine how the picture would change if the 1.1-nH inductor had to be 11 nH instead! It would dominate the IC area, and therefore the IC cost as well.

5.8   PHASE-NOISE IMPROVEMENTS OF INTEGRATED RF AND MILLIMETERWAVE OSCILLATORS[2]

5.8.1   Abstract

The generation of microwave and millimeterwave frequencies can be done by lower-frequency VCOs multiplied up in frequency, such as that provided by comb generators, by using hybrid GaAsFET-based oscillators with external resonators, or—for the best phase-noise obtainable—YIG oscillators whose physical size and cost does not always provide a practical solution. A significant improvement in oscillator noise performance can be obtained with a novel feedback circuit that uses internally generated noise and cancels the close-in noise in a bandwidth of up to 1 MHz. This results in more than 15 dB phase-noise improvement at microwave frequencies for the same topology and the feedback system can be made part of the biasing circuit. This feedback circuit can be used over a wide frequency range and works well at VHF/UHF frequencies and in the millimeterwave area.

[2]This section is based on a paper given at the Fifth International Workshop on Integrated Nonlinear Microwave and Millimeterwave Circuits, sponsored by IEEE at Gerhard-Mercator-University, Duisburg, Germany, October 1–2, 1998.

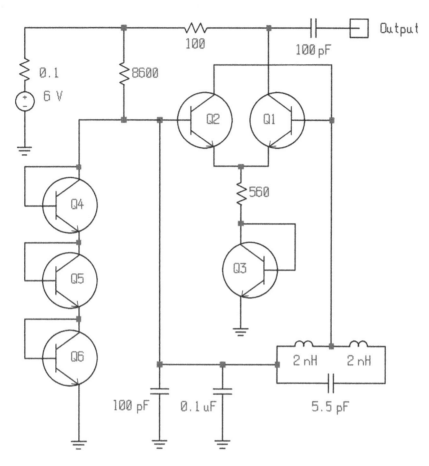

Figure 5.105 This validation circuit models the oscillator topology at the core of Motorola's MC12148 ECL oscillator IC.

5.8.2 Review of Noise Analysis

Section 5.6 of this chapter covered noise in oscillators in detail, so we will merely review oscillator noise issues here. The first linear model for a basic oscillator, without considering semiconductor noise, was developed by Leeson in 1966 [9] and has been quoted in numerous applications.

The phase noise of a VCO is completely determined by

$$\mathcal{L}(f_m) = 10\log\left\{\left[1 + \frac{f_0^2}{(2f_m q_{\text{load}})^2}\right]\left(1 + \frac{f_c}{f_m}\right)\frac{FkT}{2P_{\text{sav}}} + \frac{2kTRK_0^2}{f_m^2}\right\} \qquad (5.142)$$

where $\mathcal{L}(f_m)$ = ratio of sideband power in a 1-Hz bandwidth at f_m to total power in dB, f_m = frequency offset, f_0 = center frequency, f_c = flicker frequency, Q_{load} = loaded Q of the tuned circuit, F = noise factor, $kT = 4.1 \times 10^{-21}$ at 300 K (room temperature), P_{sav} = average power at oscillator output, R = equivalent noise resistance of tuning diode (typically 200 Ω to 10 kΩ), and K_0 = oscillator voltage gain.

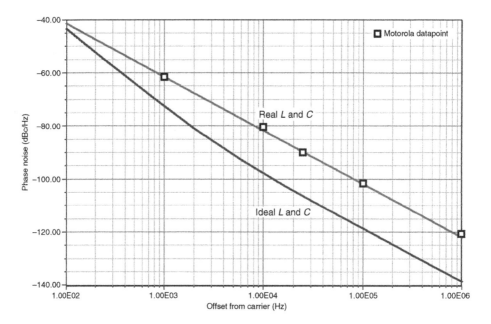

Figure 5.106 Comparison of measured versus simulated phase noise of the MC12148 differential oscillator. The simulation (1.016 GHz) predicts a phase noise of −89.54 dBc/Hz at an offset of 10 kHz (real L and C curve), which agrees well with Motorola's specification of −89.54 dBc/Hz (typical at 930 MHz) at this offset. The ideal L and C curve plots the simulation's phase-noise performance with ideal reactances, as opposed to the real L and C curve, which graphs the results with realistic Qs specified for its reactive components ($Q = 250$ for its C_s and $Q = 65$ for its L, all at 1 GHz).

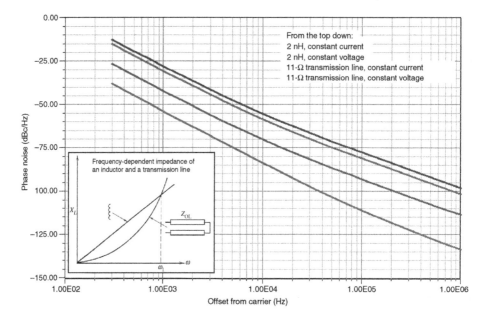

Figure 5.107 Phase-noise performance of a 2.3-GHz BJT oscillator with a resonator consisting of an inductor (2 nH) and a 1/4-λ transmission line (11 Ω, approximating the behavior of a dielectric resonator) with bias from a constant-current source and a low-impedance, resistive constant-voltage source.

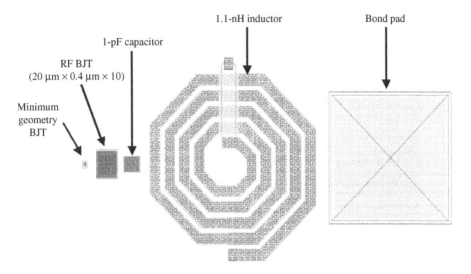

Figure 5.108 Same-scale comparison of space requirements for transistors, capacitors, inductors, and bond pads. For reference purposes, the coil diameter is 0.2 mm.

When adding an isolating amplifier, the noise of an LC oscillator is determined by

$$S_f(f_m) = \left[a_R f_0^4 + a_E (f_0/(2Q_L))^2 \right] / f_m^3$$
$$+ \left[(2GFkT/P_0)(f_0/(2Q_L))^2 \right] / f_m^2$$
$$+ (2a_R Q_L f_0^3)/f_m^2$$
$$+ a_E/f_m + 2GFkT/P_0$$

where $G =$ compressed power gain of the loop amplifier, $F =$ noise factor of the loop amplifier, $k =$ Boltzmann's constant, $T =$ temperature in Kelvins, $P_0 =$ carrier power level (in Watts) at the output of the loop amplifier, $f_0 =$ carrier frequency in Hz, $f_m =$ carrier offset frequency in Hz, $Q_L (= \pi f_0 \tau_g) =$ loaded Q of the resonator in the feedback loop, and α_R and $\alpha_E =$ flicker noise constants for the resonator and loop amplifier, respectively.

In 1978, Dieter Scherer from Hewlett-Packard added the $(1 + f_c/f_m)$ term, which addresses the $1/f$ noise or flicker corner frequency [30]. The $1/f$ frequency phenomenon is based on the surface effects inside the semiconductor and varies from 50 Hz for silicon FETs, to 5 kHz and more for silicon microwave bipolar transistors, to 1 MHz and higher for GaAsFETs. As a rule of thumb, one can state that the higher the frequency of operation, the higher the f_T for the active device, the higher the flicker corner frequency will be. Figure 5.109 shows the spectral distribution as a function of offset frequency. This linearization was first discovered by individuals from NIST, formerly the National Bureau of Standards, in Boulder, Colorado. The left corner is dominated by the flicker corner frequency and the oscillator itself shows at the output a combination of modulation and conversion noise. In addition to the $1/f$ noise, the real-life oscillator exhibits a wideband noise, which is due to nonlinearities—specifically, the AM-to-PM conversion resulting from nonlinear capacitance and transconductance of the transistor. This complete linear "noise equation" was first published in Ref. [2]. Therefore, the noise model for the transistor had to be enhanced by describing this mechanism. Essentially, it is equal to a tuning diode coupled to the resonator, which is modulated by thermal noise, and therefore, result in higher

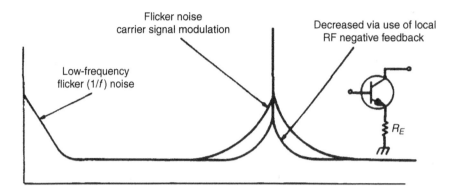

Figure 5.109   Reduction of transistor flicker-of-phase noise via use of local negative feedback (emitter resistance), showing, for a BJT, the flicker noise or corner frequency, as well as the phase noise that results from the flicker noise. The negative feedback from the unbypassed emitter resistor reduces the flicker noise by up 40 dB [31, 32].

phase-noise contribution than the oscillator itself. This noise equation also has the Q playing a major role. The phase noise, therefore, ignoring other frequencies, is proportional to $1/Q^2$. In monolithic circuits, the loss of transmission lines, which determines the Q, is related to the substrate material. One has little choice of material, particularly in microwave and millimeterwave oscillators. Printed resonators provide sadly low Q at these frequencies. A similar problem related to the material is the choices of tuning diodes. In many, if not all, cases, one uses a GaAsFET with the gate connected to the drain or to the source as a tuning diode, and the resulting Q is also disappointing.

The example in Figure 5.109 takes advantage of a microwave bipolar transistor and also is applicable for HBTs (heterojunction bipolar transistors). The transconductance of the bipolar transistor, which is approximately 39 mA/V, increases to values up to 20 times larger than exhibited by GaAsFETs. This method of using negative feedback reduces the gain of the GaAsFET too much, and is therefore, not applicable.

5.8.3   Workarounds

When looking at the modes of operation of test sets for phase-noise measurement, the one method that requires only one signal generator is most desirable. Based on the frequency discriminator method, it uses a cable of appropriate length that acts as a delay line. The oscillator under test drives a double balanced mixer (one port directly [LO port]) and drives the RF port "delayed" via the delay line cable and the mixer is, therefore, driven by the same frequency with a different phase. This setup has been used for phase-noise measurements very successfully. One can easily connect a low-noise FFT-based audio spectrum analyzer to the mixer output and measure the signal-to-noise ratio as a function of offset. With modern FFT analyzers available up to 10 MHz and higher, a fast and reliable measurement can be done. The length of the cable determines the frequency range over which an accurate measurement is possible. This phase-noise information (dc and ac voltage) at the output can be used to modulate the VCO and improve the phase noise in the frequency range over which this method is applicable. Fluke and other companies have built signal generators based on this principle to improve the phase noise. The drawback of this method is that the delay line acts like a band-limiting filter, and therefore, one cannot compensate over a large range of frequencies. The delay line in question, responsible for the phase shift, can also

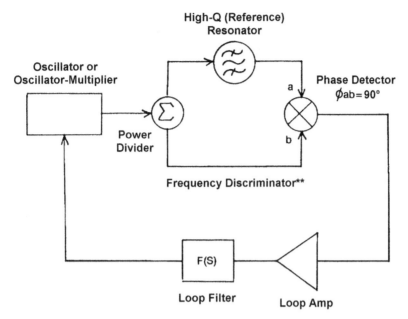

**Resonator may be operated in reflection mode as well as transmission mode.

Figure 5.110   AFC stabilization of an external oscillator.

be substituted by a high-Q resonator. By closing the loop, one can "clean up" the phase noise of the device under test (DUT). The difficulty with this is that it requires an external high-Q resonator. One might as well use the same resonator for the oscillator. Therefore, this method is not suitable for MMICs. See Figure 5.110 for more details.

As far as the actual measurement is concerned, one has to observe certain limitations of the test setup as explained in Figure 5.111. When using the delay line instead of a high-Q resonator (low loss is assumed for coaxial cable), one needs to determine the physical length of the cable. The physical length of the cable determines the delay, and therefore, the discriminator noise floor as a function of the offset frequency. Typically, 8 feet of RG-8/U cable results in a delay of 12 ns. As can be seen from Figure 5.112, it is advisable to have several lengths of cable depending upon the range of frequency of which one wants to do the measurements under different lengths of cable [33].

Since we are primarily concerned about the flicker noise, here is another, similar, approach to "cleaning up" the flicker noise, as shown in Figure 5.112. The bandwidth over which this technique works depends on the bandwidth of the phase detector system, which can be made fairly wide. A derivative of this will provide a useful solution.

5.8.4   Reduction of Flicker Noise

From inspecting a device datasheet, one is always pleasantly surprised about very low spot noise figures, while the $1/f$ corner frequency is rarely specified at all. The mechanism that transforms the noise, which is generated by the $1/f$ contribution into what we referred to as modulation noise, is the AM-to-PM conversion. If one could "linearize" the transistor, the flicker noise effect would be drastically reduced.

In the case of the bipolar transistor, you could sample the noise of the 56-Ω resistor shown in Figure 5.114a, and consistent with the previous method, feed this back into the

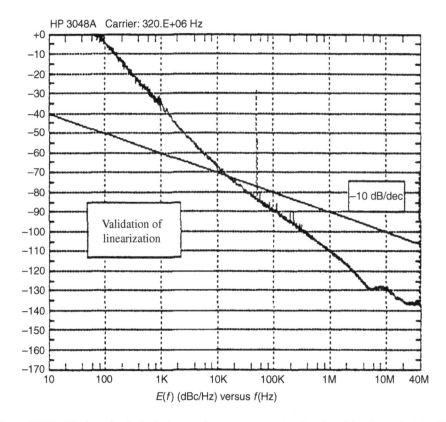

HP 3048A   Carrier: 320.E+06 Hz

$E(f)$ (dBc/Hz) versus f(Hz)

Figure 5.111 Display of a typical phase-noise measurement using the delay line principle. This method is only applicable where $x = \sin(x)$. The measurements above the solid line violate this relationship, and therefore are not valid.

oscillator system in this case and modulate the base voltage of Q_1 and Q_2. The higher-power transistor, BFR93A, is used for the oscillator, which could equally be an HBT gate "stabilizer" via Q_2 in such a manner that all the sample noise across the emitter resistor is inverted, amplified, and brought back. This noise-canceling scheme works inside the loop bandwidth determined by the circuit of Q_2. Figure 5.115 shows the Figure 5.114a circuit's phase-noise performance.

5.8.5 Applications to Integrated Oscillators

Figure 5.116 shows a silicon-based bi-CMOS technology wireless oscillator. The three big inductors can be seen easily. This design by Motorola is a good candidate for using this innovative technology. Applying this feedback technology, we measure the phase noise shown in Figure 5.117.

Moving up into the millimeterwave area, one of the target oscillators is the Ka-band MMIC voltage-controlled oscillator designed for Ka-band smart munitions applications (Figure 5.118). This voltage-controlled oscillator MMIC employs 0.25-μm gate length, double-heterojunction, pseudomorphic, high-electron-mobility transistor (PHEMT) technology. This is a custom chip developed by Martin Marietta under the U.S. Government MMIC program. Ansoft Serenade tools were used in this design.

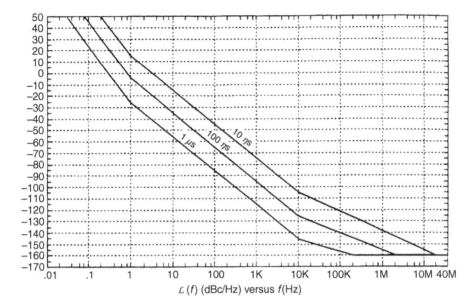

Figure 5.112 Dynamic range as a function of cable delay. 1 μs is ideal for microwave frequencies.

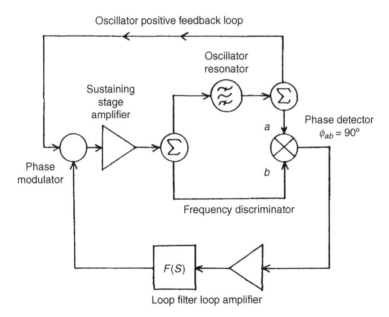

Figure 5.113 Noise reduction in an oscillator whose signal flicker-of-frequency noise is primarily due to sustaining stage flicker-of-phase noise. A phase perturbation in the sustaining stage produces a frequency change in the oscillator, which produces a change in the signal phase shift through the resonator detected as the phase detector. The resonator may be simultaneously operated in transmission line and reflection modes in the oscillator and discriminator portions of the circuit, respectively.

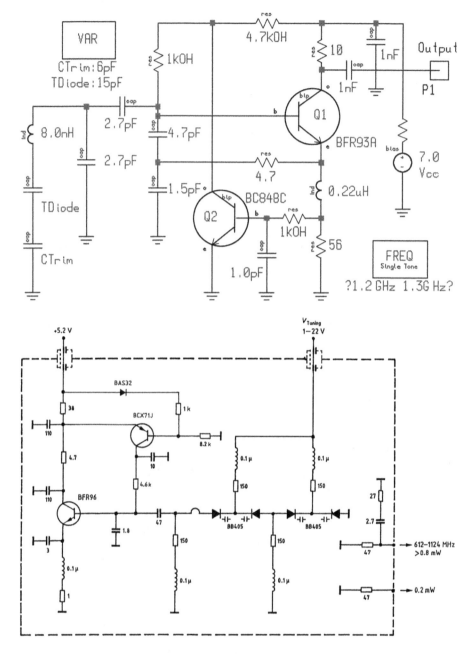

Figure 5.114 BJT-based oscillators with noise feedback. At (a), the noise sampling is done in the transistor emitter; at (b), the noise sampling is done in the collector using a variation of the active biasing circuit shown in Figure 3.206.

Features.

- Fundamental-mode, differential-ring VCO
- Electronically tunable
- Output power > 16 dBm

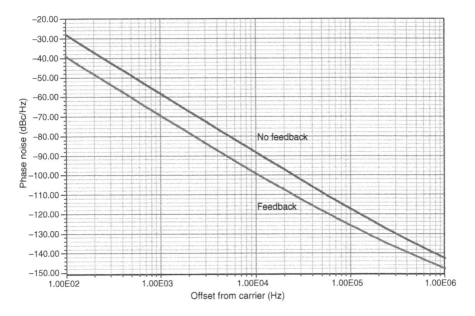

Figure 5.115 Phase noise of the Figure 5.114a oscillator with and without noise feedback.

- 13% power-added efficiency (PAE)
- Compact size for easy integration with power amplifier
- 0.25-μm pseudomorphic HEMTs

Specifications.

- Frequency range: Ka band
- Output power: 16 dBm minimum

When applying the same technology as shown in the previous bipolar example, the phase noise can be significantly reduced. Since field-effect transistors require a negative gate voltage, it is possible to sample the noise in the drain current using a resistor, amplify

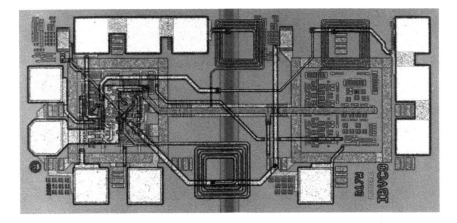

Figure 5.116 Layout of the Motorola 800-MHz monolithic differential oscillator IC in Si technology.

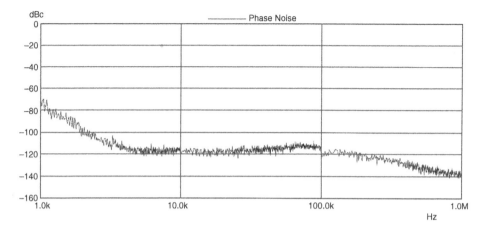

Figure 5.117 Phase noise of the oscillator of Figure 5.116. The noise floor is limited by the on-board isolation amplifier.

the noise with an appropriate loop amplifier, and modulate the gate voltage within a 1-MHz bandwidth.

A spectrum analyzer measurement of this oscillator type, as shown in Figure 5.119, indicates a phase noise of –78.6 dBc/Hz at an offset frequency of 100 kHz at a center frequency of 38 GHz, while the closed-loop phase noise of Figure 5.120 using the

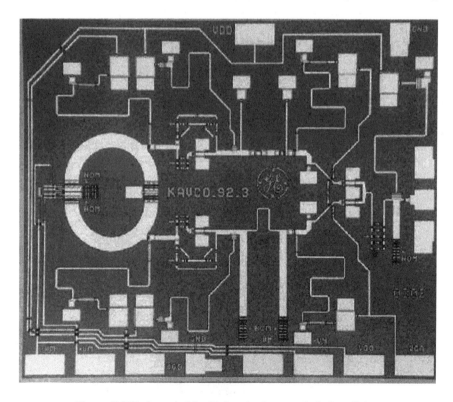

Figure 5.118 Layout of the Ka-band voltage-controlled oscillator.

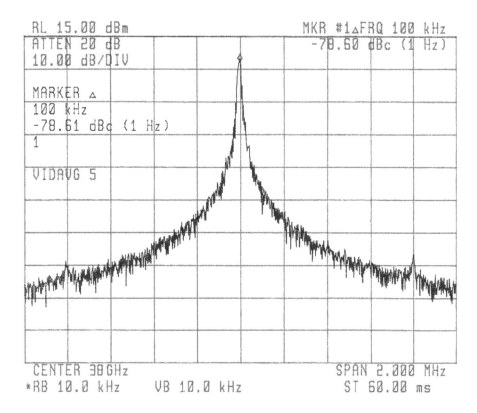

Figure 5.119 Spectrogram of the 38 GHz free-running push–pull VCO shown in Figure 5.118.

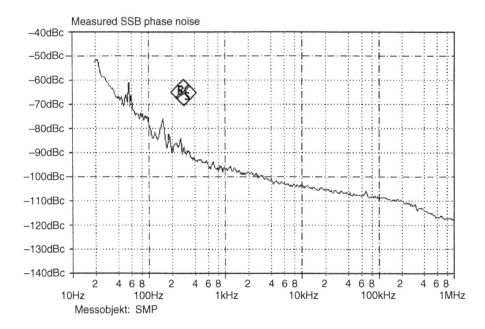

Figure 5.120 Phase-noise performance of a 38-GHz oscillator with phase-noise "clean up" system.

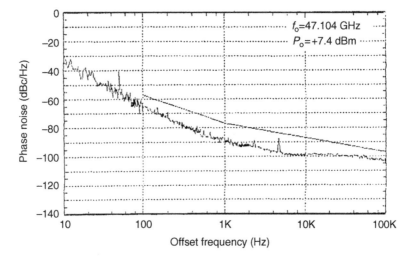

Figure 5.121   Showing a 47 GHz loop-stabilized VCO with phase-noise performance similar to that of Figure 5.120.

compensation scheme indicates a phase noise of about −108 dBc/Hz at 100 kHz. This is consistent with reported results from other researchers who have used complicated phase-locked loops to achieve similar performance as seen in Figure 5.121.

5.8.6 Summary

A novel method of improving the phase noise of oscillators has been demonstrated. This approach works well for low-frequency oscillators such as 1000 MHz up to the millimeter-wave range. It only requires an additional wideband dc biasing circuit, which applies noise feedback. The theory has been demonstrated by showing appropriate validation.

REFERENCES

1. P. Danzer, (ed.), "Chapter 14, AC/RF sources (oscillators and synthesizers)" in *The ARRL Handbook for Radio Amateurs*, 75th ed. Newington: ARRL, 1997.

2. U. L. Rohde, J. C. Whitaker, and T. T. N. Bucher, *Communications Receivers: Principles and Design*, 2nd ed. New York: McGraw-Hill, 1997, p. 380.

3. G. Vendelin, A. M. Pavio, and U. L. Rohde, *Microwave Circuit Design: Using Linear and Nonlinear Techniques*. New York, Wiley, 1990.

4. S. Y. Liao, *Microwave Devices and Circuits*, 2nd ed. Englewood Cliffs: Prentice-Hall, 1985.

5. S. C. Peterson, "900-MHz μ-strip," Philips Semiconductors application document.

6. H. Keller, "Electronic UHF tuning in TV receivers," *Radio-Fernsch-Phono-Praxis*, No. 3, 1967.

7. L. Micic, "The tuner diode," *Radio Mentor Electronic*, Vol. 32, No. 5, pp. 404–405, 1966.

8. K.-W. Yeom, "An investigation of the high-frequency limit of a miniaturized commercial voltage-controlled oscillator used in 900-MHz band mobile-communication handset, *IEEE Trans. Microwave Theory Tech.*, Vol. 46, pp. 1165–1168, August 1998.

9. D. B. Leeson, "A simple model of feedback oscillator noise spectrum," *Proc. IEEE*, pp 329–330, 1966.

10. M. Odyniec (ed.), in *RF and Microwave Oscillator Design*, "Chapter 5: Modern harmonic-balance techniques for oscillator analysis and optimization," by V. Rizzoli, A. Neri, A. Costanzo, F. Mastri, Artech House, 2002.

11. K. Kurokawa, "Noise in synchronized oscillators," *IEEE Trans. Microwave Theory Tech.*, Vol. 16, April, pp. 234–240, 1968.

12. U. L. Rohde, A. Poddar, and G. Böck, *The Design of Modern Microwave Oscillators for Wireless Applications—Theory and Optimization*. Hoboken, NJ: Wiley, 2005, Chapters 7 and 8.

13. R. J. Hawkins, "Limitations of Nielsen's and related noise equations applied to microwave bipolar transistors and a new expression for the frequency and current dependent noise figure," *Solid-State Electron.*, Vol. 20, pp. 191–196, March 1977.

14. T. H. Hsu and C. P. Snapp, "Low-noise microwave bipolar transistor with sub-half-micrometer emitter width," *IEEE Trans. Electron Dev.*, Vol. ED-25, pp. 723–730, June 1978.

15. U. L. Rohde, C. R. Chang, and J. Gerber, "Parameter extraction for large signal noise models and simulation of noise in large signal circuits like mixers and oscillators," *Proc. 23rd European Microwave Conf.*, Madrid, Spain, September 6–9, 1993.

16. C. R. Chang, "Mixer noise analysis using the enhanced microwave harmonica," *Compact Software Transmission Line News*, Vol. 6, No. 2, pp. 4–9, June 1992.

17. T. Antognetti and G. Massobrio, *Semiconductor Device Modeling with SPICE*. New York: McGraw-Hill, 1988, pp. 91.

18. R. J. Hawkins, "Limitation of Nielsen's and related noise equations applied to microwave bipolar transistors, and a new expression for the frequency and current dependent noise figure," *Solid State Electronics*, Vol. 20, pp. 191–196, 1977.

19. T.-H. Hus and C. P. Snapp, "Low noise microwave bipolar transistor sub-hall-micrometer emitter width," *IEEE Trans. Electron Devices*, Vol. ED-25, pp. 723–730, June 1978.

20. R. A. Pucel and U. L. Rohde, "An accurate expression for the noise resistance Rn of a bipolar transistor for use with the hawkins noise model," *IEEE Microwave Guided Wave Lett.*, Vol. 3, No. 2, pp. 35–37, February 1993.

21. See reference 3.

22. R. A. Pucel, W. Struble, R. Hallgren, and U. L. Rohde, "A general noise de-embedding procedure for packaged two-port linear active devices," *IEEE Trans. Microwave Theory Tech.*, Vol. 40, No. 11, pp. 2013–2024, November 1993.

23. U. L. Rohde, "Improved noise modeling of GaAs FETS: Using an enhanced equivalent circuit technique," *Microwave J.*, Part I: November 1991, pp. 87–101; Part II, December 1991, pp. 87–95.

24. V. Rizzoli, F. Mastri, and C. Cecchefti, "Computer-aided noise analysis of MESFET and HEMT mixers," *IEEE Trans. Microwave Theory Tech.*, Vol. MTT-37, pp. 1401–1410, September 1989.

25. Rizzoli and A. Lippadni, "Computer-aided noise analysis of linear multiport networks of arbitrary topology," *IEEE Trans. Microwave Theory Tech.*, Vol. MTT-33, pp. 1507–1512, December 1985.

26. V. Rizzoli, F. Mastri, and D. Masotti, "General-purpose noise analysis of forced nonlinear microwave circuits," *Military Microwave*, 1992.

27. U. L. Rohde, "Key components of modern receiver design", *VHF Conf.*, Weinheim, Germany, September 17–18, 1993.

28. C. R. Chang and U. L. Rohde, "The accurate simulation of oscillator and PLL phase noise in RF sources," *Wireless Symp.*, Santa Clara, CA, February 15–18, 1994.

29. U. L. Rohde, "Oscillator design for lowest phase noise," *Microwave Eng. Eur.*, pp. 31–40, May 1994.

30. D. Scherer, "Design principles and test methods for low phase noise RF and microwave sources," *RF Microwave Measure. Symp. Exhib.*, Hewlett-Packard.

31. D. J. Healy, III, "Flicker of frequency and phase and white frequency and phase fluctuations in frequency sources," *Proc. 26th Annu. Symp. Frequency Cont.*, Fort Monmouth, NJ, pp. 43–49, June 1972.

32. M. M. Driscoll, "Low noise oscillator design using acoustic and other high Q resonators," *44th Annu. Symp. Freq. Cont.*, Baltimore, MD, May 1990.

33. U. L. Rohde, *Microwave and Wireless Synthesizers: Theory and Design*. New York: Wiley, 1997.

INTERESTING PATENTS

M. Fache and C. Juliot, "Tunable UHF oscillator with harmonic limitation," Lignes Telegraphoques et Telephoniques (Societe), Paris Cedex, France, US patent 4,146,850, March 27, 1979.

T. M. Higgins, Jr., "Negative resistance oscillator with electronically tunable base inductance," Hewlett-Packard Company, Palo Alto, CA, US patent 5,373,264, December 13, 1994.

O. Ichiyoshi, "Voltage-control oscillator which suppresses phase noise caused by internal noise of the oscillator," NEC Corporation, Japan, US patent 5,351,014, September 27, 1994.

S. R. Joshi, U. L. Rohde, and K. Eichel, "Phase-locked loop circuits and voltage controlled oscillator circuits," Synergy Microwave Corp., Paterson, NJ, US patent 5,650,754, July 22, 1997.

J. H. Kiser, "Wide-range electronic oscillator," Vari-L Co., Denver, CO, US patent 4,621,241, November 4, 1986.

J. H. Kiser, "Oscillator voltage regulator," Vari-L Co., Denver, CO, US patent 5,675,478, October 7, 1997.

C. Lewis, "Low phase noise UHF and microwave oscillator," Z-Communications, Inc., San Diego, CA, US patent 5,748,051, May 5, 1998.

L. R. Lockwood, "Low-noise oscillator," Tektronix, Inc., Beaverton, OR, US patent 4,580,109, April 1, 1986.

B. J. Nardi, "Low phase noise reference oscillator," Watkins-Johnson Company, Palo Alto, CA, US patent 5,341,110, August 23, 1994.

L. Riebman, "Low-noise oscillator," AEL Defense Corp., Lansdale, Pennsylvania, US patent 5,166,647, November 24, 1992.

T. Saitoh and H. Hatashita, "TV tuner oscillator with feedback for more low-frequency power," Hitachi Ltd., Tokyo, Japan, US patent 4,564,822, January 14, 1986.

FURTHER READING

W. Anzill, F. X. Kärtner, and P. Russer, "Simulation of the single-sideband phase noise of oscillators", *2nd Int. Workshop Integr. Nonlinear Microwave Millimeterwave Circuits*, 1992.

Behagi, "Piece-wise linear modeling of solid-state varactor-tuned microwave oscillators," *Proc. IEEE Frequency Control Symposium*, 1992, pp. 415–519.

N. Boutin, "RF oscillator analysis and design by the loop gain method," *Appl. Microwave Wireless*, August 1999, pp. 32–48.

A. Demir, A. Mehrotra, and J. Roychowdhury, "Phase noise and timing jitter in oscillators," *Proc. IEEE Custom Integr. Circuits Conf.*, 1998, pp. 45–48.

G. Gonzalez, *Microwave Transistor Amplifiers–Analysis and Design*, 2nd ed. New York: Prentice-Hall, 1997.

A. V. Grebennikov and V. V. Nikiforov, "An analytic method of microwave transistor oscillator design," *Int. J. Electron.*, Vol. 83, pp. 849–858, December 1997.

A. V. Grebennikov, "Microwave transistor oscillators: An analytic approach to simplify computer-aided design," *Microwave J.*, May 1999, pp. 292–300.

A. Hajimiri and T. H. Lee, "A general theory of phase noise in electrical oscillators," *IEEE J. Solid-State Circuits*, Vol. 33, pp. 179–194, February 1998.

A. Hajimiri, S. Limotyrakis, and T. H. Lee, "Phase noise in multi-gigahertz CMOS ring oscillators," *Proc. IEEE Custom Integr. Circuits Conf.*, 1998, pp. 49–52.

K. M. Johnson, "Microwave varactor-tuned transistor oscillator design," *IEEE Trans. Microwave Theory Tech.*, Vol. MTT-14, No. 11, pp. 564–572, September 1966.

K. M. Johnson, "Large-signal GaAs MESFET oscillator design," *IEEE Trans. Microwave Theory Tech.*, Vol. MTT-27, No. 3, pp. 217–226, March 1979.

J. Kitchen, "Octave bandwidth varactor-tuned oscillators," *Microwave J.*, Vol. 30, No. 5, pp. 347–353, May 1987.

A. Kral, F. Behbahani, and A. A. Abidi, "RF-CMOS oscillators with switched tuning," *Proc. IEEE Custom Integr. Circuits Conf.*, 1998, pp. 555–558.

F. X. Kärtner, "Noise in oscillating systems," *2nd Int. Workshop Integr. Nonlinear Microwave Millimeterwave Circuits*, 1992.

K. Kurokawa, "Some basic characteristics of broadband negative resistance oscillator circuits," *Bell Syst. Tech. J.*, pp. 1937–1955, July 1969.

A. L. Lacaita and C. Samori, "Phase-noise performance of crystal-like LC tanks," *IEEE Trans. Circuits Syst.—II: Analog Digital Signal Process.*, Vol. 45, pp. 898–900.

M. Madihian and T. Noguchi, "An analytical approach to microwave GaAsFET oscillator design," *NEC Res. Develop.*, No. 82, pp. 49–56, July 1986.

J. O. McSpadden, L. Fan, and K. Chang, "High-efficiency Ku-band oscillators," *IEEE Trans. Microwave Theory Tech.*, Vol. 46, pp. 1566–151571, Oct. 1998.

Mini-Circuits Engineering, "Designers' considerations for VCOs used in phase-locked loops," *Microwave Products Dig.*, pp. 24, 34, 36, 37, 40, February 1999.

M. K. Nezami, "Evaluate the impact of phase noise on receiver performance," *Microwaves RF*, pp. 165–172, May 1998.

E. C. Niehenke and R. D. Hess, "A microstrip low noise X-band voltage-controlled oscillator," *IEEE Trans. Microwave Theory Tech.*, Vol. MTT-27, No. 12, pp. 1075–1079, October 1986.

C. O'Connor, "Develop trimless voltage-controlled oscillators," *Microwaves RF*, pp. 68–78, July 1999.

D. P. Owen, "Measurement of phase noise in signal generators," *Marconi Instrument.*, Vol. 15, No. 6, pp. 117–122.

D. F. Page, "A design basis for junction transistor oscillator circuits," *Proc. IRE*, Vol. 46, pp. 1271–1280, November 1958.

B.-Ha Park, "1 GHz, low-power, low-phase-noise, CMOS voltage-controlled oscillator with an integrated LC resonator," private communication with the author.

D. F. Paterson, "Varactor properties for wideband linear-tuning microwave VCOs," *IEEE Trans. Microwave Theory Tech.*, Vol. MTT-28, No. 2, pp. 110–119, February 1980.

R. A. Pucel and J. Curtis, "Near-carrier noise in FET oscillators," *IEEE MTT-S Int. Microwave Symp. Dig.*, 1983, pp. 282–284.

V. Rizzoli, F. Mastri, and D. Masotti, "A general purpose harmonic balance approach to the computation of near-carrier noise in free-running microwave oscillators," MTT-S, 1993, pp. 309–312.

C. Samori, A. L. Lacaita, F. Vill, and F. Zappa, "Spectrum folding and phase noise in LC tuned oscillators," *IEEE Trans. Circuits and Syst.—II: Analog Digital Signal Process.*, Vol. 45, pp. 781–790.

Sarafian and B. Z. Kaplan, "A new approach to the modeling of the dynamics of RF VCOs and some of its practical implications," *IEEE Trans. Circuits Syst.—Fundamental Theory Applicat.*, Vol. 40, No. 12, December 1993, pp. 895–901.

G. Sauvage, "Phase noise in oscillators: A mathematical analysis of Leeson's model," *IEEE Trans. Instrument. Measure.*, Vol. IM-26, No. 4, December 1977.

J.-S. Sun, "Design and analysis of microwave varactor-tuned oscillators," *Microwave J.*, pp. 302–310, May 1999.

R. J. Trew, "Octave-band GaAs FET YIG-tuned oscillators," *Electron. Lett.*, Vol. 13, No. 21, October 1977.

R. J. Trew, "MESFET models for microwave computer-aided design," *Microwave J.*, Vol. 33, No. 5, pp. 115–128, May 1990.

D. A. Warren, "J. M. Golio, and W. L. Seely, "Large and small signal oscillator analysis," *Microwave J.*, Vol. 32, No. 5, pp. 229–246, May 1989.

R. Winch, "Wideband varactor-tuned oscillators," *IEEE J. Solid-State Circuits*, Vol. 27, SC-17, pp. 1214–1219, December 1982.

T. Yamasaki and E. Nagata, "Microwave synthesizer for terrestrial and satellite communications", NEC Corporation, June 1989.

6

WIRELESS SYNTHESIZERS

6.1 INTRODUCTION

Increasing integration has drastically narrowed the range of component choices open to wireless synthesizer designers. The design of a high-performance synthesizer is largely reduced to selecting the most advanced synthesizer IC and, if the synthesizer uses a phase-locked loop, designing or obtaining the best VCO. This chapter covers synthesizer theory, evaluation, and design, including PLL and direct digital synthesis (DDS) techniques. Going into all the details of frequency synthesizers would be beyond the scope of this book. For depth and background on this subject, we recommend Ref. [1].

6.2 PHASE-LOCKED LOOPS

6.2.1 PLL Basics

Figure 6.1 shows a complete PLL synthesizer block diagram indicating the areas over which the designer actually has control. Once the VCO signal has been translated from analog to quasidigital (square-wave) form in circuitry similar to a line receiver, the synthesizer IC takes over. The VCO receives an analog control signal that results from the integration of digital pulses from a phase/frequency discriminator. Modern phase/frequency discriminators, which are part of the PLL IC, use edge-triggered loop locks and generate correcting pulses of either positive or negative sign. The output portion of such a phase detector is frequently referred to as a charge pump because the resulting dc control voltage, which ultimately determines the oscillator frequency, is processed by a loop filter containing at least one large capacitor that is charged or discharged as necessary to maintain the control-voltage level necessary for loop lock.

RF/Microwave Circuit Design for Wireless Applications, Second Edition. Ulrich L. Rohde and Matthias Rudolph.
© 2013 John Wiley & Sons, Inc. Published 2013 by John Wiley & Sons, Inc.

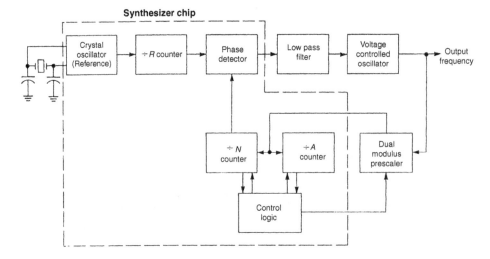

Figure 6.1 Block diagram of an integrated frequency synthesizer. In this case, the designer has control over the VCO and the loop filter; the reference oscillator is part of the chip. In most cases (up to 2.5 GHz), the dual-modulus prescaler is also inside the chip.

Figure 6.2 shows the block diagram of a single-loop synthesizer. Unless special techniques are used, such as the fractional-N-division principle, the step size or channel frequency spacing is equal to the reference frequency. When describing frequency synthesizers mathematically, we usually use a linearized model. Because most effects occurring in the phase detector are highly nonlinear, only the so-called piecewise-linear treatment allows adequate approximation.

We assume that the VCO shown in Figure 6.2 is tunable over the frequency range from 410 to 510 MHz. Its output is divided to the reference frequency in a programmable divider (i.e., divide by N) whose output is fed to one of the input of the phase/frequency detector and compared with the reference frequency supplied to the other input. The loop filter at the output of the phase detector suppresses reference-frequency components while also serving as an integrator. The dc control voltage at the output of the loop filter tunes the VCO until the divided frequency and phase equal those of the reference. In this simple

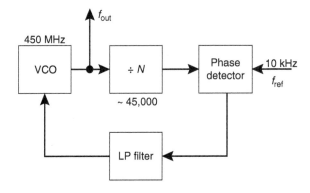

Figure 6.2 Block diagram of a single-loop synthesizer.

example with the divider set to 45,000 and the reference set to 1 kHz, the VCO is controlled to a frequency of 450 MHz. A fixed division of the frequency standard output produces the reference-frequency of the appropriate step size. Frequency standards are typically operated at 1, 5, or 10 MHz to take advantage of high crystal stability. A 5-MHz frequency standard would be divided by 500 in the example. The operating range of the PLL is determined by the maximum operating frequency of the programmable divider, by its division-range ratio, and by the tuning range of the VCO.

The PLL is nonlinear because the phase detector is nonlinear. However, it can be accurately approximated by a linear model when the loop is in lock. The response, when the loop is closed, may be expressed as

$$\frac{\theta_c(s)}{\theta_r(s)} \equiv B(s) = \frac{\text{Forward gain}}{1 + \text{open-loop gain}}$$

$$= \frac{G(s)}{1 + G(s)/N} \tag{6.1}$$

where $G(s) = G_1(s)\,G_2(s)\,F(s)\,/s$, and θ_c and θ_r are the phases of the controlled oscillator and the reference, respectively.

When the loop is locked, it is assumed that the phase-detector output voltage is proportional to the difference in phase between its inputs, that is,

$$V_\theta = K_\theta(\theta_r - \theta_i) \tag{6.2}$$

where V_θ is the output voltage of the phase detector, and θ_r and θ_i are the phases of the reference signal and the divided VCO signal, respectively. K_θ is the phase-detector gain factor and has the dimensions of volts per radian. It is also assumed that the VCO can be modeled as a linear device whose output frequency differs from its free-running frequency by an increment of frequency

$$2\pi\delta f = K_0 V_c \tag{6.3}$$

where V_c is the voltage of the output of the low-pass filter, and K_0 is the VCO gain factor with the dimensions of radians per second per volt. Because frequency is the time derivative of phase, the VCO operation can be described as

$$2\pi\delta f \equiv \frac{d\theta_c}{dt} = K_0 V_c \tag{6.4}$$

With these assumptions, the PLL may be represented by the linear model shown in Figure 6.3.

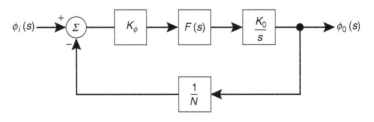

Figure 6.3 Block diagram of a linearized model of a PLL.

The linear transfer function relating $\theta_c(s)$ and $\theta_r(s)$ is

$$B(s) = \frac{\theta_c(s)}{\theta_r(s)} = \frac{K_\theta K_0 F(s)/s}{1 + K_\theta K_0 F(s)/Ns} \tag{6.5}$$

The forward gain is

$$G(s) = \frac{K_\theta K_0 F(s)}{s} \tag{6.6}$$

and the open-loop gain is

$$G(s) H(s) = \frac{K_\theta K_0 F(s)}{Ns} \tag{6.7}$$

which leads to the transfer formula of equation (6.1).

There are various choices of filter response $F(s)$. Because the VCO by itself is an integrator, we can use a simple RC filter following the phase detector. This arrangement is called a Type 1 filter. Because the components used, together with feedthrough capacitors and other stray effects, can cause excess phase shift, it is necessary to ensure than stability criteria are satisfied. If the gain of a passive loop is too small to provide adequate drift stability of the output phase, especially if a high division ratio is used, the best solution to this problem is the use of an active amplifier as an integrator. In most frequency synthesizers, the active-filter-integrator approach is preferred to the passive one. Some frequency synthesizer chips have a single-ended output. In such cases, the use of an additional integrator requires some precautions.

6.2.2 Phase-Frequency Comparators

The phase/frequency comparator can be divided into two types.

1. Phase detectors.
2. Phase-frequency comparators.

This means that the phase comparator has limited means to compare two signals and only accepts phase, not frequency, information. In this case, particular measures have to be taken to pull of the VCO into the locking range. The phase comparators require special locking help. Here we are analyzing only the performance.

6.2.2.1 Diode Rings

The diode ring is normally driven with two signals with sinusoidal waveform and also is some sort of mixer. Here it will suffice to derive the gain characteristic K_θ of the device. If the input signal is $\theta_i = A_i \sin \omega_o t$, and the reference signal is $\theta_r = A_r \sin(\omega_o t + \phi)$, where ϕ is the phase difference between the two signals, the output signal θ_e is

$$\theta_e = \theta_i \theta_r = \frac{A_i A_r}{2} K \cos\phi - \frac{A_i A_r}{2} K \cos(2\omega_o t + \phi) \tag{6.8}$$

where K is the mixer gain. One of the primary functions of the low-pass filter is to eliminate the second-harmonic term before it reaches the VCO. The second harmonic will be assumed to be filtered out and only the first term will be considered, so

$$\theta_e = \frac{A_i A_r}{2} K \cos \phi \qquad (6.9)$$

When the error signal is zero, $\phi = \pi/2$. Thus, the error signal is proportional to phase differences from 90°. For small changes in phase $\Delta\phi$,

$$\theta_e \simeq \frac{\pi}{2} + \Delta\phi = \frac{A_i A_r}{2} K \left[\cos \left(\frac{\pi}{2} + \Delta\phi \right) \right]$$
$$= \frac{A_i A_r}{2} K \sin \Delta\phi \qquad (6.10)$$

For a small phase perturbation $\Delta\phi$,

$$\theta_e \simeq \frac{A_i A_r K}{2} \Delta\phi \qquad (6.11)$$

since the phase detector output was assumed to be

$$\theta_e = K (\theta_i - \theta_o) \qquad (6.12)$$

and the phase detector scale factor K_θ is given by

$$K_\theta = \frac{A_i A_r K}{2} \qquad (6.13)$$

The phase detector scale factor K_θ depends on the input signal amplitudes; the device can be considered linear only for constant-amplitude input signals and for small deviations in phase. For larger deviations in phase,

$$\theta_e = K_\theta \sin\Delta\phi \qquad (6.14)$$

which describes a nonlinear relation between θ_e and ϕ.

In frequency synthesizers, the reference is typically generated from a reference oscillator and is lower than the VCO frequency, which is divided by a programmable divider. Both signals, therefore, are square waves rather than sine waves, and theoretically, a diode ring can be driven by those two signals.

A drawback to the diode-ring phase detector is that its output voltage is very small—several hundred millivolts at most. A post-detector dc amplifier, which will unavoidably introduce noise, is therefore required.

6.2.2.2 Edge-Triggered JK Master-Slave Flip-Flops

The fundamental idea of the sequential phase comparator we will be dealing with is that there are two outputs available, one to charge and one to discharge a capacitor. Output 1 then is high if the Signal 1 frequency is greater than the Signal 2 frequency, or of the two frequencies are equal, if Signal 1 leads Signal 2 in phase.

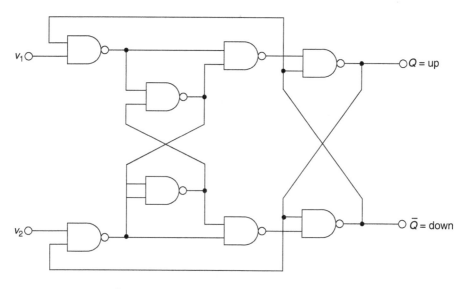

Figure 6.4 Edge-triggered JK master-slave flip-flop.

Output 2 is high if the frequency of Signal 2 is greater than that of Signal 1, or if the signal frequencies are the same and Signal 2 leads Signal 1 in phase.

Figure 6.4 shows the minimum configuration to build such a phase comparator. It can be operated from -2π to $+2\pi$, and an active amplifier is recommended as a charge pump. The Q output of the JK master-slave flip-flop is set to one by the negative edge of Signal 1, while the negative edge of Signal 2 resets it to zero. Therefore, the output $\bar{Q}$ is the complement of Q. The output voltage $\bar{V}$ is defined as the weighted duty cycle of Q and $\bar{Q}$. This means that a positive contribution is made when $Q = 1$ and a negative contribution (discharge) is made when $Q = 0$. The averaging and filtering of the unwanted ac component is done by a subsequent integrator. The integrator then is called a charge pump, as the loop capacitor is being charged and discharged depending on whether Q is high or low.

If the system using the J-K flip-flop is not in lock, and there is a large difference between frequencies f_1 and f_2 at the output, the output will not be zero, but instead will be positive or negative relative to one-half supply voltage. This is an advantage and indicates that this system is frequency sensitive. We therefore call it a phase/frequency comparator because it is capable of detecting both phase and frequency offsets. In its locking and pull-in performance, it is similar to an exclusive-OR gate.

For better understanding, let us look at a few cases where the system is in lock. It should be noted that whereas the exclusive-OR gate sensitive to the duty cycle of the input signals, the J-K flip-flop responds only to the edges, and therefore the phase/frequency comparator can be used for asymmetrical waveforms. Let us first assume that the input signals 1 and 2 have the same frequency. Figure 6.5 shows what happens if the phase error is about 0, π, and 2π. In those cases, the duty cycle at the output is about 0, 50%, or 100%, respectively. The narrow output pulses may cause spikes on the power-supply line and lines in the vicinity, and precautions must be taken to filter them.

The output voltage $\bar{V}$ is the average of the signal Q, and is a linear function of the phase error.

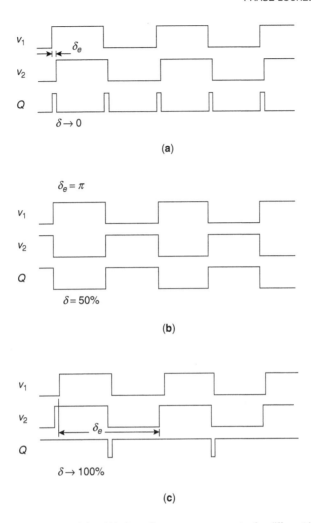

Figure 6.5   Performance of the J-K phase/frequency comparator for different input signals.

Now let us take a look at several cases where the system is not in lock. Figure 6.6 shows the case where f_1 is substantially higher than f_2. As a result, the output duty cycle is close to 100% and the VCO frequency is being pulled up to higher frequencies. If the frequency at Input 2 is much higher than that at Input 1, the opposite is true. This proves that the device is sensitive to frequency changes.

In cases where both frequencies are about the same, as shown in Figure 6.7, the crossover area is not clearly defined. The first picture shows the case where f_2 is 10% higher than f_1 and the duty cycle is changing periodically between 0 and 100%. Therefore, the ac voltages look like a sawtooth, with a rate equal to the difference of both frequencies. The same holds true if the two inputs are reversed. In the case where both frequencies are identical, the JK flip-flop behaves the same as an exclusive-OR gate. From this discussion, it can be concluded that while this phase/frequency comparator was included to explain how it works, it is not a very desirable device for practical purposes because of the uncertainty of its behavior close to lock.

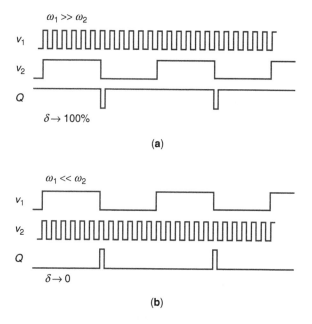

Figure 6.6   Phase detector output for two input frequencies that are substantially different.

6.2.2.3 Digital Tristate Comparators

The digital tristate phase/frequency comparator is probably the most universally used and most important next to the sample/hold comparator. Although the diode ring and exclusive-OR gate phase detectors have some applications, the tristate phase/frequency comparator can be used widely. Even in cases where a sample/hold comparator theoretically could be used, it may be inferior as far as reference attenuation or noise is concerned, but it is generally well behaved. Unfortunately, the tristate system is very complex and shows a number of unusual phenomena. Such a digital tristate comparator is shown in Figure 6.8 using two D flip-flops and a NAND gate. The Q_2 output signal is filtered with the low-pass filter. The operation of this logic circuit is readily analyzed using the state-transition diagram as shown in Figure 6.9. The D flip-flop outputs go high on the leading edge of their respective clock inputs and remain high until they are reset. The reset signal occurs when both inputs are high. When both signals are in phase and of the same frequency, both outputs will remain

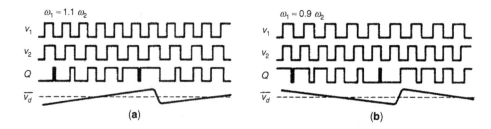

Figure 6.7   Performance of the phase detector for small frequency errors.

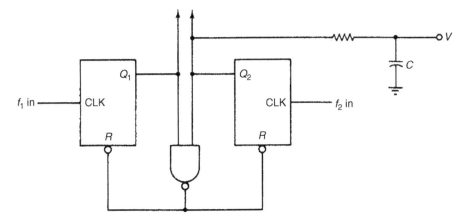

Figure 6.8 Phase detector with two D flip-flops and a NAND gate. In this text, this type of phase detector will be called a tristate comparator.

low, and no signal will applied to the operational amplifier. When the two frequencies are the same, the dc output voltage transfer characteristic will be as shown in Figure 6.10. If the two signal frequencies are not the same, the output voltage will depend on both the relative frequency difference and the phase difference. The timing diagram of Figure 6.11 illustrates the case in which $f_2 = 3f_1$. In Figure 6.11a, the leading edge of f_1 occurs just after that of f_2, so that Q_2 is high 50% of the time, and the average value of the PD output is 50%. In Figure 6.11b, the leading edge of f_1 occurs just before that of f_2, so Q_2 is high almost all the time and the average output voltage is approximately V.

The output voltage averaged over all of the phase differences is then 67% for $f_2 = 3f_1$. In general, it can be said that the average output (averaged over all the phase differences) is given by

$$V_{ave} = 1 - \frac{f_1}{f_2}V \qquad (6.15)$$

provided that f_2 is greater than f_1. This expression is plotted in Figure 6.12 together with the cases in which f_1 is greater than f_2.

The digital network used in this realization is only one of a large number of logic circuits that could be used. Many IC manufacturers now produce a quad-D circuit that functions much like the dual-D flip-flop; the main difference is that when the frequency of one signal is more than twice that of the other signal, the correspond output will be high all of the time. Therefore, a larger voltage is applied to the VCO and the loop response is faster. An example of the quad-D circuit is shown in Figure 6.13.

The most popular digital tristate phase/frequency comparator on the market is the one used in the CD4046 PLL IC, shown in Figure 6.14. It contains an additional phase comparator, an exclusive-OR gate that can be used as a lock indicator. In addition, two FETs are used to sum the two outputs. A slightly faster version in TTL technique is the Motorola MC4044. The fastest version in ECL is the MC12040, also made by Motorola, shown in Figure 6.15. Sometimes it is convenient to build the phase-frequency comparator in discrete technique to add additional features. Figure 6.16 shows an example.

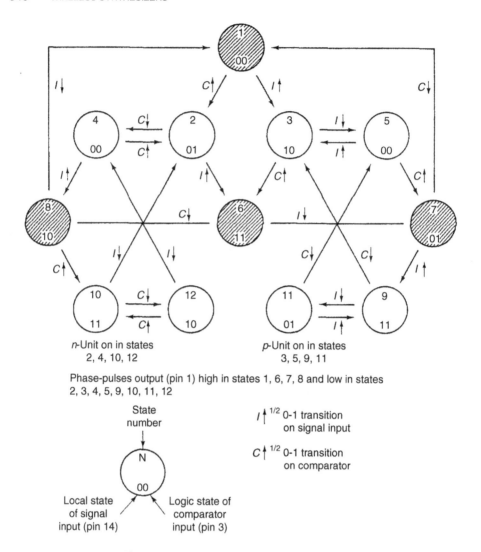

Figure 6.9   Logic diagram of the tristate detector.

This particular tristate phase-frequency comparator has a peculiarity that was first mentioned by Egan and Clark [2]. When actually building a phase-locked loop with this phase-frequency comparator, or that of the CD4046 type, by going through the normal mathematical design routine, it becomes apparent that the expected performance and the actual results differ as follows.

1. The reference suppression will be better than expected.
2. The phase error or tracking will be worse than expected.
3. The phase margin will differ and the system may not lock despite the fact that the calculation is correct.

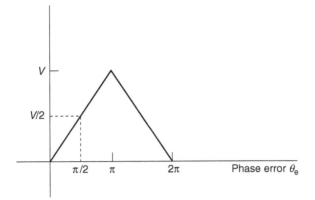

Figure 6.10 Transfer characteristic of the tristate phase/frequency comparator.

The reason for all this is due to two effects.

1. The flip-flops are not absolutely alike, and as a result of this, the output in the crossover region is not zero.
2. If there is no or very little correction voltage required, the gain of the phase detector will drop substantially.

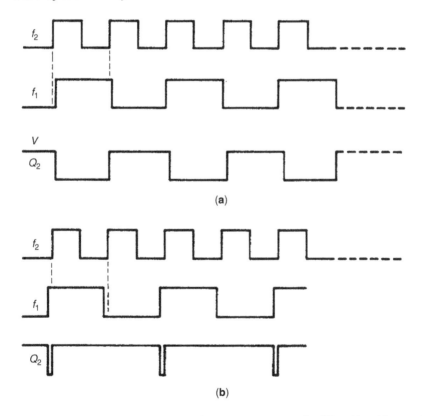

Figure 6.11 Output waveform of the tristate frequency comparator for different input frequencies.

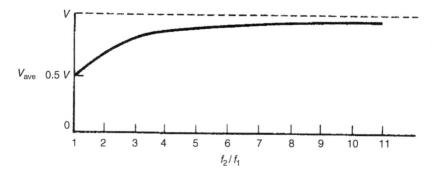

Figure 6.12 Average output voltage as a function of frequency ratio.

Let us assume the ideal situation where the output of the phase/frequency comparator feeding the charge pump does not have to correct any error, the system is drift free, and there are no leakage currents. The holding capacitor of the charge pump would maintain constant voltage and, as there is no drift, no correction voltage would be necessary.

The flip-flops, however, introduce a certain amount of jitter, and a certain amount of jitter is also introduced by the frequency dividers, both the reference divider and the programmable divider. This jitter results in an uncertainly regarding the zero crossings, and extremely narrow pulses will appear at the output of the summation amplifier used in the CD4046.

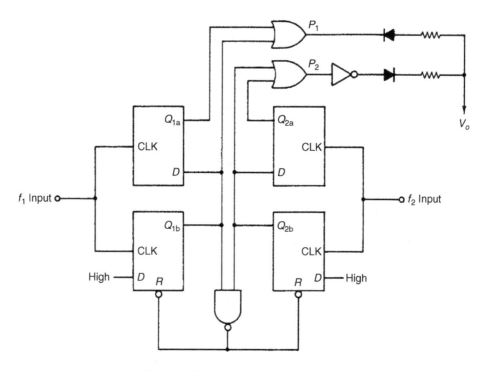

Figure 6.13 Example of a quad-D circuit.

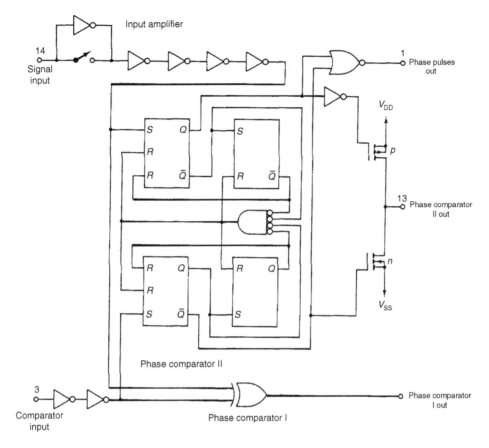

Figure 6.14 Block diagram of CD4046 phase/frequency comparator. (Courtesy of Motorola Semiconductor Products, Inc.)

Under the ideal assumption that there are no corrections required and those pulses would not exist, the reference suppression would be infinite, as there is no output. Therefore, the reference suppression—disregarding the effect of the loop filter—depends only on how well this condition is met.

The change of gain seems somewhat surprising, but as we think of it, if there is no correction and no update, there is also no gain. It is impossible to meet this condition, which is fortunate, but with regard to the temperature stability and aging characteristics of some devices, predicting actual performance may be difficult.

There are several remedies to this problem. A simple version is to introduce a controlled amount of leakage. Although the electrolytic capacitor required in the charge pump will have some leakage, it is better to set a leakage current that is independent of temperature and aging. This can be accomplished by putting a 1-MΩ resistor from the output of the CD4046 to ground. The phase/frequency comparator then has to deliver an output current, and this output current is determined by a resistor that can be independent of temperature and other effects. As a result of this, the duty cycle of the output pulses of the phase-frequency comparator will change and the pulses will become wider. The wider the pulse,

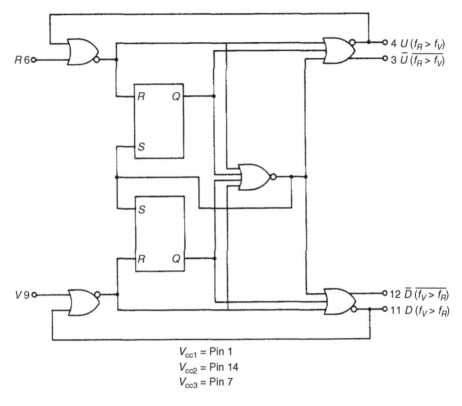

V_{cc1} = Pin 1
V_{cc2} = Pin 14
V_{cc3} = Pin 7

Figure 6.15 Block diagram of Motorola MC12040 phase/frequency comparator. (Courtesy of Motorola Semiconductor Products, Inc.)

the more energy it contains; therefore, the wider the pulse, the more the reference suppression degrades.

It is theoretically possible to put one side of the 1-MΩ resistor, instead of to ground, to the wiper of a potentiometer, and set the voltage in such a manner that this offset is compensated but again, as the phase will shift theoretically, one must adjust the potentiometer according to the actual phase error. This is not a very convenient arrangement.

A somewhat better method was proposed by Fairchild several years ago, and the hardware was possibly realized in newer ICs. It was proposed to insert a gate in one of the output arms of the phase/frequency comparator before the signal is fed to the summation amplifier and a periodic current disturbance introduced. This disturbance has the same rate as the reference frequency, and is of extremely short duration, such that the output contains only fairly high harmonics of the reference, which is easily filtered as it contains very little energy. This periodic disturbance upsets the output of the phase/frequency comparator and has an effect similar to that of a leakage resistor. The advantage of this method, however, is that this is done at a fairly high frequency and does not introduce low-frequency noise, which the 1-MΩ resistor does.

Figure 6.17 shows the circuit that accomplishes this and Figure 6.18 shows the effect on the output pulses. The charge pump output exhibits a short negative-going pulse followed immediately by a short positive-going pulse. This can also be called an antibacklash feature and it prevents operating in the dead zone. (This zone is not really a dead zone because of

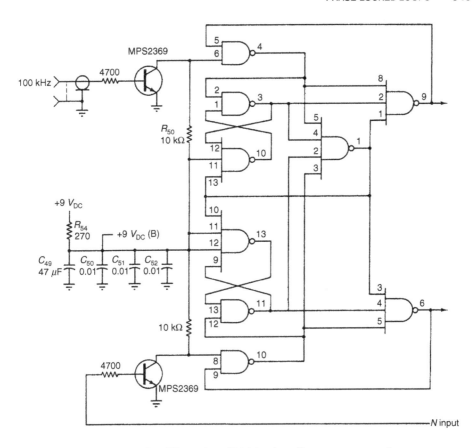

Figure 6.16   Possible version of tristate phase/frequency comparator.

leakage currents in the tuning diode.) The duration and proximity of these pulses are such that they cause no net change in the charge of the integrator. Figure 6.19 shows the response of a phase/frequency detector near loop lock, including the dead zone; this may not be true for ECL.

6.2.3   Filters for Phase Detectors Providing Voltage Output

Figure 6.20 shows the passive RC filter for the second-order loop typically used in PLL synthesizers. The transfer characteristic of the filter is

$$\frac{V_0(s)}{V_i(s)} = \frac{1 + s\tau_2}{1 + s(\tau_1 + \tau_2)} \tag{6.16}$$

where $\tau_1 = R_1 C$ and $\tau_2 = R_2 C$.

Figure 6.21 shows the schematic for the active filter for the second-order loop. Its transfer characteristic is

$$\frac{V_0(s)}{V_i(s)} = \frac{1 + s\tau_2}{s\tau_1} \tag{6.17}$$

where $\tau_1 = R_1 C$ and $\tau_2 = R_2 C$.

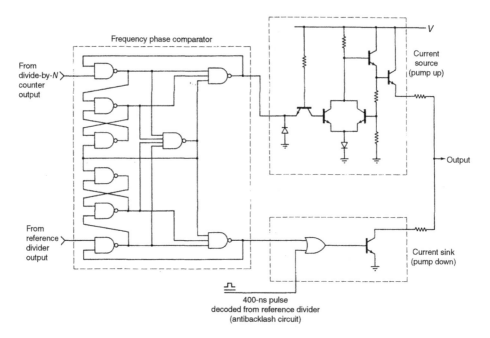

Figure 6.17   Tristate detector with antibacklash circuit included.

If only one active integrator is used, we have a type 1 PLL. If two integrators are used, as in building an active filter, we have a type 2 second-order loop. Here, second-order refers to the denominator polynomial of the transfer function. If we insert a simple low-pass filter such as the one shown in Figure 6.20, but with $R_2 = 0$, we obtain

$$F(s) = \frac{1}{1 + s\tau} \tag{6.18}$$

If we let $K = K_0 K_\theta / N$, the transfer function $B(s)$ becomes

$$B(s) = \frac{N}{s^2/\omega_n^2 + 2\zeta s/\omega_n + 1} \tag{6.19}$$

where $\omega_n = \sqrt{K/\tau}$ and $2\zeta = \omega_n/K = \sqrt{1/K\tau}$. Here, ζ is the damping factor of the loop and ω_n is the natural frequency.

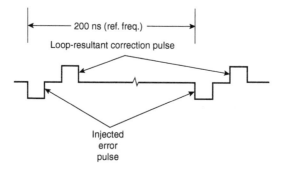

Figure 6.18   Output of frequency/phase detector with antibacklash circuit.

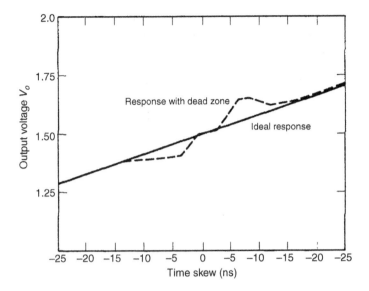

Figure 6.19 Response of frequency/phase detector near loop lock, resulting in a dead zone.

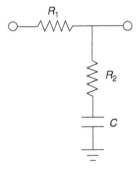

Figure 6.20 Schematic diagram of a typical passive *RC* filter.

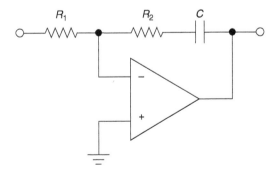

Figure 6.21 Schematic diagram of an active filter for a second-order loop.

The frequency response of the second-order transfer function is determined by ζ. For $\zeta = 0.707$, the transfer function becomes the second-order maximally flat, or Butterworth, response. For values of $\zeta < 0.707$, the gain exhibits peaking in the frequency domain. The maximum value of the frequency response can be found by setting the derivative of its maximum to zero. The frequency at which the maximum occurs is

$$\omega_p = \omega_n \sqrt{1 - 2\zeta^2} \tag{6.20}$$

The 3-dB bandwidth B is found to be

$$B = f_n \left[1 - 2\zeta^2 + \left(2 - 4\zeta^2 + 4\zeta^4 \right)^{1/2} \right]^{1/2} \tag{6.21}$$

where $f_n = \omega_n / 2\pi$.

The time required for the output to rise from 10% to 90% of its final value is the rise time t_r. It is approximately related to the system bandwidth by the relation

$$t_r = \frac{2.2}{B} \tag{6.22}$$

The RC time constant of this simple filter determines both the natural loop frequency and the damping factor ζ. To improve the performance of the filter, we need more flexibility. When the series resistor R_2 is not zero, we obtain the original RC filter of Figure 6.20. The transfer function of this filter is

$$B(s) = \frac{N \left[s\omega_n \left(2\zeta - \omega_n / K \right) + \omega_n^2 \right]}{s^2 + 2\zeta\omega_n s + \omega_n^2} \tag{6.23}$$

where $\omega_n = \sqrt{K/\tau}$ and $2\zeta = (1 + K\tau_2)/\sqrt{K\tau}$ and τ is written for $\tau_1 + \tau_2$.

The determination of the 3-dB bandwidth for this general type 1 second-order loop is somewhat more complex than the earlier computation, but after calculation, we obtain

$$B = f_n \left[a + \left(a^2 + 1 \right)^{1/2} \right]^{1/2} \tag{6.24}$$

where we have written

$$a = 2\zeta^2 + 1 - \frac{\omega_n \left(4\zeta - \omega_n / K \right)}{K} \tag{6.25}$$

The noise bandwidth of the type 1 second-order loop is

$$B_n = \pi f_n \left(\zeta + \frac{1}{4\zeta} \right) \tag{6.26}$$

In the case of the active filter, where we have two integrators, the closed-loop transfer function of the type 2 second-order PLL with a perfect integrator is

$$B(s) = \frac{N \left(2\zeta\omega_n s + \omega_n^2 \right)}{s^2 + 2\zeta\omega_n s + \omega_n^2} \tag{6.27}$$

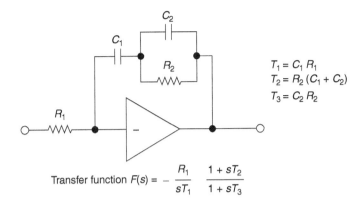

$$T_1 = C_1 R_1$$
$$T_2 = R_2 (C_1 + C_2)$$
$$T_3 = C_2 R_2$$

Transfer function $F(s) = -\dfrac{R_1}{sT_1}\dfrac{1+sT_2}{1+sT_3}$

Figure 6.22 Schematic diagram of an active filter for a third-order loop.

where $\omega_n = (KR_2/\tau_2 R_1)^{1/2}$, $2\zeta = (K\tau_2 R_2/R_1)^{1/2}$, and $K = K_\theta K_0/N$, as usual. The 3-dB bandwidth of the type 2 second-order filter is

$$B = f_n \left\{ 2\zeta^2 + 1 + \left[\left(2\zeta^2 + 1\right)^2 + 1\right]^{1/2}\right\}^{1/2} \qquad (6.28)$$

and the noise bandwidth is

$$B_n = \frac{(KR_2/R_1) + (1/\tau_2)}{4} \qquad (6.29)$$

The type 2 third-order loop is defined by the active integrator, as shown in Figure 6.22. The additional capacitor across the second resistor increases suppression of the reference frequency. The advantage of the higher-order loop is that for the same loop bandwidth, it offers more reference-frequency suppression than the second-order loop. Conversely, for the same suppression, it offers a faster lock-in time. More details are given in Ref. [1].

6.2.3.1 *Transient Response*
The Laplace transform can be used to calculate the response of the PLL to a change in frequency. Figure 6.23 shows the normalized output response of the type 1 second-order loop, and Figure 6.24 shows the normalized output response of the type 2 second-order loop. We determine from both functions that a damping ratio of 0.707 will produce a peak overshoot of less than 10% for the type 1 second-order loop and of less than 20% for the type 2 second-order loop when $\omega_n t \geq 4.5$. The settling time is therefore determined to be $t_s = 4.5/\omega_n$. For more details on the actual design of synthesizer loops, the reader should refer to Ref. [1].

The NXP UMA1018M is an example of the synthesizer implementations now available in IC form. Designed for use in portable radiotelephones, the UMA1018M contains two frequency synthesizers. One, intended for first-LO use, can operate at input frequencies from 50 MHz to 1.25 GHz; the second, intended for use as the second or "IF" LO in a double-conversion system can operate at input frequencies from 20 to 300 MHz. Both loops use the same reference signal (3–40 MHz), which must be supplied by an external crystal oscillator. Figure 6.25 shows the UMA1018M's block diagram, Figure 6.26 shows the block diagram of a sample UMA1018M application, and Figure 6.27 shows a UMA1018M application

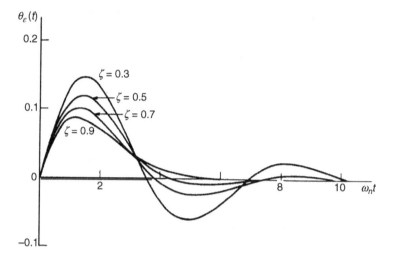

Figure 6.23 The error response of a type 1 second-order PLL to unit-step change in frequency for various damping ratios ζ with K constant. The steady-state error $2\zeta/\omega = 1/K$.

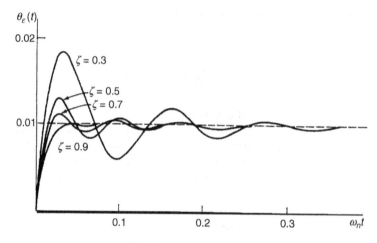

Figure 6.24 The error response of a type 2 second-order PLL to unit-step change in frequency for various damping ratios ζ with ω_n constant. The steady-state error is zero.

schematic. Finally, Figures 6.28 and 6.29 show a UMA1018M-based system's close-in phase noise and reference suppression, respectively.

6.2.4 Charge-Pump-Based Phase-Locked Loops[1]

The basic drawback of the conventional phase/frequency detector, expressed in V/rad, is its dead-zone phenomenon. The loop gain is really determined by the linearity of the phase

---

[1] Portions of this section are based on information contained in the National Semiconductor datasheet "LMX1501A/LMX1511 PLLatinum™ 1.1 GHz Frequency Synthesizer for RF Personal Communications," November 1995. Used with permission.

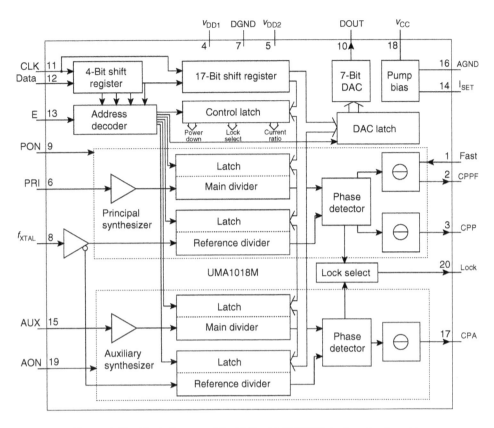

Figure 6.25 Block diagram of the Philips UMA1018M dual-synthesizer chip.

detector. It can drop to zero when no correction is needed. One way around this had been to provide an external resistance, connected between the phase detector output and common that draws current but reduces the reference suppression. A more modern way to overcome this problem is to resort to a charge pump. Figure 6.30 uses a CMOS-based charge pump, with the resistor R_L limiting the available current.

If more control over the actual current is needed, here is a recommendation by National Semiconductor that improves the flexibility.

6.2.4.1 External Charge Pump

Figure 6.31 shows one possible architecture for an external charge-pump current source. The signals ϕ_p and ϕ_r in the diagram correspond to the phase-detector outputs of the National LMX1501/1511 frequency synthesizers. These logic signals are converted into current pulses, using the circuitry shown in Figure 6.31, to enable either charging or discharging of the loop filter components to control the PLL's output frequency.

Referring to Figure 6.31, the design goal is to generate a 5-mA current that is relatively constant to within 5 V of the power-supply rail. To accomplish this, it is important to establish as large of a voltage drop across R_5 and R_8 as possible without saturating Q_2 and Q_4. Approximately 300 mV provides a good compromise, allowing the current source reference generated to be relatively repeatable in the absence of good $Q_1 - Q_2$, $Q_3 - Q_4$ matching (matched transistor pairs are recommended). The ϕ_p and ϕ_r outputs are rated for a maximum

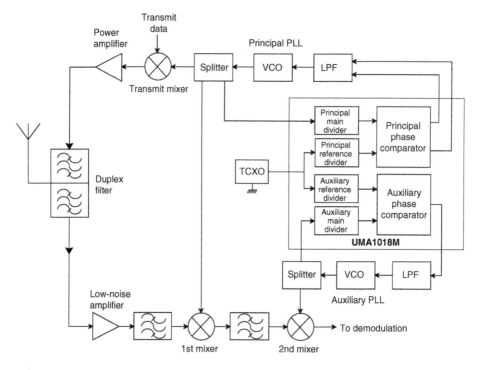

Figure 6.26   Block diagram of a typical UMA1018M application.

load current of 1 mA, while 5 mA current sources are desired. The voltages developed across R_4 and R_9 will consequently be approximately 258 mV, or 42 mV $< R_8 - R_5$, due to the current density differences [$0.026 \times \ln(5 \text{ mA}/1 \text{ mA})$] through the $Q_1 - Q_2$, $Q_3 - Q_4$ pairs.

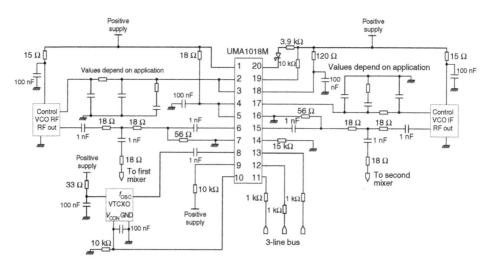

Figure 6.27   Schematic of a typical UMA1018M application.

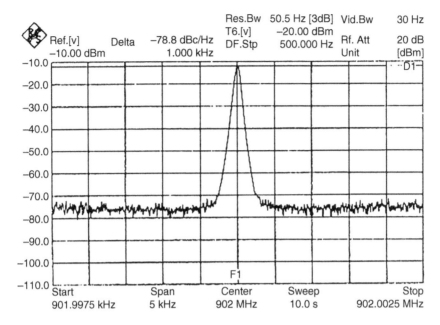

Figure 6.28 Close-in noise of a UMA1018M-based system (principal synthesizer).

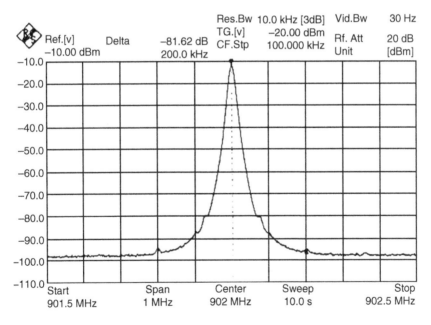

Figure 6.29 Wideband output spectrum of a UMA1018M-based system (principal synthesizer) showing reference-frequency breakthrough.

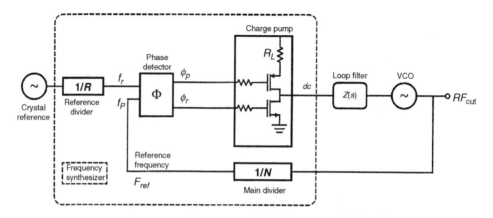

Figure 6.30 A basic charge-pump PLL.

To calculate the value of R_7, it is necessary to first estimate the forward base-to-emitter voltage drop (V_{fn}, V_{fp}) of the transistors used, the V_{OL} drop of ϕ_p, and the V_{OH} drop of ϕ_r under 1-mA loads (ϕ_p's $V_{OL} < 0.1$ V and ϕ_r's $V_{OH} < 0.1$ V).

Knowing these parameters along with the desired current allows us to design a simple external charge pump. Separating the pump-up and pump-down circuits facilitates the nodal

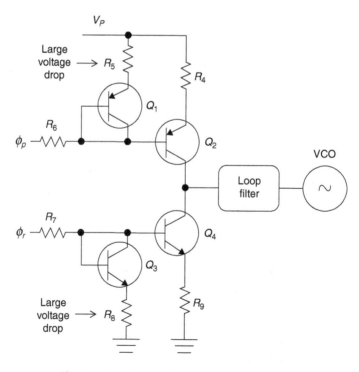

Figure 6.31 External charge pump current source.

analysis and gives the following equations:

$$R_4 = \frac{V_{R5} - V_T \times \ln\left(\frac{i_{source}}{i_{pmax}}\right)}{i_{source}} \qquad (6.30)$$

$$R_9 = \frac{V_{R8} - V_T \times \ln\left(\frac{i_{sink}}{i_{nmax}}\right)}{i_{sink}} \qquad (6.31)$$

$$R_5 = \frac{V_{R5} \times (\beta_p + 1)}{i_{pmax} \times (\beta_p + 1) - i_{source}} \qquad (6.32)$$

$$R_8 = \frac{V_{R8} \times (\beta_n + 1)}{i_{rmax} \times (\beta_n + 1) - i_{source}} \qquad (6.33)$$

$$R_6 = \frac{\left(V_p - V_{VOL\phi p}\right) - \left(V_{R5} + V_{fp}\right)}{i_{pmax}} \qquad (6.34)$$

$$R_7 = \frac{\left(V_p - V_{VOH\phi r}\right) - \left(V_{R8} + V_{fn}\right)}{i_{max}} \qquad (6.35)$$

6.2.4.2 Example

Typical device parameters: $\beta_n = 100$, $\beta_p = 50$.
Typical system parameters: $V_P = 5.0\,V$; $V_{cntl} = 0.5\,V – 4.5\,V$; $V_{\phi p} = 0.0\,V$; $V_{\phi r} = 5.0\,V$.
   Design parameters: $I_{sink} = I_{source} = 5.0\,mA$; $V_{fn} = V_{fp} = 0.8\,V$; $I_{r\,max} = I_{p\,max} = 1\,mA$; $V_{R8} = V_{R5} = 0.3\,V$; $V_{OL\phi p} = V_{OH\phi r} = 100\,mV$.

Therefore, select

$$R_4 = R_9 = \frac{0.3\,V - 0.026 \times 1n\,(5.0\,mA \div 1.0\,mA)}{5\,mA} = 51.6\,\Omega \qquad (6.36)$$

$$R_5 = \frac{0.3\,V \times (50 + 1)}{1.0\,mA \times (50 + 1) - 5.0\,mA} = 332\,\Omega \qquad (6.37)$$

$$R_8 = \frac{0.3\,V \times (100 - 1)}{1.0\,mA \times (100 + 1) - 5.0\,mA} = 315.6\,\Omega \qquad (6.38)$$

$$R_6 = R_7 = \frac{(5\,V - 0.1\,V) - (0.3\,V + 0.8\,V)}{1.0\,mA} = 3.8\,k\Omega \qquad (6.39)$$

6.2.4.3 Design Example: A Passive PLL Filter

Figure 6.32 shows the loop filter topology that we will use for this example. The advantage of a passive filter as opposed to an active filter is that the filter itself introduces no noise, while the active filter using an op-amp can frequently cause more harm than good because of its noise contribution.

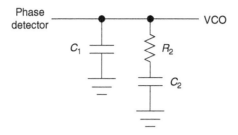

Figure 6.32 Schematic of the second-order loop filter.

The time constants that determine the pole and zero frequencies of the filter shown in Figure 6.32 are

$$\tau_1 = R_2 \frac{C_1 C_2}{C_1 + C_2} \tag{6.40}$$

and

$$\tau_2 = R_2 C_2 \tag{6.41}$$

The required values for τ_1 and τ_2 depend on the values we specify for loop bandwidth, ω_p (equal to the frequency at which the loop's gain falls to 0 dB, or unity) and phase margin, ϕ_p, which is defined as the difference between 180° and loop's phase shift at ω_p. For best loop performance, we require a phase margin of 45°. The loop bandwidth must be carefully chosen with regard to lock time, phase noise, stability, and the reference-energy suppression. In this example, we will design our loop based on 10% of the reference frequency (f_{ref}, 200 kHz), so $\omega_p = 2\pi \times 20$ kHz or 125.6 kHz. The required values for τ_1 and τ_2 can be determined from

$$\tau_1 = \frac{\sec\phi_p - \tan\phi_p}{\omega_p} \tag{6.42}$$

and

$$\tau_2 = \frac{1}{\omega_p^2 \tau_1} \tag{6.43}$$

From the time constants, τ_1 and τ_2, and the loop bandwidth ω_p, we obtain the values for C_1, C_2, and R_2 as follows:

$$C_1 = \frac{\tau_1}{\tau_2} \frac{K_\phi K_{\text{VCO}}}{\omega_p^2 \tau_1} \sqrt{\frac{1 + (\omega_p \tau_2)^2}{1 + (\omega_p \tau_1)^2}} \tag{6.44}$$

$$C_2 = C_1 \left(\frac{\tau_2}{\tau_1} - 1 \right) \tag{6.45}$$

$$R_2 = \frac{\tau_2}{C_2} \tag{6.46}$$

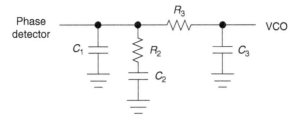

Figure 6.33   Third-order loop filter for greater reference-energy suppression.

where K_{VCO} = VCO tuning voltage constant (frequency versus voltage ratio) in MHz/V, K_ϕ = phase detector/charge pump gain constant (its ratio of current output to input phase differential) in mA, and N = main divider ratio.

If additional suppression of the loop's reference-frequency is needed, a low-pass filter section can be added, resulting in the third-order loop filter shown in Figure 6.33.

The additional reference-frequency attenuation provided by $R_3 C_3$ can be found from

$$\text{Atten} = 20 \log \left[(2\pi f_{ref} R_3 C_3)^2 + 1 \right] \tag{6.47}$$

The low-pass section contributes an additional pole, which must be low enough to provide significant additional reference attenuation and high enough ($\geq 5\omega_p$) not to compromise the loop's stability. The time constant τ_3 of the added low-pass section can be found from

$$\tau_3 = R_3 C_3 \tag{6.48}$$

We can find τ_3 for a given value of additional attenuation from

$$\tau_3 = \sqrt{\frac{10^{(\text{Atten}/20)} - 1}{(2\pi f_{ref})^2}} \tag{6.49}$$

To compensate for the added low-pass section, we recalculate the filter component values using the new loop bandwidth, ω_c, which can be found from

$$\frac{V_0(s)}{V_i(s)} = \frac{1 + s\tau_2}{1 + s(\tau_1 + \tau_2)} \tag{6.50}$$

$$\omega_c = \frac{\tan \phi (\tau_1 + \tau_3)}{[(\tau_1 + \tau_3)^2 + \tau_1 \tau_3]} \left[\sqrt{1 + \frac{(\tau_1 + \tau_3)^2 + \tau_1 \tau_3}{[\tan \phi (\tau_1 + \tau_3)]^2}} - 1 \right] \tag{6.51}$$

We then reduce the phase-margin degradation caused by $R_3 C_3$ by increasing C_1 and C_2 while slightly decreasing R_2. Calculating new values for C_1, C_2, and R_2 require that we first determine the value for τ_2 as modified by $R_3 C_3$.

$$\tau_2' = \frac{1}{[\omega_c^2 (\tau_1 + \tau_3)]} \tag{6.52}$$

We can calculate C_1 as

$$C_1 = \frac{\tau_1}{\tau_2'} \frac{K_\phi K_{\text{VCO}}}{\omega_c^2 N} \left[\frac{\left(1 + \omega_c^2 \tau_{22}'\right)}{\left(1 + \omega_c^2 \tau_1^2\right)\left(1 + \omega_c^2 \tau_3^2\right)} \right]^{1/2} \tag{6.53}$$

As with the original second-order filter, C_2 and R_2 are calculated by means of equations (6.45) and (6.46), respectively.

The values of R_3 and C_3 are somewhat arbitrary. The following rules of thumb apply. The value of C_3 should be less than that of C_1 and C_2, and preferably $\leq C_1/10$ to keep τ_3 from interacting with τ_1 and τ_2. Also, R_3 should be $\geq 2R_2$. Any capacitance already present on the VCO tuning line, including the input capacitance of the tuning diode(s), must be allowed for in selecting the value actually used for C_3 in the constructed loop filter.

The conversion from the voltage gain of a VCO to the charge current as needed here can be obtained from the formula [3]

$$K_\phi = \frac{I_{\text{charge}}}{2\pi f_r C_r} \tag{6.54}$$

which can be solved for the charge current, I_{charge}, as

$$I_{\text{charge}} = K_\phi 2\pi f_r C_r \tag{6.55}$$

where K_ϕ is the phase detector/charge pump gain factor in V/rad, f_R is the reference frequency, and C_R is the memory ("ramp") capacitance, typically 0.1 μF. The loop filters of Figures 6.32 and 6.33 include C_R as C_1.

6.2.4.4 Example
Design a third-order loop filter for a 900-MHz synthesizer with a 200-kHz reference based on the parameters:

$K_{\text{VCO}} = 20\,\text{MHz/V}$
$K_\phi = 5\,\text{mA}$
$N = 4500$
$\omega_p = 2\pi \times 20\,\text{kHz} = 1.256 \times 10^5$ Hz
$\phi_p = 45°$
Atten $= 20$ dB

$$\tau_1 = \frac{\sec\phi_p - \tan\phi_p}{\omega_p} = 3.29 \times 10^{-6}\,\text{s} \tag{6.56}$$

$$\tau_3 = \sqrt{\frac{10^{(20/20)} - 1}{\left(2\pi \cdot 200 \cdot 10^3\right)^2}} = 2.387 \times 10^{-6}\,\text{s} \tag{6.57}$$

$$\omega_c = \frac{(3.29 \cdot 10^{-6} + 2.387 \cdot 10^{-6})}{\left[(3.29 \cdot 10^{-6} + 2.387 \cdot 10^{-6})^2 + 3.29 \cdot 10^{-6} \cdot 2.387 \cdot 10^{-6}\right]}$$

$$\times \left[1 + \sqrt{\frac{(3.29 \cdot 10^{-6} + 2.387 \cdot 10^{-6})^2 + 3.29 \cdot 10^{-6} \cdot 2.387 \cdot 10^{-6}}{\left[(3.29 \cdot 10^{-6} + 2.387 \cdot 10^{-6})\right]^2}}\right] - 1$$

$$= 7.045 \cdot 10^4 \ \text{Hz} \tag{6.58}$$

$$\tau_2 = \frac{1}{(7.045 \cdot 10^4)^2 \cdot (3.29 \cdot 10^{-6} + 2.387 \cdot 10^{-6})} = 3.549 \cdot 10^{-5} \, \text{s} \tag{6.59}$$

$$C_1 = \frac{3.29 \cdot 10^{-6}}{3.549 \cdot 10^{-5}} \cdot \frac{(5.0 \cdot 10^{-3}) \cdot 20 \cdot 10^6}{(7.045 \cdot 10^4)^2 \cdot 4500}$$

$$\times \left[\frac{\left[1 + (7.045 \cdot 10^4)^2 \cdot (3.549 \cdot 10^{-5})^2\right]}{\left[1 + (7.045 \cdot 10^4)^2 \cdot (3.29 \cdot 10^{-6})^2\right]\left[1 + (7.045 \cdot 10^4)^2 \cdot (2.39 \cdot 10^{-6})^2\right]}\right]^{1/2}$$

$$= 1.085 \, \text{nF} \tag{6.60}$$

$$C_2 = 1.085 \, \text{nF} \times \left(\frac{3.55 \cdot 10^{-5}}{3.29 \cdot 10^{-6}} - 1\right) = 10.6 \, \text{nF} \tag{6.61}$$

$$R_2 = \frac{3.55 \cdot 10^{-5}}{10.6 \cdot 10^{-9}} = 3.35 \, \text{k}\Omega \tag{6.62}$$

If we choose $R_3 = 22 \, \text{k}\Omega$, then

$$C_3 = \frac{2.34 \times 10^{-6}}{22 \times 10^3} = 106 \, \text{pF} \tag{6.63}$$

Selecting the nearest standard value for each component gives $C_1 = 1000 \, \text{pF}$, $R_2 = 3.3 \, \text{k}\Omega$, $C_2 = 10 \, \text{nF}$, $R_3 = 22 \, \text{k}\Omega$, and $C_3 = 100 \, \text{pF}$.

6.3 HOW TO DO A PRACTICAL PLL DESIGN USING CAD

1. Have a modern CAD tool that give you a synthesized oscillator, such as a quarter-wave microstrip oscillator (Figure 6.34). Here we have used Compact Software's PLL Design Kit. (There are other similar programs in the market.) The loop synthesis program will ask you to specify the oscillator transistor's dc current, and others, and calculate the circuit's output power.

2. Ask your design tool for the oscillator's open- and closed-loop phase noise as a function of other noise sources (Figure 6.35). The program must consider the tuning-diode as a major noise contributor. Here the loop bandwidth has been made too wide and makes the actual phase noise worse. This plot is valid only if the dividers are the

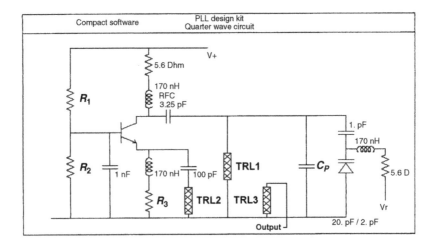

Figure 6.34 To design a PLL, begin with a suitable CAD-generated oscillator. This 900-MHz circuit is similar to that of the ceramic-resonator oscillator shown in Figure 5.33. TRL1, TRL2, and TRL3 are coupled transmission lines; the resonator, TRL1, is slightly less than λ/4 at the operating frequency. TRL2 provides feedback and TRL3 provides output coupling. R_1, R_2, and R_3 must be specified; C_P must be determined according to the desired tuned range.

only noise contributors. The phase/frequency detector also adds noise following the equation

$$\mathcal{L} = \mathcal{L}_0 + 10\log(F_r)$$

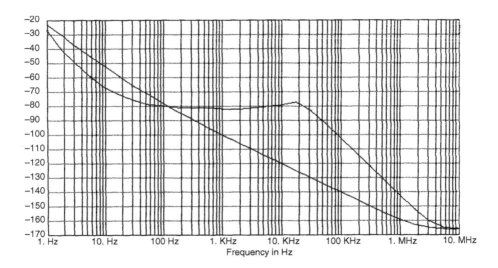

Figure 6.35 Predicted open- and closed-loop phase noise for the oscillator shown in Figure 6.28. The portion of the closed-loop curve up to 20 kHz represents the phase noise of the loop's crystal reference oscillator multiplied up. The VCO operates at 900 MHz; the reference frequency is 200 kHz. For references of 2 and 20 MHz, the phase noise would drop by 20 and 40 dB, respectively. See Table 6.1 and Figure 6.36.

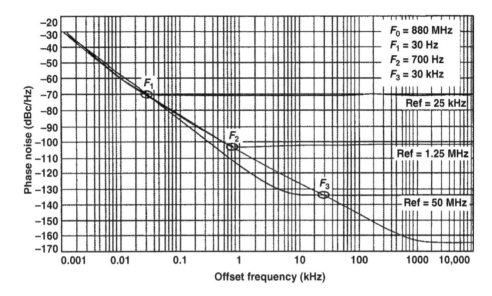

Figure 6.36 Result of dividing a 50-MHz standard oscillator down to different reference frequencies and then multiplying these frequencies to 880 MHz. The continuing line below breakpoint F_3 show the performance of the free-running 50-MHz oscillator.

where $\mathcal{L}_0$ is a constant that is equivalent to the phase/frequency detector noise with $F_r = 1\,\text{Hz}$. $\mathcal{L}$ as a function of f_r is given below for standard PLL chips:

$\mathcal{L}$ (dBc/Hz)	f_r (Hz)
-168 to -170	10 k
-164 to -168	30 k
-155 to -160	200 k
-150 to -155	1 M
-145	10 M

3. Ask your synthesis program to design a loop filter for best performance, meaning a phase margin of 45° (Figure 6.37). Here, we have arbitrarily included an additional low-pass filter with a cutoff frequency of 30 kHz to illustrate its effect on the loop performance.

4. Generate a Bode plot for the loop design and check the loop's phase margin at f_0, the frequency at which the loop gain is 0 dB. In this example, the plot (Figure 6.38) shows that because of the 30-kHz low-pass filter, we achieve only about 12° phase margin instead of the desired 45°. We can also see that the slope of the gain response through f_0 is not maintained at 6 dB/octave (20 dB/decade) over a sufficiently wide gain span. Maintaining the 6 dB/octave slope for loop-gain values of $+10$ to -10 dB is essential.

We can determine from Figure 6.38 that the loop gain is -60 dB at 200 kHz, the loop's reference frequency (f_ref). Adding this to the phase detector's reference

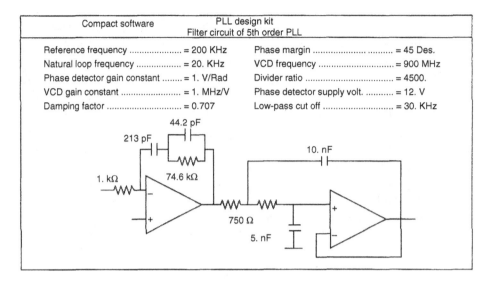

<table>
</table>

Compact software   PLL design kit
Filter circuit of 5th order PLL

Reference frequency .... = 200 KHz   Phase margin .... = 45 Des.
Natural loop frequency .... = 20. KHz   VCD frequency .... = 900 MHz
Phase detector gain constant .... = 1. V/Rad   Divider ratio .... = 4500.
VCD gain constant .... = 1. MHz/V   Phase detector supply volt. .... = 12. V
Damping factor .... = 0.707   Low-pass cut off .... = 30. KHz

Figure 6.37   CAD-generated loop-filter design based on a reference frequency of 200 kHz, a loop bandwidth of 20 kHz, and the optimum phase margin of 45°. We have arbitrarily included a 30-kHz low-pass filter to illustrate its effects on loop performance.

suppression (at least 40 dB) would appear to net us an overall reference suppression of 100 dB, but in any real implementation, crosstalk on the PLL PC board would reduce this value to between 80 and 90 dB.

5. As a result of the insufficient phase margin, the loop rings (Figure 6.39), locking in 322 μs instead of a possible 80 μs—a lock time four times as long as intended. Achievement of a phase margin of 45°, sometimes characterized merely as a rule of thumb, minimizes locking time and overshoot. Values less than 45° result in excessive overshoot and ringing, as shown here, and values greater than 45° result in an overdamped loop that crawls into lock.

Table 6.1   Phase noise vs. reference frequency derived from a high-performance 50-MHz crystal oscillator and multiplied to 880 MHz

Offset from Carrier	Phase Noise of 50-MHz Standard Oscillator (dBc/Hz)	Phase Noise in dBc/Hz After Dividing the Standard Down to		
		25 kHz	1.25 MHz	50 MHz
		and Multiplying the Result to 880 MHz		
10 Hz	−80	−55	−55	−55
100 Hz	−110	−69	−85	−85
1 kHz	−140	−69	−103	−115
10 kHz	−160	−69	−103	−135
100 kHz	−160	−69	−103	−135
1 MHz	−160	−69	−103	−135

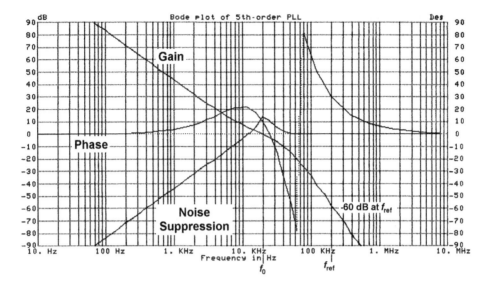

Figure 6.38 Bode plot for the PLL. The 30-kHz low-pass filter disallows a phase margin of 45° at f_0; instead, the margin is only about 20°.

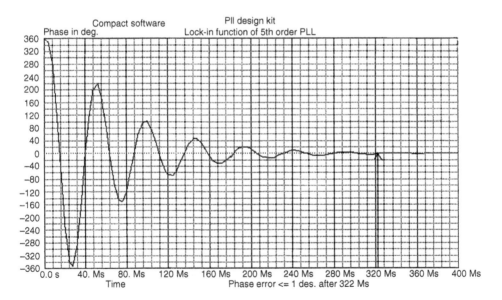

Figure 6.39 Lock-in response of the PLL. As a result of the insufficient phase margin, the loop is underdamped and takes 322 μs to achieve lock.

6.4   FRACTIONAL-*N*-DIVISION PLL SYNTHESIS

6.4.1   The Fractional-*N* Principle

The principle of the fractional-*N* PLL synthesizer has been around for a while. In the past, implementation of this has been done in an analog system. It would be ideal to be able to build a single-loop synthesizer with a 1.25-MHz or 50-MHz reference and yet obtain the desired step size resolution, such as 25 kHz. This would lead to the much smaller division ratio and much better phase noise performance.

An alternative would be for *N* to take on fractional values. The output frequency could then be changed in fractional increments of the reference frequency. Although a digital divider cannot provide a fractional division ratio, ways can be found to accomplish the same task effectively.

The most frequently used method is to divide the output frequency by $N + 1$ every M cycles and to divide by N the rest of the time. The effective division ratio is then $N + 1/M$, and the average output frequency is given by

$$f_0 = \left(N + \frac{1}{M} \right) f_r \tag{6.64}$$

This expression shows that f_0 can be varied in fractional increments of the reference frequency by varying M. The technique is equivalent to constructing a fractional divider, but the fractional part of the division is actually implemented using a phase accumulator. The phase accumulator approach is illustrated by the following example. This method can be expanded to frequencies much higher than 6 GHz using the appropriate synchronous dividers.

6.4.1.1   Example
Considering the problem of generating 899.8 MHz using a fractional-*N* loop with a 50-MHz reference frequency,

$$899.8 \, \text{MHz} = 50 \, \text{MHz} \left(N + \frac{K}{F} \right)$$

The integral part of the division N has to be set to 17 and the fractional part K/F needs to be 996/1000; (the fractional part K/F is not a integer) and the VCO output has to be divided by 996× every 1000 cycles. This can easily be implemented by adding the number 0.996 to the contents of an accumulator every cycle. Every time the accumulator overflows, the divider divides by 18 rather than by 17. Only the fractional value of the addition is retained in the phase accumulator. If we move to the lower band or try to generate 850.2 MHz, N remains 17 and K/F becomes 4/1000. This method of using fractional division was first introduced by using analog implementation and noise cancellation, but today it is implemented totally as a digital approach. The necessary resolution is obtained from the dual modulus prescaling, which allows for a well-established method for achieving a high-performance frequency synthesizer operating at UHF and higher frequencies. Dual-modulus prescaling avoids the loss of resolution in a system compared to a simple prescaler; it allows a VCO step equal to the value of the reference frequency to be obtained. This method needs an additional counter and the dual modulus prescaler then divides one or two values depending upon the state of its control. The only drawback of prescalers is the minimum division ratio of the prescaler for approximately N^2. The dual modulus divider is the key to implementing the fractional-N synthesizer principle. Although the fractional-N technique appears to have a good potential of solving the resolution limitation, it is not free of having its own complications. Typically, an overflow from the phase accumulator, which is the adder with the output feedback to the input after being latched, is used to change the instantaneous division ratio. Each overflow produces a jitter at the output frequency, caused by the fractional division, and is limited to the fractional portion of the desired division ratio.

In our case, we had chosen a step size of 200 kHz, and yet the discrete sidebands vary from 200 kHz for $K/F = 4/1000$ to 49.8 MHz for $K/F = 996/1000$. It will become the task of the loop filter to remove those discrete spurious. Although in the past the removal of the discrete spurs has been accomplished by using analog techniques, various digital methods are now available. The microprocessor has to solve the following equation:

$$N^* = \left(N + \frac{K}{F} \right) = [N(F - K) + (N + 1)K] \tag{6.65}$$

6.4.1.2 *Example*
For $f_0 = 850.2$ MHz, we obtain

$$N^* = \frac{850.2\,\text{MHz}}{50\,\text{MHz}} = 17.004$$

Following the formula above

$$N^* = \left(N + \frac{K}{f} \right) = \frac{[17(1000 - 4) + (17 + 1)4]}{1000}$$

$$= \frac{[16932 + 72]}{1000} = 17.004$$

$$f_{\text{out}} = 50\,\text{MHz} \times \frac{[16932 + 72]}{1000}$$

$$= 846.6\,\text{MHz} + 3.6\,\text{MHz}$$

$$= 850.2\,\text{MHz}$$

By increasing the number of accumulators, frequency resolution much below 1-Hz step size is possible with the same switching speed.

There is an interesting, generic problem associated with all fractional-N synthesizers. Assume for a moment that we use our 50-MHz reference and generate a 550-MHz output frequency. This means our division factor is 11. Aside from reference-frequency sidebands (± 50 MHz) and harmonics, there will be no unwanted spurious frequencies. Of course, the reference sidebands will be suppressed by the loop filter by more than 90 dB. For reasons of phase noise and switching speed, a loop bandwidth of 100 kHz has been considered. Now, taking advantage of the fractional-N principle, say we want to operate at an offset of 30 kHz (550.03 MHz). With this new output frequency, the inherent spurious-signal reduction mechanism in the fractional-N chip limits the reduction to about 55 dB. Part of the reason why the spurious-signal suppression is less in this case is that the phase-frequency detector acts as a mixer, collecting both the 50-MHz reference (and its harmonics) and

550.03 MHz. Mixing the 11th reference harmonic (550 MHz) and the output frequency (550.03 MHz) results in output at 30 kHz; since the loop bandwidth is 100 kHz, it adds nothing to the suppression of this signal. To solve this, we could consider narrowing the loop bandwidth to 10% of the offset. A 30-kHz offset would equate to a loop bandwidth of 3 kHz, at which the loop speed might still be acceptable, but for a 1-kHz offset, the necessary loop bandwidth of 100 Hz would make the loop too slow. A better way is to use a different reference frequency-one that would place the resulting spurious product considerably outside the 100-kHz loop-filter window. If, for instance, we used a 49-MHz reference, multiplication by 11 would result in 539 MHz. Mixing this with 550.03 MHz would result in spurious signals at ±11.03 MHz, a frequency so far outside the loop bandwidth that it would essentially disappear. Starting with a VHF, low-phase-noise crystal oscillator, such as 130 MHz, one can implement an intelligent reference-frequency selection to avoid these discrete spurious signals. An additional method of reducing the spurious contents is maintaining a division ratio greater than 12 in all cases. Actual tests have shown that these reference-based spurious frequencies can be repeatably suppressed by 80–90 dB.

6.4.2 Spur-Suppression Techniques

Although several methods have been proposed in the literature (see patents in Refs. [4–9]), the method of reducing the noise by using a sigma-delta modulator has shown to be most promising. The concept is to get rid of the low-frequency phase error by rapidly switching the division ratio to eliminate the gradual phase error at the discriminatory input. By changing the division ratio rapidly between different values, the phase errors occur in both polarities, positive as well as negative, and at an accelerated rate that explains the phenomenon of high-frequency noise push-up. This noise, which is converted to a voltage by the phase/frequency discriminator and loop filter, is filtered out by the low-pass filter. The main problem associated with this noise shaping technique is that the noise power rises rapidly with frequency. Figure 6.40 shows noise contributions with such a sigma-delta modulator in place.

On the other hand, we can now, for the first time, build a single-loop synthesizer with switching times as fast as 6 μs and very little phase-noise deterioration inside the loop bandwidth, as seen in Figure 6.40. Since this system maintains the good phase noise of the ceramic-resonator-based oscillator, the resulting performance is significantly better than the phase noise expected from high-end signal generators. However, this method does not allow us to increase the loop bandwidth beyond the 100-kHz limit, where the noise contribution of the sigma-delta modulator takes over.

Table 6.2 shows some of the modern spur-suppression methods. These three-stage sigma-delta methods with larger accumulators have the most potential [4–9].

Table 6.2 Modern spur-suppression methods

Technique	Feature	Problem
DAC phase estimation	Cancel spur by DAC	Analog mismatch
Pulse generation	Insert pulses	Interpolation jitter
Phase interpolation	Inherent fractional divider	Interpolation jitter
Random jittering	Randomize divider	Frequency jitter
Sigma-delta modulation	Modulate division ratio	Quantization noise

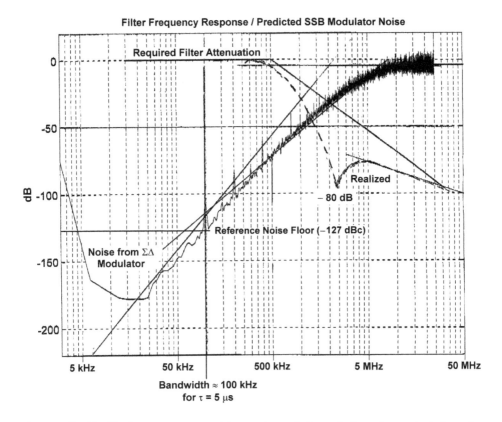

Figure 6.40 The filter frequency response/phase noise analysis graph shows the required attenuation for the reference frequency of 50 MHz and the noise generated by the sigma-delta converter (three steps) as a function of the offset frequency. It becomes apparent that the sigma-delta converter noise dominates above 80 kHz unless attenuated.

The power spectral response of the phase noise for the three-stage sigma-delta modulator is calculated from

$$\mathcal{L}(f) = \frac{(2\pi)^2}{12 \cdot f_{\mathrm{ref}}} \cdot \left[2\sin\left(\frac{\pi f}{f_{\mathrm{ref}}}\right) \right]^{2(n-1)} \mathrm{rad}^2/\mathrm{Hz} \tag{6.66}$$

where n is the number of the stage of the cascaded sigma-delta modulator [10]. Equation (6.66) shows that the phase noise resulting from the fractional controller is attenuated to negligible levels close to the center frequency, and further from the center frequency, the phase noise is increased rapidly and must be filtered out prior to the tuning input of the VCO to prevent unacceptable degradation of spectral purity. A loop filter must be used to filter the noise in the PLL loop. Figure 6.40 showed the plot of the phase noise versus the offset frequency from the center frequency. A fractional-N synthesizer with a three-stage sigma-delta modulator as shown in Figure 6.41 has been built. The synthesizer consists of a phase/frequency detector, an active low-pass filter (LPF), a voltage-controlled oscillator (VCO), a dual-modulus prescaler, a three-stage sigma-delta modulator, and a buffer. Figure 6.42 shows the inner workings of the chip in greater detail.

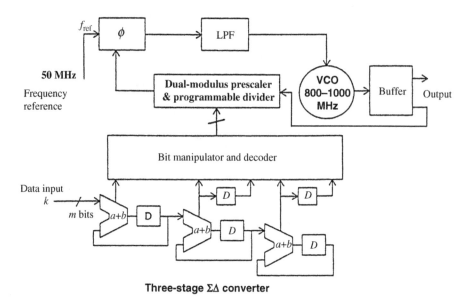

Figure 6.41   Block diagram of the fractional-*N* synthesizer built using a custom IC capable of operation at reference frequencies up to 150 MHz. The frequency is extensible up to 3 GHz using binary ($\div 2$, $\div 4$, $\div 8$, etc.) and fixed-division counters.

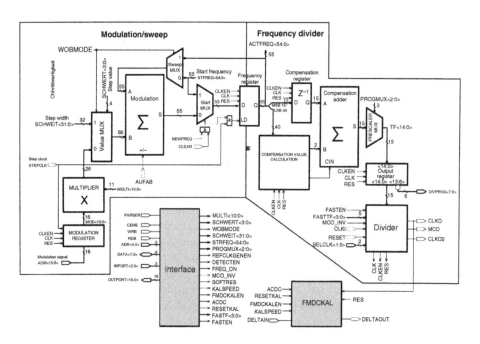

Figure 6.42   Detailed block diagram of the inner workings of the fractional-*N*-division synthesizer chip.

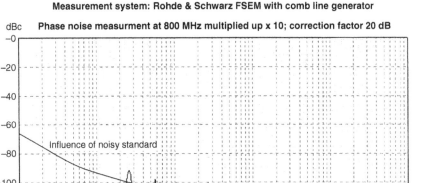

Figure 6.43 Measured phase noise of the fractional-*N*-division synthesizer using a custom-built, high-performance 50-MHz crystal oscillator as reference, with the calculated degradation due to a noisy reference plotted for comparison. Both synthesizer and spectrum analyzer use the same reference.

After designing, building, and predicting the phase noise performance of this synthesizer, it becomes clear that measuring the phase noise of such a system becomes tricky. Standard measurement techniques that use a reference synthesizer would not provide enough resolution because there are no synthesized signal generators on the market sufficiently good enough to measure such low values of phase noise. Therefore, we had to build a comb generator that would take the output of the oscillator and multiply this up 10–20 times.

Passive phase noise measurement systems, based on delay lines, are not selective, and the comb generator confuses them, however, the Rohde & Schwarz FSEM spectrum analyzer with the K-4 option has sufficient resolution to be used for phase noise measurements. All of the Rohde & Schwarz FSE-series spectrum analyzers use a somewhat more discrete fractional-division synthesizer with a 100-MHz reference. Based on the multiplication factor of 10, it turns out that there is enough dynamic range in the FSEM analyzer with the K-4 option to be used for phase noise measurement. The useful frequency range off the carrier for the system is 100 Hz to 10 MHz perfect for this measurement.

Figure 6.43 shows the measured phase noise of the final frequency synthesizer.

During the measurements, we determined that we needed a 50-MHz crystal oscillator with better phase noise. Upon examination of the measured phase noise shown in Figure 6.43, it can been seen that the oscillator used as the reference was significantly better. Otherwise, this phase noise would not have been possible. Also, the loop filter cutoff frequency of about 100 kHz can be recognized by the roll-off in Figure 6.43. This fractional-*N*-division synthesizer with a high-performance VCO has a significantly better phase noise than other example systems in this frequency range. To demonstrate this improvement, phase noise measurements were made on standard systems, using typical synthesizer chips.

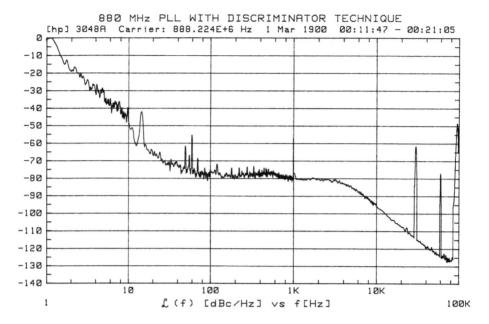

Figure 6.44 Measured phase noise of a 880-MHz synthesizer using a conventional synthesizer chip. Comparing this to Figure 6.43 shows the big improvement possible by fractional-*N*-division synthesizers as shown in this product description.

Although the phase noise by itself and the synthesizer design is quite good, it is no match for this new approach as can be seen in Figure 6.44 [11].

By combining the very best available technologies, such as using high-end VCOs with ceramic-resonator-based tuned circuits or high-Q LC arrangements including microstrips on Teflon material and modern fractional-*N*-division synthesizer blocks that can operate with a 50-MHz reference, it is possible to build an extremely high-quality system.

We have also learned that its limits are determined by the reference crystal oscillator and possibly by the phase detector; these must be specially designed to match the synthesizer's performance. (Figure 6.45 shows a block diagram of the phase detector.) Due to the very high bandwidth, switching speeds at 6 μs were made possible. The resolution of the synthesizer itself depends on the accumulator size. Step sizes of 25 kHz up to several megahertz were successfully tested. For wideband applications, some of the critical points are at 10 kHz, where −120 dBc/Hz is desired, at 800 kHz, better than −153 dBc/Hz, and at 3 MHz, better than −155 dBc/Hz.

As validation, three types of synthesizers have been built: one covering 75–105 MHz, with 1-Hz resolution, for an HF transceiver; another covering 700–2000 MHz, and a third covering 2700–3500 MHz, also with better than 1-Hz resolution.

The fractional-*N* PLL implementations described so far are intended mainly for high-performance, base-station applications. Simplified, highly integrated versions are available for mass-market wireless applications. One solution, National Semiconductor's LMX235x family, implements a fractional-*N* RF synthesizer and an integer-*N* IF synthesizer in one IC. The LMX2350's RF synthesizer can operate at input frequency up to 2.5 GHz and the LMX2352's up to 1.2 GHz. The IF synthesizer in both parts can operate from 10 to 550 MHz. Figure 6.46 shows the close-in output spectrum of an LMX2350-based synthe-

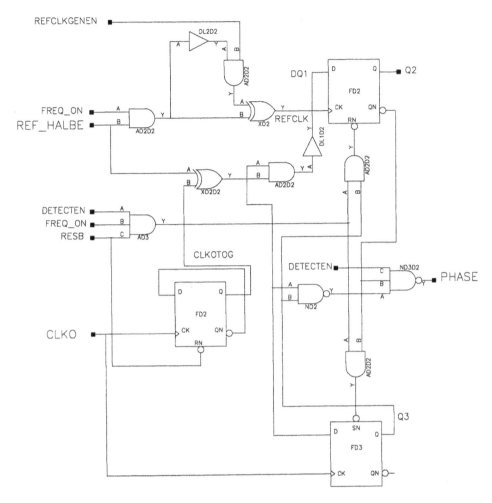

Figure 6.45 Custom-built phase detector with a noise floor of better than −168 dB. This phase detector shows extremely low-phase jitter.

sizer at 1.965 GHz, and Figure 6.47 shows its lock time. A sample LMX-2350 application is shown in Figure 6.48.

Fractional-N synthesizer chips are also available from Philips. Examples of Philips fractional-N parts include the SA7016DH (1.3 GHz), SA7025DK (main synthesizer, 1.0 GHz; auxiliary synthesizer, 150 MHz), SA7026DK (1.3 GHz/550 MHz), SA8016DH (2.5 GHz), SA8025ADK (1.8 GHz/150 MHz), and SA8026DK (2.5 GHz/550 MHz).

6.5   DIRECT DIGITAL SYNTHESIS

Direct digital frequency synthesis (DDFS), also referred to as direct digital synthesis (DDS), consists of generating a digital representation of the desired signal and then using a D/A converter to convert the digital representation to an analog waveform. Recent advances in high-speed microelectronics, particularly the microprocessor, make DDS practical at

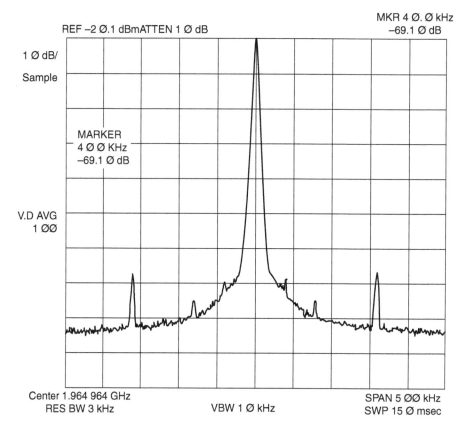

REF −2 Ø.1 dBmATTEN 1 Ø dB

MKR 4 Ø. Ø kHz
−69.1 Ø dB

1 Ø dB/

Sample

MARKER
4 Ø Ø KHz
−69.1 Ø dB

V.D AVG
1 ØØ

Center 1.964 964 GHz
RES BW 3 kHz

VBW 1 Ø kHz

SPAN 5 ØØ kHz
SWP 15 Ø msec

Figure 6.46 Block diagram of an integrated frequency synthesizer. In this case, the designer has control over the VCO and the loop filter; the reference oscillator is part of the chip. In most cases (up to 2.5 GHz), the dual-modulus prescaler is also inside the chip.

frequencies in and below the high-frequency band (as of this writing). System can be compact, use low power, and provide very fine frequency resolution with virtually instantaneous frequency changes. DDS is finding increasing application, particularly in conjunction with PLL synthesizers.

DDS uses a single-frequency source (clock) as a time reference. One method of digitally generating the values of a sine wave is to solve the digital recursion relation as follows:

$$Y_n = \left[2\cos\left(2\pi ft\right)\right] Y_{n-1} - Y_{n-2} \tag{6.67}$$

This is solved by $Y_n = \cos(2\pi f_n t)$. However, there are at least two problems with this method. The noise can increase until a limit cycle (nonlinear oscillation) occurs. Also, the finite word length used to represent $\cos(2\pi ft)$ places a limitation on the frequency resolution. Another method of DDS, direct table lookup, consists of storing the sinusoidal amplitude coefficients for successive phase increments in memory. Advances in miniaturization and the lowering cost of ROM make this the most frequently used technique.

One method of direct table lookup outputs the same N points for each cycle of the sine wave and changes the output frequency by adjusting the rate at which the points are computed. It is relatively difficult to obtain fine frequency resolution with this approach,

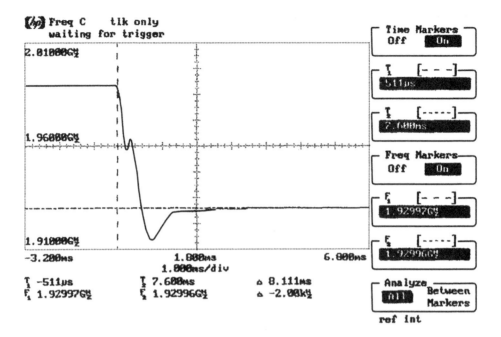

Figure 6.47   Result of lock time measurement for an LMX2350-based synthesizers.

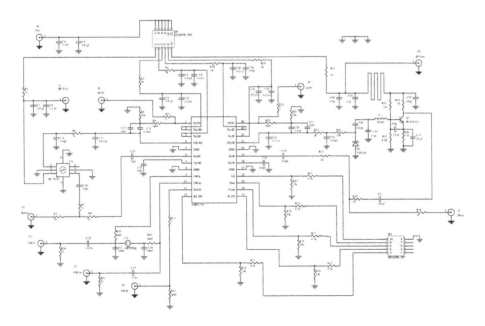

Figure 6.48   LMX2350 sample application schematic.

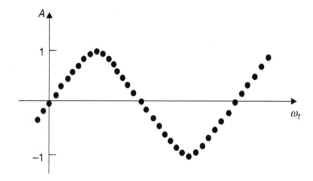

Figure 6.49 Synthesized waveform generated by direct digital synthesis.

so a modified table-lookup method is generally used. It is this method that we describe here. The function $\cos(2\pi ft)$ is approximated by outputting the function $\cos(2\pi fnT)$ for $n = 1, 2, 3, \ldots$, where T is the interval between conversions of digital words in the D/A converter and n represents the successive sample numbers. The sampling frequency, or rate, of the system is $1/T$. The lowest output frequency waveform contains N distinct points in its waveform, as illustrated in Figure 6.49. A waveform of twice the frequency can be generated, using the same sampling rate, but outputting every other data point. A waveform k times as fast is obtained by outputting every kth point at the same rate $1/T$. The frequency resolution, then, is the same as the lowest frequency, f_L.

The maximum output frequency is selected so that it is an integral multiple of f_L, that is, $f_U = kf_L$. If P points are used in the waveform of the highest frequency, $N(=kP)$ points are used in the lowest frequency waveform. The number N is limited by the available memory size. The minimum value that P can assume is usually taken to be four. With this small value of P, the output contains many harmonics of the desired frequency. These can be removed by the use of low-pass filtering in the D/A output. For $P = 4$, the period of the highest frequency is $4T$, resulting in $f_U = 4f_L$. Thus, the highest attainable frequency is determined by the fastest sampling rate possible.

In the design of this type of DDS, the following guidelines apply.

- The desired frequency resolution determines the lowest output frequency f_L.
- The number of D/A conversions used to generate f_L is $N = 4k = 4f_U/f_L$ provided that four conversions are used to generate $f_U(P = 4)$.
- The maximum output frequency f_U is limited by the maximum sampling rate of the DDS, $f_U \leq 1/4T$. Conversely, $T \leq 1/4f_U$.

The architecture of the complete DDS is shown in Figure 6.50. To generate nf_L, the integer n addresses the register and each clock cycle kn is added to the content of the accumulator so that the content of the memory address register is increased by kn. Each knth point of the memory is addressed, and the content of this memory location is transferred to the D/A converter to produce the output sampled waveform.

To complete the DDS, the memory size and length (number of bits) of the memory word must be determined. The word length is determined by system noise requirements. The amplitude of the D/A output is that of an exact sinusoid corrupted with the deterministic

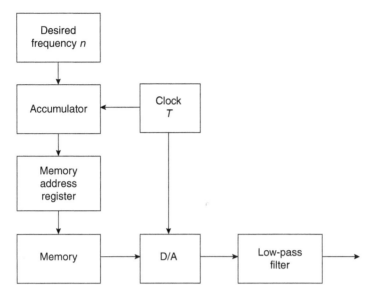

Figure 6.50   Block diagram of a direct digital frequency synthesizer.

noise due to truncation caused by the finite length of the digital words (quantization noise). If an $(n + 1)$-bit word length (including one sign bit) is used and the output of the A/D converter varies between ± 1, the mean noise from the quantization will be

$$\rho^2 = \frac{1}{12}\left(\frac{1}{2}\right)^{2n} = \frac{1}{3}\left(\frac{1}{2}\right)^{2(n+1)} \tag{6.68}$$

The mean noise is averaged over all possible waveforms. For a worst-case waveform, the noise is a square wave with amplitude $1/2\,(1/2)^n$ and $\rho^2 = 1/4\,(1/2)^{2n}$. For each bit added to the word length, the spectral purity improves by 6 dB.

The main drawback of the low-power DDS is that it is limited to relatively low frequencies. The upper frequency is directly related to the maximum usable clock frequency; today, the limit is about 1 GHz. DDS tends to be noisier than other methods, but adequate spectral purity can be obtained if sufficient low-pass filtering is used at the output. DDS systems are easily constructed using readily available microprocessors. The combination of DDS for fine frequency resolution plus other synthesis techniques to obtain higher-frequency output can provide high resolution with very rapid setting time after a frequency change. This is especially valuable for frequency-hopping spread-spectrum systems.

Figure 6.51 shows the functional block diagram of a DDS system. In analyzing both the resolution and the signal-to-noise ratio (or rather signal-to-spurious performance) of the DDS, one has to know the resolution and input frequencies. As an example, if the input frequency is approximately 35 MHz and the implementation is for a 32-bit device, the frequency resolution compared to the input frequency is $35 \cdot 10^6 \div 232 = 35 \cdot 10^6 \div 4.29496729610^9$ or 0.00815 Hz $\approx$ 0.01 Hz. Given the fact that modern shortwave radios with a first IF of about 75 MHz will have an oscillator between 75 and 105 MHz, the resolution at the output

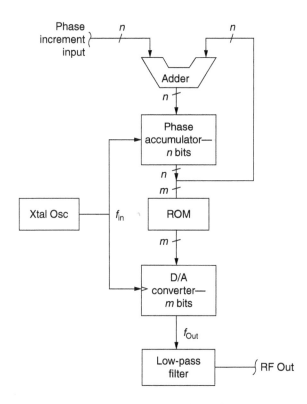

Figure 6.51 Block diagram of a DDS system.

range is more than adequate. In practice, one would use the microprocessor to round it to the next increment of 1 Hz relative to the output frequency.

As to the spurious response, the worst-case spurious response is approximately $20 \log 2R$, where R is the resolution of the digital/analog converter. For an 8-bit A/D converter, this would mean approximately 48 dB down (worst case), as the output loop would have an analog filter to suppress close-in spurious noise. In our application, we will use an 8-bit external D/A converter. However, devices such as the Analog Devices AD7008 DDS modulator have a 10-bit resolution, as shown in Figure 6.52. Ten bits of resolution can translate into $20 \log 210$ or 60 dB of suppression. The actual spurious response would be much better. The current production designs for communication applications, such as shortwave transceivers, despite the fact that they are resorting to a combination of PLLs and DDSs, still end up somewhat complicated. By using 10 MHz from the DDS and using a single-loop PLL system, one can easily extend the operation to above 1 GHz but with higher complexity and power consumption.

Figure 6.53 shows the necessary components of a single-PLL system. Some communications equipment uses this approach.

Figure 6.54 shows the combination of a standard PLL and a DDS, as implemented in the ICOM IC-736 HF/6-meter transceiver. This approach uses the DDS in a frequency range between 500 kHz and 1 MHz. This frequency is upconverted either to 60 MHz for the shortwave band or to 90 MHz for the 6-m ham band. The resulting frequency is used as an auxiliary LO to convert the frequency of the first LO (69.0415–102.0115 MHz) down to

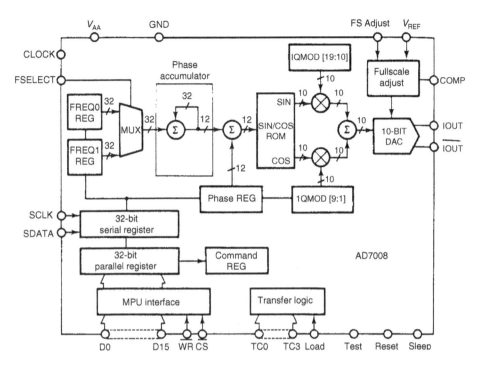

Figure 6.52   Functional block diagram of the Analog Devices AD7008 DDS modulator.

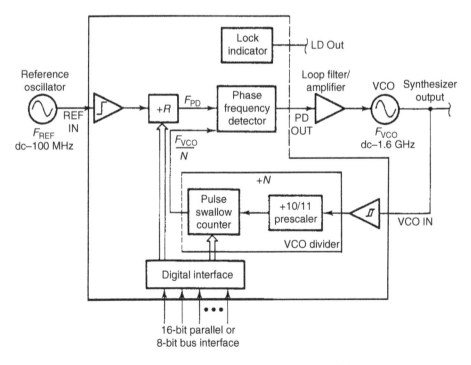

Figure 6.53   Block diagram of a single-loop PLL synthesizer showing all the necessary components for microwave and RF application.

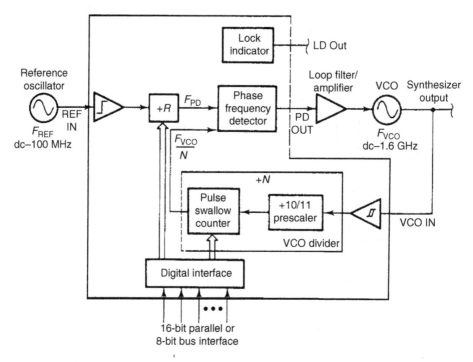

Figure 6.54 Synthesizer used in the ICOM IC-736 HF/6-m transceiver. The IC-736 combines the DDS and PLL approaches.

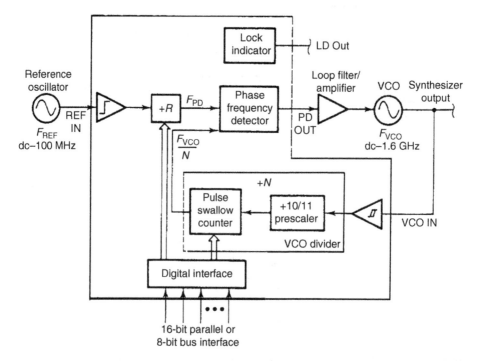

Figure 6.55 Hybrid synthesizer that provides output at about 455 kHz, and from 75 to 105 MHz at approximately 0.01-Hz resolution. This synthesizer uses a combination of a standard PLL and DDS.

the synthesizer IF between 8.5 and 41.5 MHz. There is an additional divide-by-2 stage in the loop, which therefore requires a reference frequency of 250 kHz instead of 500 kHz. This is done to extend the operating range of the synthesizer chip, including its prescaler's capability of operating at much higher frequencies, although it does not have such a hybrid DDS approach incorporated (Figure 6.55).

Although this approach obtains a fairly small division ratio, it is still a four-loop synthesizer. One loop is the DDS itself. The second is the translator loop that mixes the DDS up to 60 MHz. The third is the main loop responsible for the desired output frequency. The fourth loop, so to speak, is the generation of the auxiliary LO frequencies at 60 and 90 MHz, which are derived from the 30-MHz frequency standard. For reasons of good phase noise, it employs a total of five VCOs. Because this loop's division ratio varies between 80 and 17, its gain is subject to considerable variation.

The 10.7-MHz signal from the crystal filter goes to a single-chip PLL (U5, a Motorola MC145170) that contains all the necessary dividers and the phase/frequency discriminator. The operational amplifier (U6, an OPA27), is driven from a 28-V source, and the negative supply of the OPA27 is connected as a voltage doubler, which receives its ac voltage from the synthesizer IC. This trick allows extension of the VCO control voltage. The output from the VCO is applied to a distribution-amplifier system (Q5, Q6, Q7). Q5, a dual-gate MOSFET, drives the PLL IC's fin input; Q6, a dual-gate MOSFET, and Q7, a BJT, supply +17 dBm LO drive for the first mixer.

REFERENCES

1. U. L. Rohde, *Microwave and Wireless Synthesizers: Theory and Design*. New York: Wiley, 1997.

2. W. Egan and E. Clark, "Test your charge-pump phase detectors," *Electronic Des.*, Vol. 26, No. 12, June 7, pp. 134–137, 1978.

3. Motorola Application Note AN969, *Operation of the MC145145 PLL Frequency Synthesizer with Analog Phase Detector*.

4. A. W. Hietala and D. C. Rabe, "Latched accumulator fractional-N synthesis with residual error reduction," Motorola, Inc., Shaumburg, IL, US Patent No. 5,093,632, March 3, 1992.

5. T. A. D. Riley, "Frequency synthesizers having dividing ratio controlled sigma-delta modulator," Carleton University, Ottawa, Canada, US Patent No. 4,965,531, October 23, 1990.

6. N. J. R. King, "Phase locked loop variable frequency generator," Racal Communications Equipment Limited, England, US Patent No. 4,204,174, May 20, 1980.

7. J. N. Wells, "Frequency synthesizers," Marconi Instruments, St. Albans, Hertfordshire (GB), European Patent No. 0125790B2, July 5, 1995. See also J. N. Wells, "Frequency synthesizers," Marconi Instruments Ltd., St. Albans, England, US Patent No. 4,609,881, September 2, 1986.

8. T. Jackson, "Improvement in or relating to synthesizers," European Patent No. 0214217Bl, June 6, 1996.

9. T. Jackson, "Improvement in or relating to synthesizers," Plessey Overseas Limited, Ilford, Essex (GB), European Patent No. W086/05046, August 28, 1996.

10. B.-H. Park, "A low voltage, low power CMOS 900 MHz frequency synthesizer," Ph.D. Dissertation, Georgia Institute of Technology, Atlanta, December 1997.

11. C. E. Hill, "All digital fractional-N synthesizer for high resolution phase locked loops," *Appl. Microwave Wireless*, November/December 1997, January/February 1998.

INTERESTING PATENTS

D. A. Bradley, "Ultra-low-noise microwave synthesizer," Wiltron Company, Morgan Hill, CA, US Patent No. 5,374,902, December 20, 1994.

S. Mehrgardt, "Frequency divider circuit," Deutsche ITT Industries GmbH, Freiburg, Germany, US Patent No. 4,694,475.

E. G. Najle and R. M. Buckley, "Phase noise measurements utilization a frequency down-conversion/multiplier, direct spectrum measurement technique," Harris Corporation, Melbourne, FL, and Advanced Testing Technologies, Commach, NY, US Patent No. 5,337,014.

Fractional-N Synthesizers

A. Albarello, A. Roullet, and A. Pimentel, "Fractional-division frequency synthesizer for digital angle-modulation," Thomson-CSF, US Patent No. 4,492,936, January 8, 1985.

C. Attenborough, "Fractional-N frequency synthesizer with modulation compensation," Plessey Overseas Ltd., Ilford, Great Britain, US Patent No. 4,686,488, August 11, 1987.

R. J. Bosselaers, "PLL including an arithmetic unit," US Patent No. 3,913,928, October 1975.

R. G. Cox, "Frequency synthesizer," Hewlett-Packard Company, US Patent No. 2,976,945.

J. A. Crawford, "Enhanced analog phase interpolation for fractional-N frequency synthesis," Hughes Aircraft Company, Los Angeles, CA, US Patent No. 4,586,005, April 29, 1986.

A. T. Crowley, "PLL frequency synthesizer including fractional digital frequency divider," RCA Corporation, New York, NY, US Patent No. 4,468,632, August 28, 1984.

A. P. Edwards, "Low phase noise RF synthesizer," Hewlett-Packard Company, Palo Alto, CA, US Patent No. 4,763,083, August 9, 1988.

B.-G. Goldberg, "Digital frequency synthesizer," Sciteq Electronics, Inc., San Diego, CA, US Patent No. 4,752,902, June 21, 1988.

B.-G. Goldberg, "Digital frequency synthesizer having multiple processing paths," US Patent No. 4,958,310, September 18, 1990.

B.-G. Goldberg, "Programmable fractional-N frequency synthesizer," Sciteq Electronics, Inc., San Diego, CA, US Patent No. 5,224,132, June 29, 1993.

B.-G. Goldberg, "Analog and digital fractional-n PLL frequency synthesis: A survey and update," *Appl. Microwave Wireless*, June 1999, pp. 32–42.

W. G. Greken, "Digital frequency synthesizer," General Dynamics, US Patent No. 3,882,403.

A. W. Hietala and D. C. Rabe, "Latched accumulator fractional N synthesis with residual error reduction," Motorola, Inc., Schaumburg, IL, US Patent No. 5,093,632, March 3, 1992.

L. Jackson, "Digital frequency synthesizer," US Patent No. 3,734,269, May 1973.

T. Jackson, "Fractional-N synthesizer," Plessey Overseas Ltd., Ilford, Great Britain, US Patent No. 4,758,802, July 19, 1988.

T. Jackson, "Frequency synthesizer of the fractional type," Plessey Overseas Ltd., Ilford, Great Britain, US Patent No. 4,800,342, January 24, 1989.

T. Jackson, "Improvement in or relating to synthesizers," Plessey Overseas Ltd., Ilford, Essex, Great Britain, European Patent No. 0214217B1, June 6, 1996.

T. Jackson, "Improvement in or relating to synthesizers," Plessey Overseas Ltd., Ilford, Essex, Great Britain, European Patent No. WO86/05046, August 28, 1996.

N. J. R. King, "Phase locked loop variable frequency generator," Racal Communications Equipment Ltd., England, US Patent No. 4,204,174, May 20, 1980.

N. G. Kingsbury, "Frequency synthesizer with fractional division ratio and jitter compensation," Marconi Company Ltd., Chelmsford, Great Britain, US Patent No. 4,179,670, December 18, 1979.

C. A. Kingsford-Smith, "Device for synthesizing frequencies which are rational multiples of a fundamental frequency," Hewlett-Packard Company, Palo Alto, CA, US Patent No. 3,928,813, December 23, 1975.

F. L. Martin, "Frequency synthesizer with spur compensation," Motorola, Inc., Schaumburg, IL, US Patent No. 4,918,403, April 17, 1990.

F. L. Martin, "Frequency synthesizer with spur compensation," Motorola, US Patent No. 4,816,774, March 28, 1989.

K. D. McCann, "Frequency synthesizer having jitter compensation," US. Philips Corp., New York, NY, US Patent No. 4,599,579, July 8, 1986.

B. M. Miller, "Multiple-modulator fractional-N divider," Hewlett-Packard Company, US Patent No. 5,038,117, August 6, 1991.

E. J. Nossen, "Digitized frequency synthesizer," RCA Corp., New York, NY, US Patent No. 4,206,425, June 3, 1980.

O. Peña, "SPICE tools provide behavioral modeling of PLLs," *Microwaves RF*, November 1997, pp. 71–80.

V. S. Reinhardt and I. Shahriary, "Spurless fractional divider direct digital frequency synthesizer and method," Hughes Aircraft Company, Los Angeles, CA, US Patent No. 4,815,018, March 21, 1989.

J. Remy, "Frequency synthesizer including a fractional multiplier," Adret Electronique, Paris, France, US Patent No. 4,458,329, July 3, 1984.

T. A. D. Riley, "Frequency synthesizers having dividing ratio controlled sigma-delta modulator," Carleton University, Ottawa, Canada, US Patent No. 4,965,531, October 23, 1990.

W. P. Shepherd, D. E. Davis, and W. F. Tay, "Fractional-N synthesizer having modulation spur compensation," Motorola, Inc., Schaumburg, IL, US Patent No. 5,021,754, June 4, 1991.

W. J. Tanis, "Frequency synthesizer having fractional-N frequency divider in PLL," Engelman Microwave, US Patent No. 3,959,737, May 25, 1976.

J. N. Wells, "Frequency synthesizers," Marconi Instruments, St. Albans, Hertfordshire, Great Britain, European Patent No. 0125790B2, July 5, 1995.

J. N. Wells, "Frequency synthesizers," Marconi Instruments Ltd., St. Albans, England, US Patent No. 4,609,881, September 2, 1986.

C. E. Wheatley, III, "Digital frequency synthesizer with random jittering for reducing discrete spectral spurs," Rockwell International Corporation, El Segundo, CA, US Patent No. 4,410,954, October 18, 1983.

M. A. Wheatley, L. A. Lepper, and N. K. Webb, "Frequency modulated phase locked loop with fractional divider and jitter compensation," Racal-Dana Instruments Ltd., Berkshire, Great Britain, US Patent No. 5,038,120, August 6, 1991.

R. O. Yaeger, "Fractional frequency divider," RCA Corp., Princeton, NJ, US Patent No. 4,573,176, February 25, 1986.

FURTHER READING

Fujitsu Microelectronics, Inc., *Super PLL Application Guide*, 1998.

New Technique for Analyzing Phase-Locked Loops, Application Note 164-3, Hewlett-Packard, June 1975.

V. F. Kroupa (ed.), *Direct Digital Frequency Synthesizers*. New York: IEEE Press, 1999.

B. Sam, "Hybrid frequency synthesizer combines octave tuning range and millihertz steps," *Appl. Microwave Wireless*, May 1999, pp. 76–84.

M. Smith, "Phase noise measurement using the phase lock technique," Wireless Subscriber Systems Group (WSSG) AN1639, Motorola.

INDEX

ABCD-matrix, 375
Acceptor, 144
Access burst, 27
Active load, 496
Adjacent-channel measurement, 107
Adjacent channel power ratio (ACLR/ACPR), 86, 99, 110, 402, 410, 469, 573
Agilent E4419B power meter, 569
Agilent PNA, 336
Amplifier, 467. *See also* Automatic gain control; Cascode
  CGY96/CGY94, 398
  class E, 549
  classes A-C, 360, 544
  class F, inverse F, 548
  class F waveform, 567
  CMOS low-noise amplifier, 431
  complementary current-reuse, 436
  current conduction angle, 360, 544
  design procedure, 431
  differential, 514
  distributed, 362
  distribution, 623
  Doherty, 552
  feedback, 464
  folded cascode, 436
  gain, 423
  harmonically tuned, 548
  high gain, 467
  LNA in 130 nm CMOS, 446
  low-noise amplifier (LNA), 429
  matching, 423
  multi-stage, 499
  narrow band BFP420, 460
  NE68133, 450
  parallel and push–pull, 539
  power, 542
  single-stage feedback, 484
  specifications, 430
  stability, 423
  stability improvement, 430
  transformer-coupled LNA, 449
  turbocharged configuration, 621
  two-stages, 490
  wideband, 362
Amplitude and phase imbalance, 72
Amplitude linearity, 90
Amplitude modulator, 29
Amplitude nonlinearity, 89
Amplitude stability, 740
AM–PM conversion, 90, 98, 578, 784, 792
Angle deviation, 15
Antenna switch, 5
Antenna system noise figure, 88
Antibacklash circuit, 846
Aparicio and Hajimiri, 440
Attenuator
  diode, 678
  PIN diode, 155, 398
Automatic gain control (AGC), 153, 167, 398, 413, 524, 576, 788

Balun, 540
Band-selection filter, 5
Barkhausen, 728
Barrier height, 136, 137
Baseband predistortion, 584
Baseband processing, 65
Baseband waveforms, 34
Base-station identification code (BSIC), 28
Base transceiver station (BTS), 23

RF/Microwave Circuit Design for Wireless Applications, Second Edition. Ulrich L. Rohde and Matthias Rudolph.
© 2013 John Wiley & Sons, Inc. Published 2013 by John Wiley & Sons, Inc.

Battery-powered mobile, 436
BCR400 active bias controller, 538
BF999, 275
BFP420, 218
   matched amplifier, 457
   noise parameters, 389, 424
   s-parameters, 424
BFR380F, datasheet, 210
BFR193W, model, 351
Biasing, 524, 535
BiCMOS, 203, 275
Bipolar junction transistor (BJT). *See* BJT
Bit error rate (BER)
   vs. noise, 38, 86
      for BPSK/QPSK, 16-QAM and 64-QAM,
         43, 86
Bit synchronization, 23
BJT, 203
   base-collector capacitance, 209,
      231
   base doping densities, 223
   base-emitter time constant, 229
   base push-out, 230
   base transit time, 229
   base transport factor, 224
   base width, 220, 267
      modulation, 226
   built-in drift field, 147, 230
   collector-base breakdown, 205
   collector-base time constant, 209
   collector-emitter breakdown, 206
   collector transit time, 229
   current gain, 240
   emitter injection efficiency, 223
   Kirk effect, 230
   large-signal behavior, 204
   linear model, 239
   maximum ratings, 205
   noise model, 326
   saturation region, 232
   self-heating, 236
   transit frequency f_t, 207
   uniform-base transistor, 219
Bluetooth, 46
Bode and Fano, 601
Bode equation, 602
Bode limit, 491
Boltzmann's constant, 80
Bonding pad, 136
Breakdown, 160, 171, 186, 248
   BJT collector-base, 205
   BJT collector-emitter, 206

Bulk acoustic resonance (BAR) filter, 5
Burst
   error, 16
   structures, 22
   synchronization burst (SB), 28
   types, 27
Butterworth, 848

Cable TV distribution, 97
Cantz, 449
Cartesian loop, 579
Cascode, 436, 450, 491
Cauer, 520
Caution, 363
Cellular and cordless systems
   parameters, 59
Ceramic-resonator oscillator, 754
Ceramic resonators (CRs), 754, 756
Channel impulse response (CIR), 7, 24
Channel spacing, 18
Charge pump, 842
   external, 851
Charge-pump based PLL, 850, 854
Chebyshev, 602
Clock recovery, 55
Closed-loop transfer function, 731
CMOS, 256
   body effect, 263
   passive elements, 438
Code division multiple access (CDMA), 17, 19
Coherence bandwidth, 14
Coherent demodulation, 37
Collector efficiency η, 542
Compact Modeling Council, 291
Complex baseband signal, 32
Compression, 397, 405
Contact potential, 133
Contact resistances, 136
Convolutional encoder, 53
Coplanar waveguide (CPW), 438
Correlation
   noise sources, 377
   receiver, 36
Coupler
   3-dB hybrid, 539
   Lange coupler, 539
Cree CGH60015DE GaN HEMT, 566
Cross-correlation, 24, 36
Cross-modulation, 97, 173, 193
Crystal filter, 105
Crystal parameters, 766
CT1 standard, 20

CT2 standard, 59
Current distortion, 485
Current waveform, 546
Cyclic prefix, 49

DAB, 6, 9, 16
1-dB compression point, 93
    mixer, 650
3-dB cutoff frequency, 210
1-dB desensitization point, 651
DC offset, 654
DCS1800, 59
DC stabilization circuit, 537
Dead zone, 847
DECT, 59
Delay correction, 26
Delay spread, 9
Delay time, 26
Demodulation
    digitally modulated carriers, 35
Depletion FETs (DFETs), 298
Depletion region, 133, 147, 160, 220, 258,
        272, 274
    capacitance, 229
Detector
    diode, 133
    Schottky diodes, 657
Differential gain, 369
Differential group delay, 100
Differential phase, 369
Diffusion, 133
    capacitance, 641
    charge, 132
    current, 221
    length, 222
Digital modulation, 5
    ASK and BPSK, 34
    BPSK modulator, 676
    BPSK spectrum and constellation diagram,
        38
    constellation diagram, 34
    DBPSK, 37
    DQPSK, 37
    eye and constellation diagrams, 402
    GMSK, 27, 64
    GMSK baseband I/Q generator, 64
    I/Q modulator, 33
    linear, 62
    modulator, 30
    MPSK, 15
    MSK and GMSK, 33
    MSK, GMSK, and QPSK spectra, 413

QAM impact of amplitude compression, 94
QAM impact of AM-to-PM conversion, 101
QPSK and p/4-DQPSK constellation
        diagram, 62
QPSK modulator, 676
QPSK, MSK, and GMSK, spectra, 45
QPSK spectrum, 44
    constellation diagram, 38
spread-spectrum techniques, 45
Digital predistortion (DPD), 573
Digital tristate comparators, 838
Diode. *See* depletion region
    abrupt junction, 130, 159, 171
    abrupt junction capacitance, 136
    attenuator/switch, 678
    back-to-back, 176
    bandswitch, 202
    barrier potential, 130, 133
    breakdown, 142
    capacitance, 130, 159, 162, 166, 229, 505
        vs. frequency, 139
        vs. voltage, 140
    capacitance diode equivalent circuit, 179
    capacitance modulation, 191
    capacitance ratio, 166, 171, 178, 189,
        770, 773
        vs. breakdown voltage, 167
    capacitance temperature dependence, 180
    capacitance–voltage characteristics, 177
    chip noise equivalent circuit, 326
    current, 128
    diode ring, 834
    distortion, 171
    dynamic resistance, 136
    equivalent circuit, 148
    harmonic distortion, 175
    high breakdown, 138
    hyperabrupt, C–V curve, 164
    hyperabrupt junction, 162, 171, 508
    large-signal model, 129
    linearly graded junction, 130, 161
    linear model, 140
    LO power required for mixing, 137
    low-drive, 137
    mixer, 139, 654
        and detector, 133
    mixer diode loss, 641
    noise figure *vs.* LO power, 137
    noise model, 326
    PIN, 141, 145, 156
        attenuator, 155, 398
        diode applications, 151

Diode. *See* depletion region (*Continued*)
  diode breakdown, capacitance, Q factor, 146
  diode crossover frequency, 149
  diode I–V characteristics, 148
  diode large-signal model, 142
  diode resistance, 143
  diode reverse shunt resistance, 151
  diode series resistance, 147, 149
  diode variable resistance, 142
  planar *vs.* mesa, 165
  Q factor, 164, 182
  and loss, 168
  reverse-bias capacitance, 142
  reverse recovery, 132
  reverse-recovery time, 142
  RF Schottky diode, 657
  ring mixer, 666
  saturation current, 234
  Schottky detector diodes, 657
  Schottky diode, 133, 642
  Schottky diode band diagram, 138
  Schottky diode chip cross-section, 134
  Schottky diode reverse resistance, 136
  Schottky diode series resistance, 135
  Schottky diode small-signal parameters, 135
  Schottky junction capacitance, 136
  Schottky mixer diodes parameters, 129
  series resistance, 182
  SPICE parameters, 131
  storage time, 132
  switch, 132, 196, 200
  switch data, 198
  switch equivalent circuit, 197
  temperature dependent Schottky current, 134
  tuned resonant circuit, 186, 770
  tuning diode, 158, 771, 773, 797, 799
  tuning diode capacitance, 179
  tuning diode equivalent circuit, 178
  tuning diode parameters, 173, 182
  tuning diode physics, 159
  tuning varactors, 166
  varactor, 158, 768
  variable resistor, 144, 152
  varicap, 158
  x-band mixer diode data, 141
Direct conversion, 65
Direct digital synthesis (DDS), 33, 871
Direct-sequence spread spectrum (DSSS), 45, 47
Direct switching noise, 646
Dissipated power, 237, 238

Distortion, 405, 467
  diode, 171
Distributed effects, 440
DMOS, 270
Doherty technique, 552
Donor, 144
Doppler effect, 6, 11, 12, 17, 24, 109
Double-sideband (DSB) measurement, 396
Drain efficiency η, 542
DSP, 43, 584
Dummy burst, 28
Dynamical self-heating, 218, 291
Dynamic range (DR), 87, 96, 397, 474, 651
Dynamic stability, 193

Early
  effect, 228, 238
  voltage, 228, 479
Ebers–Moll equations, 235
Echoes, 5
Edge-triggered JK Master-Slave Flip-Flops, 835
Effective channel length, 261
Egan and Clark, 840
Electron
  drift velocity in surface channel, 267
  lifetime, 145
  velocity saturation, 268
Energy efficiency, 90
Enhancement FETs (EFETs), 298
Enhancement-mode MOSFET
  NMOS, 256
  PMOS, 422
ENR (excess noise ratio), 395
Envelope elimination and restoration, 577
Envelope feedback, 577
Equalizer, 10
Error vector magnitude (EVM), 87, 107, 581

Fading, 6, 19, 49
  simulation, 11
Fano, 490
FDD system using SA900, 65
Feedback
  active, 499
  amplifier, 464
  current, 467
  envelope, 577
  inductive source, 434
  lossless or noiseless, 489
  modulation, 577

transformation of noise parameters, 383
voltage, 467
Feedforward, 581
FET, 239
   channel charge, 266
   channel length, 267
   cold-FET and forward-biased-gate, 339
   family tree, 241
   intrinsic noise equivalent circuit, 330
   linear model, 253
   mixer, 697
   noise equivalent circuit, 255
   pinchoff voltage, 248
   short-channel effects, 267
   transconductance, 251
Film resistor above ground plane, equivalent
      circuit, 80
Filter
   attenuator, 154
   for phase detectors
Flicker noise, 327, 645, 783
   bipolar transistors, 203
   corner frequency, 292
     for different technologies, 747
   frequency dependence, 253
   GaAs diodes, 138
   upconversion, 713
FM sound broadcasting, 16
$1/f$ noise. *See* Flicker noise
Forced β, 233
Forward error correction, 109
Forward transconductance, 248
Forward transfer curve, 248
Fractional-N principle, 832, 864
Frequency correction burst (FCB), 27
Frequency division duplex (FDD), 65
Frequency division multiple access (FDMA), 17
Frequency-domain equalization, 49
Frequency doublers, 518
Frequency hopping, 17, 45
Frequency modulators, 29
Frequency shift, 193
Frequency stability, 742
Fukui, 382, 390

GaAs HEMT, 5, 241
Gain, 364, 367, 425
   circles, 390, 427
   compression, 89, 92, 583
   variation, 364
Gain-bandwidth product, 209
GaN HEMT, 241, 290

Gate oxide capacitance, 259
Gate-source cutoff voltage, 248
Gaussian distribution, 6
G1 band, 20
Gold codes, 47
GSA standards, 10
GSM, 11, 59
   GSM900, 21
   GSM1800, 21
   transmission bit rates, 23
Guard interval, 9
Guard period, 25, 49
Gummel–Poon, 327

Harmonic components, 546
Harmonic distortion
   diode, 175
   intermodulation products (HIP), 651
Harmonic output power, 803
Hartmann and Strutt, 382
Haus and Adler, 380, 391
Hawkins, 787
HBT. *See also* BJT
   band diagram, 226
   dependence of transit-time on collector
      current, 219
   InGaP/GaAs, 5, 203
   self-heating, 238
   SiGe, 203
Health effects, 2
HEMT, 290
   band diagram, 290
   InP, 290
   output conductance, 292
Heterointerface, 289
Heterojunctions, 219
Heterostructures, 225
Hewlett-Packard HP4145, 337
Hewlett-Packard HP8510, 337
Hewlett-Packard HP8640 signal generator, 748,
     793
HF/VHF voltage-controlled filter, 508
Hillbrand and Russer, 391
Hole-electron, 274
Hopf bifurcation, 629
Hot-carrier diode (HCD), 765
Hot-spot formation, 238
Hsu and Snapp, 787
Hyperabrupt junction. *See* diode
Hyperframes, 21
Hysteresis, 98
   intermodulation effects, 99

Impact ionization, 274
Impedance matching networks, 585
Information bits, 22
In-phase modulation test, 72
Integrated frequency synthesizer, 728
Interdigital capacitors, 532
Intermediate-frequency (IF), 188, 637
    filter, 15
    image, 637
Intermodulation distortion (IMD), 89, 94, 397,
        511, 637, 651, 699
    diode, 175
    improvement, 622
    PIN diodes, 153
    second order, 94, 505
    third order, 94, 576
Intermodulation intercept point (IP), 95
Interport isolation, 652
Intersymbol interference, 403
Intrinsic semiconductor, 144
Inverse current gain, 234
I/Q channel waveforms, 402
I/Q demodulator, 36
I/Q modulator, 29, 61
IS-54, 59
    Philips chipset, 66
IS-95, 59
ISM band, 46, 74
Iversen, 382

J. E. Lilienfeld, 241
JFET, 241, 248
    large-signal model, 251
    n-channel static characteristics, 248
    noise model, 329
    small-signal behavior, 251
    U310, 243
JFETs
    large-signal behavior, 248
Jitter, 842

Kahn, 577
Kirchoff, 468
Kurokawa, 786

Landscape models, 9
Large-signal S parameters, 518, 564
Large-signal validation, 341
Lazy as we are, 32
LDMOS, 241, 270, 324
Leakage, 843
    current, 146, 186

signal, 681
source-gate, 250
Leeson equation, 746, 815
Level shifting, 494
Linear distortion, 89
Linearity, 397
Linear modulation, 34
Loaded Q, 748, 759, 785, 789
Load line, 469, 518, 558, 798
Load–pull, 518, 566, 699
Load–pull/source–pull, 566
Local oscillator (LO), 637
    drive level, 652
    power required for Schottky
        diodes, 137
Lock-in, 863
    time, 849
Lossy substrate, 442
Lovelace, 777
Low-noise amplifier (LNA). *See* amplifier
Low-voltage open-collector design, 476
LTE, 29, 49, 573

Maas, 650
M-ary alphabets, 29
Matching, 588
    broadband, 490
    flexible, 482
Maximum frequency of oscillation (f_{max}), 210,
        292
Maximum likelihood estimation, 36
MC13109, 78
MC13109FB, 79
MDS, 80, 95
MESFET
    extrinsic model, 293
    I–V characteristics, 292
    noise model, 328
    small-signal model, 299
Metal-insulator-metal capacitor (MIM), 440,
        441
MFSK, 53
    baseband circuitry, 54
mHEMT, 290
Microstrip, 438
    inductor as oscillator resonator, 757
    inductors, 761
    tank, 762
    T, cross, and Y junction, 528
Microstrip inductor
    high Q, 760
Middlebrook, 249

Mikrostrip
  correction elements, 533
Miller effect, 429, 743
Minimum discernible signal (MDS),
  51
Mismatch loss, 425
Mixer
  conversion gain/loss, 640
  1 db compression, 650
  differential CMOS, 645
  diode, 133, 139, 654
  diode loss, 641
  diode ring, 666
  double-balanced, 651
  dual-gate MOS/GaAs, 701
  FET, 697
  Gilbert cell, 646, 686, 689
  image reject, 676
  MOSFET Gilbert cell, 700
  noise figure, 642, 643
  passive, 642
  quadrature IF, 676
  resistive, 701
  resistive mixer noise figure, 706
  Schottky diodes parameters, 129
  single-balanced, 646, 664
  single BJT, 685
  single diode, 655
  subharmonically pumped, 666
  transistor mixer, 685
  triple-balanced, 683
  walking stick microstrip resonator, 764
Modulated measurements, 573
Modulated RF carrier, 31
Modulation feedback, 576
MOSFET, 241, 242
  around VDS = 0, 250
  drain-substrate conductance, 275
  Gilbert cell mixer, 700
  intrinsic noise equivalent circuit, 332
  inversion, 259
  large-signal behavior, 256
  NMOS device characteristics, 263
  NMOS transfer characteristic, 272
  noise model, 331
  ohmic or triode region, 262
  saturation region, 262
  small-signal model, 263, 264, 272
  substrate flow, 274
  subthreshold conduction, 272
  transconductance, 264
  transfer characteristic, 256

voltage limitation, 262
weak inversion, 272
Motorboating phenomenon, 368
Multiframes, 21
Multipath propagation, 5, 49
Multipath reception, 13, 24, 369
Multiple-input–multiple-output (MIMO), 29

NAND gate, 838
NE71000, 341
NMT900, 20
Noise. *See also* Flicker noise; Phase noise
  atmospheric, man-made, and galactic
    noise, 88
  bandwidth, 372
  circles, 385, 390
  correlation admittance (Y_{cor}), 377
  correlation coefficient, 377
  cyclostationary sources, 713
  determine noise parameters, 396
  equivalent conductance G_n, 378
  equivalent resistance, 378
  equivalent temperature, 370
  external parasitic elements, 382
  feedback, 383
  flicker noise reduction, 819
  four noise parameters, 379
  indirect switching noise, 649
  load, 646
  measurement, 642
  noise measure, 389
  popcorn or burst noise, 256
  power ratio, 97
  resistive mixer, 706
  Schottky noise, 385
  shot noise, 326, 327
  signal-to-noise plus distortion ratio (SINAD),
    51, 86, 112
  signal-to-noise ratio, 85, 369, 371
  system noise and noise floor, 80
  two-port, 374
Noise correlation matrix, 391
  as function of noise parameters, 394
  intrinsic FET, 330, 331
  transformation, 393
Noise factor (F), 369
  mixer, 643
  vs. noise figure, 87
Noise figure (NF), 369, 370, 781
  antenna system, 88
  cascaded networks, 89, 380
  vs. LO power, 137

Noise figure (NF) (*Continued*)
  measurement, 373, 395
  mixer, 642
  passive device, 88
  vs. noise factor, 87
Noise floor, 789
Noise parameters, 384
  bias-dependence, 385
Nonlinear distortion, 89
Nonlinear generators, 584
Nonlinear modulation, 34
Norton, 476, 489
Nyquist criterion, 730
Nyquist formula, 377
Nyquist stability criterion, 625, 740
Nyquist's theorem, 644

Orthogonal, 32
Orthogonal channel, 49
Orthogonal frequency division modulation
    (OFDM), 48
Orthogonal RF carriers, 32
Oscillation condition, 750
Oscillator
  ceramic-resonator, 754
  Clapp–Gouriet, 740, 741, 745
  Colpitts, 732, 745, 749, 777, 786
  Colpitts crystal, 766
  common circuit concepts, 729
  crystal, 763
  dc-coupled, 777
  dc-stabilized, 777
  design flow chart, 753
  dielectric resonator (DRO), 752
  Hartley, 735, 744
  Hartley microstrip resonator, 763
  phase noise, 781
  push–pull, 810
  relaxation, 730
  self-limiting, 741
  specifications, 803
  tapped-inductor differential, 761
  three-reactance circuit, 733
  tuning characteristic, 808
  tuning linearity, 808
  tuning range, 189, 774, 808
  tuning speed, 808
  two-port, 738, 749
  voltage-controlled, 768
Output characteristic curves, 248
Output power
  temperature dependence, 804

Overlay capacitance, 136
Oxide capacitance, 265

Packet-switched data connections, 28
Parallel-plate capacitances, 439,
    440
Parameter extraction, 324
  database, 334
Parasitic capacitance, 265, 440, 442
Parasitic inductances, 440
Parasitic resistance, 263, 275, 326
Parasitic transistor, 219
Passivation, 146
Peak-to-average ratio, 49
Petrovic, 578
Petrovic & Smith, 578
Phase detector, 673, 833, 845
  small errors, 838
Phase error, 107
Phase-frequency comparators, 834
Phase/frequency detector noise,
    861
Phase-locked loop (PLL), 102,
    831
  block diagram, 57
  clock-recovery, 55
  linearized model, 833
  simulation, 56
Phase-locking techniques, 107
Phase noise, 101, 746, 815
  feedback model, 789
  improvement, 801, 814
  spectral density, 783
Phase response, 99
Phase stability, 742
pHEMTs, 290
Photogeneration, 151
Pierce, 746
Polar loop, 578
Poole and Paul, 388
Port VSWR, 652
Post-tuning drift, 171, 804
Power-added efficiency (PAE), 542
Power back-off, 575
Power consumption, 417, 654, 804
Power ratios–voltage ratios, 364
Preamplifier, 494
Predistortion, 583
Pseudonoise, 17
Pseudorandom binary sequence (PRBS),
    43
PTFA211801E, 275

Punch-through, 136, 147, 161, 166, 171
Push–pull, 468, 619, 640, 811
Push–push, 522

Q factor
  diode, 168
  tuning diode, 182
  unloaded Q, 788
Quarter-wave transformer, 599
Quasiparallel transistors, 621

Radial stub, 533
Radio channel, 5
  impulse response, 7
Rayleigh distribution, 6
  channel, 15
Receiver
  bandwidth, 15
  block diagram, 106
Reciprocal mixing, 104, 107
Recombination, 145, 220
Recovery, 132
Recovery time, 142
Rectifier, 151
Regrowth, 404
Rejection filter, 18
Resonance frequency, 442
RFCMOS, metal-layer stack, 438
RF feedback, 576
Richardson equation, 134
Rizzoli, 786
Rohde, 747, 786
Rohde & Schwarz subharmonically pumped
    DBM, 683
Rohde & Schwarz CMW 510, 113
Rohde & Schwarz ESH2/ESH3 test
    receiver, 776
Rohde & Schwarz ESN/ESVN40, 775
Rohde & Schwarz FSE, 869
Rohde & Schwarz SKTU, 642
Rohde & Schwarz SMDU, 748, 793
Rohde & Schwarz SME signal generator, 33
Rohde & Schwarz SMHU58, 107
Rohde & Schwarz SMY and SMIQ signal
    generators, 102
Rohde & Schwarz ZVA, 336
Rohde & Schwarz ZVR, 337
Rothe and Dahlke, 374, 387, 432, 433

SA900, 57
  TDD system, 65
  transmit modulator, 60

Saturation current, 134
  diode, 234
  Schottky diode, 136
Scalable device models, 334
Scherer, 746, 817
Selectivity, 107
Self-heating, 204
Self-limiting oscillator, 741
Sensitivity, 85
Sensitivity to load changes, 804
Series resistance, 160
Shockley, 241
Shockley equation, 128
Sideband suppression, 679
Sigma-delta modulator, 867
Signaling channel, 23
Silicon physical parameters, 144
Single-carrier frequency-domain equalization
    (SC-FDE), 49
Single-carrier frequency-domain multiple
    access (SC-FDMA), 48, 49
Single-loop synthesizer, 832
Single sideband AM (SSB-AM), 64
Single sideband (SSB)/in-phase/quadrature
    (I/Q) modulator, 678
Single-sideband (SSB) measurement,
    396
Sinusoidal oscillators, 730
Smith chart, 425
SMS, 28
Sokal, 549
S parameter-based model, 445
S parameters, 425
Spectral plots, 402
Spectral regrowth, 90, 99, 369
Spiral inductor, 443
Spurious output, 806, 866
  receiver spurious response, 52
  rejection, 51
  suppression, 866
SSB rejection, 679
SSB vs. DSB noise figure, 650
Stability, 427
  amplifier, 423
  analysis, 623
  improvement, 430
  multi-stage amplifiers, 504
  Rollett factor k, 366, 428, 568
  static, 195
Static built-in field, 226
Substrate loss, 440
Superframes, 21

Surface-acoustic wave (SAW) filter, 5,
  68, 105
Surprises, 560
Suter, 382
Switch
  diode, 132, 196, 200, 678
  diode bandswitch, 202
  diode data, 198
  diode equivalent circuit, 197
Symbol error rate, 6
Synchronization, 19, 23

TACS, 20
Tapped-C, 760
Tapped-L, 760
TDA1053, 156
TDMA timers, 22
Temperature coefficient of capacitance,
  167
Temperature effects
  compensation, 192
  drift, 808
  Schottky current, 134
Terman and Buss, 576
Termination-insensitive mixer, 673
Thermal capacitance, 237
Thermal noise sources, 370
Thermal port, 238
Thermal resistance, 237, 554
Thermal runaway, 518
Thermal subcircuit, 237
Thevenin transformation, 476
Third-order intercept, 469
Time division duplex (TDD), 65
Time-Division Multiple Access (TDMA),
  19
Time-domain solver, 445
Timing advance (TA), 26
Tracking, 190, 775
Training sequence, 23, 55
Transceiver, 3
Transconductance noise, 646
Transformer
  CMOS, 442, 444
  equivalent circuit, 445
  feedback, 489
  loss, 450
  low-noise amplifier
Transient response, 849
Transistor model
  AgilentHBT, 219
  BSIM, 324

Chalmers (Angelov), 291
Curtice–Ettenberg cubic model, 291
EEHEMT, 291
FBH-HBT, 219
FBH-HBT equivalent circuit, 237
Gummel–Poon, 218
HICUM, 219
Materka model, 291
MEXTRAM, 219, 563
modified Materka–Kacprzak, 251,
  291, 294
130-nm NMOS BSIM4 parameters,
  451
Raytheon (Statz), 291
Root, 291
Tajima, 291
TOM (TriQuint's Own Model),
  291
Transit frequency f_t, 229,
  289, 587
BFP650, 229
BJT, 267
BJT/HBT, 207
HEMT, 291
MOSFET, 266
self-heating, 231
Transit time, 132, 142
Transmission line, 528
Transmitter, 57
Trimming, 812
Triple-beat distortion, 97
Tristate detector, 840
Tuned circuit, 506
Tuned filter, 505
  VHF television tuner with electronic band
    selection, 202
Tuned resonant circuit
  diode, 186, 770
Tuning capacitance, 773
Tuning diode. See diode
Tuning ratios, 167
Tuning voltage, 195
Two-dimensional electron gas, 290

UMTS, 29
Unilateral figure of merit, 427
uPC2749TB, 499

Varactor. See diode
Varicap. See diode
VCO tank circuit, 68
VCXO, 769

Vendelin, 382
VHF band, 16
Via hole, 439, 533
Viterbi
    algorithm, 10
    decoder, 10
VMOS (vertical MOS), 270
Voltage-controlled oscillator (VCO), 768. *See also* Oscillator

W360, 337
Wideband-CDMA, 47
Wilkinson power divider, 539, 623

XOR, 46, 836

YIG oscillator, 751, 752, 800, 814

Zwischenbasis-configuration, 449